TWELFTH EDITION

METEOROLOGY TODAY

AN INTRODUCTION TO WEATHER, CLIMATE, AND THE ENVIRONMENT

C. Donald Ahrens
Emeritus, Modesto Junior College

Robert Henson
The Weather Company

 CENGAGE

Australia • Brazil • Mexico • Singapore • United Kingdom • United States

CENGAGE

Meteorology Today: An Introduction to Weather, Climate, and the Environment, **Twelfth Edition**

C. Donald Ahrens, Robert Henson

Product Director: Dawn Giovanniello

Product Manager: Rebecca Berardy Schwartz

Senior Content Developer: Brendan R. Killion

Product Assistant: Leica Lucian

Marketing Manager: Eric V. Wagner

Senior Designer: Michael C. Cook

Manufacturing Planner: Rebecca Cross

Content Project Manager: Hal Humphrey

Content Digitization Project Manager: Maya Whelan

Project Manager: Matthew Fox, SPi Global

Intellectual Property Project Manager: Sarah Shainwald

Intellectual Property Analyst: Christine Myaskovsky

Photo Researcher: Anjali Kambli, Lumina Datamatics

Text Researcher: Geolinstella Jerald, Lumina Datamatics

Copy Editor: Alyson Platt

Illustrator: Charles Preppernau

Text Designer: Diane Beasley

Cover Designer: Michael C. Cook

Cover Image: Frontpoint/iStock/Getty Images

Compositor: SPi Global

For product information and technology assistance, contact us at **Cengage Customer & Sales Support, 1-800-354-9706.**

For permission to use material from this text or product, submit all requests online at **www.cengage.com/permissions**. Further permissions questions can be e-mailed to **permissionrequest@cengage.com.**

Student Edition:
ISBN-13: 978-1-337-61666-9

Loose-leaf Student Edition:
ISBN-13: 978-1-337-61667-6

Cengage
20 Channel Center Street
Boston, MA 02210
USA

Cengage is a leading provider of customized learning solutions with employees residing in nearly 40 different countries and sales in more than 125 countries around the world. Find your local representative at **www.cengage.com**.

Cengage products are represented in Canada by Nelson Education, Ltd.

To learn more about Cengage platforms and Solutions, visit **www.cengage.com**.

Purchase any of our products at your local college store or at our preferred online store **www.cengagebrain.com**.

Printed in United States of America
Print Number: 01   Print Year: 2017

Contents in Brief

Contents

© C. Donald Ahrens

© Robert Henson

© Robert Henson

© Robert Henson

CHAPTER 13

Weather Forecasting   347

CHAPTER 14

Thunderstorms   383

NASA

© Robert Henson

CHAPTER 18

Earth's Changing Climate   503

NASA

CHAPTER 19

Air Pollution   537

CHAPTER 20

Light, Color, and Atmospheric Optics   567

APPENDIX A

APPENDIX B

APPENDIX C

APPENDIX D

APPENDIX E

APPENDIX F

APPENDIX G

APPENDIX H

© C. Donald Ahrens

Preface

The world is an ever-changing picture of naturally occurring events. From drought and famine to devastating floods, some of the greatest challenges we face come in the form of natural disasters created by weather. Yet dealing with weather and climate is an inevitable part of our lives. Sometimes it is as small as deciding what to wear for the day or how to plan a vacation. But it can also have life-shattering consequences, especially for those who are victims of a hurricane or a tornado.

Weather has always been front-page news, but in recent years, extreme weather seems to receive an ever-increasing amount of coverage. From the destruction wrought by extreme storms to the quiet, but no less devastating, impacts of severe drought, weather has enormous impact on our lives. The longer-term challenges of an evolving climate also demand our attention, whether it be rising sea levels, record global temperatures, intensified downpours, or the retreat of Arctic sea ice. Thanks in part to the rise of social media, more people than ever are sharing their weather-related observations, impressions, and photographs with the world at large. For these and many other reasons, interest in meteorology (the study of the atmosphere) continues to grow. One of the reasons that meteorology is such an engaging science to study is that the atmosphere is a universally accessible laboratory for everyone. Although the atmosphere will always provide challenges for us, as research and technology advance, our ability to understand and predict our atmosphere improves as well. We hope this book serves to assist you as you develop your own personal understanding and appreciation of our planet's dynamic, spectacular atmosphere.

About This Book

Meteorology Today is written for college-level students taking an introductory course on the atmospheric environment. As was the case in previous editions, no special prerequisites are necessary. The main purpose of the text is to convey meteorological concepts in a visual and practical manner, while simultaneously providing students with a comprehensive background in basic meteorology. This twelfth edition includes up-to-date information on important topics, including climate change, ozone depletion, air quality, and El Niño. Also included are discussions of high-profile weather events, such as droughts, heat waves, tornado outbreaks, and hurricanes of recent years.

Written expressly for the student, this book emphasizes the understanding and application of meteorological principles. The text encourages watching the weather so that it becomes "alive," allowing readers to immediately apply textbook material to the world around them. To assist with this endeavor, a color Cloud Chart appears at the end of this text. The Cloud Chart can be separated from the book and used as a learning tool in any place one chooses to observe the sky. Numerous full-color illustrations and photographs illustrate key features of the atmosphere, stimulate interest, and show how exciting the study of weather can be.

After an introductory chapter on the composition, origin, and structure of the atmosphere, the book covers energy, temperature, moisture, precipitation, and winds. Next come chapters that deal with air masses and middle-latitude cyclones, followed by weather prediction and severe storms, including a separate chapter devoted to tornadoes. Wrapping up the book are chapters on hurricanes, global climate, climate change, air pollution, and atmospheric optics.

This book is structured to provide maximum flexibility to instructors of atmospheric science courses, with chapters generally designed so they can be covered in any desired order. For example, the chapter on atmospheric optics, Chapter 20, is self-contained and can be covered before or after any chapter. Instructors, then, are able to tailor this text to their particular needs.

Each chapter contains at least two Focus sections, which expand on material in the main text or explore a subject closely related to what is being discussed. Focus sections fall into one of five distinct categories: Observations, Special Topics, Environmental Issues, Advanced Topics, and Social and Economic Impacts. Some include material that is not always found in introductory meteorology textbooks, such as temperature extremes, cloud seeding, and the weather on other planets. Others help to bridge theory and practice. Focus sections new to this edition include "GOES-16: New Windows on the Atmosphere," (Chapter 5), "Rivers in the Atmosphere" (Chapter 11), "The Weird World of Tornado Damage" (Chapter 15), and "A Forecast Challenge: The Devastating Hurricanes of 2017" (Chapter 16). Quantitative discussions of important equations, such as the geostrophic wind equation and the hydrostatic equation, are found in Focus sections on advanced topics.

Set apart as "Weather Watch" features in each chapter is weather information that may not be commonly known, yet pertains to the topic under discussion. Designed to bring the reader into the text, most of these weather highlights relate to some interesting weather fact or astonishing event.

Each chapter incorporates other effective learning aids:

- A major topic outline begins each chapter.
- Interesting introductory pieces draw the reader naturally into the main text.
- Important terms are boldfaced, with their definitions appearing in the glossary or in the text.

- Key phrases are italicized.
- English equivalents of metric units in most cases are immediately provided in parentheses.
- A brief review of the main points is placed toward the middle of most chapters.
- Each chapter ends with a summary of the main ideas.
- A list of key terms with page references follows each chapter, allowing students to review and reinforce their knowledge of key concepts.
- Questions for Review act to check how well students assimilate the material.
- Questions for Thought require students to synthesize learned concepts for deeper understanding.
- Problems and Exercises require mathematical calculations that provide a technical challenge to the student.
- References to more than 20 Concept Animations are compiled on pp. xix. These animations convey an immediate appreciation of how a process works and help students visualize the more difficult concepts in meteorology. Animations can be found in the Earth Science MindTap for Meteorology Today.

Eight appendices conclude the book. In addition, at the end of the book, a compilation of supplementary reading material is presented, as is an extensive glossary.

On the endsheet at the back of the book is a geophysical map of North America. The map serves as a quick reference for locating states, provinces, and geographical features, such as mountain ranges and large bodies of water.

Supplemental Material and Technology Support

TECHNOLOGY FOR THE INSTRUCTOR

Instructor Companion Website   Everything you need for your course in one place! This collection of book-specific lecture and class tools is available online via www.cengage.com/login. Access and download PowerPoint presentations, images, instructor's manual, and more.

Cognero Test Bank   Cengage Learning Testing Powered by Cognero is a flexible, online system that allows you to:

- Author, edit, and manage test bank content from multiple Cengage Learning solutions
- Create multiple test versions in an instant
- Deliver tests from your LMS, your classroom, or wherever you want

Global Geoscience Watch   Updated several times a day, the Global Geoscience Watch is a focused portal into GREENR— our Global Reference on the Environment, Energy, and Natural Resources—an ideal one-stop site for classroom discussion and research projects for all things geoscience! Broken into the four key course areas (Geography, Geology, Meteorology, and Oceanography), you can easily get to the most relevant content available for your course. You and your students will have access to the latest information from trusted academic journals, news outlets, and magazines. You also will receive access to statistics, primary sources, case studies, podcasts, and much more!

TECHNOLOGY FOR THE STUDENT

Earth Science MindTap for Meteorology Today   MindTap is well beyond an eBook, a homework solution or digital supplement, a resource center website, a course delivery platform, or a Learning Management System. More than 70 percent of students surveyed said that it was unlike anything they have ever seen before. MindTap is a personal learning experience that combines all of your digital assets—readings, multimedia, activities, study tools, and assessments—into a singular learning path to improve student outcomes. The twelfth edition MindTap course contains: Case Study activities with summaries and questions written by co-author Bob Henson, new Concept Animations, and auto-graded homework problems and exercises adapted from the text or newly written by the authors.

Changes in the Twelfth Edition

The authors have carried out extensive updates and revisions to this twelfth edition of *Meteorology Today*, reflecting the ever-changing nature of the field and the atmosphere itself. Dozens of new or revised color illustrations and many new photos have been added to help visualize the excitement of the atmosphere.

- Chapter 1, "Earth and Its Atmosphere," continues to serve as a broad overview of the atmosphere. Material that puts meteorology in the context of the scientific method lays the foundation for the rest of the book. Among recent events now referenced in this chapter are the severe flooding over the Southern Plains and Southeast in 2015 and the Houston flash flood of April 2016.
- Chapter 2, "Warming and Cooling Earth and Its Atmosphere," contains up-to-date statistics and background on greenhouse gases and climate change, topics covered in more detail later in the book.
- Chapter 3, "Seasonal and Daily Temperatures," includes updated details on the recently revised world high temperature record. A number of figures and tables have been updated so that they include data from the most recent reference period (1981–2010).
- Chapter 4, "Atmospheric Humidity," continues to cover essential concepts related to this important aspect of the atmosphere. A section on relative humidity and human

discomfort now stresses the danger of heat buildup in a closed vehicle, independent of humidity.

- Chapter 5, "Condensation: Dew, Fog, and Clouds," includes a new Focus section spotlighting the GOES-16 satellite and the many new capacities of the GOES-R series.

- Chapter 6, "Stability and Cloud Development," discusses atmospheric stability and instability and the resulting effects on cloud formation in a carefully sequenced manner, with numerous illustrations and several Focus sections helping to make these complex concepts understandable. A new graphic highlights the differences between absolutely stable, absolutely unstable, and conditionally unstable conditions.

- "Precipitation" (Chapter 7) includes updated information on precipitation measurement from satellites, including the new Global Precipitation Measurement (GPM) mission.

- Chapter 8, "Air Pressure and Winds," includes a recently enhanced description and revised illustrations of the interplay between the pressure gradient and Coriolis forces in cyclonic and anticylonic flow.

- Chapter 9, "Wind: Small-Scale and Local Systems," references the destructive Midwest windstorm of March 2017 and includes updated information on the continued growth of wind energy .

- Chapter 10, "Wind: Global Systems," features several new images as well as updates on a number of phenomena, including the El Niño event of 2014–2016, the California drought of 2011–2016, and recent trends in the Pacific Decadal Oscillation.

- In Chapter 11, "Air Masses and Fronts," the discussion of occluded fronts has been revised to incorporate recent perspectives, and a new Focus box illuminates the concept of atmospheric rivers.

- Chapter 12, "Middle-Latitude Cyclones," continues to provide a thorough and accessible introduction to this important topic. The Focus section on nor'easters has been revised to center around the intense storm of January 2016.

- Chapter 13, "Weather Forecasting," has undergone a major restructuring and revision. After introducing the observations used by forecasters, the chapter includes an explanation of numerical weather models and other forecast techniques, the types of forecasts that apply to various time scales, and the difference between forecast accuracy and skill. A new illustration depicts the usefulness of short-term, high-resolution mesoscale models in predicting showers and thunderstorms.

- Chapter 14, "Thunderstorms," includes updated discussions of such topics as microbursts, heat bursts, and record hailstones, featuring several new photos. The use of lightning mapping arrays to map flashes in three dimensions has also been added.

- Chapter 15, "Tornadoes," includes a new Focus section on the surprising types of damage that tornadic winds can produce. Storm chasing is discussed in the context of the VORTEX and VORTEX2 field campaigns and the tragic deaths of several storm researchers in 2013.

- Chapter 16, "Hurricanes," includes extensive background on recent and historically significant tropical cyclones, as well as new and updated illustrations depicting storm surge processes and wind-speed probabilities. A new Focus section covers the devastating hurricanes Harvey, Irma, and Maria of 2017.

- Chapter 17, "Global Climate," continues to serve as a stand-alone unit on global climatology and classification schemes. Recent updates and revisions have incorporated the 1981–2010 United States climate averages.

- Chapter 18, "Earth's Changing Climate," has undergone extensive updating to reflect recent developments and findings, including the sequence of record-setting global temperatures in 2014, 2015, and 2016; regional variations in sea-level rise; and the Paris climate agreement.

- Chapter 19, "Air Pollution," reflects a number of updates, including the vast number of deaths associated with both indoor and outdoor air pollution and the importance of the smallest airborne particulates as a health hazard.

- The book concludes with Chapter 20, "Light, Color, and Atmospheric Optics," which uses exciting photos and art to convey the beauty of the atmosphere. Several compelling new photos have been included.

Acknowledgments

Many people have contributed to this twelfth edition of *Meteorology Today*. Special thanks goes to Charles Preppernau for his care in rendering beautiful artwork and to Alyson Platt for professional and conscientious copy editing. We are indebted to the team at SPi Global, including Matthew Fox and Catherine Higginbotham, who took the photos, art, and manuscript and turned them into a beautiful end product in both print and digital forms. Special thanks go to all the people at Cengage who worked on this edition, especially Brendan Killion, Lauren Oliveira, Hal Humphrey, and Rebecca Berardy Schwartz.

Thanks to our friends who provided photos and to those reviewers who offered comments and suggestions for this edition, including:

Eric Aldrich, University of Missouri

Peter Blanken, University of Colorado Boulder

Kerry Doyle, Southern Illinois University Edwardsville

Daehyun Kim, University of Kentucky

Ryan Fogt, Ohio University

John Harrington, Kansas State University

Keeley Heise, Oklahoma State University

Cody Kirkpatrick, Indiana University Bloomington

Zachary Lebo, University of Wyoming

Kennie Leet, SUNY Broome Community College

Mark McConnaughhay, Dutchess Community College

Lou McNally, Embry–Riddle Aeronautical University

Justin Maxwell, Indiana University Bloomington

Greta Nisbet, Broward College

Keah Schuenemann, Metropolitan State University of Denver

David Schultz, University of Manchester

Bruce Sherman, Southeastern Louisiana University

Tim Wallace, Mississippi State University

Lori Weeden, University of Massachusetts Lowell

To the Student

Learning about the atmosphere can be a fascinating and enjoyable experience. This book is intended to give you some insight into the workings of the atmosphere. However, for a real appreciation of your atmospheric environment, you must go outside and observe. Although mountains take millions of years to form, a cumulus cloud can develop into a raging thunderstorm in less than an hour. The atmosphere is always producing something new for us to behold. To help with your observations, a color Cloud Chart is at the back of the book for easy reference. Remove it and keep it with you. And remember, all of the concepts and ideas in this book are out there for you to discover and enjoy. Please, take the time to look.

Donald Ahrens and Robert Henson

Explore the Concept Animations

These animations have been carefully created to bring to life key points in the chapters. They are also the perfect tool to help refresh students' memories of previous concepts, so they can keep building on knowledge already acquired. Concept Animations are accessed through the MindTap platform, which can be acquired separately or together with print or loose-leaf versions of this book. Some examples of Concept Animations are shown here.

Doppler radar images are used extensively throughout this book. To better understand Doppler radar images, watch all four parts of this *Doppler Radar* animation (Chapters 1, 11, and 15).

Learn about how air rises above an area of low atmospheric pressure and sinks above an area of high atmospheric pressure. *Converging* and *Diverging Air* (Chapters 12).

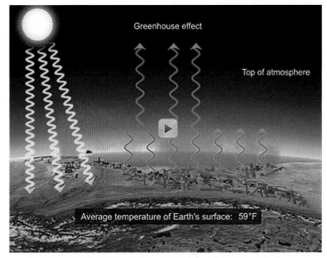

For a visual interpretation of the energy emitted by the earth without and with a *greenhouse* effect, watch the Greenhouse animation (Chapter 2).

Additional Animations

- Temperature versus Molecular Movement (Chapter 2)
- Radiant Energy and Wavelengths (Chapter 2)
- Daily Temperature Changes Above the Surface (Chapter 3)
- Air Temperature, Dew Point, and Relative Humidity (Chapter 4)
- Condensation (Chapter 4)
- Ice Crystals (Bergeron) Process (Chapter 5)
- Geostrophic Wind (Chapter 8)
- Thermally Driven Circulations (Chapter 9)
- General Circulation of the Atmosphere (Chapter 10)
- Introduction to the El Niño-Southern Oscillation (ENSO) (Chapter 10)
- Evolution of El Niño and La Niña (Chapter 10)
- Thunderstorm Evolution (Chapter 15)

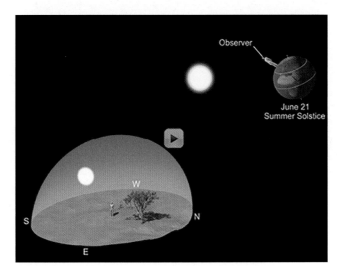

Seasons provides a complete picture of Earth revolving around the sun while it is tilted on its axis. While viewing this animation, look closely at how the sun is viewed by a mid-latitude observer at various times of the year (Chapter 3).

The concept of atmospheric stability can be a bit confusing, especially when comparing the temperature inside a rising air parcel to that of its surroundings. Watch the animations *Stable Atmosphere, Unstable Atmosphere,* and *Conditionally Unstable Atmosphere* (Chapters 6, 14, and 15).

For a visualization of the stages that a wave cyclone goes through from birth to decay, watch the animation entitled *Cyclogenesis* (Chapter 12).

To view air rising over a mountain and the formation of a rain shadow desert, watch *Air Rising Up and Over a Mountain* (Chapter 6).

For a visual presentation of the Coriolis force, watch *Coriolis Force* (Chapter 8).

For visualizations of a cold front and a warm front moving across the landscape, watch the animations *Cold Front in Winter* and *Warm Front in Winter* (Chapters 11 and 13).

For a depiction of the components of storm surge as a hurricane approaches land, watch *Storm Surge* (Chapter 16).

We live at the bottom of a turbulent ocean of air, where rising air can form into clouds composed of water and ice.

Thanongsak Kongtong/Shutterstock.com

Earth and Its Atmosphere

CONTENTS

I WELL REMEMBER A BRILLIANT RED BALLOON which kept me completely happy for a whole afternoon, until, while I was playing, a clumsy movement allowed it to escape. Spellbound, I gazed after it as it drifted silently away, gently swaying, growing smaller and smaller until it was only a red point in a blue sky. At that moment I realized, for the first time, the vastness above us: a huge space without visible limits. It was an apparent void, full of secrets, exerting an inexplicable power over all the Earth's inhabitants. I believe that many people, consciously or unconsciously, have been filled with awe by the immensity of the atmosphere. All our knowledge about the air, gathered over hundreds of years, has not diminished this feeling.

Theo Loebsack, *Our Atmosphere*

Our **atmosphere** is a delicate life-giving blanket of air that surrounds the fragile Earth. In one way or another, it influences everything we see and hear—it is intimately connected to our lives. Air is with us from birth, and we cannot detach ourselves from its presence. In the open air, we can travel for many thousands of kilometers in any horizontal direction, but should we move a mere 8 kilometers above the surface, we would suffocate. We may be able to survive without food for a few weeks, or without water for a few days, but, without our atmosphere, we would not survive more than a few minutes. Just as fish are confined to an environment of water, so we are confined to an ocean of air. Anywhere we go, air must go with us.

Earth without an atmosphere would have no lakes or oceans. There would be no sounds, no clouds, no red sunsets. The beautiful pageantry of the sky would be absent. It would be unimaginably cold at night and unbearably hot during the day. All things on Earth would be at the mercy of an intense sun beating down upon a planet utterly parched.

Living on the surface of Earth, we have adapted so completely to our environment of air that we sometimes forget how truly remarkable this substance is. Even though air is tasteless, odorless, and (most of the time) invisible, it protects us from the scorching rays of the sun and provides us with a mixture of gases that allows life to flourish. Because we cannot see, smell, or taste air, it may seem surprising that between your eyes and these words are trillions of air molecules. Some of these may have been in a cloud only yesterday, or over another continent last week, or perhaps part of the life-giving breath of a person who lived hundreds of years ago.

In this chapter, we will examine a number of important concepts and ideas about Earth's atmosphere, many of which will be expanded in subsequent chapters. These concepts and ideas are part of the foundation for understanding the atmosphere and how it produces weather. They are built on knowledge acquired and applied through the *scientific method*. This technique allows us to make informed predictions about how the natural world will behave.

The Atmosphere and the Scientific Method

For hundreds of years, the scientific method has served as the backbone for advances in medicine, biology, engineering, and many other fields. In the field of atmospheric science, the scientific method has paved the way for the production of weather forecasts that have steadily improved over time.

Investigators use the scientific method by posing a question, putting forth a hypothesis*, predicting what the hypothesis would imply if it were true, and carrying out tests to see if the prediction is accurate. Many common sayings about the weather, such as "red sky at morning, sailor take warning; red sky at night, sailor's delight," are rooted in careful observation, and there are grains of truth in some of them. However, they are not considered

---

*A hypothesis is an assertion that is subject to verification of proof.

● **FIGURE 1.1** Observing the natural world is a critical part of the scientific method. Here a vibrant red sky is visible at sunset. One might use the scientific method to verify the old proverb, "Red sky at morning, sailors take warning; red sky at night, sailor's delight."

to be products of the scientific method because they are not tested and verified in a standard, rigorous way. (See ● Fig. 1.1.)

To be accepted, a hypothesis has to be shown to be correct through a series of quantitative tests. In many areas of science, such testing is carried out in a laboratory, where it can be replicated again and again. Studying the atmosphere, however, is somewhat different, because Earth has only one atmosphere. Despite this limitation, scientists have made vast progress by studying the physics and chemistry of air in the laboratory (for instance, the way in which molecules absorb energy) and by extending those understandings to the atmosphere as a whole. Observations using weather instruments allow us to quantify how the atmosphere behaves and to determine whether a prediction is correct. If a particular kind of weather is being studied, such as hurricanes or snowstorms, a field campaign can gather additional observations to test specific hypotheses.

Over the last 60 years, computers have given atmospheric scientists a tremendous boost. The physical laws that control atmospheric behavior can be represented in software packages known as *numerical models*. Forecasts can be made and tested many times over. The atmosphere within a model can be used to depict weather conditions from the past and project them into the future. When a model can accurately simulate past weather conditions, we can have more confidence in its portrayal of tomorrow's weather. Numerical models can also provide valuable information about the types of weather and climate we may expect decades from now.

Overview of Earth's Atmosphere

The scientific method has not only illuminated our understanding of weather and climate but also provided much information about the universe that surrounds us. The universe contains billions of galaxies and each galaxy is made up of billions of stars.

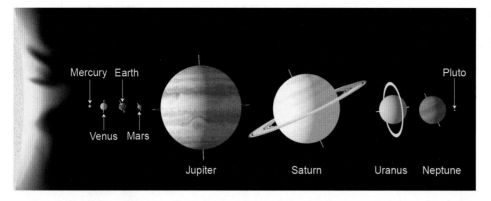

● FIGURE 1.2 The relative sizes and positions of the planets in our solar system. Pluto is included as an object called a *dwarf planet*. (Positions are not to scale.)

Stars are hot glowing balls of gas that generate energy by converting hydrogen into helium near their centers. Our sun is an average-sized star situated near the edge of the Milky Way galaxy. Revolving around the sun are Earth and seven other planets (see ● Fig. 1.2).* Our *solar system* comprises these planets, along with a host of other material (comets, asteroids, meteors, dwarf planets, etc.).

Warmth for the planets is provided primarily by the sun's energy. At an average distance from the sun of nearly 150 million kilometers (km) or 93 million miles (mi), Earth intercepts only a very small fraction of the sun's total energy output. However, it is this *radiant energy* (or *radiation*)** that drives the atmosphere into the patterns of everyday wind and weather and allows Earth to maintain an average surface temperature of about $15°C (59°F)$.† Although this temperature is mild, Earth experiences a wide range of temperatures, as readings can drop below $-85°C (-121°F)$ during a frigid Antarctic night and climb, during the day, to above $50°C (122°F)$ on the oppressively hot subtropical desert.

Earth's *atmosphere* is a relatively thin, gaseous envelope that comprises mostly nitrogen and oxygen, with small amounts of other gases, such as water vapor and *carbon dioxide* (CO_2). Nestled in the atmosphere are clouds of liquid water and ice crystals. Although our atmosphere extends upward for many hundreds of kilometers, it gets progressively thinner with altitude. Almost 99 percent of the atmosphere lies within a mere 30 km (19 mi) of Earth's surface (see ● Fig. 1.3). In fact, if Earth were to shrink to the size of a beach ball, its inhabitable atmosphere would be thinner than a piece of paper. This thin blanket of air constantly shields the surface and its inhabitants from the sun's dangerous ultraviolet radiant energy, as well as from the onslaught of material from interplanetary space. There is no definite upper limit to the atmosphere; rather, it becomes thinner and thinner, eventually merging with empty space, which surrounds all the planets.

---

*Pluto was once classified as a true planet. But recently it has been reclassified as a planetary object called a *dwarf planet*.

**Radiation is energy transferred in the form of waves that have electrical and magnetic properties. The light that we see is radiation, as is ultraviolet light. More on this important topic is given in Chapter 2.

†The abbreviation °C is used when measuring temperature in degrees Celsius, and °F is the abbreviation for degrees Fahrenheit. More information about temperature scales is given in Appendix A and in Chapter 2.

THE EARLY ATMOSPHERE The atmosphere that originally surrounded Earth was probably much different from the air we breathe today. Earth's first atmosphere (some 4.6 billion years ago) was most likely *hydrogen* and *helium*—the two most abundant gases found in the universe—as well as hydrogen compounds, such as methane (CH_4) and ammonia (NH_3). Most scientists believe that this early atmosphere escaped into space from Earth's hot surface.

A second, more dense atmosphere, however, gradually enveloped Earth as gases from molten rock within its hot interior escaped through volcanoes and steam vents. We assume that volcanoes spewed out the same gases then as they do today: mostly water vapor (about 80 percent), carbon dioxide (about 10 percent), and up to a few percent nitrogen. These gases (mostly water vapor and carbon dioxide) probably created Earth's second atmosphere.

As millions of years passed, the constant outpouring of gases from the hot interior—known as **outgassing**—provided a rich supply of water vapor, which formed into clouds. (It is also believed that when Earth was very young, some of its water may

● FIGURE 1.3 Earth's atmosphere as viewed from space. The atmosphere is the thin bluish-white region along the edge of Earth. The photo was taken from the International Space Station on April 12, 2011, over western South America.

PERMANENT GASES			VARIABLE GASES			
Gas	Symbol	Percent (by Volume) Dry Air	Gas (and Particles)	Symbol	Percent (by Volume)	Parts per Million (ppm)*
Nitrogen	N_2	78.08	Water vapor	H_2O	0 to 4	
Oxygen	O_2	20.95	Carbon dioxide	CO_2	0.041	410*
Argon	Ar	0.93	Methane	CH_4	0.00018	1.8
Neon	Ne	0.0018	Nitrous oxide	N_2O	0.00003	0.3
Helium	He	0.0005	Ozone	O_3	0.000004	0.04**
Hydrogen	H_2	0.00006	Particles (dust, soot, etc.)		0.000001	0.01–0.15
Xenon	Xe	0.000009	Chlorofluorocarbons (CFCs) and hydrofluorocarbons (HFCs)		0.00000001	0.0001

*For CO_2, 410 parts per million means that out of every million air molecules, 410 are CO_2 molecules.
**Stratospheric values at altitudes between 11 km and 50 km are about 5 to 12 ppm.

have originated from numerous collisions with small meteors that pounded Earth, as well as from disintegrating comets.) Rain fell upon Earth for many thousands of years, forming the rivers, lakes, and oceans of the world. During this time, large amounts of carbon dioxide (CO_2) were dissolved in the oceans. Through chemical and biological processes, much of the CO_2 became locked up in carbonate sedimentary rocks, such as limestone. With much of the water vapor already condensed and the concentration of CO_2 dwindling, the atmosphere gradually became dominated by molecular nitrogen (N_2), which is usually not chemically active.

It appears that molecular oxygen (O_2), the second most abundant gas in today's atmosphere, probably began an extremely slow increase in concentration as energetic rays from the sun split water vapor (H_2O) into hydrogen and oxygen during a process called *photodissociation*. The hydrogen, being lighter, probably rose and escaped into space, while the oxygen remained in the atmosphere.

It is uncertain whether this slow increase in oxygen supported the evolution of primitive plants, perhaps two to three billion years ago, or whether plants evolved in an almost oxygen-free (anaerobic) environment. At any rate, plant growth greatly enriched our atmosphere with oxygen. The reason for this enrichment is that, during the process of *photosynthesis*, plants, in the presence of sunlight, combine carbon dioxide and water to produce sugar and oxygen. Hence, after plants evolved, the atmospheric oxygen content increased more rapidly, probably reaching its present composition about several hundred million years ago.

COMPOSITION OF TODAY'S ATMOSPHERE ▼ Table 1.1 shows the various gases present in a volume of air near Earth's surface. Notice that molecular **nitrogen** (N_2) occupies about 78 percent and molecular **oxygen** (O_2) about 21 percent of the total volume of dry air. If all the other gases are removed, these percentages for nitrogen and oxygen hold fairly constant up to an elevation of about 80 km (50 mi). (For a closer look at the composition of a breath of air at Earth's surface, read Focus section 1.1.)

At the surface, there is a balance between destruction (output) and production (input) of these gases. For example, nitrogen is removed from the atmosphere primarily by biological processes that involve soil bacteria. Nitrogen is also taken from the air by tiny ocean-dwelling plankton that convert it into nutrients that help fortify the ocean's food chain. It is returned to the atmosphere mainly through the decaying of plant and animal matter. Oxygen, on the other hand, is removed from the atmosphere when organic matter decays and when oxygen combines with other substances, producing oxides. It is also taken from the atmosphere during breathing, as the lungs take in oxygen and release carbon dioxide (CO_2). The addition of oxygen to the atmosphere occurs during photosynthesis.

The concentration of the invisible gas **water vapor** (H_2O), however, varies greatly from place to place, and from time to time. Close to the surface in warm, steamy, tropical locations, water vapor may account for up to 4 percent of the atmospheric gases, whereas in colder arctic areas, its concentration may dwindle to a mere fraction of a percent (see Table 1.1). Water vapor molecules are, of course, invisible. They become visible only when they transform into larger liquid or solid particles, such as cloud droplets and ice crystals, which may grow in size and eventually fall to Earth as rain or snow. The changing of water vapor into liquid water is called *condensation*, whereas the process of liquid water becoming water vapor is called *evaporation*. The falling rain and snow is called *precipitation*. In the lower atmosphere, water is everywhere. It is the only substance that exists as a gas, a liquid, and a solid at those temperatures and pressures normally found near Earth's surface (see ● Fig. 1.4).

Water vapor is an extremely important gas in our atmosphere. Not only does it form into both liquid and solid cloud particles that grow in size and fall to Earth as *precipitation*, but it also releases large amounts of heat—called *latent heat*—when it changes from vapor into liquid water or ice. Latent heat is an important source of atmospheric energy, especially for storms, such as thunderstorms

A Breath of Fresh Air

If we could examine a breath of air, we would see that air (like everything else in the universe) is composed of incredibly tiny particles called *atoms*. We cannot see atoms individually with the naked eye. Yet, if we could see one, we would find electrons whirling at fantastic speeds about an extremely dense center, somewhat like hummingbirds darting and circling about a flower. At this center, or nucleus, are the protons and neutrons. Almost all of the atom's mass is concentrated here, in a trillionth of the atom's entire volume. In the nucleus, the proton carries a positive charge, whereas the neutron is electrically neutral. The circling electron carries a negative charge. As long as the total number of protons in the nucleus equals the number of orbiting electrons, the atom as a whole is electrically neutral (see ● Fig. 1).

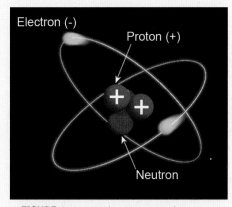

● **FIGURE 1** An atom has neutrons and protons at its center with electrons orbiting this center (or nucleus). Molecules are combinations of two or more atoms. The air we breathe is mainly molecular nitrogen (N_2) and molecular oxygen (O_2).

Most of the air particles are *molecules*, combinations of two or more atoms (such as nitrogen, N_2, and oxygen, O_2), and most of the molecules are electrically neutral. A few, however, are electrically charged, having lost or gained electrons. These charged atoms and molecules are called *ions*.

An average breath of fresh air contains a tremendous number of molecules. With every deep breath, trillions of molecules from the atmosphere enter your body. Some of these inhaled gases become a part of you, and others are exhaled.

The volume of an average size breath of air is about a liter.* Near sea level, there are roughly ten thousand million million million (10^{22})** air molecules in a liter. So,

$$1 \text{ breath of air} = 10^{22} \text{ molecules}$$

We can appreciate how large this number is when we compare it to the number of stars in the universe. Astronomers estimate that there are about 500 billion (10^{11}) stars in the Milky Way, which is considered to be an average sized galaxy, and that there may be more than 10^{11} galaxies in the universe. To determine the total number of stars in the universe, we multiply the average number of stars in a galaxy by the total number of galaxies and obtain

$$(5 \times 10^{11}) \times 10^{11} = 5 \times 10^{22} \text{ stars in the universe}$$

*One cubic centimeter is about the size of a sugar cube, and there are a thousand cubic centimeters in a liter.
**The notation, 10^{22} means the number one followed by twenty-two zeros. For a further explanation of this system of notation see Appendix A .

Therefore, just a few breaths of air contain about as many molecules as there are stars in the known universe.

In the entire atmosphere, there are nearly 10^{44} molecules. The number 10^{44} is 10^{22} squared; consequently,

$$10^{22} \times 10^{22} = 10^{44} \text{ molecules in the atmosphere}$$

We thus conclude that there are about 10^{22} breaths of air in the entire atmosphere. In other words, there are as many molecules in a single breath as there are breaths in the atmosphere.

Each time we breathe, the molecules we exhale enter the turbulent atmosphere. If we wait a long time, those molecules will eventually become thoroughly mixed with all of the other air molecules. If none of the molecules were consumed in other processes, eventually there would be a molecule from that single breath in every breath that is out there. So, considering the many breaths people exhale in their lifetimes, it is possible that in our lungs are molecules that were once in the lungs of people who lived hundreds or even thousands of years ago. In a very real way then, we all share the same atmosphere.

---

and hurricanes. Moreover, water vapor is a potent *greenhouse gas* because it strongly absorbs a portion of Earth's outgoing radiant energy (somewhat like the glass of a greenhouse prevents the heat inside from escaping and mixing with the outside air). This trapping of heat energy close to Earth's surface—called the *greenhouse effect*—keeps the average air temperature near the surface much warmer than it would be otherwise.* Thus, water vapor plays a significant role in Earth's heat-energy balance.

Carbon dioxide (CO_2), a natural component of the atmosphere, occupies a small (but important) percent of a volume of air, about 0.04 percent. Carbon dioxide enters the atmosphere mainly from the decay of vegetation, but it also comes

from volcanic eruptions, the exhalations of animal life, from the burning of fossil fuels (such as coal, oil, and natural gas), and from deforestation. The removal of CO_2 from the atmosphere takes place during photosynthesis, as plants consume CO_2 to produce green matter. The CO_2 is then stored in roots, branches, and leaves. Rain and snow can react with silicate minerals in rocks and remove CO_2 from the atmosphere through a process known as *chemical weathering*. The oceans act as a huge reservoir for CO_2, as phytoplankton (tiny drifting plants) in surface water fix CO_2 into organic tissues. Carbon dioxide that dissolves directly into surface water mixes downward and circulates through greater depths. Estimates are that the oceans hold more than 50 times the total atmospheric CO_2 content. ● Figure 1.5 illustrates important ways carbon dioxide enters and leaves the atmosphere.

*A more detailed look at the greenhouse effect is presented in Chapter 2.

● **FIGURE 1.4** Earth's atmosphere is a rich mixture of many gases, with clouds of condensed water vapor and ice crystals. Here, water evaporates from the ocean's surface. Rising air currents then transform the invisible water vapor into many billions of tiny liquid droplets that appear as puffy cumulus clouds. If the rising air in the cloud should extend to greater heights, where air temperatures are quite low, some of the liquid droplets would freeze into minute ice crystals.

● Figure 1.6 reveals that the atmospheric concentration of CO_2 has risen by around 30 percent since 1958, when regular measurements began at Mauna Loa Observatory in Hawaii. This increase means that CO_2 is entering the atmosphere at a greater

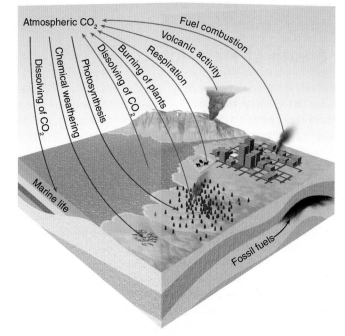

● **FIGURE 1.5** The main components of the atmospheric carbon dioxide cycle. The gray lines show processes that put carbon dioxide into the atmosphere, whereas the red lines show processes that remove carbon dioxide from the atmosphere.

rate than it is being removed. The increase is caused mainly by the burning of fossil fuels; however, deforestation also plays a role, as cut timber, burned or left to rot, releases CO_2 directly into the air. In addition, these dead trees no longer remove CO_2 from the atmosphere. Deforestation accounts for about 10 to 15 percent of the observed CO_2 increase in recent years. Measurements of CO_2 also come from ice cores. In Greenland and

● **FIGURE 1.6** (a) The solid blue line shows the average yearly measurements of CO_2 in parts per million (ppm) at Mauna Loa Observatory, Hawaii, from 1958 to mid-2017. The jagged dark line illustrates how higher readings occur in winter where plants die and release CO_2 to the atmosphere, and how lower readings occur in summer when more abundant vegetation absorbs CO_2 from the atmosphere. (b) The inset shows CO_2 values in ppm during the past 1000 years from ice cores in Antarctica (orange line) and from Mauna Loa Observatory (blue line). (Mauna Loa data courtesy of NOAA; Ice Core data courtesy of Carbon Dioxide Information Analysis Center, Oak Ridge National Laboratory)

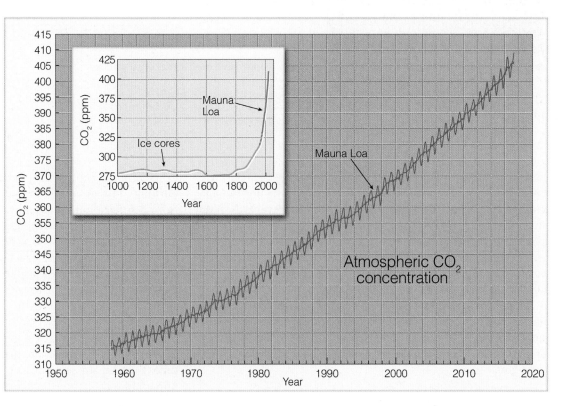

Antarctica, for example, tiny bubbles of air trapped within the ice sheets reveal that before the industrial revolution, CO_2 levels were stable at about 280 parts per million (ppm). (See the insert in Fig. 1.6.) Since the early 1800s, however, CO_2 concentrations have increased more than 40 percent. With CO_2 levels presently increasing by more than 0.5 percent annually (2.0 ppm/year), scientists now estimate that the concentration of CO_2 will likely increase from its current value of about 410 ppm to a value exceeding 550 ppm by the end of this century, assuming that fossil fuel emissions continue at or above current rates.

Like water vapor, carbon dioxide is an important greenhouse gas that traps a portion of Earth's outgoing energy. Consequently, with everything else being equal, as the atmospheric concentration of CO_2 increases, so should the average global surface air temperature. In fact, over the last 100 years or so, Earth's average surface temperature has warmed by about $1.0°C (1.8°F)$. Mathematical climate models, which predict future atmospheric conditions, estimate that if concentrations of CO_2 (and other greenhouse gases) continue to increase at or beyond their present rates, Earth's surface could warm by an additional $3°C (5.4°F)$ or more by the end of this century. As we will see in Chapter 18, the consequences of this type of *climate change* (such as rising sea levels and the rapid melting of polar ice) will be felt worldwide.

Carbon dioxide and water vapor are not the only greenhouse gases. Others include *methane* (CH_4), *nitrous oxide* (N_2O), and *chlorofluorocarbons* (CFCs). On average, methane concentrations have risen about one-half of one percent per year since the 1990s, but the pace has been uneven for reasons now being studied. Most methane appears to derive from the breakdown of plant material by certain bacteria in rice paddies, wet oxygen-poor soil, the biological activity of termites, and biochemical reactions in the stomachs of cows, although some methane is also leaked into the atmosphere by natural-gas operations. Levels of nitrous oxide—commonly known as laughing gas—have also been rising annually at the rate of about one-quarter of a percent. As well as being an industrial by-product, nitrous oxide forms in the soil through a chemical process involving bacteria and certain microbes. Ultraviolet light from the sun destroys nitrous oxide.

Chlorofluorocarbons (CFCs) represent a group of greenhouse gases that, up until the mid-1990s, had been increasing in concentration. At one time, they were the most widely used propellants in spray cans. More recently, they were used as refrigerants, as propellants for the blowing of plastic-foam insulation, and as solvents for cleaning electronic microcircuits. Although their average concentration in a volume of air is quite small (see Table 1.1, p. 6), CFCs have an important effect on our atmosphere. They not only act as greenhouse gases to trap heat but also play a part in destroying the gas ozone in the stratosphere, a region in the atmosphere located between about 11 km and 50 km above Earth's surface. CFCs have been almost completely phased out through a global agreement called the Montreal Protocol. Their main replacements, hydrofluorocarbons (HFCs), do not damage stratospheric ozone, but they are still powerful greenhouse gases.

On Earth's surface, **ozone** (O_3) is the primary ingredient of *photochemical smog*,* which irritates the eyes and throat and damages vegetation. But the majority of atmospheric ozone (about 97 percent) is found in the stratosphere, where it is formed naturally, as oxygen atoms combine with oxygen molecules. Here, the concentration of ozone averages less than 0.002 percent by volume. This small quantity is important, however, because it shields plants, animals, and humans from the sun's harmful ultraviolet rays. It is ironic that ozone, which damages plant life in a polluted environment, provides a natural protective shield in the upper atmosphere so that plants on the surface may survive.

When CFCs enter the stratosphere, ultraviolet rays break them apart, and the CFCs release ozone-destroying chlorine. Because of this effect, ozone concentration in the stratosphere decreased over parts of the Northern and Southern Hemispheres in the late twentieth century, especially over the southern polar region. ● Figure 1.7 illustrates the extent of ozone depletion above Antarctica during October 2015. Stratospheric ozone

---

*Originally the word *smog* meant the combining of smoke and fog. Today, however, the word usually refers to the type of smog that forms in large cities, such as Los Angeles, California. Because this type of smog forms when chemical reactions take place in the presence of sunlight, it is termed *photochemical smog*.

● **FIGURE 1.7** The darkest color represents the area of lowest ozone concentration, or ozone hole, over the Southern Hemisphere on October 2, 2015. Notice that the hole is larger than the continent of Antarctica. A Dobson unit (DU) is the physical thickness of the ozone layer if it were brought to Earth's surface, where 500 DU equals 5 millimeters.

● **FIGURE 1.8** Erupting volcanoes can send tons of particles into the atmosphere, along with vast amounts of water vapor, carbon dioxide, and sulfur dioxide.

concentrations plummet each year during September and October above Antarctica, to the point where so little ozone is observed that a seasonal ozone hole forms, as shown in Fig. 1.7. (We will examine stratospheric ozone and the Antarctic **ozone hole** in more detail in Chapter 19.)

Impurities from both natural and human sources are also present in the atmosphere: Wind picks up dust and soil from Earth's surface and carries it aloft; small saltwater drops from ocean waves are swept into the air (upon evaporating, these drops leave microscopic salt particles suspended in the atmosphere); smoke from forest fires is often carried high above Earth; and volcanoes spew many tons of fine ash particles and gases into the air (see ● Fig. 1.8). Collectively, these tiny solid or liquid particles of various composition, suspended in the air, are called **aerosols**.

Some natural impurities found in the atmosphere are quite beneficial. Small, floating particles, for instance, act as surfaces on which water vapor condenses to form clouds. However, most human-made impurities (and some natural ones) are a nuisance, as well as a health hazard. These we call **pollutants**. For example, many older automobile engines emit copious amounts of *nitrogen dioxide* (NO_2), *carbon monoxide* (CO), and *hydrocarbons*. In sunlight, nitrogen dioxide reacts with hydrocarbons and other gases to produce photochemical smog. Carbon monoxide is a major pollutant of city air. Colorless and odorless, this poisonous gas forms during the incomplete combustion of carbon-containing fuel. Hence, more than half of carbon monoxide in urban areas comes from road vehicles.

The burning of sulfur-containing fuels (such as coal and oil) releases sulfur gases into the air. When the atmosphere is sufficiently moist, these gases may transform into tiny dilute drops of sulfuric acid. Rain containing sulfuric acid corrodes metals and painted surfaces, and turns freshwater lakes acidic. *Acid rain* is a major environmental problem, especially downwind from major industrial areas. (More on the acid rain problem is given in Chapter 19.)

Even the tiniest pollutants are a major concern. *Particulate matter* refers to solid particles and liquid droplets that are small enough to remain suspended in the air. These particles can obscure visibility and cause respiratory and cardiovascular problems. (More information on these and other pollutants is given in Chapter 19.)

BRIEF REVIEW

Before going on to the next several sections, here is a review of some of the important concepts presented so far:

- Earth's atmosphere is a mixture of many gases. In a volume of dry air near the surface, nitrogen (N_2) occupies about 78 percent and oxygen (O_2) about 21 percent.

- Water vapor, which normally occupies less than 4 percent in a volume of air near the surface, can condense into liquid cloud droplets or transform into delicate ice crystals. Water is the only substance in our atmosphere that is found naturally as a gas (water vapor), as a liquid (water), and as a solid (ice).

- Both water vapor and carbon dioxide (CO_2) are important greenhouse gases.

- Ozone (O_3) in the stratosphere protects life from harmful ultraviolet (UV) radiation. At the surface, ozone is the main ingredient of photochemical smog.

- The majority of water on our planet is believed to have come from its hot interior through outgassing, although some of Earth's water may have come from collisions with meteors and comets.

Vertical Structure of the Atmosphere

When we examine the atmosphere in the vertical, we see that it can be divided into a series of layers. Each layer may be defined in a number of ways: by the manner in which the air temperature varies through it, by its gaseous composition, or even by its electrical properties. At any rate, before we examine these various atmospheric layers, we need to look at the vertical profile of two important atmospheric variables: air pressure and air density.

A BRIEF LOOK AT AIR PRESSURE AND AIR DENSITY Air molecules (as well as everything else) are held near Earth by *gravity*. This strong, invisible force pulling down on the air squeezes (compresses) air molecules closer together, which causes their number in a given volume to increase. The more air above a level, the greater the squeezing effect or compression.

Gravity also has an effect on the weight of objects, including air. In fact, *weight* is the force acting on an object due to gravity. Weight is defined as the mass of an object times the acceleration of gravity; thus,

$$\text{weight} = \text{mass} \times \text{gravity}$$

An object's *mass* is the quantity of matter in the object. Consequently, the mass of air in a rigid container is the same everywhere in the universe. However, if you were to instantly travel to the moon, where the acceleration of gravity is much less than that of Earth, the mass of air in the container would be the same, but its weight would decrease.

When mass is given in grams (g) or kilograms (kg), volume is given in cubic centimeters (cm^3) or cubic meters (m^3). Near sea level, air density is about 1.2 kilograms per cubic meter (nearly 1.2 ounces per cubic foot).

The **density** of air (or any substance) is determined by the masses of atoms and molecules and the amount of space between them. In other words, density tells us how much matter is in a given space (that is, volume). We can express density in a variety of ways. The molecular density of air is the number of molecules in a given volume. Most commonly, however, density is given as the mass of air in a given volume; thus

$$\text{density} = \frac{\text{mass}}{\text{volume}}$$

Because there are appreciably more molecules within the same size volume of air near Earth's surface than at higher levels, air density is greatest at the surface and decreases as we move up into the atmosphere. Notice in ● Fig. 1.9 that, because air near the surface is compressed, air density normally decreases rapidly at first, then more slowly as we move farther away from the surface.

Air molecules are in constant motion. On a mild spring day near Earth's surface, an air molecule will collide about 10 billion times each second with other air molecules. It will also bump against objects around it—houses, trees, flowers, the ground, and even people. Each time an air molecule bounces against a person,

● **FIGURE 1.9** Both air pressure and air density decrease with increasing altitude. The weight of all the air molecules above Earth's surface produces an average pressure near 14.7 lb/in.²

it gives a tiny push. This small force (push) divided by the area on which it pushes is called **pressure**; thus,

$$\text{pressure} = \frac{\text{force}}{\text{area}}$$

If we weigh a column of air 1 square inch wide, extending from the average height of the ocean surface (sea level) to the "top" of the atmosphere, it would weigh nearly 14.7 pounds (see Fig. 1.9). Thus, normal atmospheric pressure near sea level is close to 14.7 pounds per square inch $\left(14.7\ \text{lb/in.}^2\right)$. If more molecules are packed into the column, it becomes more dense, the air weighs more, and the surface pressure goes up. On the other hand, when fewer molecules are in the column, the air weighs less, and the surface pressure goes down. Thus, the surface air pressure can be changed by changing the mass of air above the surface.

Pounds per square inch is, of course, just one way to express air pressure. Presently, the most common unit found on surface weather maps is the *millibar* (mb)* although the metric equivalent, the *hectopascal* (hPa), is gradually replacing the millibar as the preferred unit of pressure on surface charts. A more traditional unit of pressure is *inches of mercury* (Hg), which is commonly used in the field of aviation and in weather reports on

---

*By definition, a *bar* is a force of 100,000 newtons (N) acting on a surface area of 1 square meter (m^2). A *newton* is the amount of force required to move an object with a mass of 1 kilogram (kg) so that it increases its speed at a rate of 1 meter per second (m/sec) each second. Because the bar is a relatively large unit, and because surface pressure changes are usually small, the unit of pressure most commonly found on surface weather maps is the *millibar*, where 1 bar = 1000 mb. The unit of pressure designated by the International System SI (Système International) of measurement is the *pascal* (Pa), where 1 pascal is the force of 1 newton acting on a surface of 1 square meter. A more common unit is the *hectopascal* (hPa), as 1 hectopascal equals 1 millibar.

television, radio, smartphones, and the Internet. At sea level, the *standard value* for atmospheric pressure is

$$1013.25 \text{ mb} = 1013.25 \text{ hPa} = 29.92 \text{ in. Hg}$$

Billions of air molecules push constantly on the human body. This force is exerted equally in all directions. We are not crushed by it because billions of molecules inside the body push outward just as hard. Even though we do not actually feel the constant bombardment of air, we can detect quick changes in it. For example, if we climb rapidly in elevation, our ears may "pop." This experience happens because air collisions outside the eardrum lessen. The popping comes about as air collisions between the inside and outside of the ear equalize. The drop in the number of collisions informs us that the pressure exerted by the air molecules decreases with height above Earth. A similar type of ear-popping occurs as we drop in elevation, and the air collisions outside the eardrum increase.

Air molecules not only take up space (freely darting, twisting, spinning, and colliding with everything around them), but—as we have seen—these same molecules have weight. In fact, air is surprisingly heavy. The weight of all the air around Earth is a staggering 5600 trillion tons, or roughly 5.08×10^{18} kg. The weight of the air molecules acts as a force upon Earth. The amount of force exerted over an area of surface is called *atmospheric pressure* or, simply, **air pressure**.* The pressure at any level in the atmosphere may be measured in terms of the total mass of air above any point. As we climb in elevation, fewer air molecules are above us; hence, *atmospheric pressure always decreases with increasing height.* Like air density, air pressure decreases rapidly at first, then more slowly at higher levels, as illustrated in Fig. 1.9.

● Figure 1.10 also illustrates how rapidly air pressure decreases with height. Near sea level, atmospheric pressure is usually close to 1000 mb. Normally, just above sea level, atmospheric pressure decreases by about 10 mb for every 100 meters (m) increase in altitude—about 1 inch of mercury (Hg) for every 1000 feet (ft) of rise. At higher levels, air pressure decreases much more slowly with height. With a sea-level pressure near 1000 mb, we can see in Fig. 1.10 that, at an altitude of only 5.5 km (3.5 mi), the air pressure is about 500 mb, or half of the sea-level pressure. This situation means that, if you were at a mere 5.5 km (about 18,000 ft) above Earth's surface, you would be above one-half of all the molecules in the atmosphere.

At an elevation approaching the summit of Mt. Everest (about 9 km, or 29,000 ft—the highest mountain peak on Earth), the air pressure would be about 300 mb. The summit is above nearly 70 percent of all the air molecules in the atmosphere. At an altitude approaching 50 km (160,000 feet), the air pressure is about 1 mb, which means that 99.9 percent of all the air molecules are below this level. Yet the atmosphere extends upwards for many hundreds of kilometers, gradually becoming thinner and thinner until it ultimately merges with outer space. (Up to now, we have concentrated on Earth's atmosphere. For a brief look at the atmospheres of the other planets, read Focus section 1.2.)

● **FIGURE 1.10** Atmospheric pressure decreases rapidly with height. Climbing to an altitude of only 5.5 km, where the pressure is 500 mb, would put you above one-half of the atmosphere's molecules.

LAYERS OF THE ATMOSPHERE Up to this point, we've looked at how both air pressure and density decrease with height above Earth—rapidly at first, then more slowly. *Air temperature*, however, has a more complicated vertical profile.**

Look closely at ● Fig. 1.11 and notice that air temperature normally decreases from Earth's surface up to an altitude of about 11 km, which is nearly 36,000 ft, or 7 mi. This decrease in air temperature with increasing height is due primarily to the fact (investigated further in Chapter 2) that sunlight warms Earth's surface, and the surface, in turn, warms the air above it. The rate at which the air temperature decreases with height is called the temperature **lapse rate**. The *average* (or *standard*) *lapse rate* in this region of the lower atmosphere is about 6.5°C for every 1000 m or about 3.6°F for every 1000-ft increase in altitude. (See ● Fig. 1.12.) Keep in mind that these values are only averages. On some days, the air becomes colder more quickly as we move upward. This would increase or steepen the lapse rate. On other days, the air temperature would decrease more slowly with height, and the lapse rate would be less. Occasionally, the air temperature may actually *increase* with height, producing a condition known as a **temperature inversion**. Thus, the lapse rate fluctuates, varying from day to day, season to season, and place to place.

Fig. 1.11 shows the region of the atmosphere from the surface up to about 11 km, which contains all of the weather we are familiar with on Earth. Also, this region is kept well stirred by rising and descending air currents, and it is common for air molecules to circulate through a depth of more than 10 km in just a few days. This region of circulating air extending upward from Earth's surface to where the air stops becoming colder with height is called the **troposphere**—from the Greek *tropein*, meaning "to turn" or "to change."

---

*Because air pressure is measured with an instrument called a *barometer*, atmospheric pressure is often referred to as *barometric pressure*.

**Air temperature* is the degree of hotness or coldness of the air and, as we will see in Chapter 2, it is also a measure of the average speed of the air molecules.

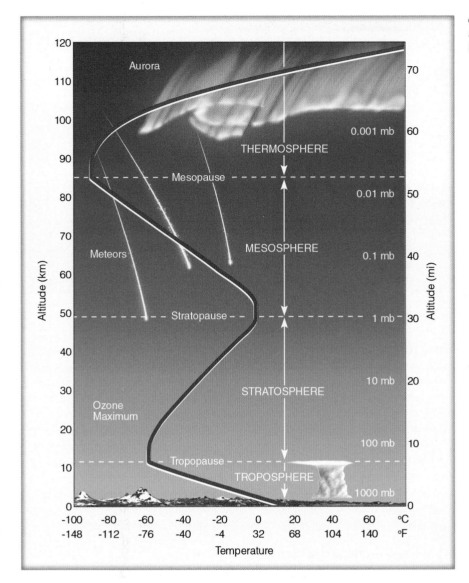

Notice in Fig. 1.11 that just above 11 km the air temperature normally stops decreasing with height. Here, the lapse rate is zero. This region, where, on average, the air temperature remains constant with height, is referred to as an *isothermal* (equal temperature) *zone*.* The bottom of this zone marks the top of the troposphere and the beginning of another layer, the **stratosphere**. The boundary separating the troposphere from the stratosphere is called the **tropopause**. The height of the tropopause varies. It is normally found at higher elevations over equatorial regions, and it decreases in elevation as we travel poleward. Generally, the tropopause is higher in summer and lower in winter at all latitudes. In some regions, the tropopause "breaks" and is difficult to locate and, here, scientists have observed tropospheric air mixing with stratospheric air and vice versa. These breaks mark the position of *jet streams*—high winds that meander in a narrow channel, like an old river, often at speeds exceeding 100 knots. (For reference, a *knot* is a nautical mile per hour, where one knot equals 1.15 miles per hour [mi/hr] or 1.85 kilometers per hour [km/hr].)

---

*In many instances, the isothermal layer is not present, and the air temperature begins to increase with increasing height.

● **FIGURE 1.12** Near Earth's surface the air temperature lapse rate is often close to 3.5°F per 1000 ft. If this temperature lapse rate is present and the air temperature at the surface (0 ft) is 46°F, the air temperature about 4000 ft above the surface would be at freezing, and snow and ice might be on the ground.

The Atmospheres of Other Planets

Earth is unique in our solar system. Not only does it lie at just the right distance from the sun so that life may flourish, it also provides its inhabitants with an atmosphere rich in both molecular nitrogen and oxygen—two gases that are not abundant in the atmospheres of either Venus or Mars, our closest planetary neighbors.

The Venusian atmosphere is mainly carbon dioxide (96.5% percent) with minor amounts of water vapor and nitrogen. An opaque acid-cloud deck encircles the planet, hiding its surface. The atmosphere is quite turbulent, as instruments reveal twisting eddies and fierce winds in excess of 320 km/hr (200 mi/hr). This thick dense atmosphere produces a surface air pressure of about 93,000 mb, which is more than 90 times greater than that on Earth. To experience such a pressure on Earth, one would have to descend in the ocean to a depth of about 900 m (2950 ft). Moreover, this thick atmosphere of CO_2 produces a strong greenhouse effect, with a scorching-hot surface temperature of around 460°C (860°F).

The atmosphere of Mars, like that of Venus, is mostly carbon dioxide, with only small amounts of other gases, dust, and water vapor. Unlike Venus, the Martian atmosphere is very thin, and heat escapes from the surface rapidly. Thus, surface temperatures on Mars are much lower, averaging around −55°C (−67°F). NASA missions have found that liquid water sometimes flows on parts of the Martian surface, but because of the thin, cold atmosphere, there is only occasional cloud cover, and the landscape is largely barren and desertlike (see ● Fig. 2). In addition, the thin atmosphere of Mars produces an average surface air pressure of only about 6 mb, which is less than one-hundredth of that experienced at the surface of Earth. Such a pressure on Earth would be observed above the surface at an altitude near 35 km (22 mi).

Occasionally, huge dust storms develop near the Martian surface. Such storms may be accompanied by winds of several hundreds of kilometers per hour. These winds carry fine dust around the entire planet. The dust gradually settles out, coating the landscape with a thin reddish veneer.

The atmosphere of the largest planet, Jupiter, is much different from that of Venus and Mars. Jupiter's atmosphere is mainly hydrogen (H_2) and helium (He), with minor amounts

● **FIGURE 2** The Martian sky and landscape, photographed by the *Spirit* robotic rover during April 2005.

of methane (CH_4) and ammonia (NH_3). A prominent feature on Jupiter is the Great Red Spot—a huge atmospheric storm about three times larger than Earth and at least 350 years old—that spins counterclockwise in Jupiter's southern hemisphere (see ● Fig. 3). Large white ovals near the Great Red Spot are similar but smaller storm systems. Not only does Jupiter have lightning, but each flash is typically several times more intense than the ones we see on Earth. Unlike Earth's weather machine, which is driven by the sun, Jupiter's massive swirling clouds

● **FIGURE 3** An image of Jupiter and the vivid cloud bands and swirls in its atmosphere, collected by the Hubble Space Telescope in April 2014.

appear to be driven by a collapsing core of hot hydrogen. Energy from this lower region rises toward the surface; then it (along with Jupiter's rapid rotation) stirs the cloud layer into more or less horizontal bands of various colors.

Swirling storms exist on other planets, too, such as on Saturn and Neptune. Studying the atmospheric behavior of other planets may give us added insight into the workings of our own atmosphere. (Additional information about size, surface temperature, and atmospheric composition of planets is given in ▼ Table 1.)

▼ **TABLE 1** **Data on Planets and the Sun**

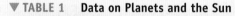

	DIAMETER	AVERAGE DISTANCE FROM SUN	AVERAGE SURFACE TEMPERATURE*		MAIN ATMOSPHERIC COMPONENTS
	Kilometers	Millions of Kilometers	°C	°F	
Sun	1,392 × 103		5,800	10,500	—
Mercury	4,880	58	260**	500	—
Venus	12,112	108	460	860	CO_2
Earth	12,742	150	15	59	N_2, O_2
Mars	6,800	228	−55	−67	CO_2
Jupiter	143,000	778	−145	−234	H_2, He
Saturn	121,000	1,427	−180	−290	H_2, He
Uranus	51,800	2,869	−225	−375	H_2, CH_4
Neptune	49,000	4,498	−210	−346	N_2, CH_4

*For the giant planets made up mostly of gas (Jupiter, Saturn, Uranus, and Neptune), the average temperature at cloud-top level is used.
**Sunlit side.

The Radiosonde

The vertical distribution of temperature, pressure, and humidity up to an altitude of about 30 km (about 19 mi) can be obtained with an instrument called a *radiosonde*.* The radiosonde is a small, lightweight box equipped with weather instruments and a radio transmitter. It is attached to a cord that has a parachute and a gas-filled balloon tied tightly at the end (see ● Fig. 4). As the balloon rises, the attached radiosonde measures air temperature with a small electrical thermometer—a thermistor—located just outside the box. The radiosonde measures humidity electrically by sending an electric current across a carbon-coated plate. Air pressure is obtained by a small barometer located inside the box. All of this information is transmitted to the surface by radio. Here, a computer rapidly reconverts the various frequencies into values of temperature, pressure, and moisture. Special tracking equipment at the surface may also be used to provide a vertical profile of winds. Radiosondes are often

equipped with Global Positioning System (GPS) receivers that lead to highly accurate wind computations. (When winds are added, the observation is called a *rawinsonde*.) When plotted on a graph, the vertical distribution of temperature, humidity, and wind is called a *sounding*. Eventually, the balloon bursts and the radiosonde returns to Earth, its descent being slowed by its parachute.

At most sites, radiosondes are released twice a day, usually at the time that corresponds to midnight and noon in Greenwich, England. Releasing radiosondes is an expensive operation because many of the instruments are never retrieved, and many of those that are retrieved are often in poor working condition. To complement the radiosonde, modern satellites (using instruments that measure radiant energy, the distortion of GPS signals due to atmospheric effects, and other variables) are providing scientists with vertical temperature profiles in inaccessible regions.

*A radiosonde that is dropped by parachute from an aircraft is called a *dropsonde*.

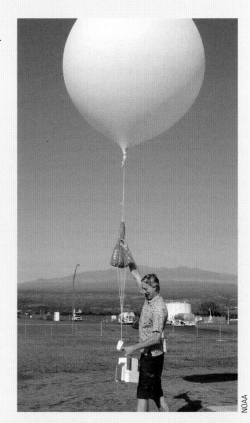

● **FIGURE 4** The radiosonde with parachute and balloon.

From Fig. 1.11 notice that in the stratosphere, the air temperature begins to increase with height, producing a *temperature inversion*. The inversion region, along with the lower isothermal layer, tends to keep the vertical currents of the troposphere from spreading into the stratosphere. The inversion also tends to reduce the amount of vertical motion in the stratosphere itself; hence, it is a stratified layer.

Even though the air temperature is increasing with height, the air at an altitude of 30 km (about 100,000 feet or 19 mi) is extremely cold, averaging less than −46°C (−51°F). At this level above polar latitudes, air temperatures can change dramatically from one week to the next, as a *sudden stratospheric warming* can raise the temperature in one week by more than 50°C. Such a rapid warming, although not well understood, is probably due to sinking air associated with circulation changes that occur in late winter or early spring as well as with the poleward displacement of strong jet stream winds in the lower stratosphere. (The instrument that measures the vertical profile of air temperature in the atmosphere up to an elevation sometimes exceeding 30 km is the **radiosonde**. More information on this instrument is given in Focus section 1.3.)

The reason for the inversion in the stratosphere is that the gas ozone plays a major part in heating the air at this altitude.

Recall that ozone is important because it absorbs energetic ultraviolet (UV) solar energy. Some of this absorbed energy warms the stratosphere from below, which explains why there is an inversion. If ozone were not present, the air probably would become colder with height, as it does in the troposphere.

Notice also in Fig. 1.11 that the level of maximum ozone concentration is observed near 25 km (at middle latitudes), yet the stratospheric air temperature reaches a maximum near 50 km. The reason for this phenomenon is that the air at 50 km is less dense than at 25 km, and so the absorption of intense solar energy at 50 km raises the temperature of fewer molecules to a much greater degree. Moreover, much of the solar energy responsible for the heating is absorbed in the upper part of the stratosphere and, therefore, does not reach down to the level of ozone maximum. And due to the low air density, the transfer of energy downward from the upper stratosphere is quite slow.

Above the stratosphere is the **mesosphere** (middle sphere). The boundary near 50 km, which separates these layers, is called the *stratopause*. The air at this level is extremely thin and the atmospheric pressure is quite low, averaging about 1 mb, which means that only one-thousandth of all the atmosphere's molecules are above this level and 99.9 percent of the atmosphere's mass is located below it.

The percentage of nitrogen and oxygen in the mesosphere is about the same as at sea level. However, given the air's low density in this region, we would not survive breathing this air for very long. Here, each breath of air would contain far fewer oxygen molecules than it would near Earth's surface. Consequently, without proper breathing equipment, the brain would soon become oxygen-starved—a condition known as *hypoxia*. Pilots who fly above 3 km (10,000 ft) for too long without oxygen-breathing apparatus may experience this. With the first symptoms of hypoxia, there is usually no pain involved, just a feeling of exhaustion. Soon, visual impairment sets in and routine tasks become difficult to perform. Some people drift into an incoherent state, neither realizing nor caring what is happening to them. Of course, if this oxygen deficiency persists, a person will lapse into unconsciousness, and death may result. In fact, in the mesosphere, we would suffocate in a matter of minutes.

Other dire effects could be experienced in the mesosphere. Exposure to ultraviolet solar energy, for example, could cause severe burns on exposed parts of the body. Also, given the low air pressure, the blood in one's veins would begin to boil at normal body temperature.

The air temperature in the mesosphere decreases with height, a phenomenon due, in part, to the fact that there is little ozone in the air to absorb solar radiation. Consequently, the molecules (especially those near the top of the mesosphere) are able to lose more energy than they absorb, which results in an energy deficit and cooling. So we find air in the mesosphere becoming colder with height up to an elevation near 85 km (53 mi). At this altitude, the temperature of the atmosphere reaches its lowest average value, $-90°C(-130°F)$.

The "hot layer" above the mesosphere is the **thermosphere**. The boundary that separates the lower, colder mesosphere from the warmer thermosphere is the *mesopause*. In the thermosphere, oxygen molecules (O_2) absorb energetic solar rays, warming the air. Because there are relatively few atoms and molecules in the thermosphere, the absorption of a small amount of energetic solar energy can cause a large increase in air temperature. Furthermore, because the amount of solar energy affecting this region depends strongly on solar activity, temperatures in the thermosphere vary from day to day (see ● Fig. 1.13). The low density of the thermosphere also means that an air molecule will move an average distance (called *mean free path*) of over one kilometer before colliding with another molecule. A similar air molecule at Earth's surface will move an average distance of less than one millionth of a centimeter before it collides with another molecule. Moreover, it is in the thermosphere where charged particles from the sun interact with air molecules to produce dazzling aurora displays. (We will look at the aurora in more detail in Chapter 2.)

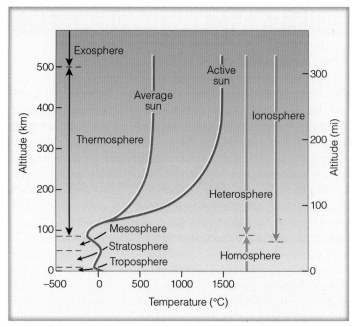

● **FIGURE 1.13** Layers of the atmosphere based on temperature (red line), composition (green line), and electrical properties (dark blue line). (An active sun is associated with large numbers of solar eruptions.)

Because the air density in the upper thermosphere is so low, air temperatures there are not measured directly. They can, however, be determined by observing the orbital change of satellites caused by the drag of the atmosphere. Even though the air is extremely tenuous, enough air molecules strike a satellite to slow it down, making it drop into a slightly lower orbit. (A number of spacecraft have fallen to Earth for this reason, including *Solar Max* in December 1989 and the Russian space station *Mir* in March 2001.) The amount of drag is related to the density of the air, and the density is related to the temperature. Therefore, by determining air density, scientists are able to construct a vertical profile of air temperature as illustrated in Fig. 1.13.

At the top of the thermosphere, about 500 km (300 mi) above Earth's surface, many of the lighter, faster-moving molecules traveling in the right direction actually escape Earth's gravitational pull. The region where atoms and molecules shoot off into space is sometimes referred to as the **exosphere**, which represents the upper limit of our atmosphere.

Up to this point, we have examined the atmospheric layers based on the vertical profile of temperature. The atmosphere, however, can also be divided into layers based on its composition. For example, the composition of the atmosphere begins to slowly change in the lower part of the thermosphere. Below the thermosphere, the composition of air remains fairly uniform (78 percent nitrogen, 21 percent oxygen) by turbulent mixing. This lower, well-mixed region is known as the **homosphere** (see Fig. 1.13). In the thermosphere, collisions between atoms and molecules are infrequent, and the air is unable to keep itself stirred. As a result, diffusion takes over as heavier atoms and molecules (such as oxygen and nitrogen) tend to settle to the bottom of the layer, while lighter gases (such as hydrogen and helium) float to the top. The region from about the base of the thermosphere to the top of the atmosphere is often called the **heterosphere**.

THE IONOSPHERE   The **ionosphere** is not really a layer, but rather an electrified region within the upper atmosphere where fairly large concentrations of ions and free electrons exist. *Ions* are atoms and molecules that have lost (or gained) one or more electrons. Atoms lose electrons and become positively charged when they cannot absorb all of the energy transferred to them by a colliding energetic particle or the sun's energy.

The lower region of the ionosphere is usually about 60 km (37 mi) above Earth's surface. From here (60 km), the ionosphere extends upward to the top of the atmosphere. Hence, as we can see in Fig. 1.13, the bulk of the ionosphere is in the thermosphere. Although the ionosphere allows TV and FM radio waves to pass on through, at night it reflects standard AM radio waves back to Earth. This situation allows AM radio waves to bounce repeatedly off the lower ionosphere and travel great distances.

BRIEF REVIEW

We have, in the last several sections, been examining our atmosphere from a vertical perspective. Following are a few of the main points:

- Atmospheric pressure at any level represents the total mass of air above that level, and atmospheric pressure always decreases with increasing height above the surface.

- The rate at which the air temperature decreases with height is called the lapse rate. A measured increase in air temperature with height is called an inversion.

- The atmosphere may be divided into layers (or regions) according to its vertical profile of temperature, its gaseous composition, or its electrical properties.

- The warmest atmospheric layer is the thermosphere; the coldest is the mesosphere. Most of the gas ozone is found in the stratosphere.

- We live at the bottom of the troposphere, which is an atmospheric layer where the air temperature normally decreases with height. The troposphere is a region that contains all of the weather with which we are familiar.

- The ionosphere is an electrified region of the upper atmosphere that normally extends from about 60 km to the top of the atmosphere.

Having looked at the composition of the atmosphere and its vertical structure, we will now turn our attention to weather events that take place in the lower atmosphere. As you read the remainder of this chapter, keep in mind that the content serves as a broad overview of material to come in later chapters, and that many of the concepts and ideas you encounter are designed to familiarize you with items you might see on TV or other electronic media, or read about on a website or in a newspaper or magazine.

Weather and Climate

When we talk about the **weather**, we are talking about the condition of the atmosphere at any particular time and place. Weather—which is always changing—includes the elements of:

1. *air temperature*—the degree of hotness or coldness of the air
2. *air pressure*—the force of the air above an area
3. *humidity*—a measure of the amount of water vapor in the air
4. *clouds*—visible masses of tiny water droplets and/or ice crystals that are above Earth's surface
5. *precipitation*—any form of water, either liquid or solid (rain or snow), that falls from clouds and reaches the ground
6. *visibility*—the greatest distance one can see
7. *wind*—the horizontal movement of air

If we measure and observe these *weather elements* over a specified interval of time, say, for many years, we would obtain the "average weather" or the **climate** of a particular region. Climate, therefore, represents the accumulation of daily and seasonal weather events (the average range of weather) over a long period of time. The concept of climate is much more than this, for it also includes the extremes of weather—the heat waves of summer and the cold spells of winter—that occur in a particular region. The *frequency* of these extremes is what helps us distinguish among climates that have similar averages.

If we were able to watch Earth for many thousands of years, even the climate would change. We would see rivers of ice moving down stream-cut valleys and huge glaciers—sheets of moving snow and ice—spreading their icy fingers over large portions of North America. Advancing slowly from Canada, a single glacier might extend as far south as Kansas and Illinois, with ice several thousands of meters thick covering the region now occupied by Chicago. Over an interval of two million years or so, we would see the ice advance and retreat many times. Of course, for this phenomenon to happen, the average temperature of North America would have to decrease and then rise in a cyclic manner.

Suppose we could photograph Earth once every thousand years for many hundreds of millions of years. In time-lapse film sequence, these photos would show that not only is the climate altering, but the whole Earth itself is changing as well: Mountains would rise up only to be torn down by erosion; isolated puffs of smoke and steam would appear as volcanoes spew hot gases and fine dust into the atmosphere; and the entire surface of Earth would undergo a gradual transformation as some ocean basins widen and others shrink.*

In summary, Earth and its atmosphere are dynamic systems that are constantly changing. While major transformations of Earth's surface are completed only after long spans of time, the state of the atmosphere can change in a matter of minutes. Hence, a watchful eye turned skyward will be able to observe many of these changes.

---

*The movement of the ocean floor and continents is explained in the theory of *plate tectonics*.

Up to this point, we have looked at the concepts of weather and climate without discussing the word *meteorology*. What does this term actually mean, and where did it originate?

METEOROLOGY—A BRIEF HISTORY **Meteorology** is the study of the atmosphere and its phenomena. The term itself goes back to the Greek philosopher Aristotle who, about 340 B.C., wrote a book on natural philosophy titled *Meteorologica*. This work represented the sum of knowledge on weather and climate at that time, as well as material on astronomy, geography, and chemistry. Some of the topics covered included clouds, rain, snow, wind, hail, thunder, and hurricanes. In those days, all substances that fell from the sky, and anything seen in the air, were called meteors, hence the term *meteorology*, which actually comes from the Greek word *meteoros*, meaning "high in the air." Today, we differentiate between those meteors that come from extraterrestrial sources outside our atmosphere (meteoroids) and particles of water and ice observed in the atmosphere (hydrometeors).

In *Meteorologica*, Aristotle attempted to explain atmospheric phenomena in a philosophical and speculative manner. Even though many of his ideas were found to be erroneous, Aristotle's work remained a dominant influence in the field of meteorology for almost two thousand years. In fact, the birth of meteorology as a genuine natural science did not take place until the invention of weather instruments, such as the hygrometer in the mid-1400s, the thermometer in the late 1500s, and the barometer (for measuring air pressure) in the mid-1600s. With the newly available observations from instruments, attempts were then made to explain certain weather phenomena employing scientific experimentation and the physical laws that were being developed at the time.

As more and better instruments were developed in the 1800s, the science of meteorology progressed. The invention of the telegraph in 1843 allowed for the transmission of routine weather observations. The understanding of the concepts of wind flow and storm movement became clearer, and in 1869 crude weather maps with *isobars* (lines of equal pressure) were drawn. Around 1920, the concepts of air masses and weather fronts were formulated in Norway. By the 1940s, daily upper-air balloon observations of temperature, humidity, and pressure gave a three-dimensional view of the atmosphere, and high-flying military aircraft discovered the existence of jet streams.

Meteorology took another step forward in the 1950s, when scientists converted the mathematical equations that describe the behavior of the atmosphere into software called *numerical models* that could be run on new high-speed computers. These calculations were the beginning of *numerical weather prediction*. Today, computers plot the observations, draw the lines on the map, and forecast the state of the atmosphere for some desired time in the future. Meteorologists evaluate the results from various numerical models and use them to issue public forecasts.

After World War II, surplus military radars became available, and many were transformed into precipitation-measuring tools. In the mid-1990s, the National Weather Service replaced these original radars with the more sophisticated *Doppler radars*, which have the ability to peer into a severe thunderstorm and unveil its winds, as well as to show precipitation intensity (see ● Fig. 1.14).

Recent upgrades to these Doppler radars make it possible for them to distinguish raindrops, snowflakes, and hailstones.

In 1960, the first weather satellite, *Tiros I*, was launched, ushering in space-age meteorology. Subsequent satellites provided a wide range of useful information, ranging from day and night time-lapse images of clouds and storms to images that depict swirling ribbons of water vapor flowing around the globe, as shown in ● Fig. 1.15. Over the last several decades, even more

● **FIGURE 1.14** Doppler radar image showing precipitation over portions of Indiana. The areas shaded light green indicate lighter rain, whereas yellow indicates heavier rain. The dark red shaded areas represent the heaviest rain and the possibility of hail and intense thunderstorms.

CRITICAL THINKING QUESTION Weather tends to move from west to east. Knowing this fact, explain why people living in Centerville would probably not be concerned about the thunderstorm just to the right of town, but very concerned about the large intense thunderstorm (large red and yellow region) just east of Indianapolis. If this intense thunderstorm is moving toward Centerville at 40 mi/hr, about how long will it take the region of hail (light purple shade just east of Indianapolis) to reach Centerville, assuming the storm is able to hold together?

● **FIGURE 1.15** This satellite image shows the dynamic nature of the atmosphere as ribbons of water vapor (gray regions) swirl counterclockwise about huge storms over the North Pacific Ocean

sophisticated satellites have been developed. These satellites are supplying forecasters and computers with a far greater network of data so that more accurate forecasts—perhaps extending up to two weeks or more—will be available in the future.

With this brief history of meteorology we are now ready to observe weather events that occur at Earth's surface.

A SATELLITE'S VIEW OF THE WEATHER A good view of the weather can be seen from a weather satellite. ● Fig. 1.16 is a satellite image showing a portion of the Pacific Ocean and the North American continent. The image was obtained from a *geostationary satellite* situated about 36,000 km (22,300 mi) above Earth. At this elevation, the satellite travels at the same rate as Earth spins, which allows it to remain positioned above the same spot so it can continuously monitor what is taking place beneath it.

The thin solid black lines running from north to south on the satellite image are called *meridians*, or lines of longitude. Because the zero meridian (or prime meridian) runs through Greenwich, England, the *longitude* of any place on Earth is simply how far east or west, in degrees, it is from the prime meridian. North America is west of Great Britain and most of the United States lies between 75°W and 125°W longitude.

The solid black lines that parallel the equator are called *parallels of latitude*. The latitude of any place is how far north or south, in degrees, it is from the equator. The latitude of the equator is 0°, whereas the latitude of the North Pole is 90°N and that of the South Pole is 90°S. Most of the United States is located between latitude 30°N and 50°N, a region commonly referred to as the **middle latitudes**.

Storms of All Sizes Probably the most prominent feature in Fig. 1.16 is the white cloud masses of all shapes and sizes. The clouds appear white because sunlight is reflected back to space from their tops. The largest of the organized cloud masses are the sprawling storms. One such storm appears as an extensive band of clouds, over 2000 km long, west of the Great Lakes. Superimposed on the satellite image is the storm's center (indicated by the large red L) and its adjoining weather fronts in red, blue, and purple. This **middle-latitude cyclonic storm system** (or *extratropical cyclone*) forms outside the tropics and, in the Northern Hemisphere, has winds spinning counterclockwise about its center, which is presently over Minnesota.

A slightly smaller but more vigorous storm is located over the Pacific Ocean near latitude 12°N and longitude 116°W. This tropical storm system, with its swirling band of rotating clouds and sustained surface winds of 65 knots* (74 mi/hr) or more, is known as a **hurricane**. The diameter of the hurricane, as measured by the presence of winds of at least 34 knots (39 mph), is about 800 km (500 mi). The tiny dot at its center is called the *eye*. Near the surface, in the eye, winds are light, skies are generally clear, and the atmospheric pressure is lowest. Around the eye, however, is an extensive region where heavy rain and high surface winds are reaching peak gusts of 100 knots.

Smaller storms are seen as white spots over the Gulf of Mexico. These spots represent clusters of towering *cumulus* clouds that have grown into **thunderstorms**, that is, tall churning clouds accompanied by lightning, thunder, strong gusty winds, and heavy rain. If you look closely at Fig. 1.16, you will see similar cloud forms in many regions. There were probably more than a thousand thunderstorms occurring throughout the world at that very moment. Although they cannot be seen individually, there are even some thunderstorms embedded in the cloud mass west of the Great Lakes. Later in the day on which this image was taken, a few of these storms spawned the most violent disturbance in the atmosphere: **tornadoes**.

A tornado is an intense rotating column of air that extends downward from the base of a thunderstorm with a circulation reaching the ground. Sometimes called *twisters*, or *cyclones*, they may appear as ropes or as a large circular cylinder. They can be more than 2 km (1.2 mi) in diameter, although most are less than a football field wide. The majority of tornadoes have sustained winds below 100 knots, but some can pack winds exceeding 200 knots. Sometimes a visibly rotating funnel cloud dips part of the way down from a thunderstorm, then rises without ever forming a tornado.

A Glimpse at a Weather Map We can obtain a better picture of the middle-latitude storm system by examining a simplified surface weather map for the same day that the satellite image was taken. The weight of the air above different regions varies and, hence, so does the atmospheric pressure. In ● Fig. 1.17, the red letter L on the map indicates a region of low atmospheric pressure, often called a *low*, which marks the center of the middle-latitude storm. (Compare the center of the storm in Fig. 1.17 with that in Fig. 1.16.) The two blue letters H on the map represent regions of high atmospheric pressure, called *highs*, or *anticyclones*. The circles on the map represent either individual weather stations or cities where observations are taken. The **wind** is the horizontal movement of air. The **wind direction**—the direction *from which* the wind is blowing*—is given by *wind barbs*, lines that parallel the wind and extend outward from the center of the station. The **wind speed**—the rate at which the air is moving past a stationary observer—is indicated by *flags*, the short lines that extend off each wind barb.

Notice how the wind blows around the highs and the lows. The horizontal pressure differences create a force that starts the air moving from higher pressure toward lower pressure. Because of Earth's rotation, the winds are deflected from their path toward the right in the Northern Hemisphere.** This deflection causes the winds to blow *clockwise* and *outward* from the center of the highs, and *counterclockwise* and *inward* toward the center of the low.

As the surface air spins into the low, it flows together and is forced upward, like toothpaste squeezed out of an upward-pointing tube. The rising air cools, and the water vapor in the air

---

*If you are facing north and the wind is blowing in your face, the wind would be called a "north wind."

**This deflecting force, known as the *Coriolis force*, is discussed more completely in Chapter 8, as are the winds.

---

*Recall from p. 15 that 1 knot equals 1.15 miles per hour.

Middle-latitude storm

Thunderstorms

Hurricane

30°

25°

20°

15°

10°

120°   110°   100°   90°

NOAA/National Weather Service

● **FIGURE 1.16** This satellite image (taken in visible reflected light) shows a variety of cloud patterns and storms in Earth's atmosphere.

condenses into clouds. Notice on the weather map that the area of precipitation (the shaded green area) in the vicinity of the low corresponds to an extensive cloudy region in the satellite image (Fig. 1.16).

Also notice by comparing Figs. 1.16 and 1.17 that, in the regions of high pressure, skies are generally clear. As the surface air flows outward away from the center of a high, air sinking from above must replace the laterally spreading surface air. Since sinking air does not usually produce clouds, we find generally clear skies and fair weather associated with the regions of high atmospheric pressure.

Areas of high and low pressure, and the swirling air around them, are the major weather producers for the middle latitudes. Look at the middle-latitude storm and the surface temperatures in Fig. 1.17 and notice that, to the southeast of the storm, southerly winds from the Gulf of Mexico are bringing warm, humid air northward over much of the southeastern portion of the nation.

On the storm's western side, cool, dry northerly breezes combine with sinking air to create generally clear weather over the Rocky Mountains. The boundary that separates the warm and cool air appears as a heavy, colored line on the map—a **front**, across which there is a sharp change in temperature, humidity, and wind direction.

Where the cool air from Canada replaces the warmer air from the Gulf of Mexico, a *cold front* is drawn in blue, with arrowheads showing the front's general direction of movement. Where the warm Gulf air is replacing cooler air to the north, a *warm front* is drawn in red, with half circles showing its general direction of movement. Where the cold front has caught up to the warm front and cold air is now replacing cool air, an *occluded front* is drawn in purple, with alternating arrowheads and half circles to show how it is moving. Along each of the fronts, warm air is rising, producing clouds and precipitation. Notice in the satellite image (Fig. 1.16) that the occluded front and the cold

● FIGURE 1.17
Simplified surface weather map that correlates with the satellite image shown in Fig. 1.16. The shaded green area represents precipitation. The numbers on the map represent air temperatures in °F. Because this map shows conditions several hours after those in Figure 1.16, the frontal system across the Midwest is farther east.

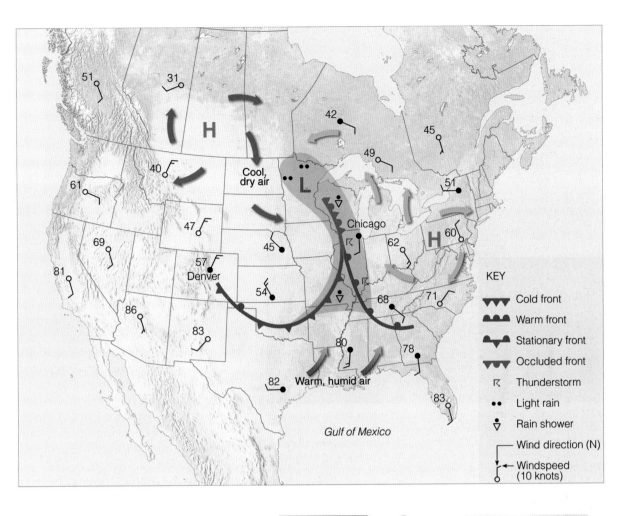

front appear as an elongated, curling cloud band that stretches from the low-pressure area over Minnesota into the northern part of Texas.

Notice in Fig. 1.17 that the frontal system is to the west of Chicago. As the westerly winds aloft push the front eastward, a person on the outskirts of Chicago might observe the approaching front as a line of towering thunderstorms similar to those in ● Fig. 1.18. On a Doppler radar image, the advancing thunderstorms might appear similar to those shown in ● Fig. 1.19. In a few hours, Chicago should experience heavy showers with thunder, lightning, and gusty winds as the front passes. All of this, however, should give way to clearing skies and surface winds from the west or northwest after the front has moved on.

Observing storm systems, we see that not only do they move but they constantly change. Steered by the upper-level westerly winds, the middle-latitude storm in Fig. 1.17 gradually weakens and moves eastward, carrying its clouds and weather with it. In advance of this system, a sunny day in Ohio will gradually cloud over and yield heavy showers and thunderstorms by nightfall. Behind the storm, cool, dry northerly winds rushing into eastern Colorado cause an overcast sky to give way to clearing conditions. Farther south, the thunderstorms presently over the Gulf of Mexico in the satellite image (Fig. 1.16) expand a little, then dissipate as new storms appear over water and land areas. To the west, the hurricane over the Pacific Ocean drifts

● FIGURE 1.18 Thunderstorms developing and advancing along an approaching cold front.

● **FIGURE 1.19** An elongated frontal system can show up on Doppler radar as a line of different colors. In this composite image, the areas shaded green and blue indicate where light-to-moderate rain is falling. Yellow indicates heavier rainfall. The red-shaded area represents the heaviest rainfall and the possibility of intense thunderstorms. Notice that a thunderstorm is approaching Chicago from the west.

NOAA/National Weather Service

northwestward and encounters cooler water. Here, away from its warm energy source, it loses its punch; winds taper off, and the storm soon turns into an unorganized mass of clouds and tropical moisture.

WEATHER AND CLIMATE IN OUR LIVES Weather and climate play a major role in our lives. Weather, for example, often dictates the type of clothing we wear, while climate influences the type of clothing we buy. Climate determines when to plant crops as well as what type of crops can be planted. Weather determines if these same crops will grow to maturity. Although weather and climate affect our lives in many ways, perhaps their most immediate effect is on our comfort. In order to survive the cold of winter and heat of summer, we build homes, heat them, air condition

WEATHER WATCH

On Saturday, April 24, 2010, a violent tornado packing winds of 150 knots roared through the town of Yazoo City, Mississippi. The tornado caused millions of dollars in damage, killed 10 people, and amazingly stayed on the ground for 149 miles—from Tallulah, Louisiana, to Oktibbeha County, Mississippi—making this one of the longest tornado paths on record.

them, insulate them—only to find that when we leave our shelter, we are at the mercy of the weather elements.

Even when we are dressed for the weather properly, wind, humidity, and precipitation can change our perception of how cold or warm it feels. On a cold, windy day the effects of *wind chill* tell us that it feels much colder than it really is, and, if not properly dressed, we run the risk of *frostbite* or even *hypothermia* (the rapid, progressive mental and physical collapse that accompanies the lowering of human body temperature). On a hot, humid day we normally feel uncomfortably warm and blame it on the humidity. If we become too warm, our bodies overheat and *heat exhaustion* or *heat stroke* may result. Those most likely to suffer these maladies are the elderly with impaired circulatory systems and infants, whose heat-regulating mechanisms are not yet fully developed.

Weather affects how we feel in other ways, too, not all of them well understood. People with arthritis often report more pain when atmospheric moisture is increasing rapidly, or when atmospheric pressure is changing abruptly. Heart attacks become more likely when the temperature decreases, perhaps because our blood thickens and blood vessels constrict. The risk of heart attack and stroke also rises during periods of stagnant, polluted air, as we inhale tiny particles that enter our bloodstream. Headaches are more common on days when we are forced to squint, often due to hazy skies or a thin, bright overcast layer of high clouds. Some people who live near mountainous regions become irritable or depressed during a warm, dry wind blowing downslope (a *chinook wind*).

When the weather turns much colder or warmer than normal, it has direct impacts on the lives and pocketbooks of many people. For example, the exceptionally warm weather observed from January to March 2012 over the United States saved people millions of dollars in heating costs. On the other side of the coin, the colder-than-normal winters of 2013–2014 and 2014–2015 over much of the northeastern United States sent utility costs soaring as demand for heating fuel escalated.

Major cold spells accompanied by heavy snow and ice can play havoc by snarling commuter traffic, curtailing airport services, closing schools, and downing power lines, thereby cutting off electricity to thousands of customers (see ● Fig. 1.20). For example, a huge ice storm during January 1998 in northern New England and Canada left millions of people without power and caused over a billion dollars in damages, and a devastating snowstorm during February 2011 produced blizzard conditions from Oklahoma to Michigan and snow drifts of up to 15 feet in the Chicago area. When frigid air settles into the Deep South, many millions of dollars' worth of temperature-sensitive fruits and vegetables may be ruined, the eventual consequence being higher produce prices for consumers.

Prolonged drought, especially when accompanied by high temperatures, can lead to a shortage of food and, in some places, widespread starvation. Parts of Africa, for example, have periodically suffered through major droughts and famine. During the summer of 2012, much of the United States experienced a severe drought with searing summer temperatures and wilting

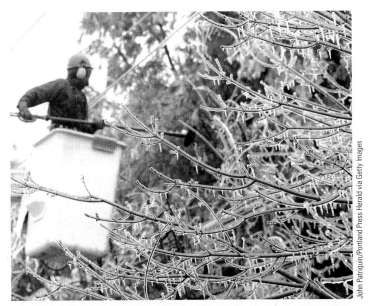

● **FIGURE 1.20** Utility workers in Maine clear off broken tree branches from power lines during a major ice storm on December 12, 2008.

crops, causing billions of dollars in crop losses. California experienced an especially destructive drought from 2011 to 2016. When the climate turns hot and dry, animals suffer too. In 1986, over 500,000 chickens perished in Georgia during a two-day period at the peak of a summer heat wave. Severe drought also has an effect on water reserves, often forcing communities to ration water and restrict its use. During periods of extended drought, vegetation often becomes tinder-dry, and, sparked by lightning or a careless human, such a dried-up region can quickly become a raging inferno. During the winter of 2005–2006, hundreds of thousands of acres in drought-stricken Oklahoma and northern Texas were ravaged by wildfire. And during

the spring and summer of 2011, wildfires destroyed millions of acres of trees and grassland over parched areas of Arizona, New Mexico, and West Texas.

Every summer, scorching *heat waves* take many lives. From 1987 to 2016, an annual average of more than 130 deaths in the United States were attributed to excessive heat exposure. In one particularly devastating heat wave that hit Chicago, Illinois, during July 1995, high temperatures coupled with high humidity claimed the lives of more than 700 people. In California, more than 100 people died during a two-week period in July 2006 as air temperatures climbed to over 46°C (115°F). Heat waves have been especially devastating in recent years across Europe, where many cities and buildings are not designed for intense heat. In the summer of 2003, tens of thousands died across Europe, including 14,000 in France alone. A record-breaking heat wave in Russia in 2010 killed nearly 11,000 people in Moscow.

Every year, the violent side of weather influences the lives of millions. Those who live along the U.S. Gulf and Atlantic coastlines keep a close watch for hurricanes during the late summer and early autumn. These large tropical systems can be among the nation's most destructive weather events. More than 250,000 people lost their homes when Hurricane Andrew struck the Miami area in 1992, and nearly 2000 people along the Central Gulf Coast were killed by Hurricane Katrina in 2005. A late-season hurricane called Sandy took a rare path in October 2012, striking the mid-Atlantic coast from the southeast. Because of its vast size and unusual path, Sandy produced catastrophic storm-surge flooding and caused more than 100 deaths across parts of New Jersey, New York, and New England (see ● Fig 1.21). In 2017, three intense hurricanes—Harvey, Irma, and Maria—caused massive destruction and took dozens of lives in Texas, Florida, Puerto Rico, and other Caribbean islands.

● **FIGURE 1.21** A resident of Long Beach, New York, digs sand out from around his car after Hurricane Sandy pushed water and sand far inland in October 2012, destroying homes, commercial businesses, and approximately 10,000 cars in this area alone.

It is amazing how many people whose family roots are in the Midwest or South know the story of someone who was severely injured or killed by a tornado. Tornadoes not only take dozens of lives each year in the United States, but annually they cause damage to buildings and property totaling in the hundreds of millions of dollars, as a single large tornado can level an entire section of a town (see ● Figs. 1.22 and 1.23).

Although the gentle rains of a typical summer thunderstorm are welcome over much of North America, the heavy downpours, high winds, and large hail of the *severe thunderstorms* are not. Cloudbursts from intense, slow-moving thunderstorms can provide too much rain too quickly, creating *flash floods* as small streams become raging rivers composed of mud and sand entangled with uprooted plants and trees. Thunderstorms dumped up to 20 inches of rain in just a few hours over parts of the Houston area in April 2016, leading to severe flash flooding (see ● Fig. 1.24). If heavy rain covers a large area, devastating *river floods* can result. Record rainfall produced both flash floods and river floods over the Southern Plains in May 2015 and across South Carolina in October 2015. On average, more people die in the United States from flooding than from either lightning strikes or tornadoes. Strong downdrafts originating inside an intense thunderstorm (a *downburst*) create turbulent winds that are capable of destroying crops and inflicting damage upon surface structures. Until a safety system implemented in the 1990s virtually eliminated such deaths, hundreds of people were killed in airline crashes attributed to turbulent *wind shear* (a rapid change in wind speed and/or wind direction) from downbursts.

● **FIGURE 1.22** Lightning flashes inside a violent tornado that tore through Joplin, Missouri, on May 22, 2011. The tornado ripped through a hospital and destroyed entire neighborhoods. (See tornado damage in Fig. 1.23.)

AP Images/tornadovideo.net

● **FIGURE 1.23** Emergency personnel walk through a neighborhood in Joplin, Missouri, damaged by a violent tornado, with winds exceeding 174 knots (200 mi/hr), on May 22, 2011. The tornado caused hundreds of millions of dollars in damage and took 159 lives, making this single tornado the deadliest in the United States since 1947.

AP Images/Mark Schiefelbein

● **FIGURE 1.24** Residents evacuate an apartment complex in the Houston area on April 18, 2016, as torrential rains from severe thunderstorms produced severe flash flooding. More than 1000 high-water rescues took place.

AP Photo/David J. Phillip

In a typical year, hail causes more than $1 billion in damage to crops and property in the United States, and lightning takes the lives of several dozen people and starts fires that destroy many thousands of acres of valuable timber (see ● Fig. 1.25).

Up to this point, we have considered the more violent side of weather and its impact on humanity. Weather- and climate-related events can have enormous economic consequences. On average, tens of billions of dollars in property damage occur each year in the United States alone. However, even the quiet side of weather has its influence. When winds die down and humid air becomes more tranquil, fog may form. Dense fog can restrict visibility at airports, causing flight delays and cancellations. Every winter, deadly fog-related auto accidents occur along our busy highways and turnpikes. But fog has a positive side, too, especially during a dry spell, as fog moisture collects on tree branches and drips to the ground, where it provides water for the tree's root system.

Weather and climate have become so much a part of our lives that the first thing many of us do in the morning is to consult our local weather forecast. For this reason, most television stations and many larger radio stations have their own "weather person" to present weather information and give daily forecasts. More and more of these people are professionally trained in meteorology, and many stations require that the weathercaster be certified by the American Meteorological Society (AMS) or hold a seal of approval from the National Weather Association (NWA). To make their weather presentation as up-to-the-minute as possible, weathercasters draw upon time-lapse satellite images, color Doppler radar displays, and other ways of illustrating current weather. Many stations work with private firms that create graphics and customized forecasts, largely based on observations and computer models from the National Weather Service (NWS). Since 1982, a staff of trained professionals at The Weather Channel have provided weather information twenty-four hours a day on cable television. (Many viewers believe the weatherperson they see on TV is a meteorologist and that all meteorologists forecast the weather. If you are interested in learning what a meteorologist or atmospheric scientist is and

A. T. Willett/Alamy Stock Photo

● **FIGURE 1.25** Estimates are that lightning strikes Earth about 40 to 50 times every second. More than 20 million lightning strikes hit the United States in a typical year. Here, lightning strikes the ground and buildings over Phoenix, Arizona.

what he or she might do for a living other than forecast the weather, read Focus section 1.4.)

The National Oceanic and Atmospheric Administration (NOAA), the parent agency of the National Weather Service, sponsors weather radio broadcasts at selected locations across the United States. Known as *NOAA Weather Radio* (and transmitted

What Is a Meteorologist?

Most people associate the term "meteorologist" with the weatherperson they see on television or hear on the radio. Many television and radio weathercasters are in fact professional meteorologists, but some are not. A professional meteorologist is usually considered to be a person who has completed the requirements for a college degree in meteorology or atmospheric science. This individual has strong, fundamental knowledge concerning how the atmosphere behaves, along with a substantial background of coursework in mathematics, physics, and chemistry.

A meteorologist uses scientific principles to explain and to forecast atmospheric phenomena. About half of the approximately 9000 meteorologists and atmospheric scientists in the United States work doing weather forecasting for the National Weather Service, the military, private firms, or for a television or radio station. The other half work mainly in research, teach atmospheric science courses in colleges and universities, or do meteorological consulting work.

Scientists who do atmospheric research may be investigating how the climate is changing, how snowflakes form, or how pollution impacts temperature patterns. Aided by supercomputers, much of the work of a research meteorologist involves simulating the atmosphere to see how it behaves (see ● Fig. 5). Researchers often work closely with such scientists as chemists, physicists, oceanographers, mathematicians, and environmental experts to determine how the atmosphere interacts with the entire ecosystem. Scientists doing work in physical meteorology may well study how radiant energy warms the atmosphere; those at work in the field of dynamic meteorology might be using the mathematical equations that describe airflow to learn more about jet streams. Scientists working in operational

© UCAR/Image courtesy Mel Shapiro

● **FIGURE 5** A three-dimensional model of Hurricane Sandy, which struck the New Jersey coast in October 2012 (see Figure 1.21). The model shows air at the surface flowing counterclockwise around the storm (arrows) and rising through it in a spiraling fashion. The height of the storm is greatly exaggerated to depict the airflow in more detail.

meteorology might be preparing a weather forecast by analyzing upper-air information over North America. A climatologist, or climate scientist, might be studying the interaction of the atmosphere and ocean to see what influence such interchange might have on planet Earth many years from now.

Meteorologists also provide a variety of services not only to the general public in the form of weather forecasts but also to city

planners, contractors, farmers, and large corporations. Meteorologists working for private weather firms create many forecasts and graphics that are found in newspapers, on television, and on the Internet. Overall, there are many exciting jobs that fall under the heading of "meteorologist"—too many to mention here. However, for more information on this topic, visit http://www.ametsoc.org/ and click on "Students."

at VHF–FM frequencies), this service provides continuous weather information and regional forecasts (as well as special weather advisories, including watches and warnings) for over 90 percent of the United States.

Although millions of people rely on weather broadcasts on radio and TV, many millions also use forecasts obtained on their smartphones or on personal computers. Websites operated by the NWS and private forecasting companies provide a wealth of local, national, and global data and forecasts. Smartphone applications can be tailored to provide conditions and forecasts for your hometown or wherever you may be traveling.

This chapter provides an overview of Earth's atmosphere. Our atmosphere is one rich in nitrogen and oxygen as well as smaller amounts of other gases, such as water vapor, carbon dioxide, and other greenhouse gases whose increasing levels are resulting in additional global warming and climate change. We examined Earth's early atmosphere and found it to be much different from the air we breathe today.

We investigated the various layers of the atmosphere: the troposphere (the lowest layer), where almost all weather events occur, and the stratosphere, where ozone protects us from a portion of the sun's harmful rays. In the stratosphere, ozone appears to be decreasing in concentration over parts of the Northern and Southern Hemispheres. Above the stratosphere lies the mesosphere, where the air temperature drops dramatically with height. Above the mesosphere lies the warmest part of the atmosphere, the thermosphere. At the top of the thermosphere is the exosphere, where collisions between gas molecules and atoms are so infrequent that fast-moving lighter molecules can actually escape Earth's gravitational pull and shoot off into space. The ionosphere represents that portion of the upper atmosphere where large numbers of ions and free electrons exist.

We looked briefly at the weather map and a satellite image and observed that dispersed throughout the atmosphere are storms and clouds of all sizes and shapes. The movement, intensification, and weakening of these systems, as well as the dynamic nature of air itself, produce a variety of weather events that we described in terms of weather elements. The sum total of weather and its extremes over a long period of time is what we call climate. Although sudden changes in weather may occur in a moment, climatic change takes place gradually over many years. The study of the atmosphere and all of its related phenomena is called *meteorology*, a term whose origin dates back to the days of Aristotle. Finally, we discussed some of the many ways weather and climate influence our lives.

KEY TERMS

The following terms are listed (with corresponding page numbers) in the order they appear in the text. Define each. Doing so will aid you in reviewing the material covered in this chapter.

atmosphere, 4
outgassing, 5
nitrogen, 6
oxygen, 6
water vapor, 6
carbon dioxide, 7
ozone, 9
ozone hole, 10
aerosol, 10
pollutant, 10
density, 11

pressure, 11
air pressure, 12
lapse rate, 12
temperature inversion, 12
troposphere, 12
stratosphere, 13
tropopause, 13
radiosonde, 15
mesosphere, 16
thermosphere, 16
exosphere, 16

homosphere, 16
heterosphere, 17
ionosphere, 17
weather, 17
climate, 17
meteorology, 18
middle latitudes, 19
middle-latitude cyclonic
    storm system, 19

hurricane, 19
thunderstorm, 19
tornado, 19
wind, 19
wind direction, 19
wind speed, 19
front, 20

QUESTIONS FOR REVIEW

1. What is the primary source of energy for Earth's atmosphere?
2. List the four most abundant gases in today's atmosphere.
3. Of the four most abundant gases in our atmosphere, which one shows the greatest variation at Earth's surface?
4. What are some of the important roles that water plays in our atmosphere?
5. Briefly explain the production and natural destruction of carbon dioxide near Earth's surface. Give two reasons for the increase of carbon dioxide over the past 100-plus years.
6. List the two most abundant greenhouse gases in Earth's atmosphere. What makes them greenhouse gases?
7. Explain how the atmosphere "protects" inhabitants at Earth's surface.
8. What are some of the aerosols in our atmosphere?
9. How has the composition of Earth's atmosphere changed over time? Briefly outline the evolution of Earth's atmosphere.
10. (a) Explain the concept of air pressure in terms of mass of air above some level.
    (b) Why does air pressure always decrease with increasing height above the surface?
11. What is standard atmospheric pressure at sea level in
    (a) inches of mercury
    (b) millibars, and
    (c) hectopascals?
12. What is the average or standard temperature lapse rate in the troposphere?
13. Briefly describe how the air temperature changes from Earth's surface to the lower thermosphere.
14. On the basis of temperature, list the layers of the atmosphere from the lowest layer to the highest.
15. What atmospheric layer contains all of our weather?
16. In what atmospheric layer do we find
    (a) the lowest average air temperature?
    (b) the highest average temperature?
    (c) the highest concentration of ozone?

17. Above what region of the world would you find the ozone hole?

18. Even though the actual concentration of oxygen is close to 21 percent (by volume) in the upper stratosphere, explain why, without proper breathing apparatus, you would not be able to survive there.

19. Define *meteorology* and discuss the origin of this word.

20. When someone says that "the wind direction today is south," does this mean that the wind is blowing *toward the south or from the south?*

21. Describe at least six features observed on a surface weather map.

22. Explain how wind blows around low- and high-pressure areas in the Northern Hemisphere.

23. How are fronts defined?

24. Rank the following storms in size from largest to smallest: hurricane, tornado, middle-latitude cyclonic storm, thunderstorm.

25. Weather in the middle latitudes tends to move in what general direction?

26. How does weather differ from climate?

27. Describe at least seven ways weather and climate influence the lives of people.

QUESTIONS FOR THOUGHT

1. Which of the following statements relate more to weather and which relate more to climate?
   (a) The summers here are warm and humid.
   (b) Cumulus clouds presently cover the entire sky.
   (c) Our lowest temperature last winter was −29°C (−18°F).
   (d) The air temperature outside is 22°C (72°F).
   (e) December is our foggiest month.
   (f) The highest temperature ever recorded in Phoenixville, Pennsylvania, was 44°C (111°F) on July 10, 1936.
   (g) Snow is falling at the rate of 5 cm (2 in.) per hour.
   (h) The average temperature for the month of January in Chicago, Illinois, is −3°C (26°F).

2. A standard pressure of 1013.25 millibars is also known as one atmosphere (1 atm).
   (a) Look at Fig. 1.10 and determine at approximately what levels you would record a pressure of 0.5 atm and 0.1 atm.
   (b) The surface air pressure on the planet Mars is about 0.007 atm. If you were standing on Mars, the surface air pressure would be equivalent to a pressure observed at approximately what altitude in Earth's atmosphere?

3. If you were suddenly placed at an altitude of 100 km (62 mi) above Earth, would you expect your stomach to expand or contract? Explain.

4. In the photo below, what are at least two kinds of weather impacts that might be occurring at this location?

● FIGURE 1.26

PROBLEMS AND EXERCISES

1. Keep track of the weather. On an outline map of North America, mark the daily position of fronts and pressure systems for a period of several weeks or more. (This information can be obtained from newspapers, the TV news, or from the Internet.) Plot the general upper-level flow pattern on the map. Observe how the surface systems move. Relate this information to the material on wind, fronts, and cyclones covered in later chapters.

2. Compose a one-week journal, including daily weather maps and weather forecasts from a newspaper or the Internet. Provide a commentary for each day regarding the correlation of actual and predicted weather.

3. Formulate a short-term climatology for your city for one month by recording maximum and minimum temperatures and precipitation amounts every day. You can get this information from television, newspapers, the Internet, or from your own measurements. Compare this data to the actual climatology for that month. How can you explain any large differences between the two?

4. Suppose a friend poses a question about how weather systems generally move in the middle latitudes. He puts forth a hypothesis that these systems generally move from east to west. How would you use the scientific method to prove his hypothesis to be incorrect?

 ONLINE RESOURCES

Visit www.cengagebrain.com to view additional resources, including video exercises, practice quizzes, an interactive eBook, and more.

Spring sunlight heats the water and melts ice in Russia's Lake Baikal.

Energy: Warming and Cooling Earth and the Atmosphere

AT HIGH LATITUDES AFTER DARKNESS HAS FALLEN a faint, white glow may appear in the sky. Lasting from a few minutes to a few hours, the light may move across the sky as a yellow green arc much wider than a rainbow; or, it may faintly decorate the sky with flickering draperies of blue, green, and purple light that constantly change in form and location, as if blown by a gentle breeze. For centuries curiosity and superstition have surrounded these eerie lights. Inuit legend says they are the lights from demons' lanterns as they search the heavens for lost souls. Nordic sagas called them a reflection of fire that surrounds the seas of the north. Even today there are those who proclaim that the lights are reflected sunlight from polar ice fields. Actually, this light show in the Northern Hemisphere is the aurora borealis—the northern lights—which is caused by invisible energetic particles bombarding our upper atmosphere. Anyone who witnesses this, one of nature's spectacular color displays, will never forget it.

Energy is everywhere. It is the basis for life. It comes in various forms: It can warm a house, melt ice, and drive the atmosphere, producing our everyday weather events. When the sun's energy interacts with our upper atmosphere we see energy at work in yet another form, a shimmering display of light from the sky—the aurora. What, precisely, is this common, yet mysterious, quantity we call "energy"? What is its primary source? How does it warm Earth and provide the driving force for our atmosphere? And in what form does it reach our atmosphere to produce a dazzling display like the aurora?

To answer these questions, we must first begin with the concept of energy itself. Then we will examine energy in its various forms and how energy is transferred from one form to another in our atmosphere. Finally, we will look more closely at the sun's energy and its influence on our atmosphere.

Energy, Temperature, and Heat

By definition, **energy** is the ability or capacity to do work on some form of matter. (Matter is anything that has mass and occupies space.) Work is done on matter when matter is either pushed, pulled, or lifted over some distance. When we lift a brick, for example, we exert a force against the pull of gravity—we "do work" on the brick. The higher we lift the brick, the more work we do. So, by doing work on something, we give it "energy," which it can, in turn, use to do work on other things. The brick that we lifted, for instance, can now do work on your toe—by falling on it.

The total amount of energy stored in any object (internal energy) determines how much work that object is capable of doing. A lake behind a dam contains energy by virtue of its position. This is called *gravitational potential energy* or simply **potential energy** because it represents the potential to do work—a great deal of destructive work if the dam were to break. The potential energy (PE) of any object is given as

$$PE = mgh$$

where m is the object's mass, g is the acceleration of gravity, and h is the object's height above the surface.

A volume of air aloft has more potential energy than the same size volume of air just above the surface. This fact is so because the air aloft has the potential to sink and warm through a greater depth of atmosphere. A substance also possesses potential energy if it can do work when a chemical change takes place. Thus, coal, natural gas, and food all contain chemical potential energy.

Any moving substance possesses energy of motion, or **kinetic energy**. The kinetic energy (KE) of an object is equal to half its mass multiplied by its velocity squared; thus

$$KE = \frac{1}{2}mv^2$$

Consequently, the faster something moves, the greater its kinetic energy; hence, a strong wind possesses more kinetic energy than a light breeze. Since kinetic energy also depends on the object's mass, a volume of water and an equal volume of air may be moving at the same speed, but, because the water has greater mass, it has more kinetic energy. The atoms and molecules that comprise all matter have kinetic energy due to their motion. This form of kinetic energy is often referred to as *heat energy*. Probably the most important form of energy in terms of weather and climate is the energy we receive from the sun—*radiant energy*.

Energy, therefore, takes on many forms, and it can change from one form into another. But the total amount of energy in the universe remains constant. *Energy cannot be created nor can it be destroyed.* It merely changes from one form to another in any ordinary physical or chemical process. In other words, the energy lost during one process must equal the energy gained during another. This is what we mean when we say that energy is conserved. This statement is known as the *law of conservation of energy*, and is also called the *first law of thermodynamics*.

We know that air is a mixture of countless billions of atoms and molecules. If they could be seen, they would appear to be moving about in all directions, freely darting, twisting, spinning, and colliding with one another like an angry swarm of bees. Close to Earth's surface, each individual molecule will travel only about a thousand times its diameter before colliding with another molecule. Moreover, we would see that all the atoms and molecules are not moving at the same speed, as some are moving faster than others. The temperature of the air (or any substance) is a measure of its average kinetic energy. Simply stated, **temperature** *is a measure of the average speed (average motion) of the atoms and molecules*, where higher temperatures correspond to faster average speeds.

Suppose we examine a volume of surface air about the size of a large flexible balloon, as shown in ● Fig. 2.1a. If we warm the air inside, the molecules will move faster, but they also will move slightly farther apart—the air becomes less dense, as illustrated in Fig. 2.1b. Conversely, if we cool the air back to its original temperature, the molecules would slow down, crowd closer together, and the air would become more dense. This molecular behavior is why, in many places throughout the book, we refer to surface air as either *warm, less-dense air* or as *cold, more-dense air*.

Along with temperature, we can also measure *internal energy*, which is the total energy (potential and kinetic) stored in a group of molecules. As we have just seen, the temperature of air and water is determined only by the *average* kinetic energy (average speed) of *all* their molecules. Since temperature only indicates how "hot" or "cold" something is relative to some set standard value, it does not always tell us the total amount of internal energy that something possesses. For example, two identical mugs, each half-filled with water and each with the same temperature, contain the same internal energy. If the water from one mug is poured into the other, the total internal energy of the filled mug has doubled because its mass has doubled. Its temperature, however, has not changed, since the average speed of all of the molecules is still the same.

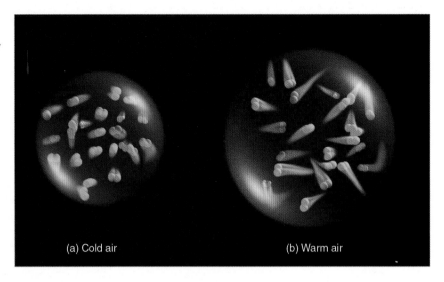

● **FIGURE 2.1** Air temperature is a measure of the average speed (motion) of the molecules. In the cold volume of air, the molecules move more slowly and crowd closer together. In the warm volume, they move faster and farther apart.

(a) Cold air   (b) Warm air

Now, imagine that you are sipping a hot cup of tea on a small raft in the middle of a lake. The tea has a much higher temperature than the lake, yet the lake contains more internal energy because it is composed of many more molecules. If the cup of tea is allowed to float on top of the water, the tea would cool rapidly. The energy that would be transferred from the hot tea to the cool water (because of their temperature difference) is called *heat*.

In essence, **heat** *is energy in the process of being transferred from one object to another because of the temperature difference between them*. After heat is transferred, it is stored as internal energy. How is this energy transfer process accomplished? In the atmosphere, heat is transferred by *conduction, convection,* and *radiation*. We will examine these mechanisms of energy transfer after we look at temperature scales and at the important concepts of *specific heat* and *latent heat*.

TEMPERATURE SCALES Suppose we take a small volume of air (such as the one shown in Fig. 2.1a) and allow it to cool. As the air slowly cools, its atoms and molecules would move more and more slowly until the air reaches a temperature of −273°C (−459°F), which is the lowest temperature possible. At this temperature, called **absolute zero**, the atoms and molecules would possess a minimum amount of energy and theoretically no thermal motion. Absolute zero is the starting point for a temperature scale called the *absolute scale*, or **Kelvin scale**, after Lord Kelvin (1824–1907), the British scientist who first introduced it. Because the Kelvin scale begins at absolute zero, it contains no negative numbers and is, therefore, quite convenient for scientific calculations.

The temperature scales most commonly used in meteorology are the Fahrenheit and Celsius (formerly centigrade) scales. The **Fahrenheit scale** was developed in the early eighteenth century by physicist G. Daniel Fahrenheit (1686–1736), who assigned the number 32 to the temperature at which water freezes, and the number 212 to the temperature at which water boils. The zero point was simply the lowest

temperature that he obtained with a mixture of ice, water, and salt. Between the freezing and boiling points are 180 equal divisions, each of which is called a *degree*. A thermometer calibrated with this scale is referred to as a Fahrenheit thermometer, for it measures an object's temperature in degrees Fahrenheit (°F).

The **Celsius scale**, named after Swedish astronomer Anders Celsius (1701–1744), was introduced later in the eighteenth century and is part of the *metric system* used around the world. The number 0 (zero) on this scale is assigned to the temperature at which pure water freezes, and the number 100 to the temperature at which pure water boils at sea level. The space between freezing and boiling is divided into 100 equal degrees. Therefore, each Celsius degree is 180/100 or 1.8 times larger than a Fahrenheit degree. Put another way, an increase in temperature of 1°C equals an increase of 1.8°F. A formula for converting °F to °C is

$$°C = \tfrac{5}{9}(°F − 32)$$

On the Kelvin scale, degrees kelvin are called *kelvins* (abbreviated K). Each degree on the Kelvin scale is exactly the same size as a degree Celsius, and a temperature of 0 K is equal to −273°C. Converting from °C to K can be done by simply adding 273 to the Celsius temperature, as

$$K = °C + 273$$

● Figure 2.2 compares the Kelvin, Celsius, and Fahrenheit scales. Converting a temperature from one scale to another can be done by simply reading the corresponding temperature from the adjacent scale. Thus, 303 K on the Kelvin scale is the equivalent of 30°C and 86°F*.

In most of the world, temperature readings are taken in °C and public weather forecasts use the Celsius scale. In the United States, however, temperatures above the surface are taken in °C while

---

*A more complete table of conversions is given in Appendix A.

● **FIGURE 2.2** Comparison of Kelvin, Celsius, and Fahrenheit scales, along with some world temperature extremes.

temperatures at the surface are typically read and reported in °F. Likewise, temperatures on upper-level maps are plotted in °C, while on surface weather maps they are in °F. Since both scales are in use in the United States, temperature readings in this book will, in most cases, be given in °C followed by their equivalent in °F.

SPECIFIC HEAT A watched pot never boils, or so it seems. The reason for this is that water requires a relatively large amount of heat energy to bring about a small temperature change. The **heat capacity** of a substance is the ratio of the amount of heat energy absorbed by that substance to its corresponding temperature rise. The heat capacity of a substance per unit mass is called **specific heat**. In other words, specific heat is the amount of heat needed to raise the temperature of 1 gram (g) of a substance 1 degree Celsius.

If we heat 1 g of liquid water on a stove, it would take about 1 calorie (cal)* to raise its temperature by 1°C. So water has a

---

*By definition, a calorie is the amount of heat required to raise the temperature of 1 g of water from 14.5°C to 15.5°C. The kilocalorie is 1000 calories and is the heat required to raise 1 kilogram (kg) of water 1°C. In the International System (Système International [SI]), the unit of energy is the joule (J), where 1 calorie = 4.186 J. (For pronunciation: *joule* rhymes with *pool*.)

specific heat of 1. If, however, we put the same amount (that is, same mass) of compact dry soil on the flame, we would see that it would take about one-fifth the heat (about 0.2 cal) to raise its temperature by 1°C. The specific heat of water is therefore five times greater than that of soil. In other words, water must absorb five times as much heat as the same quantity of soil in order to raise its temperature by the same amount. The specific heat of various substances is given in ▼ Table 2.1.

Not only does water heat slowly, it cools slowly as well. It has a much higher capacity for storing energy than other common substances, such as soil and air. A given volume of water can store a large amount of energy while undergoing only a small temperature change. Because of this attribute, water has a strong modifying effect on weather and climate. Near large bodies of water, for example, winters usually remain warmer and summers cooler than nearby inland regions—a fact well known to people who live adjacent to oceans or large lakes.

LATENT HEAT—THE HIDDEN WARMTH We know from Chapter 1 that water vapor is an invisible gas that becomes visible when it changes into larger liquid or solid (ice) particles. This process of transformation is known as a *change of state* or, simply, a *phase change*. The heat energy required to change a substance, such as water, from one state to another is called **latent heat**. But why is this heat referred to as "latent"? To answer this question, we will begin with something familiar to most of us—the cooling produced by evaporating water.

Suppose we microscopically examine a small drop of pure water. At the drop's surface, molecules are constantly escaping (evaporating). Because the more energetic, faster-moving molecules escape most easily, the average motion of all the molecules left behind decreases as each additional molecule evaporates. Because temperature is a measure of average molecular motion, the slower motion suggests a lower water temperature. *Evaporation is, therefore, a cooling process.* The energy needed to evaporate the water—that is, to change its phase from a liquid to a gas—may come from the water or other sources, including the air.

In the everyday world, we experience evaporational cooling as we step out of a shower or swimming pool into a dry

▼ **TABLE 2.1** Specific Heat of Various Substances

SUBSTANCE	SPECIFIC HEAT Cal/(g × °C)	J/(kg × °C)
Water (pure)	1	4186
Wet mud	0.6	2512
Ice (0°C)	0.5	2093
Sandy clay	0.33	1381
Dry air (sea level)	0.24	1005
Quartz sand	0.19	795
Granite	0.19	794

area. Because some of the energy used to evaporate the water comes from our skin, we may experience a rapid drop in skin temperature, even to the point where goose bumps form. In fact, on a hot, dry, windy day in Tucson, Arizona, cooling may be so rapid that we begin to shiver even though the air temperature is hovering around 38°C (100°F).

The energy lost by liquid water during evaporation can be thought of as carried away by, and "locked up" within, the water vapor molecule. The energy is thus in a "stored" or "hidden" condition and is, therefore, called *latent heat*. It is latent (hidden) in that the temperature of the substance changing from liquid to vapor is still the same. However, the heat energy will reappear as **sensible heat** (the heat we can feel, "sense," and measure with a thermometer) when the vapor condenses back into liquid water. Therefore, *condensation* (the opposite of evaporation) *is a warming process*.

The heat energy released when water vapor condenses to form liquid droplets is called *latent heat of condensation*. Conversely, the heat energy used to change liquid into vapor at the same temperature is called *latent heat of evaporation* (vaporization). Nearly 600 cal (2500 J) are required to evaporate a single gram of water at room temperature. With many hundreds of grams of water evaporating from the body, it is no wonder that after a shower we feel cold before drying off.

In a way, latent heat is responsible for keeping a cold drink with ice colder than one without ice. As ice melts, its temperature does not change. The reason for this fact is that the heat added to the ice only breaks down the rigid crystal pattern, changing the ice to a liquid without changing its temperature. The energy used in this process is called *latent heat of fusion* (melting). Roughly 80 cal (335 J) are required to melt a single gram of ice. Consequently, heat added to a cold drink with ice primarily melts the ice, while heat added to a cold drink without ice warms the beverage. If a gram of water at 0°C changes back into ice at 0°C, this same amount of heat (80 cal) would be released as sensible

heat to the environment. Therefore, when ice melts, heat is taken in; when water freezes, heat is released.

The heat energy required to change ice into vapor (a process called *sublimation*) is referred to as *latent heat of sublimation*. For a single gram of ice to transform completely into vapor at 0°C requires nearly 680 cal—80 cal for the latent heat of fusion plus 600 cal for the latent heat of evaporation. If this same vapor transformed back into ice (a process called *deposition*), approximately 680 cal (2850 J) would be released to the environment.

● Figure 2.3 summarizes the concepts examined so far. When the change of state is from left to right, heat is absorbed by the substance and taken away from the environment. The processes of melting, evaporation, and sublimation all cool the environment. When the change of state is from right to left, heat energy is given up by the substance and added to the environment. The process of freezing, condensation, and deposition all warm their surroundings.

Latent heat is an important source of atmospheric energy. Once vapor molecules become separated from Earth's surface, they are swept away by the wind, like dust before a broom. Rising to high altitudes where the air is cold, the vapor changes into liquid and ice cloud particles. During these processes, a tremendous amount of heat energy is released into the environment. This heat provides energy for storms, such as hurricanes, middle-latitude cyclones, and thunderstorms (see ● Fig. 2.4).

Water vapor evaporated from warm, tropical water can be carried into polar regions, where it condenses and gives up its heat energy. As we will see, evaporation–transportation–condensation is an extremely important mechanism for the relocation of heat energy (as well as water) in the atmosphere. We are now ready to look at other mechanisms of heat transfer in the atmosphere. (Before going on to the next section, you may wish to read Focus section 2.1, which summarizes some of the concepts considered thus far.)

● **FIGURE 2.3** Heat energy absorbed and released.

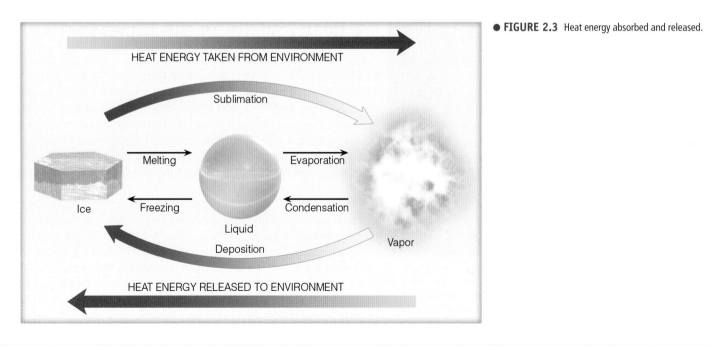

The Fate of a Sunbeam

Consider sunlight in the form of radiant energy striking a large lake, as shown in ● Fig. 1. Part of the incoming energy heats the water, causing greater molecular motion and, hence, an increase in the water's kinetic energy. This greater kinetic energy allows more water molecules to evaporate from the surface. As each molecule escapes, work is done to break it away from the remaining water molecules. This energy becomes the latent heat energy that is carried with the water vapor.

Above the lake, a large bubble* of warm, moist air rises and expands. In order for this expansion to take place, the gas molecules inside the bubble must use some of their kinetic energy to do work against the bubble's sides. This results in a slower molecular speed and a lower temperature. Well above the surface, the water vapor in the rising, cooling bubble of moist air condenses into clouds. The condensation of water vapor releases latent heat energy into the atmosphere, warming the air. The tiny suspended cloud droplets possess potential energy, which becomes kinetic energy when these droplets grow into raindrops that fall earthward.

---

*A bubble of rising (or sinking) air about the size of a large balloon is often called a parcel of air.

● **FIGURE 1** Solar energy striking a large body of water goes through many transformations.

When the drops reach the surface, their kinetic energy erodes the land. As rain-swollen streams flow into a lake behind a dam, there is a buildup of potential energy, which can be transformed into kinetic energy as water is harnessed to flow down a chute. If the moving water drives a generator, kinetic energy is converted into electrical energy, which is sent to cities. There, it heats, cools, and lights the buildings in which people work and live. Meanwhile, some of the water in the lake behind the dam evaporates and is free to repeat the cycle. Hence, the energy from the sunlight on a lake can undergo many transformations and help provide the moving force for many natural and human-made processes.

● **FIGURE 2.4** Every time a cloud forms, it warms the atmosphere. Inside this developing thunderstorm a vast amount of stored heat energy (latent heat) is given up to the air, as invisible water vapor becomes countless billions of water droplets and ice crystals. In fact, for the duration of this storm alone, more heat energy is released inside this cloud than is unleashed by a small nuclear bomb.

Heat Transfer in the Atmosphere

CONDUCTION The transfer of heat from molecule to molecule within a substance is called **conduction**. Hold one end of a metal straight pin between your fingers and place a flaming candle under the other end (see ● Fig 2.5). Because of the energy they absorb from the flame, the molecules in the pin vibrate faster. The faster-vibrating molecules cause adjoining molecules to vibrate faster. These, in turn, pass vibrational energy on to their neighboring molecules, and so on, until the molecules at the finger-held end of the pin begin to vibrate rapidly. These fast-moving molecules eventually cause the molecules of your finger to vibrate more quickly. Heat is now being transferred from the pin to your finger, and both the pin and your finger feel hot. If enough heat is transferred, your finger will become painful, and you will drop the pin. The transmission of heat from one end of the pin to the other, and from the pin to your finger, occurs by conduction. Heat transferred in this fashion always flows from *warmer to colder*

● **FIGURE 2.5** The transfer of heat from the hot end of the metal pin to the cool end by molecular contact is called *conduction*.

SUBSTANCE	HEAT CONDUCTIVITY (Watts† per meter per °C)
Still air	0.023 (at 20°C)
Wood	0.08
Dry soil	0.25
Water	0.60 (at 20°C)
Snow	0.63
Wet soil	2.1
Ice	2.1
Sandstone	2.6
Granite	2.7
Iron	80
Silver	427

*Heat (thermal) conductivity describes a substance's ability to conduct heat as a consequence of molecular motion.
†A watt (W) is a unit of power where one watt equals one joule (J) per second (J/s). One joule equals 0.24 calories.

regions. Generally, the greater the temperature difference, the more rapid the heat transfer.

When materials can easily pass energy from one molecule to another, they are considered to be good conductors of heat. How well they conduct heat depends upon how their molecules are structurally bonded together. ▼ Table 2.2 shows that solids, such as metals, are good heat conductors. It is often difficult, therefore, to judge the temperature of metal objects. For example, if you grab a metal pipe at room temperature, it will seem to be much colder than it actually is because the metal conducts heat away from the hand quite rapidly. Conversely, *air is an extremely poor conductor of heat*, which is why most insulating materials have a large number of air spaces trapped within them. Air is such a poor heat conductor that, in calm weather, the hot ground only warms a shallow layer of air a few centimeters thick by conduction. Yet, air can carry this energy rapidly from one region to another. How then does this phenomenon happen?

CONVECTION The transfer of heat by the mass movement of a fluid (such as water and air) is called **convection**. This type of heat transfer takes place in liquids and gases because they can move freely, and it is possible to set up currents within them.

Convection happens naturally in the atmosphere. On a warm, sunny day, certain areas of Earth's surface absorb more heat from the sun than others; as a result, the air near Earth's surface is heated somewhat unevenly. Air molecules adjacent to these hot surfaces bounce against them, thereby gaining some extra energy by conduction. The heated air expands and becomes less dense than the surrounding cooler air. The expanded warm air is buoyed upward and rises. In this manner, large bubbles of warm air rise and transfer heat energy upward. Cooler, heavier air flows toward the surface to replace the rising air. This cooler air becomes heated in turn, rises, and the cycle is repeated. In meteorology, this vertical exchange of

heat is called *convection*, and the rising air bubbles are known as **thermals** (see ● Fig. 2.6).

The rising air expands and gradually spreads outward. It then slowly begins to sink. Near the surface, it moves back into the heated region, replacing the rising air. In this way, a *convective circulation*, or thermal "cell," is produced in the atmosphere. In a convective circulation, the warm, rising air cools. In our atmosphere, *any air that rises will expand and cool,* and *any air that sinks is compressed and warms*. This important concept is detailed in Focus section 2.2.

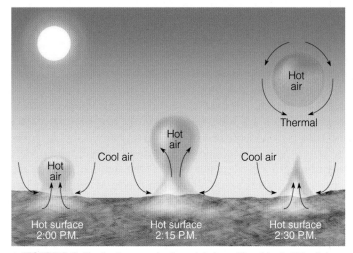

● **FIGURE 2.6** The development of a thermal. A thermal is a rising bubble of air that carries heat energy upward by *convection*.

Rising Air Cools and Sinking Air Warms

To understand why rising air cools and sinking air warms, we need to examine some air. Suppose we place air in an imaginary thin, elastic wrap about the size of a large balloon (see ● Fig. 2). This invisible balloonlike "blob" is called an *air parcel.* The air parcel can expand and contract freely, but neither external air nor heat is able to mix with the air inside. By the same token, as the parcel moves, it does not break apart but remains as a single unit.

At Earth's surface, the parcel has the same temperature and pressure as the air surrounding it. Suppose we lift the parcel. Recall from Chapter 1 that air pressure always

decreases as we move up into the atmosphere. Consequently, as the parcel rises, it enters a region where the surrounding air pressure is lower. To equalize the pressure, the parcel molecules inside push the parcel walls outward, expanding it. Because there is no other energy source, the air molecules inside use some of their own energy to expand the parcel. This energy loss shows up as slower molecular speeds, which represent a lower parcel temperature. Hence, *any air that rises always expands and cools.*

If the parcel is lowered to Earth's surface (as shown in Fig. 2), it returns to a region where

the air pressure is higher. The higher outside pressure squeezes (compresses) the parcel back to its original (smaller) size. Because air molecules have a faster rebound velocity after striking the sides of a collapsing parcel, the average speed of the molecules inside goes up. (A Ping-Pong ball moves faster after striking a paddle that is moving toward it.) This increase in molecular speed represents a warmer parcel temperature. Therefore, *any air that sinks (subsides), warms by compression.*

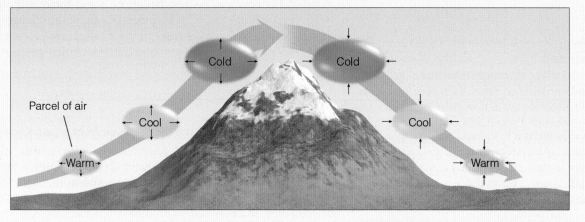

● **FIGURE 2** Rising air expands and cools; sinking air is compressed and warms.

Although the entire process of heated air rising, spreading out, sinking, and finally flowing back toward its original location is known as a convective circulation, meteorologists usually restrict the term *convection* to the process of the rising and sinking part of the circulation (see ● Fig. 2.7).

The horizontally moving part of the circulation (called *wind*) carries properties of the air in that particular area with it. The transfer of these properties by horizontally moving air is called **advection**. For example, wind blowing across a body of water will "pick up" water vapor from the evaporating surface and transport it elsewhere in the atmosphere. If the air cools, the water vapor may condense into cloud droplets and release latent heat. In a sense, then, heat is advected (carried) by the water vapor as it is swept along with the wind. Earlier we saw that this is an important way to redistribute heat energy in the atmosphere.

There is yet another mechanism for the transfer of energy—radiation, or *radiant energy*, which is what we receive from the

sun. In this method, energy may be transferred from one object to another without the space between them necessarily being heated.

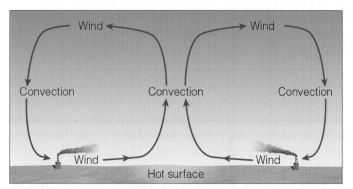

● **FIGURE 2.7** The rising of hot air and the sinking of cool air sets up a convective circulation. Normally, the vertical part of the circulation is called *convection*, whereas the horizontal part is called *wind*. Near the surface the wind is advecting smoke from one region to another.

Before moving on to the next section, here is a summary of some of the important concepts and facts we have covered:

- The temperature of a substance is a measure of the average kinetic energy (average motion) of its atoms and molecules.

- Evaporation (the transformation of liquid into vapor) is a cooling process that can cool the air, whereas condensation (the transformation of vapor into liquid) is a warming process that can warm the air.

- Heat is energy in the process of being transferred from one object to another because of the temperature difference between them.

- In conduction, which is the transfer of heat by molecule-to-molecule contact, heat always flows from warmer to colder regions.

- Air is a poor conductor of heat.

- Convection is an important mechanism of heat transfer, as it represents the vertical movement of warmer air upward and cooler air downward.

Radiant Energy

On a bright winter day, you may have noticed how warm your face feels as you stand facing the sun. Sunlight travels through the surrounding air with little effect upon the air itself. Your face, however, absorbs this energy and converts it to thermal energy. Thus, sunlight warms your face without actually warming the air. The energy transferred from the sun to your face is called **radiant energy**, or **radiation**. It travels in the form of waves that release energy when they are absorbed by an object. Because these waves have magnetic

and electrical properties, we call them **electromagnetic waves**. Electromagnetic waves do not need molecules to propagate them. In a vacuum, they travel at a constant speed of nearly 300,000 km (186,000 mi) per second—the speed of light.

● Figure 2.8 shows some of the different wavelengths of radiation. Notice that the **wavelength** (which is usually expressed by the Greek letter lambda, l) is the distance measured along a wave from one crest to another. Also notice that some of the waves have exceedingly short lengths. For example, radiation that we can see (visible light) has an average wavelength of less than one-millionth of a meter—a distance nearly one-hundredth the diameter of a human hair. To help describe these short lengths, we introduce a new unit of measurement called a **micrometer** (represented by the symbol μm), which is equal to one-millionth of a meter (m); thus

$$1 \text{ micrometer } (\mu m) = 0.000001 \text{ m} = 10^{-6} \text{ m}$$

In Fig. 2.8, notice that the average wavelength of radiation we can see (called *visible light*) is about 0.0000005 m, which is the same as 0.5 μm. To give you a common object for comparison, the average height of a letter in the printed version of this book is about 2000 μm, or 2 millimeters (2 mm), whereas the thickness of this page is about 100 μm.

TYPE OF RADIATION	RELATIVE WAVELENGTH	TYPICAL WAVELENGTH (meters)	ENERGY CARRIED PER WAVE OR PHOTON
AM radio waves	Wavelength	100	Increasing
Television waves		1	
Microwaves		10^{-3}	
Infrared waves		10^{-6}	
Visible light		5×10^{-7}	
Ultraviolet waves		10^{-7}	
X rays		10^{-9}	

● **FIGURE 2.8** Radiation characterized according to wavelength. As the wavelength decreases, the energy carried per wave increases.

We can also see in Fig. 2.8 that the longer waves carry less energy than do the shorter waves. When comparing the energy carried by various waves, it is useful to give electromagnetic radiation characteristics of particles in order to explain some of the waves' behavior. We can actually think of radiation as streams of particles or **photons** that are discrete packets of energy.*

As shown in Figure 2.8, an ultraviolet photon carries more energy than a photon of visible light. In fact, certain ultraviolet photons have enough energy to produce sunburns and penetrate skin tissue, sometimes causing skin cancer. As we discussed in Chapter 1, it is ozone in the stratosphere that protects us from the vast majority of these harmful rays.

RADIATION AND TEMPERATURE *All things (whose temperature is above absolute zero), no matter how big or small, emit radiation.* This book, your body, flowers, trees, air, Earth, and the stars are all radiating a wide range of electromagnetic waves. The energy originates from rapidly vibrating electrons, billions of which exist in every object.

The wavelengths that each object emits depend primarily on the object's temperature. The higher the temperature, the faster the electrons vibrate, and the shorter are the wavelengths of the emitted radiation. This can be visualized by attaching one end of a rope to a post and holding the other end. If the rope is shaken rapidly (high temperature), numerous short waves travel along the rope; if the rope is shaken slowly (lower temperature), longer waves appear on the rope. Although objects at a temperature of about 500°C radiate waves with many lengths, some of them are short enough to stimulate the sensation of vision. We actually see these objects glow red. Objects cooler than this radiate at wavelengths that are too long for us to see. The page of the printed version of this book, for example, is radiating electromagnetic waves. But because its temperature is only about 20°C (68°F), the waves emitted are much too long to stimulate vision. We are able to see the page, however, because light waves from other sources (such as lightbulbs or the sun) are being *reflected* (bounced) off the paper. If the book were carried into a completely dark room, it would continue to radiate, but the pages would appear black because there are no visible light waves in the room to reflect off the pages.

Objects that have a very high temperature emit energy at a greater rate or intensity than objects at a lower temperature. Thus, *as the temperature of an object increases, more total radiation is emitted each second.* This can be expressed mathematically as

$$E = \sigma T^4$$

where E is the maximum rate of radiation emitted by each square meter of surface area of the object, σ (the Greek letter sigma) is the Stefan-Boltzmann constant,** and T is the object's surface temperature in kelvins. This relationship, called the **Stefan-Boltzmann law** after Josef Stefan (1835–1893) and Ludwig Boltzmann (1844–1906), who derived it, states that all objects with temperatures above absolute zero (0 K or −273°C) emit

radiation at a rate proportional to the fourth power of their absolute temperature. Consequently, a small increase in temperature results in a large increase in the amount of radiation emitted because doubling the absolute temperature of an object increases the maximum energy output by a factor of 16, which is 2^4.

RADIATION OF THE SUN AND EARTH Most of the sun's energy is emitted from its surface, where the temperature is nearly 6000 K (10,500°F). Earth, on the other hand, has an average surface temperature of 288 K (15°C, 59°F). The sun, therefore, radiates a great deal more energy than does Earth (see ● Fig. 2.9). At what wavelengths do the sun and Earth radiate most of their energy? Fortunately, the sun and Earth both have characteristics (discussed in a later section) that enable us to use the following relationship called **Wien's law** (or *Wien's displacement law*) after the German physicist Wilhelm Wien (pronounced Ween, 1864–1928), who discovered it:

$$\lambda_{\text{max}} = \frac{\text{constant}}{T}$$

where λ_{max} is the wavelength in micrometers at which maximum radiation emission occurs, T is the object's temperature in kelvins, and the constant is 2,897 μm K. To make the numbers easy to deal with, we will round off the constant to the number 3000.

For the sun, with a surface temperature of about 6000 K, the equation becomes

$$\lambda_{\text{max}} = \frac{3000\,\mu\text{mK}}{6000\,\text{K}} = 0.5\,\mu\text{m}$$

Thus, the sun emits a maximum amount of radiation at wavelengths near 0.5 μm. The cooler Earth, with an average surface temperature of 288 K (rounded to 300 K), emits maximum radiation near wavelengths of 10 μm, since

$$\lambda_{\text{max}} = \frac{3000\,\mu\text{mK}}{300\,\text{K}} = 10\,\mu\text{m}$$

● **FIGURE 2.9** The hotter sun not only radiates more energy than that of the cooler Earth (the area under the curve), but it also radiates the majority of its energy at much shorter wavelengths. (The area under the curves is equal to the total energy emitted, and the scales for the two curves differ by a factor of roughly 1 million.)

---

*Packets of photons make up waves, and groups of waves make up a beam of radiation.

**The Stefan-Boltzmann constant σ in SI units is 5.67×10^{-8} W/m²k⁴. A watt (W) is a unit of power where 1 watt equals 1 joule (J) per second (J/s). One joule is equal to 0.24 cal. More conversions are given in Appendix A.

Thus, Earth emits most of its radiation at longer wavelengths between about 5 and 25 μm, while the sun emits the majority of its radiation at wavelengths less than 2 μm. For this reason, Earth's radiation *(terrestrial radiation)* is often called **longwave radiation**, whereas the sun's energy *(solar radiation)* is referred to as **shortwave radiation**.

Wien's law demonstrates that as the temperature of an object increases, the wavelength at which maximum emission occurs is shifted toward shorter values. For example, if the sun's surface temperature were to double to 12,000 K, its wavelength of maximum emission would be halved to about 0.25 μm. If, on the other hand, the sun's surface cooled to 3000 K, it would emit its maximum amount of radiation near 1.0 μm.

Even though the sun radiates at a maximum rate at a particular wavelength, it nonetheless emits some radiation at almost all other wavelengths. If we look at the amount of radiation given off by the sun at each wavelength, we obtain the sun's *electromagnetic spectrum*. A portion of this spectrum is shown in ● Fig. 2.10.

Given that our eyes are sensitive to radiation between 0.4 and 0.7 μm, these waves reach the eye and stimulate the sensation of color. This portion of the spectrum is referred to as the **visible region**, and the radiant energy that reaches our eye is called *visible light*. The sun emits nearly 44 percent of its radiation in this zone, with the peak of energy output found at the wavelength corresponding to the color blue-green. The color violet is the shortest wavelength of visible light. Wavelengths shorter than violet (0.4 μm) are **ultraviolet (UV)**. X-rays and gamma rays with exceedingly short wavelengths also fall into this category. The sun emits only about 7 percent of its total energy at ultraviolet wavelengths.

The longest wavelengths of visible light correspond to the color red. Wavelengths longer than red (0.7 μm) are **infrared (IR)**. These waves cannot be seen by the unaided human eye. Nearly 37 percent of the sun's energy is radiated between 0.7 μm and 1.5 μm, with only 12 percent radiated at wavelengths longer than 1.5 μm.

Whereas the hot sun emits only a part of its energy in the infrared portion of the spectrum, the relatively cool Earth emits almost all of its energy at infrared wavelengths. Although we cannot see infrared radiation, there are instruments called *infrared sensors* that can. Weather satellites that orbit the globe use these sensors to observe radiation emitted by Earth, the clouds, and the atmosphere. Since objects of different temperatures radiate their maximum energy at different wavelengths, infrared photographs can distinguish among objects of different temperatures. Clouds always radiate infrared energy; thus, cloud images using infrared sensors can be taken during both day and night.

In summary, both the sun and Earth emit radiation. The *hot sun* (6000 K) radiates nearly 88 percent of its energy at wavelengths less than 1.5 μm, with maximum emission in the *visible region* near 0.5 μm. Look again at Fig. 2.9 and notice that the *cooler Earth* (288 K) radiates nearly all its energy between 5 and 25 μm with a peak intensity in the *infrared region* near 10 μm. The sun's surface is nearly 20 times hotter than Earth's surface. From the Stefan-Boltzmann relationship, this fact means that a unit area on the sun emits nearly 160,000 (20^4) times more energy during a given time period than the same size area on Earth. And since the sun has such a huge surface area from which to radiate, the total energy emitted by the sun each minute amounts to a staggering 6 billion, billion, billion calories! (Additional information on radiation intensity and its effect on humans is given in Focus section 2.3.)

Radiation: Absorption, Emission, and Equilibrium

If Earth and all things on it are continually radiating energy, why doesn't everything get progressively colder? The answer is that all objects not only radiate energy, they absorb it as well. If an object radiates more energy than it absorbs, it gets colder; if it absorbs more energy than it emits, it gets warmer. On a sunny day, Earth's surface warms by absorbing more energy from the sun and the atmosphere than it radiates, while at night Earth cools by radiating more energy than it absorbs from its surroundings. When an object emits and absorbs energy at equal rates, its temperature remains constant.

The rate at which something radiates and absorbs energy depends strongly on its surface characteristics, such as color, texture, and moisture, as well as temperature. For example, a black object in direct sunlight is a good absorber of visible radiation. It converts energy from the sun into internal energy, and its temperature ordinarily increases. You need only walk barefoot on

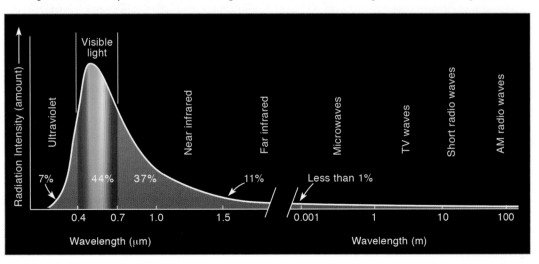

● **FIGURE 2.10** The sun's electromagnetic spectrum and some of the descriptive names of each region. The numbers underneath the curve approximate the percent of energy the sun radiates in various regions.

Wave Energy, Sunburning, and UV Rays

Standing close to a fire makes us feel warmer than we do when we stand at a distance from it. Does this mean that, as we move away from a hot object, the waves carry less energy and are, therefore, weaker? Not really. The intensity of radiation decreases as we move away from a hot object because radiant energy spreads outward in all directions.

● Figure 3 illustrates that as the distance from a radiating object increases, a given amount of energy is distributed over a larger area, so that the energy received over a given area and over a given time decreases. In fact, at twice the distance from the source, the radiation is spread over four times the area.

Another interesting fact about radiation that we learned earlier in this chapter is that shorter waves carry much more energy than do longer waves. Hence, a photon of ultraviolet light carries more energy than a photon of visible light. In fact, ultraviolet (UV) wavelengths in the range of 0.20 and 0.29 μm (known as *UVC radiation*) are harmful to living things, as certain waves can cause chromosome mutations, kill single-celled organisms, and damage the cornea of the eye. Fortunately, virtually all the ultraviolet radiation at wavelengths in the UVC range is absorbed by ozone in the stratosphere.

Ultraviolet wavelengths between about 0.29 and 0.32 μm (known as *UVB radiation*) reach Earth in small amounts. Photons in this wavelength range have enough energy to produce sunburns and penetrate skin tissues, sometimes causing skin cancer. About 90 percent of all skin cancers are linked to sun

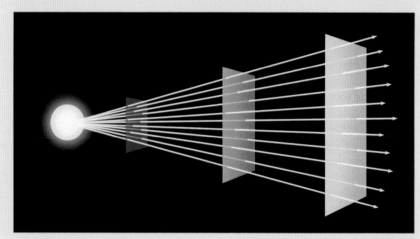

● **FIGURE 3** The intensity, or amount, of radiant energy transported by electromagnetic waves decreases as we move away from a radiating object because the same amount of energy is spread over a larger area.

exposure and UVB radiation. Oddly enough, these same wavelengths activate provitamin D in the skin and convert it into vitamin D, which is essential to health.

Longer ultraviolet waves with lengths of about 0.32 to 0.40 μm (called *UVA radiation*) are less energetic, but they are the main ones that produce skin tanning. Although UVB is mainly responsible for burning the skin, UVA can cause skin redness. It can also interfere with the skin's immune system and cause long-term skin damage that shows up years later as accelerated aging and skin wrinkling. Moreover, recent studies indicate that longer UVA exposures needed to create a tan pose about the same cancer risk as a UVB tanning dose.

Upon striking the human body, ultraviolet radiation is absorbed beneath the outer layer of

skin. To protect the skin from these harmful rays, the body's defense mechanism kicks in. Certain cells, when exposed to UV radiation, produce a dark pigment (*melanin*) that begins to absorb some of the UV radiation. (It is the production of melanin that produces a tan.) Consequently, a body that produces little melanin—one with pale skin—has little natural protection from UVB.

Additional protection can come from sunscreens that block UV rays from ever reaching the skin. Some contain chemicals (such as zinc oxide) that reflect UV radiation. (These are the white pastes once seen on the noses of lifeguards.) Others consist of a mixture of chemicals (such as dioxybenzone and para-aminobenzoic acid [PABA]) that actually absorb ultraviolet radiation. The *sun protection factor* (SPF) number on every container of sunscreen dictates how

---

a black asphalt road on a summer afternoon to experience this. At night, the blacktop road will cool quickly by emitting infrared radiation and, by early morning, it may be cooler than surrounding surfaces.

Any object that is a perfect absorber (that is, absorbs all the radiation that strikes it) and a perfect emitter (emits the maximum radiation possible at its given temperature) is called a **blackbody**. Blackbodies do not have to be colored black; they simply must absorb and emit all possible radiation. Because Earth's surface and the sun absorb and radiate with nearly 100 percent efficiency for their respective temperatures, they both behave as blackbodies. This is the reason we were able to use Wien's law and the Stefan-Boltzmann law to determine the characteristics of radiation emitted from the sun and Earth.

When we look at Earth from space, we see that half of it is in sunlight, while the other half is in darkness. The outpouring of solar energy constantly bathes Earth with radiation, while Earth, in turn, constantly emits infrared radiation. If we assume that there is no other method of transferring heat, then, when the rate of absorption of solar radiation equals the rate of emission of infrared Earth radiation, a state of *radiative equilibrium* is achieved. The average temperature at which this occurs is called the **radiative equilibrium temperature**. At this temperature, Earth (behaving as a blackbody) is absorbing solar radiation and emitting infrared radiation at equal rates, and its average temperature does not change. Because Earth is about 150 million km (93 million mi) from the sun, Earth's *radiative equilibrium temperature* is about 255 K ($-18°$C, $0°$F). But this temperature is effective the product is in protecting from UVB—the higher the number, the better the protection. Many "broad-spectrum" sunscreens protect against both UVA and UVB, although only the amount of UVB protection is considered in the SPF rating.

Protecting oneself from excessive exposure to the sun's energetic UV rays is certainly wise. Estimates are that, in a single year, more than 30,000 Americans will be diagnosed with malignant melanoma, the most deadly form of skin cancer. And in areas where the protective stratospheric ozone shield has weakened, there is an increased risk of problems associated with UVB. Using a good sunscreen and proper clothing can certainly help. The best way to protect yourself from too much sun, however, is to limit your time in direct sunlight, especially between the hours of 10 a.m. and 4 p.m. daylight saving time, when the sun is highest in the sky and its rays are most direct.

Presently, the National Weather Service makes a daily prediction of UV radiation levels for selected cities throughout the United States. The forecast, known as the *UV Index*, gives the UV level at its peak, around noon standard time or 1 p.m. daylight saving time. The index corresponds to five exposure categories set by the Environmental Protection Agency (EPA). An index value of 2 or less is considered "low," whereas a value of 11 or greater is deemed "extreme" (see ● Fig. 4). Depending on skin type, a UV index of 10 means that in direct sunlight (without sunscreen protection) a person's skin will likely begin to burn in about 6 to 30 minutes (see ● Fig. 5).

EXPOSURE CATEGORY	UV INDEX	PROTECTIVE MEASURES
Low	2 or less	Wear sunglasses on bright days; cover up and wear sunscreen.
Moderate	3 – 5	Take precautions; stay in shade near midday.
High	6 – 7	Apply SPF 15+ sunscreen; wear wide-brim hat and sunscreen; reduce time in sun between 10 a.m. and 4 p.m. (daylight savings time).
Very high	8 – 10	Take extra precautions; apply SPF 15+ sunscreen; wear wide-brim hat and sunscreen; minimize exposure between 10 a.m. and 4 p.m.
Extreme	11+	Take all precautions; apply SPF 15+ sunscreen; wear wide-brim hat and sunscreen; minimize exposure between 10 a.m. and 4 p.m. Unprotected skin can burn in minutes.

● **FIGURE 4** The UV Index.

● **FIGURE 5** If this photo were taken around 1 p.m. on a day when the UV index was 10, almost everyone on this beach without sunscreen would experience some degree of sunburning within 30 minutes.

much lower than Earth's observed average surface temperature of 288 K (15°C, 59°F). Why is there such a large difference?

The answer lies in the fact that *Earth's atmosphere absorbs and emits infrared radiation*. Unlike Earth, the atmosphere does *not* behave like a blackbody, as it absorbs some wavelengths of radiation and is transparent to others. Objects that selectively absorb and emit radiation, such as gases in our atmosphere, are known as **selective absorbers**. Let's examine this concept more closely.

SELECTIVE ABSORBERS AND THE ATMOSPHERIC GREENHOUSE EFFECT Just as some people are selective eaters of certain foods, most substances in our environment are selective absorbers; that is, they absorb only certain wavelengths of radiation. Glass is a good example of a selective absorber in that it absorbs some of the infrared and ultraviolet radiation it receives, but not the visible radiation that is transmitted through the glass. As a result, it is difficult to get a sunburn through the windshield of your car, although you can see through it.

WEATHER WATCH

The large ears of a jackrabbit are efficient emitters of infrared energy. Its ears help the rabbit survive the heat of a summer's day by radiating a great deal of infrared energy to the cooler sky above. Similarly, the large ears of the African elephant greatly increase its radiating surface area and promote cooling of its large mass.

● **FIGURE 2.11** The melting of snow outward from the trees causes small depressions to form. The melting is caused mainly by the snow's absorption of the infrared energy being emitted from the warmer tree and its branches. The trees are warmer because they are better absorbers of sunlight than is the snow.

Objects that selectively absorb radiation also selectively emit radiation at the same wavelength. This phenomenon is called **Kirchhoff's law**. This law states that *good absorbers are good emitters at a particular wavelength, and poor absorbers are poor emitters at the same wavelength.**

Snow is a good absorber as well as a good emitter of infrared energy (white snow actually behaves as a blackbody in the infrared wavelengths). The bark of a tree absorbs sunlight and emits infrared energy, which the snow around it absorbs. During the absorption process, the infrared radiation is converted into internal energy, and the snow melts outward away from the tree trunk, producing a small depression that encircles the tree, like the ones shown in ● Fig. 2.11.

● Figure 2.12 shows some of the most important selectively absorbing gases in our atmosphere. The shaded area represents the absorption characteristics of each gas at various wavelengths. Notice that both water vapor (H_2O) and carbon dioxide (CO_2) are strong absorbers of infrared radiation and poor absorbers of visible solar radiation. Other, less important, selective absorbers include nitrous oxide (N_2O) and methane (CH_4), as well as ozone (O_3), which is most abundant in the stratosphere. As these gases absorb infrared radiation emitted from Earth's surface, they gain kinetic energy (energy of motion). The gas molecules share this energy by colliding with neighboring air molecules, such as oxygen and nitrogen (both of which are poor absorbers of infrared energy). These collisions increase the average kinetic energy of the air, which results in an increase in air temperature. Thus, most of the infrared energy emitted from Earth's surface keeps the lower atmosphere warm.

Besides being selective absorbers, water vapor and CO_2 selectively emit radiation at infrared wavelengths.** This radiation travels away from these gases in all directions. A portion of this energy is radiated toward Earth's surface and absorbed, thus heating the ground. Earth, in turn, constantly radiates infrared

---

*Strictly speaking, this law only applies to gases.

**Nitrous oxide, methane, and ozone also emit infrared radiation, but their concentration in the atmosphere is much smaller than water vapor and carbon dioxide (see Table 1.1, p. 6).

● **FIGURE 2.12** Absorption of radiation by gases in the atmosphere. The dark purple shaded area represents the percent of radiation absorbed by each gas. The strongest absorbers of infrared radiation are water vapor and carbon dioxide. The bottom figure represents the percent of radiation absorbed by all of the atmospheric gases.

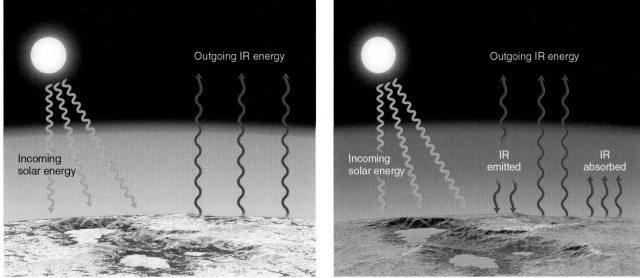

(a) Without greenhouse gases   (b) With greenhouse gases

● **FIGURE 2.13** (a) Near the surface in an atmosphere with little or no greenhouse gases, Earth's surface would constantly emit infrared (IR) radiation upward, both during the day and at night. Incoming energy from the sun would equal outgoing energy from the surface, but the surface would receive virtually no IR radiation from its lower atmosphere (i.e., there would be no atmospheric greenhouse effect). Earth's surface air temperature would be quite low, and small amounts of water found on the planet would be in the form of ice. (b) In an atmosphere with greenhouse gases, Earth's surface not only receives energy from the sun but also infrared energy from the atmosphere. Incoming energy still equals outgoing energy, but the added IR energy from the greenhouse gases raises Earth's average surface temperature to a more habitable level.

energy upward, where it is absorbed and warms the lower atmosphere. In this way, water vapor and CO_2 absorb and radiate infrared energy and act as an insulating layer around Earth, keeping part of Earth's infrared radiation from escaping rapidly into space. Consequently, Earth's surface and the lower atmosphere are much warmer than they would be if these selectively absorbing gases were not present. In fact, as we saw earlier, Earth's mean radiative equilibrium temperature without CO_2 and water vapor would be around $-18°C$ ($0°F$), or about $33°C$ ($59°F$) lower than at present.

The absorption characteristics of water vapor, CO_2, and other gases such as methane and nitrous oxide (depicted in Fig. 2.12) were, at one time, thought to be similar to the glass of a florist's greenhouse. In a greenhouse, the glass allows visible radiation to come in, but inhibits to some degree the passage of outgoing infrared radiation. For this reason, the absorption of infrared radiation from Earth by water vapor and CO_2 is popularly called the **greenhouse effect**. However, studies have shown that the warm air inside a greenhouse is probably caused more by the air's inability to circulate and mix with the cooler outside air than by the entrapment of infrared energy. Because of these findings, some scientists suggest that the greenhouse effect should be called the *atmosphere effect*. To accommodate everyone, we will usually use the term *atmospheric greenhouse effect* when describing the role that water vapor, CO_2, and other **greenhouse gases**[*] play in keeping Earth's mean surface temperature higher than it otherwise would be.

Look again at Fig. 2.12 and observe that, in the bottom diagram, there is a region between about 8 and 11 μm where neither water vapor nor CO_2 readily absorb infrared radiation. Because

---

[*]The term "greenhouse gases" derives from the standard use of "greenhouse effect." Greenhouse gases include, among others, water vapor, carbon dioxide, methane, nitrous oxide, and ozone.

these wavelengths of emitted energy pass upward through the atmosphere and out into space, the wavelength range (between 8 and 11 μm) is known as the **atmospheric window**.

Clouds can enhance the atmospheric greenhouse effect. Tiny liquid cloud droplets are selective absorbers in that they are good absorbers of infrared radiation but poor absorbers of visible solar radiation. Clouds even absorb the wavelengths between 8 and 11 μm, which are otherwise "passed up" by water vapor and CO_2. Thus, they have the effect of enhancing the atmospheric greenhouse effect by closing the atmospheric window.

Clouds—especially low, thick ones—are excellent emitters of infrared radiation. Their tops radiate infrared energy upward and their bases radiate energy back to Earth's surface where it is absorbed and, in a sense, radiated back to the clouds. This process keeps calm, cloudy nights warmer than calm, clear ones. If the clouds remain into the next day, they prevent much of the sunlight from reaching the ground by reflecting it back to space. Since the ground does not heat up as much as it would in full sunshine, cloudy, calm days are normally cooler than clear, calm days. Hence, the presence of clouds tends to keep nighttime temperatures higher and daytime temperatures lower.

In summary, the atmospheric greenhouse effect occurs because water vapor, CO_2, and other greenhouse gases are selective absorbers. They allow most of the sun's visible radiation to reach the surface, but they absorb a good portion of Earth's outgoing infrared radiation, preventing it from escaping into space. It is the atmospheric greenhouse effect, then, that keeps the temperature of our planet at a level where life can survive. The greenhouse effect is not just a "good thing"; it is essential to life on Earth, for without it, air at the surface would be extremely cold (see ● Fig. 2.13).

ENHANCEMENT OF THE GREENHOUSE EFFECT In spite of the inaccuracies that have plagued temperature measurements in the past, studies show that, during the past 120 years or so, Earth's surface air temperature has undergone a warming of about $1.0°C$ ($1.48°F$). Computer-based climate models that mathematically simulate the physical processes of the atmosphere, oceans, and ice predict that should global warming continue unabated, we would be irrevocably committed to major effects from climate change, such as a continuing rise in sea level and a shift in global precipitation patterns.

The main cause of this type of climate change is the greenhouse gas CO_2, whose concentration has been increasing primarily due to the burning of fossil fuels and to deforestation. (Look back at Fig. 1.6, p. 8.) However, increasing concentrations of other greenhouse gases, such as methane (CH_4), nitrous oxide (N_2O), and chlorofluorocarbons (CFCs),* have collectively been shown to have an effect approaching that of CO_2. In addition, as temperatures warm, more water vapor is added to the air from the world's oceans. Overall, water vapor accounts for about 60 percent of the atmospheric greenhouse effect; CO_2 accounts for about 26 percent, methane about 7 percent, and the remaining greenhouse gases about 7 percent.

Presently, the concentration of CO_2 in a volume of air near the surface is just over 0.04 percent, and it is increasing each year. Climate models predict that a continuing increase of CO_2 and other greenhouse gases will cause Earth's current average surface temperature to rise an additional 1 to $3°C$ (1.8 to $5.4°F$) or more by the end of this century. In fact, Earth's average surface temperature during the first decade of this century (2001–2010) was about $0.21°C$ ($0.38°F$) warmer than the previous decade (1991–2000). If the current warming trend were to continue at that rate, the twenty-first century would warm by $2.1°C$ ($3.8°F$). How can increasing such a small quantity of CO_2 and adding minuscule amounts of other greenhouse gases bring about such a large temperature increase?

Mathematical climate models predict that rising ocean temperatures will cause an increase in evaporation rates. The added *water vapor*—the primary greenhouse gas—will enhance the atmospheric greenhouse effect and roughly double the temperature rise in what is known as a *positive feedback* on the climate system. But there are other feedback to consider.**

The two potentially largest and least understood feedback in the climate system are the clouds and the oceans. Clouds can change area, depth, and radiation properties simultaneously with climatic changes. The net effect of all these changes is not totally clear at this time. Oceans, on the other hand, cover 70 percent of the planet. The response of ocean circulations, ocean temperatures, and sea ice to global warming will determine the global pattern and speed of climate change. Unfortunately, it is not now known how quickly each of these feedback will respond.

Satellite data and computer simulations suggest that clouds overall appear to *cool* Earth's climate, as they reflect and radiate away more energy than they retain. (Earth would be about $5°C$ ($9°F$) warmer if no clouds were present.) So an increase in global cloudiness (if it were to occur) might offset some of the global warming brought on by an enhanced atmospheric greenhouse effect. If clouds were to act on the climate system in this manner, they would provide a *negative feedback* on climate change. The actual result will also depend on what types of clouds are present, because some clouds are more reflective and have a stronger cooling effect than others. The most recent models tend to show that changes in clouds as a whole would most likely provide a small positive feedback on the climate system.

Uncertainties unquestionably exist about the impact that increasing levels of CO_2 and other greenhouse gases will have on enhancing the atmospheric greenhouse effect. Nonetheless, many independent analyses agree that climate change is occurring worldwide primarily as a result of increasing levels of greenhouse gases. The evidence for this conclusion comes from increases in global average air and ocean temperatures, as well as from the widespread melting of snow and ice, rising sea levels, and other conditions. These changes are consistent with the effects one would expect from the increase in greenhouse gases. (We will examine the important topic of climate change in more detail in Chapter 18.)

**A feedback is a process whereby an initial change in a process will tend to either reinforce the process (positive feedback) or weaken the process (negative feedback). The water vapor–greenhouse feedback is a positive feedback because the initial increase in temperature is reinforced by the addition of more water vapor, which absorbs more of Earth's infrared energy, thus strengthening the greenhouse effect and enhancing the warming.

BRIEF REVIEW

In the last several sections, we have explored examples of some of the ways radiation is absorbed and emitted by various objects. Before we continue, here are a few important facts and principles:

- All objects with a temperature above absolute zero emit radiation.

- The higher an object's temperature, the greater the amount of radiation emitted per unit surface area and the shorter the wavelength of maximum emission.

- Earth absorbs solar radiation only during the daylight hours; however, it emits infrared radiation continuously, both during the day and at night.

---

*To refresh your memory, recall from Chapter 1 that CFCs were once the most widely used propellant in spray cans.

- Earth's surface behaves as a blackbody, making it a much better absorber and emitter of radiation than the atmosphere.

- Water vapor and carbon dioxide are important atmospheric greenhouse gases that selectively absorb and emit infrared radiation, thereby keeping Earth's average surface temperature warmer than it otherwise would be.

- Cloudy, calm nights are often warmer than clear, calm nights because clouds strongly emit infrared radiation back to Earth's surface.

- It is not the greenhouse effect itself that is of concern, but the enhancement of it due to increasing levels of greenhouse gases.

- As greenhouse gases continue to increase in concentration, the average surface air temperature is projected to rise substantially by the end of this century.

With these concepts in mind, we will first examine how the air near the ground warms; then we will consider how Earth and its atmosphere maintain a yearly energy balance.

WARMING THE AIR FROM BELOW If you look back at Fig. 2.12 (p. 44), you'll notice that the atmosphere does not readily absorb radiation with wavelengths between 0.3 μm and 1.0 μm, the region where the sun emits most of its energy. Consequently, on a clear day, solar energy passes through the lower atmosphere with little effect upon the air. Ultimately it reaches the surface, warming it (see ● Fig. 2.14). Air molecules in contact with the heated surface bounce against it, gain energy by *conduction*, then shoot upward like freshly popped kernels of corn, carrying their energy with them. Because the air near the ground is very dense, these molecules only travel a short distance (about 10^{-7} m) before they collide with other molecules. During the collision, these more rapidly moving molecules share their energy with less energetic molecules, raising the average temperature of the air. But air is such a poor heat conductor that this process is only important within a few centimeters of the ground.

As the surface air warms, it actually becomes less dense than the air directly above it. The warmer air rises and the cooler air sinks, setting up thermals, or *free convection cells*, that transfer heat upward and distribute it through a deeper layer of air. The rising air expands and cools, and, if sufficiently moist, the water vapor condenses into cloud droplets, releasing latent heat that warms the air. Meanwhile, Earth constantly emits infrared energy. Some of this energy is absorbed by greenhouse gases (such as water vapor and carbon dioxide) that emit infrared energy upward and downward, back to the surface. Since the concentration of water vapor decreases rapidly above Earth, most of the absorption occurs in a layer near the surface. Hence, the lower atmosphere is mainly heated from the ground upward.

SHORTWAVE RADIATION STREAMING FROM THE SUN As the sun's radiant energy travels through space, essentially nothing interferes with it until it reaches the atmosphere. At the top of the atmosphere, solar energy received on a surface perpendicular to the sun's rays appears to remain fairly constant at nearly two

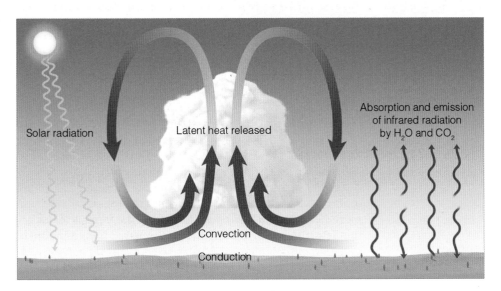

● **FIGURE 2.14** Air in the lower atmosphere is heated from the ground upward. Sunlight warms the ground, and the air above is warmed by conduction, convection, and infrared radiation. Further warming occurs during condensation as latent heat is given up to the air inside the cloud.

Blue Skies, Red Suns, and White Clouds

We know that the sky is blue because air molecules selectively scatter the shorter wavelengths of visible light—green, violet, and blue waves—more effectively than the longer wavelengths of red, orange, and yellow (see Fig. 2.15, on p. 49). When these shorter waves reach our eyes, the brain processes them as the color "blue." Therefore, on a clear day when we look up, blue light strikes our eyes from all directions, making the sky appear blue.

At noon, the sun is perceived as white because all the waves of visible sunlight strike our eyes (see ● Fig. 6). At sunrise and sunset, the white light from the sun must pass through a thick portion of the atmosphere. Scattering of light by air molecules (and particles) removes the shorter waves (blue light) from the beam, leaving the longer waves of red, orange, and yellow to pass on through. This situation often creates the image of a ruddy sun at sunrise and sunset. An observer at sunrise or sunset in Fig. 6 might see a sun similar to the one shown in ● Fig. 7.

The sky is blue, but why are clouds white? Cloud droplets are much larger than air molecules and do not selectively scatter sunlight. Instead, these larger droplets scatter all wavelengths of visible light more or less equally. Hence, clouds appear white because millions of cloud droplets scatter all wavelengths of visible light about equally in all directions.

● **FIGURE 6** At noon, the sun usually appears a bright white. At sunrise and at sunset, sunlight must pass through a thick portion of the atmosphere. Much of the blue light is scattered out of the beam (as illustrated by arrows), causing the sun to appear more red.

● **FIGURE 7** A red sunset produced by the process of scattering.

© C. Donald Ahrens

calories on each square centimeter each minute or 1361 W / m²—a value called the **solar constant**.*

When solar radiation enters the atmosphere, a number of interactions take place. For example, some of the energy is absorbed by gases, such as ozone, in the upper atmosphere. Moreover, when sunlight strikes very small objects, such as air molecules and dust particles, the light itself is deflected in all directions—forward, backward, and sideways. The distribution of light in this manner is called **scattering**. (Scattered light is also called *diffuse light*.) Because air molecules are much smaller than the wavelengths of visible light, they are more effective scatterers of the shorter (blue) wavelengths than the longer (red) wavelengths (see ● Fig. 2.15). Hence, when we look away from the direct beam of sunlight, blue light strikes our eyes from all directions, turning the daytime sky blue. (More information on the effect of scattered light and what we see is given in Focus section 2.4.)

Sunlight can be **reflected** from objects. Generally, reflection differs from scattering in that during the process of reflection more light is sent *backward*. **Albedo** is the percent of radiation returning from a given surface compared to the amount of radiation initially striking that surface. Albedo, then, represents the *reflectivity* of the surface. In ▼ Table 2.3, notice that thick clouds have a higher albedo than thin clouds. On the average, the albedo of clouds is near 60 percent. When solar energy strikes a

---

*By definition, the solar constant (which, in actuality, is not "constant") is the rate at which radiant energy from the sun is received on a surface at the outer edge of the atmosphere perpendicular to the sun's rays when Earth is at an average distance from the sun. Satellite measurements suggest the solar constant varies slightly as the sun's radiant output varies. The latest measurements and laboratory tests indicate that the solar constant is around 1361 W / m². It rises by about 1 W/m² during peaks of solar activity, which occur about every 11 years.

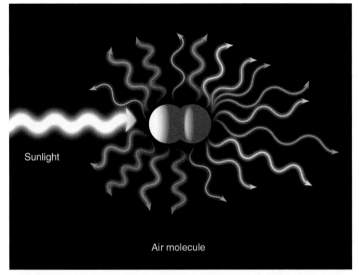

● **FIGURE 2.15** The scattering of light by air molecules. Air molecules tend to selectively scatter the shorter (violet, green, and blue) wavelengths of visible white light more effectively than the longer (orange, yellow, and red) wavelengths.

surface covered with snow, up to 95 percent of the sunlight may be reflected. Most of this energy is in the visible and ultraviolet wavelengths. Consequently, reflected radiation, coupled with direct sunlight, can produce severe sunburns on the exposed skin of unwary snow skiers, and unprotected eyes can suffer the agony of snow blindness.

Water surfaces, on the other hand, reflect only a small amount of solar energy. For an entire day, a smooth water surface will have an average albedo of about 10 percent. Water has

▼ **TABLE 2.3** **Typical Albedo of Various Surfaces**∗

SURFACE	ALBEDO (PERCENT)
Fresh snow	75 to 95
Clouds (thick)	60 to 90
Clouds (thin)	30 to 50
Venus	78
Ice	30 to 40
Sand	15 to 45
Earth and atmosphere	30
Mars	17
Grassy field	10 to 30
Dry, plowed field	5 to 20
Water	10∗
Forest	3 to 10
Moon	7

∗Daily average.

the highest albedo (and can therefore reflect sunlight best) when the sun is low on the horizon and the water is a little choppy. This may explain why people who wear brimmed hats while fishing from a boat in choppy water on a sunny day can still get sunburned during midmorning or midafternoon. Averaged for an entire year, Earth and its atmosphere (including its clouds) will redirect about 30 percent of the sun's incoming radiation back to space, which gives Earth and its atmosphere a combined albedo of 30 percent (see ● Fig. 2.16).

EARTH'S ANNUAL ENERGY BALANCE Although the average temperature at any one place may vary considerably from year to year, Earth's overall average equilibrium temperature changes only slightly from one year to the next. This fact indicates that, each year, Earth and its atmosphere combined must send off into space just as much energy as they receive from the sun. The same type of energy balance must exist between Earth's surface and the atmosphere. That is, each year, Earth's surface must return to the atmosphere the same amount of energy that it absorbs. If this did not occur, Earth's average surface temperature would change. How do Earth and its atmosphere maintain this yearly energy balance?

Suppose 100 units of solar energy reach the top of Earth's atmosphere. We can see in Fig. 2.16 that, on the average, clouds, Earth, and the atmosphere reflect and scatter 30 units back to space, and that the atmosphere and clouds together absorb 19 units, which leaves 51 units of direct and indirect solar radiation to be absorbed at Earth's surface.

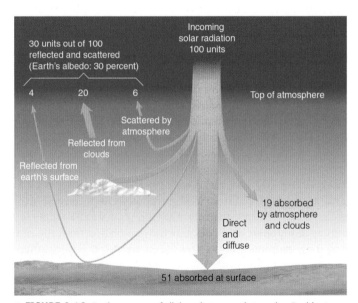

● **FIGURE 2.16** On the average, of all the solar energy that reaches Earth's atmosphere annually, about 30 percent ($^{30}/_{100}$) is reflected and scattered back to space, giving Earth and its atmosphere an albedo of 30 percent. Of the remaining solar energy, about 19 percent is absorbed by the atmosphere and clouds, and about 51 percent is absorbed at the surface.

CRITICAL THINKING QUESTION If clouds were to increase worldwide, how would this increase influence Earth's albedo? How would this increase in cloudiness affect the amount of solar energy reaching Earth's surface? Even though global cloudiness increases, why would the surface temperature not necessarily decrease?

● **FIGURE 2.17** The Earth-atmosphere energy balance. Numbers represent approximations based on surface observations and satellite data. While the actual value of each process may vary by several percent, it is the relative size of the numbers that is important.

● Figure 2.17 shows approximately what happens to the solar radiation that is absorbed by the surface and the atmosphere. Out of 51 units reaching the surface, a large amount (23 units) is used to evaporate water, and about 7 units are lost through conduction and convection, which leaves 21 units to be radiated away as infrared energy. Look closely at Fig. 2.17 and notice that Earth's surface actually radiates upward a whopping 117 units. It does so because, although it receives solar radiation only during the day, it constantly emits infrared energy both during the day and at night. Additionally, the atmosphere above only allows a small fraction of this energy (6 units) to pass through into space. The majority of it (111 units) is absorbed mainly by the greenhouse gases water vapor and CO_2 and by clouds. Much of this energy (96 units) is radiated back to Earth, producing the atmospheric greenhouse effect. Hence, Earth's surface receives nearly twice as much longwave infrared energy from its atmosphere as it does shortwave radiation from the sun. In all these exchanges, notice that the energy lost at Earth's surface (147 units) is exactly balanced by the energy gained there (147 units).

A similar balance exists between Earth's surface and its atmosphere. Again, observe in Fig. 2.17 that the energy gained by the atmosphere (160 units) balances the energy lost. Moreover, averaged for an entire year, the solar energy received at Earth's surface (51 units) and that absorbed by Earth's atmosphere (19 units) balances the infrared energy lost to space by Earth's surface (6 units) and its atmosphere (64 units).

We can see the effect that conduction, convection, and latent heat play in the warming of the atmosphere if we look at the energy balance only in radiative terms. Earth's surface receives 147 units of radiant energy from the sun and its own atmosphere, while it radiates away 117 units, producing a *surplus* of 30 units. The atmosphere, on the other hand, receives 130 units (19 units from the sun and 111 from Earth), while it loses 160 units, producing a *deficit* of 30 units. The balance (30 units) is the warming of the atmosphere produced by the heat transfer processes of conduction and convection (7 units) and by the release of latent heat (23 units).

So, Earth and the atmosphere absorb energy from the sun as well as from each other. In all of the energy exchanges, a delicate balance is maintained. Essentially, there is no yearly gain or loss of total energy, and the average temperature of Earth and its atmosphere remains fairly constant from one year to the next. This equilibrium does not imply that Earth's average temperature does not change, but that the changes are small from year to year (usually less than one-tenth of a degree Celsius) and become significant only when measured over many years.*

Even though Earth and the atmosphere together maintain an annual energy balance, such a balance is not maintained at each latitude. High latitudes tend to lose more energy to space each year than they receive from the sun, while low latitudes tend to gain more energy during the course of a year than they lose. From ● Fig. 2.18 we can see that only at middle latitudes near 38° does the amount of energy received each year balance the amount lost. From this situation, we might conclude that polar regions are growing colder each year, while tropical regions are becoming warmer. But this does not happen. To compensate for these gains and losses of energy, winds in the atmosphere and currents in the oceans circulate warm air and water toward the poles, and cold air and water toward the equator. Thus, the transfer of heat

---

*Recall that Earth's surface warmed by about 1.0°C (1.8°F) during the last century. The global average in 2001–2010 was about 0.21°C (0.38°F) warmer than in 1991–2000.

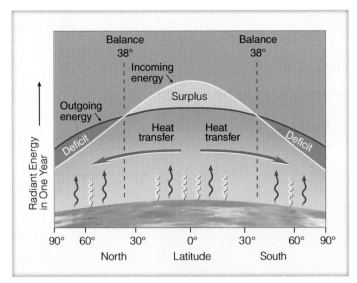

● **FIGURE 2.18** The average annual incoming solar radiation (yellow lines) absorbed by Earth and the atmosphere along with the average annual infrared radiation (red lines) emitted by Earth and the atmosphere.

CRITICAL THINKING QUESTION Climate models predict that as Earth warms, higher latitudes will warm more than lower latitudes. If lower latitudes hardly warm at all, would you expect the deficit in energy at higher latitudes to increase or decrease? How do you feel this change in energy might affect the rate at which heat is transferred poleward?

energy by atmospheric and oceanic circulations prevents low latitudes from steadily becoming warmer and high latitudes from steadily growing colder. These circulations are extremely important to weather and climate, and will be treated more completely in Chapter 10. Up to this point we have considered radiant energy of the sun and Earth. Before we turn our attention to how incoming solar energy in the form of particles produces the aurora and space weather, you may wish to read Focus section 2.5.

Solar Particles, the Aurora, and Space Weather

From the sun and its tenuous atmosphere comes a continuous discharge of particles. This discharge happens because, at extremely high temperatures, gases become stripped of electrons by violent collisions and acquire enough speed to escape the gravitational pull of the sun.

As these charged particles (ions and electrons) travel through space in the form of plasma, they are referred to as the **solar wind**. When the solar wind moves close enough to Earth, it interacts with Earth's magnetic field, sometimes producing a variety of effects known as **space weather**.

The magnetic field that surrounds Earth is much like the field around an ordinary bar magnet (see ● Fig. 2.19). Both have north and south magnetic poles, and both have invisible lines of force (field lines) that link the poles. On Earth, these field lines form closed loops as they enter near the North Magnetic Pole and leave

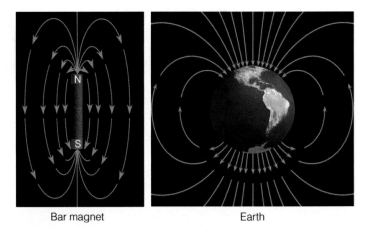

Bar magnet   Earth

● **FIGURE 2.19** A magnetic field surrounds Earth just as it does a bar magnet.

near the South Magnetic Pole. Most scientists believe that an electric current coupled with fluid motions deep in Earth's hot molten core is responsible for Earth's magnetic field. This field protects Earth, to some degree, from the onslaught of the solar wind.

Observe in ● Fig. 2.20 that when the solar wind encounters Earth's magnetic field, it severely deforms it into a teardrop-shaped cavity known as the *magnetosphere*. On the side facing the sun, the pressure of the solar wind compresses the field lines. On the opposite side, the magnetosphere stretches out into a long tail—the *magnetotail*—which reaches far beyond the moon's orbit. In a way, the magnetosphere acts as an obstacle to the solar wind by causing some of its particles to flow around Earth.

Inside Earth's magnetosphere are ionized gases. Some of these gases are solar wind particles, while others are ions from Earth's upper atmosphere that have moved upward along electric field lines into the magnetosphere.

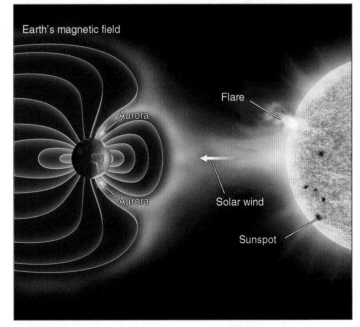

● **FIGURE 2.20** The stream of charged particles from the sun—called the *solar wind*—distorts Earth's magnetic field into a teardrop shape known as the *magnetosphere*. These particles, which spiral in along magnetic field lines, interact with atmospheric gases and produce the aurora.

Characteristics of the Sun

The sun is our nearest star. It is some 150 million km (93 million mi) from Earth. The next star, Alpha Centauri, is more than 250,000 times farther away. Even though Earth only receives about one two-billionths of the sun's total energy output, it is this energy that allows life to flourish. Sunlight determines the rate of photosynthesis in plants and strongly regulates the amount of evaporation from the oceans. It warms this planet and drives the atmosphere into the dynamic patterns we experience as everyday wind and weather. Without the sun's radiant energy, Earth would gradually cool, in time becoming encased in a layer of ice! Evidence of life on the cold, dark, and barren surface would be found only in fossils. Fortunately, the sun has been shining for billions of years, and it is likely to shine for at least several billion more.

The sun is a giant celestial furnace. Its core is extremely hot, with a temperature estimated to be near 15 million degrees Celsius. In the core, hydrogen nuclei (protons) collide at such fantastically high speeds that they fuse together to form helium nuclei. This thermonuclear process generates an enormous amount of energy, which gradually works its way to the sun's outer luminous surface—the *photosphere* ("sphere of light"). Temperatures here are much cooler than in the interior, generally between 4200°C and 5700°C. We have noted already that a body with this surface temperature emits radiation at a maximum rate in the visible region of the spectrum. The sun is, therefore, a shining example of such an object.

Dark blemishes on the photosphere, called *sunspots*, are huge, cooler regions that typically

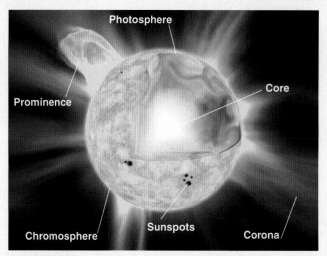

● **FIGURE 8** Various regions of the sun.

average more than five times the diameter of Earth. Although sunspots are not well understood, they are known to be regions of strong magnetic fields. They are cyclic, with the maximum number of spots occurring approximately every 11 years.

Above the photosphere are the *chromosphere* and the *corona* (see ● Fig.8). The chromosphere ("color sphere") acts as a boundary between the relatively cool photosphere (around 5000°C) and the much hotter corona, the outermost envelope of the solar atmosphere (typically 1 to 3 million degrees Celsius). During a solar eclipse, the corona is visible. It appears as a pale, milky cloud encircling the sun. Although much hotter than the photosphere, the corona radiates much less energy because its density is extremely low. This very thin solar atmosphere extends into space for many millions of kilometers.*

Violent solar activity occasionally occurs in the regions of sunspots. The most dramatic of these events are *prominences* and *flares*. Prominences are huge cloudlike jets of gas that often shoot up into the corona in the form of an arch. Solar flares are tremendous, but brief, eruptions. They emit large quantities of high-energy ultraviolet radiation, as well as energized charged particles, mainly protons and electrons, which stream outward away from the sun at extremely high speeds.

---

*You should never attempt to look at the sun's corona, either with sunglasses or through exposed negatives. Take this warning seriously. Viewing just a small area of the sun directly permits large amounts of UV radiation to enter the eye, causing serious and permanent damage to the retina. View the sun by projecting its image onto a sheet of paper, using a telescope or pinhole camera. Only when the sun is completely blocked during a total solar eclipse can one view the corona safely.

Normally, the solar wind approaches Earth at an average speed of 400 km/sec. However, during periods of high solar activity (many sunspots and flares), the solar wind is more dense, travels much faster, and carries more energy. When these energized solar particles reach Earth, they cause a variety of effects, such as changing the shape of the magnetosphere and producing auroral displays.

The aurora is not reflected light from the polar ice fields, nor is it light from demons' lanterns as they search for lost souls. The *aurora* is produced by the solar wind disturbing the magnetosphere. The disturbance involves high-energy particles within the magnetosphere being ejected into Earth's upper atmosphere, where they excite atoms and molecules. The excited atmospheric gases emit visible radiation, which causes the sky to glow like a neon light. Let's examine this process more closely.

A high-energy particle from the magnetosphere will, upon colliding with an air molecule (or atom), transfer some of its energy to the molecule. The molecule then becomes excited (see ● Fig. 2.21). Just as excited football fans leap up when their favorite team scores the winning touchdown, electrons in an excited molecule jump into a higher energy level as they orbit its center. As the fans sit down after all the excitement is over, so electrons quickly return to their lower level. When molecules de-excite,

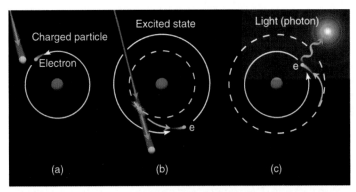

● **FIGURE 2.21** When an excited atom, ion, or molecule de-excites, it can emit visible light. (a) The electron in its normal orbit becomes excited by a charged particle and (b) jumps into a higher energy level. When the electron returns to its normal orbit, it (c) emits a photon of light.

they release the energy originally received from the energetic particle, either all at once (one big jump), or in steps (several smaller jumps). This emitted energy is given up as radiation. If its wavelength is in the visible range, we see it as visible light. In the Northern Hemisphere, we call this light show the **aurora borealis**, or *northern lights*; its counterpart in the Southern Hemisphere is the **aurora australis**, or *southern lights*.

Because each atmospheric gas has its own set of energy levels, each gas has its own characteristic color. For example,

the de-excitation of atomic oxygen can emit green or red light. Molecular nitrogen gives off red and violet light. The shades of these colors can be spectacular as they brighten and fade, sometimes in the form of waving draperies, sometimes as unmoving, yet flickering, arcs and soft coronas. On a clear, quiet night the aurora is an eerie yet beautiful spectacle. (See ● Fig. 2.22.)

The aurora is most frequently seen in polar latitudes. Energetic particles trapped in the magnetosphere move along Earth's magnetic field lines. Because these lines emerge from Earth near the magnetic poles, it is here that the particles interact with atmospheric gases to produce an aurora. Notice in ● Fig. 2.23 that the zone of most frequent auroral sightings (aurora belt) is not at the magnetic pole (marked by the flag MN), but equatorward of it, where the field lines emerge from Earth's surface. At lower

● **FIGURE 2.22** The aurora borealis is a phenomenon that forms as energetic particles from the sun interact with Earth's atmosphere.

Lindsey P. Martin Photography

● **FIGURE 2.23** The aurora belt (solid red line) represents the region where you would most likely observe the aurora on a clear night. (The numbers represent the average number of nights per year on which you might see an aurora if the sky were clear.) The flag MN denotes the North Magnetic Pole, where Earth's magnetic field lines emerge from Earth. The flag NP denotes the geographic North Pole, about which Earth rotates.

latitudes, where the field lines are oriented almost horizontal to Earth's surface, the chances of seeing an aurora diminish rapidly. How high is the aurora? The exact height varies, but it is almost always in the thermosphere, more than 80 km (50 mi) above Earth's surface.

Sometimes the northern lights are seen as far south as the southern United States. Such sightings happen only when the sun is very active—as giant flares hurl electrons and protons earthward at a fantastic rate. These particles move so fast that some of them penetrate unusually deep into Earth's magnetic field before they are trapped by it. In a process not fully understood, particles from the magnetosphere are accelerated toward Earth along electrical field lines that parallel the magnetic field lines. The acceleration of these particles gives them sufficient energy so that when they enter the upper atmosphere they are capable of producing an auroral display much farther south than usual. Auroral displays are most frequent and intense during the active part of the sun's cyclic activity, which peaks about every 11 years. During active solar periods, there are numerous sunspots (huge cooler regions on the sun's surface) and giant flares (solar eruptions) that send large quantities of solar wind particles traveling away from the sun at high speeds (hundreds of kilometers a second).

SOLAR STORMS AND SPACE WEATHER Along with producing the beauty of the aurora, the barrage of energized particles reaching our atmosphere during solar storms can also produce many negative effects. These are most common during the active phase of the 11-year solar cycle. In 1859, a huge solar storm disrupted telegraph operations around the world. Another solar storm in March 1989 caused millions of people across Quebec to lose electrical power for several hours. Airlines sometimes reroute planes away from polar regions during the most intense solar storms, which reduces the risk of interference with radio communications. Solar storms can also affect satellite operations, which is a growing concern because of the many types of satellites now used for telecommunications, navigation, and other purposes. A $630 million research satellite from Japan failed during an intense solar storm in October 2003. Because solar storms heat the outer atmosphere, they can increase the drag on satellites and reduce their lifespan in orbit.

Space weather was on the quiet side during the most recent minimum in the solar cycle (2008–2009). During this time the number of sunspots observed was at its lowest level in nearly a century. The subsequent solar maximum of the early to mid-2010s was the weakest in more than a century. Scientists cannot say for sure how long the recent trend toward lessened activity might continue, because techniques for predicting the strength of solar cycles are still being researched and tested.

In summary, energy for the aurora comes from the solar wind, which disturbs Earth's magnetosphere. This disturbance causes energetic particles to enter the upper atmosphere, where they collide with atoms and molecules. The atmospheric gases become excited and emit energy in the form of visible light. The energetic particles can also produce solar storms, which can have adverse effects on electrical systems, satellites, and radio communications.

In this chapter, we have seen how the concepts of heat and temperature differ and how heat is transferred in our environment. We learned that latent heat is an important source of atmospheric heat energy. We also learned that conduction, the transfer of heat by molecular collisions, is most effective in solids. Because air is a poor heat conductor, conduction in the atmosphere is only important in the shallow layer of air in contact with Earth's surface. A more important process of atmospheric heat transfer is convection, which involves the mass movement of air (or any fluid) with its energy from one region to another. Another significant heat transfer process is radiation—the transfer of energy by means of electromagnetic waves.

The hot sun emits most of its radiation as shortwave radiation. A portion of this energy heats Earth, and Earth, in turn, warms the air above. The cool Earth emits most of its radiation as longwave infrared radiation. Selective absorbers in the atmosphere, such as water vapor and carbon dioxide, absorb some of Earth's infrared radiation and radiate a portion of it back to the surface, where it warms the surface, producing the atmospheric greenhouse effect. Because clouds are both good absorbers and good emitters of infrared radiation, they keep calm, cloudy nights warmer than calm, clear nights. The average equilibrium temperature of Earth and the atmosphere remains fairly constant from one year to the next because the amount of energy they absorb each year is equal to the amount of energy they lose.

Finally, we examined how the sun's energy in the form of solar wind particles interacts with our atmosphere to produce auroral displays and solar storms.

KEY TERMS

The following terms are listed (with corresponding page numbers) in the order they appear in the text. Define each. Doing so will aid you in reviewing the material covered in this chapter.

QUESTIONS FOR REVIEW

1. How does the average speed (motion) of air molecules relate to the air temperature?
2. Distinguish between temperature and heat.
3. How does the Kelvin temperature scale differ from the Celsius scale?
   (a) Why is the Kelvin scale often used in scientific calculations?
   (b) Based on your experience, would a temperature of 250 K be considered warm or cold? Explain.
4. Explain how heat is transferred in Earth's atmosphere by:
   (a) conduction;
   (b) convection;
   (c) radiation.
5. How is latent heat an important source of atmospheric energy?
6. In the atmosphere, how does advection differ from convection?
7. How does the temperature of an object influence the radiation that it emits?
8. How does the amount of radiation emitted by Earth differ from that emitted by the sun?
9. How do the wavelengths of most of the radiation emitted by the sun differ from those emitted by the surface of Earth?
10. Which photon carries the most energy—infrared, visible, or ultraviolet?
11. When a body reaches a radiative equilibrium temperature, what is taking place?
12. If Earth's surface continually radiates energy, why doesn't it become colder and colder?
13. Why are carbon dioxide and water vapor called selectively absorbing greenhouse gases?
14. Explain how Earth's atmospheric greenhouse effect works.
15. What greenhouse gases are most responsible for the enhancement of Earth's greenhouse effect?
16. Why do most climate models predict that Earth's average surface temperature could increase by an additional $1°$ to $3°C$ ($1.8°$ to $5.4°F$) or more by the end of this century?

17. What processes contribute to Earth's albedo being 30 percent?
18. How is the lower atmosphere warmed from the surface upward?
19. Explain how Earth and its atmosphere balance incoming energy with outgoing energy.
20. If a blackbody is a theoretical object, why can both the sun and Earth be treated as blackbodies?
21. What is the solar wind?
22. Explain how the aurora is produced.

QUESTIONS FOR THOUGHT

1. Explain why the bridge in Fig. 2.24 (p. 57) becomes icy before the roadway.
2. Explain why the first snowfall of the winter usually "sticks" better to tree branches than to bare ground.
3. At night, why do materials that are poor heat conductors cool to temperatures less than the surrounding air?
4. Explain how, in winter, ice can form on puddles (in shaded areas) when the temperature above and below the puddle is slightly above freezing.
5. In northern latitudes, the oceans are warmer in summer than they are in winter. In which season do the oceans lose heat most rapidly to the atmosphere by conduction? Explain.

6. How is heat transferred away from the surface of the moon? (Hint: The moon has no atmosphere.)
7. Why is ultraviolet radiation more successful in dislodging electrons from air atoms and molecules than is visible radiation?
8. Why must you stand closer to a small fire to experience the same warmth you get when standing farther away from a large fire?
9. If water vapor were no longer present in the atmosphere, how would Earth's energy budget be affected?
10. Which will show the greatest increase in temperature when illuminated with direct sunlight: a plowed field or a blanket of snow? Explain.
11. Why does the surface temperature often increase on a clear, calm night as a low cloud moves overhead?
12. Which would have the greatest effect on Earth's greenhouse effect: removing all of the CO_2 from the atmosphere or removing all of the water vapor? Explain why you chose your answer.
13. Explain why an increase in cloud cover surrounding Earth would increase Earth's albedo, yet not necessarily lead to a lower Earth surface temperature.
14. If the sun's surface temperature suddenly cooled to 2000°C, explain why the sky would probably appear more red than blue.
15. Why is it that auroral displays above Colorado can be forecast up to several days in advance?

16. Why does the aurora usually occur more frequently above Maine than above Washington State, even though Maine is slightly farther south?

● **FIGURE 2.24**

PROBLEMS AND EXERCISES

1. Suppose that 500 g of water vapor condense to make a cloud about the size of an average room. If we assume that the latent heat of condensation is 600 cal/g, how much heat would be released to the air? If the total mass of air before condensation is 100 kg, how much warmer would the air be after condensation? Assume that the air is not undergoing any pressure changes. (Hint: Use the specific heat of air in Table 2.1, p. 34.)

2. Suppose planet A is exactly twice the size (in surface area) of planet B. If both planets have the same exact surface temperature (1500 K), which planet would be emitting the most radiation? Determine the wavelength of maximum energy emission of both planets, using Wien's law on p. 40.

3. Suppose, in question 2, the temperature of planet B doubles.
   (a) What would be its wavelength of maximum energy emission?
   (b) In what region of the electromagnetic spectrum would this wavelength be found?
   (c) If the temperature of planet A remained the same, determine which planet (A or B) would now be emitting the most radiation (use the Stefan-Boltzmann relationship on p. 40). Explain your answer.

4. Suppose your surface body temperature averaged 90°F. How much radiant energy in W/m² would be emitted from your body?

ONLINE RESOURCES

Visit www.cengagebrain.com to view additional resources, including video exercises, practice quizzes, an interactive eBook, and more.

Brightly colored oak and birch trees mingle with green spruce in the Czech Republic, signaling that winter is just around the corner.

Ondrej Prosicky/Shutterstock.com

CHAPTER 3

Seasonal and Daily Temperatures

THE SUN DOESN'T RISE OR FALL: it doesn't move, it just sits there, and we rotate in front of it. Dawn means that we are rotating around into sight of it, while dusk means we have turned another 180 degrees and are being carried into the shadow zone. The sun never "goes away from the sky." It's still there sharing the same sky with us; it's simply that there is a chunk of opaque earth between us and the sun which prevents our seeing it. Everyone knows that, but I really see it now. No longer do I drive down a highway and wish the blinding sun would set; instead I wish we could speed up our rotation a bit and swing around into the shadows more quickly.

Michael Collins, *Carrying the Fire*

As you sit quietly reading this book, you are part of a moving experience. Planet Earth is speeding around the sun at thousands of kilometers per hour while, at the same time, spinning on its axis. When we look down upon the North Pole, we see that the direction of spin is counterclockwise, meaning that we are moving toward the east at hundreds of kilometers per hour. We normally don't think of it in that way, but, of course, this is what causes the sun, moon, and stars to rise in the east and set in the west. It is these motions, coupled with the fact that Earth is tilted on its axis, that cause our seasons. Therefore, we will begin this chapter by examining how Earth's motions and the sun's energy work together to produce temperature variations on a seasonal basis. Later, we will examine temperature variations on a daily basis.

Why Earth Has Seasons

Earth revolves completely around the sun in an elliptical path (not quite a circle) in about 365 days and six hours (one year, plus a Leap Day every four years in February). As Earth revolves around the sun, it spins on its own axis, completing one spin in 24 hours (one day). The average distance from Earth to the sun is 150 million km (93 million mi). Because Earth's orbit is an ellipse instead of a circle and is slightly off-center from the sun, the actual distance from Earth to the sun varies during the year. Earth comes closer to the sun in January (147 million km) than it does in July (152 million km)* (see ● Fig. 3.1). From this observation we might conclude that our warmest weather should occur in January and our coldest weather in July. But in the Northern Hemisphere, we normally experience cold weather in January when we are closer to the sun and warm weather in July when we are farther away. If nearness to the sun were the primary cause

of the seasons then, indeed, January would be warmer than July. However, nearness to the sun is only a small part of the story.

Our seasons are regulated by the amount of solar energy received at Earth's surface. This amount is determined primarily by the angle at which sunlight strikes the surface, and by how long the sun shines on any latitude (daylight hours). Let's look more closely at these factors.

Solar energy that strikes Earth's surface perpendicularly (directly) is much more intense than solar energy that strikes the same surface at an angle. Think of shining a flashlight straight at a wall—you get a small, circular spot of light (see ● Fig. 3.2). Now, tip the flashlight and notice how the spot of light spreads over a larger area. The same principle holds for sunlight. Sunlight striking Earth at an angle spreads out and must heat a larger region than sunlight impinging directly on Earth. Everything else being equal, an area experiencing more direct solar rays will receive more heat than the same-size area being struck by sunlight at an angle. In addition, the more the sun's rays are slanted from the perpendicular, the more atmosphere they must penetrate. And the more atmosphere they penetrate, the more they can be scattered and absorbed. As a consequence, when the sun is high in the sky, it can heat the ground to a much higher temperature than when it is low on the horizon.

The second important factor determining how warm Earth's surface becomes is the length of time the sun shines each day, the number of daylight hours. More daylight hours, of course, mean that more energy is available from sunlight. In a given location, more solar energy reaches Earth's surface on a clear, long day than on a day that is clear but much shorter. Hence, more surface heating takes place.

From a casual observation, we know that summer days have more daylight hours than winter days. Also, the noontime summer sun is higher in the sky than is the noontime winter sun. Both of these events occur because our spinning planet is inclined on its axis (tilted) as it revolves around the sun. As ● Fig. 3.3 illustrates, the angle of tilt is $23\frac{1}{2}°$ from the perpendicular drawn to the plane of Earth's orbit. Earth's axis points to the same direction in space all year long. Thus, on one side of Earth's orbit, in summer (June), the Northern Hemisphere is tilted toward the sun; on the other side of Earth's orbit, in winter (December), the Northern Hemisphere is tilted away from the sun.

SEASONS IN THE NORTHERN HEMISPHERE Let's first discuss the *warm summer* season. Note in Fig. 3.3 that, on June 21, the northern half of the world is directed toward the sun. At noon

---

*The time around January 3rd, when Earth is closest to the sun, is called *perihelion* (from the Greek *peri*, meaning "near" and *helios*, meaning "sun"). The time when Earth is farthest from the sun (around July 4th) is called *aphelion* (from the Greek *ap*, "away from").

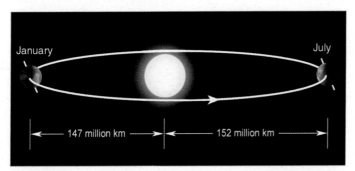

● **FIGURE 3.1** The elliptical path (highly exaggerated) of Earth about the sun brings Earth slightly closer to the sun in January than in July.

January
July
⟵ 147 million km ⟶ ⟵ 152 million km ⟶

● **FIGURE 3.2** Sunlight that strikes a surface at an angle is spread over a larger area than sunlight that strikes the surface directly. Oblique sun rays deliver less energy (are less intense) to a surface than direct sun rays.

High sun
Low sun
Earth

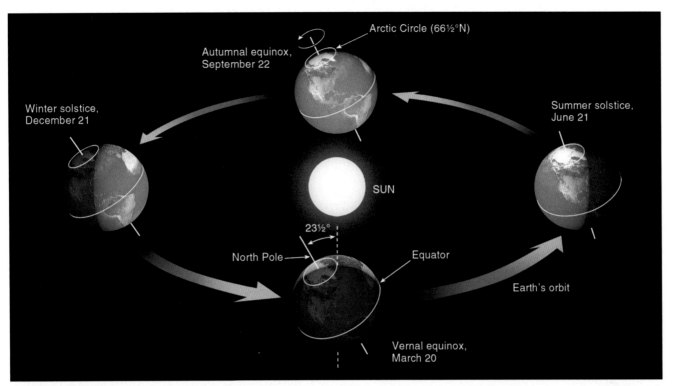

● **FIGURE 3.3** As Earth revolves about the sun, it is tilted on its axis by an angle of $23\frac{1}{2}°$. Earth's axis always points to the same area in space (as viewed from a distant star). Thus, in June, when the Northern Hemisphere is tipped toward the sun, more direct sunlight and long hours of daylight cause warmer weather than in December, when the Northern Hemisphere is tipped away from the sun. (Diagram, of course, is not to scale. The timing of each solstice and equinox varies slightly from year to year; the exact date may be a day earlier or later than shown here, depending on your time zone.)

CRITICAL THINKING QUESTION If Earth were not tilted on its axis, how would average summer temperatures and average winter temperatures change where you live?

on this day, solar rays beat down upon the Northern Hemisphere more directly than during any other time of year. The sun is at its highest position in the noonday sky, directly above $23\frac{1}{2}°$ north (N) latitude (Tropic of Cancer). If you were standing at this latitude on June 21, the sun at noon would be directly overhead. This day, called the **summer solstice**, is the astronomical first day of summer in the Northern Hemisphere.[*]

Study Fig. 3.3 closely and notice that as Earth spins on its axis, the side facing the sun is in sunshine and the other side is in darkness. Thus, half of the globe is always illuminated. If Earth's axis were not tilted, the noonday sun would always be directly overhead at the equator, and there would be 12 hours of daylight and 12 hours of darkness at each latitude every day of the year. However, Earth is tilted. Since the Northern Hemisphere faces toward the sun on June 21, each latitude in the Northern Hemisphere will have more than 12 hours of daylight. The farther north we go, the longer are the daylight hours. When we reach the Arctic Circle ($66\frac{1}{2}°$N), daylight lasts for 24 hours. Notice in Fig. 3.3 how the region above $66\frac{1}{2}°$N never gets into the "shadow" zone as Earth spins. At the North Pole, the sun actually rises above the horizon on March 20 and has six months until it

[*]As we will see later in this chapter, the seasons are reversed in the Southern Hemisphere. Hence, in the Southern Hemisphere, this same day is the winter solstice, or the astronomical first day of winter. Note that the actual date of each solstice and equinox can vary by a day or so from one year to the next because of slight variations in Earth's orbit around the sun.

sets on September 22. No wonder this region is called the "Land of the Midnight Sun"! (See ● Fig. 3.4.)

Do longer days near polar latitudes mean that the highest day-time summer temperatures are experienced there? Not really. Nearly everyone knows that New York City (41°N) "enjoys" much hotter summer weather than Barrow (Utqiaġvik), Alaska (71°N). The days in Barrow are much longer, so why isn't Barrow warmer? To figure this out, we must examine the *incoming solar radiation* (called *insolation*) on June 21. ● Figure 3.5 shows two curves: The upper curve represents the amount of insolation at the top of Earth's atmosphere on June 21, while the bottom curve represents the amount of radiation that eventually reaches Earth's surface on the same day.

The upper curve increases from the equator to the pole. This increase indicates that during the entire day of June 21, more solar radiation reaches the top of Earth's atmosphere above the

WEATHER WATCH

Does darkness (constant night) really occur at the Arctic Circle ($66\frac{1}{2}°$N) on the winter solstice? The answer is no. Due to the bending and scattering of sunlight by the atmosphere, the sky is not totally dark at the Arctic Circle on December 21. In fact, on this date, total darkness only happens north of about 82° latitude. Even at the North Pole, total darkness does not occur from September 22 through March 20, but rather from about November 5 through February 5.

● **FIGURE 3.4** Land of the Midnight Sun. A series of exposures of the sun taken before, during, and after midnight in northern Alaska during July.

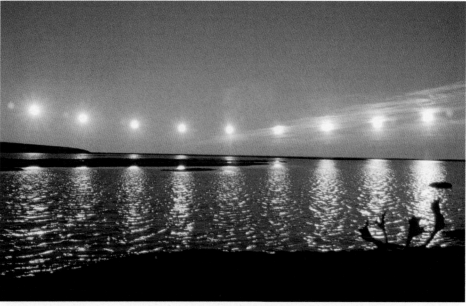

poles than above the equator. True, the sun shines on these polar latitudes at a relatively low angle, but it does so for 24 hours, causing the maximum to occur there. The lower curve shows that the amount of solar radiation eventually reaching Earth's surface on June 21 is maximum near 30°N. From there, the amount of insolation reaching the ground decreases as we move poleward.

The reason the two curves are different is that once sunlight enters the atmosphere, fine dust and air molecules scatter it, clouds reflect it, and some of it is absorbed by atmospheric gases. What remains reaches the surface. Generally, the greater the thickness of atmosphere that sunlight must penetrate, the greater are the chances that it will be either scattered, reflected, or absorbed by the atmosphere. During the summer in far northern latitudes, the sun is never very high above the horizon, so its radiant energy must pass through a thick portion of atmosphere before it reaches Earth's surface (see ● Fig. 3.6). And because of

the increased cloud cover during the arctic summer, much of the sunlight is reflected before it reaches the ground.

Solar energy that eventually reaches the surface in the far north does not heat the surface very effectively. A portion of the sun's energy is reflected by ice and snow, while some of it melts frozen soil. The amount actually absorbed is spread over a large

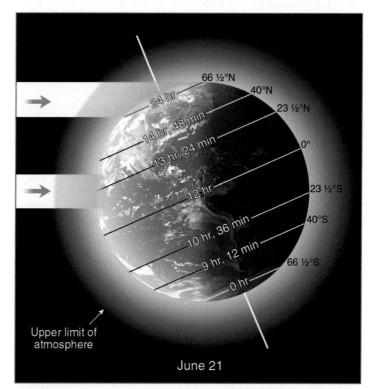

● **FIGURE 3.6** During the Northern Hemisphere summer, sunlight that reaches Earth's surface in far northern latitudes has passed through a thicker layer of absorbing, scattering, and reflecting atmosphere than sunlight that reaches Earth's surface farther south. Sunlight is lost through both the thickness of the pure atmosphere and by impurities in the atmosphere. As the sun's rays become more oblique, these effects become more pronounced.

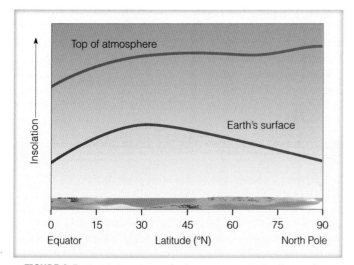

● **FIGURE 3.5** The relative amount of radiant energy received at the top of Earth's atmosphere and at Earth's surface on June 21—the summer solstice.

area. So, even though northern cities, such as Barrow, experience 24 hours of continuous sunlight on June 21, they are not warmer than cities farther south. Overall, they receive less radiation at the surface, and what radiation they do receive does not heat the surface as effectively.

In our discussion of Fig. 3.5, we saw that, on June 21, solar energy incident on Earth's surface is maximum near latitude 30°N. On this day, the sun is shining directly above latitude 23½°N. Why, then, isn't the most sunlight received here? A quick look at a world map shows that the major deserts of the world are centered near 30°N. Cloudless skies and drier air predominate near this latitude. At latitude 23½°N, the climate is more moist and cloudy, causing more sunlight to be scattered and reflected before reaching the surface. In addition, day length is longer at 30°N than at 23½°N on June 21. For these reasons, more radiation falls on 30°N latitude than at the Tropic of Cancer (23½°N).

Each day past June 21, the noon sun is slightly lower in the sky. Summer days in the Northern Hemisphere begin to shorten. June eventually gives way to September, and fall begins.

Look at Fig. 3.3 (p. 61) again and notice that, by September 22, Earth will have moved so that the sun is directly above the equator. Except at the poles, the days and nights throughout the world are of equal length. This day is called the **autumnal** (fall) **equinox**, and it marks the astronomical beginning of fall in the Northern Hemisphere. At the North Pole, the sun appears on the horizon for 24 hours, due to the bending of light by the atmosphere. The following day (or at least within several days), the sun disappears from view, not to rise again for a long, cold six months. Throughout the northern half of the world on each successive day, there are fewer hours of daylight, and the noon sun is slightly lower in the sky. Less direct sunlight and shorter hours of daylight spell cooler weather for the Northern Hemisphere. Reduced radiation, lower air temperatures, and cooling breezes stimulate the beautiful pageantry of fall colors (see ● Fig. 3.7).

In some years around the middle of autumn, there is an unseasonably warm spell, especially in the eastern two-thirds of the United States. This warm period, sometimes referred to as **Indian summer**,* may last from several days up to a week or more. It usually occurs when a large high-pressure area stalls near the southeast coast. The clockwise flow of air around this system moves warm air from the Gulf of Mexico into the central or eastern half of the nation. The warm, gentle breezes and smoke from a variety of sources respectively make for mild, hazy days. The warm weather ends abruptly when an outbreak of polar air reminds us that winter is not far away.

On December 21 (three months after the autumnal equinox), the Northern Hemisphere is tilted as far away from the sun as it will be all year (see Fig. 3.3, p. 61). Nights are long and days are short. Notice in ▼ Table 3.1 that daylight decreases from 12 hours at the equator to 0 (zero) at latitudes above 66½°N. This is the shortest day of the year, called the **winter solstice**, the

● **FIGURE 3.7** The pageantry of fall colors in New England. The weather most suitable for an impressive display of fall colors is warm, sunny days followed by clear, cool nights with temperatures dropping below 7°C(45°F), but remaining above freezing. Contrary to popular belief, it is not the first frost that causes the leaves of deciduous trees to change color. The yellow and orange colors, which are actually in the leaves, begin to show through several weeks before the first frost, as shorter days and cooler nights cause a decrease in the production of the green pigment chlorophyll.

astronomical beginning of winter in the northern world. On this day, the sun shines directly above latitude 23½°S (Tropic of Capricorn). In the northern half of the world, the sun is at its lowest position in the noon sky. Its rays pass through a thick section of atmosphere and spread over a large area on the surface.

With so little incident sunlight, Earth's surface cools quickly. A blanket of clean snow covering the ground aids in the cooling. The snow reflects much of the sunlight that reaches the surface

▼ **TABLE 3.1** **Length of Time from Sunrise to Sunset for Various Latitudes on Different Dates in the Northern Hemisphere***

LATITUDE	MARCH 20	JUNE 21	SEPT. 22	DEC. 21
0°	12 hr	12.0 hr	12 hr	12.0 hr
10°	12 hr	12.6 hr	12 hr	11.4 hr
20°	12 hr	13.2 hr	12 hr	10.8 hr
30°	12 hr	13.9 hr	12 hr	10.1 hr
40°	12 hr	14.9 hr	12 hr	9.1 hr
50°	12 hr	16.3 hr	12 hr	7.7 hr
60°	12 hr	18.4 hr	12 hr	5.6 hr
70°	12 hr	2 months	12 hr	0 hr
80°	12 hr	4 months	12 hr	0 hr
90°	12 hr	6 months	12 hr	0 hr

*Solstice and equinox dates can vary slightly from year to year.

---

*The origin of the term is uncertain, but it dates back to the eighteenth century. It may have originally referred to the good weather that allowed Native Americans time to harvest their crops. Normally, a period of cool autumn weather must precede the warm weather period for the latter to be called Indian summer.

Is December 21 Really the First Day of Winter?

On December 21 (or 22, depending on the year) after nearly a month of cold weather, and perhaps a snowstorm or two (see ● Fig. 1), someone on the radio, TV, or Internet has the audacity to proclaim that "today is the first official day of winter." If during the last several weeks it was not winter, then what season was it?

December 21 usually marks the *astronomical* first day of winter in the Northern Hemisphere (NH), just as June 21 typically marks the *astronomical* first day of summer (NH). Earth is tilted on its axis by 23½° as it revolves around the sun. This fact causes the sun (as we view it from Earth) to move in the sky from a point where it is directly above 23½° South latitude on December 21 to a point where it is directly above 23½° North latitude on June 21. The astronomical first day of spring (NH) occurs around March 20 as the sun crosses the equator moving northward and, likewise, the astronomical first day of autumn (NH) occurs around September 22 as the sun crosses the equator moving southward.

Therefore the "official" beginning of any season is simply the day on which the sun passes over a particular latitude, and has nothing to do with how cold or warm the following day will be. In fact, a period of colder or warmer than normal weather before or after a solstice or equinox is caused mainly by the upper-level winds directing cold or warm air into a region.

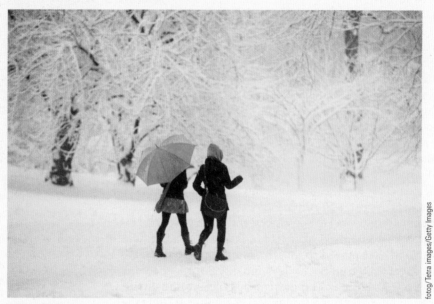

● **FIGURE 1** Snow covers Central Park in New York City on December 17, 2013. Given that the snowstorm occurred before the winter solstice, is this a late fall storm or an early winter storm?

In the middle latitudes, summer is defined as the warmest season and winter the coldest season. If the year is divided into four seasons with each season consisting of three months, then the meteorological (or *climatological*) definition of summer over much of the Northern Hemisphere would be the three warmest months of June, July, and August. Winter would be the three coldest months of December, January, and February. Autumn would be September, October, and November—the transition between summer and winter. And spring would be March, April, and May—the transition between winter and summer. The National Weather Service uses this climatological definition when reporting on seasonal temperatures.

So, the next time you hear someone remark on December 21 that "winter officially begins today," remember that this is the astronomical definition of the first day of winter. According to the climatological definition, winter has been around for several weeks.

and continually radiates away infrared energy during the long nights. In northern Canada and Alaska, the arctic air rapidly becomes extremely cold as it lies poised, ready to do battle with the milder air to the south. Periodically, this cold arctic air pushes down into the northern United States, producing a rapid drop in temperature called a *cold wave*, which occasionally reaches far into the south during the winter. Sometimes, these cold spells arrive well before the winter solstice—the "official" first day of winter—bringing with them heavy snow and blustery winds. (More information on this "official" first day of winter is given in Focus section 3.1.)

On each winter day after December 21, the sun climbs a bit higher in the midday sky. The periods of daylight grow longer until days and nights are of equal length, and we have another equinox.

The date of March 20, which marks the astronomical arrival of spring, is called the **vernal** (spring) **equinox**. At this equinox, the noonday sun is shining directly on the equator, while, at the North Pole, the sun (after hiding for six months) peeks above the horizon. Longer days and more direct solar radiation spell warmer weather for the northern world.

Three months after the vernal equinox, it is June again. The Northern Hemisphere is tilted toward the sun, which shines high in the noonday sky. The days have grown longer and warmer, and another summer season has begun.

Up to now, we have seen that the seasons are controlled by solar energy striking our tilted planet, as it makes its annual voyage around the sun. This tilt of Earth causes a seasonal variation in both the length of daylight and the intensity of

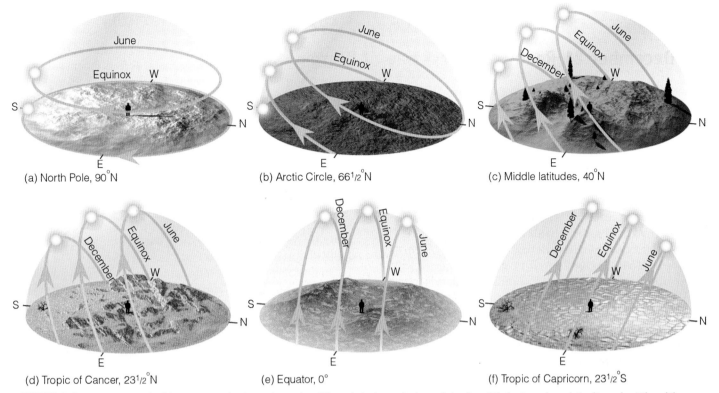

(a) North Pole, 90°N

(b) Arctic Circle, 66¹/₂°N

(c) Middle latitudes, 40°N

(d) Tropic of Cancer, 23¹/₂°N

(e) Equator, 0°

(f) Tropic of Capricorn, 23¹/₂°S

● **FIGURE 3.8** The apparent path of the sun across the sky as observed at different latitudes on the June solstice (June 21), the December solstice (December 21), and the equinox (March 20 and September 22).

sunlight that reaches the surface. These facts are summarized in ● Fig. 3.8, which shows how the sun would appear in the sky to an observer at various latitudes at different times of the year. Earlier we learned that at the North Pole the sun rises above the horizon in March and stays above the horizon for six months until September. Notice in Fig. 3.8a that at the North Pole even when the sun is at its highest point in June, it is low in the sky— only 23½° above the horizon. Farther south, at the Arctic Circle (Fig. 3.8b), the sun is always fairly low in the sky, even in June, when the sun stays above the horizon for 24 hours.

In the middle latitudes (Fig. 3.8c), notice that in December the sun rises in the southeast, reaches its highest point at noon (only about 26° above the southern horizon), and sets in the southwest. This apparent path produces little intense sunlight and short daylight hours. On the other hand, in June, the sun rises in the northeast, reaches a much higher position in the sky at noon (about 74° above the southern horizon) and sets in the northwest. This apparent path across the sky produces more intense solar heating, longer daylight hours, and, of course, warmer weather. Figure 3.8d illustrates how the tilt of Earth influences the sun's apparent path across the sky at the Tropic of Cancer (23½°). Figure 3.8e gives the same information for an observer at the Equator.

At this point it is interesting to note that although sunlight is most intense in the Northern Hemisphere on June 21, the warmest weather in middle latitudes normally occurs weeks later, usually in July or August. This situation (called the *lag in seasonal temperature*) arises because although incoming energy from the sun is greatest in June, it takes time for oceans and landmasses to release the large amounts of incoming energy they have absorbed. As a result, incoming energy still exceeds outgoing energy from Earth for a period of at least several weeks. Once the incoming solar energy and outgoing earth energy are in balance, the highest average temperature is attained. When outgoing energy exceeds incoming energy, the average temperature drops. As in the summer, there is also a seasonal temperature lag in winter. Because outgoing Earth energy exceeds incoming solar energy well past the winter solstice (December 21), we normally find our coldest weather occurring in January or February. As we will see later in this chapter, there is a similar lag in daily temperature between the time of most intense sunlight and the time of highest air temperature for the day.

SEASONS IN THE SOUTHERN HEMISPHERE On June 21, the Southern Hemisphere is adjusting to an entirely different season. Again, look back at Fig. 3.3 (p. 61), and notice that this part of the world is now tilted away from the sun. Nights are long, days are short, and solar rays come in at an angle (see Fig. 3.8f). All of these factors keep air temperatures fairly low. The June solstice marks the astronomical beginning of winter in the Southern Hemisphere. In this part of the world, summer will not "officially" begin until the sun is over the Tropic of Capricorn (23½°S)—remember that this occurs on December 21. So, when it is winter and June in the Southern Hemisphere, it is summer and June in the Northern Hemisphere. Conversely, when it is summer and December in the Southern Hemisphere, it is winter and December in the Northern Hemisphere. So, if you are tired of the cold, December weather in your Northern Hemisphere city, travel to the summer half of the world and enjoy the warmer weather. The tilt of Earth as it revolves around the sun makes all this possible.

We know Earth comes nearer to the sun in January than in July. Even though this difference in distance amounts to only about 3 percent, the energy that strikes the top of Earth's atmosphere is almost 7 percent greater on January 3 than on July 4. These statistics might lead us to believe that summer should be warmer in the Southern Hemisphere than in the Northern Hemisphere. However, this is not so. A close examination of the Southern Hemisphere reveals that nearly 81 percent of the surface is water compared to 61 percent in the Northern Hemisphere. The added solar energy due to the closeness of the sun is absorbed by large bodies of water, becoming well mixed and circulated within them. This process keeps the average summer (January) temperatures in the Southern Hemisphere cooler than average summer (July) temperatures in the Northern Hemisphere. Because of water's large heat capacity, it also tends to keep winters in the Southern Hemisphere warmer than we might expect.*

Another difference between the seasons of the two hemispheres concerns their length. Because Earth follows an elliptical path as it journeys around the sun, and because the ellipse is slightly off-center from the sun, the total number of days from the vernal (March 20) to the autumnal (September 22) equinox is about 7 days longer than from the autumnal to vernal equinox (see ● Fig. 3.9). This means that spring and summer in the Northern Hemisphere not only last about a week longer than northern fall and winter, but also about a week longer than spring and summer in the Southern Hemisphere. Hence, the shorter spring and summer of the Southern Hemisphere somewhat offset the extra insolation received due to a closer proximity to the sun.

Up to now, we have considered the seasons on a global scale. We will now shift to more local considerations.

---

* For a comparison of January and July temperatures, see Figs. 3.27 and Figs. 3.28, p. 78.

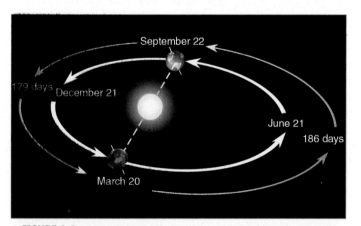

● **FIGURE 3.9** Because Earth travels more slowly when it is farther from the sun, it takes Earth a little more than 7 days longer to travel from March 20 to September 22 than from September 22 to March 20.

Local Seasonal Variations

Look back at Fig. 3.8c (p. 65), and observe that in the middle latitudes of the Northern Hemisphere, objects facing south will receive more sunlight during a year than those facing north. This fact becomes strikingly apparent in hilly or mountainous country.

Hills that face south receive more sunshine and, hence, become warmer than the partially shielded north-facing hills. Higher temperatures usually mean greater rates of evaporation and slightly drier soil conditions. Thus, south-facing hillsides are usually warmer and drier as compared to north-facing slopes at the same elevation. In many areas of the far western United States, only sparse vegetation grows on south-facing slopes, while, on the same hill, dense vegetation grows on the cooler, moister slopes that face north (see ● Fig. 3.10).

In northern latitudes, hillsides that face south usually have a longer growing season. Winemakers in western New York State do not plant grapes on the north side of hills. Grapes from vines grown on the warmer south side make better wine. Moreover, because air temperatures normally decrease with increasing height, trees found on the cooler north-facing side of mountains are often those that usually grow at higher elevations, while the warmer south-facing side of the mountain often supports trees usually found at lower elevations.

In the mountains, snow usually lingers on the ground for a longer time on north slopes than on the warmer south slopes. For this reason, ski runs are built facing north wherever possible.

● **FIGURE 3.10** In areas of the middle latitudes of the Northern Hemisphere where small temperature changes can cause major changes in soil moisture, sparse vegetation on the south-facing slopes will often contrast with lush vegetation on the north-facing slopes.

CRITICAL THINKING QUESTION Would the lush vegetation in Fig. 3.10 appear on the north-facing side or the south-facing side of the hills if this figure represented a region in the middle latitudes of the Southern Hemisphere?

Solar Heating and the Noonday Sun

The amount of solar energy that falls on a typical American home each summer day is many times the energy needed to heat the inside for a year. Thus, many people are turning to the sun as a clean, safe, and virtually inexhaustible source of energy. If solar collectors are used to heat a home, they should be placed on south-facing roofs to take maximum advantage of the energy provided. The roof itself should be constructed as nearly perpendicular to winter sun rays as possible. To determine the proper roof angle at any latitude, we need to know how high the sun will be above the southern horizon at noon.

The noon angle of the sun can be calculated in the following manner:

1. Determine the number of degrees between your latitude and the latitude where the sun is currently directly overhead.

2. Subtract the number you calculated in step 1 from 90°. This will give you the sun's elevation above the southern horizon at noon at your latitude.

For example, suppose you live in Denver, Colorado (latitude 39 ½ °N), and the date is December 21. The difference between your

● **FIGURE 2** The roof of a solar-heated home constructed in Denver, Colorado, at an angle of 45° absorbs the sun's energy in midwinter at nearly right angles.

latitude and where the sun is currently overhead is 63° (39 ½ °N to 23 ½ °S), so the sun is 27° (90° − 63°) above the southern horizon at noon. On March 20 in Denver, the angle of the sun is 50 ½ ° (90° − 39 ½ °). To determine a reasonable roof angle, we must consider the average altitude of the midwinter sun (about 39° for

Denver), building costs, and snow loads.

● Figure 2 illustrates that a roof constructed in Denver, Colorado, at an angle of 45° will be nearly perpendicular to much of the winter sun's energy. Hence, the roofs of solar-heated homes in middle latitudes are generally built at an angle between 45° and 50°.

Also, homes and cabins built on the north side of a hill usually have a steep pitched roof as well as a reinforced deck to withstand the added weight of snow from successive winter storms.

The seasonal change in the sun's position during the year can have an effect on the vegetation around the home. In winter, a large two-story home can shade its own north side, keeping it much cooler than its south side. Trees that require warm, sunny weather should be planted on the south side, where sunlight reflected from the house can even add to the warmth.

The design of a home can be important in reducing heating and cooling costs. Large windows should face south, allowing sunshine to penetrate the home in winter. To block out excess sunlight during the summer, a small eave or overhang should be built. A kitchen with windows facing east will let in enough warm morning sunlight to help heat this area. Because the west side warms rapidly in the afternoon, rooms having small windows (such as garages) can be placed here to act as a thermal buffer. Deciduous trees planted on the west or south side of a home provide shade in the summer. In winter, they drop their leaves, allowing the winter sunshine to warm the house. If you like the bedroom slightly cooler than the rest of the home, face it toward the north. Let nature help with the

heating and air conditioning. Proper house design, orientation, and landscaping can help cut the demand for electricity, as well as for natural gas and fossil fuels, which are rapidly being depleted.

From our reading of the last several sections, it should be apparent that, when using the sun to directly heat a home, proper roof angle is important in capturing much of the winter sun's energy. (The information needed to determine the angle at which sunlight will strike a roof is given in Focus section 3.2).

WEATHER WATCH

Seasonal changes can affect how we feel. For example, some people face each winter with a sense of foreboding, especially at high latitudes where days are short and nights are long and cold. If the depression is lasting and disabling, the problem is called seasonal affective disorder (SAD). People with SAD tend to sleep longer, overeat, and feel tired and drowsy during the day. One common treatment is extra doses of artificial bright light in wavelengths similar to outdoor light.

Daily Warming and Cooling of Air Near the Surface

In a way, each sunny day is like a tiny season as the air goes through a daily cycle of warming and cooling. The air warms during the morning hours, as the sun gradually rises higher in the sky, spreading a blanket of heat energy over the ground. The sun reaches its highest point around noon, after which it begins its slow journey toward the western horizon. It is around noon when Earth's surface receives the most intense solar rays. However, somewhat surprisingly, noontime is usually not the warmest part of the day. Rather, the air continues to be heated, often reaching a maximum temperature later in the afternoon. To find out why this *lag in temperature* occurs, we need to examine a shallow layer of air in contact with the ground.

DAYTIME WARMING As the sun rises in the morning, sunlight warms the ground, and the ground warms the air in contact with it by conduction. However, air is such a poor heat conductor that this process only takes place within a few centimeters of the ground. As the sun rises higher in the sky, the air in contact with the ground becomes even warmer, and there exists a thermal boundary separating the hot surface air from the slightly cooler air above. Given their random motion, some air molecules will cross this boundary: The "hot" molecules below bring greater kinetic energy to the cooler air; the "cool" molecules above bring a deficit of energy to the hot, surface air. However, on a windless day, this form of heat exchange is slow, and a substantial temperature difference usually exists between the air at ground level and the air directly above it (see ● Fig. 3.11). This explains why runners on a clear, windless, summer afternoon may experience air temperatures of over 50°C (122°F) at their feet and only 32°C (90°F) at their waist.

● **FIGURE 3.11** On a sunny, calm day, the air near the surface can be substantially warmer than the air a meter or so above the surface, where thermometers are typically located.

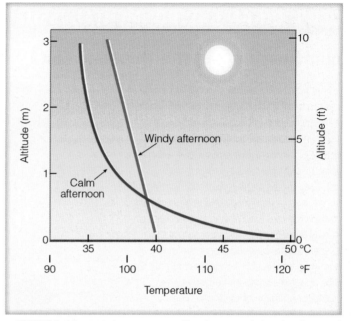

● **FIGURE 3.12** Vertical temperature profiles above an asphalt surface for a windy and a calm summer afternoon.

Near the surface, convection begins, and rising air bubbles (thermals) help to redistribute heat. In calm weather, these thermals are small and do not effectively mix the air near the surface. Thus, large vertical temperature gradients can exist. On windy days, however, turbulent eddies can mix hot surface air with the cooler air above. This form of mechanical stirring, sometimes called *forced convection*, helps the thermals to transfer heat away from the surface more efficiently. Therefore, on sunny, windy days the molecules near the surface are more quickly carried away than on sunny, calm days. ● Figure 3.12 shows a typical vertical profile of air temperature on windy days and on calm days in summer.

We can now see why the warmest part of the day is usually in the afternoon. Around noon, the sun's rays are most intense. Incoming solar radiation decreases in intensity after noon, but for a time it still exceeds outgoing heat energy from the surface for a time. This situation yields an energy surplus for two to four hours after noon and substantially contributes to a lag between the time of maximum solar heating and the time of maximum air temperature several meters above the surface (see ● Fig. 3.13).

The exact time of the highest temperature reading varies somewhat. Where the summer sky remains cloud-free all afternoon, the maximum temperature may occur sometime between 3:00 and 5:00 p.m. standard time. Where there is afternoon cloudiness or haze, the temperature maximum usually occurs an hour or two earlier. In Denver, clouds that typically build over the

WEATHER WATCH

During the summer, Death Valley, California, can sizzle at any hour of the day, even at night. On July 12, 2012, the minimum temperature dipped no lower than 107°F. This tied the record for the warmest reliably measured minimum temperature on Earth, set several days earlier—on June 27, 2012, at Khasab, Oman.

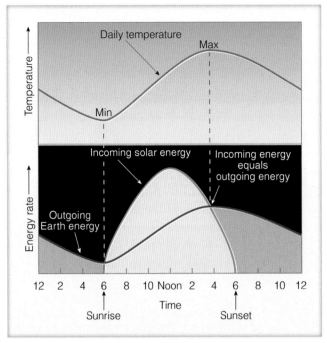

● **FIGURE 3.13** The daily variation in air temperature is controlled by incoming energy (primarily from the sun) and outgoing energy from Earth's surface. Where incoming energy exceeds outgoing energy (orange shade), the air temperature rises. Where outgoing energy exceeds incoming energy (gray shade), the air temperature falls.

mountains on warm days drift eastward early in the afternoon. These clouds reflect sunlight, sometimes causing the maximum temperature to occur as early as noon. If clouds persist throughout the day, the overall daytime temperatures are usually lower.

Adjacent to large bodies of water, cool air moving inland can modify the rhythm of temperature change such that the warmest part of the day occurs at noon or before. In winter, atmospheric storms circulating warm air northward can even cause the highest temperature to occur at night.

Just how warm the air becomes depends on such factors as the type of soil, its moisture content, and vegetation cover. When the soil is a poor heat conductor (as loosely packed sand is), heat energy does not readily transfer into the ground. This allows the surface layer to reach a higher temperature, permitting more energy to warm the air above. On the other hand, if the soil is moist or covered with vegetation, much of the available energy evaporates water, leaving less to heat the air. As you might expect, the highest summer temperatures usually occur over desert regions, where clear skies coupled with low humidities and meager vegetation permit the surface and the air above to warm up rapidly.

Where the air is humid, haze and cloudiness lower the maximum temperature by preventing some of the sun's rays from reaching the ground. In humid Atlanta, Georgia, the average maximum temperature for July is 31.7°C (89°F). In contrast, Phoenix, Arizona—in the Desert Southwest, at the same latitude as Atlanta—experiences an average July maximum of 41.1°C (106°F).

EXTREME HIGH TEMPERATURES Most people are aware of the extreme heat that exists during the summer in the Desert Southwest of the United States. But how hot does it get there? On July 10, 1913, Greenland Ranch in Death Valley, California, reported 57°C (134°F) (see ● Fig. 3.14), which is considered to be the highest temperature ever reliably observed in the world. Air temperatures in Death Valley are persistently hot throughout the summer, with the average maximum for July being 47°C (116°F).

One of the hottest urban areas in the United States is Palm Springs, California, where the average high temperature during July is 108°F. Another hot city is Yuma, Arizona. Located along the California-Arizona border, Yuma's high temperature during July also averages 108°F. In 1937, the high in Yuma reached 100°F or more for 101 consecutive days.

In a more humid climate, the maximum temperature rarely climbs above 41°C (106°F). However, during the record heat wave of 1936, the air temperature reached 121°F near Alton, Kansas. And during the heat wave of 1983, which destroyed about $7 billion in crops and increased the nation's air-conditioning bill by an estimated $1 billion, Fayetteville reported North Carolina's all-time record high temperature when the mercury hit 110°F.

Although Death Valley holds the record for the highest officially measured air temperature in the world, it is not the hottest place on earth. This distinction may belong to Dallol,

● **FIGURE 3.14** The hottest place in the world: Death Valley, California, where the air temperature reached 57°C (134°F).

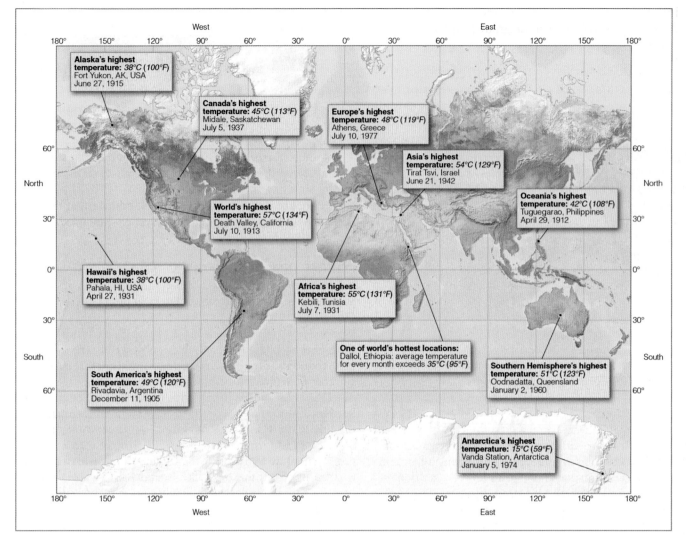

● **FIGURE 3.15** Record high temperatures throughout the world.

Ethiopia. Dallol is located near latitude 12°N, in the hot, dry Danakil Depression (see ● Fig. 3.15). A prospecting company kept weather records at Dallol from 1960 to 1966. During this time, the average daily maximum temperature exceeded 38°C (100°F) every month of the year, except during December and January, when the average maximum lowered to 98°F and 97°F, respectively. On many days, the air temperature exceeded 120°F. The average annual temperature for the six years at Dallol was 34°C (94°F). In comparison, the average annual temperature in Yuma is 23°C (74°F) and at Death Valley, 24°C (76°F).

On September 23, 1922, the temperature was reported to have reached a scorching 58°C (136°F) in Africa, northwest of Dallol at El Azizia, Libya (32°N). Until recently, this was considered to be the highest temperature recorded on Earth using standard measurement techniques. However, the reading was declared invalid in 2012 after an investigation by a panel of experts sponsored by the World Meteorological Organization (WMO). The panel found several major concerns with the El Azizia reading, including problematic instrumentation, an observer who was likely inexperienced, and asphalt-like material beneath the observing site that did not represent the native desert soil.

Because of these and other factors, the actual temperature in El Azizia may have been 7°C (11°F) cooler than the reported record. Consequently, the 1913 reading in Death Valley was declared to be the world's hottest officially measured temperature, returning it to the position it had held almost a century before. If we consider only data gathered using modern instrumentation and observing procedures, then the hottest temperature measured on Earth is 54.0°C (129.2°F), recorded at Mitzbah, Kuwait, on June 30, 2013, and at Death Valley on July 21, 2016.

NIGHTTIME COOLING We know that nights are typically much cooler than days. Nights are cooler because as the afternoon sun lowers, its energy is spread over a larger area, which reduces the heat available to warm the ground. Look at ● Fig. 3.16 and observe that sometime in late afternoon or early evening, Earth's surface and air above begin to lose more energy than they receive; hence, they start to cool.

Both the ground and air above cool by radiating infrared energy, a process called **radiational cooling**. The ground, being a much better radiator than air, is able to cool more quickly. Consequently, shortly after sunset, Earth's surface is slightly cooler than the air directly above it. The surface air transfers some energy to the ground by conduction, which the ground, in turn, quickly radiates away.

As the night progresses, the ground and the air in contact with it continue to cool more rapidly than the air a few meters higher. The warmer upper air does transfer *some* heat downward, but this process is slow due to the air's poor thermal conductivity. Therefore, by late night or early morning, the coldest air is found next to the ground, with slightly warmer air above (see ● Fig. 3.17).

This measured increase in air temperature just above the ground is known as a **radiation inversion** because it forms mainly through radiational cooling of the surface. Since radiation inversions occur on most clear, calm nights, they are also called **nocturnal inversions**.

COLD AIR AT THE SURFACE A strong radiation inversion occurs when the air near the ground is much colder than the air higher up. Ideal conditions for a strong inversion (and, hence, very low nighttime temperatures) exist when the air is calm, the night is long, and the air is fairly dry and cloud-free. Let's examine these conditions one by one.

A windless night is essential for a strong radiation inversion because a stiff breeze tends to mix the colder air at the surface with the warmer air above. This mixing, along with the cooling of the warmer air as it comes in contact with the cold ground, causes a vertical temperature profile that is almost isothermal (a constant temperature) in a layer several meters thick. In the absence of wind, the cooler, more-dense surface air does not readily mix with the warmer, less-dense air above, and the inversion is more strongly developed, as illustrated in Fig. 3.17.

A long night also contributes to a strong inversion. Generally, the longer the night, the longer the time of radiational cooling and the better are the chances that the air near the ground will

● **FIGURE 3.16** On a clear, calm night, the air near the surface can be much colder than the air above. The increase in air temperature with increasing height above the surface is called a radiation temperature inversion.

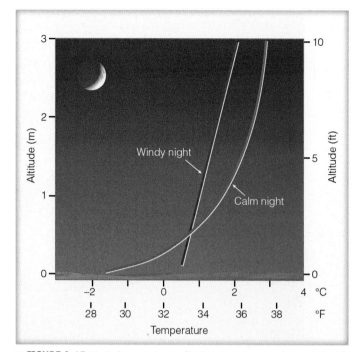

● **FIGURE 3.17** Vertical temperature profiles just above the ground on a windy night and on a calm night. Notice that the radiation inversion develops better on the calm night.

be much colder than the air above. Consequently, winter nights provide the best conditions for a strong radiation inversion, other factors being equal.

Finally, radiation inversions are more likely with a clear sky and dry air. Under these conditions, the ground is able to radiate its energy to outer space and thereby cool rapidly. With cloudy weather and moist air, much of the outgoing infrared energy is absorbed and radiated back to the surface, retarding the rate of surface cooling. Also, on humid nights, condensation in the form of fog or dew will release latent heat, which warms the air. So, radiation inversions can occur on any night. But, during long winter nights, when the air is still, cloud-free, and relatively dry, these inversions can become strong and deep.

On winter nights in middle latitudes, it is common to experience below-freezing temperatures near the ground and air 5°C (9°F) warmer at your waist. In middle latitudes, the top of the inversion—the region where the air temperature stops increasing with height—is usually not more than 100 m (330 ft) above the ground. In dry, polar regions, where winter nights are measured in months, the top of the inversion is often 1000 m (about 3300 ft) above the surface. It can, however, extend to as high as 3000 m (about 10,000 ft).

It should now be apparent that how cold the night air becomes depends primarily on the length of the night, the moisture content

WEATHER WATCH

When the surface air temperature dipped to its all-time record low of 129°F on the Antarctic Plateau of Vostok Station, a drop of saliva falling from the lips of a person taking an observation would have frozen solid before reaching the ground.

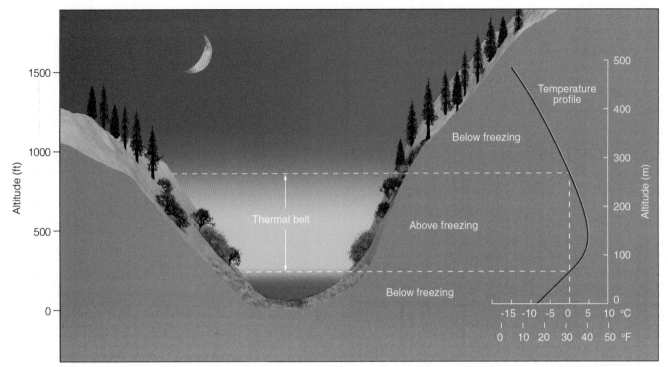

● **FIGURE 3.18** On cold, clear nights, the settling of cold air into valleys makes them colder than surrounding hillsides. The region along the side of the hill where the air temperature is above freezing is known as a *thermal belt*.

of the air, cloudiness, and the wind. Even though wind may initially bring cold air into a region, the coldest nights usually occur when the air is clear and relatively calm.

There are, however, other factors that determine how cold the night air becomes. For example, a surface that is wet or covered with vegetation can add water vapor to the air, retarding nighttime cooling. Likewise, if the soil is a good heat conductor, heat ascending toward the surface during the night adds warmth to the air, which restricts cooling. On the other hand, snow covering the ground acts as an insulating blanket that prevents heat stored in the soil from reaching the air. Snow, a good emitter of infrared energy, radiates away energy rapidly at night, which helps keep the air temperature above a snow surface quite low.

Look back at Fig. 3.13 (p. 69), and observe that the lowest temperature on any given day is usually observed around sunrise. However, the cooling of the ground and surface air may even continue beyond sunrise for a half hour or so, as outgoing energy can exceed incoming energy. This situation happens because light from the early morning sun passes through a thick section of atmosphere and strikes the ground at a low angle. Consequently, the sun's energy does not effectively heat the surface. Surface heating can be reduced further when the ground is moist and available energy is used for evaporation. (Any duck hunter lying flat in a marsh knows the sudden cooling that occurs as evaporation chills the air just after sunrise.) Hence, the lowest temperature can occur shortly after the sun has risen.

Cold, heavy surface air slowly drains downhill during the night and eventually settles in low-lying basins and valleys. Valley bottoms are thus colder than the surrounding hillsides (see ● Fig. 3.18). In middle latitudes, these warmer hillsides, called

thermal belts, are less likely to experience freezing temperatures than the valley below. This encourages farmers to plant on hillsides those trees unable to survive the valley's low temperature.

On the valley floor, the cold, dense air is unable to rise. Smoke and other pollutants trapped in this heavy air restrict visibility. Therefore, valley bottoms are not only colder, but are also more frequently polluted than nearby hillsides. Even when the land is only gently sloped, cold air settles into lower-lying areas, such as river basins and floodplains. Because the flat floodplains are agriculturally rich areas, cold air drainage often forces farmers to seek protection for their crops.

PROTECTING CROPS FROM THE COLD NIGHT AIR On cold nights, many plants may be damaged by low temperatures. To protect small plants or shrubs, cover them with straw, cloth, or plastic sheeting. This prevents ground heat from being radiated away to the colder surroundings. If you are a household gardener concerned about outside flowers and plants during cold weather, simply wrap them in plastic or cover each with a paper cup.

On cold nights, certain crops may be damaged by the low temperatures. If the cold occurs over a widespread area for a long enough time to damage certain crops, the extreme cold is called a **freeze**.* A single freeze in California, Texas, or Florida

---

*A freeze occurs over a widespread area when the surface air temperature remains below freezing for a long enough time to damage certain agricultural crops. The terms *frost* and *freeze* are often used interchangeably by various segments of society. However, to the grower of perennial crops (such as apples and citrus) who has to protect the crop against damaging low temperatures, it makes no difference if visible "frost" is present or not. The concern is whether or not the plant tissue has been exposed to temperatures equal to or below 32°F. The actual freezing point of the plant, however, can vary because perennial plants can develop hardiness in the fall that usually lasts through the winter, then wears off gradually in the spring.

can cause crop losses in the millions or even billions of dollars. In fact, citrus crop losses in Florida during the hard freeze of January 1977 exceeded $2 billion. In California, several freezes during the spring of 2001 caused millions of dollars in damages to California's north coast vineyards, which resulted in higher wine prices. And after extremely warm weather across the eastern United States in March 2012 led to early blooming of fruit trees, freezing temperatures in April destroyed nearly half of the apple crop in New York and nearly 90 percent in Michigan. Another widespread freeze across the East and Midwest in April 2007 inflicted more than $2 billion in damage to fruit and field crops.

Fruit trees are particularly vulnerable to cold weather in the spring when they are blossoming. The protection of such trees presents a serious problem to the farmer. Since the lowest temperatures on a clear, still night occur near the surface, the lower branches of a tree are the most susceptible to damage. Therefore, increasing the air temperature close to the ground may prevent damage. A traditional technique for doing this is the use of **orchard heaters**, which warm the air around the trees by setting up convection currents close to the ground (see ● Fig. 3.19). In addition, heat energy radiated from oil- or gas-fired orchard heaters is intercepted by the buds of the trees, which raises their temperature. Orchard heaters that generate smoke, known as "smudge pots," were used for many decades but are now prohibited in most areas due to their effects on local air quality.

Another way to protect trees is to mix the cold air at the ground with the warmer air above, thus raising the temperature of the air next to the ground. Such mixing can be accomplished by using **wind machines** (see ● Fig. 3.20), which are power-driven fans that resemble airplane propellers. One significant benefit of wind machines is that they can be thermostatically controlled to turn off and on at prescribed temperatures. Farmers without their own wind machines can rent air mixers in the form of helicopters. Although helicopters are effective in mixing the air, they are expensive to operate.

If sufficient water is available, trees can be protected by irrigation. On potentially cold nights, farmers might flood the orchard. Because water has a high heat capacity, it cools more slowly than dry soil, and so the surface does not become as cold as it would if it were dry. Furthermore, wet soil has a higher thermal conductivity than dry soil. In wet soil, heat is conducted upward from subsurface soil more rapidly, which helps to keep the surface warmer.

If the air temperature both at the surface and above fall below freezing, farmers are left with a difficult situation. Wind machines won't help because they would only mix cold air at the surface with the colder air above. Orchard heaters and irrigation are of little value, as they would only protect the branches just above the ground. Fortunately, there is one form of protection that does work: An orchard's sprinkling system can be turned on so that it emits a fine spray of water. In the cold air, the water freezes around the branches and buds, coating them with a thin veneer of ice. As long as the spraying continues, the latent heat—given off as the water changes into ice—keeps the ice temperature at 0°C (32°F). The ice acts as a protective coating against the subfreezing air by keeping the buds (or fruit) at a temperature higher than their damaging point (see ● Fig. 3.21). Care must be taken since too much ice can cause the branches to break. The fruit can be saved from the cold air, but the tree itself may be damaged by

● **FIGURE 3.19** Orchard heaters circulate the air by setting up convection currents.

● **FIGURE 3.20** Wind machines mix cooler surface air with warmer air above.

too much protection. Sprinklers work well when the air is fairly humid. They do not work well when the air is dry, as a good deal of the water may be lost through evaporation.

RECORD LOW TEMPERATURES One city in the United States that experiences very low temperatures is International Falls, Minnesota, where the average temperature for January is −15°C (4°F). Located several hundred miles to the south of International Falls, Minneapolis–St. Paul is the coldest major urban area in the nation, with an average temperature of −8°C (17°F) for January. For duration of extreme cold,

● **FIGURE 3.21** Ice covers citrus trees in Clermont, Florida, that were sprayed with water during the early morning to protect them from damaging low temperatures that dipped into the 20s (°F) on December 15, 2010.

● **FIGURE 3.22** Antarctica, the coldest continent on Earth, where air temperatures often drop below −100°F.

Minneapolis reported 186 consecutive hours of temperatures below 0°F during the winter of 1911–1912. Within the forty-eight adjacent states, however, the record for the longest duration of severe cold belongs to Langdon, North Dakota, where the thermometer remained below 0°F for 41 consecutive days (January 11 to February 20, 1936).

The most extensive cold wave in the United States occurred in February 1899. Temperatures during this cold spell fell below 0°F in every existing state, including Florida. This extreme cold event was the first and only one of its kind in recorded history. Record temperatures set during this extremely cold outbreak still stand today in many cities of the United States. The official record for the lowest temperature in the forty-eight adjacent states, however, belongs to Rogers Pass, Montana, where on the morning of January 20, 1954, the mercury dropped to −57°C (−70°F). The lowest official temperature for Alaska, −62°C (−80°F), occurred at Prospect Creek on January 23, 1971.

The coldest areas in North America are found in the Yukon and Northwest Territories and Nunavut in Canada. Resolute, Canada (latitude 75°N), has an average temperature of −32°C (−26°F) for the month of January.

The lowest temperatures and coldest winters in the Northern Hemisphere are found in the interior of Siberia and Greenland. For example, the average January temperature in Yakutsk, Siberia (latitude 62°N), is −39°C (−38°F). There, the mean temperature for the entire year is a bitterly cold −9°C (16°F). At Eismitte, Greenland, the average temperature for February (the coldest month) is −47°C (−53°F), with the mean annual temperature being a frigid −30°C (−22°F). Even though these temperatures are extremely low, they do not come close to the coldest area of the world: the Antarctic (see ● Fig. 3.22).

At the geographical South Pole, over 9000 feet above sea level, where the Amundsen-Scott scientific station has been keeping records for more than sixty years, the average temperature for the month of July (winter) is −56°C (−69°F) and the mean annual temperature is −49°C (−56°F). The lowest temperature ever recorded there (−83°C or −177°F) occurred under clear skies with a light wind on the morning of June 23, 1983. Cold as it was, it was not the record low for the world. That belongs to the Russian station at Vostok, Antarctica (latitude 78°S), where the temperature plummeted to −89°C (−129°F) on July 21, 1983. (● Figure 3.23 provides more information on record low temperatures throughout the world.)

BRIEF REVIEW

Up to this point we have examined temperature variations on a seasonal and daily basis. Before going on, here is a review of some of the important concepts and facts we have covered:

● The seasons are caused by Earth being tilted on its axis as it revolves around the sun. The tilt causes annual variations in the amount of sunlight that strikes the surface as well as variations in the length of time the sun shines at each latitude.

● During the day, Earth's surface and air above will continue to warm as long as incoming energy (mainly sunlight) exceeds outgoing energy from the surface.

● At night, Earth's surface cools, mainly by giving up more infrared radiation than it receives—a process called radiational cooling.

● The coldest nights of winter normally occur when the air is calm, fairly dry (low water-vapor content), and cloud-free.

● The highest temperatures during the day and the lowest temperatures at night are normally observed at Earth's surface.

● Radiation inversions exist usually at night when the air near the ground is colder than the air above.

● Farmers use a variety of techniques to protect crops or fruit from damaging low temperatures, including heating the air, mixing the air, irrigating, and spraying water onto trees in below-freezing weather.

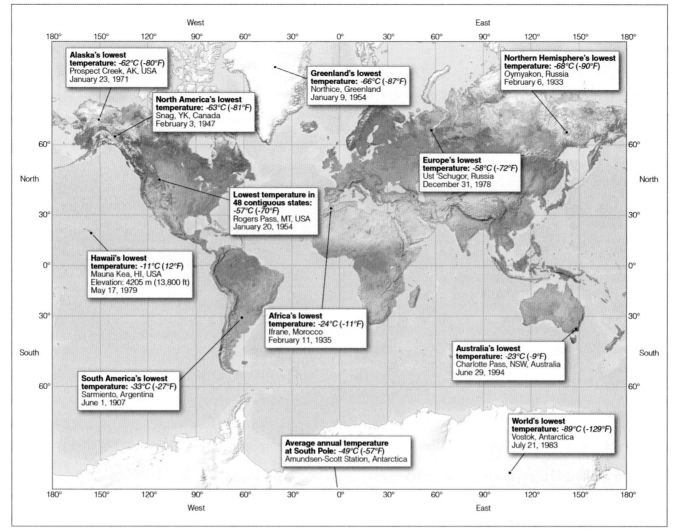

● **FIGURE 3.23** Record low temperatures throughout the world.

DAILY TEMPERATURE VARIATIONS So far, we have looked at how and why the air temperature near the ground changes during the course of a 24-hour day. We saw that during the day the air near Earth's surface can become quite warm, whereas at night it can cool off dramatically. ● Figure 3.24 summarizes these observations by illustrating how the average air temperature above the ground can change over a span of 24 hours. Notice in the figure that although the air several feet above the surface both cools and warms, it does so at a slower rate than air at the surface.

In fact, the greatest variation in daily temperature occurs at Earth's surface. The difference between the daily maximum and minimum temperature—called the **daily (diurnal) range of temperature**—is greatest next to the ground and becomes progressively smaller as we move away from the surface (see ● Fig. 3.25). This daily variation in temperature is also much larger on clear days than on cloudy ones.

The largest diurnal range of temperature occurs on high deserts, where the air is fairly dry, often cloud-free, and there is little water vapor to radiate much infrared energy back to the surface. By day, clear summer skies allow the sun's energy to quickly warm the ground which, in turn, warms the air above to a temperature often exceeding 38°C (100°F). At night, the ground cools rapidly by radiating infrared energy to space, and the minimum temperature in these regions occasionally dips below 7°C (45°F), thus giving an extremely high daily temperature range of more than 31°C (55°F).

A good example of a location with a large diurnal temperature range is the town of Winnemucca, Nevada, which is located at an elevation of 1310 m (4300 ft) above sea level. Here, in the dry, thin summer air, the average daily maximum temperature for July is 34°C (93°F)—short-sleeve weather, indeed. But don't lose

WEATHER WATCH

One of the greatest temperature ranges ever recorded in the Northern Hemisphere (100°F) occurred at Browning, Montana, on January 23, 1916, when the air temperature plummeted from 44°F to −56°F in less than 24 hours.

● **FIGURE 3.24** An idealized distribution of air temperature above the ground during a 24-hour day. The temperature curves represent the variations in average air temperature above a grassy surface for a mid-latitude city during the summer under clear, calm conditions.

● **FIGURE 3.25** The daily range of temperature decreases as we climb away from Earth's surface. Hence, there is less day-to-night variation in air temperature near the top of a high-rise apartment complex than at the ground level.

your shirt in Winnemucca. You will need it at night, as the average daily minimum temperature for July is 11°C (52°F). Winnemucca has a typical daily range in July of 23°C (41°F)!

Clouds can have a large effect on the daily range in temperature. As we saw in Chapter 2, clouds (especially low, thick ones) are good reflectors of incoming solar radiation, and so they prevent much of the sun's energy from reaching the surface. This effect tends to lower daytime temperatures (see ● Fig. 3.26a). If the clouds persist into the night, they tend to keep nighttime temperatures higher, as clouds are excellent absorbers and emitters of infrared radiation—the clouds actually emit a great deal of

infrared energy back to the surface. Clouds, therefore, have the effect of lowering the daily range of temperature. In clear weather (Fig. 3.26b), daytime air temperatures tend to be higher as the sun's rays impinge directly upon the surface, while nighttime temperatures are usually lower due to rapid radiational cooling. Therefore, clear days and clear nights combine to promote a large daily range in temperature.

Humidity can also have an effect on diurnal temperature ranges. For example, in humid regions, the diurnal temperature range is usually small. Here, haze and clouds lower the maximum temperature by preventing some of the sun's energy from reaching the surface.

(a) Small daily temperature range

(b) Large daily temperature range

● **FIGURE 3.26** (a) Clouds tend to keep daytime temperatures lower and nighttime temperatures higher, producing a small daily range in temperature. (b) In the absence of clouds, days tend to be warmer and nights cooler, producing a larger daily range in temperature.

When It Comes to Temperature, What's Normal?

When the weathercaster reports that "the normal high temperature for today is 68°F," does this mean that the high temperature on this day is usually 68°F? Or does it mean that we should expect a high temperature near 68°F? Actually, we should expect neither one.

Remember that the word *normal*, or *norm*, refers to weather data averaged over a period of 30 years. For example, ● Fig. 3 shows the high temperatures measured each May 6 for 30 years (1981 to 2010) in Salt Lake City, Utah. The average (mean) high temperature for this period is 68°F; hence, the normal high temperature for this date is 68°F (dashed line). Notice, however, that only on two days during this 30-year period did the high temperature actually measure 68°F (large red dots). In fact, the most common high temperature (called the *mode*) was 58°F, which occurred on three days (blue dots).

So what would be considered a typical high temperature for this date? Actually, any high temperature that lies between about 46°F and 90°F (two standard deviations* on either side of 68°F in Salt Lake City) would be considered in the range of typical high temperatures for this day. It

---
*A standard deviation is a statistical measure of the spread of the data. Two standard deviations for this set of data mean that 95 percent of the time the high temperature occurs between 46°F and 90°F.

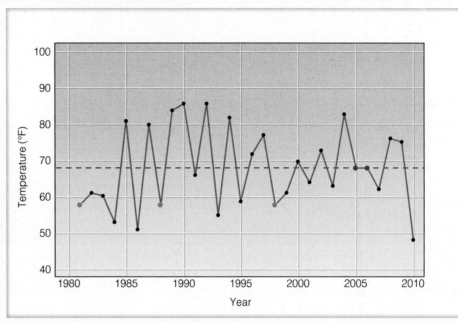

● **FIGURE 3** High temperatures measured on May 6 each year from 1981 to 2010 in Salt Lake City, Utah. The dashed line represents the *normal* (average) temperature for the 30-year period.

would be truly noteworthy, and "abnormal," if the high temperature happened to be 68°F on May 6 every single year, even though 68°F is the *normal* (average) high temperature for this 30-year period. While a high temperature of 86°F may be quite warm and a high temperature of 51°F may be on the cool side, they are both

no more uncommon (unusual) in this 30-year period than a high temperature of 68°F, which is the *normal* (average) high temperature for the 30-year period. This same type of reasoning applies to *normal rainfall*, as the actual amount of precipitation will likely be greater or less than the 30-year average.

At night, the moist air keeps the minimum temperature high by absorbing Earth's infrared radiation and radiating a portion of it to the ground. An example of a humid city with a small summer diurnal temperature range is Charleston, South Carolina, where the average July maximum temperature is 33°C (91°F), the average minimum is 23°C (73°F), and the diurnal range is only 10°C (18°F).

Cities near large bodies of water typically have smaller diurnal temperature ranges than cities farther inland. This phenomenon is caused in part by the additional water vapor in the air and by the fact that water warms and cools much more slowly than land.

Moreover, cities whose temperature readings are obtained at airports often have larger diurnal temperature ranges than those whose readings are obtained in downtown areas, because nighttime temperatures in cities tend to be warmer than those in outlying rural areas. This nighttime city warmth—called the *urban heat island*—forms as the sun's energy is absorbed by urban structures and concrete; then, during the night, this heat energy is slowly released into the city air.

The average of the highest and lowest temperature for a 24-hour period is known as the **mean (average) daily temperature**. The average of the mean daily temperatures for a particular date averaged for a 30-year period gives the average (or "*normal*") temperatures for that date. Currently, the National Weather Service uses the period 1981–2010 to calculate average temperatures, including the mean, maximum, and minimum for each day of the year. These readings are updated every ten years. Sometimes weather websites, TV weathercasts, or daily newspapers will provide these averages, along with the highest and lowest temperatures observed on the preceding day. (More information on the concept of "normal" temperature is given in Focus section 3.3.)

REGIONAL TEMPERATURE VARIATIONS The main factors that cause variations in temperature from one place to another are called the controls of temperature. Earlier we saw that the greatest factor in determining temperature is the amount of solar radiation that reaches the surface. This amount, of course, is determined by the length of daylight hours and the intensity

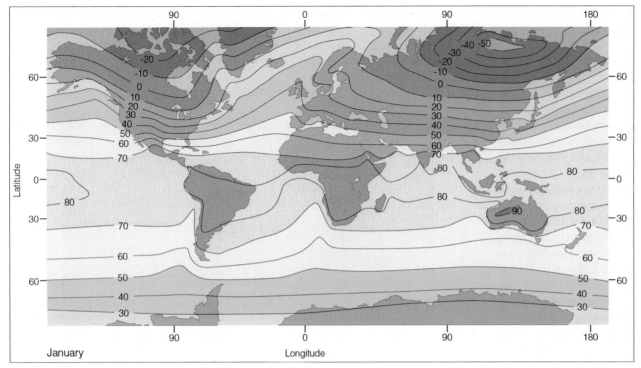

● **FIGURE 3.27** Average air temperature near sea level in January (°F). Temperatures in Central Antarctica are not visible on this map.

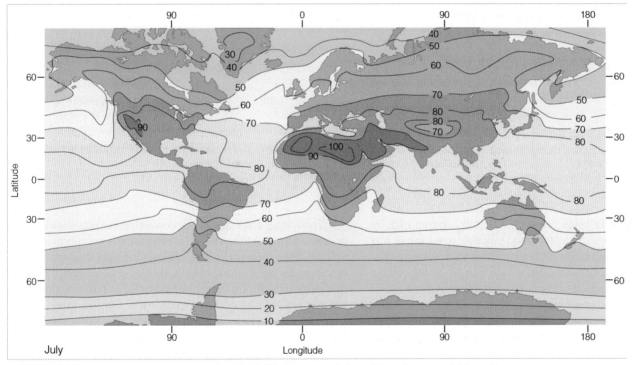

● **FIGURE 3.28** Average air temperature near sea level in July (°F). Temperatures in Central Antarctica are not visible on this map.

of incoming solar radiation. Both of these factors are a function of latitude; hence, latitude is considered an important control of temperature. The main controls are:

1. latitude
2. land and water distribution
3. ocean currents
4. elevation

We can obtain a better picture of these controls by examining ● Fig. 3.27 and ● Fig. 3.28, which show the average monthly temperatures throughout the world for January and July. (The average temperature for each month is the average of the daily mean temperatures for that month.) The lines on the map are **isotherms**—lines connecting places that have the same temperature. Because air temperature normally decreases with height, cities at very high elevations are much colder than their sea-level counterparts. Consequently, the isotherms in Fig. 3.27 and Fig. 3.28 are corrected to read at the same horizontal level (sea level) by adding to each station above sea level the equivalent average temperature change with height.*

Figures 3.27 and 3.28 show the importance of latitude on temperature. Notice that on both maps and in both hemispheres the isotherms are oriented east-west, indicating that locations at the same latitude receive nearly the same amount of solar energy. In addition, the annual solar heat that each latitude receives decreases from lower to higher latitudes; hence, average temperatures in January and July tend to decrease from lower to higher latitudes. But there is a greater variation in solar radiation between low and high latitudes in winter than in summer. Thus, the isotherms in January (during the Northern Hemisphere winter) are closer together (a tighter gradient—which represents a larger change in temperature over a given distance) than they are in July. If you travel from New Orleans to Chicago in January, you are more likely to experience greater temperature variations than if you make the same trip in July.

Even though average temperatures tend to decrease from low latitudes toward high latitudes, notice on the July map (Fig. 3.28) that the highest average temperatures do not occur in the tropics, but rather in the subtropical deserts of the Northern Hemisphere. Here, sinking air associated with high-pressure areas generally produces clear skies and low humidity. These conditions, along with a high sun beating down upon a relatively barren landscape, produce scorching heat.

The most extreme cold over land areas in January is typically across the interior of Siberia (Fig. 3.27), where the average January temperature dips below $-50°F$. Even colder readings, on average, occur in Antarctica during its dark winter months of June and July, as relatively dry air, high elevations, and snow-covered surfaces allow for rapid radiational cooling. Although not shown in Fig. 3.28, the average temperature for the coldest month at the South Pole is below $-70°F$. And for absolute cold, the lowest average temperature for any month $(-100°F)$ was recorded at the Plateau Station during July 1968.

So far we've seen that January temperatures in the Northern Hemisphere are much lower in the middle of continents than they are at the same latitude near the oceans. Notice on the July map that the reverse is true. One reason for these temperature differences is the unequal heating and cooling properties of land and water, as discussed earlier in this chapter. For one thing, solar energy reaching land is absorbed in a thin layer of soil; reaching water, it penetrates deeply. Because water circulates, it can distribute its heat through a much deeper layer. In addition, some of the solar energy striking the water evaporates it rather than heats it.

Another reason for the sharp temperature difference between oceans and interior locations is that it takes a great deal more heat to raise the temperature of a given amount of water by one degree than it does to raise the temperature of the same amount of land

by one degree.* Water not only heats more slowly than land, it cools more slowly as well, allowing the oceans to act like huge heat reservoirs. Thus, mid-ocean surface temperatures change relatively little from summer to winter compared to the much larger annual temperature changes over the middle of continents.

As a result of the warming and cooling properties of water, even large lakes can modify the temperature around them. In summer, for example, the Great Lakes remain cooler than the land and refreshing breezes blow inland, bringing relief from the sometimes sweltering heat. As winter approaches, the water cools more slowly than the land. The first blast of cold air from Canada is modified as it crosses the lakes, and so the first freeze is delayed on the eastern shores of Lake Michigan.

Look closely at Figs. 3.27 and 3.28 and notice that in many places the isotherms on both maps tend to bend when they approach an ocean-continent boundary. Such bending of the isotherms along the margin of continents is due in part to the unequal heating and cooling properties of land and water, and in part to *ocean currents*. For example, along the eastern margins of continents warm ocean currents transport warm water toward the poles, whereas, along the western margins, they transport cold water toward the equator. As we will see in Chapter 10, some coastal areas also experience upwelling, which brings cold water from below to the surface.

At any location, the difference in average temperature between the warmest month (often July in the Northern Hemisphere) and coldest month (often January) is called the **annual range of temperature**. As we would expect, annual temperature ranges are largest over interior continental landmasses and much smaller over larger bodies of water (see ● Fig. 3.29). Moreover, inland cities have larger annual temperature ranges than do coastal cities. Near the equator (because daylight length varies little and the sun is always high in the noon sky), annual temperature ranges are small, usually less than $3°C$ $(5°F)$. Quito, Ecuador—on the equator at an elevation of 2850 m (9350 ft)—experiences an annual range of less than $1°C$. In middle and high latitudes, annual ranges are large, especially in the middle of a continent. Yakutsk, in northeastern Siberia near the Arctic Circle, has an extremely large annual temperature range of $58°C$ $(104°F)$.

The average temperature of any station for the entire year is the **mean (average) annual temperature**, which represents the average of the 12 monthly average temperatures.** When two cities have the same mean annual temperature, it might first seem that their temperatures throughout the year are quite similar. However, often this is not the case. For example, San Francisco, California, and Richmond, Virginia, are situated at the same latitude $(37°N)$. Both have similar hours of daylight during the year; both have a mean annual temperature near $15°C$ $(59°F)$. But here

---

*The amount of change is usually less than the standard temperature lapse rate of $3.6°F$ per 1000 feet ($6.5°C$ per 1000 meters). The reason is that the standard lapse rate is computed for altitudes above Earth's surface in the "free" atmosphere. In the less-dense air at high elevations, the absorption of solar radiation by the ground causes an overall slightly higher temperature than that of the free atmosphere at the same level.

---

*Recall from Chapter 2 that the amount of heat needed to raise the temperature of one gram of a substance by one degree Celsius is called *specific heat*, and that water has a higher specific heat than does land.

**The mean annual temperature may be obtained by multiplying each of the 12 monthly means by the number of days in that month, adding the 12 numbers, and dividing that total by 12; or by obtaining the sum of the daily means and dividing that total by 365.

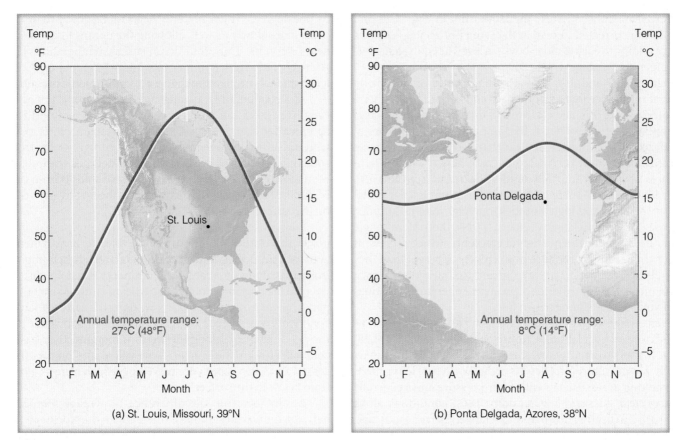

● **FIGURE 3.29** Monthly temperature data and annual temperature range for (a) St. Louis, Missouri, a city located near the middle of the North American continent, and (b) Ponta Delgada, a city located in the Azores in the Atlantic Ocean. Notice that the annual temperature range is much higher in St. Louis, even though both cities are at the same latitude.

the similarities end. The temperature differences between the two cities are apparent to anyone who has traveled to San Francisco during the summer with a suitcase full of clothes suitable for summer weather in Richmond.

● Figure 3.30 summarizes the average temperatures for San Francisco and Richmond. Notice that the coldest month for both cities is January. Even though January in Richmond averages only about 8°C (14°F) colder than January in San Francisco, people in Richmond awaken to an average January minimum temperature of −2°C (28°F), only 1°F above the lowest temperature ever recorded in San Francisco. Trees that thrive in San Francisco's weather would find it difficult to survive a winter in Richmond. So, even though San Francisco and Richmond have the same mean annual temperature, the behavior and range of their temperatures differ greatly.

Applications of Air Temperature Data

There are a variety of applications for the mean daily temperature. An application developed by heating engineers in estimating energy needs is the **heating degree day**. The heating degree day is based on the assumption that people will begin to use their furnaces when the mean daily temperature drops below 65°F.

Therefore, heating degree days are determined by subtracting the mean temperature for the day from 65°F. Thus, if the mean temperature for a day is 64°F, the heating degree day for this day is 1.*

On days when the mean temperature is above 65°F, there are no heating degree days. Hence, the lower the average daily temperature, the more heating degree days and the greater the predicted consumption of fuel. When the number of heating degree days for a whole year is calculated, the heating fuel requirements for any location can be estimated. ● Figure 3.31 shows the yearly average number of heating degree days in various locations throughout the United States.

As the mean daily temperature climbs above 65°F, people begin to cool their indoor environment. Consequently, an index called the **cooling degree day** is used during warm weather to estimate the energy needed to cool indoor air to a comfortable level. The forecast of mean daily temperature is converted to cooling degree days by subtracting 65°F from the mean. The remaining value is the number of cooling degree days for that day. For example, a day with a mean temperature of 70°F would correspond to (70 − 65), or 5 cooling degree days. High values indicate warm weather and high power production for cooling (see ● Fig. 3.32).

*In the United States, the National Weather Service and the Department of Agriculture use degrees Fahrenheit in their computations.

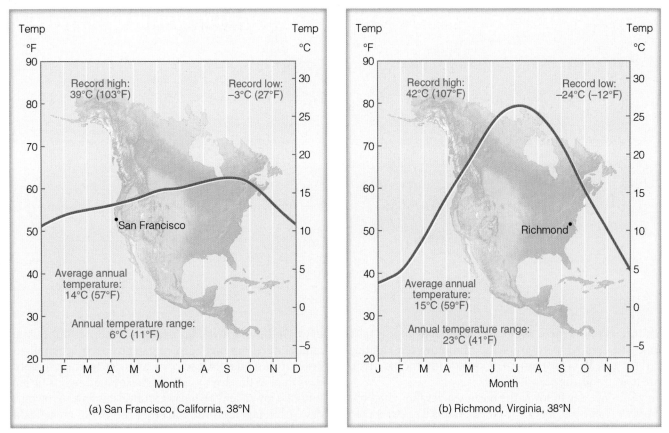

● FIGURE 3.30 Temperature data for (a) San Francisco, California (38°N) and (b) Richmond, Virginia (38°N)—two cities with the same mean annual temperature.

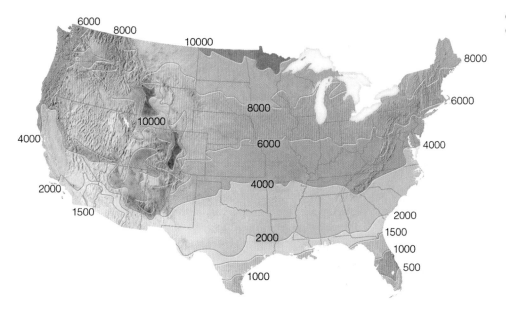

● FIGURE 3.31 Mean annual total heating degree days across the United States (base 65°F).

Knowledge of the number of cooling degree days in an area allows a builder to plan the size and type of equipment that should be installed to provide adequate air conditioning. Also, the forecasting of cooling degree days during the summer gives power companies a way of predicting the energy demand during peak energy periods. A composite of heating plus cooling degree days gives a practical indication of the energy requirements over the year.

Farmers use an index called **growing degree days** as a guide to planting and for determining the approximate dates when a crop will be ready for harvesting. A growing degree day for a particular crop is defined as a day on which the mean daily

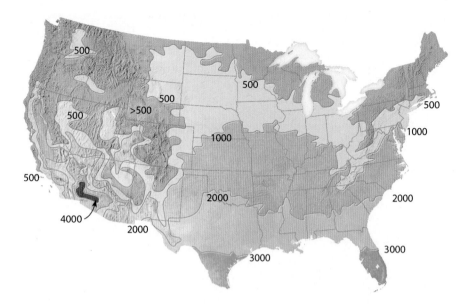

temperature is one degree above the *base temperature* (also known as the *zero temperature*)—the minimum temperature required for growth of that crop. For sweet corn, the base temperature is 50°F and, for peas, it is 40°F.

On a summer day in Iowa, the mean temperature might be 80°F. From ▼ Table 3.2, we can see that, on this day, sweet corn would accumulate (80 − 50), or 30 growing degree days. Theoretically, sweet corn can be harvested when it accumulates a total of 2200 growing degree days. For example, if sweet corn is planted in early April and each day thereafter averages about 20 growing degree days, the corn would be ready for harvest about 110 days later, or around the middle of July.*

At one time, corn varieties were rated in terms of "days to maturity." This rating system was unsuccessful because, in actual practice, corn took considerably longer in some areas than in others. This discrepancy was the reason for defining "growing degree

days." In humid Iowa, for example, where summer nighttime temperatures are high, growing degree days accumulate much faster. Consequently, the corn matures in considerably fewer days than in the drier west, where summer nighttime temperatures are lower, and each day accumulates fewer growing degree days. Although moisture and other conditions are not taken into account, growing degree days nevertheless serve as a useful guide in forecasting approximate dates of crop maturity.

AIR TEMPERATURE AND HUMAN COMFORT Probably everyone realizes that the same air temperature can feel different on different occasions. For example, a temperature of 20°C (68°F) on a clear, windless March afternoon in New York City can feel almost balmy after a long hard winter. Yet, this same temperature can feel uncomfortably cool on a summer afternoon in a stiff breeze. The human body's perception of temperature obviously changes with varying atmospheric conditions. The reason for these changes is related to how we exchange heat energy with our environment.

The body stabilizes its temperature primarily by converting food into heat (*metabolism*). To maintain a constant temperature, the heat produced and absorbed by the body must be equal to the heat it loses to its surroundings. There is, therefore, a constant exchange of heat—especially at the surface of the skin—between the body and the environment.

One way the body loses heat is by emitting infrared energy. But we not only emit radiant energy, we absorb it as well. Another way the body loses and gains heat is by conduction and convection, which transfer heat to and from the body by air motions. On a cold day, a thin layer of warm air molecules forms close to the skin, protecting it from the surrounding cooler air and from the rapid transfer of heat. In cold weather, when the air is calm, the temperature we perceive—called the **sensible temperature**—is often higher than a thermometer might indicate. (Could the opposite effect occur where the air temperature is very high and a person might feel exceptionally cold? If you are not sure how to answer this question, read the Focus section 3.4.)

▼ TABLE 3.2 **Estimated Growing Degree Days for Certain Naturally Grown Agricultural Crops to Reach Maturity**

CROP (VARIETY, LOCATION)	BASE TEMPERATURE (°F)	GROWING DEGREE DAYS TO MATURITY
Beans (Snap/South Carolina)	50	1200–1300
Corn (Sweet/Indiana)	50	2200–2800
Cotton (Delta Smooth Leaf/Arkansas)	60	1900–2500
Peas (Early/Indiana)	40	1100–1200
Rice (Vegold/Arkansas)	60	1700–2100
Wheat (Indiana)	40	2100–2400

*When the air temperature climbs above 86°F in the Corn Belt of the Midwest, the hot air puts added stress on the growth of the corn. Consequently, the corn grows more slowly. Because of this fact, any maximum temperature over 86°F is reduced to 86°F when computing the mean air temperature for growing degree days.

A Thousand Degrees and Freezing to Death

Is there somewhere in our atmosphere where the air temperature can be exceedingly high (say above 1000°C or 1800°F) yet a person might feel extremely cold? There is such a region, but it's not at Earth's surface.

You may recall from Chapter 1 (Fig. 1.13, p. 16) that in the upper reaches of our atmosphere (in the middle and upper thermosphere), air temperatures can exceed 1000°C. However, a thermometer shielded from the sun in this region of the atmosphere would indicate an extremely low temperature. This apparent discrepancy lies in the meaning of air temperature and how we measure it.

In Chapter 2, we learned that the air temperature is directly related to the average speed at which the air molecules are moving—faster speeds correspond to higher temperatures. In the middle and upper thermosphere (at altitudes approaching 300 km, or 200 mi) air molecules are zipping about at speeds corresponding to extremely high

● **FIGURE 4** How can an astronaut survive when the "air" temperature is 1000°C?

temperatures. However, in order to transfer enough energy to heat something up by conduction (exposed skin or a thermometer bulb),

an extremely large number of molecules must collide with the object. In the "thin" air of the upper atmosphere, air molecules are moving extraordinarily fast, but there are simply not enough of them bouncing against the thermometer bulb for it to register a high temperature. In fact, when properly shielded from the sun, the thermometer bulb loses far more energy than it receives and indicates a temperature near absolute zero. This explains why an astronaut, when spacewalking, will not only survive temperatures exceeding 1000°C, but will also feel a profound coldness when shielded from the sun's radiant energy. At these high altitudes, the traditional meaning of air temperature (that is, regarding how "hot" or "cold" something feels) is no longer applicable.

Once the wind starts to blow, the insulating layer of warm air is swept away, and heat is rapidly removed from the skin by the constant bombardment of cold air. When all other factors are the same, the faster the wind blows, the greater the heat loss, and the colder we feel. How cold the wind makes us feel is usually expressed as a **wind-chill index (WCI)**.

The most recent version of the wind-chill index (see ▼ Table 3.3) was formulated in 2001 by a joint action group of the National Weather Service and other agencies. The index takes into account the wind speed at about 1.5 m (5 ft) above the ground (close to where an adult's upper body would be) instead of the 10 m (33 ft) where "official" readings are usually taken. In addition, the index translates the capacity of the air to take heat away from a person's face (the air's cooling power) into a wind-chill equivalent temperature.* For example, notice in Table 3.3 that an air temperature of 10°F with a wind speed of 10 mi/hr produces a wind-chill equivalent temperature of −4°F. In other words, the skin of a person's exposed face would lose as much heat in one minute in air with a temperature of 10°F and a wind speed of 10 mi/hr as it would in calm air with a temperature of −4°F.

Of course, how cold we feel actually depends on a number of factors, including the fit and type of clothing we wear, the amount of sunshine striking the body, and the actual amount of exposed skin.

High winds, in below-freezing air, can remove heat from exposed skin so quickly that the skin may actually freeze and discolor. The freezing of skin, called **frostbite**, usually occurs on the body extremities first because they are the greatest distance from the source of body heat.

In cold weather, wet skin can be a factor in how cold we feel. A cold rainy day (drizzly, or even foggy) often feels colder than a "dry" one because water on exposed skin conducts heat away from the body better than air does. In fact, in cold, wet, and windy weather a person may actually lose body heat faster than the body can produce it. This may even occur in relatively mild weather with air temperatures as high as 10°C (50°F). The rapid loss of body heat may lower the body temperature below its normal level and bring on a condition known as **hypothermia**—the rapid, progressive mental and physical collapse that accompanies the lowering of human body temperature.

The first symptom of hypothermia is exhaustion. If exposure continues, judgment and reasoning power begin to disappear. Prolonged exposure, especially at temperatures near or below freezing, produces stupor, collapse, and death when the internal body temperature drops to 26°C (79°F). Most cases of hypothermia occur when the air temperature is between 0°C and 10°C

---

*The wind-chill equivalent temperature formulas are as follows: Wind chill (°F) = $35.74 + 0.6215T - 35.75(V^{0.16}) + 0.4275T(V^{0.16})$, where T is the air temperature in (°F) and V is the wind speed in mi/hr. Wind chill (°C) = $13.12 + 0.6215T - 11.37(V^{0.16}) + 0.3965T(V^{0.16})$, where T is the air temperature in (°C), and V is the wind speed in km/hr.

▼ **TABLE 3.3** **Wind-Chill Equivalent Temperature (°F). A 20-mi/hr Wind Combined with an Air Temperature of 20°F Produces a Wind-Chill Equivalent Temperature of 4°F.***

	AIR TEMPERATURE (°F)																
Calm	40	35	30	25	20	15	10	5	0	−5	−10	−15	−20	−25	−30	−35	−40
5	36	31	25	19	13	7	1	−5	−11	−16	−22	−28	−34	−40	−46	−52	−57
10	34	27	21	15	9	3	−4	−10	−16	−22	−28	−35	−41	−47	−53	−59	−66
15	32	25	19	13	6	0	−7	−13	−19	−26	−32	−39	−45	−51	−58	−64	−71
20	30	24	17	11	4	−2	−9	−15	−22	−29	−35	−42	−48	−55	−61	−68	−74
25	29	23	16	9	3	−4	−11	−17	−24	−31	−37	−44	−51	−58	−64	−71	−78
30	28	22	15	8	1	−5	−12	−19	−26	−33	−39	−46	−53	−60	−67	−73	−80
35	28	21	14	7	0	−7	−14	−21	−27	−34	−41	−48	−55	−62	−69	−76	−82
40	27	20	13	6	−1	−8	−15	−22	−29	−36	−43	−50	−57	−64	−71	−78	−84
45	26	19	12	5	−2	−9	−16	−23	−30	−37	−44	−51	−58	−65	−72	−79	−86
50	26	19	12	4	−3	−10	−17	−24	−31	−38	−45	−52	−60	−67	−74	−81	−88
55	25	18	11	4	−3	−11	−18	−25	−32	−39	−46	−54	−61	−68	−75	−82	−89
60	25	17	10	3	−4	−11	−19	−26	−33	−40	−48	−55	−62	−69	−76	−84	−91

(WIND SPEED (MI/HR))

*Dark-shaded areas represent conditions where frostbite occurs in 30 minutes or less.

(between 32°F and 50°F). This may be because many people apparently do not realize that wet clothing in windy weather greatly enhances the loss of body heat, even when the temperature is well above freezing.

In cold weather, heat is more easily dissipated through the skin. To counteract this rapid heat loss, the peripheral blood vessels of the body constrict, cutting off the flow of blood to the outer layers of the skin. In hot weather, the blood vessels enlarge, allowing a greater loss of heat energy to the surroundings. Perspiration is also a factor. As water evaporates from our skin, the skin cools because it supplies the large latent heat of vaporization (about 560 cal/g). When the air contains a great deal of water vapor (very humid), and it is close to being saturated, perspiration does not readily evaporate from the skin. Less evaporational cooling causes most people to feel hotter than it really is, and a number of people start to complain about the "heat and humidity." A closer look at how we feel in hot, humid weather will be given in Chapter 4 after we examine the concepts of relative humidity, dew point, and wet-bulb temperature.

MEASURING AIR TEMPERATURE Thermometers were developed to measure air temperature. Each thermometer has a definite scale and is calibrated so that a thermometer reading of 0°C will denote the same temperature whether it is in Vermont or North Dakota. If a particular temperature were to represent different degrees of hot or cold, depending on location, thermometers would be useless.

A very common thermometer for measuring surface air temperature is the **liquid-in-glass thermometer**. This type of thermometer has a glass bulb attached to a sealed, graduated tube about 25 cm (10 in.) long. A very small opening, or bore, extends from the bulb to the end of the tube. A liquid in the bulb (usually mercury or red-colored alcohol) is free to move from the bulb up through the bore and into the tube. The length of the liquid in the tube represents the air temperature. When the air temperature increases, the liquid in the bulb expands and rises up the tube. When the air temperature decreases, the liquid contracts, and moves down the tube. Hence, the length of the liquid in the tube represents the air temperature. Because the bore is very narrow, a small temperature change shows up as a relatively large change in the length of the liquid column.

Maximum and minimum thermometers are liquid-in-glass thermometers used for determining daily maximum and minimum temperatures. The **maximum thermometer** looks like any other liquid-in-glass thermometer with one exception: It has a small constriction within the bore just above the bulb (see ● Fig. 3.33). As the air temperature increases, the mercury expands and freely moves past the constriction up the tube, until the maximum temperature occurs. However, as the air temperature begins to drop, the small constriction prevents the mercury from flowing back into the bulb. Thus, the end of the stationary mercury column indicates the maximum temperature for the day. The mercury will stay at this position until either the air warms to a higher reading or the thermometer is reset by whirling it on a special holder and pivot. Usually, the whirling is sufficient to push the mercury back into the bulb past the constriction until the end of the column indicates the present air temperature.*

*Liquid-in-glass thermometers that measure body temperature are maximum thermometers, which is why they are shaken both before and after you take your temperature.

● **FIGURE 3.33** A section of a maximum thermometer.

● **FIGURE 3.34** A section of a minimum thermometer showing both the current air temperature and the minimum temperature in °F.

A **minimum thermometer** measures the lowest temperature reached during a given period. Most minimum thermometers use alcohol as a liquid, since it freezes at a temperature of −130°C compared to −39°C for mercury. The minimum thermometer is similar to other liquid-in-glass thermometers except that it contains a small barbell-shaped index marker in the bore (see ● Fig. 3.34). The small index marker is free to slide back and forth within the liquid. It cannot move out of the liquid because the surface tension at the end of the liquid column (the *meniscus*) holds it in.

A minimum thermometer is mounted horizontally. As the air temperature drops, the contracting liquid moves back into the bulb and brings the index marker down the bore with it. When the air temperature stops decreasing, the liquid and the index marker stop moving down the bore. As the air warms, the alcohol expands and moves freely up the tube past the stationary index marker. Because the index marker does not move as the air warms, the minimum temperature is read by observing the upper end of the marker.

To reset a minimum thermometer, simply tip it upside down. This allows the index marker to slide to the upper end of the alcohol column, which is indicating the current air temperature. The thermometer is then remounted horizontally, so that the marker will move toward the bulb as the air temperature decreases.

Highly accurate temperature measurements can be made with **electrical thermometers**. One type of electrical thermometer is the *electrical resistance thermometer*. This does not directly measure air temperature; rather, it measures the resistance of a wire, usually platinum or nickel, whose resistance increases as the temperature increases. An electrical meter measures the resistance and is calibrated to represent air temperature.

Electrical resistance thermometers are the type of thermometers used in the measurement of air temperature at more than 900 fully automated surface weather stations (known as *ASOS* for *A*utomated *S*urface *O*bserving *S*ystem) that exist at airports and military facilities throughout the United States (see ● Fig. 3.35). They have replaced many of the liquid-in-glass thermometers formerly in use.

The replacement of liquid-in-glass thermometers with electrical thermometers has some implications for the long-term observational record. For one thing, the response of the electrical thermometers to temperature change is faster. Thus, electrical thermometers might reach a brief extreme reading that could have been missed by the slower-responding liquid-in-glass thermometer. In addition, many temperature readings that were previously taken at airport weather offices are now taken at ASOS locations situated near or between runways at the airport. This change

in instrumentation and relocation of the measurement site can sometimes introduce a small, but significant, temperature change at the reporting station. To reduce the impact of such changes, the United States has created a Climate Reference Network of about 100 weather stations. These stations are carefully placed, calibrated, and maintained to produce a consistent, accurate, long-term reading of air temperature.

Thermistors are another type of electrical thermometer. They are made of ceramic material whose resistance increases as the temperature decreases. A thermistor is the temperature-measuring device of the radiosonde—the instrument that measures air temperature from the surface up to an altitude near 30 kilometers. (For additional information on radiosondes, read Focus section 1.3 in Chapter 1, on p. 15.)

Another electrical thermometer is the *thermocouple*. This device operates on the principle that the temperature difference between the junction of two dissimilar metals sets up a weak electrical current. When one end of the junction is maintained at a temperature different from that of the other end, an electrical

● **FIGURE 3.35** The instruments that comprise the ASOS system. The max-min temperature shelter is the middle white box.

Why Thermometers Should Be Read in the Shade

When we measure air temperature with a common liquid thermometer, an incredible number of air molecules bombard the bulb, transferring energy either to or away from it. When the air is warmer than the thermometer, the liquid gains energy, expands, and rises up the tube; the opposite will happen when the air is colder than the thermometer. The liquid stops rising (or falling) when equilibrium between incoming and outgoing energy is established. At this point, we can read the temperature by observing the height of the liquid in the tube.

It is *impossible* to measure air temperature accurately in direct sunlight, because the thermometer absorbs radiant energy from the sun in addition to energy from the air molecules. The thermometer gains energy at a much faster rate than it can radiate it away, and the liquid keeps expanding and rising until there is equilibrium between incoming and outgoing

energy. Because of the direct absorption of solar energy, the level of the liquid in the thermometer indicates a temperature *much* higher than the actual air temperature. Thus, a claim that "today the air temperature measured 100 degrees in the sun" has no meaning; a thermometer must be kept in a shady place to measure the temperature of the air accurately.

● **FIGURE 5** Instrument shelters such as the one shown here serve as a shady place for thermometers. Thermometers inside shelters measure the temperature of the air, whereas thermometers held in direct sunlight do not.

Ross DePaola

---

current will flow in the circuit. This current is proportional to the temperature difference between the junctions.

Air temperature can also be obtained with instruments called *infrared sensors*, or **radiometers**. Radiometers do not measure temperature directly; rather, they measure emitted radiation (usually infrared). By measuring both the intensity of radiant energy and the wavelength of maximum emission of a particular gas, radiometers in orbiting satellites are now able to obtain temperature measurements at selected levels in the atmosphere.

A **bimetallic thermometer** consists of two different pieces of metal (usually brass and iron) welded together to form a single strip. As the temperature changes, the brass expands more than the iron, causing the strip to bend. The small amount of bending is amplified through a system of levers to a pointer on a calibrated scale. The bimetallic thermometer is usually the temperature-sensing part of the **thermograph**, an instrument that measures and records temperature (see ● Fig. 3.36).

Mechanical thermographs are gradually being replaced with *data loggers*. These small instruments have a thermistor connected to a circuit board inside the logger. A computer programs the interval at which readings are taken. The loggers are not only more responsive to air temperature than are thermographs—they are also less expensive.

Chances are, you may have heard someone exclaim something like, "Today the thermometer measured a hundred degrees in the shade!" Does this mean that the air temperature is sometimes measured in the sun? If you are unsure of the answer,

read the Focus section 3.5 before reading the next section on instrument shelters.

Thermometers and other instruments are usually housed in an **instrument shelter**. The shelter completely encloses the instruments, protecting them from rain, snow, and the sun's direct rays. It is painted white to reflect sunlight, faces north to avoid direct exposure to sunlight, and has louvered sides, so that air is free to flow through it. This construction helps to keep the air inside the shelter at the same temperature as the air outside.

The thermometers inside a standard shelter are mounted about 1.5 to 2 m (5 to 6 ft) above the ground. As we saw in an earlier section, on a clear, calm night the air at ground level may be much colder than the air at the level of the shelter. As a result,

● **FIGURE 3.36** The thermograph with a bimetallic thermometer.

on clear winter mornings it is possible to see ice or frost on the ground even though the minimum thermometer in the shelter did not reach the freezing point.

Many older instrument shelters (such as the one shown in the Focus section 3.5, Fig. 5) have been replaced by the *Max-Min Temperature Shelter* of the ASOS system (the middle white box in Fig. 3.35). The shelter is mounted on a pipe, and wires from the electrical temperature sensor inside are run to a building. A readout inside the building displays the current air temperature and stores the maximum and minimum temperatures for later retrieval.

Because air temperatures vary considerably above different types of surfaces, shelters are placed over grass where possible, to ensure that the air temperature is measured at the same elevation over the same type of surface. Unfortunately, some shelters are placed on asphalt, others sit on concrete, while others are located on the tops of tall buildings, making it difficult to compare air temperature measurements from different locations. In fact, if either the maximum or minimum air temperature in your area seems suspiciously different from those of nearby towns, find out where the instrument shelter is situated.

SUMMARY

Earth has seasons because Earth is tilted on its axis as it revolves around the sun. The tilt of Earth causes a seasonal variation in both the length of daylight and the intensity of sunlight that reaches the surface. When the Northern Hemisphere is tilted toward the sun, the Southern Hemisphere is tilted away from the sun. Longer hours of daylight and more intense sunlight produce summer in the Northern Hemisphere, while, in the Southern Hemisphere, shorter daylight hours and less intense sunlight produce winter. In a more local setting, Earth's inclination influences the amount of solar energy received on the north and south side of a hill, as well as around a home.

The daily variation in air temperature near Earth's surface is controlled mainly by the input of energy from the sun and the output of energy from the surface. On a clear, calm day, the surface air warms, as long as heat input (mainly sunlight) exceeds heat output (mainly convection and radiated infrared energy). The surface air cools at night, as long as heat output exceeds input. Because the ground at night cools more quickly than the air above, the coldest air is normally found at the surface where a radiation inversion usually forms. When the air temperature in agricultural areas drops to dangerously low readings, fruit trees and grape vineyards can be protected from the cold by a variety of means, from mixing the air to spraying the trees and vines with water.

The greatest daily variation in air temperature occurs at Earth's surface. Both the diurnal and annual ranges of temperature are greater in dry climates than in humid ones. Even though two cities may have similar average annual temperatures, the range and extreme of their temperatures can differ greatly. Temperature information impacts our lives in many ways, from influencing decisions on what clothes to take on a trip to providing critical information for energy-use predictions and agricultural planning. We reviewed some of the many types of thermometers in use. Those designed to measure air temperatures near the surface are housed in instrument shelters to protect them from direct sunlight and precipitation.

KEY TERMS

The following terms are listed (with corresponding page numbers) in the order they appear in the text. Define each. Doing so will aid you in reviewing the material covered in this chapter.

summer solstice, 61
autumnal equinox, 63
Indian summer, 63
winter solstice, 63
vernal equinox, 64
radiational cooling, 70
radiation inversion, 71
nocturnal inversion, 71
thermal belts, 72
freeze, 72
orchard heaters, 73
wind machines, 73
daily (diurnal) range of temperature, 75
mean (average) daily temperature, 77
isotherms, 78
annual range of temperature, 79

mean (average) annual temperature, 79
heating degree day, 80
cooling degree day, 80
growing degree days, 81
sensible temperature, 82
wind-chill index (WCI), 83
frostbite, 83
hypothermia, 83
liquid-in-glass thermometer, 84
maximum thermometer, 84
minimum thermometer, 85
electrical thermometers, 85
radiometers, 86
bimetallic thermometer, 86
thermograph, 86
instrument shelter, 86

QUESTIONS FOR REVIEW

1. In the Northern Hemisphere, why are summers warmer than winters, even though Earth is actually closer to the sun in January?
2. What are the main factors that determine seasonal temperature variations?
3. During the Northern Hemisphere's summer, the daylight hours in northern latitudes are longer than in middle latitudes. Explain why northern latitudes are not warmer.
4. If it is winter and January in New York City, what is the season in Sydney, Australia?
5. Explain why Southern Hemisphere summers are not warmer on average than Northern Hemisphere summers, even though Earth is closer to the sun in January than in July.
6. Explain why the vegetation on the north-facing side of a hill is frequently different from the vegetation on the south-facing side of the same hill.

7. Look at Figures 3.12 (p. 68) and 3.17 (p. 71), which show vertical profiles of air temperature during different times of the day. Explain why the temperature curves are different.

8. What are some of the factors that determine the daily fluctuation of air temperature just above the ground?

9. Explain how incoming energy and outgoing energy regulate the daily variation in air temperature.

10. On a calm, sunny day, why is the air next to the ground normally much warmer than the air just above?

11. Explain why the warmest time of the day is usually in the afternoon, even though the sun's rays are most direct at noon.

12. Explain how radiational cooling at night produces a radiation temperature inversion.

13. What weather conditions are best suited for the formation of a cold night and a strong radiation inversion?

14. Explain why thermal belts are found along hillsides at night.

15. List some of the measures farmers use to protect their crops against the cold. Explain the physical principle behind each method.

16. Why are the lower tree branches most susceptible to damage from low temperatures?

17. Describe each of the controls of temperature.

18. Look at Fig. 3.27 (temperature map for January), p. 78, and explain why the isotherms dip southward (equatorward) over the Northern Hemisphere continents.

19. Explain why the daily range of temperature is normally greater
    (a) in dry regions than in humid regions and
    (b) on clear days than on cloudy days.

20. Why is the largest annual range of temperatures normally observed over continents away from large bodies of water?

21. Two cities have the same mean annual temperature. Explain why this fact does not mean that their temperatures throughout the year are necessarily similar.

22. During a cold, calm, sunny day, why do we usually feel warmer than a thermometer indicates?

23. What atmospheric conditions can bring on hypothermia?

24. During the winter, white frost can form on the ground when the minimum thermometer in an instrument shelter indicates a low temperature above freezing. Explain.

25. Why do daily temperature ranges decrease as you increase in altitude?

26. Why do the first freeze in autumn and the last freeze in spring occur in low-lying areas?

27. Someone says, "The air temperature today measured 99°F in the sun." Why does this statement have no meaning?

28. Briefly describe how the following thermometers measure air temperature:
    (a) liquid-in-glass
    (b) bimetallic
    (c) electrical
    (d) radiometer

QUESTIONS FOR THOUGHT

1. Explain (with the aid of a diagram) why the morning sun shines brightly through a south-facing bedroom window in December, but not in June.

2. If the tilt of Earth's axis suddenly increases to 45°, give two reasons why average winter temperatures in the middle latitudes of the Northern Hemisphere would decrease.

3. In New York City, on October 21 and on February 21, the intensity of sunlight and the number of daylight hours are almost identical. Why, then, in New York City is it usually much colder on February 21?

4. At the top of Earth's atmosphere during the early summer (Northern Hemisphere), above what latitude would you expect to receive the most solar radiation in one day? During the same time of year, where would you expect to receive the most solar radiation at the surface? Explain why the two locations are different. (If you are having difficulty with this question, refer to Fig. 3.5, p. 62.)

5. If a construction company were to build a solar-heated home in middle latitudes in the Southern Hemisphere, in

which direction should the solar panels on the roof be directed for maximum daytime heating?

6. Aside from the aesthetic appeal (or lack of such), explain why painting the outside north-facing wall of a middle-latitude house one color and the south-facing wall another color is not a bad idea.

7. How would the lag in daily temperature experienced over land compare to the daily temperature lag over water?

8. Where would you expect to experience the smallest variation in temperature from year to year and from month to month? Why?

9. The average temperature in San Francisco, California, for December, January, and February is 52°F. During the same three-month period the average temperature in Richmond, Virginia, is 40°F. Yet, San Francisco and Richmond have nearly the same yearly total of heating degree days. Explain why. (Hint: See Fig. 3.30, p. 81.)

10. On a warm summer day, one city experienced a daily range of 22°C (40°F), while another had a daily range of 10°C (18°F). One of these cities is located in New Jersey and the other in New Mexico. Which location most likely had the highest daily range, and which one had the smallest? Explain.

11. Minimum thermometers are usually read during the morning, yet they are reset in the afternoon. Explain why.

12. If clouds arrive at 2 a.m. in the middle of a calm, clear night, it is quite common to see temperatures rise after 2 a.m. How does this happen?

13. In the Northern Hemisphere, south-facing mountain slopes normally have a greater diurnal range in temperature than north-facing slopes. Why?

14. If the poles have 24 hours of sunlight during the summer, why is the average summer temperature there still below 0°F?

15. In Pennsylvania and New York, wine grapes are planted on the side of hills rather than in valleys. Explain why this practice is so common in these areas.

PROBLEMS AND EXERCISES

1. Draw a graph similar to Fig. 3.5 p. 62. Include in it the amount of solar radiation reaching Earth's surface in the Northern Hemisphere on the equinox.

2. Each day past the winter solstice the noon sun is a little higher above the southern horizon.
   (a) Determine how much change takes place each day at your latitude.
   (b) Does the same amount of change take place at each latitude in the Northern Hemisphere? Explain.

3. On approximately what dates will the sun be overhead at noon at latitudes:
   (a) 10°N?
   (b) 15°S?

4. Design a solar-heated home that sits on the north side of an east-west running street. If the home is located at 40°N, draw a proper roof angle for maximum solar heating. Design windows, doors, overhangs, and rooms with the intent of reducing heating and cooling costs. Place trees around the home that will block out excess summer sunlight and yet let winter sunlight inside. Choose a paint color for the house that will add to the home's energy efficiency.

5. Suppose peas are planted in Indiana on May 1. If the peas need 1200 growing degree days before they can be picked, and if the average maximum temperature for May and June is 80°F and the average minimum is 60°F, on about what date will the peas be ready to pick? (Assume a base temperature of 55°F.)

6. What is the wind-chill equivalent temperature when the air temperature is 5°F and the wind speed is 35 mi/hr? (Use Table 3.3, p. 84.)

 ONLINE RESOURCES

Visit www.cengagebrain.com to view additional resources, including video exercises, practice quizzes, an interactive eBook, and more.

On this summer morning, sunlight is diffused by saturated air whose temperature has dropped to the dew point.

Efimenko Alexander/Shutterstock.com

Atmospheric Humidity

IT'S 9 A.M. ON APRIL 26, 2005, IN BANGKOK, THAILAND, one of the hottest and most humid major cities in the world. The streets are clogged with traffic and on this hot, muggy morning perspiration streams down the faces of anxious people struggling to get to work. What makes this day so eventful is that a rare weather event is occurring: Presently, the air temperature is 91°F, the relative humidity is 94 percent, and the heat index, which tells us how hot it really feels, is a staggering 130°F.

We know from Chapter 1 that in our atmosphere, the concentration of the invisible gas water vapor is normally less than a few percent of all the atmospheric molecules. Yet water vapor is exceedingly important, for it transforms into cloud particles—particles that grow in size and fall to Earth as precipitation. The term *humidity* can describe the amount of water vapor in the air. To most of us, a moist day suggests high humidity. However, there is usually more water vapor in the hot, "dry" air of the Sahara Desert than in the cold, "damp" winter air of New England, which raises an interesting question: Does the desert air have a higher humidity? As we will see later in this chapter, the answer to this question is both yes and no, depending on the type of humidity we mean.

So that we may better understand the concept of humidity, we will begin this chapter by examining the circulation of water in the atmosphere. Then, we will look at different ways to express humidity. At the end of the chapter, we will investigate various ways to measure humidity.

Circulation of Water in the Atmosphere

Within the atmosphere, there is an unending circulation of water. Since the oceans occupy over 70 percent of Earth's surface, we can think of this circulation as beginning over the ocean. Here, the sun's energy transforms enormous quantities of liquid water into water vapor in a process called **evaporation**. Winds then transport the moist air to other regions, where the water vapor changes back into liquid (or ice), forming clouds, in a process called **condensation**. Under certain conditions, the liquid cloud particles (or solid ice crystals) may grow in size and fall to the surface as **precipitation**—rain, snow, or hail.* If the precipitation falls into an ocean, the water begins its cycle again. If, on the other hand, the precipitation falls on a continent, a great deal of the water returns to the ocean only after a complex journey. This cycle of moving and transforming water molecules from liquid to vapor and back to liquid again is called the **hydrologic (water) cycle**. In the form with which we are most concerned, water molecules travel from ocean to atmosphere to land and then back to the ocean.

● Figure 4.1 illustrates the complexities of the hydrologic cycle. For example, before falling rain ever reaches the ground, a portion of it evaporates back into the air. Some of the precipitation may be intercepted by vegetation, where it evaporates or drips to the ground long after a storm has ended. Once on the surface, a portion of the water soaks into the ground by percolating downward through small openings in the soil and rock, forming groundwater that can be tapped by wells. What does not soak into the ground collects in puddles of standing water or runs off into streams and rivers, which find their way back to the ocean. Even the underground water moves slowly and eventually surfaces, only to evaporate or be carried seaward by rivers.

Over land, a considerable amount of water vapor is added to the atmosphere through evaporation from the soil, lakes, and streams. Even plants give up moisture by a process called *transpiration*. The water absorbed by a plant's root system moves upward through the stem and emerges from the plant through numerous small openings on the underside of the leaf. In all, evaporation and transpiration from continental areas amount to

---

*Precipitation is any form of water that falls from a cloud and reaches the ground.

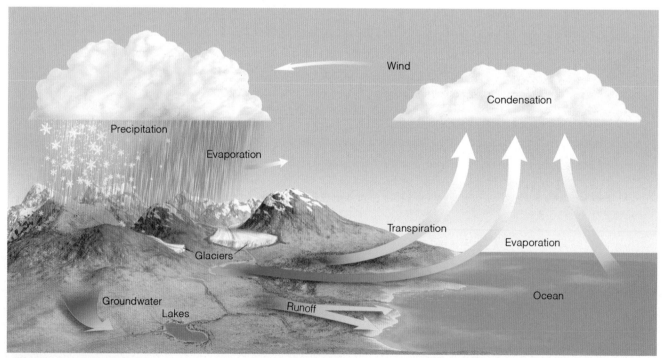

● **FIGURE 4.1** The hydrologic cycle.

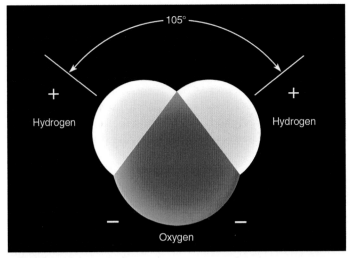

● **FIGURE 4.2** The water molecule.

● **FIGURE 4.3** The three states of matter: water as a gas, as a liquid, and as a solid.

only about 15 percent of the nearly 1.5 quintillion (10^{18}) gallons of water vapor that annually evaporate into the atmosphere; the remaining 85 percent evaporates from the oceans. If all of this water vapor were to suddenly condense and fall as rain, it would be enough to cover the entire globe with about 2.5 centimeters (or 1 inch) of water.* The total mass of water vapor stored in the atmosphere at any moment adds up to only a little over a week's supply of the world's precipitation. Since this amount varies only slightly from day to day, the hydrologic cycle is exceedingly efficient in circulating water in the atmosphere.

The Many Phases of Water

If we could see individual water molecules, we would find that, in the lower atmosphere, water is everywhere. If we could observe just one single water molecule by magnifying it billions of times, we would see an H_2O molecule in the shape of a tiny head that somewhat resembles Mickey Mouse (see ● Fig. 4.2). The bulk of the "head" of the molecule is the oxygen atom. The "mouth" is a region of excess negative charge. The "ears" are partially exposed protons of the hydrogen atom, which are regions of excess positive charge.

When we look at many H_2O molecules, we see that, as a gas, water vapor molecules move about quite freely, mixing well with neighboring atoms and molecules (see ● Fig. 4.3). As we learned in Chapter 2, the higher the temperature of the gas, the faster the molecules move. In the liquid state, the water molecules are closer together, constantly jostling and bumping into one another. If we lower the temperature of the liquid, water molecules will move slower and slower until, when cold enough, they arrange themselves into an orderly pattern with each molecule more or less locked into a rigid position, able to vibrate but not able to move

about freely. In this solid state called *ice*, the shape and charge of the water molecule helps arrange the molecules into six-sided (hexagonal) crystals.

As we observe the ice crystal in freezing air, we see an occasional molecule gain enough energy to break away from its neighbors and enter into the air above. The molecule changes from an ice molecule directly into a vapor molecule without passing through the liquid state. This ice-to-vapor phase change is called **sublimation**. If a water vapor molecule should attach itself to the ice crystal, the vapor-to-ice phase change is called **deposition**. And if we apply warmth to the ice crystal, its molecules will vibrate faster. In fact, some of the molecules will actually vibrate out of their rigid crystal pattern into a disorderly condition—that is, the ice melts.

And so water vapor is a gas that becomes visible to us only when millions of molecules join together to form tiny cloud droplets or ice crystals. In this process—known as a *change of state* or, simply, *phase change*—water only changes its disguise, not its identity.

Evaporation, Condensation, and Saturation

Suppose we were able to observe individual water molecules in a beaker, as illustrated in ● Fig. 4.4a. What we would see are water molecules jiggling, bouncing, and moving about. However, we would also see that the molecules are not all moving at the same speed—some are moving much faster than others. At the surface, molecules with enough speed (and traveling in the right direction) would occasionally break away from the liquid surface and enter into the air above. These molecules, changing from the *liquid state into the vapor state*, are *evaporating*. While some water molecules are leaving the liquid, others are returning. Those returning are *condensing*, as they are changing from a *vapor state to a liquid state*.

---

*Precipitable water refers to the amount of water that would accumulate if all of the water vapor in a column of air condensed and fell to Earth as rain.

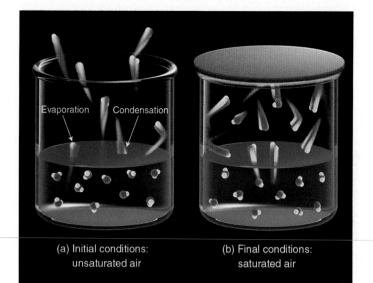

FIGURE 4.4 (a) Water molecules at the surface of the water are evaporating (changing from liquid into vapor) and condensing (changing from vapor into liquid). Since more molecules are evaporating than condensing, net evaporation is occurring. (b) When the number of water molecules escaping from the liquid (evaporating) balances those returning (condensing), the air above the liquid is saturated with water vapor. (For clarity, only water molecules are illustrated.)

When a cover is placed over the beaker (see Fig. 4.4b), after a while the total number of molecules escaping from the liquid (evaporating) is balanced by the number returning (condensing). When this condition exists, the air is said to be **saturated** with water vapor. Under saturated conditions, for every molecule that evaporates, one must condense, and no net loss of liquid or vapor molecules results.

If we remove the cover and blow across the top of the water, some of the vapor molecules already in the air above will be blown away, creating a difference between the actual number of vapor molecules and the total number required for *saturation*. This helps prevent saturation from occurring and allows for a greater amount of evaporation, just as if wind were blowing atop the water surface. Wind, therefore, enhances evaporation.

The temperature of the water also influences evaporation. All else being equal, warm water will evaporate more readily than cool water, because when water molecules are heated, they will speed up. At higher temperatures, a greater fraction of the molecules have sufficient speed to break through the surface tension of the water and zip off into the air above. In other words, the warmer the water, the greater the rate of evaporation.

If we could examine the air above the water in Fig. 4.4b, we would observe the water vapor molecules freely darting about and bumping into each other as well as into neighboring molecules of oxygen and nitrogen. When these gas molecules collide, they tend to bounce off one another, constantly changing in speed and direction. However, the speed lost by one molecule is gained by another, and so the average speed of all the molecules does not change. Consequently, the temperature of the

air does not change. Mixed in with all of the air molecules are microscopic bits of dust, smoke, salt, and other particles called *condensation nuclei* (so called because water vapor condenses on them). In the warm air above the water, fast-moving water vapor molecules strike the nuclei with such impact that they simply bounce away (see ● Figure 4.5a). However, if the air is chilled (Fig. 4.5b), the molecules move more slowly and are more apt to stick and condense to the nuclei. When many billions of these water vapor molecules condense onto the nuclei, tiny liquid cloud droplets form.

We can see, then, that condensation is more likely to happen as the air cools and the speed of the water vapor molecules decreases. As the air temperature increases, condensation is less likely because most of the water vapor molecules have sufficient speed (sufficient energy) to remain as a vapor. As we will see in this and other chapters, *condensation occurs primarily when the air is cooled.*[*]

Even though condensation is more likely to occur when the air cools, it is important to note that no matter how cold the air becomes, there will always be a few water vapor molecules with sufficient speed (sufficient energy) to remain as a vapor. It should be apparent, then, that with the same number of water vapor molecules in the air, saturation is more likely to occur in cool air than in warm air. This fact often leads to the statement that "warm air can hold more water vapor molecules before becoming saturated than can cold air" or, simply, "warm air has a greater capacity for water vapor than does cold air." At this point, it is important to realize that although these statements are correct, the use of such words as "hold" and "capacity" are misleading when describing water vapor content, as air does not really "hold" water vapor in the sense of making "room" for it.

---

[*]As we will see later, another way of explaining why cooling produces condensation is that the saturation vapor pressure decreases with lower temperatures.

FIGURE 4.5 Condensation is more likely to occur as the air cools. (a) In the warm air, fast-moving water vapor molecules tend to bounce away after colliding with nuclei. (b) In the cool air, slow-moving vapor molecules are more likely to join together on nuclei. The condensing of many billions of water molecules produces tiny liquid water droplets.

Humidity

We are now ready to look more closely at the concept of **humidity**, which may refer to any one of a number of ways of specifying the amount of water vapor in the air. Most people have heard of *relative humidity*, which we will examine later, but there are several other ways to express atmospheric water vapor content, and there are several meanings for the concept of humidity. The first one we'll take a look at is *absolute humidity*.

ABSOLUTE HUMIDITY   Suppose we enclose a volume of air in an imaginary thin elastic container—a *parcel*—about the size of a large balloon, as illustrated in ● Fig. 4.6. With a chemical drying agent, we can extract the water vapor from the air, weigh it, and obtain its mass. If we then compare the vapor's mass with the volume of air in the parcel, we will have determined the **absolute humidity** of the air—that is, the mass of water vapor in a given volume of air, which can be expressed as

$$\text{Absolute humidity} = \frac{\text{mass of water vapor}}{\text{volume of air}}$$

Absolute humidity represents the water *vapor density* (mass/volume) in the parcel and, normally, is expressed as grams of water vapor in a cubic meter of air. For example, if the water vapor in 1 cubic meter of air weighs 25 grams, the absolute humidity of the air is 25 grams per cubic meter ($25\,\text{g/m}^3$).

We learned in Chapter 2 that a rising or descending parcel of air will experience a change in its volume because of the changes in surrounding air pressure. Consequently, when a volume of air fluctuates, the absolute humidity changes—even though the air's vapor content has remained constant (see ● Fig. 4.7).

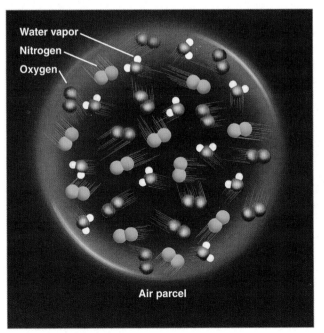

● **FIGURE 4.6** The water vapor content (humidity) inside this air parcel can be expressed in a number of ways.

Parcel Size	Mass of Parcel	Mass of H₂o Vapor	Absolute Humidity
2 m³	2 kg	10 g	5 g/m³
1 m³	1 kg	10 g	10 g/m³

● **FIGURE 4.7** With the same amount of water vapor in a parcel of air, an increase in volume decreases absolute humidity, whereas a decrease in volume increases absolute humidity.

Parcel Size	Mass of Parcel	Mass of H₂o Vapor	Specific Humidity
2 m³	1 kg	1 g	1 g/kg
1 m³	1 kg	1 g	1 g/kg

● **FIGURE 4.8** The specific humidity does not change as air rises and descends.

For this reason, the absolute humidity is not commonly used in atmospheric studies.

SPECIFIC HUMIDITY AND MIXING RATIO   Humidity, however, can be expressed in ways that are not influenced by changes in air volume. When the mass of the water vapor in the air parcel in Fig. 4.6 is compared with the mass of all the air in the parcel (including vapor), the result is called the **specific humidity**; thus,

$$\text{Specific humidity} = \frac{\text{mass of water vapor}}{\text{total mass of air}}$$

Another convenient way to express humidity is to compare the mass of the water vapor in the parcel to the mass of the remaining dry air. Humidity expressed in this manner is called the **mixing ratio**; thus,

$$\text{Mixing ratio} = \frac{\text{mass of water vapor}}{\text{mass of dry air}}$$

Both specific humidity and mixing ratio are expressed as grams of water vapor per kilogram of air (g/kg).

The specific humidity and mixing ratio of an air parcel remain constant *as long as water vapor is not added to or removed from the parcel.* This happens because the total number of molecules (and, hence, the mass of the parcel) remains constant, even as the parcel expands or contracts (see ● Fig. 4.8). Since changes in parcel size do not affect specific humidity and mixing ratio, these two concepts are used extensively in the study of the atmosphere.

Figure 4.9 shows how specific humidity varies with latitude. The average specific humidity is highest in the warm, muggy tropics. As we move away from the tropics, it decreases, reaching its lowest average value in the polar latitudes. Although the

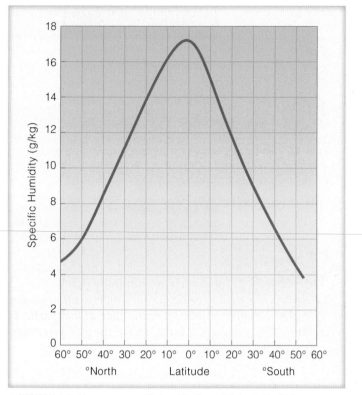

● **FIGURE 4.9** The average specific humidity for each latitude. The highest average values are observed in the tropics and the lowest values in polar regions.

major deserts of the world are located near latitude 30°, Fig. 4.9 shows that, at this latitude, the average air contains nearly twice the water vapor as does the air at latitude 50°N. Hence, the air of a desert is certainly not "dry," nor is the water vapor content extremely low. Since the hot, desert air of the Sahara often contains more water vapor than the cold, polar air farther north, we can say that *summertime Sahara air has a higher specific humidity.* (We will see later in what sense we consider desert air to be "dry.")

VAPOR PRESSURE The air's moisture content may also be described by measuring the pressure exerted by the water vapor in the air. Suppose the air parcel in Fig. 4.6 (p. 95) is near sea level. The total pressure inside the parcel is due to the collision of all the molecules against the inside surface of the parcel. In other words, the total pressure inside the parcel is equal to the sum of the pressures of the individual gases. (This phenomenon is known as *Dalton's law of partial pressure.*) If the total pressure inside the parcel is 1000 millibars (mb),* and the gases inside include nitrogen (78 percent), oxygen (21 percent), and water vapor (1 percent), then the partial pressure exerted by nitrogen would be 780 mb and by oxygen, 210 mb. The partial pressure of water vapor, called the **actual vapor pressure**, would be only 10 mb (1 percent of

1000).** It is evident, then, that because the number of water vapor molecules in any volume of air is small compared to the total number of air molecules in the volume, the actual vapor pressure is normally a small fraction of the total air pressure.

Everything else being equal, the more air molecules in a parcel, the greater the total air pressure. When you blow up a balloon, you increase its pressure by putting in more air. Similarly, an increase in the number of water vapor molecules will increase the total vapor pressure. Hence, the actual vapor pressure is a fairly good measure of the total amount of water vapor in the air: *High actual vapor pressure indicates large numbers of water vapor molecules, whereas low actual vapor pressure indicates comparatively small numbers of vapor molecules.**

In summer across North America, the highest vapor pressures are observed along the humid Gulf Coast, whereas the lowest values are experienced over the drier Great Basin, especially Nevada. In winter, the highest average vapor pressures are again observed along the Gulf Coast with lowest values over the northern Great Plains into Canada.

Actual vapor pressure indicates the air's total water vapor content, whereas **saturation vapor pressure** describes how much water vapor is necessary to make the air saturated at any given temperature. Put another way, *saturation vapor pressure is the pressure that the water vapor molecules would exert if the air were saturated with vapor at a given temperature.**

We can obtain a better picture of the concept of saturation vapor pressure by imagining molecules evaporating from a water surface. Look back at Fig. 4.4b (p. 94) and recall that when the air is saturated, the number of molecules escaping from the water's surface equals the number returning. Since the number of "fast-moving" molecules increases as the temperature increases, the number of water molecules escaping per second increases also. To maintain equilibrium, this situation causes an increase in the number of water vapor molecules in the air above the liquid. Consequently, at higher air temperatures, it takes more water vapor to saturate the air. And more vapor molecules exert a greater pressure. *Saturation vapor pressure, then, depends primarily on the air temperature.* From the graph in ● Fig. 4.10, we can see that at 10°C, the saturation vapor pressure is about 12 mb, whereas at 30°C it is about 42 mb.

The inset in Fig. 4.10 shows that, when both water and ice exist at the same temperature below freezing, *the saturation vapor pressure just above the water is greater than the saturation vapor pressure over the ice.* In other words, at any temperature below freezing, it takes more vapor molecules to saturate air directly above water than it does to saturate air directly above ice. This situation occurs because it is harder for molecules to escape an ice surface than a water surface. Consequently, fewer molecules escape the ice surface at a given temperature, requiring fewer in the vapor phase to maintain equilibrium. Likewise, salts in solution bind water molecules, reducing the number escaping. These

---

*You may recall from Chapter 1 that the millibar is the unit of pressure most commonly found on surface weather maps, and that it expresses atmospheric pressure as a force over a given area.

**When we use the percentages of various gases in a volume of air, Dalton's law only gives us an approximation of the actual vapor pressure. The point here is that, near Earth's surface, the actual vapor pressure is often close to 10 mb.

˙Remember that actual vapor pressure is only an approximation of the total vapor content. A change in total air pressure will affect the actual vapor pressure even though the total amount of water vapor in the air remains the same.

**When the air is saturated, the amount of water vapor is the maximum possible at the existing temperature and pressure.

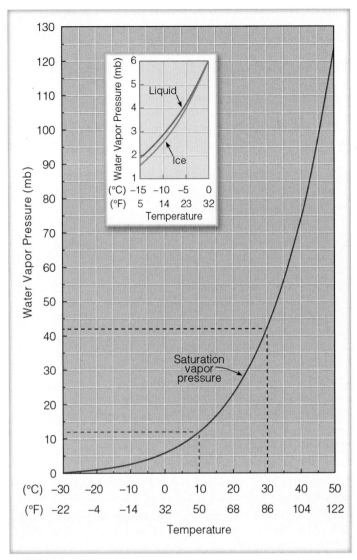

● FIGURE 4.10 Saturation vapor pressure increases with increasing temperature. At a temperature of 10°C, the saturation vapor pressure is about 12 mb, whereas at 30°C it is about 42 mb. The inset illustrates that the saturation vapor pressure over water is greater than the saturation vapor pressure over ice.

CRITICAL THINKING QUESTION Why would you *not* expect the saturation vapor pressure to fall to zero, even at extremely low temperatures?

concepts are important and (as we will see in Chapter 7) play a role in the process of rain formation.

So far, we've discussed the amount of moisture actually in the air and how much more would be needed to saturate the air. We have several options for describing the amount of moisture in the air:

1. *Absolute humidity* tells us the *mass* of water vapor in a fixed volume of air, or the *water vapor density*.

2. *Specific humidity* measures the *mass* of water vapor in a fixed *total mass* of air, and the *mixing ratio* describes the mass of water vapor in a fixed mass of the remaining dry air.

3. The *actual vapor pressure* of air expresses the amount of water vapor in terms of the amount of *pressure* that the water vapor molecules exert.

We also have an option for describing how much moisture could be added to the atmosphere:

4. The *saturation vapor pressure* is the pressure that the water vapor molecules would exert if the air were saturated with vapor at a given temperature.

Each of these measures has its uses but, as we will see, the concepts of vapor pressure and saturation vapor pressure are critical to an understanding of the sections that follow. (Before looking at the most commonly used moisture variable—relative humidity—you may wish to read Focus section 4.1 on vapor pressure and boiling.)

RELATIVE HUMIDITY While relative humidity is the most common way of describing atmospheric moisture, it is also, unfortunately, the most misunderstood. The concept of relative humidity may at first seem confusing because it does not indicate the actual amount of water vapor in the air. Instead, it tells us how close the air is to being saturated. The **relative humidity (RH)** *is the ratio of the amount of water vapor actually in the air to the maximum amount of water vapor required for saturation at that particular temperature (and pressure).* It is the *ratio* of the air's water vapor *content* to its *capacity*; thus,

$$RH = \frac{\text{water vapor content}}{\text{water vapor capacity}}$$

We can think of the actual vapor pressure as a measure of the air's actual water vapor content, and the saturation vapor pressure as a measure of air's total capacity for water vapor. Hence, the relative humidity can be expressed as

$$RH = \frac{\text{actual vapor pressure}}{\text{saturation vapor pressure}} \times 100 \text{ percent}^{*}$$

Relative humidity is given as a percent. Air with a 50 percent relative humidity actually contains one-half the amount of water vapor required for saturation. Air with a 100 percent relative humidity is said to be *saturated* because it is filled to capacity with water vapor. Air with a relative humidity greater than 100 percent is said to be **supersaturated**, a condition that does not tend to occur often or last long. Since relative humidity is used so much in the everyday world, let's examine it more closely.

A change in relative humidity can be brought about in two primary ways:

1. by changing the air's water vapor content
2. by changing the air temperature

In ● Fig. 4.11a, we can see that an increase in the water vapor content of the air (with no change in air temperature) increases the air's relative humidity. The reason for this increase resides in the fact

---

* Relative humidity may also be expressed as

$$RH = \frac{\text{actual mixing ratio}}{\text{saturation mixing ratio}} \times 100$$

where the actual mixing ratio is the mixing ratio of the air, and the saturation mixing ratio is the mixing ratio of saturated air at that particular temperature.

Vapor Pressure and Boiling—The Higher You Go, the Longer Cooking Takes

If you camp in the mountains, you may have noticed that the higher you camp, the longer it takes vegetables to cook in boiling water. To understand this observation, we need to examine the relationship between vapor pressure and boiling. As water boils, bubbles of vapor rise to the top of the liquid and escape. For this to occur, the saturation vapor pressure exerted by the bubbles must equal the pressure of the atmosphere; otherwise, the bubbles would collapse. Boiling, therefore, occurs when the saturation vapor pressure of the escaping bubbles is equal to the total atmospheric pressure.

Because the saturation vapor pressure is directly related to the temperature of the liquid, higher water temperatures produce higher vapor pressures. Hence, any change in atmospheric pressure will change the temperature at which water boils: An increase in air pressure raises the boiling point, while a decrease in air pressure lowers it. Notice in ● Fig. 1 that, to make pure water boil at sea level, the water must be heated to a temperature of 100°C (212°F). At Denver, Colorado, which is situated about 1500 m (5000 ft) above sea level, the air pressure is near 850 millibars, and water boils at 95°C (203°F).

Once water starts to boil, its temperature remains constant, even if you continue to heat it.

● **FIGURE 1** The lower the air pressure, the lower the saturation vapor pressure and, hence, the lower the boiling point temperature.

This happens because energy supplied to the water is used to convert the liquid to a gas (steam). Now we can see why vegetables take longer to cook in the mountains. To be thoroughly cooked, they must boil for a longer time because the boiling water is cooler than at lower levels. In New York City, which is near sea level, it takes about five minutes to hard-boil an egg. An egg boiled for five minutes in the "mile high city" of Denver, Colorado, turns out to be runny.

---

that as more water vapor molecules are added to the air, there is a greater likelihood that some of the vapor molecules will stick together and condense. Condensation takes place in saturated air. So, as more and more water vapor molecules are added to the air, the air gradually approaches saturation, and the relative humidity of the air increases.* Conversely, removing water vapor from the air decreases the likelihood of saturation, which lowers the air's relative humidity. In summary, with no change in air temperature, adding water vapor to the air increases the relative humidity; removing water vapor from the air lowers the relative humidity.

Figure 4.11b illustrates that as the air temperature increases (with no change in water vapor content), the relative

humidity decreases. This decrease in relative humidity occurs because in the warmer air the water vapor molecules are zipping about at such high speeds they are unlikely to join together and condense. The higher the temperature, the faster the molecular speed, the less likely saturation will occur, and the lower the relative humidity.* As the air temperature lowers, the water vapor molecules move more slowly, condensation becomes more likely as the air approaches saturation, and the relative humidity increases. In summary, with no change in water vapor content, *an increase in air temperature lowers the relative humidity, while a decrease in air temperature raises the relative humidity.*

---

*We can also see in Fig. 4.11a that as the total number of vapor molecules increases (at a constant temperature), the actual vapor pressure increases and approaches the saturation vapor pressure at 20°C. As the actual vapor pressure approaches the saturation vapor pressure, the air approaches saturation and the relative humidity rises.

*Another way to look at this concept is to realize that, as the air temperature increases, the air's saturation vapor pressure also increases. As the saturation vapor pressure increases, with no change in water vapor content, the air moves farther away from saturation, and the relative humidity decreases.

In many places, the air's total vapor content varies only slightly during an entire day, and so it is the changing air temperature that primarily regulates the daily variation in relative humidity (see ● Fig. 4.12). As the air cools during the night, the relative humidity increases. Normally, the highest relative humidity occurs in the early morning, during the coolest part of the day. As the air warms

● **FIGURE 4.11** (a) At the same air temperature, an increase in the water vapor content of the air increases the relative humidity as the air approaches saturation. (b) With the same water vapor content, an increase in air temperature causes a decrease in relative humidity as the air moves farther away from being saturated.

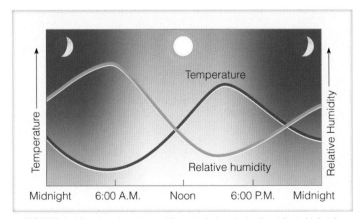

● **FIGURE 4.12** When the air is cool (morning), the relative humidity is high. When the air is warm (afternoon), the relative humidity is low. These conditions exist in clear weather when the air is calm or of constant wind speed.

during the day, the relative humidity decreases, with the lowest values usually occurring during the warmest part of the afternoon.

These changes in relative humidity are important in determining the amount of evaporation from vegetation and wet surfaces. If you water your lawn on a hot afternoon, when the relative humidity is low, much of the water will evaporate quickly from the lawn, instead of soaking into the ground. Watering the same lawn in the evening or during the early morning, when the relative humidity is higher, will cut down the evaporation and increase the effectiveness of the watering.

RELATIVE HUMIDITY AND DEW POINT Suppose it is early morning and the outside air is saturated. The air temperature is 10°C (50°F) and the relative humidity is 100 percent. We know from the previous section that relative humidity can be expressed as

$$RH = \frac{\text{actual vapor pressure}}{\text{saturation vapor pressure}} \times 100 \text{ percent}$$

Looking back at Fig. 4.10 (p. 97), we can see that air with a temperature of 10°C has a saturation vapor pressure of 12 mb. Since the air is saturated and the relative humidity is 100 percent, the actual vapor pressure *must* be the same as the saturation vapor pressure (12 mb), since

$$RH = \frac{12 \text{ mb}}{12 \text{ mb}} \times 100\% = 100 \text{ percent}$$

Suppose during the day the air warms to 30°C (86°F), with no change in water vapor content (or air pressure). Because there is no change in water vapor content, the actual vapor pressure must be the same (12 mb) as it was in the early morning when the air was saturated. The saturation vapor pressure, however, has increased because the air temperature has increased. From Fig. 4.10, note that air with a temperature of 30°C has a saturation vapor pressure of 42 mb. The relative humidity of this unsaturated, warmer air is now much lower, as

$$RH = \frac{12 \text{ mb}}{42 \text{ mb}} \times 100\% = 29 \text{ percent}$$

To what temperature must the outside air, with a temperature of 30°C, be cooled so that it is once again saturated? The answer, of course, is 10°C. For this amount of water vapor in the air, 10°C is called the **dew-point temperature** or, simply, the **dew point**. It represents *the temperature to which air would have to be cooled (with no change in air pressure or moisture content) for saturation to occur.* The dew point is determined with respect to a flat

| Dew-point temperature, 70°F | Dew-point temperature, 60°F | Dew-point temperature, 50°F |

● **FIGURE 4.13** On a calm, clear night, the lower the dew-point temperature, the lower the expected minimum temperature. With the same initial evening air temperature 80°F and with no change in weather conditions during the night, as the dew point lowers, the expected minimum temperature lowers. This situation occurs because a lower dew point means that there is less water vapor in the air to absorb and radiate infrared energy back to the surface. More infrared energy from the surface is able to escape into space, producing more rapid radiational cooling at the surface. (Dots in each diagram represent the amount of water vapor in the air. Red wavy arrows represent infrared [IR] radiation).

surface of water. When the dew point is determined with respect to a flat surface of ice, it is called the **frost point**.

The dew point is an important measurement used in predicting the formation of dew, frost, fog, and even the minimum temperature (see ● Fig. 4.13). It is also the main quantity used to indicate moisture on weather maps. When used with an empirical formula (as illustrated in Chapter 6 on p. 157), the dew point can help determine the height of the base of a cumulus cloud. Since atmospheric pressure varies only slightly at Earth's surface, *the dew point is a good indicator of the air's actual water vapor content. High dew points indicate high water vapor content;* *low dew points, low water vapor content.* Adding water vapor to the air increases the dew point; removing water vapor lowers it.

● Figure 4.14 shows the average dew-point temperatures across the United States and southern Canada for January. Notice that the dew points are highest (the greatest amount of water vapor in the air) over the Gulf Coast states and lowest over the interior. Compare New Orleans with Fargo. Cold, dry winds from northern Canada flow relentlessly into the Central Plains during the winter, keeping this area dry. But warm, moist air from the Gulf of Mexico helps maintain a higher dew-point temperature in the southern states.

● **FIGURE 4.14** Average surface dew-point temperatures (°F) across the United States and Canada for January.

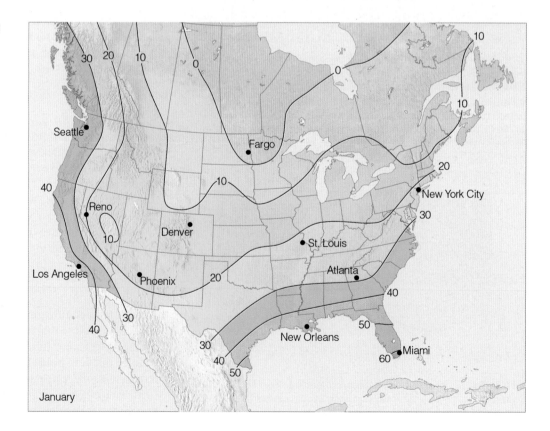

Figure 4.15 is a similar diagram showing the average dew-point temperatures for July. Again, the highest dew points are observed along the Gulf Coast, with some areas experiencing average dew-point temperatures near 75°F. In fact, most people consider it to be "humid" when the dew-point temperature exceeds 65°F, and "oppressive" when it equals or exceeds 75°F. Note, too, in Fig. 4.15 that the dew points over the eastern and central portion of the United States are much higher in July, meaning that the July air contains between three and six times more water vapor than the January air. The reason for the high dew points is that, in summertime, this region is almost constantly receiving humid air from the warm Gulf of Mexico. The lowest dew point, and hence the driest air, is found in the West, with the lowest values observed in Nevada—a region surrounded by mountains that effectively shield it from significant amounts of moisture moving in from the southwest and northwest.

The difference between air temperature and dew point can indicate whether the relative humidity is low or high. When the air temperature and dew point are far apart, the relative humidity is low; when they are close to the same value, the relative humidity is high. When the air temperature and dew point are equal, the air is *saturated* and the relative humidity is 100 percent.

Under certain conditions, the air can be considered "dry" even though the relative humidity may be 100 percent.

Observe, for example, in Fig. 4.16a that, because the air temperature and dew point are the same in the polar air, the air is saturated and the relative humidity is 100 percent. On the other hand, the desert air (Fig. 4.16b), with a large separation between air temperature and dew point, has a much lower relative humidity—21 percent.* However, since dew point is a measure of the amount of water vapor in the air, the desert air (with a higher dew point) must contain *more* water vapor. So even though the polar air has a higher relative humidity, the desert air that contains more water vapor has a higher water vapor density, or *absolute humidity,* and a higher specific humidity and mixing ratio as well.

Now we can see why polar air is often described as being "dry" when the relative humidity is high (often close to 100 percent). In cold polar air, the dew point and air temperature are normally close together. But the low dew-point temperature means there is little water vapor in the air. Consequently, the air is said to be "dry" even though the relative humidity is quite high.

There is a misconception that if it is raining (or snowing), the outside relative humidity must be 100 percent. Look at Fig. 4.17 and observe that inside the cloud the relative humidity is 100 percent, but at the ground the relative humidity is much less than 100 percent. As the rain falls into the drier air near the surface, some of the drops evaporate, a process that chills the

*The relative humidity can be computed from Fig. 4.10 (p. 97). The desert air with an air temperature of 35°C has a saturation vapor pressure of about 56 mb. A dew-point temperature of 10°C gives the desert air an actual vapor pressure of about 12 mb. These values produce a relative humidity of 12/56 × 100, or 21 percent.

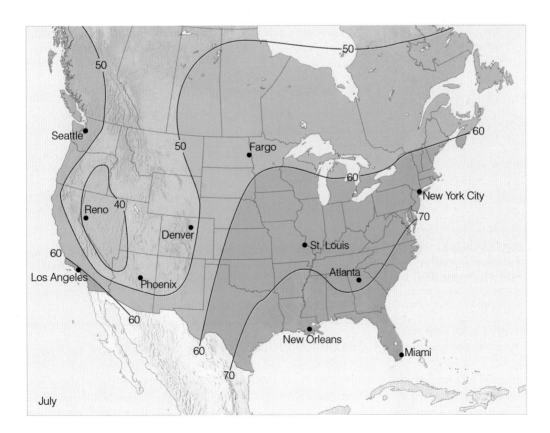

● FIGURE 4.15 Average surface dew-point temperatures across the United States and Canada (°F) for July.

(a) POLAR AIR: Air temperature –2°C (28°F)
Dew point –2°C (28°F)
Relative humidity 100 percent

(b) DESERT AIR: Air temperature 35°C (95°F)
Dew point 10°C (50°F)
Relative humidity 21 percent

© C. Donald Ahrens

● **FIGURE 4.16** In this example, the polar air has the higher relative humidity, whereas the desert air, with the higher dew point, contains more water vapor.

$T = 40°F$
$T_d = 40°F$
RH = 100%

$T = 70°F$
$T_d = 65°F$
RH = 84%

● **FIGURE 4.17** Inside the cloud the air temperature (T) and dew point (T_d) are the same, the air is saturated, and the relative humidity (RH) is 100 percent. However, at the surface where the air temperature and dew point are not the same, the air is *not* saturated (even though it is raining), and the relative humidity is considerably less than 100 percent.

CRITICAL THINKING QUESTION If the cloud remains in the same spot for some time, why would you expect the cloud base to begin to lower?

air and increases the air's water vapor content. The lowering air temperature and rising dew point cause the relative humidity to rise. If the falling rain persists, the air at the surface may become saturated and the relative humidity may reach 100 percent.

BRIEF REVIEW

Up to this point we have looked at the different ways of describing humidity. Before going on, here is a review of some of the important concepts and facts we have covered:

● Relative humidity tells us how close the air is to being saturated.

● Relative humidity can change when the air's water-vapor content changes, or when the air temperature changes.

● With a constant amount of water vapor, cooling the air raises the relative humidity and warming the air lowers it.

● The dew-point temperature is a good indicator of the air's water-vapor content: High dew points indicate high water-vapor content; and low dew points, low water-vapor content.

● Where the air temperature and dew point are close together, the relative humidity is high; when they are far apart, the relative humidity is low.

● Dry air can have a high relative humidity when the air is very cold and the air temperature and dew point are close together.

COMPARING HUMIDITIES ● Figure 4.18 shows how the average relative humidity varies from the equator to the poles. High relative humidities are normally found in the tropics and near the poles, where there is little separation between air temperature and dew point. The average relative humidity is low near latitude 30°—a latitude where we find the deserts of the world girdling the globe.

Of course, not all locations near 30°N are deserts. Take, for example, humid New Orleans, Louisiana. During July, the air in New Orleans with an average dew-point temperature of 22°C (72°F) contains a great deal of water vapor—nearly 50 percent more than does the air along the southern California coast. Since both locations are adjacent to large bodies of water, why is New Orleans much more humid?

Figure 4.19 shows a summertime situation where air from the Pacific Ocean is moving into southern California and air from the Gulf of Mexico is moving into the southeastern states. Notice that the Pacific water is much cooler than the Gulf water. Westerly winds, blowing across the Pacific, cool to just about the same temperature as the water. Likewise, air over the warmer Gulf reaches a temperature near that of the water below it. Over the water, at both locations, the air is nearly saturated with water vapor. This means that the dew-point temperature of the air over the cooler Pacific Ocean is much lower than the dew-point temperature over the warmer Gulf. Consequently, the air from the Gulf of Mexico contains a great deal more water vapor than the Pacific air.

As the air moves inland, away from the source of moisture, the air temperature in both cases increases. But the amount of water vapor in the air (and, hence, the dew-point temperature) hardly changes. Therefore, as the humid air moves into the southeastern states, high air temperatures along with high dew-point temperatures produce high relative humidities, often greater than 75 percent during the hottest part of the day. On the other hand, over the southwestern part of the nation, high air temperatures and low dew-point temperatures produce low relative humidities, often less than 25 percent during the hottest part of the afternoon. Much of this inland area over the Southwest is a desert. However, keep in mind that although considered "dry," this area, with a dew-point temperature above freezing, still contains far more water vapor than does the cold, arctic air in polar regions.

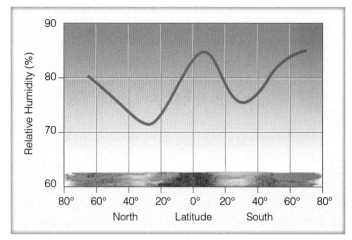

● **FIGURE 4.18** Relative humidity averaged for latitudes north and south of the equator.

(For more information on the computation of relative humidity and dew point, read Focus section 4.2.)

RELATIVE HUMIDITY IN THE HOME During the winter, the relative humidity inside a home can drop to an extremely low value and the inhabitants are usually unaware of it. When cold polar air is brought indoors and heated, its relative humidity decreases dramatically. Notice in ● Fig 4.20 that when outside air with a temperature and dew point of −15°C (5°F) is brought indoors and heated to 20°C (68°F), the relative humidity of the heated air drops to 8 percent—a value lower than what you would normally experience in a desert during the hottest time of the day.*

---

* $RH = \dfrac{1.9\,mb}{23.4\,mb} \times 100 = 8\%$

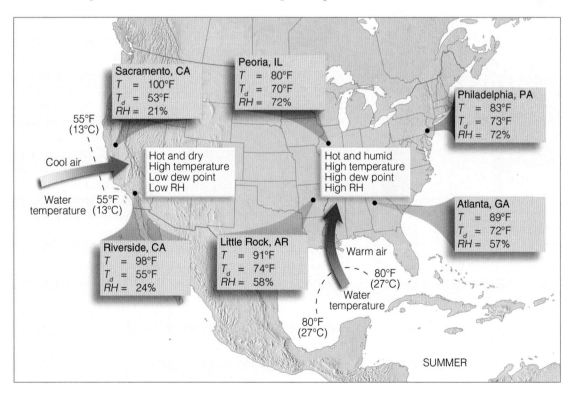

● **FIGURE 4.19** Air from the Pacific Ocean is hot and dry over land, whereas air from the Gulf of Mexico is hot and muggy over land. For each city, *T* represents the air temperature, T_d the dew point, and *RH* the relative humidity. (All data represent conditions during a July afternoon at 3 p.m. local time.)

Sacramento, CA
T = 100°F
T_d = 53°F
RH = 21%

Peoria, IL
T = 80°F
T_d = 70°F
RH = 72%

Philadelphia, PA
T = 83°F
T_d = 73°F
RH = 72%

55°F (13°C)

Cool air

Hot and dry
High temperature
Low dew point
Low RH

Hot and humid
High temperature
High dew point
High RH

Water temperature 55°F (13°C)

Atlanta, GA
T = 89°F
T_d = 72°F
RH = 57%

Riverside, CA
T = 98°F
T_d = 55°F
RH = 24%

Little Rock, AR
T = 91°F
T_d = 74°F
RH = 58%

Warm air

80°F (27°C)
Water temperature

80°F (27°C)

SUMMER

Very low relative humidity in a house can have an adverse effect on living things inside. For example, house plants have a difficult time surviving because the moisture from their leaves and the soil evaporates rapidly. Thus, they usually need watering more frequently in winter than in summer. People suffer, too, when the relative humidity is quite low. The rapid evaporation of moisture from exposed flesh causes skin to crack, dry, flake, or itch. These low humidities also irritate the mucous membranes in the nose and throat, producing an "itchy" or "scratchy" throat. Similarly, dry nasal passages permit inhaled bacteria to incubate, causing persistent infections. The remedy for most of these problems is simply to increase the relative humidity. Inside the home, the relative humidity can be increased simply by heating water and allowing it to evaporate into the air. The added water vapor raises the relative humidity to a more comfortable level. In modern homes, a humidifier, installed near the furnace, adds moisture to the air at a rate of about one gallon per room per day. This air, with its increased water vapor, is circulated throughout the home by a forced-air heating system. In this way, all rooms get their fair share of moisture—not just the room where the vapor is added.

To lower the air's moisture content, as well as the air temperature, many homes are air-conditioned. Outside air cools as it passes through a system of cold coils located in the air-conditioning unit. The cooling increases the air's relative humidity, and the air reaches saturation. The water vapor condenses into liquid water, which is carried away. The cooler, dehumidified air is now forced into the home.

In hot regions, where the relative humidity is low, *evaporative cooling systems* can be used to cool the air. These systems operate by having a fan blow hot, dry outside air across pads that are saturated with water. Evaporation cools the air, which is forced into the home, bringing some relief from the hot weather.

Evaporative coolers, also known as "swamp coolers," work best when the relative humidity is low and the air is warm. They do not work well in hot, muggy weather because a high relative humidity greatly reduces the rate of evaporation. Besides, swamp coolers add water vapor to the air—something that is not needed when the air is already uncomfortably humid. That is why swamp coolers may be found on homes in Arizona, but not on homes in Alabama.

RELATIVE HUMIDITY AND HUMAN DISCOMFORT On a hot, muggy day when the relative humidity is high, it is common to hear someone exclaim (often in exasperation), "It's not the heat, it's the humidity!" Actually, this statement has validity. In warm weather, the main source of body cooling is through evaporation of perspiration. Recall from Chapter 2 that evaporation is a cooling process, so when the air temperature is high and the relative humidity is low, perspiration on the skin evaporates quickly, often making us feel that the air temperature is lower than it really is. However, when both the air temperature and relative humidity are high and the air is nearly saturated with water vapor, body moisture does not readily evaporate; instead, it collects on the skin as beads of perspiration. Less evaporation means less cooling, and so we usually feel warmer than we did with a similar air temperature, but a lower relative humidity.

A good measure of how cool the skin can become is the **wet-bulb temperature**—*the lowest temperature that can be reached by evaporating water into the air.** On a hot day when the wet-bulb temperature is low, rapid evaporation (and, hence, cooling) takes place at the skin's surface. As the wet-bulb temperature approaches the air temperature, less cooling occurs, and the skin temperature may begin to rise. When the wet-bulb temperature exceeds the skin's temperature, no net evaporation occurs, and the body temperature can rise quite rapidly. Fortunately, the wet-bulb temperature is almost always considerably below the temperature of the skin.

When the weather is hot and muggy, a number of heat-related problems can occur. For example, in hot weather when the human body temperature rises, the *hypothalamus* gland (a gland in the brain that regulates body temperature) activates the body's heat-regulating mechanism, and more than 10 million sweat glands wet the body with as much as two liters of liquid per hour. As this perspiration evaporates, rapid loss of water and salt can result in a chemical imbalance that may lead to painful *heat cramps*. Excessive water loss through perspiring coupled with an increasing body temperature can result in *heat exhaustion*— fatigue, headache, nausea, and even fainting. If one's

● **FIGURE 4.20** When outside air with an air temperature and a dew point of −15°C (5°F) is brought indoors and heated to a temperature of 20°C (68°F) (without adding water vapor to the air), the relative humidity drops to 8 percent, placing stress on plants, animals, and humans living inside. (*T* represents temperature; T_d, dew point; and *RH*, relative humidity.)

INSIDE AIR

$T = 20°C (68°F)$
$T_d = -15°C (5°F)$
$RH = 8\%$

OUTSIDE AIR

$T = -15°C (5°F)$
$T_d = -15°C (5°F)$
$RH = 100\%$

© C. Donald Ahrens

CRITICAL THINKING QUESTION How would boiling potatoes inside this home raise the dew-point temperature and make the people inside feel more comfortable?

---

*The wet-bulb temperature and the dew-point temperature are different. The wet-bulb temperature is attained by *evaporating water* into the air, whereas the dew-point temperature is reached by *cooling* the air. Also, recall from an earlier discussion that the dew-point temperature is a measure of the amount of water vapor in the air and that most people begin to feel that it is humid when the dew point exceeds 65°F and oppressive when it equals or exceeds 75°F.

Computing Relative Humidity and Dew Point

Suppose we want to compute the air's relative humidity and dew point from ▼ Table 1. Earlier, we learned that relative humidity may be expressed as the actual vapor pressure divided by the saturation vapor pressure times 100 percent. If the actual vapor pressure is designated by the letter e, and the saturation vapor pressure by e_s, then the expression for relative humidity becomes

$$RH = \frac{e}{e_s} \times 100\%\,{}^*$$

Let's look at a practical example of using vapor pressure to measure relative humidity and obtain dew point. Suppose the air temperature in a room is 27°C (80°F). Because the saturation vapor pressure (e_s) is dependent on the temperature of the air, to obtain e_s from Table 1 we simply read the value adjacent to the air temperature much like we did in Fig. 4.10 on p. 97. Hence, air with a temperature of 27°C has a saturation vapor pressure of 35 mb.

Now, suppose that the air in the room is cooled suddenly with no change in moisture content. At successively lower temperatures, the saturation vapor pressure decreases. As the lowering saturation vapor pressure (e_s) approaches the actual vapor pressure (e), the relative humidity increases. If the air reaches saturation when the air cools to 21°C (70°F), then the dew-point temperature in the room must be 21°C. Since the air is saturated, we can see from Table 1 that both e_s and e must be equal to 25 millibars. If, then, we know the actual vapor pressure in a room, we can determine the dew point by using Table 1 to locate the temperature at which air will be saturated with what amount of water vapor. Similarly, if we are told what the dew point in the room is, we can look up that temperature in Table 1 and find the actual vapor pressure.

*Relative humidity may also be expressed as $RH = w/w_s \times 100\%$, where w is the actual mixing ratio and W_s is the saturation mixing ratio. Relative humidity computations using mixing ratio and adiabatic charts are given in Chapter 6, p. 161.

▼ TABLE 1   Saturation Vapor Pressure Over Water for Various Air Temperatures*

AIR TEMPERATURE		SATURATION VAPOR PRESSURE	AIR TEMPERATURE		SATURATION VAPOR PRESSURE
(°C)	(°F)	(MB)	(°C)	(°F)	(MB)
−18	(0)	1.5	18	(65)	21.0
−15	(5)	1.9	21	(70)	25.0
−12	(10)	2.4	24	(75)	29.6
−9	(15)	3.0	27	(80)	35.0
−7	(20)	3.7	29	(85)	41.0
−4	(25)	4.6	32	(90)	48.1
−1	(30)	5.6	35	(95)	56.2
2	(35)	6.9	38	(100)	65.6
4	(40)	8.4	41	(105)	76.2
7	(45)	10.2	43	(110)	87.8
10	(50)	12.3	46	(115)	101.4
13	(55)	14.8	49	(120)	116.8
16	(60)	17.7	52	(125)	134.2

*The data in this table can be obtained in Fig. 4.10, p. 97, by reading where the air temperature intersects the saturation vapor pressure curve.

In essence, we can use Table 1 to obtain the saturation vapor pressure (e_s) and the actual vapor pressure (e) if the air temperature and dew point of the air are known. With this information we can calculate relative humidity. For example, what is the relative humidity of air with a temperature of 29°C and a dew point of 18°C?

Answer: At 29°C, Table 1 shows $e_s = 41$mb. For a dew point of 18°C, the actual vapor pressure (e) is 21 mb; therefore, the relative humidity is

$$RH = \frac{e}{e_s} = \frac{21}{41} \times 100\% = 51\%$$

If we know the air temperature is 27°C and the relative humidity is 60 percent, what is the

dew-point temperature of the air? From Table 1, an air temperature of 27°C produces a saturation vapor pressure (e_s) of 35 mb. To obtain the actual vapor pressure (e), we simply plug the numbers into the formula

$$RH = \frac{e}{e_s} \times 100\%;\ 60\% = \frac{e}{35}$$
$$e = 21mb$$

As we saw in the previous example, an actual vapor pressure of 21 mb yields a dew-point temperature of 18°C.

body temperature rises above about 41°C (106°F), **heat stroke** can occur, resulting in complete failure of the circulatory functions. If the body temperature continues to rise, death may result. In fact, each year across North America, hundreds of people die from heat-related maladies. Even strong, healthy individuals can succumb to heat stroke, as did the Minnesota Vikings' all-pro offensive lineman Korey Stringer, who collapsed after practice in Mankato, Minnesota, on July 31, 2001, and died 15 hours later. Before Stringer fainted, temperatures on the practice field were in the 90s (°F) with the relative humidity above 55 percent.

In an effort to draw attention to this serious weather-related health hazard, an index called the **heat index (HI)** is used by the National Weather Service. The index combines air temperature with relative humidity to determine an **apparent temperature**— what the air temperature "feels like" to the average person for various combinations of air temperature and relative humidity. For example, in ● Fig. 4.21 an air temperature of 100°F and a relative humidity of 60 percent produce an apparent temperature of 129°F. As we can see in Fig. 4.21, heat stroke is highly likely when the index reaches this level. However, as we saw in the preceding paragraph,

deaths from heat stroke can occur when the heat index value is considerably lower than 129°F. Also, the values shown in Fig. 4.21 are calculated assuming light winds and shade. Under full sunshine, the effective heat index can climb as much as 15°F higher than indicated, and strong winds can increase the danger even further.

Tragically, hundreds and even thousands of people can be killed by a single heat wave. One example is the devastating Chicago heat wave of July 1995. Dew-point temperatures were extremely high, which led to several days of unusually warm overnight readings that kept residents from gaining relief from the heat. On July 13, the afternoon air temperature reached 106°F at Midway Airport, followed by an overnight low temperature of only 84°F. Many residents either had no air conditioning or could not afford to use it. More than 700 deaths occurred over five days. Since the time of that disaster, Chicago and many other cities in the United States have added neighborhood "cooling centers" and taken other steps to address the danger of heat waves.

In a closed vehicle, temperatures can soar far above outdoor readings in a matter of minutes. Since 1998, more than 600 children in the United States have died after being left in parked vehicles. When sunshine is strong, temperatures need not be blistering outside to cause such a tragedy. On an 80°F day in full sun, the air temperature inside a closed car can rise to 99°F in just 10 minutes and to 114°F in half an hour. On a 100°F day, interior temperatures can top 140°F within an hour. Unfortunately, "cracking" the windows does little to reduce the buildup of heat.

At this point it is important to dispel a common myth about hot, humid weather. Often people will recall a particularly sultry day as having been "90 degrees with 90 percent humidity" or even "95 degrees with 95 percent humidity." We see in Fig. 4.21 that a temperature of 90°F with 90 percent relative humidity would

● **FIGURE 4.21** Air temperature (°F) and relative humidity are combined to determine an apparent temperature or heat index (HI). An air temperature of 96°F with a relative humidity of 55 percent produces an apparent temperature (HI) of 112°F.

Temperature (°F)	Relative humidity (%)																	Possible heat-related risks
	20	25	30	35	40	45	50	55	60	65	70	75	80	85	90	95	100	
120	130	138	148															**Extreme danger** Heat stroke or sunstroke highly likely
118	126	134	142															
116	122	129	137	146														
114	119	125	132	140	148													
112	116	121	127	134	142	150												
110	112	117	122	129	136	143												**Danger** Sunstroke muscle cramps and/or heat exhaustion likely
108	109	113	118	123	130	137	144											
106	106	109	114	119	124	130	137	145										
104	103	106	110	114	119	124	131	137	145									
102	100	103	106	110	114	119	124	130	137	144								
100	97	100	102	106	109	114	118	124	129	136	143	150						**Extreme Caution** Sun stroke muscle cramps and/or heat exhaustion possible
98	95	97	99	102	105	109	113	117	123	128	134	141	148					
96	93	94	96	98	101	104	108	112	116	121	126	132	138	145				
94	90	91	93	95	97	100	103	106	110	114	119	124	129	135	141	148		
92	88	89	90	92	94	96	99	101	105	108	112	116	121	126	131	137	143	
90	86	87	88	89	91	93	95	97	100	103	106	109	113	117	122	127	132	
88	85	85	86	87	88	89	91	93	95	98	100	103	106	110	113	117	121	**Caution** Fatigue possible
86	83	83	84	85	85	87	88	89	91	93	95	97	100	102	105	108	112	
84	81	82	82	83	83	84	85	86	88	89	90	92	94	96	98	100	103	
82	80	80	80	81	81	82	83	84	84	85	86	88	89	90	92	93	95	
80	79	79	79	80	80	80	81	81	82	82	83	84	84	85	86	86	87	

© C. Donald Ahrens

produce a heat index of 122°F. Although this weather situation is remotely possible, it is *extremely unlikely*, as a temperature of 90°F and a relative humidity of 90 percent can occur only if the dew-point temperature is incredibly high (nearly 87°F), and a dew point this high rarely occurs anywhere in the United States, even on the muggiest of days.

Similarly, in hot, muggy weather, there are people who will remark about how "heavy" or how dense the air feels. Is hot, humid air really more dense than hot, dry air? If you are interested in the answer, read Focus section 4.3.

MEASURING HUMIDITY Humidity is most often measured through automated instruments (though some measurements are still taken manually at some observing sites). One common instrument used to obtain dew point and relative humidity is the **psychrometer**, which consists of two liquid-in-glass thermometers mounted side by side and attached to a piece of metal that has either a handle or chain at one end (see Fig. 4.22). The thermometers are exactly alike except that one has a piece of cloth (wick) covering the bulb. The wick-covered thermometer—called the *wet bulb*—is dipped in clean (usually distilled) water, while the other thermometer is kept dry. Both thermometers are ventilated for a few minutes, either by being whirled *(sling psychrometer)*, or by having air drawn past it with an electric fan *(aspirated psychrometer)*. Water evaporates from the wick and the thermometer cools. The drier the air, the greater the amount of evaporation and cooling. After a few minutes, the wick-covered thermometer will cool to the lowest value possible. Recall from an earlier section that this is the *wet-bulb temperature*—the lowest temperature that can be attained by evaporating water into the air.

The dry thermometer (commonly called the *dry bulb*) gives the current air temperature, or *dry-bulb temperature*. The temperature difference between the dry bulb and the wet bulb is known as the *wet-bulb depression*. A large depression indicates that a great deal of water can evaporate into the air and that the relative humidity is low. A small depression indicates that little evaporation of water vapor is possible, so the air is close to saturation and the relative humidity is high. If there is no depression, the dry bulb, the wet bulb, and the dew point are the same; the air is saturated and the relative humidity is 100 percent. (Tables used to compute relative humidity and dew point are given in Appendix C.)

Instruments that measure humidity are commonly called **hygrometers**. One type—called the **hair hygrometer**—is constructed on the principle that the length of human hair increases by 2.5 percent as the relative humidity increases from 0 to 100 percent. This instrument uses human or horse hair, or a synthetic fiber, to measure relative humidity. A number of strands of hair (with oils removed) are attached to a system of levers. A small change in hair length is magnified by a linkage system and transmitted to a dial (see ● Fig. 4.23) calibrated to show relative humidity, which can then be read directly or recorded on a chart. (Often, the chart is attached to a clock-driven rotating drum that gives a continuous record of relative humidity.) Because the hair hygrometer is not as accurate as the psychrometer (especially at very high and very low relative humidities and

very low temperatures), it requires frequent calibration, principally in areas that experience large daily variations in relative humidity.

An automated instrument that measured humidity is the *electrical hygrometer*. It consists of a flat plate coated with a film of carbon. An electric current is sent across the plate. As water vapor is absorbed, the electrical resistance of the carbon coating changes. These changes are translated into relative humidity. This instrument is commonly used in the radiosonde, which gathers atmospheric data at various levels above Earth. Still another instrument—the *infrared hygrometer*—measures atmospheric humidity by measuring the amount of infrared energy absorbed by water vapor in a sample of air, and the *dew cell* determines the amount of water vapor in the air by measuring the air's actual water vapor pressure. Finally, the *dew-point hygrometer* measures the dew-point temperature by cooling the surface of a mirror until condensation (dew) forms. This sensor is the type that measures dew-point temperature in the hundreds of fully automated weather stations—*Automated Surface Observing System* (ASOS)—that exist throughout the United States. (A picture of ASOS is shown in Fig. 3.35, on p. 85.)

● **FIGURE 4.22** The sling psychrometer.

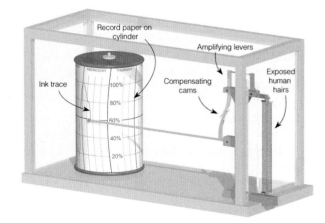

● **FIGURE 4.23** The hair hygrometer measures relative humidity by amplifying and measuring changes in the length of human (or horse) hair.

Which Is "Heavier"—Humid Air or Dry Air?

Does a volume of hot, humid air weigh more than a similarly sized volume of hot, dry air? The answer is no! At the same temperature and at the same level, humid air weighs *less* than dry air. (Keep in mind that we are referring strictly to water vapor—a gas—and not suspended liquid droplets.) To understand why, we must first see what determines the weight of atoms and molecules.

Almost all of the weight of an atom is concentrated in its nucleus, where the protons and neutrons are found. Neutrons weigh nearly the same as protons. To get some idea of how heavy an atom is, we simply add up the number of protons and neutrons in the nucleus. (Electrons are so light that we ignore them in comparing weights.) The larger this total, the heavier the atom. Now, we can compare one atom's weight with another's. For example, hydrogen, the lightest known atom, has only 1 proton in its center (no neutrons). Thus, it has an *atomic weight* of 1. Nitrogen, with 7 protons and 7 neutrons in its nucleus, has an atomic weight of 14. Oxygen, with 8 protons and 8 neutrons, weighs in at 16.

A molecule's weight is the sum of the atomic weights of its atoms. For example, molecular oxygen, with two oxygen atoms (O_2), has a molecular weight of 32. The most abundant atmospheric gas, molecular nitrogen (N_2), has a molecular weight of 28.

When we determine the weight of air, we are dealing with the weight of a mixture. As you might expect, a mixture's weight is a little more complex. We cannot just add the weights of all its atoms and molecules because the mixture might contain more of one kind than another. Air, for example, has far more nitrogen (78 percent) than oxygen (21 percent). We allow for this by multiplying the molecule's weight by its share in the mixture. Since dry air is essentially composed of N_2 and O_2 (99 percent), we ignore the other parts of air for the rough average shown in ▼ Table 2.

The symbol ≈ means "is approximately equal to." Therefore, dry air has a molecular weight of about 29. How does this compare with humid air?

Water vapor is composed of two atoms of hydrogen and one atom of oxygen (H_2O). It is an invisible gas, just as oxygen and nitrogen are invisible. It has a molecular weight; its two atoms of hydrogen (each with atomic weight of 1) and one atom of oxygen (atomic weight 16) give water vapor a molecular weight of 18. Obviously, air, at nearly 29, weighs appreciably *more* than water vapor.

Suppose we take a given volume of completely dry air and weigh it, then take exactly the same amount of water vapor at the same temperature and weigh it. We will find that the dry air weighs slightly more. If we replace dry air molecules one for one with water vapor molecules, the total number of molecules remains the same, but the total weight of the drier air decreases. Since density is mass per unit volume, *hot, humid air at the surface is less dense (lighter) than hot, dry air*.

This fact can have an important influence on our weather. The lighter the air becomes, the more likely it is to rise. All other factors being equal, hot, humid (less-dense) air will rise more readily than hot, dry (more-dense) air. It is of course the water vapor in the rising air that changes into liquid cloud droplets and ice crystals, which, in turn, grow large enough to fall to Earth as precipitation (see ● Fig. 2).

Of lesser importance to weather but of greater importance to sports is the fact that a baseball will "carry" farther in less-dense air. Consequently, without the influence of wind, a ball will travel slightly farther on a hot, humid day than it will on a hot, dry day. So when the sports announcer proclaims "the air today is heavy because of the high humidity," remember that this statement is not true and, in fact, a 404-foot home run on this humid day might simply be a 400-foot out on a very dry day.

▼ TABLE 2

GAS	WEIGHT		NUMBER OF ATOMS		MOLECULAR WEIGHT		PERCENT BY VOLUME
Oxygen	16	×	2	=	32	×	21% = 7
Nitrogen	14	×	2	=	28	×	78% = 22

Molecular weight of dry air ≈ 29

● **FIGURE 2** On this summer afternoon in Maryland, lighter (less-dense) hot, humid air rises and its water vapor condenses into towering cumulus clouds.

© C. Donald Ahrens

SUMMARY

This chapter examines the concept of atmospheric humidity. The chapter begins by looking at the hydrologic cycle and the circulation of water in our atmosphere. It then looks at the different phases of water, showing how evaporation, condensation, and saturation occur at the molecular level. The next several sections look at the many ways of describing the amount of water vapor in the air. Here we learn that there are many ways of describing humidity. The absolute humidity represents the density of water vapor in a given volume of air. Specific humidity measures the mass of water vapor in a fixed mass of air, while the mixing ratio expresses humidity as the mass of water vapor in the fixed mass of remaining dry air. The actual vapor pressure indicates the air's total water vapor content by expressing the amount of water vapor in terms of the amount of pressure that the water vapor molecules exert. The saturation vapor pressure describes how much water vapor the air could hold at any given temperature in terms of how much pressure the water vapor molecules would exert if the air were saturated at that temperature. A good indicator of the air's actual water vapor content is the dew point—the temperature to which air would have to be cooled (at constant pressure) for saturation to occur.

Relative humidity is a measure of how close the air is to being saturated. Air with a high relative humidity does not necessarily contain a great deal of water vapor; it is simply close to being saturated. With a constant water-vapor content, cooling the air causes the relative humidity to increase, while warming the air causes the relative humidity to decrease. When the air temperature and dew point are close together, the relative humidity is high, and, when they are far apart, the relative humidity is low. High relative humidity in hot weather makes us feel hotter than it really is by retarding the evaporation of perspiration. The heat index is a measure of how hot it feels to an average person for various combinations of air temperature and relative humidity. Although relative humidity can be confusing (because it can change with either air temperature or moisture content), it is nevertheless the most widely used way of describing the air's moisture content.

The chapter concludes by examining the various instruments that measure humidity, such as the psychrometer and hair hygrometer.

KEY TERMS

The following terms are listed (with page number) in the order they appear in the text. Define each. Doing so will aid you in reviewing the material covered in this chapter.

evaporation, 92
condensation, 92
precipitation, 92
hydrologic (water)
   cycle, 92
sublimation, 93
deposition, 93

saturated (air), 94
humidity, 95
absolute humidity, 95
specific humidity, 95
mixing ratio, 95
actual vapor
   pressure, 96

saturation vapor
   pressure, 96
relative humidity
   (RH), 97
supersaturated (air), 97
dew-point temperature
   (dew point), 99
frost point, 100

wet-bulb
   temperature, 104
heat stroke, 106
heat index (HI), 106
apparent temperature, 106
psychrometer, 107
hygrometer, 107
hair hygrometer, 107

QUESTIONS FOR REVIEW

1. Briefly explain the movement of water in the hydrologic cycle.
2. Basically, how do the three states of water differ?
3. What are the primary factors that influence evaporation?
4. Explain why condensation occurs primarily when the air is cooled.
5. How are evaporation and condensation related to saturated air above a flat water surface?
6. How does condensation differ from precipitation?
7. Why are specific humidity and mixing ratio more commonly used in representing atmospheric moisture than absolute humidity? What is the only way to change the specific humidity or mixing ratio of an air parcel?
8. In a volume of air, how does the actual vapor pressure differ from the saturation vapor pressure? When are they the same?
9. Upon what does saturation vapor pressure primarily depend?
10. Explain why it takes longer to cook vegetables in the mountains than at sea level.
11. (a) What does the relative humidity represent?
   (b) When the relative humidity is given, why is it also important to know the air temperature?
   (c) Explain two ways the relative humidity can be changed.
12. Explain why, during a summer day, the relative humidity will change as shown in Fig. 4.12 (p. 99).
13. Why do hot, humid summer days usually feel hotter than hot, dry summer days?
14. Why is the wet-bulb temperature a good measure of how cool human skin can become?
15. Explain why the air on a hot, humid day is less dense than on a hot, dry day.
16. (a) What is the dew-point temperature?
   (b) How is the difference between dew point and air temperature related to the relative humidity?
17. Why is cold polar air described as "dry" even when the relative humidity of that air is very high?

18. How can a region have a high specific humidity and a low relative humidity? Give an example.

19. Why is the air from the Gulf of Mexico so much more humid than air from the Pacific Ocean at the same latitude?

20. How are the dew-point temperature and wet-bulb temperature different? Can they ever read the same? Explain.

21. When outside air is brought indoors on a cold winter day, the relative humidity of the heated air inside often drops below 25 percent. Explain why this situation occurs.

22. Describe how a sling psychrometer works. What does it measure? Does it give you dew point and relative humidity? Explain.

23. Why are human hairs often used in a hair hygrometer?

QUESTIONS FOR THOUGHT

1. Would you expect water in a glass to evaporate more quickly on a windy, warm, dry summer day or on a calm, cold, dry winter day? Explain.

2. How can frozen clothes "dry" outside in subfreezing weather? What exactly is taking place?

3. Explain how and why each of the following will change as a parcel of air with an unchanging amount of water vapor rises, expands, and cools:
   (a) absolute humidity;
   (b) relative humidity;
   (c) actual vapor pressure; and
   (d) saturation vapor pressure.

4. Where in the United States and Canada would you go to experience the *least* variation in dew point (actual moisture content) from January to July? (Hint: Look at Fig. 4.14 and Fig. 4.15.)

5. After completing a grueling semester of meteorological course work, you get in touch with a friend from Arizona to discuss a much-needed summer vacation. When your friend suggests heading to the desert, you decline because of a concern that the dry air will make your skin feel uncomfortable. Your friend's mother, who is a travel agent, assures you that almost daily "desert relative humidities are above 90 percent." Could your friend's mother be correct? Explain.

6. On a clear, calm morning, water condenses on the ground in a thick layer of dew. As the water slowly evaporates into the air, you measure a slow increase in dew point. Explain why.

7. (a) Two cities have exactly the same amount of water vapor in the air. The 6:00 a.m. relative humidity in one city is 93 percent, while the 3:00 p.m. relative humidity in the other city is 28 percent. Explain how this can come about.

   (b) If two cities have exactly the same amount of water vapor in the air, how can one city have a relative humidity of 90 percent while the other city has a relative humidity of only 40 percent?

8. Suppose the dew point of cold outside air is the same as the dew point of warm air indoors. If the door is opened, and cold air replaces some of the warm inside air, would the new relative humidity indoors be (a) lower than before, (b) higher than before, or (c) the same as before? Explain your answer.

9. On a warm, muggy day, the air is described as "close." What are several plausible explanations for this expression?

10. Outside, on a very warm day, you swing a sling psychrometer for about a minute and read a dry-bulb temperature of 38°C and a wet-bulb temperature of 24°C. After swinging the instrument again, the dry bulb is still 38°C, but the wet bulb is now 26°C. Explain how this could happen.

11. Why are evaporative coolers used in Arizona, Nevada, and California but not in Florida, Georgia, or Indiana?

12. Devise a way of determining elevation above sea level if all you have is a thermometer and a pot of water.

13. A large family lives in northern Minnesota. This family gets together for a huge dinner three times a year: on Thanksgiving, on Christmas, and on the March equinox. The Thanksgiving and Christmas dinners consist of turkey, ham,

mashed potatoes, and lots of boiled vegetables. The equinox dinner is pizza. The air temperature inside the home is about the same for all three meals (70°F), yet everyone remarks on how "warm, cozy, and comfortable" the air feels during the Thanksgiving and Christmas dinners, and how "cool" the inside air feels during the March meal. Explain to the family members why they might feel "warmer" inside the house during Thanksgiving and Christmas, and "cooler" during the March equinox. (The answer has nothing to do with the amount or type of food consumed.)

PROBLEMS AND EXERCISES

1. On a bitterly cold, snowy morning, the air temperature and dew point of the outside air are both −7°C. If this air is brought indoors and warmed to 21°C, with no change in vapor content, what is the relative humidity of the air inside the home? (Hint: See Table 1, p. 105.)

2. (a) With the aid of Fig. 4.15 (p. 101), determine the average July dew points in St. Louis, Missouri; New Orleans, Louisiana; and Los Angeles, California.
   (b) If the high temperature on a particular summer day in all three cities is 32°C (90°F), and the dew points are the same as the average July values, then calculate the afternoon relative humidity at each of the three cities. (Hint: Either Fig. 4.10, p. 97, or Table 1, p. 105, will be helpful.)

3. Suppose with the aid of a sling psychrometer you obtain an air temperature of 30°C and a wet-bulb temperature of 25°C. What is
   (a) the wet-bulb depression,
   (b) the dew point, and
   (c) the relative humidity of the air?
   (Use the tables in Appendix C.)

4. If the air temperature is 35°C and the dew point is 21°C, determine the relative humidity using
   (a) Table 1, p. 105;
   (b) Fig. 4.10, p. 97.

5. Suppose the average vapor pressure in Nevada is about 8 mb.
   (a) Use Table 1 (p. 105) to determine the average dew point of this air.
   (b) Much of the state is above an elevation of 1500 m (5000 ft). At 1500 m, the normal pressure is about 12.5 percent less than at sea level. If the air over Nevada were brought down to sea level, without any change in vapor content, what would be the new water vapor pressure of the air?

6. In Yellowstone National Park, there are numerous ponds of boiling water. If Yellowstone is about 2200 m (7200 ft) above sea level (where the air pressure is normally about 775 mb), what is the normal boiling point of water in Yellowstone? (Hint: See Fig. 1, p. 98.)

7. Three cities have the following temperature (T) and dew point (T_d) during a July afternoon:
   Atlanta, Georgia, $T = 90°F$; $T_d = 75°F$
   Baltimore, Maryland, $T = 80°F$; $T_d = 70°F$
   Norman, Oklahoma, $T = 70°F$; $T_d = 65°F$
   (a) Which city appears to have the highest relative humidity?
   (b) Which city appears to have the lowest relative humidity?
   (c) Which city has the *most* water vapor in the air?
   (d) Which city has the *least* water vapor in the air?
   (e) For each city use Table 1 on p. 105 and the information on the same page to calculate the relative humidity for each city.
   (f) Using both the relative humidity calculated in (e) and the air temperature, determine the heat index for each city using Fig. 4.21 (p. 106).

 ONLINE RESOURCES

Visit www.cengagebrain.com to view additional resources, including video exercises, practice quizzes, an interactive eBook, and more.

Skyscrapers peek out from a layer of thick fog enshrouding Dubai, United Arab Emirates.

Zohaib Anjum/Shutterstock.com

Condensation: Dew, Fog, and Clouds

THE WEATHER IS AN EVER-PLAYING DRAMA before which we are a captive audience. With the lower atmosphere as the stage, air and water as the principal characters, and clouds for costumes, the weather's acts are presented continuously somewhere about the globe. The script is written by the sun; the production is directed by Earth's rotation; and, just as no theater scene is staged exactly the same way twice, each weather episode is played a little differently, each is marked with a bit of individuality.

Clyde Orr, Jr., *Between Earth and Space*

Have you walked barefoot across a lawn on a summer morning and felt the wet grass under your feet? Did you ever wonder how those glistening droplets of dew could form on a clear summer night? Or why they formed on grass but not on bushes several meters above the ground? In this chapter, we will investigate the formation of dew and frost. Then we will examine the different types of fog. The chapter concludes with the identification and observation of clouds.

The Formation of Dew and Frost

On clear, calm nights, objects near Earth's surface cool rapidly by emitting infrared radiation. The ground and objects on it often become much colder than the surrounding air. Air that comes in contact with these cold surfaces cools by conduction. Eventually, the air cools to the *dew point*—the temperature at which saturation occurs. As surfaces such as twigs, leaves, and blades of grass cool below this temperature, water vapor begins to condense upon them, forming tiny visible specks of water called **dew** (see ● Fig. 5.1). If the air temperature should drop to freezing or below, the dew will freeze, becoming tiny beads of ice called **frozen dew**. Because the coolest air is usually at ground level, dew is more likely to form on blades of grass than on objects several meters above the surface. This thin coating of dew dampens bare feet, of course, but more importantly, it is a valuable source of moisture for many plants during periods of low rainfall. Averaged for an entire year in middle latitudes, dew yields a blanket of water between 12 and 50 mm (0.5 and 2 in.) thick.

Dew is more likely to form on nights that are clear and calm than on nights that are cloudy and windy. Clear nights allow objects near the ground to cool rapidly by emitting infrared radiation, and calm winds mean that the coldest air will be located at ground level. These atmospheric conditions are usually associated with large fair-weather, high-pressure systems. On the other hand, the cloudy, windy weather that inhibits rapid cooling near the ground and the forming of dew often signifies the approach of a rain-producing storm system. These observations inspired the following folk rhyme:

> When the dew is on the grass,
> rain will never come to pass.
> When grass is dry at morning light,
> look for rain before the night!

Visible white frost forms on cold, clear, calm mornings when the dew-point temperature is at or below freezing. When the air temperature cools to the dew point (now called the *frost point*) and further cooling occurs, water vapor can change directly to ice without becoming a liquid first—a process called *deposition*.* The delicate, white crystals of ice that form in this manner are called *hoarfrost, white frost,* or simply **frost**. Frost has a treelike branching pattern that easily distinguishes it from the nearly spherical beads of frozen dew.

***

*Recall that when the ice changes back into vapor without melting, the process is called *sublimation*.

On cold winter mornings, frost may form on the inside of a windowpane in much the same way as it does outside, except that the cold glass chills the indoor air adjacent to it. When the temperature of the inside of the window drops below freezing, water vapor in the room forms a light, feathery deposit of frost (see ● Fig. 5.2).

In very dry weather, the air temperature may become quite cold and drop below freezing without ever reaching the frost point, and no visible frost forms. *Freeze* and *black frost* are words denoting this situation, one that can severely damage crops (see Chapter 3, pp. 72-73).

So, dew, frozen dew, and frost form in the rather shallow layer of air near the ground on clear, calm nights. But what happens to air as a deeper layer adjacent to the ground is cooled? We've seen in Chapter 4 that if air cools without any change in water vapor content, the relative humidity increases. When air cools to the dew point, the relative humidity becomes 100 percent and the air is saturated. Continued cooling condenses some of the vapor into tiny cloud droplets.

● **FIGURE 5.1** Dew forms on clear nights when objects on the surface cool to a temperature below the dew point. If these beads of water should freeze, they would become frozen dew.

● **FIGURE 5.2** These are the delicate ice-crystal patterns that frost exhibits on a window during a cold winter morning.

Condensation Nuclei

Actually, the condensation process that produces clouds is not quite so simple. Just as dew and frost need a surface to form on, there must be airborne particles on which water vapor can condense to produce cloud droplets.

Although the air may look clean, it never really is. On an ordinary day, a volume of air about the size of your index finger contains between 1000 and 150,000 particles. Because many of these serve as surfaces on which water vapor can condense, they are called **condensation nuclei**. Without them, relative humidities of several hundred percent would be required before condensation could begin.

Some condensation nuclei are quite small and have a radius of less than 0.1 μm; these are referred to as *Aitken nuclei*, after the British physicist who discovered that water vapor condenses on nuclei. Particles ranging in size from 0.1 to 1 μm are called *large nuclei*, while others, called *giant nuclei*, are much larger and have radii exceeding 1 μm (see ▼ Table 5.1). The condensation nuclei most favorable for producing clouds (called *cloud condensation nuclei*) have radii of 0.1 μm or more. Usually, between 100 and 1000 nuclei of this size exist in a cubic centimeter of air. These particles enter the atmosphere in a variety of ways: dust, volcanoes, factory smoke, forest fires, salt from ocean spray, and even sulfate particles emitted by phytoplankton in the oceans. In fact, studies show that sulfates provide the major source of cloud condensation nuclei in the marine atmosphere. Because most particles are released into the atmosphere near the ground, the largest concentrations of nuclei are observed in the lower atmosphere near Earth's surface.

Condensation nuclei are extremely light (many have a mass less than one-trillionth of a gram), so they can remain suspended in the air for many days. They are most abundant over industrial cities, where highly polluted air may contain nearly 1 million particles per cubic centimeter. They decrease in cleaner "country" air and over the oceans, where concentrations may dwindle to only a few nuclei per cubic centimeter.

Some particles are **hygroscopic** ("water-seeking"), and water vapor condenses upon these surfaces when the relative humidity is considerably lower than 100 percent. Ocean salt is hygroscopic, as is common table salt. In humid weather, it is difficult to pour salt from a shaker because water vapor condenses onto the salt crystals, sticking them together. Moreover, on a humid day, salty potato chips left outside in an uncovered bowl turn soggy. Other hygroscopic nuclei include sulfuric and

● **FIGURE 5.3** Hygroscopic nuclei are "water-seeking," and water vapor rapidly condenses on their surfaces. Hydrophobic nuclei are "water-repelling" and resist condensation.

nitric acid particles. Not all particles serve as good condensation nuclei. Some are **hydrophobic*** ("water-repelling")—such as oils, gasoline, and paraffin waxes—and resist condensation even when the relative humidity is above 100 percent (see ● Fig. 5.3). As we can see, condensation may begin on some particles when the relative humidity is well below 100 percent and on others only when the relative humidity is much higher than 100 percent. However, at any given time there are usually many nuclei present, so that haze, fog, and clouds will form at relative humidities near or below 100 percent.

Haze

Suppose you visit an area that has a layer of **haze** (that is, a layer of dust or salt particles) suspended above the region. There, you may notice that distant objects are usually more visible in the afternoon than in the morning, even when the concentration of particles in the air has not changed. Why? During the warm afternoon, the relative humidity of the air is often below the point where water vapor begins to condense, even on active hygroscopic nuclei. Therefore, the floating particles remain small—usually no larger than about one-tenth of a micrometer. These tiny *dry haze* particles selectively scatter some rays of sunlight, while allowing others to penetrate the air. The scattering effect

---

*A synthetic hydrophobic is PTFE, or Teflon—the material used in rain-repellent fabric.

▼ **TABLE 5.1** Characteristic Sizes and Concentration of Condensation Nuclei and Cloud Droplets

TYPE OF PARTICLE	APPROXIMATE RADIUS (MICROMETERS)	NO. OF PARTICLES (PER CM³)	
		Range	Typical
Small (Aitken) condensation nuclei	<0.1	1000 to 10,000	1000
Large condensation nuclei	0.1 to 1.0	1 to 1000	100
Giant condensation nuclei	>1.0	<1 to 10	1
Fog and cloud droplets	>10	10 to 1000	300

● **FIGURE 5.4** On a still winter morning, the high relative humidity of the cold air above the lake is causing a layer of haze to form.

© C. Donald Ahrens

of dry haze produces a bluish color when viewed against a dark background and a yellowish tint when viewed against a light-colored background.

As the air cools during the night, the relative humidity increases. When the relative humidity reaches about 75 percent, condensation may begin on the most active hygroscopic nuclei, producing a *wet haze*. As water collects on the nuclei, their size increases and the particles, although still small, become large enough to scatter light much more efficiently. In fact, as the relative humidity increases from about 60 percent to 80 percent, the scattering effect increases by a factor of nearly 3. Given that relative humidities are normally high during cool mornings, much of the light from distant objects is scattered away by the wet haze particles before reaching you; hence, it is difficult to see these distant objects.

Not only does wet haze restrict visibility more than dry haze, it also appears dull gray or white (see ● Fig. 5.4). Near seashores and in clean air over the open ocean, large salt particles suspended in air with a high relative humidity often produce a thin white veil across the horizon.

Fog

By now, it should be apparent that condensation is a continuous process beginning when water vapor condenses onto hygroscopic nuclei at relative humidities as low as 75 percent. As the relative humidity of the air increases, the visibility decreases, and the landscape becomes masked with a grayish tint. As the relative humidity gradually approaches 100 percent, the haze particles grow larger, and condensation begins on the less-active nuclei. Now water is condensing onto a large fraction of the available nuclei. As water condenses onto them, the droplets grow even bigger, until eventually they become visible to the naked eye. The increasing size and concentration of droplets further restrict visibility. When the visibility lowers to less than 1 km (0.62 mi), and the air is wet with countless millions of tiny floating water droplets, the wet haze becomes a cloud resting near the ground, which we call **fog**.*

With the same water content, fog that forms in dirty city air often is thicker than fog that forms over the ocean. Normally, the smaller number of condensation nuclei over the middle of the ocean produces fewer, but larger, fog droplets. City air with its abundant nuclei produces many tiny fog droplets, which greatly increases the thickness (or opaqueness) of the fog and reduces visibility. A dramatic example of a thick fog forming in air with abundant nuclei occurred in London, England, during the early 1950s. The fog became so thick, and the air so laden with smoke particles, that sunlight could not penetrate the smoggy air, requiring that street lights be left on at midday. Moreover, fog that forms in polluted air can turn acidic as the tiny liquid droplets combine with gaseous impurities, such as oxides of sulfur and nitrogen. **Acid fog** poses a threat to human health, especially to people with preexisting respiratory problems. We'll examine in more detail the health problems associated with acid fog and other forms of pollution in Chapter 19.

As tiny fog droplets grow larger, they become heavier and tend to fall toward Earth. A fog droplet with a diameter of 25 μm settles toward the ground at about 5 cm (2 in.) each second. At this rate, most of the droplets in a fog layer 180 m (about 600 ft) thick would reach the ground in less than one hour. Therefore, two questions arise: How does fog form? And how is fog maintained once it does form?

Fog, like any cloud, usually forms in one of two ways:

1. by cooling—air is cooled below its saturation point (dew point)
2. by evaporation and mixing—water vapor is added to the air by evaporation, and the moist air mixes with relatively dry air

---

*This is the official international definition of fog. The U.S. National Weather Service reports fog as a restriction to visibility when fog restricts the visibility to 6 miles or less and the spread between the air temperature and dew point is 5°F or less. When the visibility is less than one-quarter of a mile, the fog is considered dense.

Once fog forms it is maintained by new fog droplets, which constantly form on available nuclei. In other words, the air must maintain its degree of saturation either by continual cooling or by evaporation and mixing of vapor into the air. Let's examine both processes.

RADIATION FOG How can the air cool so that a cloud will form near the surface? Radiation and conduction are the primary means for cooling nighttime air near the ground. Fog produced by Earth's radiational cooling is called **radiation fog**, or *ground fog*. It forms best on clear nights when a shallow layer of moist air near the ground is overlain by drier air. Because the moist layer is shallow, it does not absorb much of Earth's outgoing infrared radiation. The ground, therefore, cools rapidly and so does the air directly above it, and a surface inversion forms, with cooler air at the surface and warmer air above. The moist lower layer (chilled rapidly by the cold ground) quickly becomes saturated, and fog forms. The longer the night, the longer the time of cooling and the greater the likelihood of fog. Hence, radiation fogs are most common over land in late fall and winter.

Another factor promoting the formation of radiation fog is a light breeze of less than 5 knots. Although radiation fogs may form in calm air, slight air movement brings more of the moist air in direct contact with the cold ground, and the transfer of heat occurs more rapidly. A strong breeze tends to prevent radiation fog from forming by mixing the air near the surface with the drier air above. The ingredients of clear skies and light winds are associated with large high-pressure areas (anticyclones). Consequently, during the winter, when an anticyclone becomes stagnant over an area, radiation fog may form on many consecutive days.

Because cold, heavy air drains downhill and collects in valley bottoms, we normally see radiation fog forming in low-lying areas. Hence, radiation fog is frequently called *valley fog*. The cold air and high moisture content in river valleys make them susceptible to radiation fog. Since radiation fog normally forms in lowlands, hills may be clear all day long, while adjacent valleys are fogged in (see ● Fig. 5.5).

Radiation fogs form upward from the ground as the night progresses and are usually deepest around sunrise. However, fog may occasionally form after sunrise, especially when evaporation and mixing take place near the surface. This usually occurs at the end of a clear, calm night as radiational cooling brings the air temperature close to the dew point in a rather shallow layer above the ground. At the surface, the air becomes saturated, forming a thick blanket of dew on the grass. At daybreak, the sun's rays evaporate the dew, adding water vapor to the air. A light breeze then stirs the moist air with the drier air above, causing saturation (and, hence, fog) to form in a shallow layer near the ground.

Often a shallow fog layer will dissipate or *burn off* by the afternoon. Of course, the fog does not "burn"; rather, sunlight penetrates the fog and warms the ground, causing the air temperature in contact with the ground to increase. The warm air rises and mixes with the foggy air above, which increases the temperature of the foggy air. In the slightly warmer air, some of the fog droplets evaporate, allowing more sunlight to reach the ground, which produces more heating, and soon the fog completely evaporates and disappears.

Satellite images show that a blanket of radiation fog tends to evaporate ("burn off") first around its periphery, where the fog is usually thinnest. Sunlight rapidly warms this region, causing the fog to dissipate as the warmer air mixes in toward the denser foggy area.

If the fog is thick, with little sunlight penetrating it, and there is little mixing along the outside edges, the fog may not dissipate. This is often the case in the Central Valley area of California during the late fall and winter. A fog layer over 500 m (1700 ft) thick settles between two mountain ranges, while a strong inversion normally keeps the warmest air above the top of the fog. This is often called *tule fog*, after a type of plant commonly found in the Central Valley where it forms. During the day, much of the light from the low winter sun reflects off the top of tule fog, allowing only a small amount of sunlight to penetrate the fog and warm the ground. As the air warms from below, the fog dissipates upward from the surface in a rather shallow layer less than 150 m (500 ft), creating the illusion that the fog is lifting. Because the fog no longer touches the ground and a strong inversion exists above it, the fog is called a *high inversion fog*. (The low cloud above the ground is also called *stratus*, or, simply, *high fog*.) As soon as the sun sets, radiational cooling lowers the air temperature, and the fog once again forms on the ground. This daily lifting and lowering of the fog without the sun ever breaking through it may last for many days or even weeks during winter in California's Central Valley (see ● Fig. 5.6).

ADVECTION FOG Cooling surface air to its saturation point may be accomplished by warm moist air moving over a cold surface. The surface must be sufficiently cooler than the air above so that the transfer of heat from air to surface will cool the air to its dew point and produce fog. Fog that forms in this manner is called **advection fog**.

A good example of advection fog can be observed along the Pacific Coast during summer. The main reason fog forms in this region is that the surface water near the coast is much

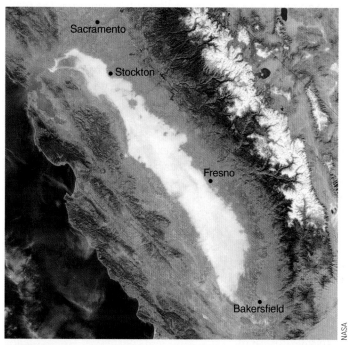

● **FIGURE 5.6** Visible satellite image of dense radiation fog in the southern half of California's Central Valley on the morning of November 20, 2002. The white region to the east (right) of the fog is the snow-capped Sierra Nevada range. During the late fall and winter, the fog, nestled between two mountain ranges, can last for many days without dissipating. The fog on this day was responsible for several auto accidents, including a 14-car pileup near Fresno.

CRITICAL THINKING QUESTION Notice in this satellite image that the fog mainly covers the western (left) side of the valley and that the cities of Stockton and Fresno are not in the fog. Based on what you've learned so far about radiation fog, come up with at least one possible reason why, at the time of this image, fog did not cover the eastern side of the valley.

colder than the surface water farther offshore. Warm moist air from the Pacific Ocean is carried (advected) by westerly winds over the cold coastal waters. Chilled from below, the air temperature drops to the dew point, and fog forms. Advection fog, unlike radiation fog, always involves the movement of air, so when there is a stiff summer breeze in San Francisco, it's common to watch advection fog roll in past the Golden Gate Bridge (see ● Fig. 5.7). It is also more common to see advection fog forming at headlands that protrude seaward than in the mouths of bays. If you are curious as to why, read Focus section 5.1.

As summer winds carry the fog inland over the warmer land, the fog near the ground dissipates, leaving a sheet of low-lying gray clouds that block out the sun. Farther inland, the air is sufficiently warm so that even these low clouds evaporate and disappear. Given that the fog is more likely to burn off during the warmer part of the day, a typical summertime weather forecast for coastal areas would read, "Fog and low cloudiness along the coast, extending locally inland both nights and mornings, with sunny afternoons."

Because they provide moisture to the coastal redwood trees, advection fogs are important to the scenic beauty of the Pacific Coast. The needles and branches of the redwoods absorb moisture from the fog, allowing the trees to grow very tall without having to draw moisture from their roots far below. Additional moisture drips to the ground (*fog drip*), where it is utilized by the tree's shallow root system. Without the summer fog, the coast's redwood trees would have trouble surviving the dry California summers. Hence, we find them nestled in the fog belt along the coast.

● **FIGURE 5.7** Advection fog forms as the wind moves moist air over a cooler surface. Here, advection fog, having formed over the cold, coastal water of the Pacific Ocean, is rolling inland past the Golden Gate Bridge in San Francisco. As fog moves inland, the air warms and the fog lifts above the surface. Eventually, the air becomes warm enough to totally evaporate the fog.

Why Are Headlands Usually Foggier Than Beaches?

If you drive along a highway that parallels an irregular coastline, you may have observed that advection fog is more likely to form in certain regions. For example, headlands that protrude seaward usually experience more fog than do beaches that are nestled in the mouths of bays. Why?

As air moves onshore, it crosses the coastline at nearly a right angle. This causes the air to flow together or converge in the vicinity of the headlands (see ● Fig. 1). This area of weak convergence causes the surface air to rise and cool just a little. If the rising air is close to being saturated, it will cool to its dew point, and fog will form.

Meanwhile, near the beach area, notice in Fig. 1 that the surface air spreads apart or diverges as it crosses the coastline. This area of weak divergence creates sinking and slightly warmer air. Because the sinking of air increases the separation between air temperature and

● **FIGURE 1** Along an irregular coastline, advection fog is more likely to form at the headland (the region of land extending seaward) where moist surface air converges and rises than at the beach where air diverges and sinks.

dew point, fog is less likely to form in this region. Hence, the headlands can be shrouded in fog while the beaches are basking in sunshine.

Advection fogs also prevail where two ocean currents with different temperatures flow next to one another. Such is the case in the Atlantic Ocean off the coast of Newfoundland, where the cold southward-flowing Labrador Current lies almost parallel to the warm northward-flowing Gulf Stream. Warm southerly air moving over the cold water produces fog in that region—so frequently that fog occurs on about two out of three days during summer.

Advection fog also forms over land. In winter, warm moist air from the Gulf of Mexico moves northward over progressively colder and slightly elevated land in the southern or central United States. As the air cools to its saturation point, fog will form. Because the cold ground is often the result of radiational cooling, fog that forms in this manner is sometimes called **advection-radiation fog**.

During this same time of year, air moving across the warm Gulf Stream encounters the colder land of the British Isles and produces the thick fogs of England. Similarly, fog forms as marine air moves over an ice or snow surface. In extremely cold arctic air, ice crystals form instead of water droplets, producing an *ice fog*.

Keep in mind that advection fog forms when wind blows moist air over a cooler surface, whereas radiation fog forms under relatively calm conditions. ● Figure 5.8 visually summarizes the formation of these two types of fog.

UPSLOPE FOG   Fog that forms as moist air flows up along an elevated plain, hill, or mountain is called **upslope fog**. Typically, upslope fog forms during the winter and spring on the eastern

(a) Radiation fog
(b) Advection fog

● **FIGURE 5.8** (a) Radiation fog tends to form on clear, relatively calm nights when cool, moist surface air is overlain by drier air and rapid radiational cooling occurs. (b) Advection fog forms when the wind moves moist air over a cold surface and the moist air cools to its dew point.

Fog That Forms by Mixing

How can unsaturated bodies of air mix together to produce fog (or a cloud)? To answer this question, let's first examine two unsaturated air parcels. (Later, we will look at the parcels mixed together.) The two parcels in ● Fig. 2 are essentially the same size and have a mass of 1 kg. Yet each has a different temperature and a different relative humidity. (We will assume that the parcels are near sea level where the atmospheric pressure is close to 1000 mb.)

In Chapter 4, we used both the actual vapor pressure and the saturation vapor to obtain the air's relative humidity (see p. 97). Here we will use another formula to express relative humidity:

$$RH = \frac{\text{actual mixing ratio } (w)}{\text{saturation mixing ratio } (w_s)} \times 100\%*$$

Look closely at Fig. 2 and observe that parcel A has an air temperature (T) of 20°C and a dew-point temperature (T_d) of 15°C. To obtain the *saturation mixing ratio*, we look at ▼ Table 1 and read the value that corresponds to the parcel's air temperature. For a temperature of 20°C, the saturation mixing ratio is 15.0 g/kg. Likewise,

---

*Recall from Chapter 4 that the actual mixing ratio (w) is the mass of water vapor per kilogram (kg) of dry air, usually expressed as grams per kilogram (g/kg), and the saturation mixing ratio (w_s) is the mixing ratio of saturated air, also expressed as g/kg.

PARCEL A
T = 20°C
T_d = 15°C
w = 10.8 g/kg
RH = 72%
1.0 kg

PARCEL B
T = -10°C
T_d = -15°C
w = 1.2 g/kg
RH = 67%
1.0 kg

MIXING

MIXED PARCEL
T = 5°C
w_s = 5.5 g/kg
w = 6.0 g/kg
RH = 109%
2.0 kg

● **FIGURE 2** Mixing two unsaturated air parcels can produce fog. Notice that the mixed parcel has an actual mixing ratio (W) that is greater than the saturation mixing ratio (W_s). This produces an RH of 109% and results in fog. (As the mixed parcel cools below its dew-point temperature, water vapor will condense onto condensation nuclei, producing the liquid droplets that create fog.)

the *actual mixing ratio* is obtained in Table 1 by reading the value that corresponds to the parcel's dew-point temperature. For a dew-point temperature of 15°C, the actual mixing ratio is 10.8 g/kg. Consequently, the relative humidity of the air in parcel A is 72 percent because

$$RH = \frac{w}{w_s} = \frac{10.8}{15.0} \times 100\% = 72\%$$

Air parcel B in Fig. 2 is considerably colder than parcel A with a temperature of −10°C, and considerably drier with a dew-point temperature of −15°C. These temperatures yield a relative humidity of 67 percent as:

$$RH = \frac{w}{w_s} = \frac{1.2}{1.8} \times 100\% = 67\%$$

Suppose we now thoroughly mix the two parcels in Fig. 2. After mixing, the new parcel's temperature will be close to the average of parcel A and parcel B, or about 5°C. The total water vapor content (the actual mixing ratio) of the mixed parcel will be the sum of the mixing ratios of parcel A and parcel B, or

$$10.8 \tfrac{g}{kg} + 1.2 \tfrac{g}{kg} = \frac{12.0 \tfrac{g}{kg}}{2} = 6.0 \tfrac{g}{kg}$$

Look at Table 1 and observe that the saturation mixing ratio for a saturated parcel at 5°C

---

side of the Rockies, where the eastward-sloping plains are nearly a kilometer higher than the land farther east. Occasionally, cold air moves from the lower eastern plains westward. The air gradually rises, expands, becomes cooler, and—if sufficiently moist—a fog forms (see ● Fig. 5.9). Upslope fogs that form over an extensive area can last for days.

Up to now, we have seen how the cooling of air produces fog. But remember that fog can also form from the mixing of two unsaturated masses of air. Fog that forms in this manner is usually called *evaporation fog* because evaporation initially enriches the air with water vapor. Probably a more appropriate name for the fog is **evaporation (mixing) fog**. (For a better understanding of how mixing can produce fog, read Focus section 5.2.)

EVAPORATION (MIXING) FOG On a cold day, you may have unknowingly produced evaporation fog. When moist air from your mouth or nose meets the cold air and mixes with it, the air becomes saturated, and a tiny cloud forms with each exhaled breath.

Upslope fog

Moist air

● **FIGURE 5.9** Upslope fog forms as moist air slowly rises, cools, and condenses over elevated terrain.

A common form of evaporation (mixing) fog is **steam fog**, which forms when cold air moves over warm water. This type of fog forms above a heated outside swimming pool in winter. As long as

▼ **TABLE 1** Saturation Mixing Ratios of Water Vapor for Various Air Temperatures (Assuming Air Pressure Is 1000 mb)

AIR TEMPERATURE (°C)	SATURATION MIXING RATIO (g/kg)
20	15
15	10.8
10	7.8
5	5.5
0	3.8
−5	2.6
−10	1.8
−15	1.2
−20	0.8

is only 5.5 g/kg. This means that the water vapor content of the mixed parcel, at 6.0 g/kg, is *above* that required for saturation and that the parcel is *supersaturated* with a relative humidity of 109 percent. Of course, such a high relative humidity is almost impossible to obtain, as water vapor would certainly condense on condensation nuclei, producing liquid water droplets as the two parcels mix together and the relative humidity approaches 100 percent. Hence, mixing two initially unsaturated masses of air can produce fog or a cloud.

Another way to look at this mixing process is to place the two unsaturated air parcels into Fig. 3, which is a graphic representation of Table 1. The solid blue line in Fig. 3 represents the saturation mixing ratio. Any air parcel with an air temperature and actual mixing ratio that falls on the blue line is saturated with a relative humidity of 100 percent. If an air parcel is located to the right of the blue line, the air parcel is unsaturated. If an air parcel lies to the left of the line, the parcel is supersaturated, and condensation will occur.

Notice that when parcel A and parcel B from Fig. 2 are plotted in Fig. 3, both unsaturated air parcels fall to the right of the blue line. However, when parcel A and parcel B are mixed, the mixed air parcel (with an air temperature of 5°C and an actual mixing ratio of 6.0 g/kg) falls to the left of the blue line, indicating that water vapor inside the mixed parcel will condense into either fog or a cloud. Fog that forms in this manner is called evaporation (mixing) fog (p. 120). As you read that section, keep in mind that although evaporation has a part in fog formation, mixing plays the dominant role.

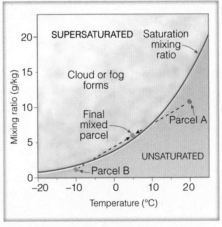

● **FIGURE 3** The blue line is the saturation mixing ratio. The mixing of two unsaturated air parcels (A and B) can produce a saturated air parcel and fog.

the water is warmer than the unsaturated air above, water will evaporate from the pool into the air. The increase in water vapor raises the dew point, and, if mixing is sufficient, the air above becomes saturated. The colder air directly above the water is heated from below and becomes warmer than the air directly above it. This warmer air rises and, from a distance, the rising condensing vapor appears as "steam."

It is common to see steam fog forming over lakes on autumn mornings, as cold air settles over water still warm from the long summer. On occasion, over the Great Lakes and other warm bodies of water, columns of condensed vapor rise from the fog layer, forming whirling *steam devils*, which appear similar to the dust devils observed on land. If you travel to Yellowstone National Park, you will see steam fog forming above thermal ponds all year long (see ● Fig. 5.10). Over the ocean in polar regions, steam fog is referred to as *arctic sea smoke*.

Steam fog may form above a wet surface on a sunny day. This type of fog is commonly observed after a rain shower as

● **FIGURE 5.10** Even in summer, warm air rising above thermal pools in Yellowstone National Park condenses into a type of steam fog.

sunlight shines on a wet road, heats the asphalt, and quickly evaporates the water. This added vapor mixes with the air above, producing steam fog. Fog that forms in this manner is short-lived and disappears as the road surface dries.

A warm rain falling through a layer of cold moist air can produce fog. Remember from Chapter 4 that the saturation vapor pressure depends on temperature: Higher temperatures correspond to higher saturation vapor pressures. When a warm raindrop falls into a cold layer of air, the saturation vapor pressure over the raindrop is greater than that of the air. This vapor-pressure difference causes water to evaporate from the raindrop into the air. This process may saturate the air and, if mixing occurs, fog forms. Fog of this type is often associated with warm air riding up and over a mass of colder surface air. The fog usually develops in the shallow layer of cold air just ahead of an approaching warm front or behind a cold front, which is why this type of evaporation fog is also known as *precipitation fog*, or **frontal fog**. Snow covering the ground is an especially favorable condition for the formation of frontal fog. The melting snow extracts heat from the environment, thereby cooling the already rain-saturated air.

Foggy Weather

The foggiest regions in the United States are shown in ● Fig. 5.11. Notice that dense fog is more prevalent in coastal margins (especially those regions lapped by cold ocean currents and near the Great Lakes and Appalachian Mountains) than in the center of the continent. In fact, the foggiest spot near sea level in the United States is Cape Disappointment, Washington. Located at the mouth of the Columbia River, it averages 2556 hours (or the equivalent of 106.5 twenty-four-hour days) of dense fog each year. Anyone hoping to enjoy the sun during August and September by traveling to this spot would find its name appropriate indeed.

Notice in Fig. 5.11 that the coast of Maine is also foggy. In fact, Moose Peak Lighthouse on Mistake Island averages 1580 hours (66 equivalent days) of dense fog. To the south, Nantucket Island has on average 2040 hours (85 equivalent days) of fog each year.

Although fog is basically a nuisance, it has many positive aspects. For example, the California Central Valley fog that many residents scorn is extremely important to the economy of

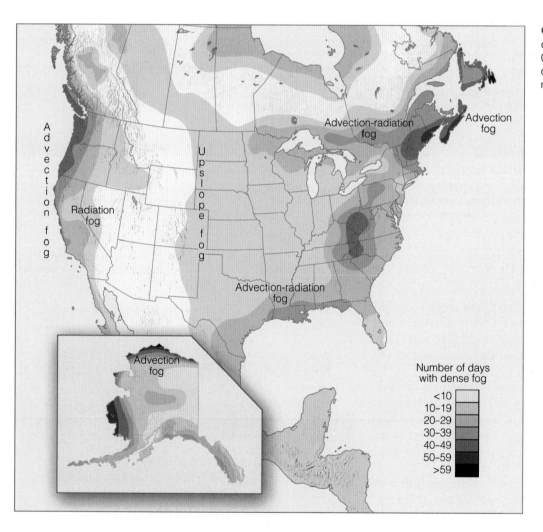

● **FIGURE 5.11** Average annual number of days with dense fog (visibility less than 0.25 miles) across North America. (Dense fog observed in small mountain valleys and on mountain tops is not shown.)

that area.* Fruit and nut trees that have finished growing during the summer and fall require **winter chilling**—a large number of hours with the air temperature below 7°C (45°F) before trees will begin to grow again. The winter fog blocks out the sun and helps keep daytime temperatures quite cool, while keeping nighttime temperatures above freezing: The more continuous the fog, the more effective the chilling. Consequently, the agricultural economy of the region depends heavily on the fog, for without it and the winter chill it stimulates, many of the fruit and nut trees would not grow well. During the spring, when trees are in bloom, fog prevents nighttime air temperatures from dipping to dangerously low readings by trapping infrared energy that is radiated by Earth and releasing latent heat to the air as fog droplets form.

Unfortunately, fog also has many negative aspects. Along a gently sloping highway, the elevated sections may have excellent visibility, while in lower regions—only a few kilometers away—fog may cause poor visibility. Driving from a clear area into the fog on a major freeway can be extremely dangerous. In fact, every winter many people are involved in fog-related auto accidents. These usually occur when a driver enters fog and, because of the reduced visibility, the driver puts on the brakes to slow down. The car behind then slams into the slowed vehicle, causing a chain-reaction accident with many cars involved. One such accident actually occurred near Fresno, California, in February 2002, when 87 vehicles smashed into each other along a stretch of foggy Highway 99. The accident left dozens of people injured, three people dead, and a landscape strewn with cars and trucks twisted into heaps of jagged steel.

Extremely limited visibility exists while driving at night in thick fog with the high-beam lights on. The light scattered back to the driver's eyes from the fog droplets makes it difficult to see very far down the road. However, even in thick fog, there is usually a drier and therefore clearer region extending about 35 cm (14 in.) above the road surface. People who drive a great deal in foggy weather take advantage of this by installing extra head lamps—called *fog lamps*—just above the front bumper. These lights are directed downward into the clear space where they provide improved visibility.

Fog-related problems are not confined to land. Even with sophisticated electronic equipment, dense fog in the open sea hampers navigation. A Swedish liner rammed the luxury liner *Andrea Doria* in thick fog off Nantucket Island on July 25, 1956, causing 52 deaths. On a fog-covered runway in the Canary Islands, two 747 jet airliners collided, taking the lives of more than 570 people on March 27, 1977.

Airports suspend flight operations when fog causes visibility to drop below a prescribed minimum. The resulting delays and cancellations become costly to the airline industry and irritate passengers. With fog-caused problems such as these, it is no wonder that scientists have explored ways to disperse, or at least "thin," fog. *Cold fog*—which forms when the air temperature is below freezing, and most of the fog droplets remain as supercooled liquid—can be cleared by injecting large amounts of dry ice into it.

_____

*For reference, look back at Fig. 5.6 (p. 118), and see how the fog can cover a vast region. Also, note that the fog can last for many days on end.

This approach has been tried at various airports, but as much as several hundred pounds of dry ice are needed to clear a small area. Moreover, this technique does not work in *warm fog*—that is, fog that forms when the air temperature is above freezing. One approach to clearing warm fogs is to warm the air enough so that the fog droplets evaporate and visibility improves. However, tests in the 1950s showed that the procedure cleared the runway only for a short time. Helicopters have also been flown across fog layers so that the helicopter blades produce a turbulent downwash, bringing drier air above the fog into contact with the moist fog layer. This method can work well for shallow radiation fog with a relatively low liquid water content. But many fogs are thick, have a high liquid water content, and form by other means. Given all of these challenges, no single method of fog dispersal has proven versatile and affordable enough to be used widely.

Up to this point, we have looked at the different forms of condensation that occur on or near Earth's surface. In particular, we learned that fog is simply many billions of tiny liquid droplets (or ice crystals) that form near the ground. In the following sections, we will see how these same particles, forming well above the ground, are classified and identified as clouds.

BRIEF REVIEW

Before going on to the section on clouds, here is a brief review of some of the facts and concepts we have covered so far:

- Dew, frost, and frozen dew generally form on clear nights when the temperature of objects on the surface cools below the air's dew-point temperature.

- Visible white frost forms in saturated air when the air temperature is at or below freezing. Under these conditions, water vapor can change directly to ice, in a process called deposition.

- Condensation nuclei act as surfaces on which water vapor condenses. Those nuclei that have an affinity for water vapor are called hygroscopic.

- Fog is a cloud resting on the ground. It can be composed of water droplets, ice crystals, or a combination of both.

- Radiation fog, advection fog, and upslope fog all form as the air cools. The cooling for radiation fog is mainly radiational cooling at Earth's surface; for advection fog, the cooling is mainly warmer air moving over a colder surface; for upslope fog, the cooling occurs as moist air gradually rises and expands along sloping terrain.

- Evaporation (mixing) fog, such as steam fog and frontal fog, forms as water evaporates and mixes with drier air.

Clouds

Clouds are aesthetically appealing and add excitement to the atmosphere. Without them, there would be no rain or snow, thunder or lightning, rainbows or halos. How monotonous it would be if there were only a clear blue sky to look at.

A *cloud* is a visible aggregate of tiny water droplets or ice crystals suspended in the air. Some are found only at high elevations, whereas others nearly touch the ground (or are classified as fog if they do touch the ground). Clouds can be thick or thin, big or little—they exist in a seemingly endless variety of forms. To impose order on this variety, we divide clouds into 10 basic types. With a careful and practiced eye, you can become reasonably proficient in correctly identifying them.

CLASSIFICATION OF CLOUDS Although ancient astronomers named the major stellar constellations about 2000 years ago, clouds were not formally identified and classified until the early nineteenth century. The French naturalist Jean-Baptiste Lamarck (1744–1829) proposed the first system for classifying clouds in 1802; however, his work did not receive wide acclaim. One year later, Luke Howard, an English naturalist, developed a cloud classification system that found general acceptance. In essence, Howard's innovative system employed Latin words to describe clouds as they appear to a ground observer. He named a sheetlike cloud *stratus* (Latin for "layer"); a puffy cloud *cumulus* ("heap"); a wispy cloud *cirrus* ("curl of hair"); and a rain cloud *nimbus* ("violent rain"). In Howard's system, these were the four basic cloud forms. Other clouds could be described by combining the basic types. For example, nimbostratus is a rain cloud that shows layering, whereas cumulonimbus is a rain cloud having pronounced vertical development.

In 1887, Ralph Abercromby and Hugo Hildebrandsson expanded Howard's original system and published a classification system that, with only slight modification, is still in use today. Ten principal cloud forms are divided into four primary cloud groups. Each group is identified by the height of the cloud's base above the surface: high clouds, middle clouds, and low clouds. The fourth group contains clouds showing more vertical than horizontal development. Within each group, cloud types are identified by their appearance. ▼ Table 5.2 lists these four groups and their cloud types.

The approximate base height of each cloud group is given in ▼ Table 5.3. Note that the altitude separating the high and middle cloud groups overlaps and varies with latitude. Large temperature changes cause most of this latitudinal variation. For example, high cirriform clouds are composed almost entirely of ice crystals. In tropical regions, air temperatures low enough to freeze all liquid water usually occur only above 6000 m (about 20,000 ft). In polar regions, however, these same temperatures may be found at altitudes as low as 3000 m (about 10,000 ft). Although you may observe cirrus clouds at 3600 m (about 12,000 ft) over northern Alaska, you will not see them at that elevation above southern Florida.

Clouds cannot be accurately identified strictly on the basis of elevation. Other visual clues are necessary. Some of these are explained in the following section.

CLOUD IDENTIFICATION

High Clouds High clouds in middle and low latitudes generally form above 5000 m (16,000 ft). Because the air at these elevations is quite cold and "dry," high clouds are composed almost exclusively of ice crystals and are also rather thin.* High clouds usually appear white, except near sunrise and sunset, when the unscattered (red, orange, and yellow) components of sunlight are reflected from the underside of the clouds.

The most common high clouds are **cirrus** (Ci), which are thin, wispy clouds blown by high winds into long streamers called *mares' tails*. Notice in ● Fig. 5.12 that they can look like a white, feathery patch with a faint wisp of a tail at one end. Cirrus clouds usually move across the sky from west to east, indicating the prevailing winds at their elevation, and they generally occur during periods of fair, pleasant weather.

Cirrocumulus (Cc) clouds, seen less frequently than cirrus, appear as small, rounded, white puffs that may occur individually

*Small quantities of liquid water in cirrus clouds at temperatures as low as −36°C (−33°F) were discovered during research conducted above Boulder, Colorado.

▼ **TABLE 5.2** **The Four Major Cloud Groups and Their Types**

HIGH CLOUDS	LOW CLOUDS
Cirrus (Ci)	Stratus (St)
Cirrostratus (Cs)	Stratocumulus (Sc)
Cirrocumulus (Cc)	Nimbostratus (Ns)
MIDDLE CLOUDS	CLOUDS WITH VERTICAL DEVELOPMENT
Altostratus (As)	Cumulus (Cu)
Altocumulus (Ac)	Cumulonimbus (Cb)

▼ **TABLE 5.3** **Approximate Height of Cloud Bases Above the Surface for Various Locations**

CLOUD GROUP	TROPICAL REGION	MID-LATITUDE REGION	POLAR REGION
High (Ci, Cs, Cc)	20,000 to 60,000 ft	16,000 to 43,000 ft	10,000 to 26,000 ft
	(6000 to 18,000 m)	(5000 to 13,000 m)	(3000 to 8000 m)
Middle (As, Ac)	6500 to 26,000 ft	6500 to 23,000 ft	6500 to 13,000 ft
	(2000 to 8000 m)	(2000 to 7000 m)	(2000 to 4000 m)
Low (St, Sc, Ns)	surface to 6500 ft	surface to 6500 ft	surface to 6500 ft
	(0 to 2000 m)	(0 to 2000 m)	(0 to 2000 m)

● FIGURE 5.12 Cirrus clouds. Notice the silky "mare's tail" appearance.

● FIGURE 5.14 Cirrostratus clouds. Notice the faint halo encircling the sun. The sun is the bright white area in the center of the circle.

or in long rows (see ● Fig. 5.13). When in rows, the cirrocumulus cloud has a rippling appearance that distinguishes it from the silky look of cirrus and the sheetlike cirrostratus. Cirrocumulus seldom cover more than a small portion of the sky. The dappled cloud elements that reflect the red or yellow light of a setting sun make this one of the most beautiful of all clouds. The small ripples in the cirrocumulus strongly resemble the scales of a fish; hence, the expression *mackerel sky* commonly describes a sky full of cirrocumulus clouds.

The thin, sheetlike, high clouds that often cover the entire sky are **cirrostratus** (Cs) (see ● Fig. 5.14), which are so thin that

the sun and the moon can be clearly seen through them. The ice crystals in these clouds bend the light passing through them and will often produce a *halo*—a ring of light that encircles the sun or the moon. In fact, the veil of cirrostratus may be so thin that a halo is the only clue to its presence. Thick cirrostratus clouds give the sky a glary white appearance and frequently form ahead of an advancing cyclonic storm; hence, they can be used to predict rain or snow within 12 to 24 hours, especially if they are followed by middle-type clouds.

Middle Clouds The middle clouds have bases between about 2000 and 7000 m (6500 and 23,000 ft) in the middle latitudes. These clouds are composed of water droplets and—when the temperature becomes low enough—some ice crystals. Precipitation can form in middle clouds if they become thick enough.

Altocumulus (Ac) clouds are middle clouds that are composed mostly of water droplets and are rarely more than 1 km (about 3300 ft) thick. They appear as gray, puffy masses, sometimes rolled out in parallel waves or bands (see ● Fig. 5.15). Usually, one part of each cloud element is darker than another, which helps to distinguish it from the higher cirrocumulus. Also, the individual puffs of the altocumulus appear larger than those of the cirrocumulus. A layer of altocumulus can sometimes be confused with altostratus; in case of doubt, clouds are called altocumulus if there are rounded masses or rolls present. Altocumulus clouds that look like "little castles" *(castellanus)* in the sky indicate the presence of rising air at cloud level. The appearance of these clouds on a warm, humid summer morning often portends thunderstorms by late afternoon.

Altostratus (As) are gray or blue-gray clouds composed of ice crystals and water droplets. Altostratus clouds often cover the entire sky across an area that extends over many hundreds of square kilometers. In the thinner section of the cloud, the sun (or the moon) may be *dimly visible* as a round disk, as if the sun were shining through ground glass. This appearance is sometimes

● FIGURE 5.13 Cirrocumulus clouds.

● **FIGURE 5.15** Altocumulus clouds. Notice the dark-to-light contrasting patterns that distinguish these clouds from cirrocumulus clouds.

referred to as a "watery sun" (see ● Fig. 5.16). Thick cirrostratus clouds are occasionally confused with thin altostratus clouds. The gray color, height, and dimness of the sun are good clues to identifying an altostratus. The fact that halos occur only with cirriform clouds also helps one to distinguish them. Another way to separate the two is to look at the ground for shadows. If there are none, it is a good bet that the cloud is altostratus because cirrostratus are usually transparent enough to produce them. Altostratus clouds often form ahead of mid-latitude cyclonic storms having widespread and relatively continuous precipitation. If precipitation falls from altostratus, the cloud base usually lowers, and the precipitation is steady and not showery as found with cumuliform clouds. If the precipitation reaches the ground, the cloud is then classified as *nimbostratus*.

Low Clouds   Low clouds, with their bases lying below 2000 m (6500 ft), are almost always composed of water droplets, although in cold weather they may contain ice particles.

Nimbostratus (Ns) are dark-gray, "wet"-looking cloud layers associated with more or less continuously falling rain or snow (see ● Fig. 5.17). The intensity of this precipitation is usually light or moderate—it is never of the heavy, showery variety, unless well-developed cumulus clouds are embedded within the nimbostratus cloud. Precipitation often makes the base of the nimbostratus cloud impossible to identify clearly. The distance from the cloud's base to its top can be over 3 km (10,000 ft). Nimbostratus is easily confused with altostratus. Thin nimbostratus is usually darker gray than thick altostratus, and you normally cannot see the sun or the moon through a layer of nimbostratus. Visibility below a nimbostratus cloud deck is usually quite poor because rain will evaporate and mix with the air in this region. If this air becomes

● **FIGURE 5.16** Altostratus clouds. The appearance of a dimly visible "watery sun" through a deck of light gray clouds is usually a good indication that the clouds are altostratus.

● **FIGURE 5.17** The nimbostratus is the sheetlike cloud from which light rain is falling. The ragged-appearing clouds beneath the nimbostratus are stratus fractus, or scud.

● **FIGURE 5.18** Stratocumulus clouds forming along the south coast of Florida. Notice that the rounded masses are larger than those of altocumulus.

Altocumulus cloud elements will generally be about the size of your thumbnail, whereas stratocumulus will usually be about the size of your fist. Although precipitation rarely falls from stratocumulus, light rain showers or winter snow flurries can occur if the cloud develops vertically into a much thicker cloud with a top colder than about −5°C (23°F).

Stratus (St) is a uniform grayish cloud that often covers the entire sky. It resembles a fog that does not reach the ground (see ● Fig. 5.19). Actually, when a thick fog "lifts," the resulting cloud is a deck of low stratus. Normally, no precipitation falls from stratus, but sometimes it is accompanied by a light mist or drizzle. This cloud commonly occurs over Pacific and Atlantic coastal waters in summer. A thick layer of stratus might be confused with nimbostratus, but the distinction between them can be made by observing the low base of the stratus cloud and remembering that light-to-moderate precipitation occurs with nimbostratus. Moreover, stratus often has a more uniform base than does nimbostratus. Also, a deck of stratus may be confused with a layer of altostratus. However, if you remember that stratus are lower and darker gray and that the sun normally appears "watery" through altostratus, the distinction can be made.

saturated, a lower layer of clouds or fog can form beneath the original cloud base. Because these lower clouds drift rapidly with the wind, they form irregular shreds with a ragged appearance that are called *stratus fractus*, or *scud* (see Fig. 5.17).

Stratocumulus (Sc) are low, lumpy clouds that appear in rows, in patches, or as rounded masses with blue sky visible between individual cloud elements (see ● Fig. 5.18). Often they appear near sunset as the spreading remains of a much larger cumulus cloud. Occasionally, the sun will shine through the cloud breaks, producing bands of light (called *crepuscular rays*) that appear to reach down to the ground. The color of stratocumulus ranges from light to dark gray. It differs from altocumulus in that it has a lower base and larger individual cloud elements. (Compare Fig. 5.15 with Fig. 5.18.) To distinguish between the two, hold your hand at arm's length and point toward one of these clouds.

Clouds with Vertical Development Familiar to almost everyone, the puffy **cumulus** (Cu) cloud takes on a variety of shapes, but most often it looks like a piece of floating cotton with sharp outlines and a flat base (see ● Fig. 5.20). The base appears white to light gray, and, on a humid day, may be only 1000 m (3300 ft) above the ground and a kilometer or so wide. The top of the cloud—often in the form of rounded towers—denotes the limit of rising air and is usually not very high. These clouds can be distinguished from stratocumulus by the fact that cumulus clouds are detached (usually with a great deal of blue sky between each cloud), while stratocumulus usually occur in groups or patches. Also, the cumulus has a dome- or tower-shaped top as opposed to the generally flat tops of the stratocumulus. Cumulus clouds that show only slight vertical growth are called *cumulus humilis* and are associated with fair weather; therefore, we call these clouds *"fair-weather*

● **FIGURE 5.19** A layer of low-lying stratus clouds hides these mountains in Iceland.

● **FIGURE 5.20** Cumulus clouds. Small cumulus clouds such as these are sometimes called *fair weather cumulus,* or *cumulus humilis.*

cumulus." Ragged-edge cumulus clouds that are smaller than cumulus humilis and scattered across the sky are called *cumulus fractus.*

Harmless-looking cumulus often develop on warm summer mornings and, by afternoon, become much larger and more vertically developed. When the growing cumulus resembles a head of cauliflower, it becomes a *cumulus congestus,* or *towering cumulus* (Tcu). Most often, it is a single large cloud, but, occasionally, several grow into each other, forming a line of towering clouds, as shown in ● Fig. 5.21. Precipitation that falls from a cumulus congestus is always showery with frequent changes in intensity.

If a cumulus congestus continues to grow vertically, it develops into a giant **cumulonimbus** (Cb)—a thunderstorm cloud (see ● Fig. 5.22). While its dark base may be no more than 600 m (2000 ft) above Earth's surface, its top can extend upward to the

tropopause, over 12,000 m (39,000 ft) higher. A cumulonimbus can occur as an isolated cloud or as part of a line or "wall" of clouds.

The tremendous amounts of energy released by the condensation of water vapor within a cumulonimbus result in the development of violent updrafts and downdrafts, which can exceed 70 knots. The lower (warmer) part of the cloud is usually composed of only water droplets. Higher up in the cloud, water droplets and ice crystals both abound, while toward the cold top, there are only ice crystals. Swift winds at these higher altitudes can reshape the top of the cloud into a huge flattened anvil* *(cumulonimbus incus)*. These great thunderheads can contain all forms of

---

*An anvil is a heavy block of iron or steel with a smooth, flat top on which metals are shaped by hammering.

● **FIGURE 5.21** Cumulus congestus clouds are frequently called towering cumulus. These clouds are taller than cumulus clouds and are more likely to produce showers. Here a line of cumulus congestus clouds is building along Maryland's eastern shore.

● FIGURE 5.22 A cumulonimbus cloud (thunderstorm). Strong upper-level winds blowing from right to left produce a well-defined anvil. Sunlight scattered by falling ice crystals produces the white (bright) area beneath the anvil. Notice the heavy rain shower falling from the base of the cloud.

precipitation—large raindrops, snowflakes, snow pellets, and sometimes hailstones—all of which can fall to Earth in the form of heavy showers. Lightning, thunder, and even tornadoes are associated with cumulonimbus. (More information on thunderstorms and tornadoes is given in Chapters 14 and 15.)

Cumulus congestus and cumulonimbus frequently look alike, making it difficult to distinguish between them. However, you can usually do so by looking at the top of the cloud. If the sprouting upper part of the cloud is sharply defined and not fibrous, it is usually a cumulus congestus; conversely, if the top of the cloud loses its sharpness and becomes fibrous in texture, it is usually a cumulonimbus. Compare Fig. 5.21 with Fig. 5.22. The weather associated with these clouds also differs: Lightning, thunder, and large hail typically occur with cumulonimbus.

So far, we have discussed the 10 primary cloud forms, summarized pictorially in ● Fig. 5.23. This figure, along with the

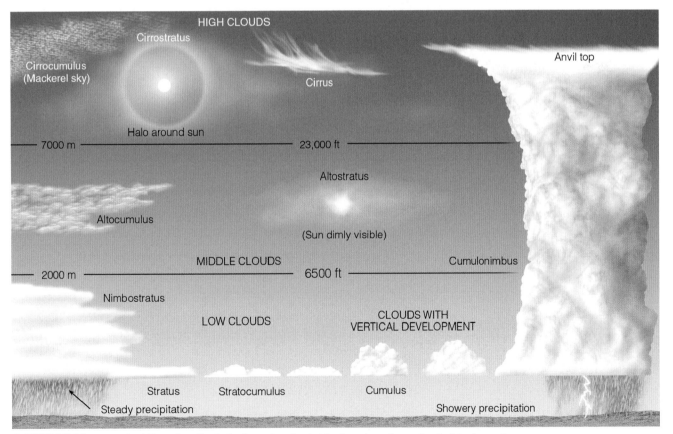

● FIGURE 5.23 A generalized illustration of basic cloud types based on height above the surface and vertical development.

cloud photographs and descriptions (and the cloud chart at the back of the book), should help you identify the more common cloud forms. Don't worry if you find it hard to estimate cloud heights. This is a difficult procedure, requiring much practice. You can use local objects (hills, mountains, and tall buildings) of known height as references on which to base your height estimates.

To better describe a cloud's shape and form, a number of descriptive words may be used in conjunction with its name. We mentioned a few in the previous section; for example, a stratus cloud with a ragged appearance is a stratus fractus, and a cumulus cloud with marked vertical growth is a cumulus congestus. ▼ Table 5.4 lists some of the more common terms that are used in cloud identification.

SOME UNUSUAL CLOUDS Although the 10 basic cloud forms are the most frequently seen, there are some unusual clouds that deserve mentioning. For example, moist air crossing a mountain barrier often forms into waves. The clouds that form in the wave crest usually have a lens shape and are, therefore, called **lenticular clouds** (see ● Fig. 5.24). Frequently, they form one above the other like a stack of pancakes, and at a distance they can resemble a hovering spacecraft. Hence, it is no wonder that lenticular clouds may be reported as UFOs, especially in areas where these clouds are uncommon. When a cloud forms over and extends downwind of an isolated mountain peak, as shown in ● Fig. 5.25, it is called a **banner cloud**.

Similar to the lenticular is the *cap cloud,* or **pileus**, that usually resembles a silken scarf capping the top of a sprouting cumulus cloud (see ● Fig. 5.26). Pileus clouds form when moist winds are deflected up and over the top of a building cumulus congestus or cumulonimbus. If the air flowing over the top of the cloud condenses, a pileus often forms.

Most clouds form in rising air, but the mammatus forms in sinking air. **Mammatus clouds** derive their name from their appearance—baglike sacs that hang beneath the cloud and resemble a cow's udder (see ● Fig. 5.27). Although mammatus most frequently forms on the underside of cumulonimbus, it may develop beneath cirrocumulus, altostratus, altocumulus, and stratocumulus. For mammatus to form, the sinking air must

▼ **TABLE 5.4** **Common Terms Used in Identifying Clouds**

TERM	LATIN ROOT AND MEANING	DESCRIPTION
Lenticularis	(*lens, lenticula,* lentil)	Clouds having the shape of a lens or an almond, often elongated and usually with well-defined outlines. This term applies mainly to cirrocumulus, altocumulus, and stratocumulus
Fractus	(*frangere,* to break or fracture)	Clouds that have a ragged or torn appearance; applies only to stratus and cumulus
Humilis	(*humilis,* of small size)	Cumulus clouds with generally flattened bases and slight vertical growth
Congestus	(*congerere,* to bring together; to pile up)	Cumulus clouds of great vertical extent that from a distance may resemble a head of cauliflower
Calvus	(*calvus,* bald)	Cumulonimbus in which at least some of the upper part is beginning to lose its cumuliform outline
Capillatus	(*capillus,* hair; having hair)	Cumulonimbus characterized by the presence in the upper part of cirriform clouds with fibrous or striated structure
Undulatus	(*unda,* wave; having waves)	Clouds in patches, sheets, or layers showing undulations
Translucidus	(*translucere,* to shine through; transparent)	Clouds that cover a large part of the sky and are sufficiently translucent to reveal the position of the sun or the moon
Incus	(*incus,* anvil)	The smooth cirriform mass of cloud in the upper part of a cumulonimbus that is anvil shaped
Mammatus	(*mamma,* mammary)	Baglike clouds that hang like a cow's udder on the underside of a cloud; may occur with cirrus, altocumulus, altostratus, stratocumulus, and cumulonimbus
Pileus	(*pileus,* cap)	A cloud in the form of a cap or hood above or attached to the upper part of a cumuliform cloud, particularly during its developing stage
Castellanus	(*castellum,* a castle)	Clouds that show vertical development and produce towerlike extensions, often in the shape of small castles

● **FIGURE 5.24** Lenticular clouds forming downwind of the Front Range of the Rocky Mountains, near Boulder, Colorado.

be cooler than the air around it and have a high liquid water or ice content. As saturated air sinks, it warms, but the warming is retarded because of the heat taken from the air to evaporate the liquid or melt ice particles. If the sinking air remains saturated and cooler than the air around it, the sinking air can extend below the cloud base, appearing as the rounded masses called mammatus clouds.

Jet aircraft flying at high altitudes often produce a cirrus-like trail of condensed vapor called a *condensation trail* or **contrail** (see ● Fig. 5.28). Contrails evaporate rapidly when the relative humidity of the surrounding air is low. If the relative humidity is high, however, contrails may persist for many hours. Contrails can form directly from the water vapor added to the air from engine exhaust. In this case, there must be sufficient mixing of the hot exhaust gases with the cold air to produce saturation. The release of particles in the exhaust can even provide nuclei on which ice crystals form. They may also form by a cooling process. The reduced pressure produced by air flowing over the wing causes the air to cool. This cooling can supersaturate the air, producing an *aerodynamic contrail*. This type of trail usually disappears quickly in the turbulent wake of the aircraft.

Aside from the cumulonimbus cloud that sometimes penetrates into the stratosphere, all of the clouds described so far are observed in the lower atmosphere—in the troposphere. Occasionally, however, clouds can be seen above the troposphere. For example, soft, pearly looking clouds called **nacreous** or **polar stratospheric clouds** (also called *mother-of-pearl clouds*) form in the stratosphere at altitudes above 30 km (see ● Fig. 5.29). They are best viewed in polar latitudes during the winter months when the sun, being just below the horizon, is able to illuminate them because of their high altitude. Their exact composition is not known, although they appear to be composed of water in either solid or liquid (supercooled) form.

Wavy bluish-white clouds, so thin that stars shine brightly through them, can sometimes develop in the upper mesosphere, at altitudes above 75 km (46 mi). These clouds are at such a high altitude that they appear bright against a dark background. For this reason, they are called **noctilucent clouds**, meaning "luminous

● **FIGURE 5.25** The cloud forming over and downwind of Mount Rainier is called a banner cloud.

● **FIGURE 5.26** A pileus cloud forming above a developing cumulus cloud.

● **FIGURE 5.27** Mammatus clouds forming beneath a thunderstorm anvil.

● **FIGURE 5.28** A contrail forming behind a jet aircraft.

● **FIGURE 5.29** The clouds in this photograph are nacreous or polar stratospheric clouds. They form in the stratosphere and are most easily seen at high latitudes.

night clouds" (see ● Fig. 5.30). They are best seen at twilight, during the summer at latitudes poleward of 50°, although they have been observed in recent years in Utah and Colorado. Studies reveal that these clouds are composed of tiny ice crystals. The

water to make the ice may originate in meteoroids that disintegrate when entering the upper atmosphere or from the chemical breakdown of methane gas at high levels in the atmosphere.

Our system of classifying clouds may never be completely finalized. One intriguing formation, *asperitas*, has only recently been recognized as a specific cloud type (see Weather Watch on p. 134 and ● Fig. 5.31).

CLOUD OBSERVATIONS

Determining Sky Conditions Often, a daily weather forecast will include a phrase such as, "overcast skies with clouds becoming scattered by evening." To the average person, this means that the cloudiness will diminish, but to the meteorologist, the terms *overcast* and *scattered* have a more specific meaning. In meteorology, descriptions of sky conditions are defined by the fraction of sky covered by clouds. A *clear* sky, for example, is one where no clouds are present.* When there are between one-eighth and two-eighths clouds covering the sky, there are a *few* clouds present. When cloudiness increases to between three-eighths and four-eighths, the sky is described as being *scattered* with clouds. "Partly cloudy" also describes these sky conditions. Clouds covering between five-eighths and seven-eighths of the sky denote a sky with *broken* clouds ("mostly cloudy"), and *overcast* conditions exist when the sky is covered (eight-eighths) with clouds. ▼ Table 5.5 presents a summary of sky cover conditions.

Observing sky conditions far away can sometimes fool even the trained observer. A broken cloud deck near the horizon usually appears as overcast because the open spaces between the clouds are less visible at a distance. Therefore, cloudiness is usually overestimated when clouds are near the horizon. Viewed from afar, clouds not normally associated with precipitation can appear darker and thicker than they actually are. The reason for

*In automated station usage, the phrase "clear sky" means that no clouds are reported whose bases are at or below 12,000 ft.

● **FIGURE 5.30** The wavy clouds in this photograph are noctilucent clouds. They are usually observed at high latitudes, at altitudes between 75 and 90 km above Earth's surface.

Description	OBSERVATION		Meaning
	ASOS*	Human	
Clear (CLR or SKC)	0 to 5%	0	No clouds
Few	>5 to ≤25%	0 to ⅜	Few clouds visible
Scattered (SCT)	>25 to ≤50%	⅜ to ½	Partly cloudy
Broken (BKN)	>50 to ≤87%	⅝ to ⅞	Mostly cloudy
Overcast (OVC)	>87 to 100%	⅞	Sky is covered by clouds
Sky obscured	—	—	Sky is hidden by surface-based phenomena, such as fog, blowing snow, or smoke, rather than by cloud cover

*Automated Surface Observing System. Symbol > means greater than; < means less than; ≥ means equal to or greater than. ASOS does not report clouds above 12,000 ft.

this observation is that light from a distant cloud travels through more atmosphere and is more attenuated than the light from the same type of cloud closer to the observer. (Information on measuring the height of cloud bases is given in Focus section 5.3.)

Up to this point, we have seen how clouds look from the ground. We will now look at clouds from a different vantage point—the satellite view.

SATELLITE OBSERVATIONS   The weather satellite is a cloud-observing platform in Earth's orbit. It provides extremely valuable images of clouds over areas where there are no ground-based observations. Because water covers over 70 percent of Earth's surface, there are vast regions where few (if any) surface cloud observations are made. Before weather satellites were available, tropical cyclones, such as hurricanes and typhoons, often went undetected until they moved dangerously near inhabited areas. Residents of the regions affected had little advance warning. Today, satellites spot these storms while they are still far out in the ocean and track them accurately.

Two primary types of weather satellites are used for viewing clouds. The first type are called **geostationary satellites** (or *geosynchronous satellites*) because they orbit the equator at the same rate Earth spins and, hence, remain above a fixed spot on Earth's surface, at an altitude of nearly 36,000 km (22,300 mi) (see ● Fig. 5.32). This positioning allows continuous monitoring of a specific region.

Geostationary satellites are also important because they use a real-time data system, meaning that the satellites transmit images to the receiving system on the ground as soon as the image is taken. Successive cloud images from these satellites can be put into a time-lapse movie sequence to show the cloud movement, dissipation, or development associated with weather fronts and storms. This information is a great help in monitoring and forecasting the progress of large weather systems. Wind directions and speeds at various levels may also be approximated by monitoring cloud movement with the geostationary satellite.

Complementing the geostationary satellites are **polar-orbiting satellites**, which closely parallel Earth's meridian lines. These satellites pass over the north and south polar regions on each revolution. As Earth rotates to the east beneath the satellite, each pass monitors an area to the west of the previous pass (see ● Fig. 5.33). Eventually, the satellite covers the entire Earth.

Courtesy of Gary McArthur

● **FIGURE 5.31** Asperitas over Burnie, Tasmania, Australia.

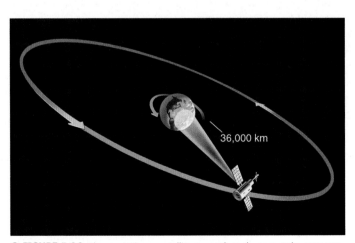

36,000 km

● **FIGURE 5.32** The geostationary satellite moves through space at the same rate that Earth rotates, so it remains above a fixed spot on the equator and monitors one area constantly.

Measuring Cloud Ceilings

In addition to knowing about sky conditions, it is usually important to have a good estimate of the height of cloud bases. Aircraft could not operate safely without accurate cloud height information, particularly at lower elevations.

The term *ceiling* is defined as the height of the lowest layer of clouds above the surface that are either broken or overcast, but not thin. Direct information on cloud height can be obtained from pilots who report the altitude at which they encounter the ceiling. Less directly, *ceiling balloons* can measure the height of clouds. A small balloon filled with a known amount of hydrogen or helium rises at a fairly constant and known rate. The ceiling is determined by measuring the time required for the balloon to enter the lowest cloud layer.* Ceiling balloon observations can be made at night simply by attaching a small battery-operated light to the balloon.

For many years, the *optical drum ceilometer* (or *rotating beam ceilometer*) provided information on cloud ceiling, especially at airports. This instrument consists of a ground-based projector that rotates vertically from

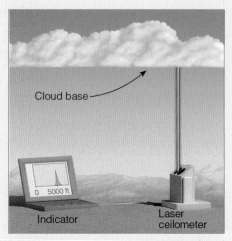

● **FIGURE 4** The laser ceilometer sends pulses of infrared radiation up to the cloud. Part of this beam is reflected back to the ceilometer. The interval of time between pulse transmission and return is a measure of cloud height, as displayed on the indicator screen.

horizon to horizon. As it rotates, it sends out a powerful light beam that moves along the base of the cloud. A light-sensitive detector, some known distance from the projector, points upward and picks up the light from the cloud base. By knowing the projector angle and its distance from the detector, the cloud height is determined mathematically.

Most of the optical drum ceilometers have been phased out and replaced with

laser ceilometers. The laser ceilometer is a fixed-beam type whose transmitter and receiver point straight up at the cloud base (see ● Fig. 4). Short, intense pulses of infrared radiation from the transmitter strike the cloud base, and a portion of this radiation is reflected back to the receiver. The time interval between pulse transmission and its return from the cloud determines the cloud-base height.

The Automated Surface Observing System (ASOS) uses a laser ceilometer to measure cloud height. The ceilometer measures the cloud height and then infers the amount of cloud cover by averaging the amount of clouds that have passed over the sensor for a duration of 30 minutes. The ASOS laser ceilometer is unable to measure clouds that are not above the sensor. To help remedy this situation, a second laser ceilometer may be located nearby. Another limitation of the ASOS ceilometer is that it does not report clouds above 12,000 ft.* A new laser ceilometer that can provide cloud height information up to 25,000 ft is being developed.

---

*For example, if the balloon rises 125 m (about 400 ft) each minute, and it takes three minutes to enter a broken layer of stratocumulus, the ceiling would be 375 m (about 1200 ft).

---

*The latest geostationary satellites above North America are equipped to measure cloud heights above 12,000 ft over ASOS stations.

Polar-orbiting satellites have the advantage of scanning clouds directly beneath them as they move. They can gather sharp images from polar regions, where images from a geostationary satellite are distorted because of the low angle at which the satellite "sees" this region. Polar orbiters circle Earth at a much lower altitude (about 850 km or 530 mi) than geostationary satellites. The lower altitude allows them to provide more detailed images of clouds and storms. However, because polar orbiters are constantly moving, they cannot produce images of a given area as frequently and consistently as geostationary satellites do.

Continuously improved detection devices make weather observation by satellites more versatile than ever. Early satellites, such as *TIROS I*, launched in 1960, used television cameras to photograph clouds. Contemporary satellites use radiometers, which can observe clouds during both day and night by detecting radiation that emanates from the top of the clouds. Additionally, satellites have the capacity to obtain vertical profiles of atmospheric temperature and moisture by detecting emitted radiation from atmospheric gases, such as water vapor. In modern satellites,

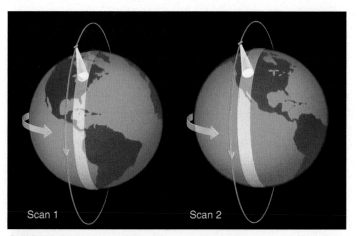

● **FIGURE 5.33** Polar-orbiting satellites scan from north to south, and on each successive orbit the satellite scans an area farther to the west.

GOES-16: New Windows on the Atmosphere

A new era in monitoring weather across North America began on November 19, 2016, with the launch of a satellite from Cape Canaveral, Florida. This satellite is GOES-16, the first entry in the fourth generation of the Geostationary Operational Environmental Satellite. Forecasters will benefit greatly from GOES-16 as well as its three sibling satellites that are in the works over the next few years.

Since 1975, GOES has been the workhorse of the round-the-clock United States weather satellite fleet. Each GOES satellite collects images of the atmosphere from fixed points about 22,000 mi (35,400 km) high. One GOES satellite is typically positioned over the equator at about 75°W longitude, allowing it to sample the United States from an eastern perspective. Another GOES satellite over the equator is perched at around 135°W, giving it a western perspective on the United States. Together, the two units can view nearly half of the globe, providing a "full disk" perspective. A third satellite is typically stationed elsewhere, ready to pinch-hit as needed, while another satellite may remain in storage, available to be launched if major problems develop with any of the units in space.

GOES images provide an immediate boost to forecasters who need to identify where dangerous weather may be brewing. The satellite data are also fed into computer forecast models, giving them a more complete starting point when they project our weather days into the future.

One of the big improvements of the new generation over prior GOES satellites is speed. Earlier GOES units collected images of the contiguous United States every 30 minutes. The new satellites will gather such data every 5 minutes—and as often as every 30 seconds during periods of severe weather. Because a tornado or a burst of damaging wind may come and go in less than a half hour, the new rapid-response images will be enormously helpful in giving forecasters advance notice of these fast-changing conditions. The new GOES satellites also provide crisper images than their predecessors. The spacing between data points has been cut in half in both the east-west and north-south directions. The closest spacing is now 500 m (about 0.3 mi), which is enough to distinguish features such as cloud elements atop severe thunderstorms.

Along with its sharper, more frequent images, the new GOES satellite can view the atmosphere in more dimensions with the help of its Advanced Baseline Imager. The ABI senses radiation at 16 different wavelengths: two visible, four near-infrared, and ten infrared (see ● Fig. 5). Each wavelength provides a different type of detail on the evolution of clouds and storms.

GOES-16 is the world's first satellite to include a Geostationary Lightning Mapper (GLM), another key advance. The GLM can detect lightning within storms as well as lightning striking the ground. This will be especially helpful over the ocean, where ground-based lightning detection networks cannot be deployed. For the first time, it will be possible to keep close tabs on lightning within a developing hurricane, where bursts of thunderstorm activity are often associated with rapid strengthening.

Several more launches of the next-generation GOES satellites are scheduled to take place through the mid-2020s. If all goes according to plan, the new GOES satellites will be providing forecasters with invaluable data until at least the mid-2030s.

● **FIGURE 5** A collection of four images gathered by the GOES-16 satellite showing the development of a mid-latitude cyclonic storm system across the south central United States at 12:07 pm CST on January 15, 2017. Each of the four images was measured by sensing radiation at a different wavelength.

a special type of advanced radiometer (called an *imager*) provides satellite images with much better resolution than did previous imagers. The latest U.S. *Geostationary Operational Environmental Satellite* (*GOES*) series debuted in 2016–2017. This series includes many enhancements, including better resolution in both space and time and the ability to map lightning activity from space. For more on the new *GOES* series, read Focus section 5.4.

Information on cloud thickness and height can be deduced from satellite images. Visible images show the sunlight reflected

WEATHER WATCH

Cloud lovers have a newly identified type to watch for. Called *asperitas* (see Fig. 5.31, p. 133), this dramatic cloud appears as a rolling, very turbulent, choppy wave cloud that looks very ominous, but doesn't produce stormy weather. In 2017, the World Meteorological Organization (WMO) added it to the International Cloud Atlas, which made asperitas the first new cloud formation to be recognized by the WMO since 1951.

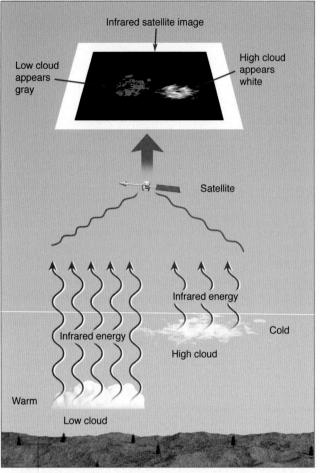

● **FIGURE 5.34** Generally, the lower the cloud, the warmer its top. Warm objects emit more infrared energy than do cold objects. Thus, an infrared satellite picture can distinguish warm, low (gray) clouds from cold, high (white) clouds.

● **FIGURE 5.35** A 3-D visible satellite image of Hurricane Rita over the Gulf of Mexico on September 21, 2005.

● **FIGURE 5.36** A visible satellite image of the eastern Pacific Ocean taken at just about the same time on the same day as the image in Fig. 5.37. Notice that the clouds in this visible image appear white. Superimposed on the image are the weather fronts of the mid-latitude cyclonic storm.

from a cloud's upper surface. Because thick clouds have a higher albedo (reflectivity) than thin clouds, they appear brighter on a visible satellite image. However, middle and low clouds have just about the same albedo, so it is difficult to distinguish among them simply by viewing them in visible light. To make this distinction, *infrared cloud images* are used. Such pictures produce a better image of the actual radiating surface because they do not show the strong visible reflected light. Because warm objects radiate more energy than cold objects, high-temperature regions can be artificially made to appear darker on an infrared image. Because the tops of low clouds are warmer than those of high clouds, cloud observations made in the infrared can distinguish between warm low clouds (dark) and cold high clouds (light) (see ● Fig. 5.34). Moreover, cloud temperatures can be converted by a computer into a three-dimensional (3-D) image of the cloud. These are the 3-D cloud photos presented on television by many weathercasters (see ● Fig. 5.35).

● Figure 5.36 shows a visible satellite image (from a geostationary satellite) of a mid-latitude cyclonic storm system in the eastern Pacific. Notice that all of the clouds in the image appear white. However, in the infrared image (see ● Fig. 5.37), taken on the same day (and just about the same time), the clouds appear to have many shades of gray. In the visible image (Fig. 5.36), the clouds covering part of Oregon and northern California appear relatively thin compared to the thicker, bright clouds to the west. Furthermore, these thin clouds must be high because they also appear bright in the infrared image (Fig. 5.37). Along the elongated cloud band, associated with the cyclonic

● **FIGURE 5.37** Infrared satellite image of the eastern Pacific Ocean taken at just about the same time on the same day as the image in Fig. 5.36. Notice that the low clouds in this infrared image appear in various shades of gray.

● **FIGURE 5.38** An enhanced infrared image of the eastern Pacific Ocean taken on the same day as the images shown in Figs. 5.36 and 5.37.

storm and the occluded front and cold front extending from it, the clouds appear white and bright in both images, indicating a zone of thick, heavy clouds. Behind the front, the lumpy clouds are probably cumulus because they appear gray in the infrared image, indicating that their tops are low and relatively warm.

When temperature differences are small, it is difficult to directly identify significant cloud and surface features on an infrared image. Some way must be found to increase the contrast between features and their backgrounds. This can be accomplished by a process called *computer enhancement*. Certain temperature ranges in the infrared image are assigned specific shades of gray—grading from black to white. Normally, clouds with cold tops and those tops near freezing are assigned the darkest gray color.

● Figure 5.38 is an *enhanced infrared image* for the same day as shown in Figs. 5.36 and 5.37. Often in this type of image, dark blue or red is assigned to clouds with the coldest (highest) tops to distinguish them from lower-topped clouds. Hence, the dark red areas embedded along the front in Fig. 5.38 represent the region where the coldest and, therefore, highest and thickest clouds are found. It is here where the stormiest weather is probably occurring. Also notice that, near the southern tip of the image, the dark

red blotches surrounded by areas of white are thunderstorms that have developed over warm tropical waters. They show up clearly as thick white clouds in both the visible and infrared images. By examining the movement of these clouds on successive satellite images, forecasters can predict the arrival of clouds and storms, as well as the passage of weather fronts.

In regions where there are no clouds, it is difficult to observe the movement of the air. To help with this situation, geostationary satellites are equipped with water vapor sensors that can profile the distribution of atmospheric water vapor in the middle and upper troposphere (see ● Fig. 5.39). In time-lapse films, the swirling patterns of moisture clearly show wet regions and dry regions, as well as middle tropospheric swirling wind patterns and jet streams.

Specialized satellites have gathered data on clouds and precipitation for more than 20 years. From 1997 to 2015, the long-lived *TRMM* (*Tropical Rainfall Measuring Mission*) satellite provided information on clouds and precipitation from about 35°N to 35°S. A joint venture of NASA and the Japan Aerospace Exploration Agency (JAXA), this satellite orbited Earth at an altitude of about 400 km (250 mi), sensing individual cloud features as small as

Satellites Do More Than Observe Clouds

The use of satellites to monitor weather is not restricted to observing clouds. For example, there are satellites that relay data communications and television signals, and provide military surveillance. Moreover, satellites measure radiation from Earth's surface and atmosphere, giving us information about the Earth-atmosphere energy balance, discussed in Chapter 2. The infrared radiation measurements, obtained by an atmospheric *sounder,* are transformed into vertical profiles of temperature and moisture, which are fed into National Weather Service computer forecast models.

Radiation intensities from the ocean surface are translated into sea-surface temperature readings (see ● Fig. 6). This information is valuable to the fishing industry, as well as to the meteorologist. In fact, the *Tropical Rainfall Measuring Mission (TRMM)* satellite obtains sea-surface temperatures with a microwave scanner, even through clouds and atmospheric particles.

Satellites also monitor the amount of snow cover in winter, the extent of ice fields in the Arctic and Antarctic, the movement of large icebergs that drift into shipping lanes, and the height of the ocean's surface. One polar-orbiting satellite actually carries equipment that can detect faint distress signals anywhere on the globe, and relay them to rescue forces on the ground.

Infrared sensors on polar-orbiting satellites are able to assess conditions of crops, areas of deforestation, regions of extensive drought, and any changes in salt content in the upper levels of the ocean. Satellites are also able to detect volcanic eruptions and follow the movement of ash clouds. The *Global Positioning System (GPS)* consists of 24 polar-orbiting satellites that transmit radio signals to ground receivers, which then use the signals for navigation and relative positioning on earth. The signal the satellites send to Earth is slowed and bent by variations in atmospheric density. Because of this effect, a GPS receiver can be used to estimate the atmosphere's *precipitable water vapor* (the total atmospheric water vapor contained in a vertical column of air). Networks of GPS receivers, such as the *Constellation Observing System for Meteorology, Ionosphere, and Climate (COSMIC),* can be launched into orbit to monitor upper-level temperature, water vapor, and pressure around the globe.

Geostationary satellites, such as *GOES,* are equipped with systems that receive environmental information from remote data-collection platforms on the surface. These platforms include instrumented buoys, river gauges, automatic weather stations, seismic and tsunami ("tidal" wave) stations, and ships. This information is transmitted to the satellite, which relays it to a central receiving station.

Normally, a network of five geostationary satellites positioned over the equator gives nearly complete global coverage from about latitude 60°N to 60°S. Along with monitoring clouds and the atmosphere, the latest *GOES* series provides forecasters and researchers with data from Doppler radars and the network of automated surface-observing stations. Geostationary satellites detect pollution and haze, and provide accurate cloud-height measurements during the day. They even have the capacity to monitor the seasonal and daily trend in atmospheric ozone.

Satellites specifically designed to monitor the natural resources of Earth *(LandSat)* circle Earth 14 times a day in a near-polar circular orbit that stays within sunlight. Photographs taken in several wavelength bands provide valuable information about our planet's geology, hydrology, oceanography, and ecology. *LandSat* also collects data transmitted from remote ground stations in North America. These stations monitor a variety of environmental data, with water quality, rainfall amount, and snow depth of particular interest to the meteorologist and hydrologist.

Satellite information is not confined to the lower atmosphere. There are satellites that monitor the concentrations of ozone, air temperature, and winds in the upper atmosphere. And both geostationary and polar orbiting satellites carry instruments that monitor solar activity. Even with all this information available, there are more sophisticated satellites on the drawing board that will provide more and improved data in the future.

● **FIGURE 6** Sea-surface temperatures for July 1, 2014. Temperatures are derived mainly from satellites, but temperature information also comes from buoys and ships.

● **FIGURE 5.39** Infrared water vapor image. The darker areas represent dry air aloft; the brighter the gray, the more moist the air in the middle or upper troposphere. Bright white areas represent dense cirrus clouds or the tops of thunderstorms. The area in color represents the coldest cloud tops. The swirl of moisture off the West Coast represents a well-developed mid-latitude cyclonic storm.

● **FIGURE 5.41** *(Top)* Visible satellite image of Super Typhoon Choi-Wan over the tropical eastern Pacific Ocean on September 15, 2009. *(Bottom) CloudSat* vertical radar profile through Super Typhoon Choi-Wan. The location of the profile is shown by the red line in the top view.

● **FIGURE 5.40** Precipitation in Hurricane Arthur as sensed by the *Global Positioning Mission* (*GPM*) satellite on July 3, 2014, off the South Carolina coast. Colors on the surface ranging from light green to dark red to purple indicate areas ranging from low to high rainfall. The violet areas aloft indicate frozen precipitation.

2.4 km (1.5 mi) in diameter. *TRMM* gathered 3-D images of clouds and storms, along with the intensity and distribution of precipitation and data on Earth's energy budget and lightning discharges in storms. *TRMM* has been succeeded by another NASA/JAXA project called the *Global Precipitation Mission* (*GPM*), whose core observatory was launched in 2014. *GPM* covers a much broader swath than *TRMM*—from about 65°S to 65°N—and it includes advanced sensors that can distinguish precipitation intensity and type as well as cloud characteristics (see ● Fig. 5.40).

Another specialized satellite has also provided enhanced detail on clouds and precipitation. Launched in 2006, the satellite *CloudSat* circles Earth in an orbit about 700 km above the surface. Onboard *CloudSat*, a very sensitive radar (called Cloud Profiling Radar [CPR]) uses microwave radiation to peer into a cloud and unveil its very fine structures, including the altitude of the cloud's top and base, its thickness, optical properties, the abundance of liquid and ice particles, along with the intensity of precipitation inside the cloud. *CloudSat* provides this information in a vertical view, as shown in ● Fig. 5.41. Such vertical profiling of a cloud's makeup will hopefully provide scientists with a better understanding of precipitation processes that go on inside the cloud and the role that clouds play in Earth's global climate system.

At this point, it should be apparent that today's satellites do much more than observe clouds. More information on satellites and the information they provide is given in Focus section 5.5.

SUMMARY

In this chapter, we examined the different forms of condensation. We saw that dew forms when the air temperature cools to the dew point in a shallow layer of air near the surface. If the dew should freeze, it produces tiny beads of ice called frozen dew. Frost forms when the air cools to a dew point that is at freezing or below.

As the air cools in a deeper layer near the surface, the relative humidity increases and water vapor begins to condense on hygroscopic condensation nuclei, forming wet haze. As the relative humidity approaches 100 percent, condensation occurs on most nuclei, and the air becomes filled with tiny liquid droplets (or ice crystals) called fog.

Fog forms in two primary ways: cooling of air and evaporating and mixing water vapor into the air. Radiation fog, advection fog, and upslope fog form by the cooling of air, while steam fog and frontal fog are two forms of evaporation (mixing) fog. Although fog has some beneficial effects—providing winter chilling for fruit trees and water for thirsty redwoods—in many places it is a nuisance, for it disrupts air traffic and is the primary cause of a number of auto accidents.

Condensation above Earth's surface produces clouds. When clouds are classified according to their height and physical appearance, they are divided into four main groups: high, middle, low, and clouds with vertical development. Given that each cloud has physical characteristics that distinguish it from all the others, careful cloud observations normally lead to correct identification.

Satellites enable scientists to obtain a bird's-eye view of clouds on a global scale. Polar-orbiting satellites obtain data covering Earth from pole to pole, while geostationary satellites located above the equator continuously monitor a desired portion of Earth. Both types of satellites use radiometers (imagers) that detect emitted radiation. As a consequence, clouds can be observed both day and night.

Visible satellite images, which show sunlight reflected from a cloud's upper surface, can distinguish thick clouds from thin clouds. Infrared images show an image of the cloud's radiating top and can distinguish low clouds from high clouds. To increase the contrast between cloud features, infrared photographs are enhanced. Specialized instruments aboard satellites can be used to analyze cloud characteristics and precipitation.

Satellites do a great deal more than simply photograph clouds. They provide us with a wealth of physical information about Earth and the atmosphere.

KEY TERMS

The following terms are listed (with corresponding page number) in the order they appear in the text. Define each. Doing so will aid you in reviewing the material covered in this chapter.

dew, 114
frozen dew, 114
frost, 114
condensation nuclei, 115
hygroscopic nuclei, 115
hydrophobic nuclei, 115
haze, 115
fog, 116
acid fog, 116
radiation (ground) fog, 117
advection fog, 118
advection-radiation fog, 119
upslope fog, 119
evaporation (mixing) fog, 120
steam fog, 120
frontal fog, 122
winter chilling, 123
cirrus, 124
cirrocumulus, 124
cirrostratus, 125
altocumulus, 125
altostratus, 125
nimbostratus, 126
stratocumulus, 127
stratus, 127
cumulus, 127
cumulonimbus, 128
lenticular clouds, 130
banner cloud, 130
pileus clouds, 130
mammatus clouds, 130
contrail, 131
nacreous clouds, 131
polar stratospheric clouds, 131
noctilucent clouds, 131
geostationary satellites, 133
polar-orbiting satellites, 133

QUESTIONS FOR REVIEW

1. Explain how dew, frozen dew, and visible frost form.
2. Distinguish among dry haze, wet haze, and fog.
3. How can fog form when the air's relative humidity is less than 100 percent?
4. Name and describe four types of fog.
5. Why do ground fogs usually "burn off" by early afternoon?
6. List as many positive consequences of fog as you can.
7. List and describe three methods of fog dispersal.
8. What atmospheric conditions are necessary for the development of advection fog?
9. How does evaporation (mixing) fog form?
10. Clouds are most generally classified by height above Earth's surface. List the major height categories and the cloud types associated with each.
11. List at least two distinguishable characteristics of each of the 10 basic clouds.
12. Why are high clouds normally thin? Why are they composed almost entirely of ice crystals?
13. How can you distinguish altostratus from cirrostratus?
14. Which clouds are normally associated with each of the following characteristics?
    (a) lightning
    (b) heavy rain showers
    (c) mackerel sky
    (d) mares' tails
    (e) halos
    (f) light continuous rain or snow
    (g) hailstones
    (h) anvil top
15. How do geostationary satellites differ from polar-orbiting satellites?

16. Explain why visible and infrared images can be used to distinguish:
    (a) high clouds from low clouds
    (b) thick clouds from thin clouds
17. Why are infrared images enhanced?
18. Name two clouds that form above the troposphere.
19. List and explain the various types of environmental information obtained from satellites.

QUESTIONS FOR THOUGHT

1. Explain the reasoning behind the wintertime expression, "Clear moon, frost soon."
2. Explain why icebergs are frequently surrounded by fog.
3. During a summer visit to New Orleans, you stay in an air-conditioned motel. One afternoon, you put on your sunglasses, step outside, and within no time your glasses are "fogged up." Explain what has caused this.
4. While driving from cold air (well below freezing) into much warmer air (well above freezing), frost forms on the windshield of the car. Does the frost form on the inside or outside of the windshield? How can the frost form when the air is so warm?
5. Why are extremely clean and extremely dirty atmospheres undesirable?
6. Why do relative humidities seldom reach 100 percent in polluted air?
7. Why are advection fogs rare over tropical water?
8. A January snowfall covers central Arkansas with 5 inches of snow. The following day, a south wind brings heavy fog to this region. Explain what has apparently happened.
9. If all fog droplets gradually settle earthward, explain how fog can last (without disappearing) for many days at a time.
10. Explain why steam fog is more likely to form during the autumn, and advection fog in early spring, near the shore of an extremely large lake.
11. The air temperature during the night cools to the dew point in a deep layer, producing fog. Before the fog formed, the air temperature cooled each hour about 2°C. After the fog formed, the air temperature cooled by only 0.5°C each hour. Give *two* reasons why the air cooled more slowly after the fog formed.
12. On a winter night, the air temperature cooled to the dew point and fog formed. Before the formation of fog, the dew point remained almost constant. After the fog formed, the dew point began to decrease. Explain why.

13. Why can you see your breath on a cold morning? Does the air temperature have to be below freezing for this to occur?
14. Explain why altocumulus clouds might be observed at 6400 m (21,000 ft) above the surface in Mexico City, Mexico, but never at that altitude above Fairbanks, Alaska.
15. The sky is overcast and it is raining. Explain how you could tell if the cloud above you is a nimbostratus or a cumulonimbus.
16. Suppose it is raining lightly from a deck of nimbostratus clouds. Beneath the clouds are small, ragged, puffy clouds that are moving rapidly with the wind. What would you call these clouds? How did they probably form?
17. You are sitting inside your house on a sunny afternoon. The shades are drawn and you look out the window and notice the sun disappears for about 10 seconds. The alternating light and dark period lasts for nearly 30 minutes. Are the clouds passing in front of the sun cirrocumulus, altocumulus, stratocumulus, or cumulus? Give a reasonable explanation for your answer.

PROBLEMS AND EXERCISES

1. The data in ▼ Table 5.6 represent the dew-point temperature and expected minimum temperature near the ground for various clear winter mornings in a southeastern U.S. city. Assume that the dew point remains constant throughout the night. Answer the following questions about the data.
    (a) On which morning would there be the greatest likelihood of observing visible frost? Explain why.
    (b) On which morning would frozen dew most likely form? Explain why.
    (c) On which morning would there be black frost with no sign of visible frost, dew, or frozen dew? Explain.
    (d) On which morning would you probably only observe dew on the ground? Explain why.
2. If a ceiling balloon rises at 120 m (about 400 ft) each minute, what is the ceiling of an overcast deck of stratus clouds 1500 m (about 5000 ft) thick if the balloon disappears into the clouds in 5 minutes?

ONLINE RESOURCES

Visit www.cengagebrain.com to view additional resources, including video exercises, practice quizzes, an interactive eBook, and more.

▼ **TABLE 5.6**

	MORNING 1	MORNING 2	MORNING 3	MORNING 4	MORNING 5
Dew-point temperature	2°C (35°F)	−7°C (20°F)	1°C (34°F)	−4°C (25°F)	3°C (38°F)
Expected minimum temperature	4°C (40°F)	−3°C (27°F)	0°C (32°F)	−4.5°C (24°F)	2°C (35°F)

The setting sun illuminates the top of a shallow layer of clouds as viewed from an airplane.

Stability and Cloud Development

IN JULY AND AUGUST ON THE HIGH DESERT the thunderstorms come. Mornings begin clear and dazzling bright, the sky as blue as the Virgin's cloak, unflawed by a trace of cloud in all that emptiness. . . . By noon, however, clouds begin to form over the mountains, coming it seems out of nowhere, out of nothing, a special creation.

The clouds multiply and merge, cumulonimbus piling up like whipped cream, like mashed potatoes, like sea foam, building upon one another into a second mountain range greater in magnitude than the terrestrial range below.

The massive forms jostle and grate, ions collide, and the sound of thunder is heard over the sun-drenched land. More clouds emerge from empty sky, anvil-headed giants with glints of lightning in their depths. An armada assembles and advances, floating on a plane of air that makes it appear, from below, as a fleet of ships must look to the fish in the sea.

Edward Abbey, *Desert Solitaire—A Season in the Wilderness*

Clouds, spectacular features in the sky, add beauty and color to the natural landscape. Yet clouds are important for nonaesthetic reasons, too. As they form, vast quantities of heat are released into the atmosphere. Clouds help regulate Earth's energy balance by reflecting and scattering solar radiation and by absorbing Earth's infrared energy. And, of course, without clouds there would be no precipitation. But clouds are also significant because they visually indicate the physical processes taking place in the atmosphere; to a trained observer, they are signposts in the sky. This chapter examines the atmospheric processes these signposts point to, the first of which is atmospheric stability.

Atmospheric Stability

Most clouds form as air rises and cools. Why does air rise on some occasions and not on others? And why do the size and shape of clouds vary so much when the air does rise? Let's see how knowing about the air's stability will help us to answer these questions.

When we speak of atmospheric stability, we are referring to a condition of equilibrium. For example, rock A resting in the depression in ● Fig. 6.1 is in *stable* equilibrium. If the rock is pushed up along either side of the hill and then let go of, it will quickly return to its original position. On the other hand, rock B, resting on the top of the hill, is in a state of *unstable* equilibrium, as a slight push will set it moving away from its original position. Applying these concepts to the atmosphere, we can see that air is in stable equilibrium when, after being lifted or lowered, it tends to return to its original position—it resists upward and downward air motions. Air that is in unstable equilibrium will, when given a little push, move farther away from its original position—it favors upward and downward motion.

To explore the behavior of rising and sinking air, we must first put some air in an imaginary thin elastic wrap. This small volume of air is referred to as a **parcel of air**.* Although the air parcel can expand and contract freely, it does not break apart, but remains as a single unit. At the same time, neither external air nor heat can mix with the air inside the parcel. The space occupied by the air molecules within the parcel defines the air density. The average speed of the molecules is directly related to the air temperature, and the molecules colliding against the parcel walls determine the air pressure inside.

---

*An air parcel is an imaginary body of air about the size of a basketball. The concept of an air parcel is illustrated in several places in the text, including in Fig. 4.6, p. 95.

At Earth's surface, the parcel has the same temperature and pressure as the air surrounding it. Suppose we lift the air parcel up into the atmosphere. We know from Chapter 1 that air pressure decreases with height. Consequently, the air pressure surrounding the parcel lowers. The lower pressure outside allows the air molecules inside to push the parcel walls outward, expanding the parcel. Because there is no other energy source, the air molecules inside must use some of their own energy to expand the parcel. This shows up as slower average molecular speeds, which result in a lower parcel temperature. If the parcel is lowered to the surface, it returns to a region where the surrounding air pressure is higher. The higher pressure squeezes (compresses) the parcel back into its original (smaller) volume. This squeezing increases the average speed of the air molecules and the parcel temperature rises. Hence, *a rising parcel of air expands and cools, while a sinking parcel is compressed and warms.*

If a parcel of air expands and cools, or compresses and warms, and there is no interchange of heat with its surroundings, this situation is called an **adiabatic process**. As long as the air in the parcel is unsaturated (the relative humidity is less than 100 percent), the rate of adiabatic cooling or warming remains constant. This rate of heating or cooling is about 10°C for every 1000 m of change in elevation (5.5°F per 1000 ft). Because this rate applies only to unsaturated air, it is called the **dry adiabatic rate*** (see ● Fig. 6.2).

As the rising air cools, its relative humidity increases as the air temperature approaches the dew-point temperature. If the rising air cools to its dew-point temperature, the relative humidity becomes 100 percent. Further lifting results in condensation; a cloud forms, and latent heat is released inside the rising air parcel. Because the heat added during condensation offsets some of the cooling due to expansion, the air no longer cools at the dry adiabatic rate but at a lesser rate called the **moist adiabatic rate**. If a saturated parcel containing water droplets were to sink, it would compress and warm at the moist adiabatic rate because evaporation of the liquid droplets

---

*For aviation purposes, the dry adiabatic rate is sometimes expressed as 3°C per 1000 ft.

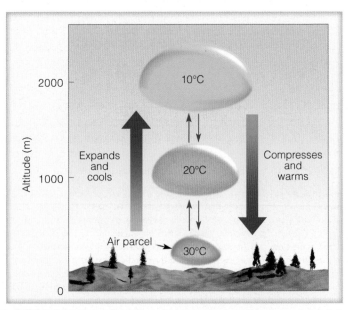

● **FIGURE 6.2** The dry adiabatic rate. As long as the air parcel remains unsaturated, it expands and cools by 10°C per 1000 m; the sinking parcel compresses and warms by 10°C per 1000 m.

● **FIGURE 6.1** When rock A is disturbed, it will return to its original position; rock B, however, will accelerate away from its original position.

Pressure (mb)	TEMPERATURE (°C)					TEMPERATURE (°F)				
	−40	−20	0	20	40	−40	−5	30	65	100
1000	9.5	8.6	6.4	4.3	3.0	5.2	4.7	3.5	2.4	1.6
800	9.4	8.3	6.0	3.9		5.2	4.6	3.3	2.2	
600	9.3	7.9	5.4			5.1	4.4	3.0		
400	9.1	7.3				5.0	4.0			
200	8.6					4.7				

would offset the rate of compressional warming. Hence, the rate at which rising or sinking saturated air changes temperature—the moist adiabatic rate—is less than the dry adiabatic rate.*

Unlike the dry adiabatic rate, the moist adiabatic rate is not constant, but varies greatly with temperature and, hence, with moisture content, because warm saturated air produces more liquid water than cold saturated air. The added condensation in warm, saturated air liberates more latent heat. Consequently, the moist adiabatic rate is much less than the dry adiabatic rate when the rising air is warm; however, the two rates are nearly the same when the rising air is very cold (see ▼ Table 6.1). Although the moist adiabatic rate does vary, to make the numbers easy to deal with, we will use an average of 6°C per 1000 m (3.3°F per 1000 ft) in most of our examples and calculations.

Determining Stability

We determine the stability of the air by comparing the temperature of a rising parcel to that of its surroundings. If the rising air is colder than its environment, it will be more dense** (heavier) and tend to sink back to its original level. In this case, the air is *stable* because it resists upward movement. If the rising air is warmer and, therefore, less dense (lighter) than the surrounding air, it will continue to rise until it reaches the same temperature as its environment. This is an example of *unstable* air. To figure out the air's stability, we need to measure the temperature both of the rising air and of its environment at various levels above Earth.

A STABLE ATMOSPHERE Suppose we release a balloon-borne instrument called a *radiosonde*. (A photo of a radiosonde is found in Focus section 1.3.) As the balloon carries the radiosonde upward, the radiosonde sends back temperature data, as shown in ● Fig. 6.3 (such a vertical profile of temperature is called a *sounding*). Notice that the radiosonde-measured air temperature

---

*Consider an air parcel initially at rest. Suppose the air parcel rises and cools, and a cloud forms. Further suppose that no precipitation (rain or snow) falls from the cloud (leaves the parcel). If the parcel should descend to its original level, the latent heat released inside the parcel during condensation will be the same amount that is absorbed as the cloud evaporates. This process is called a *reversible adiabatic process*. If, on the other hand, rain or snow falls from the cloud during uplift and leaves the parcel, the sinking parcel will not recover during evaporation the same amount of latent heat released during condensation because the parcel's water content is lower. This process is known as an *irreversible pseudoadiabatic process*.

**When, at the same level in the atmosphere, we compare parcels of air that are equal in size but vary in temperature, we find that cold air parcels are more dense than warm air parcels; that is, in the cold parcel, there are more molecules that are crowded closer together.

in Fig. 6.3 decreases by 4°C for every 1000 m rise in altitude. Remember from Chapter 1 that the rate at which the air temperature changes with altitude is called the *lapse rate*. Because this is the rate at which the air temperature surrounding us will be changing if we were to climb upward into the atmosphere, we will refer to it as the **environmental lapse rate**.

Now suppose in Fig. 6.3a that a parcel of unsaturated air with a temperature of 30°C is lifted from the surface. As it rises, it cools at the dry adiabatic rate (10°C per 1000 m), and the temperature inside the parcel at 1000 m would be 20°C, or 6°C lower than the air surrounding it. Look at Fig. 6.3a closely and notice that, as the air parcel rises higher, the temperature difference between it and the surrounding air becomes even greater. Even if the parcel is initially saturated (see Fig. 6.3b), it will cool at the moist rate—6°C per 1000 m—and will be colder than its environment at all levels. In both cases, the rising air is colder and heavier than the air surrounding it. In this example, the atmosphere is **absolutely stable**. *The atmosphere is always absolutely stable when the environmental lapse rate is less than the moist adiabatic rate.*

Because air in an absolutely stable atmosphere strongly resists upward vertical motion, it will, *if forced to rise*, tend to spread out horizontally. If clouds form in this rising air, they, too, will spread horizontally in relatively thin layers and usually have flat tops and bases. We might expect to see stratiform clouds—such as cirrostratus, altostratus, nimbostratus, or stratus—forming in a stable atmosphere.

What conditions are necessary to bring about a stable atmosphere? As we have just seen, the atmosphere is stable when the environmental lapse rate is small; that is, when the difference in temperature between the surface air and the air aloft is relatively small. Consequently, the atmosphere tends to become more stable—that is, it stabilizes—as the air aloft warms or the surface air cools. If the air aloft is being replaced by warmer air (warm advection), and the surface air is not changing appreciably, the environmental lapse rate decreases and the atmosphere becomes more stable. Similarly, the environmental lapse rate decreases and the atmosphere becomes more stable when the lower layer cools (see ● Fig. 6.4). The *cooling* of the *surface air* can be due to:

1. nighttime radiational cooling of the surface
2. an influx of cold surface air brought in by the wind (cold advection)
3. air moving over a cold surface

So, on a typical day, the atmosphere is usually most stable in the early morning around sunrise, when the lowest surface air temperature is recorded. If the surface air becomes saturated in a stable atmosphere, a persistent layer of haze or fog can form (see ● Fig. 6.5).

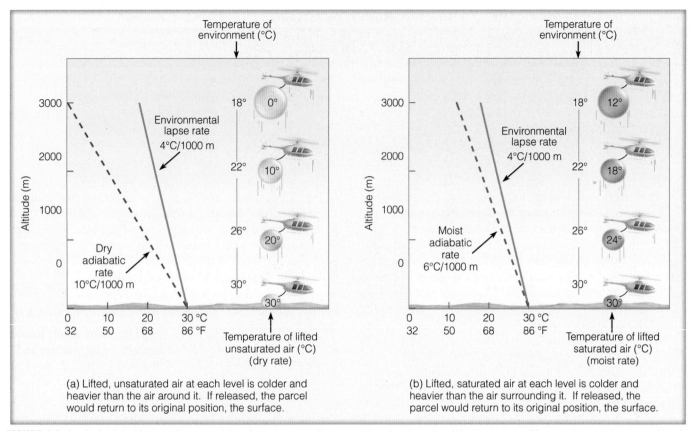

● **FIGURE 6.3** An absolutely stable atmosphere occurs when the environmental lapse rate is less than the moist adiabatic rate. In a stable atmosphere, a rising air parcel is colder and more dense than the air surrounding it, and, if given the chance (in other words, if it is released), it will return to its original position. (In both situations, the helicopter shows that the air is being lifted. In the real world, this type of parcel lifting, of course, would be impossible.)

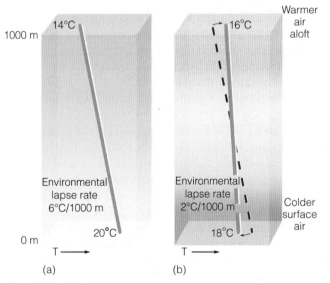

● **FIGURE 6.4** The initial environmental lapse rate in diagram (a) will become more stable (stabilize) as the air aloft warms and the surface air cools, as illustrated in diagram (b).

Another way the atmosphere becomes more stable is when an entire layer of air sinks. For example, if a layer of unsaturated air over 1000 m thick and covering a large area subsides, the entire layer will warm by adiabatic compression. As the layer subsides, it becomes compressed by the weight of the atmosphere and shrinks vertically. The upper part of the layer sinks farther, and, hence, warms more

than the bottom part. This phenomenon is illustrated in ● Fig. 6.6. After subsiding, the top of the layer is actually warmer than the bottom, and an inversion is formed. (Recall from Chapter 3 that an inversion represents an atmospheric condition where the air becomes warmer with height.) Inversions that form as air slowly sinks over a large area are called **subsidence inversions**. They sometimes occur at the surface, but more frequently, they are observed aloft and are often associated with large high-pressure areas because of the sinking air motions associated with these systems.

An inversion represents an atmosphere that is absolutely stable. Why? Within the inversion, warm air overlies cold air, and, if air rises into the inversion, it is becoming colder, while the air around it is getting warmer. Obviously, the colder air would tend to sink. Inversions, therefore, act as lids on vertical air motion. When an inversion exists near the ground, stratus clouds, fog, haze, and pollutants are all kept close to the surface. In fact, as we will see in Chapter 19, most air pollution episodes occur with subsidence inversions. (For additional information on subsidence inversions, read Focus section 6.1.)

At this point, we can see why extremely high surface air temperatures can occur when the air aloft is sinking. Notice in ● Fig. 6.7 that the air inside an air parcel at an altitude of about 3000 m (10,000 ft), where the air pressure is 700 mb, has a moderately cool temperature of 10°C (50°F). If this air parcel sinks all the way to the surface (0 meters), where the pressure is 1000 mb, the final temperature inside the parcel having warmed at the dry adiabatic rate will be a whopping 40°C (104°F)! Thus, the air

● **FIGURE 6.5** On this morning near Boulder, Colorado, cold surface air has produced a stable atmosphere that inhibits vertical air motions and allows the fog and haze to linger close to the ground.

parcel at 700 mb has the potential of being very warm when brought to the surface.*

Before we turn our attention to an unstable atmosphere, let's first examine a condition known as **neutral stability**. If the lapse rate is exactly equal to the dry adiabatic rate, rising or sinking unsaturated air will cool or warm at the same rate as the air around it. At each level, it would have the same temperature and density as the surrounding air. Because this air tends neither to continue rising nor sinking, the atmosphere is said to be neutrally stable. For saturated air, *neutral stability* exists when the environmental lapse rate is equal to the moist adiabatic rate.

AN UNSTABLE ATMOSPHERE   Suppose a radiosonde sends back air temperatures above Earth as plotted in ● Fig. 6.8a. Once again, we determine the atmosphere's stability by comparing the environmental lapse rate to the moist and dry adiabatic rates. In this case, the environmental lapse rate is 11°C per 1000 m (6°F per 1000 ft). A rising parcel of unsaturated surface air will cool at the dry adiabatic rate. Because the dry adiabatic rate is less than the environmental lapse rate, the parcel will be warmer than the surrounding air and will continue to rise, constantly moving upward, away from its original

_____
*The temperature an air parcel would have if lowered at the dry adiabatic rate to a pressure of 1000 millibars is called *potential temperature*. Moving parcels to the same level allows them to be observed under identical conditions so it can be determined which parcels are potentially warmer than others. In this example, the potential temperature of the parcel is 40°C, or 313 K.

position. The atmosphere is unstable. Of course, a parcel of saturated air cooling at the lower moist adiabatic rate will be even warmer than the air around it (see Fig. 6.8b). In both cases, the air parcels, once they start upward, will continue to rise on their own because the rising air parcels are warmer and less dense than the air around them. The atmosphere in this example is said to be **absolutely unstable**. An air parcel in an unstable atmosphere, being warmer and less dense than its surroundings, has an upward-directed force (called *buoyant force*) acting on it. The warmer the air parcel compared to its surroundings, the greater the buoyant force, and the more rapidly the air parcel rises. *Absolute instability results when the environmental lapse rate is greater than the dry adiabatic rate.*

It should be noted, however, that deep layers in the atmosphere are seldom, if ever, absolutely unstable. Absolute instability is usually limited to a very shallow layer near the ground on hot, sunny days. Here the environmental lapse rate can exceed the dry adiabatic rate, and the lapse rate is called *superadiabatic*. On rare occasions when the environmental lapse rate exceeds about 3.4°C per 100 m (the *autoconvective lapse rate*), convection becomes spontaneous, resulting in the automatic overturning of the air.

So far, we have seen that the atmosphere is absolutely stable when the environmental lapse rate is less than the moist adiabatic rate and absolutely unstable when the environmental lapse rate is greater than the dry adiabatic rate. However, a typical type of atmospheric instability exists when the lapse rate lies between the moist and dry adiabatic rates.

● **FIGURE 6.6** The layer x–y is initially 1400 m thick. If the entire layer slowly subsides, it shrinks in the more-dense air near the surface. As a result of the shrinking, the top of the layer warms more than the bottom, and the entire layer (x'–y') becomes more stable, and in this example forms an inversion.

● **FIGURE 6.7** An air parcel initially with a temperature of 10°C (50°F) at an altitude where the pressure is 700 mb (about 3000 m or 10,000 ft) has the potential to warm to 40°C (104°F) if it were to sink all the way to the surface, where the pressure is 1000 mb.

A CONDITIONALLY UNSTABLE ATMOSPHERE

The environmental lapse rate in ● Fig. 6.9 is 7°C per 1000 m (4°F per 1000 ft). When a parcel of unsaturated air rises, it cools dry adiabatically and is colder at each level than the air around it (see Fig. 6.9a). It will, therefore, tend to sink back to its original level because it is in a stable atmosphere. Now, suppose the rising parcel is saturated. As we can see in Fig. 6.9b, the rising air is warmer than its environment at each level. Once the parcel is given a push upward, it will tend to move in that direction; the atmosphere is unstable for the saturated parcel. In this example, the atmosphere is said to be **conditionally unstable**. This type of stability depends on whether or not the rising air is saturated. When the rising parcel of air is unsaturated, the atmosphere is stable; when the parcel of air is saturated, the atmosphere is unstable. Conditional instability means that, if unsaturated air can be lifted to a level where it becomes saturated, instability will result.

Conditional instability occurs whenever the environmental lapse rate is between the moist adiabatic rate and the dry adiabatic rate. Recall from Chapter 1 that the average lapse rate in the troposphere is about 6.5°C per 1000 m (3.6°F per 1000 ft). As this value lies between the dry adiabatic rate and the average moist rate, *the atmosphere is ordinarily in a state of conditional instability.* ● Figure 6.10 summarizes how the three categories of instability (absolutely stable, conditionally unstable, and absolutely unstable) relate to the dry and moist adiabatic rates.

CAUSES OF INSTABILITY

What causes the atmosphere to become more unstable? The atmosphere becomes more unstable

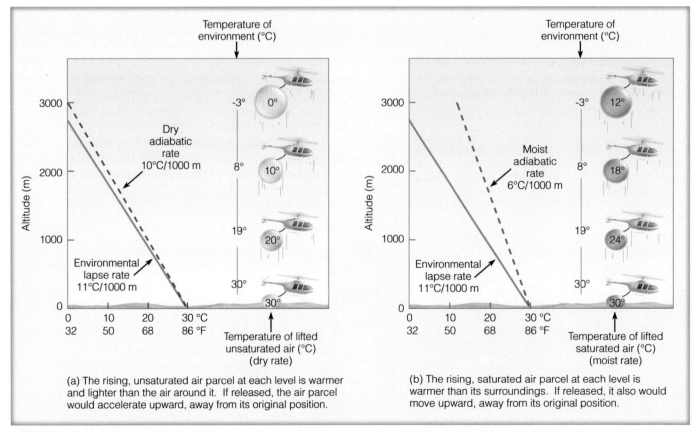

(a) The rising, unsaturated air parcel at each level is warmer and lighter than the air around it. If released, the air parcel would accelerate upward, away from its original position.

(b) The rising, saturated air parcel at each level is warmer than its surroundings. If released, it also would move upward, away from its original position.

● **FIGURE 6.8** An absolutely unstable atmosphere occurs when the environmental lapse rate is greater than the dry adiabatic rate. In an unstable atmosphere, a rising air parcel will continue to rise because it is warmer and less dense than the air surrounding it.

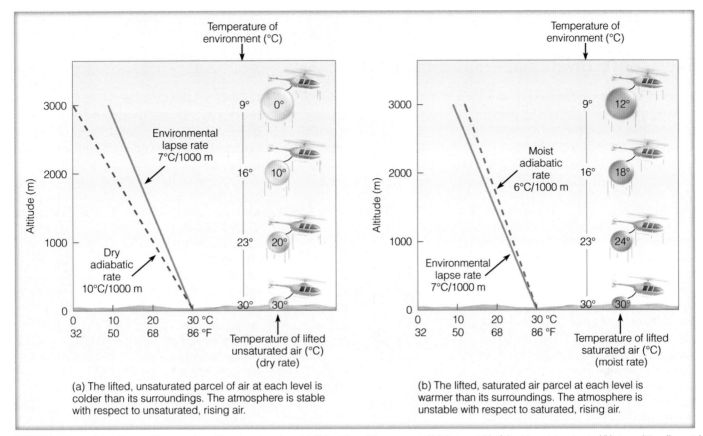

● FIGURE 6.9 Conditionally unstable atmosphere. The atmosphere is *stable* if the rising air is *unsaturated* (a), but *unstable* if the rising air is *saturated* (b). A conditionally unstable atmosphere occurs when the environmental lapse rate is between the moist adiabatic rate and the dry adiabatic rate. (As in Fig. 6.3, the lifting of air parcels by a helicopter would be impossible.)

(a) The lifted, unsaturated parcel of air at each level is colder than its surroundings. The atmosphere is stable with respect to unsaturated, rising air.

(b) The lifted, saturated air parcel at each level is warmer than its surroundings. The atmosphere is unstable with respect to saturated, rising air.

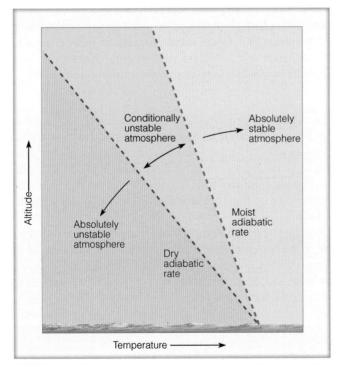

● FIGURE 6.10 When the environmental lapse rate is greater than the dry adiabatic rate (red region), the atmosphere is absolutely unstable. When the environmental lapse rate is less than the moist adiabatic rate (blue region), the atmosphere is absolutely stable. And when the environmental lapse rate lies between the dry adiabatic rate and the moist adiabatic rate (green region), the atmosphere is conditionally unstable.

as the environmental lapse rate steepens; that is, as the air temperature drops rapidly with increasing height. This circumstance may be brought on by either air aloft becoming colder or the surface air becoming warmer (see ● Fig. 6.11).

The *cooling of the air aloft* may be due to:

1. winds bringing in colder air (cold advection)
2. clouds (or the air) emitting infrared radiation to space (radiational cooling)

The *warming of the surface air* may be due to:

1. daytime solar heating of the surface
2. an influx of warm air brought in by the wind (warm advection)
3. air moving over a warm surface

The combination of cold air aloft and warm surface air can produce a steep lapse rate and atmospheric instability (see ● Fig. 6.12).

At this point, we can see that the stability of the atmosphere changes during the course of a day. In clear, calm weather around sunrise, surface air is normally colder than the air above it, a radiation inversion exists, and the atmosphere is quite stable, as indicated by smoke or haze lingering close to the ground. As the day progresses, sunlight warms the surface and the surface warms the air above. As the air temperature near the ground increases, the lower atmosphere gradually becomes more unstable—it *destabilizes*—with maximum instability usually occurring during the hottest part of the day.

Up to now, we have seen that a layer of air may become more unstable by either cooling the air aloft or warming the air at the

Subsidence Inversions—Put a Lid on It

● Figure 1 shows a typical summertime vertical profile of air temperature and dew point measured with a radiosonde near the coast of California. Notice that the air temperature decreases from the surface up to an altitude of about 300 m (1000 ft). Notice also that, where the air temperature reaches the dew point, a cloud forms.

Above about 300 m, the air temperature increases rapidly up to an altitude near 900 m (about 3000 ft). This region of increasing air temperature with increasing height marks the region of the *subsidence inversion*. Within the inversion, air from aloft warms by compression. The sinking air at the top of the inversion is not only warm (about 24°C or 75°F) but also dry with a low relative humidity, as indicated by the large spread between air temperature and dew point. The subsiding air, which does not reach the surface, is associated with a large high-pressure area, located to the west of California.

Immediately below the base of the inversion lies cool, moist air. The cool air is unable to penetrate the inversion because a lifted parcel of cool marine air within the inversion would be much colder and heavier than the air surrounding it. Because the colder air parcel would fall back to its original position, the atmosphere is absolutely stable within the inversion. The subsidence inversion, therefore, acts as a lid on the

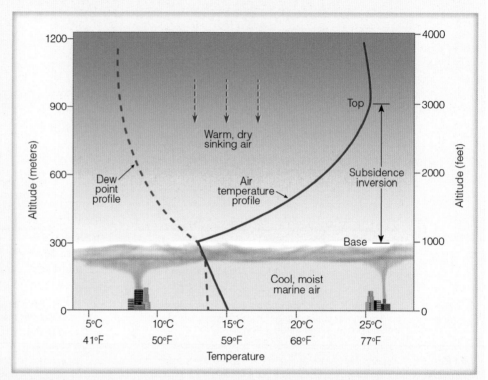

● **FIGURE 1** A strong subsidence inversion along the coast of California. The base of the stable inversion acts as a cap or lid on the cool marine air below. An air parcel rising into the inversion layer would sink back to its original level because the rising air parcel would be colder and more dense than the air surrounding it.

air below, preventing the air from mixing vertically into the inversion. And so the marine air with its pollution and clouds is confined to a relatively shallow region near Earth's surface. It

is this trapping of air near the surface, associated with a strong subsidence inversion, that helps to give West Coast cities such as Los Angeles frequent periods of pollution.

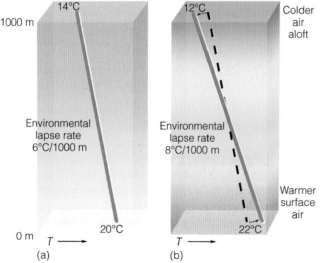

● **FIGURE 6.11** The initial environmental lapse rate in diagram (a) will become more unstable (that is, destabilize) as the air aloft cools and the surface air warms, as illustrated in diagram (b).

surface. A layer of air may also be made more unstable by either mixing or lifting. Let's look at mixing first. In ● Fig. 6.13, the environmental lapse rate before mixing is less than the moist rate, and the layer is stable (A). Now, suppose the air in the layer is mixed either by convection or by wind-induced turbulent eddies. Air is cooled adiabatically as it is brought up from below and heated adiabatically as it is mixed downward. The up and down motion in the layer redistributes the air in such a way that the temperature at the top of the layer decreases, while, at the base, it increases. This steepens the environmental lapse rate and makes the layer more unstable. If this mixing continues for some time, and the air remains unsaturated, the vertical temperature distribution will eventually be equal to the dry adiabatic rate (B).

WEATHER WATCH

The cumulus cloud that is forming in Fig 6.12 is called a pyrocumulus. If the pyrocumulus builds into a rain-producing cloud it may act as a natural fire-extinguisher by dumping rainwater onto the raging fire beneath it.

● **FIGURE 6.12** ● **FIGURE 6.12** The warmth from this forest fire in Idaho during August 2003 heats the air, causing instability near the surface. Warm, less-dense air (and smoke) bubbles upward, expanding and cooling as it rises. Eventually the rising air cools to its dew point, condensation begins, and a cumulus cloud forms (top of photo). If the rising air parcels are large and strong enough, the resulting clouds (sometimes called *pyrocumulus*) may produce lightning and precipitation.

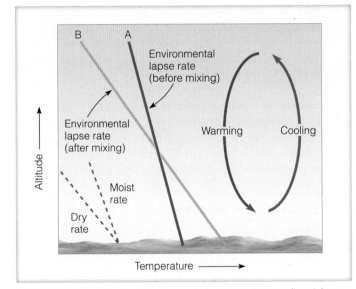

● **FIGURE 6.13** Mixing tends to steepen the lapse rate. Rising, cooling air lowers the temperature toward the top of the layer, while sinking, warming air increases the temperature near the bottom.

Just as lowering an entire layer of air makes it more stable, the lifting of a layer makes it more unstable. In ● Fig. 6.14, the air lying between 1000 and 900 mb is initially absolutely stable because the environmental lapse rate of layer x–y is less than the moist adiabatic rate. The layer is lifted, and, as it rises, the rapid decrease in air density aloft causes the layer to stretch out vertically. If the layer remains unsaturated, the entire layer cools at the dry adiabatic rate. Due to the stretching effect, however, the top of the layer cools more than the bottom. This steepens the environmental lapse rate. Note that the absolutely stable layer x–y, after rising, has become conditionally unstable between 500 and 600 mb (layer x′–y′).

A very stable air layer may be converted into an absolutely unstable layer when the lower portion of a layer is moist and the upper portion is quite dry. In ● Fig. 6.15, the inversion layer between

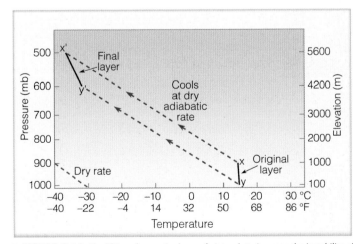

● **FIGURE 6.14** The lifting of an entire layer of air tends to increase the instability of the layer. The initial stable layer (x–y) after lifting is now a conditionally unstable layer (x′–y′).

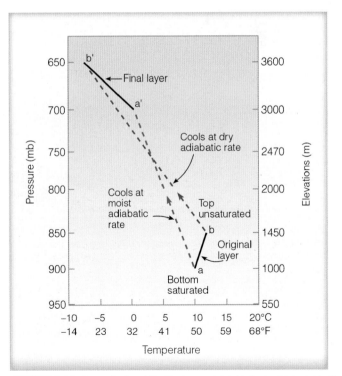

● **FIGURE 6.15** Convective instability. The layer a–b is initially absolutely stable. The lower part of the layer is saturated, and the upper part is "dry." After lifting, the entire layer (a′–b′) becomes absolutely unstable.

900 and 850 mb is absolutely stable. Suppose the bottom of the layer is saturated while the air at the top is unsaturated. If the layer is forced to rise, even a little, the upper portion of the layer cools at the dry adiabatic rate and grows cold quite rapidly, while the air near the bottom cools more slowly at the moist adiabatic rate. It does not take much lifting before the upper part of the layer is much colder than the bottom part; the environmental lapse rate steepens and the entire layer becomes absolutely unstable (layer a–b). The potential instability brought about by the lifting of a stable layer whose surface is humid and whose top is "dry" is called *convective instability*. Convective instability is associated with the development of severe storms, such as thunderstorms and tornadoes, which are investigated more thoroughly in Chapters 14 and 15.

BRIEF REVIEW

Before going on to the next section, here is a brief review of some of the facts and concepts concerning atmospheric stability.

- The air temperature in a rising parcel of unsaturated air decreases at the dry adiabatic rate, whereas the air temperature in a rising parcel of saturated air decreases at the moist adiabatic rate.

- The dry adiabatic rate and moist adiabatic rate of cooling are different due to the fact that latent heat is released in a rising parcel of saturated air.

- In a stable atmosphere, a lifted parcel of air will be colder (heavier) than the air surrounding it. Because of this fact, the lifted parcel will tend to sink back to its original position.

- In an unstable atmosphere, a lifted parcel of air will be warmer (lighter) than the air surrounding it, and thus will continue to rise upward, away from its original position.

- The atmosphere becomes more stable (stabilizes) as the surface air cools, the air aloft warms, or a layer of air sinks (subsides) over a vast area.

- The atmosphere becomes more unstable (destabilizes) as the surface air warms, the air aloft cools, or a layer of air is either mixed or lifted.

- A conditionally unstable atmosphere exists when the environmental lapse rate is between the moist adiabatic rate and the dry adiabatic rate.

- The atmosphere is normally most stable in the early morning and most unstable in the afternoon.

- Layered clouds tend to form in a stable atmosphere, whereas cumuliform clouds tend to form in a conditionally unstable atmosphere.

Up to this point we've looked at how the stability of the atmosphere can change during the course of a day. It's interesting to note that these changes can play a role in making afternoons windier than mornings. So, before going on to the next section, which describes how stability is responsible for the development of individual cloud types, you may wish to read Focus section 6.2, which explains why clear afternoons are often windy.

Cloud Development

We know that most clouds form as air rises and cools and its water vapor condenses. Given that air normally needs a "trigger" to start it moving upward, what is it that causes the air to rise so that clouds are able to form? The following mechanisms are primarily responsible for the development of the majority of clouds we observe:

1. surface heating and free convection
2. uplift along topography
3. widespread ascent due to convergence of surface air
4. uplift along weather fronts (see ● Fig. 6.16)

(a) Convection
—5 km—
Heated surface

(b) Lifting along topography
—150 km—

(c) Convergence of air
—500 km—
Low pressure

(d) Lifting along weather fronts
—1500 km—
Cold air   Warm air

● **FIGURE 6.16** The primary ways clouds form: (a) surface heating and convection; (b) forced lifting along topographic barriers; (c) convergence of surface air; (d) forced lifting along weather fronts.

Atmospheric Stability and Windy Afternoons—Hold On to Your Hat

On warm days when the weather is clear or partly cloudy, you may have noticed that the windiest time of the day is usually in the afternoon. Windy afternoons occur because of several factors all working together, including surface heating, convection, and atmospheric stability.

We know that in the early morning the atmosphere is most stable, meaning that the air resists upward and downward motions. As an example, consider the flow of air in the early morning as illustrated in ● Fig. 2a. Notice that weak winds exist near the surface with much stronger winds aloft. Because the atmosphere is stable, there is little vertical mixing between the surface air and the air higher up.

As the day progresses, and the sun rises higher in the sky, the surface heats up and the lower atmosphere becomes more unstable. Over hot surfaces, the air begins to rise in the form of thermals that carry the slower-moving air with them (see Fig. 2b). At some level above the surface, the rising air links up with the faster-moving air aloft. If the air begins to sink as part of a convective circulation, it may pull some of the stronger winds aloft downward

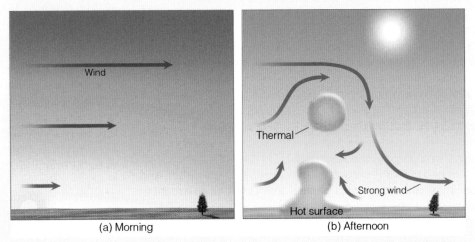

(a) Morning   (b) Afternoon

FIGURE 2 (a) During the early morning, there is little exchange between the surface winds and the winds aloft. (b) In the afternoon, when the atmosphere is usually most unstable, convection in the form of rising thermals links surface air with the air aloft, causing strong winds from aloft to reach the ground and produce strong, gusty surface winds.

with it. If this sinking air should reach the surface, it produces a momentary gust of strong wind. This exchange of air also increases the average wind speed at the surface. Because this type of air exchange is greatest on a clear day in the afternoon when the atmosphere is most unstable, we tend to experience the strongest, most gusty winds in the afternoon. At night, when the atmosphere stabilizes, the interchange between the surface air and the air aloft is at a minimum, and the winds at the surface tend to die down.

The first mechanism that can cause the air to rise is *convection*. Although we briefly looked at convection in Chapter 2 when we examined rising thermals and how they transfer heat upward into the atmosphere, we will now look at convection from a slightly different perspective—how rising thermals are able to form into cumulus clouds.

CONVECTION AND CLOUDS   Some areas of Earth's surface are better absorbers of sunlight than others and, therefore, heat up more quickly. The air in contact with these "hot spots" becomes warmer than its surroundings. A hot "bubble" of air—a *thermal*—breaks away from the warm surface and rises, expanding and cooling as it ascends. As the thermal rises, it mixes with the cooler, drier air around it and gradually loses its identity. Its upward movement now slows. Before the thermal is completely diluted, subsequent rising thermals often penetrate it and help the air rise a little higher. If the rising air cools to its saturation point, the moisture will condense, and the thermal becomes visible to us as a cumulus cloud.

Observe in ● Fig. 6.17 that the air motions are downward on the outside of the cumulus cloud. The downward motions are caused in part by evaporation around the outer edge of the cloud, which cools the air, making it heavy (more dense). Another reason for the downward motion is the completion of the convection

FIGURE 6.17 Cumulus clouds form as warm, invisible air bubbles detach themselves from the surface, then rise and cool to the condensation level. Below and within the cumulus clouds, the air is rising. Around the cloud, the air is sinking.

current started by the thermal. Cool air slowly descends to replace the rising warm air. Therefore, we have rising air in the cloud and sinking air around it. Because subsiding air greatly inhibits the growth of thermals beneath it, small cumulus clouds usually have a great deal of blue sky between them (see ● Fig. 6.18).

As the cumulus clouds grow, they shade the ground from the sun. This, of course, cuts off surface heating and upward convection. Without the continual supply of rising air, the cloud begins to erode as its droplets evaporate. Unlike the sharp outline of a growing cumulus, the cloud now has indistinct edges, with cloud fragments extending from its sides. As the cloud dissipates (or moves along with the wind), surface heating begins again and regenerates another thermal, which becomes a new cumulus. This is why you often see cumulus clouds form, gradually disappear, then reform in the same spot.

Suppose that it is a warm, humid summer afternoon and the sky is full of cumulus clouds. The cloud bases are all at nearly the same level above the ground and the cloud tops extend only about a thousand meters higher. The development of these clouds depends primarily upon the air's stability and moisture content. To illustrate how these factors influence the formation of a convective cloud, we will examine the temperature and moisture characteristics within a rising bubble of air. Because the actual air motions that go into forming a cloud are rather complex, we will simplify matters by making these assumptions:

1. No mixing takes place between the rising air and its surroundings.
2. Only a single thermal produces the cumulus cloud.
3. The cloud forms when the relative humidity is 100 percent.
4. The rising air in the cloud remains saturated.

The environmental lapse rate on this particular day is plotted in ● Fig. 6.19 and is represented as a dark gray line on the far left of the illustration. The changing environmental air temperature indicates changes in the atmosphere's stability. The environmental lapse rate in layer A is greater than the dry adiabatic rate, so the layer is absolutely unstable. The air layers above it—layer B and layer C—are both absolutely stable since the environmental lapse rate in each layer is less than the moist adiabatic rate. However, the overall environmental lapse rate from the surface up to the base of the inversion (2000 m) is 7.5°C per 1000 m (4.1°F per 1000 ft), which indicates a conditionally unstable atmosphere.

Now, suppose that a very warm, moist bubble of air with an air temperature and dew-point temperature of 35°C and 27°C (95°F and 80.5°F), respectively, breaks away from the surface and begins to rise (which is illustrated in the middle of Fig. 6.19). Notice that, a short distance above the ground, the air inside the bubble is warmer than the air around it, so it is buoyant and rises freely. This level in the atmosphere where the rising air becomes warmer than the surrounding air is called the *level of free convection*. The rising bubble will continue to rise as long as it is warmer than the air surrounding it.

The rising air cools at the dry adiabatic rate and the dew point falls, but not as rapidly.* The rate at which the dew point

---

*The decrease in dew-point temperature is caused by the rapid decrease in air pressure within the rising air. Because the dew point is directly related to the actual vapor pressure of the rising air, a decrease in total air pressure causes a corresponding decrease in vapor pressure and, hence, a lowering of the dew-point temperature. If the air should sink, the increase in total pressure will cause an increase in vapor pressure and a corresponding increase in the dew-point temperature.

● **FIGURE 6.18** Cumulus clouds building on a warm summer afternoon. Each cloud represents a region where thermals are rising from the surface. The clear areas between the clouds are regions where the air is sinking.

© UCAR

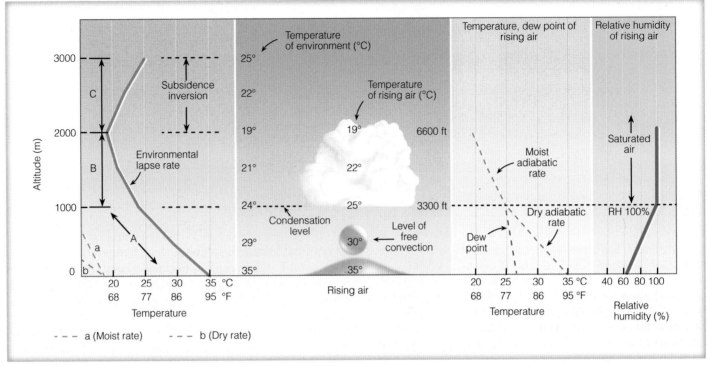

● **FIGURE 6.19** The development of a cumulus cloud.

drops varies with the moisture content of the rising air, but an approximation of 2°C per 1000 m (1°F per 1000 ft) is commonly used. So, as unsaturated rising air cools, the air temperature and dew point approach each other at the rate of 8°C per 1000 m (4.5°F per 1000 ft). This process causes an increase in the air's relative humidity (illustrated in the far right-hand side of Fig. 6.19, by the dark green line).

At an elevation of 1000 m (3300 ft), the air has cooled to the dew point, the relative humidity is 100 percent, condensation begins, and a cloud forms. The elevation where the cloud forms is called the **condensation level**. Above the condensation level the rising air is saturated and cools at the moist adiabatic rate. Condensation continues to occur, and since water vapor is transforming into liquid cloud droplets, the dew point within the cloud now drops more rapidly with increasing height than before. The air remains saturated as both the air temperature and dew point decrease at the moist adiabatic rate (illustrated in the area of Fig. 6.19 shaded tan).

Notice that inside the cloud the rising air remains warmer than the air surrounding it and continues its spontaneous rise upward through layer B. The top of the bulging cloud at 2000 m (about 6600 ft) represents the top of the rising air, which has now cooled to a temperature equal to its surroundings. The air would have a difficult time rising much above this level because of the stable subsidence inversion directly above it. The subsidence inversion, associated with the downward air motions of a high-pressure system, prevents the clouds from building very high above their bases. Hence, an afternoon sky full of flat-base cumuli with little vertical growth indicates fair weather. (Recall from Chapter 5 that the proper name of these fair-weather cumulus clouds is *cumulus humilis*.)

As we can see, the stability of the atmosphere above the condensation level plays a major role in determining the vertical growth of a cumulus cloud. Notice in ● Fig. 6.20 that, when a deep stable layer begins a short distance above the cloud base, only cumulus humilis are able to form. If a deep conditionally unstable layer exists above the cloud base, cumulus congestus are likely to grow, with billowing cauliflowerlike tops. When the conditionally unstable layer is extremely deep—usually greater than 4 km (2.5 mi)—the cumulus congestus may even develop into a cumulonimbus.

Seldom do cumulonimbus clouds extend very far above the tropopause. The stratosphere is quite stable, so once a cloud penetrates the tropopause, it usually stops growing vertically and spreads horizontally. The low temperature at this altitude produces ice crystals in the upper section of the cloud. In the middle latitudes, high winds near the tropopause blow the ice crystals laterally, producing the flat anvil-shaped top characteristic of cumulonimbus clouds (see ● Fig. 6.21).

The vertical development of a convective cloud also depends upon the mixing that takes place around its periphery. The rising, churning cloud mixes cooler air into it. Such mixing is called **entrainment**. If the environment around the cloud is very dry, the cloud droplets quickly evaporate. The effect of entrainment, then, is to increase the rate at which the rising air cools by the injection of cooler air into the cloud and the subsequent evaporation of the cloud droplets. If the rate of cooling approaches the dry adiabatic rate, the air stops rising and the cloud no longer builds, even though the lapse rate may indicate a conditionally unstable atmosphere.

● **FIGURE 6.20** Variations in the air's stability, as indicated by the environmental lapse rate, greatly influence the growth of cumulus clouds.

Up to now, we have looked at convection over land. Convection and the development of cumulus clouds also occur over large bodies of water. As cool air flows over a body of relatively warm water, the lowest layer of the atmosphere becomes warm and moist. This induces instability—convection begins and cumulus clouds form. If the air moves over progressively warmer water, as is sometimes the case over the open ocean, more active convection occurs and a cumulus cloud can build into cumulus congestus and finally into cumulonimbus. This sequence of cloud development is observed from satellites as cold northerly winds move southward over the northern portions of the Atlantic and Pacific oceans (see ● Fig. 6.22).

Once a convective cloud forms, stability, humidity, and entrainment all play a part in its vertical development. The level at which the cloud initially forms, however, is determined primarily by the surface temperature and moisture content of the original thermals. (Focus section 6.3 uses this information and a simple formula to determine the bases of convective clouds.)

TOPOGRAPHY AND CLOUDS Horizontally moving air obviously cannot go through a large obstacle, such as a mountain, so the air must go over it. Forced lifting along a topographic barrier is called **orographic uplift**. Often, large masses of air rise when they approach a long chain of mountains like the Sierra Nevada or Rockies. This lifting produces cooling, and, if the air is humid, clouds form. Clouds produced in this manner are called *orographic clouds*. The type of cloud that forms will depend on the air's stability and moisture content. On the leeward (downwind) side of the mountain, as the air moves downhill, it warms. This sinking air is now drier, given that much of its moisture was

● **FIGURE 6.21** Cumulus clouds developing into thunderstorms in a conditionally unstable atmosphere over the Great Plains. Notice that, in the distance, the cumulonimbus with the flat anvil-shaped top has reached a stable layer of the atmosphere.

● **FIGURE 6.22** Satellite view of stratocumulus clouds forming in rows over the Atlantic Ocean as cold, dry arctic air sweeps over Canada, then out over warmer water. Notice that the clouds are absent over the landmass and directly along the coast, but form and gradually thicken as the surface air warms and destabilizes farther offshore.

Determining Convective Cloud Bases

The bases of cumulus clouds that form by convection on warm, sunny afternoons can be estimated quite easily when the surface air temperature and dew point are known. If the air is not too windy, we can assume that entrainment of air will not change the characteristics of a rising thermal. Because the rising air cools at the dry adiabatic rate of about 10°C per 1000 m, and the dew point drops at about 2°C per 1000 m, the air temperature and dew point approach each other at the rate of 8°C for every 1000 m of rise. Rising surface air with an air temperature and dew point spread of 8°C would produce saturation and a cloud at an elevation of 1000 m. Put another way, a 1°C difference between the surface air temperature and the dew point produces a cloud base at 125 m. Therefore, by finding the difference between surface air temperature (T) and dew point (T_d), and multiplying this value by 125, we can estimate the base of the convective cloud forming overhead, as

$$H_{meter} = 125(T - T_d),*$$ (1)

where H is the height of the base of the cumulus cloud in meters above the surface, with both T and T_d measured in degrees Celsius. If T and T_d are in °F, H can be calculated with the formula

$$H_{feet} = 228(T - T_d)$$ (2)

To illustrate the use of formula (1), let's determine the base of the cumulus cloud in Fig. 6.19, p. 155. Recall that the surface air temperature and dew point were 35°C and 27°C, respectively. The difference, $T - T_d$, is 8°C. This value multiplied by 125 gives us a cumulus cloud with a base at 1000 above the ground. This agrees with the condensation level we originally calculated.

---

*The formula works best when the air is well mixed from the surface up to the cloud base, such as in the afternoon on a sunny day. The formula does not work well at night or in the early morning.

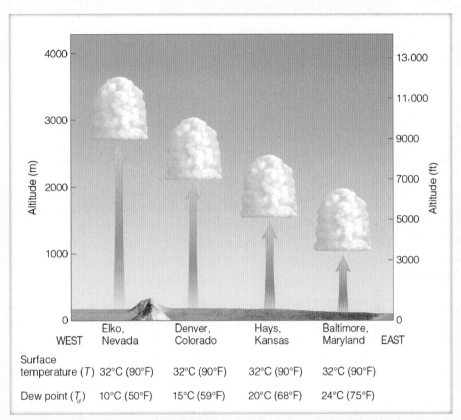

● **FIGURE 3** During the summer, cumulus cloud bases typically increase in elevation above the ground as one moves westward into the drier air of the Central Plains. Note that cloud altitudes are shown relative to ground-level altitude, which increases more rapidly from east to west than shown here.

	Elko, Nevada	Denver, Colorado	Hays, Kansas	Baltimore, Maryland
Surface temperature (T)	32°C (90°F)	32°C (90°F)	32°C (90°F)	32°C (90°F)
Dew point (T_d)	10°C (50°F)	15°C (59°F)	20°C (68°F)	24°C (75°F)

Along the East Coast in summer, when the air is warm and muggy, the separation between air temperature and dew point may be smaller than 9°C (16°F). The bases of afternoon cumulus clouds over cities such as Philadelphia and Baltimore are typically about 1000 m (3300 ft) above the ground (see ● Fig. 3). Farther west, in the Central Plains, where the air is drier and the spread between surface air temperature and dew point is greater, the cloud bases are higher. For example, west of Salina, Kansas, the cumulus cloud bases are generally greater than 1500 m (about 5000 ft) above the surface. On a summer afternoon in central Nevada, it is not uncommon to observe cumulus forming at 2400 m (about 8000 ft above the surface). In the Central Valley of California, where the summer afternoon spread between air temperature and dew point usually exceeds 22°C (40°F), the air must rise to almost 2700 m (about 9000 ft above the surface) before a cloud forms. Due to sinking air aloft, thermals in this area are unable to rise to that elevation, and afternoon cumulus clouds are seldom observed forming overhead.

---

removed in the form of clouds and precipitation on the windward side. This region on the leeward side, where precipitation is noticeably less, is called a **rain shadow**.

Clouds and Orographic Uplift   An example of orographic uplift and cloud development is given in ● Fig. 6.23. Before rising up and over the barrier, the air at the base of the mountain (0 m) on the windward side has an air temperature of 20°C (68°F) and a dew-point temperature of 12°C (54°F). Notice that the atmosphere is conditionally unstable, as indicated by the environmental lapse rate of 8°C per 1000 m. (Remember from our earlier discussion that the atmosphere is conditionally unstable when

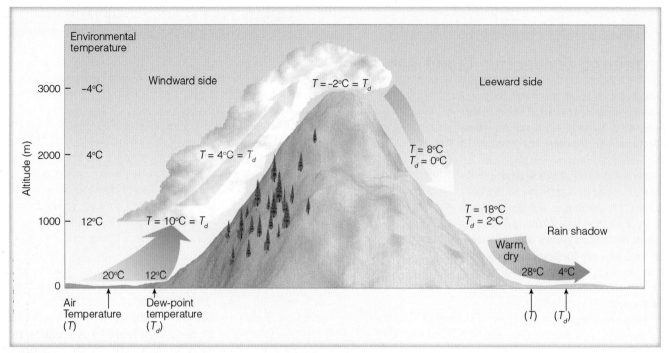

Environmental temperature

Windward side

$T = -2°C = T_d$

Leeward side

Altitude (m)

3000 — −4°C

2000 — 4°C $T = 4°C = T_d$ $T = 8°C$ $T_d = 0°C$

1000 — 12°C $T = 10°C = T_d$ $T = 18°C$ $T_d = 2°C$

Rain shadow

Warm, dry

0 —   20°C   12°C   28°C   4°C

Air Temperature (T)

Dew-point temperature (T_d)

(T)   (T_d)

● **FIGURE 6.23** Orographic uplift, cloud development, and the formation of a rain shadow.

CRITICAL THINKING QUESTION   Suppose the cloud in Fig. 6.23 forms at an altitude of 2000 meters (instead of 1000 meters). How would this change in cloud base influence the dew-point temperature of the air after it crosses the mountain and reaches 0 meters on the leeward side?

the environmental lapse rate falls between the dry adiabatic rate and the moist adiabatic rate.)

As the unsaturated air rises, the air temperature decreases at the dry adiabatic rate $(10°C$ per 1000 m$)$ and the dew-point temperature decreases at 2°C per 1000 m. Notice that the rising, cooling air reaches its dew point and becomes saturated at 1000 m. This level (called the **lifting condensation level [LCL]**) marks the base of the cloud that has formed as air is lifted (in this case by the mountain). As the rising saturated air condenses into many billions of liquid cloud droplets, and as latent heat is liberated by the condensing vapor, both the air temperature and dew-point temperature decrease at the moist adiabatic rate.

At the top of the mountain, the air temperature and dew point are both −2°C. Note in Fig. 6.23 that this temperature is higher than that of the surrounding air (−4°C). Consequently, the rising air at this level is not only warmer but unstable with respect to its surroundings. Therefore, the rising air should continue to rise and build into a much larger cumuliform cloud.

Suppose, however, that the air at the top of the mountain (temperature and dew point of −2°C) is forced to descend to the base of the mountain (0 m) on the leeward side. If we assume that the cloud remains on the windward side and does not extend beyond the mountaintop, the temperature of the sinking air will increase at the dry adiabatic rate $(10°C$ per 1000 m$)$ all the way down to the base of the mountain. (The dew-point temperature increases at a much lower rate of 2°C per 1000 m.)

We can see in Fig. 6.23 that on the leeward side, after descending 3000 m, the air temperature is 28°C (82°F) and the dew-point temperature is 4°C (39°F). The air is now 8°C (14°F) warmer than it was before being lifted over the barrier. The higher air temperature on the leeward side is the result of latent heat being converted into sensible heat during condensation on the windward side. (In fact, the rising air at the *top* of the mountain is considerably warmer than it would have been had condensation not occurred.) The lower dew-point temperature and, hence, drier air on the leeward side are the result of water vapor condensing and then remaining as liquid cloud droplets and precipitation on the windward side.

In summary, there are several important concepts to come away with after reading the previous section on air rising up and over a mountain:

1. Air descending a mountain warms by compressional heating and, upon reaching the surface, can be much warmer than the air at the same level on the windward side, especially when condensation occurs and latent heat is released on the windward side.

2. Air on the leeward side of a mountain is normally drier (has a lower dew point) than the air on the windward side because water in the form of clouds and precipitation often remains on the windward side. The lower dew point and higher air temperature on the leeward side produce a lower relative humidity, a greater potential for evaporation of water, and a rain-shadow desert.

CLOUDS THAT FORM DOWNWIND OF MOUNTAINS Although clouds are more prevalent on the windward side of mountains, they may, under certain atmospheric conditions, form on the leeward side as well. For example, stable air flowing over a mountain often moves in a series of waves that may extend for several hundred kilometers on the leeward side (see ● Fig. 6.24). These waves resemble the waves that form in a river downstream from a large boulder. Recall from Chapter 5 that *wave clouds* often have a characteristic lens shape and are commonly called *lenticular clouds*.

The formation of lenticular clouds is shown in ● Fig. 6.25. As moist air rises on the upwind side of the wave, it cools and condenses, producing a cloud. On the downwind side of the wave, the air sinks and warms, and the cloud evaporates. Viewed from the ground, the clouds appear motionless as the air rushes through them; hence, they are often referred to as *standing wave clouds*. Since they most frequently form at altitudes where middle clouds form, they are also called *altocumulus standing lenticulars*.

When the air between the cloud-forming layers is too dry to produce clouds, lenticular clouds will form one above the other. Actually, when a strong wind blows almost perpendicular to a high mountain range, mountain waves may extend into the stratosphere, producing a spectacular display, sometimes resembling a fleet of hovering spacecraft (see ● Fig. 6.26, p. 162).

Notice in Fig. 6.25 that beneath the lenticular cloud downwind of the mountain range, a large swirling eddy forms. The rising part of the swirling air may cool enough to produce a visible cloud called a **rotor cloud**. The air in the rotor is extremely turbulent and presents a major hazard to aircraft in the vicinity. Dangerous flying conditions also exist near the lee side of the mountain, where strong downwind air motions are present. (These types of winds will be treated in more detail in Chapter 9.) Now, having examined the concept of stability and the formation of clouds, we are ready to see what role stability might play in changing a cloud from one type into another. (Before going on to the next section, you may wish to read the Focus section 6.4, which provides you with a graphic representation of air rising up and over a mountain using an *adiabatic chart*.)

CHANGING CLOUD FORMS Under certain conditions, a layer of altostratus may change into altocumulus. This happens if the top of the original cloud deck cools while the bottom warms. Because clouds are such good absorbers and emitters of infrared radiation, the top of the cloud will often cool as it radiates infrared energy to space more rapidly than it absorbs solar energy. Meanwhile, the bottom of the cloud will warm as it absorbs infrared energy from below more quickly than it radiates this energy away. This process makes the cloud layer conditionally unstable to the point that small convection cells begin within the cloud itself. The up and down motions in a layered cloud produce globular elements that give the cloud a lumpy appearance. The cloud forms in the rising part of a cell, and clear spaces appear where descending currents occur.*

Cirrocumulus and stratocumulus may form in a similar way. When the wind is fairly uniform throughout a cloud layer, these new cloud elements appear evenly distributed across the sky. However, if the wind speed or direction changes with height,

---

*An example of an altocumulus cloud with a lumpy appearance is given in Fig. 5.15 on p. 126.

● **FIGURE 6.24** Satellite view of wave clouds forming many kilometers downwind of the mountains in Scotland and Ireland.

NASA/GSFC

● **FIGURE 6.25** Lenticular clouds that form in the wave directly over the mountain are called *mountain wave clouds,* whereas those that form downwind of the mountain are called *lee wave clouds.* On the underside of the lee wave's crest a turbulent rotor cloud may form.

Adiabatic Charts

The adiabatic chart is a valuable tool for anyone who studies the atmosphere. The chart itself, shown on p. 161, is a graph that shows how various atmospheric elements change with altitude. At first glance, the chart appears complicated because of its many lines. We will, therefore, construct these lines on the chart step by step.

● Figure 4 shows horizontal lines of pressure decreasing with altitude, and vertical lines of temperature in °C increasing toward the right. The height values along the right-hand side of Fig. 4 are approximate elevations that have been computed assuming that the air temperature decreases at a standard rate of 6.5°C per kilometer.

In ● Fig. 5, the slanted solid red lines are called *dry adiabats*. They show how the air temperature would change inside a rising or descending *unsaturated* air parcel. Suppose, for example, that an unsaturated air parcel at the surface (pressure 1013 mb) with a temperature of 10°C rises and cools at the dry adiabatic rate (10°C per km). What would be the parcel temperature at a pressure of 900 mb? To find out, simply follow the dry adiabat from the surface temperature of 10°C up to where it crosses the 900-mb line. Answer: about 0°C. If the same parcel returns to the surface, follow the dry adiabat back to the surface and read the temperature, 10°C.

On some charts, the dry adiabats are expressed as a potential temperature in Kelvin. The *potential temperature* is the temperature an air parcel would have if it were moved dry adiabatically to a pressure of 1000 mb. Moving parcels to the same level allows them to be observed under identical conditions. Thus, it can be determined which parcels are potentially warmer than others.

● FIGURE 4

● FIGURE 5

● FIGURE 6

● FIGURE 7

The sloping dashed blue lines in ● Fig. 6 are called *moist adiabats*. They show how the air temperature would change inside a rising or descending parcel of *saturated* air. In other words, they represent the moist adiabatic rate for a rising or sinking saturated air parcel, such as in a cloud.

The sloping gray lines in ● Fig. 7 are lines of constant *mixing ratio*. At any given temperature and pressure, they show how much water vapor the air could hold if it were saturated—the

saturation mixing ratio (w_s) in grams of water vapor per kilogram of dry air (g/kg). At a given dew-point temperature, they show how much water vapor the air is actually holding—the *actual mixing ratio (w)* in g/kg. Hence, given the air temperature and dew-point temperature at some level, we can compute the relative humidity of the air.* For example, suppose at the

---

*The relative humidity (RH) of the air can be expressed as RH = w/w_s × 100%.

---

the horizontal axes of the convection cells align with the average direction of the wind. The new cloud elements then become arranged in rows and are given the name **cloud streets** (see ● Fig. 6.27). When the changes in wind speed and direction reach a critical value, and an inversion caps the cloud-forming layers, wavelike clouds called **billow clouds** may form along the top of the cloud layer as shown in ● Fig. 6.28.

WEATHER WATCH

The first widely reported "sighting" of a flying saucer in the United States came in 1947 from a pilot who flew near Mt. Rainier, Washington, a region where lenticular clouds commonly form.

surface (pressure 1013 mb) the air temperature and dew-point temperature are 29°C and 15°C, respectively. In Fig. 7, observe that at 29°C the saturation mixing ratio (w_s) is 26 g/kg, and with a dew-point temperature of 15°C, the actual mixing ratio (w) is 11 g/kg. This produces a relative humidity of $^{11}/_{26} \times 100$ percent, or 42 percent.

The mixing ratio lines also show how the dew-point temperature changes in a rising or sinking unsaturated air parcel. If an unsaturated air parcel with a dew point of 15°C rises from the surface (pressure 1013 mb) up to where the pressure is 700 mb (approximately 3 km), notice in Fig. 7 that the dew-point temperature inside the parcel would have dropped to a temperature near 10°C.

● Figure 8 shows all of the lines described thus far on a single chart. We have already seen that the chart can be used to obtain graphically a number of atmospheric mathematical relationships. Therefore, let's use the chart to obtain information on air that rises up and over a mountain range.

Suppose we use the example given in Fig. 6.23 on p. 158 . Air at an elevation of 0 m (pressure 1013 mb), with a temperature of 20°C (T_1) and a dew-point temperature of 12°C (D_1), first ascends, then descends a 3000-meter high mountain range. Look at Fig. 8 closely and observe that the surface air with a temperature of 20°C indicates a saturation mixing ratio of about 15 g/kg, and at 12°C the dew-point temperature indicates an actual mixing ratio of about 9 g/kg. Hence, the relative humidity of the air before rising over the mountain is $^9/_{15}$, or 60 percent.

Now, as the unsaturated air rises (as indicated by arrows in Fig. 8), the air temperature follows a dry adiabat (solid red line), and the

● **FIGURE 8** The adiabatic chart. The arrows illustrate the example given in the text. The cloud on the right side represents the base and height of the cloud given in the example.

dew-point temperature follows a line of constant mixing ratio (gray line). Carefully follow the mixing ratio line in Fig. 8 from 12°C up to where it intersects the dry adiabat that slopes upward from 20°C. Notice that the intersection occurs at an elevation near 1 km. This, of course, marks the base of the cloud—the *lifting condensation level* (*LCL*)—where the relative humidity is 100 percent and condensation begins. Above this level, the rising air is saturated. Consequently, the air temperature and dew-point temperature together follow a moist adiabat (dashed blue line) to the top of the mountain.

Notice in Fig. 8 that, at the top of the mountain (at 3 km or about 700 mb), both the air temperature and dew point are −2°C. If we

assume that the cloud stays on the windward side, then from 3 km (700 mb) the descending air follows a dry adiabat all the way to the surface (1013 mb). Notice that, after descending, the air has a temperature of 28°C (T_2). From the mountaintop, the dew-point temperature follows a line of mixing ratio and reaches the surface (1013 mb) with a temperature of 4°C (D_2). Observe in Fig. 8 that, with an air temperature of 28°C, the saturation mixing ratio is about 25 g/kg and, with a dew point of 4°C, the actual mixing ratio is about 5 g/kg. Thus, the relative humidity of the air after descending is about $^5/_{25}$, or 20 percent.

Occasionally, altocumulus show vertical development and produce towerlike extensions. The clouds often resemble floating castles and, for this reason, they are called *altocumulus castellanus* (see ● Fig. 6.29). They form when rising currents within the cloud extend into conditionally unstable air above the cloud. Apparently, the buoyancy for the rising air comes from the latent heat released during condensation within the cloud. This process can occur in cirrocumulus clouds, producing *cirrocumulus castellanus*. When altocumulus castellanus appear, they indicate that the middle level of the troposphere is becoming more unstable (destabilizing). This destabilization is often the precursor to shower activity. So a morning sky full of

● **FIGURE 6.26** Lenticular clouds tend to form over and downwind of mountains. They also tend to remain in one place as air rushes through them. Here, lenticular clouds are forming over mountainous terrain in Argentina's Los Glaciares National Park.

● **FIGURE 6.27** Satellite view of cloud streets, rows of stratocumulus clouds forming over the warm Georgia landscape.

● **FIGURE 6.28** Billow clouds forming in a region where wind speed changes rapidly with altitude. This is a region of strong vertical wind shear. (More on the topic of *wind shear* and the formation of these clouds is given in Chapter 9, p. 234.)

altocumulus castellanus will likely become afternoon showers and even thunderstorms.

Occasionally, the stirring of a moist layer of stable air will produce a deck of stratocumulus clouds. In ● Fig. 6.30a, the air is stable and close to saturation. Suppose a strong wind mixes the layer from the surface up to an elevation of 600 m (2000 ft) (Fig. 6.30b). As we saw earlier, the lapse rate will steepen as the

upper part of the layer cools and the lower part warms. At the same time, mixing will make the moisture distribution in the layer more uniform. The warmer temperature and decreased moisture content cause the lower part of the layer to dry out. On the other hand, the decrease in temperature and increase in moisture content saturate the top of the mixed layer, producing a layer of stratocumulus clouds. Notice in Fig. 6.30b that the air above the region of mixing is still stable and inhibits further mixing. In some cases, an inversion may actually form above the clouds. However, if the surface warms substantially, rising thermals may penetrate the stable region and the stratocumulus clouds may change into more widely separated clouds, such as cumulus or cumulus congestus. A stratocumulus layer changing to a sky dotted with growing cumulus clouds often occurs as surface heating increases on a warm, humid summer day.

● **FIGURE 6.29** An example of altocumulus castellanus.

CRITICAL THINKING QUESTION What do you feel would be the most important factor(s) in predicting how high these altocumulus castellanus clouds will rise into the atmosphere?

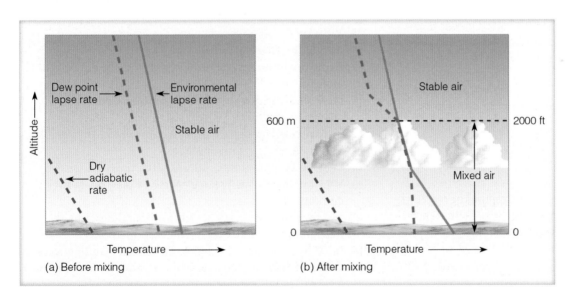

(a) Before mixing

(b) After mixing

● **FIGURE 6.30** The mixing of a moist layer of air near the surface can produce a deck of stratocumulus clouds.

SUMMARY

In this chapter, we tied together the concepts of stability and the formation of clouds. We learned that rising unsaturated air cools at the dry adiabatic rate and, due to the release of latent heat, rising saturated air cools at the moist adiabatic rate. In a stable atmosphere, a lifted parcel of air will be colder (heavier) than the air surrounding it at each new level, and it will sink back to its original position. Because stable air tends to resist upward vertical motions, clouds forming in a stable atmosphere often spread horizontally and have a stratified appearance, such as cirrostratus and altostratus. A stable atmosphere may be caused by either cooling the surface air, warming the air aloft, or the sinking (subsidence) of an entire layer of air, in which case a very stable subsidence inversion usually forms.

In an unstable atmosphere, a lifted parcel of air will be warmer (lighter) than the air surrounding it at each new level, and it will continue to rise upward away from its original position. In a conditionally unstable atmosphere, an unsaturated parcel of air can be lifted to a level where condensation begins, latent heat is released, and instability results, as the temperature inside the rising parcel becomes warmer than the air surrounding it. In a conditionally unstable atmosphere, rising air tends to form clouds that develop vertically, such as cumulus congestus and cumulonimbus. Instability can be caused by warming the surface air, cooling the air aloft, or by the lifting or mixing of an entire layer of air.

The development of most clouds results from either surface heating, uplift along topography (orographic uplift), convergence of surface air, or lifting along weather fronts. On warm, humid days, the instability generated by surface heating can produce cumulus clouds at a height determined by the temperature and moisture content of the surface air. Instability may cause changes in existing clouds as convection changes an altostratus into an altocumulus. Also, mixing can change a clear day into a cloudy one.

KEY TERMS

The following terms are listed (with corresponding page numbers) in the order they appear in the text. Define each. Doing so will aid you in reviewing the material covered in this chapter.

parcel of air, 144	neutral stability, 147
adiabatic process, 144	absolutely unstable
dry adiabatic rate, 144	atmosphere, 147
moist adiabatic rate, 144	conditionally unstable
environmental lapse rate, 145	atmosphere, 148
absolutely stable atmosphere, 145	condensation level, 155
subsidence inversion, 146	entrainment 155

orographic uplift, 156	rotor clouds, 159
rain shadow, 157	cloud streets, 160
lifting condensation level (LCL), 158	billow clouds, 160

QUESTIONS FOR REVIEW

1. What is an adiabatic process?
2. Why are moist and dry adiabatic rates of cooling different?
3. Under what conditions would the moist adiabatic rate of cooling be almost equal to the dry adiabatic rate?
4. Explain the difference between environmental lapse rate and dry adiabatic rate.
5. How would one normally obtain the environmental lapse rate?
6. What is a stable atmosphere and how can it form?
7. Describe the general characteristics of clouds associated with stable and unstable atmospheres.
8. List and explain several processes by which a stable atmosphere can be made unstable.
9. If the atmosphere is conditionally unstable, what condition is necessary to bring on instability?
10. Explain why cumulus clouds are conspicuously absent over a cool water surface.
11. Why are cumulus clouds more frequently observed during the afternoon than at night?
12. Explain why an inversion represents an absolutely stable atmosphere.
13. How and why does lifting or lowering a layer of air change its stability?
14. List and explain several processes by which an unstable atmosphere can be made stable.
15. Why do cumulonimbus clouds often have flat tops?
16. Why are there usually large spaces of blue sky between cumulus clouds?
17. List four primary ways clouds form, and describe the formation of one cloud type by each method.
18. (a) Why are lenticular clouds also called standing wave clouds? (b) On which side of a mountain (windward or leeward) would lenticular clouds most likely form?
19. Explain why rain shadows form on the leeward side of mountains.
20. How can a layer of altostratus change into one of altocumulus?
21. Describe the conditions necessary to produce stratocumulus clouds by mixing.

22. Briefly describe how each of the following clouds forms:
- (a) lenticular
- (b) rotor
- (c) billow
- (d) castellanus

QUESTIONS FOR THOUGHT

1. How is it possible for a layer of air to be convectively unstable and absolutely stable at the same time?

2. Are the bases of convective clouds generally higher during the day or the night? Explain.

3. Where would be the safest place to build an airport in a mountainous region? Why?

4. Use Fig. 4.15, p. 101 (Chapter 4) to help you explain why the bases of cumulus clouds, which form from rising thermals during the summer, often increase in height above the surface as you move from east to west across Kansas.

5. For least polluted conditions, what would be the best time of day for a farmer to burn agricultural debris?

6. Suppose that surface air on the windward side of a mountain rises and descends on the leeward side. Recall from Chapter 4 that the dew-point temperature is a measure of the amount of water vapor in the air. Explain, then, why the relative humidity of the descending air drops as the dew-point temperature of the descending air increases.

7. Usually when a cumulonimbus cloud begins to dissipate, the bottom half of the cloud dissipates first. Give an explanation as to why this situation might happen.

PROBLEMS AND EXERCISES

1. Under which set of conditions would a cumulus cloud base be observed at the highest level above the surface? Surface air temperatures and dew points are as follows: (a) 35°C, 14°C; (b) 30°C, 19°C; (c) 34°C, 9°C; (d) 29°C, 7°C; (e) 32°C, 6°C.

2. If the height of the base of a cumulus cloud is 1000 m above the surface, and the dew point at Earth's surface beneath the cloud is 20°C, determine the air temperature at Earth's surface beneath the cloud.

3. The condensation level over New Orleans, Louisiana, on a warm, muggy afternoon is 2000 ft. If the dew-point temperature of the rising air at this level is 73°F, what is the approximate dew-point temperature and air temperature at the surface? Determine the surface relative humidity. (Hint: See Chapter 4, p. 105.)

4. Suppose the air pressure outside a conventional jet airliner flying at an altitude of 10 km (about 33,000 ft) is 250 mb. Further, suppose the air inside the aircraft is pressurized to 1000 mb. If the outside air temperature is −50°C (−58°F), what would be the temperature of this air if brought inside the aircraft and compressed at the dry adiabatic rate to a pressure of 1000 mb? (Assume that a pressure of 1000 mb is equivalent to an altitude of 0 m.)

5. In ● Fig. 6.31, a radiosonde is released and sends back temperature data as shown in the diagram. (This is the environment temperature.)
- (a) Calculate the environmental lapse rate from the surface up to 3000 m.
- (b) What type of atmospheric stability (stable or unstable) does the sounding indicate?

Suppose the wind is blowing from the west and a parcel of surface air with a temperature of 10°C and a dew point of 2°C begins to rise upward along the western (windward) side of the mountain.
- (c) What is the parcel's relative humidity at 0 m (pressure 1013 mb) before rising? (Hint: See Chapter 4, p. 105.)
- (d) As the air parcel rises, at approximately what altitude would condensation begin and a cloud start to form?
- (e) What is the air temperature and dew point of the rising air at the base of the cloud?
- (f) What is the air temperature and dew point of the rising air inside the cloud at an altitude of 3000 m? (Use the moist adiabatic rate of 6°C per 1000 m.)
- (g) At an altitude of 3000 m, how does the air temperature inside the cloud compare with the temperature outside the cloud, as measured by the radiosonde? What type of atmospheric stability does this suggest? Explain.
- (h) At an altitude of 3000 m, would you expect the cloud to continue to develop vertically? Explain.

● **FIGURE 6.31** The information in this illustration is to be used in answering question 5 in the Problems and Exercises section.

ONLINE RESOURCES

Visit www.cengagebrain.com to view additional resources, including video exercises, practice quizzes, an interactive eBook, and more.

Even a leafless forest in winter can be
transformed by a coating of snow or ice.

Precipitation

BY AN UNFORTUNATE COINCIDENCE, AS I WRITE, the New Jersey countryside around me is in the grip of an ice storm—"the worst ice storm in a generation" so the papers tell me, and a look at my garden suffices to convince me. A 150-year-old tulip tree has already lost enough limbs to keep us in firewood for the rest of the winter; a number of black locusts stand beheaded; the silver birches are bent double to the ground; and almost every twig of every bush and tree is encased in a translucent cylinder of ice one to two inches in diameter. There is beauty in the sight, to be sure, for the sun has momentarily transmuted the virginal whites and grays into liquid gold. And there is hope, too, for some of the trees are still unbowed and look as though they had every intention of living to tell the tale.

George H. T. Kimble, *Our American Weather*

The young boy pushed his nose against the cold window-pane, hoping to see snowflakes glistening in the light of the street lamp across the way. Perhaps if it snowed, he thought, accumulations would be deep enough to cancel school—maybe for a day, possibly a week, or, perhaps, forever. But a full moon with a halo gave little hope for snow on this evening, as did the voice from the back room that insisted, "Don't even think about snow. You know it won't snow tonight—it's too cold to snow."

Is it ever "too cold to snow"? Although many believe in this expression, the fact remains that it is *never* too cold to snow. True, colder air cannot "hold" as much water vapor as warmer air, but snow can fall into extremely cold surface air from less-cold air above. And no matter how cold the air becomes, it always contains some water vapor that could produce snow. At Fort Yellowstone, Wyoming, for example, 3 in. of snow fell on February 2, 1899, when the maximum temperature reached only −28°C (−18°F). In fact, tiny ice crystals have been observed falling at temperatures as low as −47°C (−53°F). We usually associate extremely cold air with "no snow" because the coldest winter weather usually occurs on clear, calm nights—conditions that normally prevail with strong high-pressure areas that have few, if any, clouds.

This chapter raises a number of interesting questions to consider regarding **precipitation**.* Why, for example, does the largest form of frozen precipitation—hail—often fall during the warmest time of the year? Why does it sometimes rain on one side of the street but not on the other? What is sleet, and how does it differ from hail? First, we will examine the processes that produce rain and snow; then, we will look closely at the other forms of precipitation. Our discussion will conclude with a section on how precipitation is measured.

Precipitation Processes

As we all know, cloudy weather does not necessarily mean that it will rain or snow. In fact, clouds may form, linger for many days, and never produce precipitation. In Eureka, California, the August daytime sky is overcast more than 50 percent of the time, yet the town's average precipitation for August is merely one-tenth of an inch. We know that clouds form by condensation, yet apparently condensation alone is not sufficient to produce rain. Why not? To answer this question we need to closely examine the tiny world of cloud droplets.

HOW DO CLOUD DROPLETS GROW LARGER? An ordinary cloud droplet is extremely small, having an average diameter of 20 micrometers (μm)** or 0.002 cm. Notice in ● Fig. 7.1 that a typical cloud droplet is 100 times smaller in diameter than a typical raindrop. If a cloud droplet is in equilibrium with its surroundings, the size of the droplet does not change because the water molecules condensing onto the droplet will be exactly balanced by those evaporating from it. If, however, it is not in equilibrium,

---

*Recall from Chapter 4 that precipitation is any form of water (liquid or solid) that falls from a cloud and reaches the ground.

**Remember from Chapter 2 that one micrometer (μm) equals one-millionth of a meter.

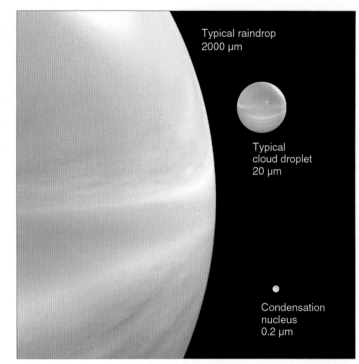

● **FIGURE 7.1** Relative sizes of raindrops, cloud droplets, and condensation nuclei, with diameters shown in micrometers (μm).

Typical raindrop 2000 μm

Typical cloud droplet 20 μm

Condensation nucleus 0.2 μm

the droplet size will either increase or decrease, depending on whether condensation or evaporation predominates.

Consider a cloud droplet in equilibrium with its environment. The total number of vapor molecules around the droplet remains fairly constant and defines the droplet's *saturation vapor pressure*. Since the droplet is in equilibrium, the saturation vapor pressure is also called the **equilibrium vapor pressure**. ● Figure 7.2 shows a cloud droplet and a flat water surface, both of which are in equilibrium. Because more vapor molecules surround the droplet, it has a greater equilibrium vapor pressure. The reason for this fact is that water molecules are less strongly attached to a curved (convex) water surface; hence, they evaporate more readily.

To keep the droplet in equilibrium, more vapor molecules are needed around it to replace those molecules that are constantly evaporating from its surface. Smaller cloud droplets exhibit a greater curvature relative to the size of a water molecule, which causes a more rapid rate of evaporation. As a result of this

● **FIGURE 7.2** At equilibrium, the vapor pressure over a curved droplet of water (right) is greater than that over a much larger volume of water (left), whose surface would be relatively flat compared to the size of the molecules.

Temperature 10°C

Flat water surface   Cloud droplet

process (called the **curvature effect**), smaller droplets require an even greater vapor pressure to keep them from evaporating away. Therefore, *when air is saturated with respect to a flat surface, it is unsaturated with respect to a curved droplet of pure water*, and the droplet evaporates. So, to keep tiny cloud droplets in equilibrium with the surrounding air, the air must be *supersaturated*; that is, the relative humidity must be greater than 100 percent. The smaller the droplet, the greater its curvature relative to the size of water molecules, and the higher the supersaturation needed to keep the droplet in equilibrium.

● Figure 7.3 shows the curvature effect for pure water. The dark blue line represents the relative humidity needed to keep a droplet with a given diameter in equilibrium with its environment. Note that when the droplet's size is less than 2 μm, the relative humidity (measured with respect to a flat surface) must be above 100.1 percent for the droplet to survive. As droplets become larger, the effect of curvature lessens; for a droplet whose diameter is greater than 20 μm, the curvature effect is so small that the droplet behaves as if its surface were flat.

Just as relative humidities less than that required for equilibrium permit a water droplet to evaporate and shrink, those greater than the equilibrium value allow the droplet to grow by condensation. From Fig. 7.3, we can see that a droplet whose diameter is 1 μm will grow larger as the relative humidity approaches 101 percent. But relative humidities, even in clouds, rarely become greater than 101 percent. How, then, do tiny cloud droplets of less than 1 μm grow to the size of an average cloud droplet?

Recall from Chapter 5 (fog formation discussion) that condensation begins on tiny particles called *cloud condensation nuclei*. Because many of these nuclei are *hygroscopic* (that is, they have an affinity for water vapor), condensation may begin on such particles when the relative humidity is well below 100 percent. When condensation begins on hygroscopic salt particles, for example, they dissolve, forming a solution. Because the salt ions in solution bind closely with water molecules, it is more difficult for the water molecules to evaporate. This condition reduces the equilibrium vapor pressure, an effect known as

the **solute effect**. Due to the solute effect, once an impurity (such as a salt particle) replaces a water molecule in the lattice structure of the droplet, the equilibrium vapor pressure surrounding the droplet is lowered. As a result of the solute effect, a droplet containing salt can be in equilibrium with its environment when the atmospheric relative humidity is much lower than 100 percent. Should the relative humidity of the air increase, water vapor molecules would attach themselves to the droplet at a faster rate than they would leave, and the droplet would grow larger in size.

As the droplet grows larger, however, the solution becomes more dilute, and the solute effect diminishes. So the curvature effect and the solute effect are in direct competition: The curvature effect acts to inhibit the growth of small droplets, whereas the solute effect acts to enhance their growth. The combination of these two effects results in varying sizes of individual cloud droplets.

Imagine that we place cloud condensation nuclei of varying sizes into moist but unsaturated air. As the air cools, the relative humidity increases. When the relative humidity reaches a value near 78 percent, condensation occurs on the majority of nuclei. As the air cools further, the relative humidity increases, with the droplets containing the most salt reaching the largest sizes. And since the smaller nuclei are more affected by the curvature effect, only the larger nuclei are able to become cloud droplets.

Over landmasses where large concentrations of nuclei exist, there may be many hundreds of droplets per cubic centimeter, all competing for the available supply of water vapor. Over the oceans where the concentration of nuclei is less, there are normally fewer (typically less than 100 per cubic centimeter) but larger cloud droplets. So, in a given volume we tend to find more cloud droplets in clouds that form over land and fewer, but larger, cloud droplets in clouds that form over the ocean.

We now have a cloud composed of many small droplets—too small to fall as rain. These minute droplets require only slight upward air currents to keep them suspended. Those droplets that do fall descend slowly and evaporate in the drier air beneath the cloud. It is evident, then, that most clouds cannot produce precipitation. The condensation process by itself is entirely too slow to produce rain. Even under ideal conditions, it would take several days for this process alone to create a raindrop. However, observations show that clouds can develop and begin to produce rain in less than an hour. Because it takes about 1 million average-size (20 μm) cloud droplets to make an average-size (2000 μm) raindrop, there must be some other process by which cloud droplets grow large and heavy enough to fall as precipitation. Even though all of the intricacies of how rain is produced are not yet fully understood, two important processes stand out: (1) the collision-coalescence process and (2) the ice-crystal (Bergeron) process.

COLLISION AND COALESCENCE PROCESS In clouds with tops warmer than −15°C (5°F), the **collision-coalescence process** can play a significant role in producing precipitation. To produce the many collisions necessary to form a raindrop, some cloud droplets must be larger than others. Larger drops can form on large condensation nuclei, such as salt particles, or they may form through random collisions of droplets. Studies suggest that turbulent mixing between the cloud and its drier environment can play a role in producing larger droplets.

● **FIGURE 7.3** The curved line represents the relative humidity needed to keep a droplet in equilibrium with its environment. For a given droplet size, the droplet will evaporate and shrink when the relative humidity is less than that given by the curve. The droplet will grow by condensation when the relative humidity is greater than the value on the curve.

▼ **TABLE 7.1** **Terminal Velocity of Different Size Particles Involved in Condensation and Precipitation Processes**

Diameter (μm)	TERMINAL VELOCITY		Type of Particle
	m/sec	ft/sec	
0.2	0.0000001	0.0000003	Condensation nuclei
20	0.01	0.03	Typical cloud droplet
100	0.27	0.9	Large cloud droplet
200	0.7	2.3	Large cloud droplet or drizzle
1000	4	13.1	Small raindrop
2000	6.5	21.4	Typical raindrop
5000	9	29.5	Large raindrop

As cloud droplets fall, air slows them down. The amount of air resistance depends on the size of the drop and on its rate of fall: The greater its speed, the more air molecules the drop encounters each second. The speed of the falling drop increases until the air resistance equals the pull of gravity. At this point, the drop continues to fall, but at a constant speed, which is called its **terminal velocity**. Because larger drops have a smaller surface-area-to-weight ratio, they must fall faster before reaching their terminal velocity. Thus *larger drops fall faster than smaller drops* (see ▼ Table 7.1). Note in Table 7.1 that, in calm air, a typical raindrop falls over 600 times faster than a typical cloud droplet!

Eventually, droplets overtake and collide with smaller drops in their path. This merging of cloud droplets by collision is called **coalescence**. Within a cloud, there can be both updrafts and downdrafts. If an updraft is especially strong, then droplets of various sizes can all be pushed upward. Coalescence may now occur as smaller droplets are pushed upward more quickly, colliding with larger droplets in their path. Laboratory studies show that collision does not always guarantee coalescence; sometimes the droplets actually bounce apart during collision. For example, the forces that hold a tiny droplet together (surface tension) are so strong that if one tiny droplet were to collide with another, chances are the two would not stick together (coalesce) (see ● Fig. 7.4). Coalescence appears to be enhanced if colliding droplets have opposite (and, hence, attractive) electrical charges.* An important factor influencing cloud droplet growth by the collision process is the amount of time the droplet spends in the cloud. A very large cloud droplet of 200 μm falling in still air takes about 12 minutes to travel through a cloud 500 m (1640 ft) thick and over an hour if the cloud thickness is 2500 m (8200 ft). Rising air currents in a forming cloud slow the rate at which droplets fall toward the ground. Consequently, a thick cloud with strong updrafts maximizes the time cloud droplets spend in the cloud and, hence, the size to which they can grow.

A warm stratus cloud is typically less than 500 m thick and has slow upward air movement (generally less than 0.1 m/sec). Under these conditions, a large droplet would be in the cloud for a relatively short time and grow by coalescence to only about 200 μm. If the air beneath the cloud is moist, the droplets may reach the ground as *drizzle*, the lightest form of rain. If, however, the stratus cloud base is fairly high above the ground, the drops will evaporate before reaching the surface, even when the relative humidity is 90 percent.

Clouds that have above-freezing temperatures at all levels are called *warm clouds*. In such clouds, precipitation forms by the collision and coalescence process. For example, in tropical regions, where warm cumulus clouds build to great heights, convective updrafts of at least 1 m/sec (and some exceeding many tens of meters per second) occur. Look at the warm cumulus cloud in ● Fig. 7.5. Suppose a cloud droplet of 100 μm is caught in an updraft whose velocity is 6.5 m/sec (about 15 mi/hr). As the droplet rises, it collides with and captures smaller drops in its path and grows until it reaches a size of about 1000 μm. At this point, the updraft in the cloud is just able to balance the pull of gravity on the drop. Here, the drop remains suspended until it grows just a little bigger. Once the fall velocity of the drop is greater than the updraft velocity in the cloud, the drop slowly descends. As the drop falls, larger cloud droplets are captured by the falling drop, which then grows larger. By the time this drop reaches the bottom of the cloud, it will be a large raindrop with a diameter of over 5000 μm (5 mm). Because raindrops of this size fall faster and reach the ground first, they typically occur at the beginning of a rain shower originating in these warm, convective cumulus clouds.

Raindrops that reach Earth's surface are seldom much larger than about 5 mm. The collisions between raindrops (whether

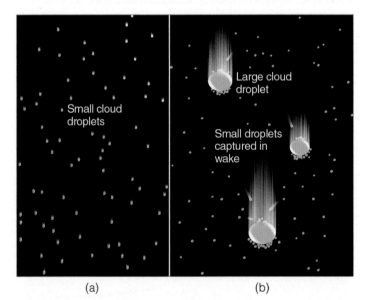

● **FIGURE 7.4** Collision and coalescence. (a) In a warm cloud composed only of small cloud droplets of uniform size, the droplets are less likely to collide as they all fall very slowly at about the same speed. Those droplets that do collide, frequently do not coalesce because of the strong surface tension that holds together each tiny droplet. (b) In a cloud composed of droplets of varying sizes, larger droplets fall faster than smaller droplets. Although some tiny droplets are swept aside, some collect on the larger droplet's forward edge, while others (captured in the wake of the larger droplet) coalesce on the droplet's back side.

---

*It was once thought that atmospheric electricity played a significant role in the production of rain. Today, many scientists feel that the difference in electrical charge that exists between cloud droplets results from the bouncing collisions between them. It is felt that the weak separation of charge and the weak electrical fields in developing, relatively warm clouds are not significant in initiating precipitation. However, studies show that coalescence is often enhanced in thunderstorms where strongly charged droplets exist in a strong electrical field.

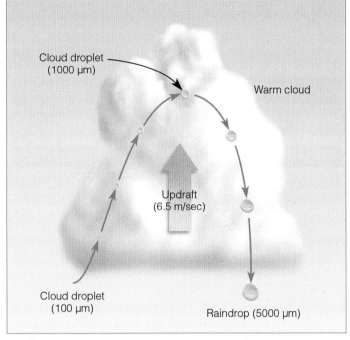

● **FIGURE 7.5** A cloud droplet rising then falling through a warm cumulus cloud can grow by collision and coalescence, and emerge from the cloud as a large raindrop.

glancing or head-on) tend to break them up into many smaller drops. Additionally, a large drop colliding with another large drop may result in oscillations within the combined drop. As the resultant drop grows, these oscillations may tear the drop apart into many fragments, all smaller than the original drop.

So far, we have examined the way cloud droplets in warm clouds (that is, those clouds with temperatures above freezing) grow large enough by the collision-coalescence process to fall as raindrops. Rain that falls from warm clouds is sometimes called *warm rain*. The most important factor in the production

of raindrops is the cloud's *liquid water content*. In a cloud with sufficient water, other significant factors are:

1. the range of droplet sizes
2. the cloud thickness
3. the updrafts of the cloud
4. the electric charge of the droplets and the electric field in the cloud

Relatively thin stratus clouds with slow, upward air currents are, at best, only able to produce drizzle, whereas the towering cumulus clouds associated with rapidly rising air can cause heavy showers. Now, Let's turn our attention to seeing how clouds with temperatures below freezing are able to produce precipitation.

ICE-CRYSTAL (BERGERON) PROCESS The **ice-crystal** (or **Bergeron) process*** of rain formation is extremely important in middle and high latitudes, where clouds extend upward into regions where the air temperature is well below freezing. Such clouds are called *cold clouds*. ● Figure 7.6 illustrates a typical cold cloud that has formed over the Great Plains, where the "cold" part is well above the 0°C isotherm.

Suppose we take an imaginary balloon flight up through the cumulonimbus cloud in Fig. 7.6. Entering the cloud, we observe cloud droplets growing larger by processes described in the previous section. As expected, only water droplets exist here, for the base of the cloud is warmer than 0°C. Surprisingly, in the cold air just above the 0°C isotherm, almost all of the cloud droplets are still composed of liquid water. Water droplets existing at temperatures below freezing are referred to as **supercooled droplets**. Even at higher levels, where the air temperature is −10°C (14°F), there is only one ice crystal for every million liquid droplets. Near 5500 m (18,000 ft), where the temperature becomes −20°C (−4°F),

---

*The ice-crystal process is also known as the *Bergeron process* after the Swedish meteorologist Tor Bergeron, who proposed that essentially all raindrops begin as ice crystals.

● **FIGURE 7.6** The distribution of ice and water in a typical cumulonimbus cloud.

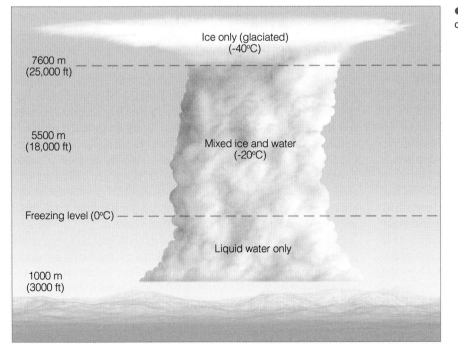

The Freezing of Tiny Cloud Droplets

Over large bodies of fresh water, ice ordinarily forms when the air temperature drops slightly below 0°C. Yet, a cloud droplet of pure water about 25 μm in diameter will not freeze spontaneously until the air temperature drops to about −40°C (−40°F) or below.

The freezing of pure water (without the benefit of some nucleus) is called *spontaneous or homogeneous freezing*. For this type of freezing to occur, enough molecules within the water droplet must join together in a rigid pattern to form a tiny ice structure, or ice *embryo*. When the ice embryo grows to a critical size, it acts as a nucleus. Other molecules in the droplet then attach themselves to the nucleus of ice and the water droplet freezes.

Tiny ice embryos form in water at temperatures just below freezing, but at these temperatures thermal agitations are large enough to weaken their structure. The ice embryos simply form and then break apart. At lower temperatures, thermal motion is reduced, making it

● **FIGURE 1** This cirrus cloud is probably composed entirely of ice crystals, because any liquid water droplet, no matter how small, must freeze spontaneously at the very low temperature (below −40°C) found at this altitude, 9 km (29,500 ft).

easier for bigger ice embryos to form. Hence, freezing is more likely.

The chances of an ice embryo growing large enough to freeze water before the embryo is broken up by thermal agitation increase with larger volumes of water. Consequently, only larger cloud droplets will freeze by homogeneous freezing at air temperatures higher than

−40°C. In air colder than −40°C, however, it is almost certain that an ice embryo will grow to critical size in even the smallest cloud droplet. Thus, any cloud that forms in extremely cold air (below −40°C), such as cirrus clouds (see ● Fig. 1), will almost certainly be composed of ice, since any cloud droplets that form will freeze spontaneously.

---

ice crystals become more numerous, but are still outnumbered by water droplets.* The distribution of ice crystals, however, is not uniform, as the downdrafts contain more ice than the updrafts. Not until we reach an elevation of 7600 m (25,000 ft), where temperatures drop below −40°C (also −40°F), do we find *only* ice crystals. The region of a cloud where only ice particles exist is called *glaciated*. Why are there so few ice crystals in the middle of the cloud, even though temperatures there are well below freezing? Laboratory studies reveal that the smaller the amount of pure water, the lower the temperature at which water freezes. Given that cloud droplets are extremely small, it takes very low temperatures to turn them into ice. (More on this topic is given in Focus section 7.1.)

Just as liquid cloud droplets form on condensation nuclei, ice crystals may form in subfreezing air on particles called **ice nuclei**. The number of ice-forming nuclei available in the atmosphere is small, especially at temperatures above −10°C (14°F). However, as the temperature decreases, more particles become active and promote freezing. Although some uncertainty exists regarding the principal source of ice nuclei, it is known that clay minerals, such as kaolinite, become effective nuclei at temperatures near −15°C (5°F). Some types of bacteria in decaying plant leaf material are also effective ice nuclei, as are ice crystals themselves and

other particles whose geometry resembles that of an ice crystal. However, it is difficult to find substances in nature that have a lattice structure similar to ice, given that there are so many possible lattice structures. In the atmosphere, it is easy to find hygroscopic ("water-seeking") particles. Consequently, ice-forming nuclei are rare compared to cloud condensation nuclei.

In a cold cloud, there may be several types of ice-forming nuclei present. For example, certain ice nuclei allow water vapor to deposit as ice directly onto their surfaces in cold, saturated air. These are called *deposition nuclei* because, in this situation, water vapor changes directly into ice without going through the liquid phase. Ice nuclei that promote the freezing of supercooled liquid droplets are called *freezing nuclei*. Some freezing nuclei cause freezing after they are immersed in a liquid drop; some promote condensation, then freezing; still others cause supercooled droplets to freeze if they collide with them. This last process is called **contact freezing**, and the particles involved are called *contact nuclei*. Studies suggest that contact nuclei can be just about any substance and that contact freezing may be the dominant force in the production of ice crystals in some clouds.

We can now understand why there are so few ice crystals in the cold mixed region of some clouds. Cloud droplets may freeze spontaneously, but only at the very low temperatures usually found at high altitudes. Ice nuclei may initiate the growth of ice crystals, but they do not abound in nature. Because there are many more cloud condensation nuclei than ice nuclei, we are left with a cold cloud that contains many more liquid droplets than

---

*In continental clouds, such as the one shown in Fig. 7.6, where there are many small cloud droplets less than 20 μm in diameter, ice-crystal formation begins at temperatures between −9°C and −15°C. In clouds where larger but fewer cloud droplets are present, ice crystals begin to form at temperatures between −4°C and −8°C.

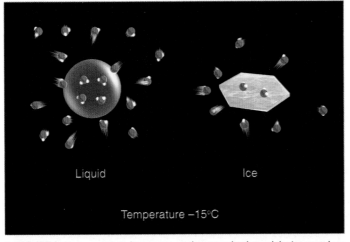

● FIGURE 7.7 In a saturated environment, the water droplet and the ice crystal are in equilibrium, as the number of molecules leaving the surface of each droplet and ice crystal equals the number returning. There are more water vapor molecules above the droplet than above the ice, which produces a greater vapor pressure above the droplet. At saturation, then, the pressure exerted by the water molecules is greater over the water droplet than over the ice crystal.

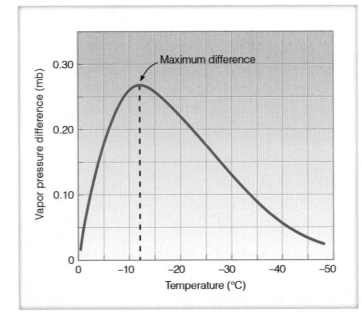

● FIGURE 7.8 The difference in saturation vapor pressure between supercooled water and ice at different temperatures.

ice particles, even at temperatures as low as −10°C (14°F). Neither the tiny liquid nor solid particles are large enough to fall as precipitation. How, then, does the ice-crystal (Bergeron) process produce rain and snow?

In the subfreezing air of a cloud, many supercooled liquid droplets will surround each ice crystal. Suppose that the ice crystal and liquid droplet in ● Fig. 7.7 are part of a cold (−15°C), supercooled saturated cloud. Because the air is saturated, both the liquid droplet and the ice crystal are in equilibrium, meaning that the number of molecules leaving the surface of both the droplet and the ice crystal must equal the number of molecules returning. Observe, however, that there are more vapor molecules above the liquid. The reason for this fact is that molecules escape the surface of water much more easily than they escape the surface of ice. Consequently, more molecules escape the water surface at a given temperature, requiring more in the vapor phase to maintain saturation. This situation reflects the important fact discussed briefly in Chapter 4: At the same subfreezing temperature, *the saturation vapor pressure just above a water surface is greater than the saturation vapor pressure above an ice surface*. This difference in saturation vapor pressure between water and ice is illustrated in ● Fig. 7.8.

This difference in saturation vapor pressure causes water vapor molecules to move (diffuse) from the water droplet toward the ice crystal. The removal of vapor molecules reduces the vapor pressure above the water droplet. Since the droplet is now out of equilibrium with its surroundings, it evaporates to replenish the diminished supply of water vapor above it. This process provides a continuous source of moisture for the ice crystal, which absorbs the water vapor and grows rapidly (see ● Fig. 7.9). Hence, during the *ice-crystal (Bergeron) process, ice crystals grow larger at the expense of the surrounding water droplets*.

The constant supply of moisture to the ice crystal allows it to enlarge rapidly. At some point, the ice crystal becomes heavy enough to overcome updrafts in the cloud and begins to fall. But a single falling ice crystal does not comprise a snowstorm; consequently, other ice crystals must quickly form.

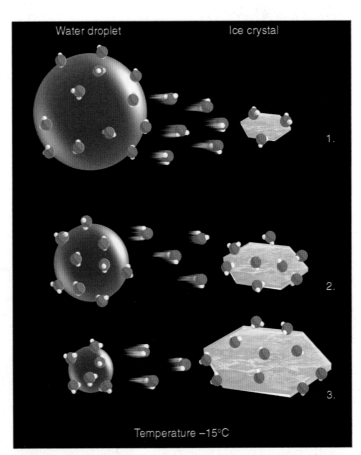

● FIGURE 7.9 The ice-crystal (Bergeron) process. (1) The greater number of water vapor molecules around the liquid droplet causes water molecules to diffuse from the liquid droplet toward the ice crystal. (2) The ice crystal absorbs the water vapor and grows larger, while (3) the water droplet grows smaller.

CRITICAL THINKING QUESTION Would you expect the ice-crystal (Bergeron) process to be more effective at a temperature near freezing, rather than at −15°C as shown here? If not, why not?

In some clouds, especially those with relatively warm tops, ice crystals might collide with supercooled droplets. Upon contact, the liquid droplets freeze into ice and stick together. This process of ice crystals growing larger as they collide with supercooled cloud droplets is called **accretion**. The icy matter that forms is called **graupel** (or *snow pellets*). As the graupel falls, it may fracture or splinter into tiny ice particles when it collides with cloud droplets. These splinters may grow to become new graupel, which, in turn, may produce more splinters.

In colder clouds, the delicate ice crystals may collide with other crystals and fracture into smaller ice particles, or tiny seeds, which freeze hundreds of supercooled droplets on contact. In both cases a chain reaction can develop, producing many ice crystals. As the ice crystals fall, they may collide and stick to one another. The process of ice crystals colliding and then sticking together is called **aggregation**.* The end product of this clumping together of ice crystals is a **snowflake** (see ● Fig. 7.10). If the snowflake melts before reaching the ground, it continues its fall as a raindrop. Much of the rain falling in middle and high latitudes—even in summer—actually begins as snow.

For ice crystals to grow large enough to produce precipitation there must be many, many times more water droplets than ice crystals. Generally, the ratio of ice crystals to water droplets must be on the order of 1:100,000 to 1:1,000,000. When there are too few ice crystals in the cloud, each crystal grows large and falls out of the cloud, leaving the majority of the cloud behind (unaffected). Because there are very few ice crystals, there is very little precipitation. If, on the other hand, there are too many ice crystals

(such as an equal number of crystals and droplets), then each ice crystal receives the mass of one droplet. This would create a cloud of many tiny ice crystals, each too small to fall to the ground, and no precipitation. Now, if the ratio of crystals to droplets is on the order of 1:100,000, then each ice crystal would receive the mass of 100,000 droplets. Most of the cloud would convert to precipitation, as the majority of ice crystals would grow large enough to fall to the ground as precipitation.

The first person to formally propose the theory of ice-crystal growth due to differences in the vapor pressure between ice and supercooled water was Alfred Wegener (1880–1930), a German climatologist who also proposed the geological theory of continental drift. In the early 1930s, important additions to this theory of ice-crystal growth were made by the Swedish meteorologist Tor Bergeron. Several years later, the German meteorologist Walter Findeisen made additional contributions to Bergeron's theory; hence, the ice-crystal theory of rain formation has come to be known as the *Wegener-Bergeron-Findeisen process,* or, simply, the *Bergeron process.*

CLOUD SEEDING AND PRECIPITATION   The primary goal in many **cloud seeding** experiments is to inject (or seed) a cloud with small particles that will act as nuclei, so that the cloud particles will grow large enough to fall to the surface as precipitation. The first ingredient in any seeding project is, of course, the presence of clouds, as seeding does not generate clouds. However, not just any cloud will do. For optimum results, the cloud must be cold; that is, at least a portion of it (preferably the upper part) must be supercooled, because cloud seeding uses the ice-crystal (Bergeron) process to cause the cloud particles to grow.

---

*Significant aggregation seems possible only when the air is relatively warm, usually warmer than −10°C (14°F).

(a) Falling ice crystals may freeze supercooled droplets on contact (accretion), producing larger ice particles.

(b) Falling ice particles may collide and fracture into many tiny (secondary) ice particles.

(c) Falling ice crystals may collide and stick to other ice crystals (aggregation), producing snowflakes.

● **FIGURE 7.10** Ice particles in clouds.

The idea in cloud seeding is to first find clouds that have *too low* a ratio of ice crystals to droplets and then to add enough artificial ice nuclei so that the ratio of crystals to droplets is about 1:100,000. However, it should be noted that the natural ratio of ice nuclei to cloud condensation nuclei in a typical cold cloud is about 1:100,000, just about optimal for producing precipitation.

Some of the first experiments in cloud seeding were conducted by Vincent Schaefer and Irving Langmuir during the late 1940s. To seed a cloud, they dropped crushed pellets of *dry ice* (solid carbon dioxide) from a plane. Because dry ice has a temperature of −78°C (−108°F), it acts as a cooling agent. As the extremely cold, dry ice pellets fall through the cloud, they quickly cool the air around them. This cooling causes the air around the pellet to become supersaturated. In this supersaturated air, water vapor directly forms many tiny cloud droplets. In the very cold air created by the falling pellets (below −40°C), the tiny droplets instantly freeze into tiny ice crystals. The newly formed ice crystals then grow larger by deposition as the water vapor molecules attach themselves to the ice crystals at the expense of the nearby liquid droplets and, upon reaching a sufficiently large size, fall as precipitation.

In 1947, Bernard Vonnegut demonstrated that silver iodide (AgI) could be used as a cloud-seeding agent. Because silver iodide has a crystalline structure similar to an ice crystal, it acts as an effective ice nucleus at temperatures of −4°C (25°F) and lower. Silver iodide causes ice crystals to form in two primary ways:

1. Ice crystals form when silver iodide crystals come in contact with supercooled liquid droplets.
2. Ice crystals grow in size as water vapor is deposited onto the silver iodide crystal.

Silver iodide is much easier to handle than dry ice, as it can be supplied to the cloud from burners located either on the ground or on the wing of a small aircraft. Although other substances, such as lead iodide and cupric sulfide, are also effective ice nuclei, silver iodide still remains the most commonly used substance in cloud-seeding projects. (Additional information on the controversial topic of cloud seeding's effectiveness is given in Focus section 7.2.)

Under certain conditions, clouds may be seeded naturally. For example, when cirriform clouds lie directly above a lower cloud deck, ice crystals may descend from the higher cloud and seed the cloud below (see ● Fig. 7.11). As the ice crystals mix into the lower cloud, supercooled droplets are converted to ice crystals, and the precipitation process is enhanced. Sometimes the ice crystals in the lower cloud may settle out, leaving a clear area or "hole" in the cloud (see Fig. 2 in the Focus section 7.2). When the cirrus clouds form waves downwind from a mountain chain, bands of precipitation often form—producing heavy precipitation in some areas and practically no precipitation in others (see ● Fig. 7.12).

PRECIPITATION IN CLOUDS
In cold, strongly convective clouds, precipitation may begin only minutes after the cloud forms and may be initiated by either the collision-coalescence or

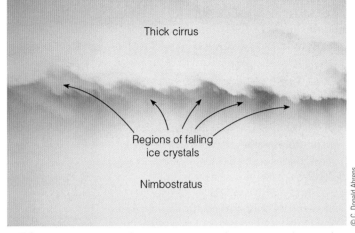

● **FIGURE 7.11** Ice crystals falling from a dense cirriform cloud into a lower nimbostratus cloud. This photo was taken at an altitude near 6 km (19,700 ft) above western Pennsylvania. At the surface, moderate rain was falling over the region.

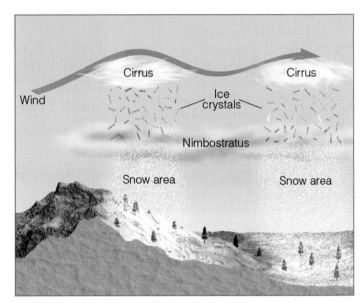

● **FIGURE 7.12** Natural seeding by cirrus clouds can form bands of precipitation downwind of a mountain chain. Notice that heavy snow is falling only in the seeded areas.

the ice-crystal (Bergeron) process. Once either process begins, most precipitation growth is by accretion, as supercooled liquid droplets freeze on impact with snowflakes and ice crystals. Although precipitation is commonly absent in warm-layered clouds, such as stratus, it is often associated with such cold-layered clouds as nimbostratus and altostratus. This precipitation is thought to form principally by the ice-crystal (Bergeron) process because the liquid water content of these clouds is generally lower than that in convective clouds, thus making the collision-coalescence process much less effective. Nimbostratus clouds are normally thick enough to extend to levels where air temperatures are quite low, and they usually last long enough for the ice-crystal (Bergeron) process to initiate precipitation. ● Figure 7.13 illustrates how ice crystals produce precipitation in clouds of both low and high liquid water content.

Does Cloud Seeding Enhance Precipitation?

How effective is artificial seeding with silver iodide or other substances to increase precipitation? In any given year, dozens of cloud-seeding projects are taking place around the world. However, the ability of cloud seeding to reliably increase precipitation is a much-debated question among meteorologists.

First of all, it is difficult to evaluate the results of any cloud-seeding experiment. When a seeded cloud produces precipitation, the question always remains as to how much precipitation would have fallen had the cloud not been seeded. Other factors must be considered when evaluating cloud-seeding experiments: the type of cloud, its temperature and moisture content, the droplet size distribution, and the updraft velocities in the cloud.

Some experiments suggest that cloud seeding in some areas, *under the right conditions*, may enhance precipitation by anywhere from 5 percent to 20 percent or more. However, the results vary from project to project, and it can be difficult to confirm or disprove claims of success.

Some cumulus clouds show an "explosive" growth after being seeded. The latent heat given off when the droplets freeze acts to warm the cloud, causing it to become more buoyant. It grows rapidly and becomes a longer-lasting cloud, which may produce more precipitation.

The business of cloud seeding can be a bit tricky, because overseeding can produce too many ice crystals. When this phenomenon occurs, the cloud becomes glaciated (all liquid droplets become ice) and the ice particles, being very small, do not fall as precipitation. Since few liquid droplets exist, the ice crystals cannot

● **FIGURE 2** When an aircraft flies through a layer of altocumulus clouds composed of supercooled droplets, a hole in the cloud layer may form. The cirrus-type cloud in the center is probably the result of inadvertent cloud seeding by the aircraft.

grow by the ice-crystal (Bergeron) process; rather, they evaporate, leaving a clear area in a thin, stratified cloud (see ● Fig. 2). Because dry ice can produce the most ice crystals in a super-cooled cloud, it is the substance most suitable for deliberate overseeding. Hence, it has been the substance most commonly used to dissipate cold fog at airports.

Warm clouds with temperatures above freezing have also been seeded in an attempt to produce rain. Tiny water drops and particles of hygroscopic salt are injected into the base (or top) of the cloud. These particles (called *seed drops*),

when carried into the cloud by updrafts, create large cloud droplets, which grow even larger by the collision-coalescence process. Apparently, the seed drop size plays a major role in determining the effectiveness of seeding with hygroscopic particles. To date, however, the results obtained using this method are inconclusive.

In summary, cloud seeding in certain instances can lead to more precipitation; in others, to less precipitation, and, in still others, to no change in precipitation amounts. Many of the questions about cloud seeding have yet to be resolved.

BRIEF REVIEW

In the last few sections we encountered a number of important concepts and ideas about how cloud droplets can grow large enough to fall as precipitation. Before examining the various types of precipitation, here is a summary of some of the important ideas presented so far:

● Cloud droplets are very small, much too small to fall as rain.

● The smaller the cloud droplet, the greater its curvature relative to water molecules, and the more likely it will evaporate.

● Cloud droplets form on cloud condensation nuclei. Hygroscopic nuclei, such as salt, allow condensation to begin when the relative humidity is less than 100 percent.

● Cloud droplets, in above-freezing air, can grow larger as faster-falling, bigger droplets collide and coalesce with smaller droplets in their path.

● In the ice-crystal (Bergeron) process of rain formation, both ice crystals and liquid cloud droplets must coexist at below-freezing temperatures. The difference in saturation vapor pressure between liquid and ice causes water vapor to diffuse from the liquid droplets (which shrink) toward the ice crystals (which grow).

● Most of the rain that falls over middle latitudes results from melted snow that formed from the ice-crystal (Bergeron) process.

● Cloud seeding with silver iodide can only be effective in coaxing precipitation from a cloud if the cloud is supercooled and the correct ratio of cloud droplets to ice crystals exists.

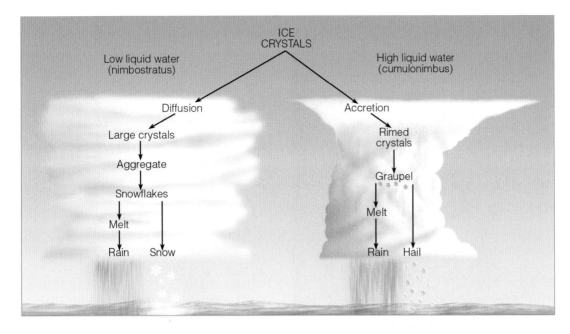

● **FIGURE 7.13** How ice crystals grow and produce precipitation in clouds with a low liquid water content (left) and a high liquid water content (right). The nimbostratus cloud at the left has been expanded vertically to show internal processes; in the actual atmosphere, the nimbostratus cloud at left is shallower than the cumulonimbus at right.

Precipitation Types

Up to now, we have seen how cloud droplets can grow large enough to fall to the ground as rain or snow. While falling, raindrops and snowflakes can be altered by atmospheric conditions encountered beneath the cloud and transformed into other forms of precipitation that can profoundly influence our environment.

RAIN Most people consider **rain** to be any falling drop of liquid water. To the meteorologist, however, that falling drop must have a diameter equal to, or greater than, 0.5 mm to be considered rain. Fine, uniform drops of water with diameters are smaller than 0.5 mm (which is a diameter about one-third the width of the letter "o" in the print version of this page) are called **drizzle**. Most drizzle falls from stratus clouds; however, small raindrops may fall through air that is unsaturated, partially evaporate, and reach the ground as drizzle. Surfaces can be moistened by fog or mist, especially in windy conditions, even though the droplets in fog and mist are too tiny to fall to the ground as precipitation. In various parts of the world, where the air is typically moist but rainfall is scarce, *fog collectors* are used to gather fresh water.

Occasionally, the rain falling from a cloud never reaches the surface because the low humidity causes rapid evaporation. As the drops become smaller, their rate of fall decreases, and they appear to hang in the air as a rain streamer. These evaporating streaks of precipitation are called **virga*** (see ● Fig. 7.14).

Raindrops can also fall from a cloud and not reach the ground if they encounter rapidly rising air. Large raindrops have a terminal velocity of about 9 m/sec (20 mi/hr), and, if they encounter rising air whose speed is greater than 9 m/sec, they will not reach the surface. If the updraft weakens or changes direction and becomes a downdraft, the suspended drops will fall to the ground as a sudden rain **shower**. The showers falling from cumuliform clouds, such as cumulonimbus or cumulus congestus, are usually brief and sporadic, as the cloud moves overhead and then drifts on by. If the shower is excessively heavy, it may be informally called a *cloudburst*. Beneath a cumulonimbus cloud, which normally contains strong, deep convection currents of rising and descending air, it is entirely possible for one side of a street to be dry (updraft side), while a heavy shower is occurring across the street (downdraft side) (see ● Fig. 7.15). Continuous rain, on the other hand, usually falls from a layered cloud that covers a large area and has weaker, more shallow vertical air currents. These are the conditions normally associated with nimbostratus clouds.

As we saw earlier in this chapter, raindrops that reach Earth's surface are seldom much larger than about 5 mm (0.2 in.), the reason being that the collisions (whether glancing or head-on) between raindrops tend to break them up into many smaller drops. Additionally, when raindrops grow too large they become

---

*Studies suggest that the "rain streamer" is actually caused by ice (which is more reflective) changing to water (which is less reflective). Apparently, most evaporation occurs below the virga line.

Ross DePaola

● **FIGURE 7.14** The streaks of falling precipitation that evaporate before reaching the ground are called *virga*.

unstable and break apart. What is the shape of the falling raindrop? Is it tear-shaped, or is it round? You may be surprised at the answer, which is given in Focus section 7.3.

After a rainstorm, visibility usually improves primarily because precipitation removes (scavenges) many of the suspended particles. When rain combines with gaseous pollutants, such as oxides of sulfur and nitrogen, it becomes acidic. *Acid rain*, which has an adverse effect on plants and water resources, has become a major problem in many industrialized regions of the world over the past few decades. We will investigate the acid rain problem more thoroughly in Chapter 19, which emphasizes air pollution.

It is important to know the interval of time over which rain falls. Did it fall over several days, gradually soaking into the soil?

Or did it come all at once in a cloudburst, rapidly eroding the land, clogging city gutters, and causing floods along creeks and rivers unable to handle the sudden increased flow? The *intensity* of rain is the amount that falls in a given period; intensity of rain is always based on the accumulation during a certain interval of time (see ▼ Table 7.2).

SNOW We know that much of the precipitation reaching the ground actually begins as **snow**. In summer, the freezing level is usually above 3600 m (12,000 ft), and the snowflakes falling from a cloud melt before reaching the ground. However, in winter, the freezing level is much lower, and falling snowflakes have a better chance of survival. Snowflakes can generally fall about 300 m (1000 ft) below the freezing level before completely melting.

Occasionally, you can spot the melting level when you look in the direction of the sun, if it is near the horizon. Because snow scatters incoming sunlight better than rain, the darker region beneath the cloud contains falling snow, while the lighter region is falling rain. The melting zone, then, is the transition between the light and dark areas (see ● Fig. 7.16).

The sky will look different, however, if you are looking directly up at the precipitation. Because snowflakes are such effective scatterers of light, they redirect the light beneath the cloud in all directions—some of it eventually reaching your eyes,

● **FIGURE 7.15** Strong updrafts and downdrafts of a cumulonimbus cloud can cause rain to fall on one side of a street but not on the other.

▼ TABLE 7.2 **Rainfall Intensity**

RAINFALL DESCRIPTION	RAINFALL RATE (IN./HR)*
Light	0.01 to 0.10
Moderate	0.11 to 0.30
Heavy	>0.30

*In the United States, the National Weather Service measures rainfall in inches.

Are Raindrops Tear Shaped?

As rain falls, the drops take on a characteristic shape. Choose the shape in ● Fig. 3 that you think most accurately describes that of a falling raindrop. Did you pick number 1? The tear-shaped drop has been depicted by artists for many years. Unfortunately, *raindrops are not tear shaped*. Actually, the shape depends on the drop size. Raindrops less than 2 mm in diameter are nearly spherical and look like rain-drop number 2. The attraction among the mole-cules of the liquid (surface tension) tends to squeeze the drop into a shape that has the smallest surface area for its total volume—a sphere.

● **FIGURE 3** Which of the three drops shown here represents the real shape of a falling raindrop?

Large raindrops, with diameters exceed-ing 2 mm, take on a different shape as they fall. Surprisingly, they look like the number 3, slightly elongated, flattened on the bottom, and rounded on top. As the larger drop falls, the air pressure against the drop is greatest on the bottom and least on the sides. The pres-sure of the air on the bottom flattens the drop, while the lower pressure on its sides allows it to expand a little. This mushroom shape has been described as resembling everything from a falling parachute to a loaf of bread, or even a hamburger bun. You may call it what you wish, but remember: A raindrop is not tear shaped.

---

making the region beneath the cloud appear a lighter shade of gray. Falling raindrops, on the other hand, scatter very little light toward you, and the underside of the cloud appears dark. It is this change in shading that enables some observers to predict with uncanny accuracy whether precipitation reaching the ground in cold weather will be in the form of rain or snow.

When ice crystals and snowflakes fall from high cirrus clouds they are called **fallstreaks**. Fallstreaks behave in much the same way as virga. As the ice particles fall into drier air, they usually sublimate (that is, change from ice into vapor). Because the winds at higher levels move the cloud and ice particles hori-zontally more quickly than do the slower winds at lower levels, fallstreaks appear as dangling white streamers (see ● Fig. 7.17). Moreover, fallstreaks descending into lower, supercooled clouds may actually seed them.

WEATHER WATCH

Does rain have an odor? Often before it rains the air has a distinc-tive, somewhat earthy smell to it. This odor may originate from soil bacteria that produce aromatic gases. As rain falls onto the soil, it pushes these gases into the air, where winds carry them out ahead of the advancing rain shower.

SNOWFLAKES AND SNOWFALL Snowflakes that fall through moist air that is slightly above freezing slowly melt as they descend. A thin film of water forms on the edge of a flake, act-ing like glue when other snowflakes come in contact with it. In this way, several flakes can join to produce giant snowflakes often measuring several inches or more in diameter. These

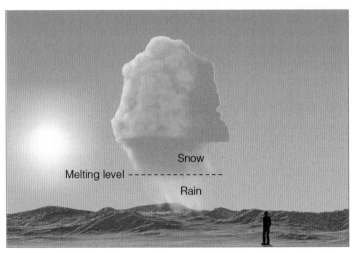

● **FIGURE 7.16** Snow scatters sunlight more effectively than rain. Consequently, when you look toward the sun, the region of falling precipitation looks darker above the melting level than below it.

© C. Donald Ahrens

● **FIGURE 7.17** The dangling white streamers of ice crystals beneath these cirrus clouds are known as *fallstreaks*. The bending of the streaks is due to the changing wind speed with height.

Snowing When the Air Temperature Is Well Above Freezing

In the beginning of this chapter, we learned that it is never too cold to snow. So when is it too warm to snow? A person who has never been in a snowstorm might answer, "When the air temperature rises above freezing." However, anyone who lives in a climate that experiences cold winters knows that snow can fall when the air temperature is considerably above freezing (see ● Fig. 4). In fact, in some areas, snowstorms often begin with a surface air temperature near 2°C (36°F). Why doesn't the falling snow melt in this air? Actually, it does melt, at least to some degree. Let's examine this in more detail.

For falling snowflakes to survive in air with temperatures much above freezing, the air must be unsaturated (relative humidity is less than 100 percent), and the wet-bulb temperature must be at freezing or below. You may recall from our discussion on humidity in Chapter 4 that the wet-bulb temperature is the lowest temperature that can be attained by evaporating water into the air. Consequently, it is a measure of the amount of cooling that can occur in the atmosphere as water evaporates into the air. When rain falls into a layer of dry air with a low wet-bulb temperature, rapid evaporation and cooling occurs, which is why the air temperature often decreases when it begins to rain. During the winter, as raindrops evaporate in this dry air, rapid cooling may actually change a rainy day into a snowy one. This same type of cooling allows snowflakes to survive above freezing (melting) temperatures.

Suppose it is winter and the sky is overcast. At the surface, the air temperature is 2°C (36°F), the dew point is −6°C (21°F), and

● **FIGURE 4** It is snowing at 40°F during the middle of July near the summit of Beartooth Mountain, Montana.

© C. Donald Ahrens

the wet-bulb temperature is 0°C (32°F).* The air temperature drops sharply with height, from the surface up to the cloud deck. Soon, flakes of snow begin to fall from the clouds into the unsaturated layer below. In the above-freezing temperatures, the snowflakes begin to partially melt. The air, however, is dry, so the water quickly evaporates, cooling the air. In addition, evaporation cools the falling snowflake to the wet-bulb temperature, which retards the flake's rate of melting. As snow continues to fall, evaporative cooling causes the air temperature to continue to drop. The addition of water vapor to the air increases the dew point, while the wet-bulb temperature remains essentially unchanged. Eventually, the entire layer of air cools to the wet-bulb temperature and becomes saturated at 0°C. As long as the wind

---

*The wet-bulb temperature is always higher than the dew point, except when the air is saturated. At that point, the air temperature, wet-bulb temperature, and dew point are all the same.

doesn't bring in warmer air, the precipitation remains as snow.

We can see that when snow falls into warmer air (say at 8°C or 46°F), the air must be extremely dry to have a wet-bulb temperature at freezing or below. In fact, with an air temperature of 8°C (46°F) and a wet-bulb temperature of 0°C (32°F), the dew point would be −23°C (−9°F) and the relative humidity 11 percent. Conditions such as these are extremely unlikely at the surface before the onset of precipitation. Actually, the highest air temperature possible with a below-freezing wet-bulb temperature is about 10°C (50°F). Hence, snowflakes will melt rapidly in air with a temperature above this value. However, it is still possible to see flakes of snow at temperatures greater than 10°C (50°F), especially if the snowflakes are swept rapidly earthward by the cold, relatively dry downdraft of a thunderstorm.

---

large, soggy snowflakes are associated with moist air and temperatures near freezing. However, when snowflakes fall through extremely cold air with a low moisture content, small, powdery flakes of "dry" snow accumulate on the ground. (To understand how snowflakes can survive in air that is above freezing, read Focus section 7.4, "Snowing When the Air Temperature Is Well Above Freezing.")

If you catch falling snowflakes on a dark object and examine them closely, you will see that the most common snowflake form is a fernlike branching star shape called a *dendrite*. Since many types of ice crystals grow (see ● Fig. 7.18), why is the dendrite crystal the most common shape for snowflakes? The

type of crystal that forms, as well as its growth rate, depends on the air temperature and relative humidity (the degree of supersaturation between water and ice). ▼ Table 7.3 summarizes the crystal forms (habits) that develop when supercooled water and ice coexist in a saturated environment. Note that dendrites are common at temperatures between −12°C and −16°C. The maximum growth rate of ice crystals depends on the difference in saturation vapor pressure between water and ice, and this difference reaches a maximum in the temperature range where dendrite crystals are most likely to grow. (Look back at Fig. 7.8, on p. 173.) Therefore, this type of crystal grows more rapidly than the other crystal forms. As ice crystals fall through a cloud, they

▼ **TABLE 7.3** **Ice Crystal Habits That Form at Various Temperatures***

ENVIRONMENTAL TEMPERATURE		CRYSTAL HABIT
(°C)	(°F)	
0 to −4	32 to 25	Thin plates
−4 to −6	25 to 21	Needles
−6 to −10	21 to 14	Columns
−10 to −12	14 to 10	Plates
−12 to −16	10 to 3	Dendrites, plates
−16 to −22	3 to −8	Plates
−22 to −40	−8 to −40	Hollow columns

*Note that at each temperature, the type of crystal that forms (e.g., hollow columns versus solid columns) will depend on the difference in saturation vapor pressure between ice and supercooled water.

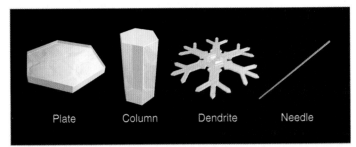

● **FIGURE 7.18** Common ice crystal forms (habits).

Plate Column Dendrite Needle

are constantly exposed to changing temperature and moisture conditions. Because many ice crystals can join together (aggregate) to form a much larger snowflake, snow crystals may assume many complex patterns (see ● Fig. 7.19).

Snow falling from developing cumulus clouds is often in the form of **flurries**. These are usually light showers that fall intermittently for short durations and produce only light accumulations. A more intense snow shower is called a **snow squall**. These brief but heavy falls of snow are comparable to summer rain showers and, like snow flurries, usually fall from cumuliform clouds. If the updrafts are strong enough, the snow may even be accompanied by thunder and lightning. If the snow falls from intense cumuliform clouds producing thunder and lightning, the snow is often referred to as **thundersnow**. A more continuous snowfall (sometimes persisting for several hours) accompanies nimbostratus and altostratus clouds. The intensity of snow is based on its reduction of horizontal visibility at the time of observation (see ▼ Table 7.4). However, the intensity of snow as measured by visibility does not tell us how much water the snow is bringing

▼ **TABLE 7.4** **Snowfall Intensity**

SNOWFALL DESCRIPTION	VISIBILITY
Light	Greater than ½ mile*
Moderate	Greater than ¼ mile, less than or equal to ½ mile
Heavy	Less than or equal to ¼ mile

*In the United States, the National Weather Service determines visibility (the greatest distance you can see) in miles.

WEATHER WATCH

Residents of New England experienced a mishmash of seasons during an intense snowfall on February 9, 2017. Updrafts were strong enough to generate widespread thundersnow, including more than 120 cloud-to-ground lightning strikes. One of those bolts struck a home and destroyed part of a tree in Warwick, Rhode Island.

Scott Cumazine/Science Source

● **FIGURE 7.19** Color-enhanced image of dendrite snowflakes.

to the surface. A moderate snow made up of dense, small flakes can deposit more frozen liquid than a heavy snow consisting of large, fluffy flakes.

When a strong wind is blowing at the surface, snow can be picked up and deposited into huge drifts. Drifting snow is usually accompanied by *blowing snow*; that is, snow lifted from the surface by the wind and blown about in such quantities that horizontal visibility is greatly restricted. The combination of drifting and blowing snow, after falling snow has ended, is called a *ground blizzard*. A true **blizzard** is a weather condition characterized by low temperatures and strong winds (greater than 30 knots) bearing large amounts of fine, dry, powdery particles of snow that reduces visibility to less than one-quarter mile (and sometimes to as little as a few feet) for at least three hours (see ● Fig. 7.20).

WEATHER WATCH

Snowfalls do not have to be enormous to cause major trouble. At midday on January 28, 2014, snow began falling in the Atlanta metropolitan area. Although less than 3 in. accumulated, the snow fell into surface air that was well below freezing, which allowed ice to form quickly on roads just as schools and businesses were closing early. Thousands of people ended up stuck in gridlock for many hours, in some cases overnight. More than 2000 students were forced to spend the night at school, and at least three babies were born in cars that were stuck on area roads.

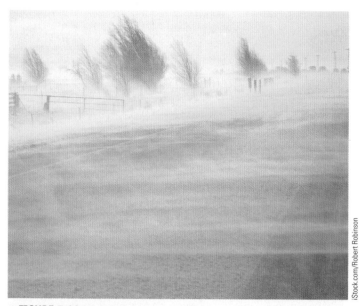

● **FIGURE 7.20** High winds, blowing and falling snow, and cold temperatures accompanied this blizzard over the Great Plains.

A BLANKET OF SNOW A mantle of snow covering the landscape is much more than a beautiful setting—it is a valuable resource provided by nature. A blanket of snow is a good insulator (poor heat conductor). In fact, the more air spaces there are between the individual snowflake crystals, the better insulator this blanket becomes. A light, fluffy covering of snow protects sensitive plants and their root systems from damaging low temperatures by retarding the loss of ground heat.

Radiational cooling from a deep snowpack can lead to extremely cold surface air temperatures, but underneath, the ground maintains a higher temperature than if it were directly exposed to cold air. In this way, snow can prevent the ground from freezing downward to great depths. In cold climates that receive little snow, it is often difficult to grow certain crops because the frozen soil makes spring cultivation almost impossible. Because frozen ground also prevents early spring rains from percolating downward into the soil, it can lead to rapid water runoff and flooding. If subsequent rains do not fall, the soil could even become moisture-deficient. The accumulation of snow in mountains provides for winter recreation, and the melting snow in spring and summer is of great economic value in that it supplies streams and reservoirs with much-needed water. However, as spring approaches, rapid melting of the snowpack may flood low-lying areas.

There are other hardships and potential hazards associated with heavy snowfall. Too much snow on the side of a steep hill or mountain may lead to an *avalanche*. The added weight of snow on the roof of a building may cause it to collapse, leading to costly repairs and even loss of life.

Each winter, heavy snows clog streets and disrupt transportation. To keep traffic moving, streets must be plowed and sanded, or treated with various types of salts or special compounds to lower the temperature at which the snow freezes (melts). This effort can be expensive, especially if the snow is heavy and wet. Cities unaccustomed to snow are usually harder hit by a moderate snowstorm than cities that frequently experience snow. A January snowfall of several inches in New Orleans, Louisiana, can bring traffic to a standstill, while the same amount of snow in Buffalo, New York, would go practically unnoticed. ● Figure 7.21 gives the annual average snowfall across the United States and southern Canada. (A blanket of snow also has an effect on the way sounds are transmitted. More on this subject is given in Focus section 7.5.)

● **FIGURE 7.21** Average annual snowfall over the United States and Canada for 1981–2010.

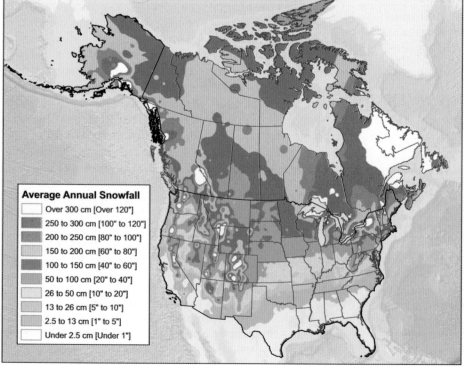

Average Annual Snowfall

- Over 300 cm [Over 120"]
- 250 to 300 cm [100" to 120"]
- 200 to 250 cm [80" to 100"]
- 150 to 200 cm [60" to 80"]
- 100 to 150 cm [40" to 60"]
- 50 to 100 cm [20" to 40"]
- 26 to 50 cm [10" to 20"]
- 13 to 26 cm [5" to 10"]
- 2.5 to 13 cm [1" to 5"]
- Under 2.5 cm [Under 1"]

Sounds and Snowfalls

A blanket of snow is not only beautiful, but it can affect what we hear. You may have noticed that after a snowfall, it seems quieter than usual: Freshly fallen snow can absorb sound—just like acoustic tiles. As the snow gets deeper, this absorption increases. Anyone who has walked through the woods on a snowy evening knows the quiet created by a thick blanket of snow. As snow becomes older and more densely packed, its ability to absorb sound is reduced. That's why sounds you couldn't hear right after a snowstorm become more audible several days later.

New snow covering a pavement will sometimes squeak as you walk in it. The sound produced is related to the snow's temperature. When the air and snow are only slightly below freezing, the pressure from the heel of a boot partially melts the snow. The snow can then

● **FIGURE 5** This freshly fallen blanket of snow absorbs sound waves so effectively that even the water flowing in the tiny stream is difficult to hear.

© C. Donald Ahrens

flow under the weight of the boot and no sound is made. However, on cold days when the snow temperature drops below −10°C (14°F), the heel of the boot will not melt the snow, and the ice crystals are crushed. The crunching of the crystals produces the creaking sound.

SLEET AND FREEZING RAIN Consider the falling snowflake in ● Fig. 7.22. As it falls into warmer air, it begins to melt. When it falls through the deep subfreezing surface layer of air, the partially melted snowflake or cold raindrop turns back into ice, not as a snowflake, but as a *tiny ice pellet* called **sleet**.* Generally, these ice pellets are transparent (or translucent), with diameters of 5 mm (0.2 in.) or less. They bounce when striking the ground

_____

* Occasionally, news media in the United States will use the term sleet to describe a mixture of rain and snow. While this is not the American definition, it is correct in some other nations, including the United Kingdom.

and produce a tapping sound when they hit a window or piece of metal.

The cold surface layer beneath a cloud may be too shallow to freeze raindrops as they fall. In this case, they reach the surface as supercooled liquid drops. Upon striking a cold object, the drops spread out and almost immediately freeze, forming a thin veneer of ice. This form of precipitation is called **freezing rain**, or **glaze**. If the drops are small (less than 0.5 mm in diameter), the precipitation is called **freezing drizzle**. When small, supercooled cloud or fog droplets strike an object whose temperature is below freezing, the tiny droplets freeze, forming an accumulation of white or milky granular ice called **rime** (see ● Fig. 7.23).

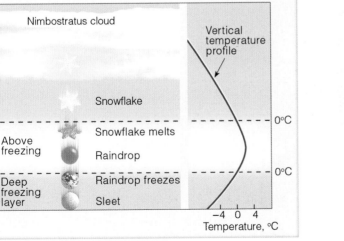

● **FIGURE 7.22** Sleet forms when a partially melted snowflake or a cold raindrop freezes into a pellet of ice before reaching the ground.

NCAR/UCAR/NSF

● **FIGURE 7.23** An accumulation of rime forms on tree branches as supercooled fog droplets freeze on contact in the below-freezing air.

● **FIGURE 7.24** A heavy coating of freezing rain (glaze) covers Syracuse, New York, causing tree limbs to break and power lines to sag. This massive ice storm in January 1998 left millions without power and caused over $1 billion in damage in northern New England and southeast Canada.

Occasionally, light rain, drizzle, or supercooled fog droplets will come in contact with surfaces such as bridges and overpasses that have cooled to a temperature below freezing. The tiny liquid droplets freeze on contact to road surfaces or pavements, producing a sheet of ice that often appears relatively dark. Such ice, usually called **black ice**, can produce extremely hazardous driving conditions.

Freezing rain can create a beautiful winter wonderland by coating everything with silvery, glistening ice. At the same time, highways turn into skating rinks for automobiles, and the destructive weight of the ice—which can be many tons on a single tree—breaks tree branches, power lines, and telephone cables. When there is a substantial accumulation of freezing rain, the result is an **ice storm** (see ● Fig. 7.24).

A case in point is the huge ice storm of January 1998, which left millions of people without power in northern New England and Canada and caused over $1 billion in damages. Another catastrophic ice storm struck an area from Oklahoma to West Virginia in January 2009. More than 2 million people lost power, and 65 people died. Many of the deaths were attributed to the use of emergency heaters without proper ventilation, which led to carbon monoxide poisoning. The area most frequently hit by these storms extends over a broad region from Texas into Minnesota and eastward into the middle Atlantic states and New England. Such storms are extremely rare in most of California and Florida (see ● Fig. 7.25). (For additional information on freezing rain and its effect on aircraft, read Focus section 7.6.)

In summary, ● Fig. 7.26 shows various winter temperature profiles and the type of precipitation associated with each. In profile (a), the air temperature is below freezing at all levels, and snowflakes reach the surface. In (b), a zone of above-freezing air causes snowflakes to partially melt; then, in the deep subfreezing air at the surface, the liquid freezes into sleet. In the shallow subfreezing surface air in (c), the melted snowflakes, now supercooled liquid drops, freeze on contact, producing freezing rain. In (d), the air temperature is above freezing in a sufficiently deep layer so that precipitation reaches the surface as rain. (Weather symbols for these and other forms of precipitation are presented in Appendix B.)

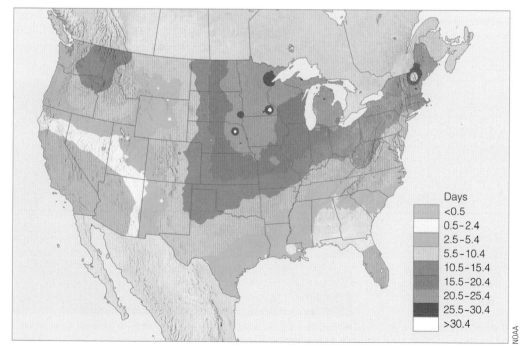

● **FIGURE 7.25** Average annual number of days with freezing rain and freezing drizzle over the United States.

Days
<0.5
0.5–2.4
2.5–5.4
5.5–10.4
10.5–15.4
15.5–20.4
20.5–25.4
25.5–30.4
>30.4

Aircraft Icing

The formation of ice on an aircraft—called *aircraft icing*—can be extremely dangerous, sometimes leading to tragic accidents. Icing is believed to be the probable cause for the crash-landing of a passenger plane as it approached the Detroit airport on January 9, 1997. All 29 people aboard the flight were killed. Fortunately, the number of icing-related U.S. aircraft fatalities has dropped in recent years with increased safety measures.

Consider an aircraft flying through an area of freezing rain or through a region of large supercooled droplets in a cumuliform cloud. As the large, supercooled drops strike the leading edge of the wing, they break apart and form a film of water, which quickly freezes into a solid sheet of ice. This smooth, transparent ice—called *clear ice*—is similar to the freezing rain or glaze that coats trees during ice storms. Clear ice can build up quickly; it is heavy and difficult to remove, even with modern de-icers.

When an aircraft flies through a cloud composed of tiny, supercooled liquid droplets, *rime ice* may form. Rime ice forms when some of the cloud droplets strike the wing and freeze before they have time to spread, thus leaving a rough and brittle coating of ice on the wing. Because the small, frozen droplets trap air between them, rime ice usually appears white (see Fig. 7.23, p. 183). Even though rime ice redistributes the flow of air over the wing more than clear ice does, it is lighter in weight and is more easily removed with de-icers.

● **FIGURE 6** An aircraft undergoing de-icing during inclement winter weather.

Because the raindrops and cloud droplets in most clouds vary in size, a mixture of clear and rime ice usually forms on aircraft. Also, because concentrations of liquid water tend to be greatest in warm air, icing is usually heaviest and most severe when the air temperature is between 0°C and −10°C (32°F and 14°F).

A major hazard to aviation, icing reduces aircraft efficiency by increasing weight. Icing has other adverse effects, depending on where it forms. On a wing or fuselage, ice can disrupt the airflow and decrease the plane's flying capability. When ice forms in the air intake of the engine, it robs the engine of air, causing a reduction in power. Icing may also affect the operation of brakes, landing gear, and instruments. Because of the hazards of ice on an aircraft, its wings are usually sprayed with a type of antifreeze before taking off during cold, inclement weather (see ● Fig. 6). In a snowstorm, the amount of liquid in the snow is the primary factor in determining how much de-icing fluid is needed for aircraft.

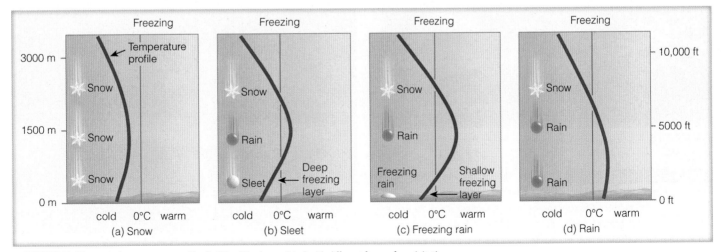

● **FIGURE 7.26** Vertical temperature profiles (solid red line) associated with different forms of precipitation.

SNOW GRAINS AND SNOW PELLETS **Snow grains** are small, opaque grains of ice, the solid equivalent of drizzle. They are fairly flat or elongated, with diameters generally less than 1 mm (0.04 in.). They fall in small quantities from stratus clouds, and never in the form of a shower. Upon striking a hard surface, they neither bounce nor shatter. **Snow pellets**, on the other hand, are white, opaque grains of ice, with diameters less than 5 mm (0.2 in.). They are sometimes confused with snow grains. The distinction is easily made, however, by remembering that, unlike snow grains, snow pellets are brittle, crunchy, and bounce (or break apart) upon hitting a hard surface. They usually fall as showers, especially from cumulus congestus clouds.

To understand how snow pellets form, consider the cumulus congestus cloud with a high liquid water content in ● Fig. 7.27. The freezing level is near the surface and, since the atmosphere is conditionally unstable, the air temperature drops quickly with height. An ice crystal falling into the cold (−23°C) middle region of the cloud would be surrounded by many supercooled cloud droplets and ice crystals. In the very cold air, the crystals tend to rebound after colliding rather than sticking to one another. However, when the ice crystals collide with the supercooled water droplets, they immediately freeze the droplets, producing a spherical accumulation of icy matter (rime) containing many tiny air spaces. These small air bubbles have two effects on the growing ice particle: (1) They keep its density low; and (2) they scatter light, making the particle opaque. By the time the ice particle reaches the lower half of the cloud, it has grown in size and its original shape is gone. When the ice particle accumulates a heavy coating of rime, it is called *graupel*. Since the freezing level is at a low elevation, the graupel reaches the surface as a light, round clump of snowlike ice called a *snow pellet* (see ● Fig. 7.28).

On the surface, the accumulation of snow pellets sometimes gives the appearance of tapioca pudding; hence, it can be referred to as *tapioca snow*. In a thunderstorm, when the freezing level is well above the surface, graupel that reaches the ground is sometimes called *soft hail*. During summer, graupel may melt and reach the surface as large raindrops. In vigorously convective clouds, however, graupel may develop into full-fledged hailstones.

HAIL **Hailstones** are pieces of ice, either transparent or partially opaque, ranging in size from that of small peas to that of golf balls or larger (see ● Fig. 7.29). Some are round; others take on irregular shapes. In the United States, the hailstone with the greatest measured circumference (18.7 in.) fell on Aurora, Nebraska, on June 22, 2003. This giant hailstone, almost as large as a soccer ball, had a measured diameter of 7 in. and probably weighed over 1.75 lbs. The hailstone with the largest diameter (8 in.) and the largest weight (1.94 lbs.) ever measured in the United States fell on Vivian, South Dakota, on July 23, 2010 (see ● Fig. 7.30). The world's heaviest confirmed hailstone, at 2.25 lbs., fell in a Bangladesh thunderstorm on April 14, 1986.

Needless to say, large hailstones are quite destructive. They can break windows, dent cars, and batter roofs of homes. Even small hail can injure livestock and cause extensive damage to

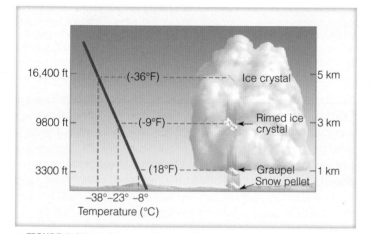

● **FIGURE 7.27** The formation of snow pellets. In the cold air of a convective cloud, with a high liquid water content, ice particles collide with supercooled cloud droplets, freezing them into clumps of icy matter called *graupel*. Upon reaching the relatively cold surface, the graupel is classified as snow pellets.

● **FIGURE 7.28** A snowflake becoming a rimed snowflake, then finally graupel (a snow pellet).

CRITICAL THINKING QUESTION Of all the hailstones you see in this photo (see Fig. 7.29), which probably reached the greatest speed before hitting the surface? Take a guess as to how fast it fell to Earth. Devise a method of obtaining the diameter of large hailstones in a remote area, even though you might not visit the region for a long time after the hailstorm occurred.

● **FIGURE 7.29** Hailstones of varying sizes fell over west Texas during an intense thunderstorm.

© 2009 Warren Faidley

● **FIGURE 7.30** This whopping hailstone fell on Vivian, South Dakota, on July 23, 2010. It had a record diameter of 8 in., weighed a record 1.94 lbs., and had a circumference of 18.6 in.

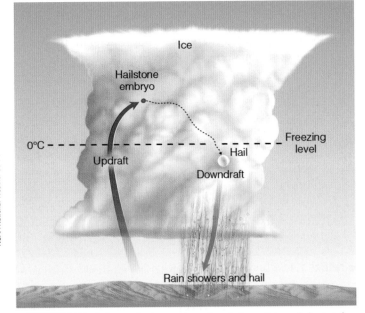

● **FIGURE 7.31** Hailstones begin as embryos (usually ice particles called *graupel*) that remain suspended in the cloud by violent updrafts. The updrafts (or a single broad updraft, as illustrated here) can sweep the ice particles horizontally through the cloud, producing the optimal trajectory for hailstone growth. Along their path, the ice particles collide with supercooled liquid droplets, which freeze on contact. The ice particles eventually grow large enough and heavy enough to fall toward the ground as hailstones.

crops if it falls heavily, especially in a high wind. A single hailstorm can destroy a farmer's crop in a matter of minutes, which is why farmers sometimes call it "the white plague."

Estimates are that, in the United States alone, hail damage amounts to hundreds of millions of dollars annually. The costliest hailstorm in U.S. history swept along Interstate 70 from eastern Kansas into Illinois on May 18, 2001. Hailstones as large as baseballs caused a total of more than $2 billion in damage to homes, businesses, and crops. Although hailstones are potentially lethal, only a few human fatalities due to falling hail have been documented in the United States since 1900. Deaths due to hail are somewhat more frequent in parts of Asia, where 92 people were killed in the Bangladesh thunderstorm of April 1986.

Hail forms in a cumulonimbus cloud—usually an intense thunderstorm—when graupel, or large frozen raindrops, or just about any particles (even insects) act as *embryos* that grow by *accretion*, the accumulation of supercooled liquid droplets. It takes about a million cloud droplets to form a single raindrop, but it takes about 10 billion cloud droplets to form a golf ball–sized hailstone. For a hailstone to grow to this size, it must remain in the cloud between 5 and 10 minutes. Upsurging air currents within the storm carry small ice particles high above the freezing level, where the ice particles grow by colliding with supercooled liquid cloud droplets. Violent, rotating updrafts in severe thunderstorms (especially the long-lived storms called *supercells*) are even capable of sweeping the growing ice particles laterally through the cloud. In fact, it appears that the best trajectory for hailstone growth is one that is nearly horizontal through the storm (see ● Fig. 7.31).

As growing ice particles pass through regions of varying liquid water content, a coating of ice forms around them, causing them to grow larger and larger. In a strong updraft, the larger hailstones ascend very slowly and may appear to "float" in the updraft, where they continue to grow rapidly by colliding with numerous supercooled liquid droplets. When winds aloft carry the large hailstones away from the updraft or when the hailstones reach appreciable size, they become too heavy to be supported by the rising air, and they begin to fall.

In the warmer air below the cloud, the hailstones begin to melt. Small hail often completely melts before reaching the ground, but in the violent thunderstorms of late spring and summer, hailstones often grow large enough to reach the surface before completely melting. Strangely, then, we find the largest form of frozen precipitation occurring during the warmest time of the year.

● Figure 7.32 shows a cut section of a very large hailstone. Notice that it has distinct concentric layers of milky white and clear ice. We know that a hailstone grows by accumulating supercooled water droplets. If the growing hailstone enters a region inside the storm where the liquid water content is relatively low (called the *dry growth regime*), supercooled droplets will freeze immediately on the stone, producing a coating of white or opaque rime ice containing many air bubbles. Should the hailstone get swept into a region of the storm where the liquid water content is higher (called the *wet growth regime*), supercooled water droplets will collect so rapidly on the stone that, due to the release of latent heat, the stone's surface temperature remains at freezing, even though the surrounding air may be much colder. Now the supercooled droplets no longer freeze on impact; instead, they spread a coating of water around the hailstone, filling in the porous regions, which leaves a layer of clear ice around the stone. Therefore, as a hailstone passes through a thunderstorm of changing liquid water content (the dry and wet growth regimes), alternating layers of opaque and clear ice form, as illustrated in Fig. 7.32.

As a thunderstorm moves along, it may deposit its hail in a long narrow band (often over a mile wide and about 6 mi long) known as a **hailstreak**. If the storm should remain almost stationary for a period of time, substantial accumulation of hail is possible. A rare British hailstorm in the town of Ottery St. Mary dropped 9 to 10 inches of hail in two hours in October 2008, and

● **FIGURE 7.32** A large hailstone cut and then photographed under regular light (left) and polarized light (right). This procedure reveals its layered structure.

drifts of up to 6 feet of hail occurred during an intense storm in Cheyenne, Wyoming, in August 1985. During November 2003, an unusual hailstorm dumped more than 5 inches of hail over sections of Los Angeles, California, causing gutters to clog and floods to occur. In addition to its destructive effect, accumulation of hail on a roadway is a hazard to traffic, as when, for example, four people lost their lives near Soda Springs, California, in a 15-vehicle pileup on a hail-covered freeway during September 1989.

Because hailstones are so damaging, various methods have been tried to prevent them from forming in thunderstorms. One method employs the seeding of clouds with large quantities of silver iodide. These nuclei freeze supercooled water droplets and convert them into ice crystals. The ice crystals grow larger as they come in contact with additional supercooled cloud droplets. In time, the ice crystals grow large enough to be called graupel, which then becomes a hailstone embryo. Large numbers of embryos are produced by seeding in hopes that competition for the remaining supercooled droplets can be so great that none of the embryos can grow into large and destructive hailstones. In the United States, the results of most hail-suppression experiments have been inconclusive, although such cloud seeding has been carried out in many areas for a number of years.

Up to this point, we have examined the various types of precipitation. The different types (from drizzle to hail) are summarized in ▼ Table 7.5.

▼ **TABLE 7.5** **Summary of Precipitation Types**

PRECIPITATION TYPE	WEATHER SYMBOL	DESCRIPTION
Drizzle	ˏˏ (light)	Tiny water drops with diameters less than 0.5 mm that fall slowly, usually from a stratus cloud
Rain	•• (light)	Falling liquid drops that have diameters greater than 0.5 mm
Snow	✱✱ (light)	White (or translucent) ice crystals in complex hexagonal (six-sided) shapes that often join together to form snowflakes
Sleet (ice pellets)	△	Frozen raindrops that form as cold raindrops (or partially melted snowflakes) refreeze while falling through a relatively deep subfreezing layer
Freezing rain	∿ (light)	Supercooled raindrops that fall through a relatively shallow subfreezing layer and freeze upon contact with cold objects at the surface
Snow grains (granular snow)	⏦	White or opaque particles of ice less than 1 mm in diameter that usually fall from stratus clouds and are the solid equivalent of drizzle
Snow pellets (graupel)	⧊ (light showers)	Brittle, soft white (or opaque), usually round particles of ice with diameters less than 5 mm that generally fall as showers from cumuliform clouds; they are softer and larger than snow grains
Hail	⧊ (moderate or heavy showers)	Transparent or partially opaque ice particles in the shape of balls or irregular lumps that range in size from that of a pea to that of a softball or larger; the largest form of precipitation. *Severe hail* has a diameter of 1 in. or greater. Almost all hail is produced in thunderstorms

Measuring Precipitation

INSTRUMENTS   Any instrument that can collect and measure rainfall is called a *rain gauge*. A **standard rain gauge** consists of a funnel-shaped collector attached to a long measuring tube (see ● Fig. 7.33). The cross-sectional area of the collector is 10 times that of the tube. Hence, rain falling into the collector is amplified tenfold in the tube, permitting measurements of great precision. A wooden scale, calibrated to allow for the vertical exaggeration, is inserted into the tube and withdrawn. The wet portion of the scale indicates the depth of water. So, 10 in. of water in the tube would be measured as 1 in. of rainfall. This amplification permits rainfall measurements to be as precise as one-hundredth (0.01) of an inch. An amount of rainfall less than one-hundredth of an inch is called a **trace**.

The measuring tube can only collect 2 in. of rain. Rainfall of more than this amount causes an overflow into an outer cylinder. Here, the excess rainfall is stored and protected from appreciable evaporation. When the gauge is emptied, the overflow is carefully poured into the tube and measured.

Another instrument that measures rainfall is the **tipping bucket rain gauge**. In ● Fig. 7.34, notice that this gauge has a receiving funnel leading to two small metal collectors (buckets) attached to each other and mounted on a pivot. The bucket beneath the funnel collects the rain water. When it accumulates the equivalent of one-hundredth of an inch of rain, the weight of the water causes it to tip and empty itself. As the first bucket turns on the pivot, the second bucket immediately moves under the funnel to catch the water. When it fills, it also tips and empties itself, moving the other direction on the pivot, and the original bucket moves back beneath the funnel. Each time a bucket tips, an electric contact is made, causing a pen to register a mark on a remote recording chart. Adding up the total number of marks gives the rainfall for a certain time period. A problem with the tipping bucket rain gauge is that during each "tip" it loses some rainfall and, therefore, undermeasures rainfall amounts, especially during heavy downpours. The tipping bucket is the rain gauge used in the automated (ASOS) weather stations.

Remote recording of precipitation can also be made with a **weighing-type rain gauge**. With this gauge, precipitation is caught in a cylinder and accumulates in a bucket. The bucket sits on a sensitive weighing platform. Special gears translate the accumulated weight of rain or snow into millimeters or inches of precipitation. The precipitation totals are recorded by a pen on chart paper, which covers a clock-driven drum. By using special electronic equipment, this information can be transmitted from rain gauges in remote areas to satellites or land-based stations, thus providing precipitation totals from previously inaccessible regions.

Snow is challenging to measure, since accumulations can vary greatly from one spot to the next, especially when winds are strong. Traditionally, the depth of snow in a region is determined by measuring its depth at three or more representative areas. The amount of snowfall is defined as the average of these measurements. Snow accumulation rates may be measured at a single site by using a *snow measuring board*, a small wooden platform resting near the ground that is cleared off after each measurement (normally every six hours). Snow depth may also be measured by removing the collector and inner cylinder of a standard rain gauge and allowing snow to accumulate in the outer tube. Automated gauges are now used in many areas to collect snowfall and measure the amount of liquid water it contains. Typically, these gauges are surrounded by one or more octagonal fences that help to block wind and produce more accurate readings.

Remote sensing techniques are becoming a popular way to measure snow depth, especially where harsh winter conditions make it difficult to reach observing stations. Laser beams and pulses of ultrasonic energy can be sent from a transmitter to a snowpack, where the energy bounces off the snowpack back to the transmitter. This procedure allows the height of the snow accumulation to be measured in much the same way that radar measures the distance of falling rain from a transmitter. (The topic of radar and precipitation is discussed in the following

● **FIGURE 7.33** Components of the standard rain gauge.

Measuring scale
Collector
Measuring tube
Overflow cylinder

50 cm

20 cm

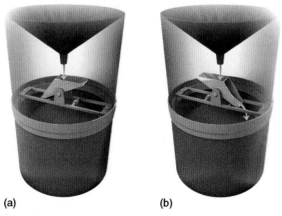

(a)   (b)

● **FIGURE 7.34** The tipping bucket rain gauge. (a) Each time one of the two buckets mounted on a pivot fills with one-hundredth of an inch of rain, it tips (b), sending an electric signal to the remote recorder.

section.) Snowpack height can also be obtained by measuring how long it takes a GPS signal to travel from a satellite to a snow field and then be reflected upward to a receiver located in the snow field.

On average, about 10 in. of snow will melt down to about 1 in. of water, giving a typical fresh snowpack a **water equivalent** * ratio of 10:1. This ratio, however, will vary greatly, depending on the texture and packing of the snow. Very wet snow falling in air near freezing may have a water equivalent of 6:1. On the other hand, in dry powdery snow, the ratio may be as high as 30:1. Toward the end of the winter, large compacted drifts representing the accumulation of many storms may have a water equivalent of less than 2:1.

Determining the water equivalent of snow is a fairly straightforward process: The snow accumulated in a rain gauge is melted and its depth is measured. Another method uses a long, hollow tube pushed into the snow to a desired depth. This snow sample is then melted and poured into a rain gauge for measuring its depth. Knowing the water equivalent of snow can provide valuable information about spring runoff and the potential for flooding, especially in mountain areas.

---

* *Water equivalent is* the depth of water that would result from the melting of a snow sample.

Precipitation is a highly variable weather element. A huge thunderstorm may drench one section of a town while leaving another section completely dry. Given this variability, it should be apparent that a single rain gauge on top of a building cannot represent the total precipitation for any particular region.

DOPPLER RADAR AND PRECIPITATION Radar (*r*adio *d*etection *a*nd *r*anging) has become an essential tool of the atmospheric scientist, because it gathers information about storms and precipitation in otherwise inaccessible regions. Atmospheric scientists use radar to examine the inside of a cloud in much the same way that physicians use X rays to examine the inside of a human body. Essentially, the radar unit consists of a transmitter that sends out short, powerful microwave pulses. When this energy encounters a foreign object—called a *target*—a fraction of the energy is scattered back toward the transmitter and is detected by a receiver (see ● Fig. 7.35). The returning signal is amplified and displayed on a screen, producing an image, or *echo*, from the target. The elapsed time between transmission and reception indicates the target's distance.

Smaller targets require detection by shorter wavelengths. Cloud droplets are detected by radar using wavelengths of 1 cm, whereas longer wavelengths (between 3 and 10 cm) are only weakly scattered by tiny cloud droplets, but are strongly scattered by larger

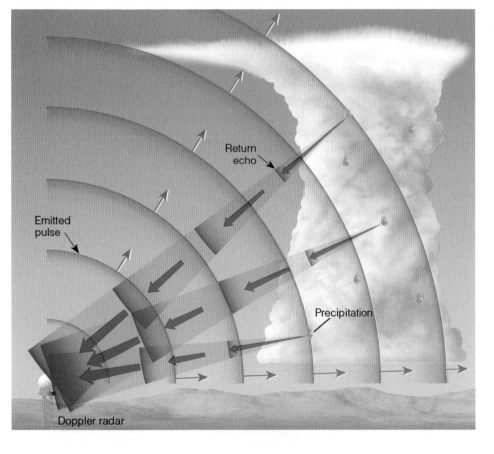

● **FIGURE 7.35** A microwave pulse is sent out from the radar transmitter. The pulse strikes raindrops and a fraction of its energy is reflected back to the radar unit, where it is detected and displayed, as shown in Fig. 7.36.

Return echo

Emitted pulse

Precipitation

Doppler radar

(a) Precipitation intensity (reflectivity)

(b) Precipitation total (estimated 1-hour rainfall)

NOAA/NWS

● **FIGURE 7.36** (a) Doppler radar display showing precipitation intensity over North Carolina for August 27, 2011, as Hurricane Irene moves onshore. (b) Doppler radar display showing one-hour rainfall estimates over North Carolina for August 27, 2011. Notice that in some places Doppler radar estimated that more than 1.50 in. of rain had fallen in one hour.

precipitation particles. The brightness of the echo is directly related to the amount (intensity) of rain, snow, or both falling in the cloud. So, the radar screen shows not only *where* precipitation is occurring, but also how *intense* it is. The radar image typically is displayed using various colors to denote the intensity of precipitation, with light blue or green representing the lightest precipitation and orange or red representing the heaviest precipitation.

During the 1990s, **Doppler radar** replaced the conventional radar units that were put into service shortly after World War II. Doppler radar is like conventional radar in that it can detect areas of precipitation and measure rainfall intensity (see ● Fig. 7.36a). Using special computer programs called *algorithms*, the rainfall intensity, over a given area for a given time, can be computed and displayed as an estimate of total rainfall over that particular area (see Fig. 7.36b). But the Doppler radar can do more than conventional radar.

Because the Doppler radar uses the principle called *Doppler shift*,[*] it has the capacity to measure the speed at which falling precipitation is moving horizontally toward or away from the radar antenna. Falling rain moves with the wind. Consequently, Doppler radar allows scientists to peer into a tornado-generating thunderstorm and observe its wind. We will investigate these ideas further in Chapters 14 and 15, when we consider the formation of severe thunderstorms and tornadoes.

In some instances, radar displays indicate precipitation where there is none reaching the surface. This situation happens because the radar beam travels in a nearly straight line while Earth's surface curves away from it. Hence, the return echo is

not necessarily that of precipitation reaching the ground, but is that of raindrops in the cloud. So, if Doppler radar indicates that it's raining in your area, and outside you observe that it is not, remember that it is raining, but the raindrops are probably evaporating before reaching the ground.

The most recent improvement for Doppler radar is *polarimetric radar*, also referred to as *dual-polarization radar*. This form of Doppler radar, which was installed around the United States in the early 2010s, transmits both a vertical and horizontal pulse, which makes it easier to determine whether falling precipitation is in the form of rain or snow. ● Fig. 7.37 shows an example

NOAA/NWS

● **FIGURE 7.37** Doppler radar with dual-polarization technology shows precipitation and cloud particles inside a thunderstorm near Wichita, Kansas, on May 30, 2012. This technology was added to the NWS radars to more clearly identify different types of precipitation and to more precisely identify tornado circulations.

---

[*]The Doppler shift (or effect) is the change in the frequency of waves that occurs when the emitter or the observer is moving toward or away from the other. As an example, suppose a high-speed train is approaching you. The higher-pitched (higher-frequency) whistle you hear as the train approaches will shift to a lower pitch (lower frequency) after the train passes.

of a dual-polarization radar display. Another technology being explored is *phased-array radar*, which can gather a much greater amount of data using a grid of small transmitters instead of a single large one.

MEASURING PRECIPITATION FROM SPACE   As we discussed in Chapter 5 (p. 133), specialized satellites can also be used to observe precipitation from space. Up through 2015, the *Tropical Rainfall Measuring Mission* (*TRMM*) satellite gathered a massive amount of three-dimensional data on precipitation. The newer *Global Precipitation Mission* (*GPM*) satellite has taken the place of TRMM, gathering precipitation data over a much larger area. GPM is able to gather more detail on lighter precipitation than TRMM, and it includes a dual-polarization radar that can deduce the sizes of raindrops, hailstones, and other precipitation elements (see Fig. 7.38). Another satellite that gathers three-dimensional data on precipitation intensity, NASA's *CloudSat* satellite, was launched in 2006. An image from *CloudSat* is shown in Fig. 5.41 on p. 138 . It is hoped that such vertical profiling of liquid water and ice will provide scientists with a better understanding of precipitation processes that go on inside the cloud and the role that clouds play in Earth's global climate system.

● **FIGURE 7.38**   A three-dimensional image of Hurricane Jimena over the Central Pacific Ocean, collected on September 1, 2015, by the Global Precipitation Mission satellite. Reds and oranges indicate the heaviest rainfall at the surface. The eye of Jimena is visible as the hollow area in the middle of intense showers and thunderstorms.

SUMMARY

In this chapter, we learned that cloud droplets are too small and light to reach the ground as rain. Cloud droplets do grow larger by condensation, but this process by itself is much too slow to produce substantial precipitation. Because larger cloud droplets fall faster and farther than smaller ones, they grow larger as they fall by coalescing with drops in their path. If the air temperature in a cloud drops below freezing, then ice crystals play an important role in producing precipitation. Some ice crystals can form directly on ice nuclei, or they may result when an ice nucleus makes contact with and freezes a supercooled water droplet. Because of differences in vapor pressures between water and ice, an ice crystal surrounded by water droplets grows larger at the expense of the droplets. As an ice crystal begins to fall, it can grow larger by colliding with supercooled liquid droplets, which freeze on contact. In an attempt to coax more precipitation from them, some clouds are seeded with silver iodide or other substances.

Precipitation can reach the surface in a variety of forms. In winter, raindrops may freeze on impact, producing freezing rain that can disrupt electrical service by downing power lines. Raindrops can freeze into tiny pellets of ice above the ground and reach the surface as sleet. Depending on conditions, snow may fall as pellets, grains, or flakes, all of which can influence how far we see and hear. Strong updrafts in a cumulonimbus cloud may keep ice particles suspended above the freezing level, where they can acquire an additional coating of ice and form destructive hailstones. Although the rain gauge is still the most commonly used method of measuring precipitation, Doppler radar has become an important instrument for determining precipitation intensity and estimating rainfall amount. In some regions, rainfall estimates can be obtained from radar and microwave scanners on board satellites.

KEY TERMS

The following terms are listed (with corresponding page number) in the order they appear in the text. Define each. Doing so will aid you in reviewing the material covered in this chapter.

precipitation, 168	curvature effect,
equilibrium vapor pressure,	169
168	solute effect, 169

QUESTIONS FOR REVIEW

1. What is the primary difference between a cloud droplet and a raindrop?
2. Why do typical cloud droplets seldom reach the ground as rain?
3. Describe how the process of collision and coalescence produces precipitation.
4. Would the collision-coalescence process work better at producing rain in (a) a warm, thick nimbostratus cloud or (b) a warm, towering cumulus congestus cloud? Explain.
5. List and describe three ways in which ice crystals can form in a cloud.
6. When the temperature in a cloud is $-30°C$, are larger cloud droplets more likely to freeze than smaller cloud droplets? Explain.
7. In a cloud where the air temperature is $-10°C$, why are there many more cloud droplets than ice crystals?
8. How does the ice-crystal (Bergeron) process produce precipitation? What is the *main* premise describing this process?
9. Why do heavy showers usually fall from cumuliform clouds? Why does steady precipitation normally fall from stratiform clouds?
10. Why are large snowflakes most often observed when the air temperature near the ground is just below freezing?
11. In a cloud composed of water droplets and ice crystals, is the saturation vapor pressure greater over the droplets or over the ice?
12. Why would it be foolish to seed a clear sky with silver iodide?
13. When seeding a cloud to promote rainfall, is it possible to overseed the cloud so that it prevents rainfall? Explain.
14. Explain how clouds can be seeded naturally.
15. What atmospheric conditions are necessary for snow to fall when the air temperature is considerably above freezing?
16. List the advantages and disadvantages of heavy snowfall.
17. How do the atmospheric conditions that produce sleet differ from those that produce hail?
18. What is the difference between freezing rain and sleet?
19. Describe how hail might form in a cumulonimbus cloud.
20. Why is hail more common in summer than in winter?
21. List the common precipitation gauges that measure rain and snow.
22. (a) What is Doppler radar?
    (b) How does Doppler radar measure the intensity of precipitation?

QUESTIONS FOR THOUGHT

1. Ice crystals that form by accretion are fairly large. Explain why they fall slowly.
2. Why is a warm, tropical cumulus cloud more likely to produce precipitation than a cold, stratus cloud?
3. Explain why very small cloud droplets of pure water evaporate even when the relative humidity is 100 percent.
4. Suppose a thick nimbostratus cloud contains ice crystals and cloud droplets all about the same size. Which precipitation process will be most important in producing rain from this cloud? Why?
5. Clouds that form over water are usually more efficient in producing precipitation than clouds that form over land. Why?
6. On many summer days, a blizzard occurs somewhere over the Great Plains. Explain where and why.
7. During a recent snowstorm, Denver, Colorado, received 3 in. of snow. Sixty kilometers east of Denver, a city received no measurable snowfall, while 150 km east of Denver another city received 4 in. of snow. Given that Denver is located to the east of the Rockies, and the upper-level winds were westerly during the snowstorm,

give an explanation as to what *could* account for this snowfall pattern.

8. Raindrops rarely grow larger than 5 mm. Two reasons were given on pp. 169–171. Can you think of a third? (Hint: See Focus section 7.3, and look at the shape of a large drop.)

9. Lead iodide is an effective ice-forming nucleus. Why do you think it has not been used for that purpose?

10. When cirrus clouds are above a deck of altocumulus clouds, occasionally a clear area, or "hole," will appear in the altocumulus cloud layer. What do you suppose could cause this to happen?

11. It is −12°C (10°F) in Albany, New York, and freezing rain is falling. Can you explain why? Draw a vertical profile of the air temperature (a sounding) that illustrates why freezing rain is occurring at the surface.

12. When falling snowflakes become mixed with sleet, why is this condition often followed by the snowflakes changing into rain?

13. A major snowstorm occurred in a city in northern New Jersey. Three volunteer weather observers measured the snowfall. Observer 1 measured the depth of newly fallen snow every hour. At the end of the storm, Observer 1 added up the measurements and came up with a total of 12 in. of new snow. Observer 2 measured the depth of new snow twice: once in the middle of the storm and once at the end, and came up with a total snowfall of 10 in. Observer 3 measured the new snowfall only once, after the storm had stopped, and reported 8.4 in. Which of the three observers do you think has the correct snowfall total? List *at least five* possible reasons why the snowfall totals were different.

PROBLEMS AND EXERCISES

1. In the daily newspaper, a city is reported as receiving 0.52 in. of precipitation over a 24-hour period. If all the precipitation fell as snow, and if we assume a normal water equivalent ratio of 10:1, how much snow did this city receive?

2. How many times faster does a large raindrop (diameter 5000 μm) fall than a cloud droplet (diameter 20 μm), if both are falling at their terminal velocity in still air?

3. (a) How many minutes would it take drizzle with a diameter of 200 μm to reach the surface if it falls at its terminal velocity from the base of a cloud 1000 m (about 3300 ft) above the ground? (Assume that the air is saturated beneath the cloud, the drizzle does not evaporate, and the air is still.)

   (b) Suppose that the drizzle in problem 3a evaporates on its way to the ground. If the drop size is 200 μm for the first 450 m of descent, 100 μm for the next 450 m, and 20 μm for the final 100 m, how long will it take the drizzle to reach the ground if it falls in still air?

4. Suppose that a large raindrop (diameter 5000 μm) falls at its terminal velocity from the base of a cloud 1500 m (about 5000 ft) above the ground.

   (a) If we assume that the raindrop does not evaporate, how long would it take the drop to reach the surface?

   (b) What would be the shape of the falling raindrop just before it reaches the ground?

   (c) What type of cloud would you expect this raindrop to fall from? Explain.

5. In ● Fig. 7.39, a drawing of a large hailstone, explain how the areas of clear ice and rime ice could have formed.

● **FIGURE 7.39**

 ONLINE RESOURCES

Visit www.cengagebrain.com to view additional resources, including video exercises, practice quizzes, an interactive eBook, and more.

Contestants in a yachting competition take advantage of a stiff breeze.

Air Pressure and Winds

DECEMBER 19, 1980, WAS A COOL DAY IN LYNN, Massachusetts, but not cool enough to dampen the spirits of more than 2000 people who gathered in Central Square—all hoping to catch at least one of the 1500 dollar bills that would be dropped from a small airplane at noon. Right on schedule, the aircraft circled the city and dumped the money onto the people below. However, to the dismay of the onlookers, a westerly wind caught the currency before it reached the ground and carried it out over the cold Atlantic Ocean. Had the pilot or the sponsoring leather manufacturer examined the weather charts beforehand, it might have been possible to predict that the wind would ruin the advertising scheme.

The scenario on the previous page raises two questions: (1) Why does the wind blow? and (2) How can one tell its direction by looking at weather charts? Chapter 1 has already answered the first question: Air moves in response to horizontal differences in pressure. This is what happens when we open a vacuum-packed can: air rushes from the higher-pressure region outside the can toward the region of lower pressure inside. In the atmosphere, the wind blows in an attempt to equalize imbalances in air pressure. Does this mean that the wind always blows directly from high to low pressure? Not really, because the movement of air is controlled not only by pressure differences but by other forces as well.

In this chapter, we will first consider how and why atmospheric pressure varies, then we will look at the forces that influence atmospheric motions aloft and at the surface. Through studying these forces, we will be able to tell how the wind should blow in a particular region by examining surface and upper-air charts.

Atmospheric Pressure

In Chapter 1, we learned several important concepts about atmospheric pressure. One is that **air pressure** is simply the mass of air above a given level. As we climb in altitude above Earth's surface, there are fewer air molecules above us; hence, *atmospheric pressure always decreases with increasing height*. Another concept we learned is that our atmosphere is highly compressible. This means that the weight of higher layers compresses the atmosphere beneath, and so most of the molecules in our atmosphere are crowded close to Earth's surface. Hence, air pressure decreases with height, rapidly at first, then more slowly at higher altitudes.

So one way to change air pressure is to simply move up or down in the atmosphere. But what causes the air pressure to change in the horizontal? And why does the air pressure change at the surface?

HORIZONTAL PRESSURE VARIATIONS: A TALE OF TWO CITIES   To answer these questions, we eliminate some of the complexities of the atmosphere by constructing *models*. ● Fig. 8.1 shows a simple atmospheric model—a column of air, extending well up into the atmosphere. In the column, the dots represent

air molecules. Our model assumes: (1) that the air molecules are not crowded close to the surface and, unlike the real atmosphere, the air density remains constant from the surface up to the top of the column, (2) that the width of the column does not change with height, and (3) that the air is unable to freely move into or out of the column.

Suppose we somehow force more air into the column in Fig. 8.1. What would happen? If the air temperature in the column does not change, the added air would make the column more dense, and the added weight of the air in the column would increase the surface air pressure. Likewise, if a great deal of air were removed from the column, the surface air pressure would decrease. Consequently, to change the surface air pressure, we need to change the mass of air in the column above the surface. But how can this feat be accomplished?

Look at the air columns in ● Fig. 8.2a.* Suppose both columns are located at the same elevation, both have the same air temperature, and both have the same surface air pressure. There must therefore be the same number of molecules (same mass of air) in each column above both cities. Further suppose that the surface air pressure for both cities remains the same, while the air above city 1 cools and the air above city 2 warms (see Fig. 8.2b).

As the air in column 1 cools, the molecules move more slowly and crowd closer together, so the air becomes more dense. In the warm air above city 2, the molecules move faster and spread farther apart, and the air becomes less dense. Because the width of the columns does not change (and if we assume an invisible barrier exists between the columns), the total number of molecules above each city remains the same, and the surface pressure does not change. Therefore, in the more-dense, colder air above city 1, the height of the column decreases, while it increases in the less-dense, warmer air above city 2.

We now have a colder, shorter, more-dense column of air above city 1 and a warmer, taller, less-dense air column above city 2. From this situation, we can conclude that *it takes a shorter column of cold, more-dense air to exert the same surface pressure as a taller column of warm, less-dense air*. This concept has a great deal of meteorological significance.

Atmospheric pressure decreases more rapidly with height in the cold column of air. In the cold air above city 1 (Fig. 8.2b), move up the column and observe how quickly you pass through the densely packed molecules. This activity indicates a rapid change in pressure. In the warmer, less-dense air, the pressure does not decrease as rapidly with height simply because you climb above fewer molecules in the same vertical distance.

In Fig. 8.2c, move up the warm, tan column until you come to the letter *H*. Now move up the cold, blue column the same distance until you reach the letter *L*. Notice there are more molecules above the letter *H* in the warm column than above the letter *L* in the cold column. The fact that the number of molecules above any level is a measure of the atmospheric pressure leads to an important concept: *Warm air aloft is normally associated with high atmospheric pressure, and cold air aloft is associated with low atmospheric pressure.*

Air column

● **FIGURE 8.1** A model of the atmosphere where air density remains constant with height. The air pressure at the surface is related to the number of molecules above. When air of the same temperature is stuffed into the column, the surface air pressure rises. When air is removed from the column, the surface pressure falls. (In the actual atmosphere, unlike this model, density decreases with height.)

---

*We will keep our same assumption as in Fig. 8.1; that is, (1) the air molecules are not crowded close to the surface, (2) the width of the columns does not change, and (3) air is unable to move into or out of the columns.

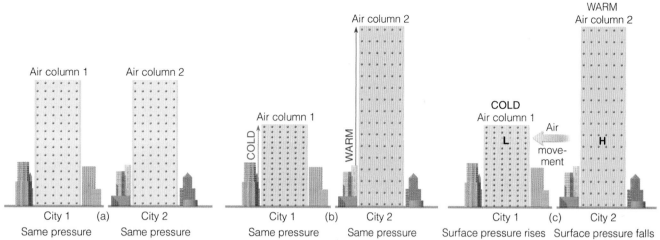

● **FIGURE 8.2** Illustration of how variations in temperature can produce horizontal pressure forces. (Note that for simplicity, this model assumes that the air density is constant with height, whereas in the actual atmosphere, density decreases with height.) (a) Two air columns, each with identical mass, have the same surface air pressure. (b) Because it takes a shorter column of cold air to exert the same surface pressure as a taller column of warm air, as column 1 cools, it must shrink, and as column 2 warms, it must rise. (c) Because at the same level in the atmosphere there is more air above the H in the warm column than above the L in the cold column, warm air aloft is associated with high pressure and cold air aloft with low pressure. The pressure differences aloft create a force that causes the air to move from a region of higher pressure toward a region of lower pressure. The removal of air from column 2 causes its surface pressure to drop, whereas the addition of air into column 1 causes its surface pressure to rise. (The difference in height between the two columns is greatly exaggerated.)

In Fig. 8.2c, the horizontal difference in temperature creates a horizontal difference in pressure. The pressure difference establishes a force (called the *pressure-gradient force*) that causes the air to move from higher pressure toward lower pressure. Consequently, if we remove the invisible barrier between the two columns near the top of column 1 and allow the air aloft to move horizontally, the air will move from column 2 toward column 1. As the air aloft leaves column 2, the mass of the air in the column decreases, and so does the surface air pressure. Meanwhile, the accumulation of air in column 1 causes the surface air pressure to increase.

Higher air pressure at the surface in column 1 and lower air pressure at the surface in column 2 causes the surface air to move from city 1 toward city 2 (see ● Fig. 8.3). As the surface air moves out away from city 1, the air aloft slowly sinks to replace this outwardly spreading surface air. As the surface air flows into city 2, it slowly rises to replace the depleted air aloft. In this manner, a complete circulation of air is established due to the heating and cooling of air columns. As we will see in Chapter 9, this type of thermal circulation is the basis for a wide range of wind systems throughout the world.

In summary, we can see how heating and cooling columns of air can establish horizontal variations in air pressure both aloft and at the surface. It is these horizontal differences in air pressure that cause the wind to blow. Before moving on to the next section, you may wish to look at Focus section 8.1, which describes how air pressure, air density, and air temperature are interrelated.

DAILY PRESSURE VARIATIONS From what we have learned so far, we might expect to see the surface pressure dropping as the air temperature rises, and vice versa. Over large continental areas, especially the southwestern United States in summer, hot surface air is accompanied by surface low pressure. Likewise, bitter cold arctic air in winter is often accompanied by surface high

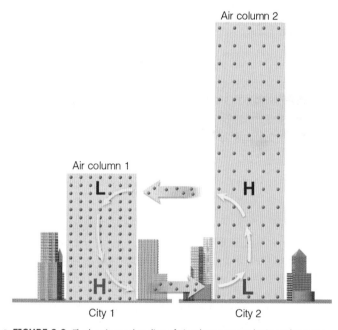

● **FIGURE 8.3** The heating and cooling of air columns causes horizontal pressure variations aloft and at the surface. These pressure variations force the air to move from areas of higher pressure toward areas of lower pressure. In conjunction with these horizontal air motions, the air slowly sinks above the surface high and rises above the surface low.

pressure. Yet, on a daily basis, any cyclic change in surface pressure brought on by daily temperature changes is concealed by the pressure changes created by the warming and cooling of the upper atmosphere.

In the tropics, for example, pressure rises and falls in a regular pattern twice a day (see ● Fig. 8.4). Maximum pressures occur around 10:00 a.m. and 10:00 p.m., minimum near 4:00 a.m. and 4:00 p.m. The largest pressure difference, about 2.5 mb, occurs

● **FIGURE 8.4** Diurnal surface pressure changes in the middle latitudes and in the tropics.

Sea-level pressure changes in middle latitudes on a given day

Average sea-level pressure changes in the tropics

near the equator. It also shows up in higher latitudes, but with a much smaller amplitude. This daily (*diurnal*) fluctuation of pressure appears to be due primarily to the absorption of solar energy by ozone in the upper atmosphere and by water vapor in the lower atmosphere. The warming and cooling of the air creates density oscillations known as *thermal* (or *atmospheric*) tides that show up as small pressure changes near Earth's surface.

In middle latitudes, surface pressure changes are primarily the result of large high- and low-pressure areas that move toward or away from a region. Generally, when an area of high pressure approaches a city, surface pressure usually rises. When it moves away, pressure usually falls. Likewise, an approaching low causes the air pressure to fall, and one moving away causes surface pressure to rise.

PRESSURE MEASUREMENTS   Instruments that detect and measure pressure changes are called **barometers**, which literally means an instrument that measures bars. You may recall from Chapter 1 that a *bar* is a unit of pressure that describes a force over a given area.* Because the bar is a relatively large unit, and because surface pressure changes are normally small, the unit of pressure commonly found on surface weather maps is, as we saw in Chapter 1, the **millibar (mb)**, where 1 mb = 1/1000 bar or

$$1 \, bar = 1000 \, mb$$

---

*A bar is a force of 100,000 newtons acting on a surface area of 1 square meter. A *newton* (N) is the amount of force required to move an object with a mass of 1 kilogram so that it increases its speed at a rate of 1 meter per second each second. Additional pressure units and conversions are given in Appendix A.

**Standard atmospheric pressure at sea level is the pressure extended by a column of mercury 29.92 in. (760 mm) high, having a density of 1.36×10^4 kg/m^3, and subject to an acceleration of gravity of 9.80 m/sec^2.

A common pressure unit used in aviation is *inches of mercury* (Hg). At sea level, **standard atmospheric pressure**** is

$$1013.25 \, mb = 29.92 \, in. \quad Hg = 76 \, cm$$

As a reference, ● Fig. 8.5 compares pressure readings in inches of mercury and millibars.

The unit of pressure designated by the International System (SI) of measurement is the *pascal*, named in honor of Blaise Pascal (1632−1662), whose experiments on atmospheric pressure greatly increased our knowledge of the atmosphere. A pascal (Pa) is the force of 1 newton acting on a surface area of 1 square meter. Thus, 100 pascals equals 1 millibar. The scientific community often uses the *kilopascal* (kPa) as the unit of pressure, where 1 kPa = 10 mb. However, a more convenient unit is the **hectopascal (hPa)**, as

$$1 \, hPa = 1 \, mb$$

The hectopascal is gradually replacing the millibar as the preferred unit of pressure on surface and upper-air weather maps. (No conversion is needed in moving between millibars and hectopascals.)

Because we measure atmospheric pressure with an instrument called a *barometer*, atmospheric pressure is also referred to as *barometric pressure*. Evangelista Torricelli, a student of Galileo, invented the **mercury barometer** in 1643. His barometer, similar to those in use today, consisted of a long glass tube open at one end and closed at the other (see ● Fig. 8.6). Removing air from the tube and covering the open end, Torricelli immersed the lower portion into a dish of mercury. He removed the cover, and the mercury rose up the tube to nearly 76 cm (or about 30 in.) above the level in the dish. Torricelli correctly concluded that the column of mercury in the tube was balancing the weight of the air above the dish, and hence its height was a measure of

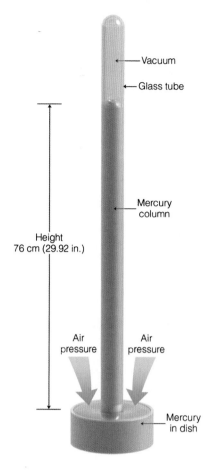

FIGURE 8.6 The mercury barometer. The height of the mercury column is a measure of atmospheric pressure.

Labels on figure: Vacuum, Glass tube, Mercury column, Height 76 cm (29.92 in.), Air pressure, Air pressure, Mercury in dish

---

Scale labels for FIGURE 8.5:

in. Hg	mb
32.78	1110
32.48	1100
32.19	1090
31.89	1080
31.60	1070
31.30	1060
31.00	1050
30.71	1040
30.42	1030
30.12	1020
29.82	1010
29.53	1000
29.24	990
28.94	980
28.64	970
28.35	960
28.05	950
27.76	940
27.46	930
27.17	920
26.87	910
26.58	900
26.28	890
25.99	880
25.69	870
25.40	860
25.10	850

1085 mb (32.03 in.) Highest recorded sea-level pressure: Tongsontsen, Mongolia (December 2001)

1064 mb (31.42 in.) Highest recorded sea-level pressure in the continental United States (excluding Alaska): Miles City, Montana (December 1983)

Strong high-pressure system

1013.25 mb (29.92 in.) Standard sea-level pressure

Deep low-pressure system

920 mb (27.17 in.) Hurricane Katrina during landfall (August 2005)

882 mb (26.04 in.) Hurricane Wilma (October 2005)

870 mb (25.70 in.) Lowest recorded sea-level pressure: Typhoon Tip (October 1979)

FIGURE 8.5 Atmospheric pressure in inches of mercury and in millibars.

WEATHER WATCH

Although 1013.25 mb (29.92 in.) is the standard atmospheric pressure at sea level, it is not the *average* sea-level pressure. Earth's average sea-level pressure is 1011.0 mb (29.85 in.). Because much of Earth's surface is above sea level, Earth's annual average surface pressure is estimated to be 984.43 mb (29.07 in.).

---

atmospheric pressure. Mercury barometers are still used today in many settings, although in some areas, including Europe, they are no longer manufactured or sold because mercury use can pose a risk to human health.

Why is mercury rather than water used in barometers? The primary reason is convenience. (Also, water can evaporate in the tube.) Mercury seldom rises to a height above 80 cm (31.5 in.). A water barometer, however, presents a problem. Because water is 13.6 times less dense than mercury, an atmospheric pressure of 76 cm (30 in.) of mercury would be equivalent to 1034 cm (408 in.) of water. A water barometer resting on the ground near sea level would have to be read from a ladder over 10 m (33 ft) tall. Water is sometimes used as a more general indicator of atmospheric pressure in smaller devices called "storm glasses."

The most common type of home barometer—the **aneroid barometer**—contains no fluid. Inside this instrument is a small, flexible metal box called an *aneroid cell*. Before the cell is tightly sealed, air is partially removed, so that small changes in external air pressure cause the cell to expand or contract. The size of the cell is calibrated to represent different pressures, and any change in its size is amplified by levers and transmitted to an indicating arm, which points to the current atmospheric pressure (see ● Fig. 8.7).

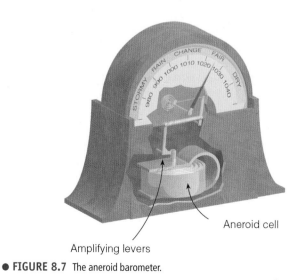

Labels on figure: STORMY, RAIN, CHANGE, FAIR, DRY, Amplifying levers, Aneroid cell

FIGURE 8.7 The aneroid barometer.

The Atmosphere Obeys the Gas Law

Air temperature, air pressure, and air density are all interrelated. If one of these variables changes, the other two usually change as well.

The relationship among the pressure, temperature, and density of air can be expressed by

Pressure = temperature × density × constant

This simple relationship, often referred to as the *gas law (or equation of state)*, tells us that the pressure of a gas is equal to its temperature times its density times a constant. When we ignore the constant and look at the gas law in symbolic form, it becomes

$$p \sim T \times \rho$$

where, of course, *p* is pressure, *T* is temperature, and ρ (the Greek letter rho, pronounced "row") represents air density.* The line ~ is a symbol meaning "is proportional to." A change in one variable causes a corresponding change in the other two variables. Thus, it will be easier to understand the behavior of a gas if we keep one variable from changing and observe the behavior of the other two.

Suppose, for example, we hold the temperature constant. The relationship then becomes

$$p \sim \rho$$
(*assuming temperature is constant*)

This expression says that the pressure of the gas is proportional to its density, as long as its temperature does not change. Consequently, if the temperature of a gas (such as air) is held constant, as the pressure increases the density increases, and as the pressure decreases the density decreases. In other words, *at the same temperature, air at a higher pressure is more dense than air at a lower pressure*. If we apply this concept to the atmosphere, then with nearly the same temperature and elevation, air above a region of surface high pressure is more dense than air above a region of surface low pressure (see ● Fig. 1).

We can see, then, that for surface high-pressure areas (anticyclones) and surface

● **FIGURE 1** Air above a region of surface high pressure is more dense than air above a region of surface low pressure (at the same temperature). (The dots in each column represent air molecules.)

low-pressure areas (mid-latitude cyclones) to form, the air density (mass of air) above these systems must change. As we will see later in this chapter, as well as in other chapters, surface air pressure increases when the wind causes more air to move into a column of air than is able to leave (called *net convergence*), and surface air pressure decreases when the wind causes more air to move out of a column of air than is able to enter (called *net divergence*).

Earlier, we considered how pressure and density are related when the temperature is not changing. What happens to the gas law when the pressure of a gas remains constant? In shorthand notation, the relationship becomes

$$(\text{constant pressure}) \sim T \times \rho$$

This relationship tells us that when the pressure of a gas is held constant, the gas becomes less dense as the temperature goes up, and more dense as the temperature goes down. Therefore, *at a given atmospheric pressure, air that is cold is more dense than air that is warm*. Keep in mind that the idea that cold air is more dense than warm air applies only when we compare volumes of air at the same level, where pressure changes are small in any horizontal direction compared to the vertical.

We can use the gas law to obtain information about the atmosphere. For example, at an altitude of about 5600 m (18,400 ft) above sea level, the atmospheric pressure is normally close to 500 millibars. If we obtain the average density at this level, with the aid of the gas law we can calculate the average air temperature.

Recall that the gas law is written as

$$p = T \times \rho \times C$$

With the pressure *(p)* in millibars (mb), the temperature *(T)* in kelvin, and the density (ρ) in kilograms per cubic meter (kg / m³), the numerical value of the constant *(C)* is about 2.87.*

At an altitude of 5600 m above sea level, where the average (or standard) air pressure is about 500 mb and the average air density is 0.690 kg / m³, the average air temperature becomes

$$
\begin{aligned}
p &= T \times \rho \times C \\
500 &= T \times 0.690 \times 2.87 \\
\frac{500}{0.690 \times 2.87} &= T \\
252.5\ K &= T
\end{aligned}
$$

To convert kelvins into degree Celsius, we subtract 273 from the Kelvin temperature and obtain a temperature of −20.5°C, which is the same as −5°F.

If we know the numerical values of temperature and density, with the aid of the gas law we can obtain the air pressure. For example, in Chapter 1 we saw that the average global temperature near sea level is about 15°C (59°F), which is the same as 288 K. If the average air density at sea level is 1.226 kg/m³, what would be the standard (average) sea-level pressure?

Using the gas law, we obtain

$$
\begin{aligned}
p &= T \times \rho \times C \\
p &= 288 \times 1.226 \times 2.87 \\
p &= 1013\ \text{mb}
\end{aligned}
$$

Because the air pressure is related to both temperature and density, a small change in either or both of these variables can bring about a change in pressure.

---

*This gas law may also be written as $p \times v = T \times$ constant. Consequently, pressure and temperature changes are also related to changes in volume.

---

*The constant is usually expressed as 2.87 × 10⁶ erg/gK, or, in the SI system, as 287 J/kg K. (See Appendix A for information regarding the units used here.)

Notice that the aneroid barometer often has descriptive weather-related words printed above specific pressure values. These descriptions indicate the most likely weather conditions when the needle is pointing to that particular pressure reading. Generally, the higher the reading, the more likely clear weather will occur, and the lower the reading, the better are the chances for inclement weather. This situation occurs because surface high-pressure areas are associated with sinking air and normally fair weather, whereas surface low-pressure areas are associated with rising air and usually cloudy, wet weather. A steady rise in atmospheric pressure (a rising barometer reading) usually indicates clearing weather or fair weather, whereas a steady drop in atmospheric pressure (a falling barometer reading) often signals the approach of a cyclonic storm with inclement weather.

The *altimeter* and *barograph* are two types of aneroid barometers. Altimeters are aneroid barometers that measure pressure, but are calibrated to indicate altitude. Barographs are recording aneroid barometers. Basically, the barograph consists of a pen attached to an indicating arm that marks a continuous record of pressure on chart paper. The chart paper is attached to a drum rotated slowly by an internal mechanical clock (see ● Fig. 8.8).

Digital barometers are becoming more common. This type of barometer uses a device called a transducer that detects the change in pressure exerted by the atmosphere on a precisely engineered surface. The change in pressure is then converted into an electrical signal. Some digital barometers are small enough to be included in smartphones, while others are designed for research settings.

PRESSURE READINGS Obtaining the correct air pressure from a mercury barometer involves more than simply reading the height of the mercury column. Being a fluid, mercury is sensitive to changes in temperature; it will expand when heated and contract when cooled. Consequently, to obtain accurate pressure readings without the influence of temperature, all mercury barometers are corrected as if they were read at the same temperature. Also, because Earth is not a perfect sphere, the force of gravity is not a constant. Because small gravity differences influence the height of the mercury column, they must be taken into account when reading the barometer. Finally, each mercury barometer has its own "built-in" error, called *instrument error*, which is caused, in part, by the surface tension of the mercury against the glass tube. After being corrected for temperature, gravity, and instrument error, the barometer reading at a particular location and elevation is termed **station pressure**.

● Figure 8.9a gives the station pressure measured at four locations only a few hundred kilometers apart. The different station pressures of the four cities are due primarily to the cities being at different elevations above sea level. This fact becomes even clearer when we realize that atmospheric pressure changes much more quickly when we move upward than it does when we move sideways. As an example, the vertical change in air pressure from the base to the top of the Empire State Building—a distance of a little more than $^1/_2$ km—is typically much greater than the horizontal difference in air pressure from New York City to Miami, Florida—a distance of over 1600 km. Therefore, we can see that a small vertical difference between two observation sites can yield a large difference in station pressure. Thus, to properly monitor horizontal changes in pressure, barometer readings must be corrected for altitude.

Altitude corrections are made so that a barometer reading taken at one elevation can be compared with a barometer reading taken at another. Station pressure observations are normally adjusted to a level of *mean sea level*—the level representing the average surface of the ocean. The adjusted reading is called **sea-level pressure**. The size of the correction depends primarily on how high the station is above sea level.

Surface and Upper-Level Charts

Near Earth's surface, atmospheric pressure decreases on the average by about 10 millibars for every 100 meters of increase in elevation (about 1 in. of mercury for each 1000-ft rise).* Notice in Fig. 8.9a that city A has a station pressure of 952 millibars. Notice also that city A is 600 meters above sea level. Adding 10 millibars per 100 meters to its station pressure yields a sea-level pressure of 1012 mb (Fig. 8.9b). After all the station pressures are adjusted to sea level (Fig. 8.9c), we are able to see the horizontal variations in sea-level pressure—something we were not able to see from the station pressures alone in Fig. 8.9a.

When more pressure data are added (see Fig. 8.9c), the chart can be analyzed and the pressure pattern visualized. **Isobars** (lines connecting points of equal pressure) are drawn at intervals of 4 mb,** with 1000 mb being the base value. Note that the isobars do not pass through each point, but, rather, between many

---

*This decrease in atmospheric pressure with height (10 mb/100 m) occurs when the air temperature decreases at the standard lapse rate of 6.5°C/1000 m. Because atmospheric pressure decreases more rapidly with height in cold (more-dense) air than it does in warm (less-dense) air, the vertical rate of pressure change is typically greater than 10 mb per 100 m in cold air and less than that in warm air.

**An interval of 2 mb would put the lines too close together, and an 8-mb interval would spread them too far apart.

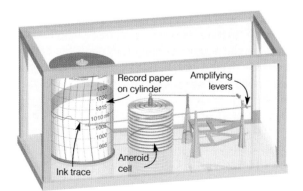

● **FIGURE 8.8** A recording barograph.

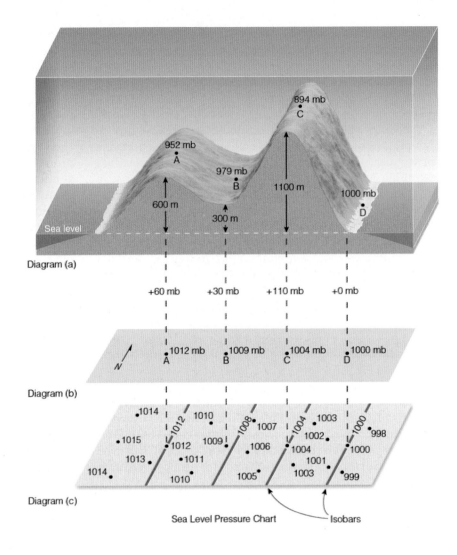

● **FIGURE 8.9** The top diagram (a) shows four cities (A, B, C, and D) at varying elevations above sea level, all with different station pressures. The middle diagram (b) represents sea-level pressures of the four cities plotted on a sea-level chart. The bottom diagram (c) shows sea-level pressure readings of the four cities plus other sea-level pressure readings at locations not shown in (a) and (b). Isobars are drawn on the chart (gray lines) at intervals of 4 millibars.

of them, with the exact values being interpolated from the data given on the chart. For example, follow the 1008-mb line from the top of the chart southward and observe that there is no plotted pressure of 1008 millibars. The 1008-millibars isobar, however, comes closer to the station with a sea-level pressure of 1007 mb than it does to the station with a pressure of 1010 mb. With its isobars, the bottom chart (Fig. 8.9c) is now called a *sea-level pressure chart* or simply a **surface map**. When weather data are plotted on the map it becomes a *surface weather map*.

The isobars in ● Fig. 8.10 have been smoothed to eliminate small-scale wiggles produced by data collected at high-altitude stations and at stations that might have small observational errors. Otherwise, the isobars might be significantly distorted. An extreme case of this type of error occurs at Leadville, Colorado (elevation 3096 m), the highest city in the United States. Here, the station pressure is typically near 700 mb. This means that nearly 300 mb must be added to obtain a sea-level pressure reading! A mere 1 percent error in estimating the exact correction would result in a 3-mb error in sea-level pressure.

The sea-level pressure chart described so far is called a *constant height chart* because it represents the atmospheric pressure at a constant level—in this case, sea level. The same type of chart could be drawn to show the horizontal variations in pressure at any level in the atmosphere; for example, at 3000 meters (see ● Fig. 8.11).

Another type of chart more commonly used in studying the weather is the *constant pressure chart*, or **isobaric chart**. Instead of showing pressure variations at a constant altitude, these charts are constructed to show height variations along a constant pressure (*isobaric*) surface. Constant pressure charts are convenient to use because the height variables they show are easier to deal with in meteorological equations than the variables of pressure. Given that isobaric charts are routinely used by meteorologists, let's examine them in detail.

Imagine that the dots inside the air column in ● Fig. 8.12 represent tightly packed air molecules from the surface up to the tropopause. Assume that the air density is constant throughout the entire air layer and that all of the air molecules are squeezed

● **FIGURE 8.10** Sea-level isobars, drawn every 4 millibars and smoothed to account for small-scale variations related to high-altitude stations and potential observational errors.

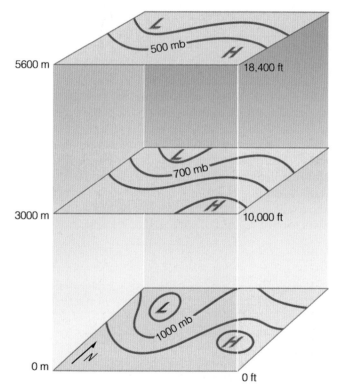

● **FIGURE 8.11** Each map shows isobars on a constant height chart. The isobars represent variations in horizontal pressure at that altitude. An average isobar at sea level would be about 1000 mb; at 3000 m, about 700 mb; and at 5600 m, about 500 mb.

into this layer. If we climb halfway up the air column and stop, then draw a sheetlike surface representing this level, we will have made a constant height surface. This altitude (5600 m) is where we would, under standard conditions, measure a pressure of 500 millibars. Observe that everywhere along this surface (shaded tan in the diagram) there are an equal number of molecules above and below it. This condition means that the level of constant

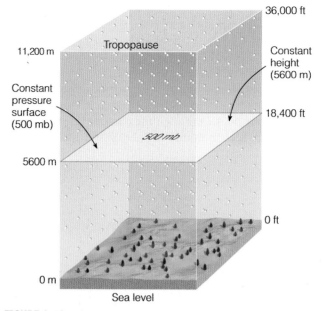

● **FIGURE 8.12** When there are no horizontal variations in pressure, constant pressure surfaces are parallel to constant height surfaces. In the diagram, a measured pressure of 500 millibars is 5600 meters above sea level everywhere. The actual atmosphere extends well above the tropopause level shown here. (Dots in the diagram represent air molecules.)

height also represents a level of constant pressure. At every point on this *isobaric surface*, the height is 5600 meters above sea level and the pressure is 500 millibars. Within this simplified air column, we could cut any number of horizontal slices, each one at a different altitude, and each slice would represent both an isobaric and constant height surface. A contour map of any one of these surfaces would be blank, as there are no horizontal variations in either pressure or altitude.

If the air temperature should change in any portion of the column, the air density and pressure would change along with it. Notice in ● Fig. 8.13 that we have colder air to the north and warmer air to the south. To simplify this situation, we will assume that the atmospheric pressure at Earth's surface remains constant. Hence, the total number of molecules in the column above each region must remain constant.

In Fig. 8.13, the area shaded gray at the top of the column represents a constant pressure (isobaric) surface, where the atmospheric pressure at all points along this surface is 500 millibars. Notice that in the warmer, less-dense air the 500-mb pressure surface is found at a higher (than average) level, while in the colder, more-dense air, it is observed at a much lower (than average) level. From these observations, we can see that when *the air aloft is warm, constant pressure surfaces are typically found at higher elevations than normal, and when the air aloft is cold, constant pressure surfaces are typically found at lower elevations than normal.*

The variations in height of the isobaric surface in Fig. 8.13 are shown in ● Fig. 8.14. Note that where the constant altitude lines intersect the 500-mb pressure surface, **contour lines** (lines connecting points of equal elevation) are drawn on the 500-mb map. Each contour line, of course, tells us the altitude above sea level at which we can obtain a pressure reading of 500 mb. In the warmer air to the south, the elevations are high, while in the cold air to the north, the elevations are low. The contour lines are crowded together in the middle of the chart, where the pressure surface dips rapidly due to the changing air temperatures. Where there is little horizontal temperature change, there are also few contour lines. Although the contour lines are lines of constant height, keep in mind that they illustrate pressure as do isobars, in that *contour lines of low height represent a region of lower pressure and contour lines of high height represent a region of higher pressure.*

Because cold air aloft is normally associated with low heights and warm air aloft with high heights, on upper-air charts representing the Northern Hemisphere, contour lines (and isobars) usually decrease in value from south to north because the air is typically warmer to the south and colder to the north. The lines, however, are not straight; they bend and turn, indicating **ridges** (*elongated highs*) where the air is warm and indicating depressions, or **troughs** (*elongated lows*), where the air is cold. In ● Fig. 8.15, we can see how the wavy contours on the map relate to the changes in altitude of the isobaric surface.

Although we have examined only the 500-mb chart, other isobaric charts are commonly used. ▼ Table 8.1 lists these charts and their approximate heights above sea level.

Upper-level charts are a valuable tool. As we will see, they can be used to analyze wind flow patterns that are extremely important in forecasting the weather. They can also be used to determine the movement of weather systems and to predict the behavior of surface pressure areas. To the pilot of a small aircraft, a constant pressure chart can help determine whether the plane is flying at an altitude either higher or lower than its altimeter indicates. (For more information on this topic, read Focus section 8.2.)

● Figure 8.16a is a simplified surface map that shows areas of high and low pressure and arrows that indicate *wind direction*—the direction from which the wind is blowing. The large blue H's on the map indicate the centers of high pressure, which are also called **anticyclones**. The large L's represent centers of low pressure, also known as *depressions* or **mid-latitude cyclonic storms** because they form in the middle latitudes, outside of the tropics. The solid dark lines are isobars with units in millibars. Notice that the surface winds tend to blow across the isobars toward regions of lower pressure. In fact, as we briefly

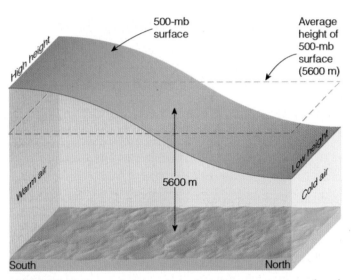

● **FIGURE 8.13** The area shaded gray in the above diagram represents a surface of constant pressure, or isobaric surface. Because of the changes in air density, the isobaric surface rises in warm, less-dense air and lowers in cold, more-dense air. Where the horizontal temperature changes most quickly, the isobaric surface changes elevation most rapidly.

● **FIGURE 8.14** Changes in altitude of an isobaric surface (500 mb) show up as contour lines on an isobaric (500 mb) map. Where the isobaric surface dips most rapidly, the contour lines are closer together on the 500-mb map.

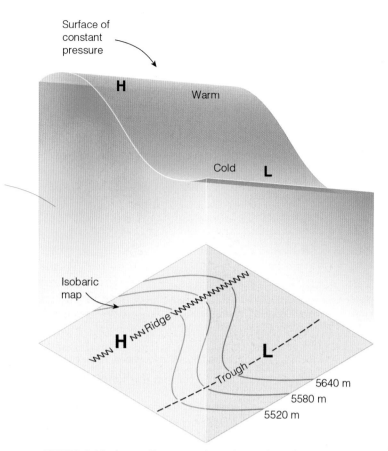

● **FIGURE 8.15** The wavelike patterns of an isobaric surface reflect the changes in air temperature. An elongated region of warm air aloft shows up on an isobaric map as higher heights and a ridge; the colder air shows as lower heights and a trough.

ISOBARIC SURFACE (MB) CHARTS	APPROXIMATE ELEVATION	
	(M)	(FT)
1000	120	400
850	1460	4800
700	3000	9800
500	5600	18,400
300	9180	30,100
200	11,800	38,700
100	16,200	53,200

Figure 8.16b shows a simplified upper-air chart (a 500-mb isobaric map) for the same day as the idealized surface map in Fig. 8.16a. The solid gray lines on the map are contour lines given in meters above sea level. The difference in elevation between each contour line (called the *contour interval*) is 60 meters. Superimposed on this map are dashed red lines, which represent lines of equal temperature (isotherms). Observe how the contour lines tend to parallel the isotherms. As we would expect, the contour lines tend to decrease in value from south to north.

The arrows on the 500-mb map show the wind direction. Notice that, unlike the surface winds that cross the isobars in Fig. 8.16a, the winds on the 500-mb chart tend to flow *parallel* to the contour lines in a wavy west-to-east direction. Why does the wind tend to cross the isobars on a surface map, yet blow parallel to the contour lines (or isobars) on an upper-air chart? To answer this question we will now examine the forces that affect winds.

observed in Chapter 1, in the Northern Hemisphere the winds blow counterclockwise and inward toward the center of the lows and clockwise and outward from the center of the highs.

(a) Surface map

(b) Upper-air map (500 mb)

● **FIGURE 8.16** (a) Surface map showing areas of high and low pressure. The solid lines are isobars drawn at 4-mb intervals. The arrows represent wind direction. Notice that the wind blows *across* the isobars. (b) The upper-level (500-mb) map for the same day as the surface map. Solid lines on the map are contour lines in meters above sea level. Dashed red lines are isotherms in °C. Arrows show wind direction. Notice that, on this upper-air map, the wind blows *parallel* to the contour lines.

Flying on a Constant Pressure Surface—High to Low, Look Out Below

Aircraft that use pressure altimeters typically fly along a constant pressure surface rather than a constant altitude surface. They do this because the *altimeter*, as we saw earlier, is simply an aneroid barometer calibrated to convert atmospheric pressure to an approximate altitude. The altitude indicated by an altimeter assumes a standard atmosphere where the air temperature decreases at the rate of 65°C/100 m (3.6°F/1000 ft). Given that the air temperature seldom, if ever, decreases at exactly this rate, altimeters generally indicate an altitude different from their true elevation.

● Figure 2 shows a standard column of air bounded on each side by air with a different temperature and density. On the left side, the air is warm; on the right, it is cold. The orange line represents a constant pressure surface of 700 mb as seen from the side. In the standard air, the 700-mb surface is located at 10,000 ft above sea level.

In the warm air, the 700-mb surface rises; in the cold air, it descends. An aircraft flying along the 700-mb surface would be at an altitude less than 10,000 ft in the cold air, equal to 10,000 ft in the standard air, and greater than 10,000 ft in the warmer air. With no corrections for temperature, the altimeter would indicate the same altitude at all three positions because the air pressure does not change. We can see that, if no temperature corrections are made, an aircraft flying into warm air will increase in altitude and fly higher than its altimeter indicates. Put another way: The altimeter inside the plane will read an altitude lower than the plane's true elevation.

Flying from standard air into cold air represents a potentially dangerous situation. As an aircraft flies into cold air, it flies along a lowering pressure surface. If no correction for temperature is made, the altimeter shows no change in elevation even though the aircraft is losing altitude; hence, the plane will be flying lower than the altimeter indicates. This problem can be serious, especially for planes flying above mountainous terrain with poor visibility and where high winds and turbulence can reduce the air pressure drastically. To ensure adequate clearance under these conditions, pilots fly their aircraft higher than

● **FIGURE 2** An aircraft flying along a surface of constant pressure (orange line) may change altitude as the air temperature changes. Without being corrected for the temperature change, a pressure altimeter will continue to read the same elevation.

they normally would, consider air temperature, and compute a more realistic altitude by resetting their altimeters to reflect these conditions.

Even without sharp temperature changes, pressure surfaces may dip suddenly (see ● Fig. 3). An aircraft flying into an area of decreasing pressure will lose altitude unless corrections are made. For example, suppose a pilot has set the altimeter for sea-level pressure above station A. At this location, the plane is flying along an isobaric surface at a true altitude of 500 ft. As the plane flies toward station B, the pressure surface (and the plane) dips but the altimeter continues to read 500 ft, which is too high. To correct for such changes in pressure, a pilot can obtain a current altimeter setting from ground control.

With this additional information, the altimeter reading will more closely match the aircraft's actual altitude.

Because of the inaccuracies inherent in the pressure altimeter, most high-performance and commercial aircraft are equipped with a *radio altimeter*, also known as a *radar altimeter*. This device measures the altitude of the aircraft by sending out radio waves that bounce off the terrain below. The time it takes these waves to reach the surface and return is a measure of the aircraft's altitude. When this device is used in conjunction with a pressure altimeter, a pilot can determine the variations in a constant pressure surface simply by flying along that surface and observing how the true elevation measured by the radio altimeter changes.

● **FIGURE 3** In the absence of horizontal temperature changes, pressure surfaces can dip toward Earth's surface. An aircraft flying along the pressure surface will either lose or gain altitude, depending on the direction of flight.

Newton's Laws of Motion

Our understanding of why the wind blows stretches back through several centuries, with many scientists contributing to our knowledge. When we think of the movement of air, however, one great scholar stands out—Isaac Newton (1642–1727), who formulated several fundamental laws of motion.

Newton's first law of motion states that *an object at rest will remain at rest and an object in motion will remain in motion (and travel at a constant velocity along a straight line) as long as no force is exerted on the object.* For example, a baseball in a pitcher's hand will remain there until a force (a push) acts upon the ball. Once the ball is pushed (thrown), it would continue to move in that direction forever if it were not for the force of air friction (which slows it down), the force of gravity (which pulls it toward the ground), and the catcher's mitt (which exerts an equal but opposite force to bring it to a halt). Similarly, to start air moving, to speed it up, to slow it down, or even to change its direction requires the action of an external force. This brings us to Newton's second law.

Newton's second law states that *the force exerted on an object equals its mass times the acceleration produced.* In symbolic form, this law is written as

$$F = ma$$

From this relationship we can see that, when the mass of an object is constant, the force acting on the object is directly related to the acceleration that is produced. A force in its simplest form is a push or a pull. *Acceleration is the speeding up, the slowing down, and/or the changing of direction of an object.* (More precisely, acceleration is the change in velocity* over a period of time.)

Because more than one force may act upon an object, Newton's second law always refers to the *net*, or total, force that results. An object will always accelerate in the direction of the total force acting on it. Therefore, to determine in which direction the wind will blow, we must identify and examine all of the forces that affect the horizontal movement of air. These forces include:

1. pressure-gradient force
2. Coriolis force
3. friction

We will first study the forces that influence the flow of air aloft. Then we will see which forces modify winds near the ground.

Forces That Influence the Winds

We already know that horizontal differences in atmospheric pressure cause air to move and, hence, the wind to blow. Given that air is an invisible gas, it may be easier to see how pressure differences cause motion if we examine a visible fluid, such as water.

---

*Velocity specifies both the speed of an object and its direction of motion.

In ● Fig. 8.17, the two large tanks are connected by a pipe. Tank A is two-thirds full and tank B is only one-half full. Since the water pressure at the bottom of each tank is proportional to the weight of water above, the pressure at the bottom of tank A is greater than the pressure at the bottom of tank B. Moreover, since fluid pressure is exerted equally in all directions, there is a greater pressure in the pipe directed from tank A toward tank B than from B toward A.

Because pressure is force per unit area, there must also be a net force directed from tank A toward tank B. This force causes the water to flow from left to right, from higher pressure toward lower pressure. The greater the pressure difference, the stronger the force, and the faster the water moves. In a similar way, horizontal differences in atmospheric pressure cause air to move.

PRESSURE-GRADIENT FORCE ● Figure 8.18 shows a region of higher pressure on the map's left side, lower pressure on the right. The isobars show how the horizontal pressure is changing. If we compute the amount of pressure change that occurs over a given distance, we have the **pressure gradient**; thus

$$\text{Pressure gradient} = \frac{\text{difference in pressure}}{\text{distance}}$$

If we let the symbol delta (Δ) mean "a change in," we can simplify the expression and write the pressure gradient as

$$PG = \frac{\Delta p}{d}$$

where Δp is the pressure difference between two places some horizontal distance (d) apart. In Fig. 8.18 the pressure gradient between points 1 and 2 is 4 millibars per 100 kilometers.

● **FIGURE 8.17** The higher water level creates higher fluid pressure at the bottom of tank A and a net force directed toward the lower fluid pressure at the bottom of tank B. This net force causes water to move from higher pressure toward lower pressure.

Suppose the pressure in Fig. 8.18 were to change and the isobars become closer together. This condition would produce a rapid change in pressure over a relatively short distance, or what is called a *steep* (or *strong*) *pressure gradient*. However, if the pressure were to change such that the isobars spread farther apart, then the difference in pressure would be small over a relatively large distance. This condition is called a *gentle* (or *weak*) *pressure gradient*.

Notice in Fig. 8.18 that when differences in horizontal air pressure exist there is a net force acting on the air. This force, called the **pressure-gradient force (PGF)**, *is directed from higher toward lower pressure at right angles to the isobars*. The magnitude of the force is directly related to the pressure gradient. Steep pressure gradients correspond to strong pressure-gradient forces and vice versa. • Figure 8.19 shows the relationship between pressure gradient and pressure-gradient force.

The *pressure-gradient force is the force that causes the wind to blow*. Because of this effect, closely spaced isobars on a weather map indicate steep pressure gradients, strong forces, and high winds. On the other hand, widely spaced isobars indicate gentle pressure gradients, weak forces, and light winds. An example of a steep pressure gradient and strong winds associated with Hurricane (Superstorm) Sandy is given in • Fig. 8.20.

If the pressure-gradient force were the only force acting upon air, we would always find winds blowing directly from higher toward lower pressure. However, the moment air starts to move, it is deflected in its path by the *Coriolis force*.

CORIOLIS FORCE   The **Coriolis force** describes an apparent force that is due to the rotation of Earth. To understand how it works, consider two people playing catch as they sit opposite one another on the rim of a merry-go-round (see • Fig. 8.21, platform A). If the merry-go-round is not moving, each time the ball is thrown, it moves in a straight line to the other person.

Suppose the merry-go-round starts turning counterclockwise—the same direction Earth spins as viewed from above the North Pole. If we watch the game of catch from above, we see that the ball moves in a straight-line path just as before. However, to the people playing catch on the merry-go-round, the ball seems to veer to its right each time it is thrown, always landing

• **FIGURE 8.19** The closer the spacing of the isobars, the greater the pressure gradient. The greater the pressure gradient, the stronger the pressure-gradient force. The stronger the PGF, the greater the wind speed. The red arrows represent the relative magnitude of the force, which is always directed from higher toward lower pressure.

to the right of the point intended by the thrower (see Fig. 8.21, platform B). This perception is accounted for by the fact that while the ball moves in a straight-line path, the merry-go-round rotates beneath it; by the time the ball reaches the opposite side, the catcher has moved. To anyone on the merry-go-round, it seems as if there is some force causing the ball to deflect to the right. This apparent force is called the *Coriolis force* after Gaspard Coriolis, a nineteenth-century French scientist who worked it out mathematically. (Because it is an *apparent force* due to the rotation of Earth, it is also called the *Coriolis effect*.) This effect occurs not only on merry-go-rounds but on rotating Earth, too. All free-moving objects, such as ocean currents, aircraft, artillery projectiles, and even air molecules seem to deflect from a straight-line path because Earth rotates under them.

The Coriolis force *appears to deflect the wind to the right of its intended path in the Northern Hemisphere and to the left of its intended path in the Southern Hemisphere*. To illustrate this, consider a satellite in polar circular orbit. If Earth were not rotating, the path of the satellite would be observed to move directly from north to south, parallel to Earth's meridian lines. However, Earth *does* rotate, carrying us and meridians eastward with it. Because of this rotation in the Northern Hemisphere, we see the satellite moving southwest instead of due south; it seems to veer off its path and move toward *its right*. In the Southern Hemisphere, Earth's direction of rotation is clockwise as viewed from above the South Pole. Consequently, a satellite moving northward from the South Pole would appear to move northwest and, hence, would veer to the *left* of its path.

The magnitude of the Coriolis force varies with the speed of the moving object and the latitude. • Figure 8.22 shows this variation for various wind speeds at different latitudes. In each case, as the wind speed increases, the Coriolis force increases; hence, *the stronger the wind speed, the greater the deflection*. Also, note that the Coriolis force increases for all wind speeds from a value of *zero at the equator to a maximum at the poles*. We can see this latitude effect better by examining • Fig. 8.23.

• **FIGURE 8.18** The pressure gradient between point 1 and point 2 is 4 mb per 100 km. The net force directed from higher toward lower pressure is the *pressure-gradient force*.

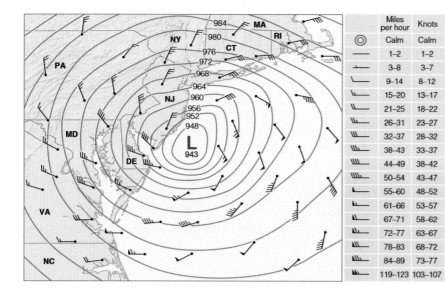

● **FIGURE 8.20** Surface map for 4 p.m. (EST) Monday, October 29, 2012, as Hurricane (Superstorm) Sandy approaches the New Jersey shore from the east. Isobars are dark gray lines with units in millibars. The interval between isobars is 4 mb. Sandy's central pressure is 943 mb or 27.85 in. The tightly packed isobars are associated with a strong pressure gradient, a strong pressure gradient force, and high winds, gusting to over 80 knots over portions of Long Island, New York. Wind directions are given by lines that parallel the wind. Wind speeds are indicated by barbs and flags, where a wind indicated by the symbol would be a wind from the northwest at between 23 and 27 knots. (See blue insert.) Wind speeds over the ocean, where no observations are available, are based on computer model calculations.

● **FIGURE 8.21** On nonrotating platform A, the thrown ball moves in a straight line. On platform B, which rotates counterclockwise, the ball continues to move in a straight line. However, platform B is rotating while the ball is in flight; thus, to the person throwing the ball on platform B, the ball appears to deflect to the right of its intended path.

CRITICAL THINKING QUESTION Suppose the rotating platform in diagram B is rotating in a clockwise direction, as viewed from above the platform. How would the ball appear to deflect (bend) to the person throwing the ball on the platform? How would the ball appear to move to a person not on the platform?

Imagine in Fig. 8.23 that there are three aircraft, each at a different latitude and each flying along a straight-line path, with no external forces acting on them. The destination of each aircraft is due east and is marked on the diagram (see Fig. 8.23a). Each plane travels in a straight path relative to an observer positioned at a fixed spot in space. Earth rotates beneath the moving planes, causing the destination points at latitudes 30° and 60° to

WEATHER WATCH

An intense low in November 1998 produced extremely high winds that gusted over 90 knots in Wisconsin. The winds caused blizzard conditions over the Dakotas, closed many interstate highways, shut down airports, and overturned trucks. The winds pushed a school bus off the road near Albert Lea, Minnesota, injuring two children, and blew the roofs off homes in Wisconsin. This notorious storm set an all-time record low pressure for Minnesota of 963 mb (28.43 in.). That record was broken on October 26, 2010, by an even stronger storm that deepened to a reading at Bigfork, Minnesota, of 955 mb (28.21 in.)—the lowest surface pressure measured anywhere in the contiguous United States up to that time, apart from the Atlantic coast.

change direction slightly when seen by an observer in space (see Fig. 8.23b). To an observer standing on Earth, however, it is the plane that appears to deviate. The amount of deviation is greatest toward the pole and nonexistent at the equator. Therefore, the Coriolis force has a far greater effect on the plane at high latitudes (large deviation) than on the plane at low latitudes (small deviation). On the equator, it has no effect at all. The same, of course, is true of its effect on winds.

In summary, to an observer on Earth, objects moving in *any direction* (north, south, east, or west) are deflected to the *right* of their intended path in the Northern Hemisphere and to the *left* of their intended path in the Southern Hemisphere.

The amount of deflection depends upon:

1. the rotation of Earth
2. the latitude
3. the object's speed*

---

*These three factors are grouped together and shown in the expression Coriolis force = $2m\Omega V \sin\phi$, where m is the object's mass, Ω is Earth's angular rate of spin (a constant), V is the speed of the object, and ϕ is the latitude.

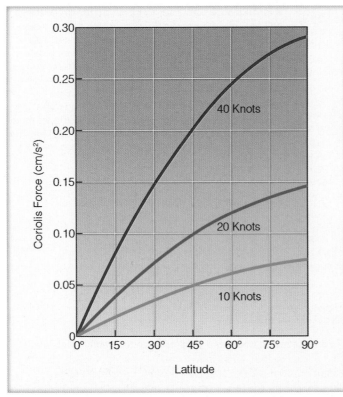

● **FIGURE 8.22** The relative variation of the Coriolis force at different latitudes with different wind speeds.

In addition, the *Coriolis force acts at right angles to the wind, only influencing wind direction and never wind speed.*

The Coriolis force is present in all motions relative to Earth's surface. However, because its strength is directly related to the speed of motion, the impact of the Coriolis force tends to be smaller when looking at small areas. In most of our everyday experiences, the Coriolis force is so small (compared to other forces involved in those experiences) that it is negligible. Contrary to popular belief, it does not cause water to turn clockwise or counterclockwise when draining from a sink. The Coriolis force is also minimal on small-scale winds, such as those that blow inland along coasts in summer. Here, the Coriolis force itself might be strong because of high winds, but the force cannot produce much deflection over the relatively short distances. Only where winds blow over long distances is the Coriolis effect significant.

BRIEF REVIEW

In summary, we know that:

● Atmospheric (air) pressure is the pressure exerted by the mass of air above a region.

● A change in surface air pressure can be brought about by changing the mass (amount of air) above the surface.

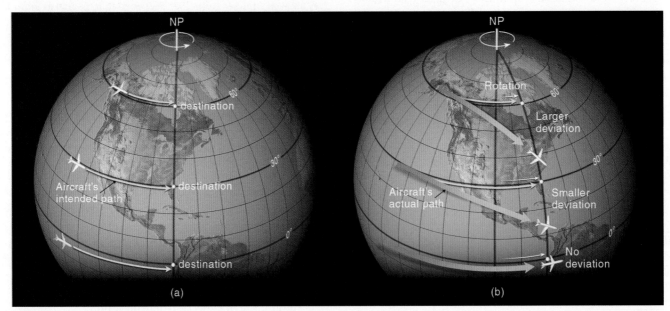

● **FIGURE 8.23** Except at the equator, a free-moving object heading either east or west (or any other direction) will appear from Earth to deviate from its path as Earth rotates beneath it. The deviation (Coriolis force) is greatest at the poles and decreases to zero at the equator. Notice that the aircraft's deviation from its intended destination is greatest at high latitudes and nonexistent at the equator.

CRITICAL THINKING QUESTION If Earth were to stop rotating in diagram (b), how would the path of the aircraft at 60°N, 30°N, and at the equator appear to change to a person standing on Earth's surface?

- Heating and cooling columns of air can establish horizontal variations in atmospheric pressure aloft and at the surface.

- A difference in horizontal air pressure produces a horizontal pressure-gradient force.

- The pressure-gradient force is always directed from higher pressure toward lower pressure, and it is the pressure-gradient force that causes the air to move and the wind to blow.

- Steep pressure gradients (tightly packed isobars on a weather map) indicate strong pressure-gradient forces and high winds; gentle pressure gradients (widely spaced isobars) indicate weak pressure-gradient forces and light winds.

- Once the wind starts to blow, the Coriolis force causes it to bend to the right of its intended path in the Northern Hemisphere and to the left of its intended path in the Southern Hemisphere.

With this information in mind, we will first examine how the pressure-gradient force and the Coriolis force produce straight-line winds aloft, above the frictional influence of Earth's surface. We will then look at what other forces come into play as winds blow along a curved path.

STRAIGHT-LINE FLOW ALOFT—GEOSTROPHIC WINDS

Earlier in this chapter, we saw that the winds aloft on an upper-level chart blow more or less parallel to the isobars or contour lines. We can see why this phenomenon happens by carefully looking at ● Fig. 8.24, which shows a map in the Northern Hemisphere, above Earth's frictional influence at an altitude of about 1 kilometer (3300 ft) above Earth's surface. Horizontal pressure changes are shown by isobars. The evenly spaced isobars indicate a constant pressure-gradient force directed from south toward north as indicated by the red arrow at the left. Why, then, does the map show a wind blowing from the west? We can answer this question by placing a parcel of air at position 1 in the diagram and watching its behavior.

At position 1, the PGF acts immediately upon the air parcel, accelerating it northward toward lower pressure. However, the instant the air begins to move, the Coriolis force deflects the air toward its right, curving its path. As the parcel of air increases in speed (positions 2, 3, and 4), the magnitude of the Coriolis force increases (as shown by the longer arrows), bending the wind more and more to its right. Eventually, the wind speed increases to a point where the Coriolis force just balances the PGF. At this point (position 5), the wind no longer accelerates because the net force is zero. Here the wind flows in a straight path, parallel to the isobars at a constant speed.* This flow of air is called a **geostrophic** (*geo*: "earth"; *strophic*: "turning") **wind**. Notice that the geostrophic wind blows in the Northern Hemisphere with lower pressure to its left and higher pressure to its right.

---

*At first, it may seem odd that the wind blows at a constant speed with no net force acting on it. But when we remember that the net force is necessary only to accelerate ($F = ma$) the wind, it makes more sense. For example, it takes a considerable net force to push a car and get it rolling from rest. But once the car is moving, it only takes a force large enough to counterbalance friction to keep it going. There is no net force acting on the car, yet it rolls along at a constant speed.

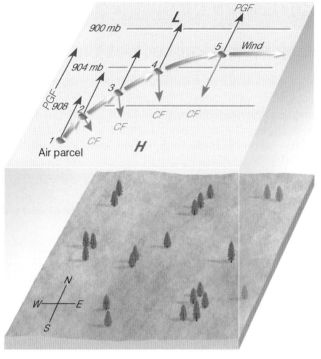

● **FIGURE 8.24** Above the level of friction, air initially at rest will accelerate until it flows parallel to the isobars at a steady speed with the PGF balanced by the CF. Wind blowing under these conditions is called *geostrophic*.

WEATHER WATCH

As you drive your car along a highway (at the speed limit), the Coriolis force would "pull" your vehicle to the right about 1500 feet for every 100 miles you travel if it were not for the friction between your tires and the road surface.

When the flow of air is purely geostrophic, the isobars (or contours) are straight and evenly spaced, and the wind speed is constant. In the atmosphere, isobars are rarely straight or evenly spaced, and the wind normally changes speed as it flows along. So, the geostrophic wind is usually only an approximation of the real wind. However, the approximation is generally close enough to help us more clearly understand the behavior of the winds aloft.

As we would expect from our previous discussion of winds, the speed of the geostrophic wind is directly related to the pressure gradient. In ● Fig. 8.25, we can see that a geostrophic wind flowing parallel to the isobars is similar to water in a stream flowing parallel to its banks. At position 1, the wind is blowing at a low speed; at position 2, the pressure gradient increases and the wind speed picks up. Notice also that at position 2, where the wind speed is greater, the Coriolis force is greater and balances the stronger pressure-gradient force. (A more mathematical approach to the concept of geostrophic wind is given in Focus section 8.3.)

In ● Fig. 8.26, we can see that the geostrophic wind direction can be determined by studying the orientation of the

A Mathematical Look at the Geostrophic Wind

We know from an earlier discussion that the geostrophic wind gives us a good approximation of the real wind above the level of friction, about 500 to 1000 m above Earth's surface. Above the friction layer, the winds tend to blow parallel to the isobars, or contours. We know that, for any given latitude, the speed of the geostrophic wind is proportional to the pressure gradient. This may be represented as

$$V_g \sim \frac{\Delta p}{d}$$

where V_g is the geostrophic wind and Δp is the pressure difference between two places some horizontal distance (d) apart. From this, we can see that the greater the pressure gradient, the stronger the geostrophic wind.

When we consider a unit mass of moving air, we must take into account the air density (mass per unit volume) expressed by the symbol ρ. The geostrophic wind is now directly proportional to the pressure-gradient force; thus

$$V_g \sim \frac{1}{\rho} \frac{\Delta p}{d}$$

We can see from this expression that, with the same pressure gradient (at the same latitude), the geostrophic wind will increase with increasing altitude because air density decreases with height.

In a previous section, we saw that the geostrophic wind represents a balance of forces between the Coriolis force and the pressure-gradient force. Here, it should be noted that the Coriolis force (per unit mass) can be expressed as

$$\text{Coriolis force} = 2\,\Omega\, V \sin \phi$$

where Ω is Earth's angular spin (a constant), V is the speed of the wind, and ϕ is the latitude. The $\sin \phi$ is a trigonometric function that takes into account the variation of the Coriolis force with latitude. At the equator (0°), $\sin \phi$ is 0; at 30° latitude, $\sin \phi$ is 0.5, and, at the poles (90°), $\sin \phi$ is 1.

This balance between the Coriolis force and the pressure-gradient force can be written as

$$CF = PGF$$

$$2\Omega V_g \sin \phi = \frac{1}{\rho} \frac{\Delta p}{d}$$

Solving for V_g, the geostrophic wind, the equation becomes

$$V_g = \frac{1}{2\,\Omega\, \sin \phi\, \rho} \frac{\Delta p}{d}$$

Customarily, the rotational (2Ω) and latitudinal ($\sin \phi$) factors are combined into a single value f, called the *Coriolis parameter*. Thus, we have the geostrophic wind equation written as

$$V_g \sim \frac{1}{f\rho} \frac{\Delta p}{d}$$

$$500$$
40°N 200 km $V_g \longrightarrow$
$$504$$
0   200 km
N   Scale

● **FIGURE 4** A portion of an upper-air chart for part of the Northern Hemisphere at an altitude of 5600 meters above sea level. The lines on the chart are isobars, where 500 equals 500 millibars. The air temperature is −25°C and the air density is 0.70 kg/m³.

Suppose we compute the geostrophic wind for the example given in ● Fig. 4. Here the wind is blowing parallel to the isobars in the Northern Hemisphere at latitude 40°. The spacing between the isobars is 200 km and the pressure difference is 4 mb. The altitude is 5600 m above sea level, where the air temperature is −25°C (−13°F) and the air density is 0.70 kg/m³. First, we list our data and put them in the proper units, as

$$\Delta p = 4 \text{ mb} = 400 \text{ Newtons/m}^2$$
$$d = 200 \text{ km} = 2 \times 10^5 \text{ m}$$
$$\sin \phi = \sin(40°) = 0.64$$
$$\rho = 0.70 \text{ kg/m}^3$$
$$2\,\Omega = 14.6 \times 10^{-5} \text{ radians/sec *}$$

When we use equation (1) to compute the geostrophic wind, we obtain

$$V_g = \frac{1}{2\,\Omega\, \sin \phi\, \rho} \frac{\Delta p}{d}$$

$$V_g = \frac{400}{14.6 \times 10^{-5} \times 0.64 \times 0.70 \times 2 \times 10^5}$$

$$V_g = 30.6 \text{ m/sec, or } 59.4 \text{ knots}$$

*The rate of Earth's rotation (Ω) is 360° in one day, actually a sidereal day consisting of 23 hr, 56 min, 4 sec, or 86,164 seconds. This gives a rate of rotation of 4.18×10^{-3} degrees per second. Most often, Ω is given in radians, where 2π radians equals 360° ($\pi = 3.14$). Therefore, the rate of Earth's rotation can be expressed as 2π radians/ 86,164 sec, or 7.29×10^{-5} radians/sec, and the constant 2Ω becomes 14.6×10^{-5} radians/sec.

isobars; its speed can be estimated from the spacing of the isobars. On an isobaric chart, the geostrophic wind direction and speed are related in a similar way to the contour lines. Therefore, if we know the isobar or contour patterns on an upper-level chart, we also know the direction and relative speed of the geostrophic wind, even for regions where no direct wind measurements have been made. Similarly, if we know the geostrophic wind direction and speed, we can estimate the orientation and spacing of the isobars, even if we don't have a current weather map. (It is also possible to estimate the wind flow and pressure patterns aloft by watching the movement of clouds. Focus section 8.4 illustrates these phenomena further.)

We know that the winds aloft do not always blow in a straight line; frequently, they curve and bend into meandering loops as they tend to follow the patterns of the isobars. In the Northern Hemisphere, winds blow counterclockwise around lows and clockwise around highs. The next section explains why.

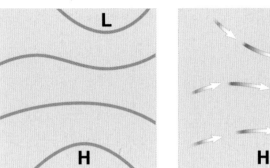

● **FIGURE 8.25** The isobars and contours on an upper-level chart are like the banks along a flowing stream. When they are widely spaced, the flow is weak; when they are narrowly spaced, the flow is stronger. The increase in winds on the chart results in a stronger CF, which balances a larger PGF.

(a) Isobar or contour pattern   (b) Wind pattern

● **FIGURE 8.26** By observing the orientation and spacing of the isobars (or contours) in diagram (a), the geostrophic wind direction and speed can be determined in diagram (b).

CURVED WINDS AROUND LOWS AND HIGHS ALOFT— GRADIENT WINDS Because lows are also known as cyclones, the flow of air around them (which is counterclockwise in the Northern Hemisphere) is often called *cyclonic flow.* Likewise, the flow of air around a high, or anticyclone (which is clockwise in the Northern Hemisphere), is called *anticyclonic flow.* Look at the wind flow around the upper-level low (Northern Hemisphere) in ● Fig. 8.27. At first, it appears as though the wind is defying the Coriolis force by bending to the left as it moves counterclockwise around the system. Let's see why the wind blows in this manner.

Suppose we consider a parcel of air initially at rest at position 1 in Fig. 8.27a. The pressure-gradient force accelerates the air inward toward the center of the low and the Coriolis force deflects the moving air to its right, until the air is moving parallel to the isobars at position 2. If the isobars were straight lines ahead of this point, the wind would move northward at a constant speed, bringing it to position 3. In reality, the isobars are curved, and the wind follows that curvature (position 4). A wind that blows at a constant speed parallel to *curved isobars* above the level of frictional influence is termed a **gradient wind**. Why does the wind follow this curved path?

Look closely at position 2 (Fig. 8.27a), where a parcel is moving northward, and observe if the wind were in geostrophic balance, the inward-directed pressure-gradient force (would be in balance with the outward-directed *Coriolis force (CF)* and the wind should continue blowing in a straight line toward

position 3. Suppose the parcel reaches position 3. Notice in Fig. 8.27b that here the PGF would now be angled toward the southwest, because the PGF is always directed toward lower pressure at right angles to the isobars. This means that part of the PGF will be directed southward, against the northward motion of the parcel. This situation slows the wind down just a bit. Recall that the Coriolis force is directly related to wind speed, so the slower wind weakens the CF. As a result of the weaker CF, the PGF causes the wind to bend the parcel toward the left and thus move in a circular, counterclockwise path, parallel to curved isobars around the low, as illustrated in Fig. 8.27c.

Look at ● Fig. 8.28a and notice that above the level of frictional influence, the winds blow clockwise around an area of high pressure. The same spacing of the isobars tells us that the magnitude of the PGF is the same as in Fig. 8.27a. Suppose at position 2 in Fig. 8.28a the wind is in geostrophic balance with the PGF directed outward and the Coriolis force directed inward. If the wind was in geostrophic balance, the air would move in a straight line parallel to straight-line isobars. But the isobars are curved. Let's further suppose that the air parcel moves from position 2 directly southward to position 3. We can see in Fig. 8.28b that at position 3 the PGF crosses the isobars toward the southeast, producing a southerly component of the PGF, which increases the wind speed just a bit. The increase in wind speed increases the magnitude of the Coriolis force, thereby bending the parcel to the right, which causes the wind to blow clockwise, parallel to the isobars in a circular path around the high.*

The greater Coriolis force around the high as compared to the low results in an interesting relationship. Because the CF (at any given latitude) can increase only when the wind speed

---

*Earlier in this chapter we learned that an object accelerates when there is a change in its speed or direction (or both). Therefore, the gradient wind blowing *around* the low-pressure center is constantly accelerating because it is constantly changing direction. This acceleration, called the *centripetal acceleration,* is directed at right angles to the wind, inward toward the low center. Remember from Newton's second law that, if an object is accelerating, there must be a net force acting on it. In this case, the *net force* acting on the wind must be directed toward the center of the low, so that the air will keep moving in a circular path. This inward-directed force is called the *centripetal force (centri:* center; *petal:* to push toward). In some cases, it is more convenient to express the centripetal force (and the centripetal acceleration) as the centrifugal force, an apparent force that is equal in magnitude to the *centripetal force,* but directed outward from the center of rotation.

increases, we can see that for the same pressure gradient (the same spacing of the isobars), the winds around a high-pressure area (or a ridge) must be greater than the winds around a low-pressure area (or a trough). Typically, however, this difference in speed is overshadowed by the stronger winds that occur around low-pressure areas (cyclonic storms) because of the closer spacing of the isobars and stronger pressure gradients associated with them.

In the Southern Hemisphere, the pressure-gradient force starts the air moving, and the Coriolis force deflects the moving air to the *left*, thereby causing the wind to blow *clockwise around lows* and *counterclockwise around highs*. • Figure 8.29 shows a satellite image of clouds and wind flow (dark arrows) around a low-pressure area in the Northern Hemisphere (Fig. 8.29a) and in the Southern Hemisphere (Fig. 8.29b).

So far we have seen how winds blow in theory, but how do they appear on an actual map?

WINDS ON UPPER-LEVEL CHARTS

On the upper-level 500-mb map (• Figure 8.30), notice that, as we would expect, the winds tend to parallel the contour lines in a wavy west-to-east direction. Notice also that the contour lines tend to decrease in height from south to north. This situation occurs because the air at this level is warmer to the south and colder to the north. On the map, where horizontal temperature contrasts are large there is also a large height gradient—the contour lines are close together and the winds are strong. Where the horizontal temperature contrasts are small, there is a small height gradient—the contour lines are spaced farther apart and the winds are weaker. In general, on maps such as this we find stronger north-to-south temperature contrasts in winter than in summer, which is why the winds aloft are usually stronger in winter.

In Fig. 8.30, the wind is geostrophic where it blows in a straight path parallel to evenly spaced lines; it is gradient where it blows parallel to curved contour lines. There are two additional ways of classifying wind. Where the wind flows in large, looping meanders, following a more or less north-south trajectory (as it does in Fig. 8.30 off the west coast of North America), the wind-flow pattern is called **meridional**. Where the winds are blowing

in a west-to-east direction (as seen in Fig. 8.30 over the eastern third of the United States), the flow is termed **zonal**.

Because the winds aloft in middle and high latitudes generally blow from west to east, planes flying in this direction have a beneficial tail wind, which explains why a flight from San Francisco to New York City takes, on average, about 30 to 45 minutes less than the return flight. If the flow aloft is zonal, clouds, storms, and surface anticyclones tend to move more rapidly from west to east. However, where the flow aloft is meridional, as we will see in Chapter 12, surface storms tend to move more slowly, often intensifying into major cyclonic storm systems.

We know that because of the pressure-gradient force and the Coriolis force (which, as we have seen, bends moving air to the right in the Northern Hemisphere), the winds aloft in the middle latitudes of the Northern Hemisphere tend to blow in a west-to-east pattern. Because the Coriolis force bends moving air to the

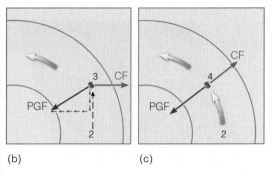

• **FIGURE 8.27** Winds and related forces around an area of low pressure above the friction level in the Northern Hemisphere. Notice that the pressure-gradient force is in red, while the Coriolis force is in blue.

• **FIGURE 8.28** Winds and related forces around an area of high pressure above the friction level in the Northern Hemisphere. Notice that the pressure-gradient force is in red, while the Coriolis force is in blue.

Estimating Wind Direction and Pressure Patterns Aloft by Watching Clouds

Both the wind direction and the orientation of the isobars aloft can be estimated by observing middle- and high-level clouds from Earth's surface. Suppose, for example, we are in the Northern Hemisphere watching clouds directly above us move from southwest to northeast at an elevation of about 3000 m or 10,000 ft (see ● Fig. 5a). This indicates that the geostrophic wind at this level is southwesterly. Looking downwind, the geostrophic wind blows parallel to the isobars with lower pressure on the left and higher pressure on the right. Thus, if we *stand with our backs to the direction from which the clouds are moving, lower pressure aloft will always be to our left and higher pressure to our right.* From this observation, we can draw a rough upper-level chart (see Fig. 5b), which shows isobars and wind direction for an elevation of approximately 10,000 ft.

The isobars aloft will not continue in a southwest-northeast direction indefinitely;

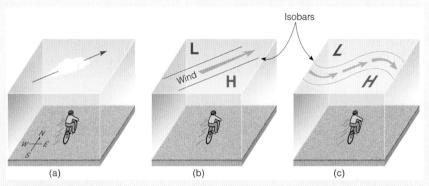

rather, they will often bend into wavy patterns. We may carry our observation one step farther, then, by assuming a bending of the lines (Fig. 5c). Thus, with a southwesterly wind aloft, a trough of low pressure will be found to our

west and a ridge of high pressure to our east. What would be the pressure pattern if the winds aloft were blowing *from* the northwest? Answer: A trough would be to the east and a ridge to the west.

● **FIGURE 5** The movement of clouds can help determine the geostrophic wind direction and orientation of the isobars. Upper-level clouds moving from the southwest (a) indicate isobars and winds aloft (b). When extended horizontally, the upper-level chart appears as in (c), where a trough of low pressure is to the west and a ridge of high pressure is to the east.

(a) Northern Hemisphere (b) Southern Hemisphere

● **FIGURE 8.29** Clouds and related wind flow patterns (purple arrows) around low-pressure areas. (a) In the Northern Hemisphere, winds blow counterclockwise around an area of low pressure. (b) In the Southern Hemisphere, winds blow clockwise around an area of low pressure.

● **FIGURE 8.30** An upper-level 500-mb map showing wind direction, as indicated by lines that parallel the wind. Wind speeds are indicated by barbs and flags. (See the blue insert.) Solid gray lines are contours in meters above sea level. Dashed red lines are isotherms in °C.

left in the Southern Hemisphere, does this situation mean that the winds aloft in the Southern Hemisphere blow from east to west? If you are unsure of the answer to this question, read Focus section 8.5.

Take a minute and look at Fig. 8.20 on p. 211 and observe that the surface winds around Superstorm Sandy tend to cross the isobars, blowing from higher pressure toward lower pressure. Observe also that the tightly packed isobars are indicating strong winds. Yet this same distribution of pressure (same pressure gradient) and same temperature on an upper-level chart would produce even stronger winds. Why, then, do surface winds normally cross the isobars, and why do they blow more slowly than the winds aloft? The answer to both of these questions is *friction*.

SURFACE WINDS The frictional drag of the ground slows the wind down. Because the effect of friction decreases as we move away from Earth's surface, wind speeds tend to increase with height above the ground. The atmospheric layer that is influenced by friction, called the **friction layer** (or *planetary boundary layer*), usually extends upward to an altitude near 1000 m (3300 ft) above the surface, but this altitude can vary due to strong winds or irregular terrain. (We will examine the planetary boundary layer winds more thoroughly in Chapter 9.)

In ● Fig. 8.31a, the wind aloft is blowing at a level above the frictional influence of the ground. At this level, the wind is approximately geostrophic and blows parallel to the isobars, with the horizontal *PGF* on its left balanced by the CF on its right. At Earth's surface, the same pressure gradient will not produce the same wind speed, and the wind will not blow in the same direction.

Near Earth's surface, *friction reduces the wind speed, which in turn reduces the Coriolis force*. Consequently, the weaker Coriolis force no longer balances the pressure-gradient force, and the wind blows across the isobars toward lower pressure. The angle (α) at which the wind crosses the isobars varies, but averages about 30°.* As we can see in Fig. 8.31a, at the surface the pressure-gradient force is now balanced by the sum of the frictional force and the Coriolis force. Therefore, in the Northern Hemisphere, we find surface winds blowing counterclockwise and *into* a low; they flow clockwise and *out* of a high (see Fig. 8.31b). In the Southern Hemisphere, winds blow clockwise and inward around surface lows; counterclockwise and outward around surface highs (see ● Fig. 8.32). ● Figure 8.33 illustrates a surface weather map and the general wind flow pattern on a particular day in South America.

We know that, because of friction, surface winds move more slowly than do the winds aloft with the same pressure gradient. Surface winds also blow across the isobars toward lower pressure. The angle at which the winds cross the isobars depends upon surface friction, wind speed, and the height above the surface. Aloft, however, the winds blow parallel to contour lines, with lower pressure (in the Northern Hemisphere) to their left.

---

*The angle at which the wind crosses the isobars to a large degree depends upon the roughness of the terrain. Everything else being equal, the rougher the surface, the larger the angle. Over hilly land, the angle might average between 35° and 40°, while over an open body of relatively smooth water it may average between 10° and 15°. Taking into account all types of surfaces, the average is near 30°. This angle also depends on the wind speed. Typically, the angle is smallest for high winds and largest for gentle breezes. As we move upward through the friction layer, the wind becomes more and more parallel to the isobars.

Winds Aloft in the Southern Hemisphere

In the Southern Hemisphere, just as in the Northern Hemisphere, the winds aloft blow because of horizontal differences in pressure. The pressure differences, in turn, are due to variations in temperature. Recall from an earlier discussion of pressure that warm air aloft is associated with high pressure and cold air aloft with low pressure. Look at ● Fig. 6. It shows an upper-level chart that extends from the Northern Hemisphere into the Southern Hemisphere. Over the equator, where the air is warmer, the pressure aloft is higher. North and south of the equator, where the air is colder, the pressure aloft is lower.

Let's assume, to begin with, that there is no wind on the chart. In the Northern Hemisphere, the pressure-gradient force directed northward starts the air moving toward lower pressure. Once the air is set in motion, the Coriolis force bends it to the right until it is a *west wind*, blowing parallel to the isobars.

● **FIGURE 6** Upper-level chart that extends over the Northern and Southern Hemispheres. Solid gray lines on the chart are isobars.

In the Southern Hemisphere, the pressure-gradient force directed southward starts the air moving south. But notice that the Coriolis force in the Southern Hemisphere bends the moving air to its *left*, until the wind is blowing parallel to the isobars *from the west*. Hence, in the middle and high latitudes of both hemispheres, we generally find westerly winds aloft.

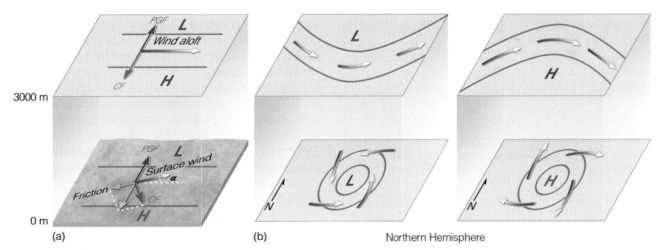

(a)   (b)   Northern Hemisphere

● **FIGURE 8.31** (a) The effect of surface friction is to slow down the wind so that, near the ground, the wind crosses the isobars and blows toward lower pressure. (b) This phenomenon at the surface produces an inflow of air around a low and an outflow of air around a high. Aloft, the winds blow parallel to the lines, usually in a wavy west-to-east pattern. Both diagrams (a) and (b) are in the Northern Hemisphere.

Consequently, because of this fact, if you (in the Northern Hemisphere) stand with the wind aloft to your back, lower pressure will be to your left and higher pressure to your right (see ● Fig. 8.34a). The same rule applies to the surface wind, but with a slight modification due to the fact that here the wind crosses the isobars. Look at Fig. 8.34b and notice that, at the surface, *if you stand with your back to the wind, then turn clockwise about 30°, the center of lowest pressure will be to your left.** This relationship between wind and pressure is often called **Buys Ballot's law**, after the Dutch meteorologist Christophorus Buys Ballot (1817–1890), who formulated it.

---

*In the Southern Hemisphere, stand with your back to the wind, then turn counterclockwise about 30°—the center of lowest pressure will then be to your right.

● FIGURE 8.33 Surface weather map showing isobars and winds on a day in December in South America.

Winds and Vertical Air Motions

Up to this point, we have seen that surface winds blow in toward the center of low pressure and outward away from the center of high pressure. Notice in ● Fig. 8.35 that as air moves inward toward the center of low pressure, it must go somewhere. Since this converging air cannot go into the ground, it slowly rises. Above the surface low (at about 6 km or so), the air begins to diverge (spread apart).

As long as the upper-level diverging air balances the converging surface air, the central pressure in the surface low does not change. However, the surface pressure *will change* if upper-level divergence and surface convergence are not in balance. For example, as we saw earlier in this chapter (when we examined the air pressure above two cities), the surface pressure will change if the mass of air above the surface changes. Consequently, if upper-level divergence exceeds surface convergence (that is, more air is removed at the top than is taken in at the surface), the air pressure at the center of the surface low will decrease, and isobars around the low will become more tightly packed. This situation increases the pressure gradient (and, hence, the pressure-gradient force), which, in turn, increases the surface winds.

Surface winds move outward (diverge) away from the center of a high-pressure area. Observe in Fig. 8.35 that to replace this laterally spreading air, the air aloft converges and slowly descends. Again, as long as upper-level converging air balances surface diverging air, the air pressure in the center of the high will not change. (Convergence and divergence of air are so important to the development or weakening of surface pressure systems that we will examine this topic again when we look more closely at the vertical structure of pressure systems in Chapter 12.)

The rate at which air rises above a low or descends above a high is small compared to the horizontal winds that spiral about these systems. Generally, the vertical motions are usually only about several centimeters per second, or about 1.5 km (or 1 mi) per day.

Earlier in this chapter we learned that air moves in response to pressure differences. Because air pressure decreases rapidly with increasing height above the surface, there is always a strong pressure-gradient force directed upward, much stronger than in the horizontal. Why, then, doesn't the air rush off into space?

Air does not rush off into space because the upward-directed pressure-gradient force is nearly always exactly balanced by the downward force of gravity. When these two forces are in exact balance, the air is said to be in **hydrostatic equilibrium**. When air is in hydrostatic equilibrium, there is no net vertical force acting on it, and so there is no net vertical acceleration. (Remember the example of the moving car in the footnote at the bottom of

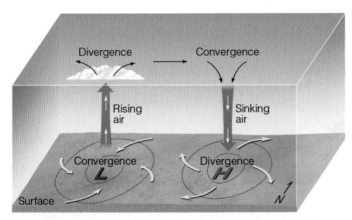

● FIGURE 8.34 (a) In the Northern Hemisphere, if you stand with the wind aloft at your back, lower pressure aloft will be to your left and higher pressure to your right. (b) At the surface, the center of lowest pressure will be to your left if, with your back to the surface wind, you turn clockwise about 30°.

● FIGURE 8.35 Winds and air motions associated with surface highs and lows in the Northern Hemisphere.

p. 213). Most of the time, the atmosphere approximates hydrostatic balance, even when air slowly rises or descends at a constant speed. However, this balance does not exist in violent thunderstorms and tornadoes, where the air shows appreciable vertical acceleration. But these occur over relatively small vertical distances, considering the total vertical extent of the atmosphere. (A more mathematical look at hydrostatic equilibrium, expressed by the *hydrostatic equation*, is given in Focus section 8.6.)

The Hydrostatic Equation

Air is in hydrostatic equilibrium when the upward-directed pressure-gradient force is exactly balanced by the downward force of gravity.

● Figure 7 shows air in hydrostatic equilibrium. Because there is no net vertical force acting on the air, there is no net vertical acceleration, and the sum of the forces is equal to zero, all of which is represented by

$$PGF_{vertical} + g = 0$$

$$\frac{1}{\rho}\frac{\Delta p}{\Delta z} + g = 0$$

where ρ is the air density, Δp is the decrease in pressure along a small change in height (Δz), and g is the force of gravity. This expression is usually given as

$$\frac{\Delta p}{\Delta z} = -\rho g$$

This equation is called the *hydrostatic equation*. The hydrostatic equation tells us that the rate at which the pressure decreases with height is equal to the air density times the acceleration of gravity (where ρg is actually the force of gravity per unit volume). The minus sign indicates that, as the air pressure decreases, the height increases. When the hydrostatic equation is given as

$$\Delta p = -\rho g\, \Delta z$$

● **FIGURE 7** When the vertical PGF is in balance with the force of gravity *(g)*, the air is in hydrostatic equilibrium.

it tells us something important about the atmosphere that we learned earlier: The air pressure decreases more rapidly with height in cold (more-dense) air than it does in warm (less-dense) air. In addition, we can use the hydrostatic equation to determine how rapidly the air pressure decreases with increasing height above the surface. For example, suppose at the surface a 1000-meter-thick layer of air (under standard conditions) has an average density of 1.1 kg/m³ and an acceleration of gravity of 9.8 m/sec². Therefore, we have

$$\rho = 1.1 \text{ kg}/\text{m}^3$$
$$g = 9.8 \text{ m}/\text{sec}^2$$
$$\Delta z = 1000 \text{ m}$$

(This value is the height difference from the surface [0 m] to an altitude of 1000 m.)

Using the hydrostatic equation to compute Δp, the difference in pressure in a 1000-meter-thick layer of air, we obtain

$$\Delta p = \rho g\, \Delta z$$
$$\Delta p = (1.1)(9.8)(1000)$$
$$\Delta p = 10,780 \text{ Newtons}/\text{m}^2$$

Since 1 mb = 100 Newtons/m²,

$$\Delta p = 108 \text{ mb}$$

Hence, air pressure decreases by about 108 mb in a standard 1000-meters layer of air near the surface. This closely approximates the pressure change of 10 mb per 100 meters we used in converting station pressure to sea-level pressure earlier in this chapter.

SUMMARY

This chapter gives us a broad view of how and why the wind blows. We examined constant pressure charts and found that low heights correspond to low pressure and high heights to high pressure. In regions where the air aloft is cold, the air pressure is normally lower than average; where the air aloft is warm, the air pressure is normally higher than average. Where horizontal variations in temperature exist, there is a corresponding horizontal change in pressure. The difference in pressure establishes a force, the pressure-gradient force, which starts the air moving from higher toward lower pressure.

Once the air is set in motion, the Coriolis force bends the moving air to the right of its intended path in the Northern Hemisphere and to the left in the Southern Hemisphere. Above the level of surface friction, the wind is bent enough so that it blows nearly parallel to the isobars, or contours. Where the wind blows in a straight-line path, and a balance exists between the pressure-gradient force and the Coriolis force, the wind is termed *geostrophic*. Where the wind blows parallel to curved isobars (or contours), the centripetal acceleration becomes important, and the wind is called a gradient wind. When the wind flow pattern aloft is more west-to-east, the flow is called *zonal*; where the wind flow aloft is more north-to-south or south-to-north, the flow is called *meridional*.

The interaction of the forces causes the wind in the Northern Hemisphere to blow clockwise around regions of high pressure and counterclockwise around areas of low pressure.

In the Southern Hemisphere, the wind blows counterclockwise around highs and clockwise around lows. The effect of surface friction is to slow down the wind. This causes the surface air to blow across the isobars from higher pressure toward lower pressure. Consequently, in both hemispheres, surface winds blow outward, away from the center of a high, and inward, toward the center of a low.

When the upward-directed pressure-gradient force is in balance with the downward force of gravity, the air is in hydrostatic equilibrium. As there is no net vertical force acting on the air, it does not rush off into space.

KEY TERMS

The following terms are listed (with corresponding page number) in the order they appear in the text. Define each. Doing so will aid you in reviewing the material covered in this chapter.

air pressure, 198
barometer, 200
millibar (mb), 200
standard atmospheric pressure, 200
hectopascal (hPa), 200
mercury barometer, 200
aneroid barometer, 201
station pressure, 203
sea-level pressure, 203
isobars, 203
surface map, 204
isobaric chart, 204
contour lines (on isobaric charts), 205
ridges, 206

troughs, 206
anticyclones, 206
mid-latitude cyclonic storms, 206
pressure gradient, 209
pressure-gradient force (PGF), 210
Coriolis force (CF), 210
geostrophic wind, 213
gradient wind, 215
meridional flow, 216
zonal flow, 216
friction layer, 218
Buys Ballot's law 219
hydrostatic equilibrium, 220

QUESTIONS FOR REVIEW

1. Why does air pressure decrease with height more rapidly in cold air than in warm air?
2. What can cause the air pressure to change at the bottom of a column of air?
3. What is considered standard sea-level atmospheric pressure in millibars? In inches of mercury? In hectopascals?
4. How does an aneroid barometer differ from a mercury barometer?
5. How does sea-level pressure differ from station pressure? Can the two ever be the same? Explain.
6. On an upper-level chart, is cold air aloft generally associated with low or high pressure? What about warm air aloft?
7. What do Newton's first and second laws of motion tell us?
8. Explain why, in the Northern Hemisphere, the average height of contour lines on an upper-level isobaric chart tends to decrease northward.

9. What is the force that initially sets the air in motion?
10. What does the Coriolis force do to moving air (a) in the Northern Hemisphere? (b) in the Southern Hemisphere?
11. Explain how each of the following influences the Coriolis force: (a) rotation of Earth; (b) wind speed; (c) latitude.
12. How does a steep (or strong) pressure gradient appear on a weather map?
13. Explain why on a map, closely spaced isobars (or contours) indicate strong winds, and widely spaced isobars (or contour lines) indicate weak winds.
14. What is a geostrophic wind? Why would you *not* expect to observe a geostrophic wind at the equator?
15. Why do upper-level winds in the middle latitudes of both hemispheres generally blow from the west?
16. Describe how the wind blows around highs and lows aloft and near the surface (a) in the Northern Hemisphere and (b) in the Southern Hemisphere.
17. What are the forces that affect the horizontal movement of air?
18. What factors influence the angle at which surface winds cross the isobars?
19. How does zonal flow differ from meridional flow?
20. Describe the type of vertical air motions associated with surface high- and low-pressure areas.
21. Because there is always an upward-directed pressure-gradient force, why doesn't the air rush off into space?
22. How does Buys Ballot's Law help to locate regions of high and low pressure aloft and at the surface?
23. Explain the effect surface friction has on wind speed and direction.
24. Explain how on a 500-mb chart you would be able to distinguish a trough from a ridge.

QUESTIONS FOR THOUGHT

1. Explain why, on a sunny day, an aneroid barometer would indicate "stormy" weather when carried to the top of a mountain.
2. The gas law states that pressure is proportional to temperature times density. Use the gas law to explain why a balloon will deflate when placed inside a refrigerator. Use the gas law to explain why the same balloon will inflate when removed from the refrigerator and placed in a warm room.
3. In ● Fig. 8.36 suppose the air column above city Q is completely saturated with water vapor, and the air column above city T is completely dry. If the temperature of the air in both columns is the same, which column will have the highest atmospheric pressure at the surface? Explain. (Hint: Refer back to the Focus section 4.3 on p. 108 in Chapter 4, "Which Is 'Heavier'—Humid Air or Dry Air?")

● **FIGURE 8.36**

4. Could station pressure ever *exceed* sea-level pressure? Explain.

5. Suppose you are in the Northern Hemisphere watching altocumulus clouds 4000 m (13,000 ft) above you drift from the northeast. Draw the orientation of the isobars above you. Locate and mark regions of lowest and highest pressure on this map. Finish the map by drawing isobars and the upper-level wind flow pattern hundreds of kilometers in all directions from your position. Would this type of flow be zonal or meridional? Explain.

6. Pilots often use the expression "high to low, look out below." In terms of upper-level temperature and pressure, explain what this can mean.

7. Suppose an aircraft using a pressure altimeter flies along a constant pressure surface from standard temperature into warmer-than-standard air without any corrections. Explain why the altimeter would indicate an altitude lower than the aircraft's true altitude.

8. If Earth were not rotating, how would the wind blow with respect to centers of high and low pressure?

9. Why are surface winds that blow over the ocean closer to being geostrophic than those that blow over the land?

10. If the wind aloft is blowing parallel to curved isobars, with the horizontal pressure-gradient force being of greater magnitude than the Coriolis force, would the wind flow be cyclonic or anticyclonic?

11. With your present outside surface wind, use Buys-Ballot's Law to determine where regions of surface high- and low-pressure areas are located. If clouds are moving overhead, use the relationship to locate regions of higher and lower pressure aloft.

12. If you live in the Northern Hemisphere and a region of surface low pressure is directly west of your location, what would probably be the surface wind direction at your home? If an upper-level low is also directly west of you, describe the probable wind direction aloft and the direction in which middle-type clouds would move. How would the wind direction and speed change from the surface to where the middle clouds are located?

13. In the Northern Hemisphere, you observe surface winds shift from N to NE to E, then to SE. From this observation, you determine that a west-to-east moving high-pressure area (anticyclone) has passed north of your location. Describe with the aid of a diagram how you were able to come to this conclusion.

14. The CF causes winds to deflect to the right of their intended path in the Northern Hemisphere, yet around a surface low-pressure area, winds blow counterclockwise, appearing to bend to their left. Explain why.

15. Why is it that, on the equator, winds may blow either counterclockwise or clockwise with respect to an area of low pressure?

16. Use the gas law in the Focus section 8.1 on p. 202 to explain why a car with tightly closed windows will occasionally have a window "blow out" or crack when exposed to the sun on a hot day.

17. Consider wind blowing over a land surface that crosses a coastline and then blows over a lake. How will the wind speed and direction change as it moves from the land surface to the lake surface?

18. As a cruise ship crosses the equator, the entertainment director exclaims that water in a tub will drain in the opposite direction now that the ship is in the Southern Hemisphere. Give *two* reasons to the entertainment director why this assertion is not so.

PROBLEMS AND EXERCISES

1. A station 300 m above sea level reports a station pressure of 994 mb. What would be the sea-level pressure for this station, assuming standard atmospheric conditions? If the observation were taken on a hot summer afternoon, would the sea-level pressure be greater or less than that obtained during standard conditions? Explain.

2. Figure 8.37 is a sea-level pressure chart (Northern Hemisphere), with isobars drawn every 4 mb. Answer the following questions, which refer to this map.

   (a) What is the lowest possible pressure in whole millibars that can be in the center of the closed low? What is the highest pressure possible?

   (b) Place a dashed line through the ridge and a dotted line through the trough.

   (c) What would be the wind direction at point A and at point B?

   (d) Where would the stronger wind be blowing, at point A or B? Explain.

   (e) Compute the pressure gradient between points 1 and 2, and between points 3 and 4. How do the computed pressure gradients relate to the pressure-gradient force?

   (f) If point A and point B are located at 30°N, and if the air density is 1.2 kg/m³, use the geostrophic wind equation in the Focus section 8.3 to compute the geostrophic wind at point A and point B. (Hint: Be sure to convert km to m and mb to Newtons/m², where 1 mb = 100 Newtons/m².)

   (g) Would the actual winds at point A and point B be greater than, less than, or equal to the wind speeds computed in problem (f)? Explain.

● FIGURE 8.37

3. (a) Suppose the atmospheric pressure at the bottom of a deep air column 5.6 km thick is 1000 mb. If the average air density of the column is 0.91 kg/m³, and the acceleration of gravity is 9.8 m/ sec², use the hydrostatic equation on p. 222 to determine the atmospheric pressure at the top of the column. (Hint: Be sure to convert km to m and mb to Newtons/m², where 1 mb = 100 N/m².)

   (b) If the air in the column discussed in problem (a) becomes much colder than average, would the atmospheric pressure at the top of the new column be greater than, less than, or equal to the pressure computed in problem (a)? Explain.

   (c) Determine the atmospheric pressure at the top of the air column in problem (a) if the air in the column is quite cold and has an average density of 0.97 kg/m³.

4. Use the gas law in the Focus section 8.1 to calculate the air pressure in millibars when the air temperature is −23°C and the air density is 0.700 kg/m³. (Hint: Be sure to use the Kelvin temperature.) At approximately what altitude would you expect to observe this pressure?

5. Suppose air in a closed container has a pressure of 1000 mb and a temperature of 20°C.

   (a) Use the gas law to determine the air density in the container.

   (b) If the density in the container remains constant, but the pressure doubles, what would be the new temperature?

6. A large balloon is filled with air so that the air pressure inside just equals the air pressure outside. The volume of the filled balloon is 3 m³, the mass of air inside is 3.6 kg, and the temperature inside is 20°C. What is the air pressure? (Hint: Density = mass/volume.)

7. If the clouds overhead are moving from north to south, would the upper-level center of low pressure be to the east or west of you? Draw a simplified map to explain.

▣ ONLINE RESOURCES

Visit www.cengagebrain.com to view additional resources, including video exercises, practice quizzes, an interactive eBook, and more.

High winds blow sand along Folly Beach,
South Carolina.

Wind: Small-Scale and Local Systems

ON DECEMBER 28, 1997, A UNITED AIRLINES BOEING 747 carrying 374 passengers was over the Pacific Ocean en route to Hawaii from Japan. About two hours into the flight, the aircraft was at a cruising altitude of 31,000 feet when suddenly, east of Tokyo, this routine, uneventful flight turned harrowing. Seat-belt signs were turned on because of reports of severe air turbulence nearby. The plane hurtled upward, then quickly dropped by about 30 meters (100 feet) before stabilizing. Screaming, terrified passengers not fastened to their seats were flung against the walls of the aircraft, then dropped. Bags, serving trays, and luggage that slipped out from under the seats were flying about inside the plane. Within seconds, thztire ordeal was over. A total of 160 people were injured. Tragically, there was one fatality: A 32-year-old woman, who had been hurled against the ceiling of the plane, died of severe head injuries. What sort of atmospheric phenomenon could cause such turbulence?

The aircraft in our opening vignette on p. 227 encountered a turbulent eddy—an "air pocket"—in perfectly clear air. Such violent eddies are not uncommon, especially in the vicinity of jet streams. In this chapter, we will examine a variety of eddy circulations. First, we will look at the different scales of motion found within our atmosphere, then we will see how eddies form and how eddies and other small-scale winds interact with our environment. Next, we will examine slightly larger circulations—local winds—such as the sea breeze and the chinook, describing how they form and the type of weather they generally bring. In Chapter 10, we will examine winds and circulations as they unfold on the global scale.

The air in motion—what we commonly call *wind*—is a powerful phenomenon. It is invisible, yet we see evidence of it nearly everywhere we look. It sculpts rocks, moves leaves, blows smoke, and lifts water vapor upward to where it can condense into clouds. The wind is with us wherever we go. On a hot day, it can cool us off; on a cold day, it can make us shiver. A breeze can sharpen our appetite when it blows the aroma from the local bakery or a food truck in our direction. Wind is the workhorse of weather, moving storms and large fair-weather systems around the globe. It transports heat, moisture, dust, insects, bacteria, and pollen from one area to another.

Circulations of all sizes exist within the atmosphere. Little whirls form inside bigger whirls, which encompass even larger whirls—one huge mass of turbulent, twisting *eddies*.* For clarity, meteorologists arrange circulations according to their size. This hierarchy of motion from tiny gusts to giant storms is called the **scales of motion**.

Scales of Atmospheric Motion

Consider smoke rising from a chimney into the otherwise clean air in an industrial section of a large city (see ● Fig. 9.1a). Within the smoke, small chaotic motions—tiny eddies—cause it to tumble and turn. These eddies constitute the smallest scale of atmospheric motion—the **microscale**. At the microscale level, eddies with diameters of a few meters or less not only disperse smoke, but they also sway branches and swirl dust and papers into the air. They form by convection or by the wind blowing past obstructions and are usually short-lived, lasting only a few minutes at best.

In Fig. 9.1b observe that, as the smoke rises, it drifts toward the center of town. Here the smoke rises even higher and is carried many kilometers downwind. This circulation of city air constitutes the next larger scale—the **mesoscale** (meaning middle scale). Typical mesoscale circulations range from a few kilometers to about a hundred kilometers in diameter. Generally, they last longer than microscale motions, often many minutes, hours, or in some cases, as long as a day. Mesoscale circulations include local winds (which form along shorelines and mountains), as well as thunderstorms, tornadoes, and small tropical cyclones.

When we look at the smokestack on a surface weather map (see Fig. 9.1c), neither the smokestack nor the circulation of city air shows up. All that we see are the circulations around high- and low-pressure areas—the cyclone and anticyclones of the middle latitudes, as well as the large tropical cyclones of lower latitudes. We are now looking at the **synoptic scale**, or weather-map scale. Circulations of this magnitude dominate regions of hundreds to even thousands of square kilometers and, although the life spans of these features vary, they typically last for days and sometimes weeks. These include large hurricanes and typhoons as well as the frequent mid-latitude storm systems that bring rain, snow, and wind.

The largest wind patterns are seen at the **global scale**, or **planetary scale**. Here, we have wind patterns ranging over the entire earth. Together, the synoptic and global scales are referred to as the **macroscale**—the largest scale of atmospheric motion. ● Figure 9.2 summarizes the various scales of motion and their average life span. Having looked at the different scales of

*Eddies are spinning globs of air that have a life history of their own.

(a) Microscale   (b) Mesoscale   (c) Synoptic scale

● **FIGURE 9.1** Scales of atmospheric motion. The tiny microscale motions constitute a part of the larger mesoscale motions, which, in turn, are part of the much larger synoptic scale. Notice that as the scale becomes larger, motions observed at the smaller scale are no longer visible.

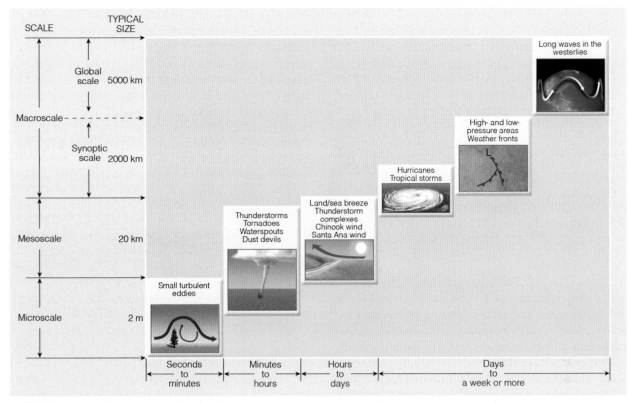

● **FIGURE 9.2** The scales of atmospheric motion with the phenomenon's average size and life span. (Because the actual size of certain features can vary, some of the features fall into more than one category.)

atmospheric motion, we turn our attention to see the effect that microscale winds can have on our environment.

Small-Scale Winds Interacting with the Environment

We begin our discussion of microscale winds by examining the important topic of *turbulent flow*, called **turbulence**, which represents any disturbed flow of air that produces wind gusts and eddies.

FRICTION AND TURBULENCE IN THE BOUNDARY LAYER We are all familiar with friction. If we rub our hand over the top of a table, friction tends to slow its movement because of irregularities in the table's surface. On a microscopic level, friction arises as atoms and molecules of the two surfaces seem to adhere, then snap apart, as the hand slides over the table. Friction is not restricted to solid objects; it occurs in moving fluids as well. Consider, for example, a steady flow of water in a stream. When a paddle is placed in the stream, turbulent whirls *(eddies)* form behind it. These eddies create fluid friction by draining energy from the main stream flow, slowing it down. Let's examine the idea of fluid friction in more detail.

The friction of fluid flow is called **viscosity**. When the slowing of a fluid—such as air—is due to the random motion of the gas molecules, the viscosity is referred to as *molecular viscosity*.

Consider a mass of air gliding horizontally and smoothly *(laminar flow)* over a stationary mass of air. Even though the molecules in the stationary air are not moving horizontally, they are darting about and colliding with each other. At the boundary separating the air layers, there is a constant exchange of molecules between the stationary air and flowing air. The overall effect of this molecular exchange is to slow down the moving air. If molecular viscosity were the only type of friction acting on moving air, the effect of friction would disappear in a thin layer just above the surface. There is, however, another frictional effect that is far more important in reducing wind speeds.

When laminar flow gives way to irregular turbulent motion, there is an effect similar to molecular viscosity, but which occurs throughout a much larger portion of the moving air. The internal friction produced by turbulent whirling eddies is called *eddy viscosity*. Near the surface, it is related to the roughness of the ground. As wind blows over a landscape dotted with trees and buildings, it breaks into a series of irregular, twisting eddies that can influence the airflow for hundreds of meters above the surface. Within each eddy, the wind speed and direction fluctuate rapidly, producing the irregular air motion known as *wind gusts*. Eddy motions created by obstructions are commonly referred to as **mechanical turbulence**. Mechanical turbulence creates a drag on the flow of air, one far greater than that caused by molecular viscosity.

The frictional drag of the ground normally decreases as we move away from Earth's surface. Because of the reduced friction, wind speeds tend to increase with height above the ground. In fact, at a height of only 10 m (33 ft), the wind is often moving twice as fast as at the surface. As we saw in Chapter 8, the atmospheric layer near the surface that is influenced by friction (turbulence) is called the *friction layer* or **planetary** *(atmospheric)* **boundary layer**. The top of the boundary layer is usually near 1000 m (3300 ft), but this height may vary somewhat as both strong winds and rough terrain extend the region of frictional influence.

Surface heating and instability also cause turbulence to extend to greater altitudes. As Earth's surface heats, thermals rise and convection cells form. The resulting vertical motion creates **thermal turbulence**, which increases with the intensity of surface heating and the degree of atmospheric instability. During the early morning, when the air is most stable, thermal turbulence is normally at a minimum. As surface heating increases, instability is induced and thermal turbulence becomes more intense. If this heating produces convective clouds that rise to great heights, there may be turbulence from Earth's surface to the base of the stratosphere, more than 10 km (6.2 mi) above the ground.

Although we have treated thermal and mechanical turbulence separately, they occur together in the atmosphere—each magnifying the influence of the other. Let's consider a simple example: the eddy forming behind the barn in ● Fig. 9.3. In stable air with weak winds, the eddy is nonexistent or small. As wind speed and surface heating increase, instability develops, and the eddy becomes larger and extends through a greater depth. The rising side of the eddy carries slow-moving surface air upward, causing a frictional drag on the faster flow of air aloft. Some of

the faster-moving air is brought down with the descending part of the eddy, producing a momentary gust of wind. Because of the increased depth of circulating eddies in unstable air, strong, gusty surface winds are more likely to occur when the atmosphere is unstable. Greater instability also leads to a greater exchange of faster-moving air from upper levels with slower-moving air at lower levels. In general, this exchange increases the average wind speed near the surface and decreases it aloft, producing the distribution of wind speed with height shown in ● Fig. 9.4.

We can now see why surface winds are usually stronger in the afternoon. Vertical mixing during the middle of the day links

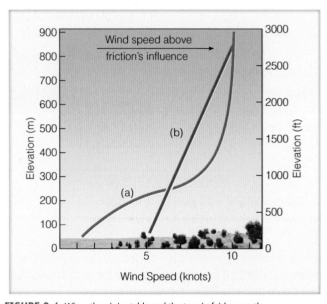

● **FIGURE 9.4** When the air is stable and the terrain fairly smooth (a), vertical mixing is at a minimum, and the effect of surface friction only extends upward a relatively short distance above the surface. When the air is unstable and the terrain rough (b), vertical mixing is at a maximum, and the effect of surface friction extends upward through a much greater depth of atmosphere. Within the region of frictional influence, vertical mixing increases the wind speed near the ground and decreases it aloft. (Wind at the surface is normally measured at 10 m [33 ft] above ground level.)

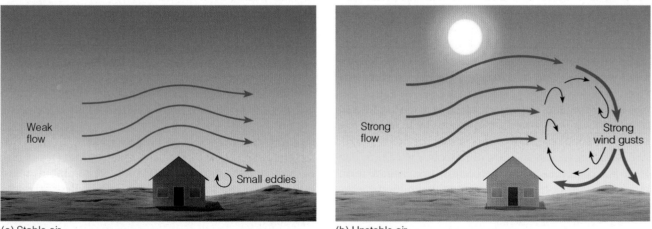

(a) Stable air

(b) Unstable air

● **FIGURE 9.3** Winds flowing past an obstacle. (a) In stable air, light winds produce small eddies and little vertical mixing. (b) Greater winds in unstable air create deep, vertically mixing eddies that produce strong, gusty surface winds.

surface air with the faster-moving air aloft. The result is that the surface air is pulled along more quickly. At night, when convection is reduced, the interchange between the air at the surface and the air aloft is at a minimum. Hence, the wind near the ground is less affected by the faster wind flow above, and so it blows more slowly.

In summary, the friction of airflow (viscosity) is a result of the exchange of air molecules moving at different speeds. The exchange brought about by random molecular motions (molecular viscosity) is quite small in comparison with the exchange brought about by turbulent motions (eddy viscosity). Therefore, the frictional effect of the surface on moving air depends largely on mechanical and thermal turbulent mixing. The depth of mixing and, hence, frictional influence (in the boundary layer) depend primarily on three factors:

1. surface heating—producing a steep lapse rate and strong thermal turbulence
2. strong wind speeds—producing strong mechanical turbulent motions
3. rough or hilly landscape—producing strong mechanical turbulence

When these three factors occur simultaneously, the frictional effect of the ground is transferred upward to considerable heights, and the wind at the surface is typically strong and gusty.

EDDIES—BIG AND SMALL When the wind encounters a solid object, a whirl of air—an eddy—forms on the object's leeward side (see ● Fig. 9.5). The size and shape of the eddy often depend on the size and shape of the obstacle and on the speed of the wind. Light winds produce small stationary eddies. Wind moving past trees, shrubs, and even your body produces small eddies. (You may have had the experience of dropping a piece of paper on a windy day only to have it carried away by a swirling eddy as you bend down to pick it up.) Air flowing over a building produces larger eddies that will, at best, be about the size of the building. Strong winds blowing past an open sports stadium can produce eddies that may rotate in such a way as to create surface winds on the playing field that move in a direction opposite to the wind flow above the stadium. Wind blowing over a fairly smooth surface produces few eddies, but when the surface is rough, many eddies form.

The eddies that form downwind from obstacles can produce a variety of interesting effects. For instance, wind moving over a mountain range in a stable atmosphere with a speed greater than 40 knots usually produces waves and eddies, such as those shown in ● Fig. 9.6. We can see that eddies form both close to the mountain and beneath each wave crest. These are called *roll eddies*, or **rotors**, and have violent vertical motions that produce extreme turbulence and hazardous flying conditions. Strong winds blowing over a mountain in stable air sometimes provide a *mountain wave eddy* on the downwind side, with a reverse flow near the ground.

The largest atmospheric eddies form as the flow of air becomes organized into huge spiraling whirls—the cyclones and anticyclones of middle latitudes—which can have diameters

● **FIGURE 9.5** Satellite image of eddies forming on the leeward (downwind) side of the Cape Verde Islands during April 2004. As trade winds blow from the northeast past the islands, the air breaks into a variety of swirls toward the southwest, as indicated by the cloud pattern. (The islands are situated in the Atlantic Ocean, off Africa's western coast.)

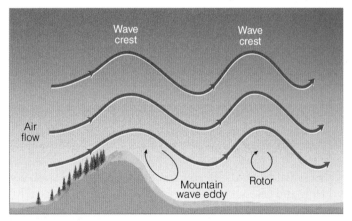

● **FIGURE 9.6** Under stable conditions, air flowing past a mountain range can create eddies many kilometers downwind of the mountain itself.

greater than 1000 km (600 mi). Since it is these migrating systems that make our middle latitude weather so changeable, we will examine the formation and movement of these systems in Chapters 11 and 12.

Turbulent eddies form aloft as well as near the surface. Turbulence aloft can occur suddenly and unexpectedly, especially where the wind changes its speed or direction (or both) abruptly. Such a change is called **wind shear**. The shearing creates forces that produce eddies along a mixing zone. If the eddies form in clear air, this form of turbulence is called **clear air turbulence (CAT)**. When an airplane is flying through such turbulence, the bumpiness can range from small vibrations to violent up-and-down motions that force passengers against their seats and toss objects throughout the cabin. (Additional information on CAT is given in Focus section 9.1.)

THE STRONG FORCE OF THE WIND The force of the wind on an object is proportional to the wind speed squared, which means that a small increase in wind speed can greatly increase the force of the wind acting on an object. So, strong winds may blow down trees, overturn mobile homes, and even move railroad cars. For example, in February 1965 the wind presented people in North Dakota with a "ghost train" as it pushed five railroad cars from Portal to Minot (about 125 km/77 mi) without a locomotive. On May 2, 2009, while the Dallas Cowboys rookie football players were going through workouts at the indoor practice facilities near Dallas, Texas, a strong wind—estimated at over 60 knots—ripped the roof off the facility, injuring 12 people. And on March 8, 2017, more than a million people lost power when a vast swath of high wind uprooted trees and knocked down utility lines across the Great Lakes states, all under bright blue skies.

Wind blowing with sufficient force to rip the roof off buildings is uncommon. However, wind blowing with enough force to move an automobile is very common, especially when the automobile is exposed to a strong crosswind. On a normal road, the force of a crosswind is usually insufficient to move a car sideways because of the reduced wind flow near the ground. However, when the car crosses a high bridge, where the frictional influence of the ground is reduced, the increased wind speed can be felt by the driver. Near the top of a high bridge, where the wind flow is typically strongest, complicated eddies pound against the car's side as the air moves past obstructions, such as guard railings and posts. In a strong wind, these eddies may even break into extremely turbulent whirls that buffet the car, causing difficult handling as it moves from side to side. If there is a wall on the bridge, the wind may swirl around and strike the car from the side opposite the wind direction, producing hazardous driving conditions.

A similar effect occurs where the wind moves over low hills paralleling a highway (see ● Fig. 9.7). When the vehicle moves by the obstruction, a wind gust from the opposite direction can suddenly and without warning push it to the opposite side. This wind hazard is a special problem for trucks, campers, and trailers. Consequently, highway signs warning of gusty wind areas are often posted.

Up to now we've seen that, when the wind meets a barrier, it exerts a force upon it. If the barrier doesn't move, the wind moves around, up, and over it. When the barrier is long and low like a water wave, the slight updrafts created on the windward side support the wings of birds, allowing them to skim the water in search of food without having to flap their wings. Elongated hills and cliffs that face into the wind create upward air motions that can support a hang glider in the air for a long time. The cliffs in

the Hawaiian Islands and along the California coast, with their steep escarpments, are especially fine hang-gliding areas (see ● Fig. 9.8). Wind speeds greater than about 15 knots blowing over a smooth yet moderately sloping ridge may provide excellent ridge-soaring for the sailplane enthusiast.

As stable air flows over a ridge, it increases in speed. Thus, winds blowing over mountains tend to be stronger than winds blowing at the same level on either side. In fact, one of the greatest wind speeds ever recorded near the ground occurred at the summit of Mt. Washington, New Hampshire, elevation 1909 m (6262 ft), where the wind gusted to 201 knots (231 mi/hr) on April 12, 1934. A similar increase in wind speed occurs where air accelerates as it funnels through a narrow constriction, such as a low pass or saddle in a mountain crest.

● **FIGURE 9.7** Strong winds flowing past an obstruction, such as these hills, can produce a reverse flow of air that strikes an object from the side opposite the general wind direction.

● **FIGURE 9.8** With the prevailing wind blowing from off the ocean, the steep cliffs along the coast of Southern California promote rising air and good hang-gliding conditions.

Eddies and "Air Pockets"

To better understand how eddies form along a zone of wind shear, imagine that, high in the atmosphere, there is a stable layer of air having vertical wind speed shear (changing wind speed with height) as depicted in ● Fig. 1a. The top half of the layer slowly slides over the bottom half, and the relative speed of both halves is low. As long as the wind shear between the top and bottom of the layer is small, few if any eddies form. However, if the shear and the corresponding relative speed of these layers increases (Figs. 1b and 1c), wave-like undulations may form. When the shearing exceeds a certain value, the waves break into large swirls, with significant vertical movement (Fig. 1d). Eddies such as these often form in the upper troposphere near jet streams, where large wind speed shears exist.

Turbulent eddies also occur in conjunction with mountain waves, which may extend upward into the stratosphere (see ● Fig. 2). As we learned earlier, when these huge eddies develop in clear air, this form of turbulence is referred to as *clear air turbulence (CAT)*.

The eddies that form in clear air may have diameters ranging from a couple of meters to several hundred meters. An unsuspecting aircraft entering such a region may be in for more than just a bumpy ride. If the aircraft flies into a zone of descending air, it may drop suddenly, producing the sensation that there is no air to support the wings. Consequently, these regions have come to be known as *air pockets*.

Commercial aircraft entering air pockets have dropped hundreds of meters, injuring passengers and flight attendants not strapped into their seats. Five passengers and crew members had to be hospitalized in February 2014 after severe turbulence struck a Boeing 737 jetliner as it approached Billings, Montana. In April 1981, a DC-10 jetliner flying at 11,300 m (37,000 ft) over central Illinois encountered a region of severe clear air turbulence and reportedly plunged about 600 m (2000 ft) toward Earth before stabilizing. Twenty-one of the 154 people aboard were injured; one person sustained a fractured hip and another person, after hitting the ceiling, jabbed himself in the nose with a fork, then landed in the seat in front of him.* Clear air turbulence has occasionally caused structural damage to aircraft by breaking off vertical stabilizers and tail structures. Fortunately, the effects are usually not this dramatic.

The potential adverse effects of clear air turbulence is one important reason why passengers are frequently told to "fasten your seat belts" while flying, even when there are no thunderstorms or obvious hazards in sight.

---

*Another example of an aircraft that experienced severe turbulence as it flew into an air pocket is given in the opening vignette on p. 227.

(a) Small shear   (b) Increasing shear, boundary deforms   (c) Waves appear   (d) Turbulent eddies break

● **FIGURE 1** The formation of clear air turbulence (CAT) along a boundary of increasing wind speed shear. The wind in the top layer increases in speed from (a) through (d) as it flows over the bottom layer.

● **FIGURE 2** Turbulent eddies forming downwind of a mountain chain in a wind shear zone produce these waves called *Kelvin-Helmholtz waves*. The visible clouds that form are called *billow clouds*.

Brooks Martner—www.cloudphotos.net

WIND AND SOIL Where the wind blows over exposed soil, it takes an active role in shaping the landscape. This is especially noticeable in deserts. Tiny, loose particles of sand, silt, and dust are lifted from the surface and carried away by the wind, leaving the surface lower than it once was. These same winds may also help to move desert rocks across wet ground, as shown in ● Fig. 9.9.

Blowing sand eventually comes to rest *behind* obstacles, which can be anything from a rock to a clump of vegetation. As the sand grains accumulate, they pile into a larger heap that, when high enough, acts as an obstacle itself. If the wind speed is strong and continues to blow in the same direction for a sufficient time, the sand piles up higher and eventually becomes a *sand dune*. On the dune's surface, the sand rolls, slides, and gradually creeps along, producing wavelike patterns called *sand ripples*. Each ripple forms perpendicular to the wind direction, with a gentle slope on the upwind side and a steeper slope on the downwind side. (If the wind direction frequently changes, the ripple becomes more symmetric.) On a larger scale, the dune itself may take on a more symmetric shape. Sand is carried forward and up the dune until it reaches the top. Here, the airflow is strongest, and the sand

continues its forward movement and cascades down the backside of the dune into quieter air. The effect of this migration is to create a dune whose windward slope is more gentle than its leeward slope. Therefore, the shape of a sand dune reveals the prevailing wind direction during its formation (see ● Fig. 9.10).

WIND AND SNOW Wind blowing over a snow-covered landscape may also create wavelike patterns several centimeters high and oriented at right angles to the wind. These *snow ripples* are similar to sand ripples. On a larger scale, winds may create *snow drifts* and even *snow dunes*, which are quite similar to sand dunes. Irregularities at the surface can cause a strong wind (40 knots) to break into turbulent eddies. If the snow on the ground is moist and sticky, some of it may be picked up by the wind and sent rolling. As it rolls along, it collects more snow and grows bigger. If the wind is sufficiently strong, the moving clump of snow becomes cylindrical, often with a hole extending through it lengthwise. These **snow rollers** range from the size of eggs to that of small barrels. The tracks they make in the snow are typically less than 1 centimeter deep and several meters long. Snow rollers are rare, but, when they occur, they create a striking winter scene (see ● Fig. 9.11). In populated areas, they may escape notice as they are often mistaken as having been made by children rather than by nature.

Strong winds blowing over a vast region of open plains can alter the landscape in a different way. Consider, for example, a light snowfall several centimeters deep covering a large portion of central South Dakota. After the snow stops falling, strong winds may whip it into the air, leaving fields barren of snow. The cold, dry wind also robs the soil of any remaining moisture and freezes it solid. Meanwhile, the snow settles out of the air when the wind encounters obstacles. Because the greatest density of such obstructions is normally in towns, municipal snowfall measurements may show an accumulation of many centimeters, while the surrounding countryside, which may desperately need the snow, has practically none.

To help remedy this situation, *snow fences* are constructed in open spaces (see ● Fig. 9.12). Behind the snow fence, the wind speed is reduced because the air is broken into small eddies,

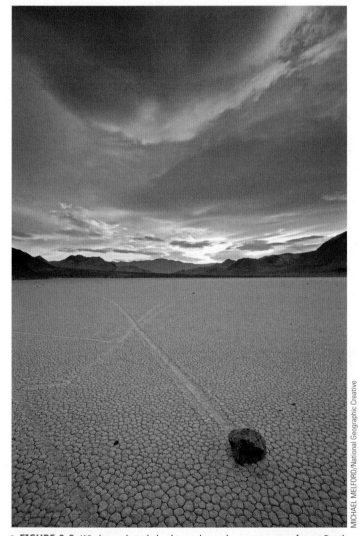

● **FIGURE 9.9** Winds may have helped to push a rock across a wet surface at Death Valley National Park, California.

MICHAEL MELFORD/National Geographic Creative

● **FIGURE 9.10** The shape of this sand dune reveals that the wind was blowing from left to right when it formed. Note also the shape of the sand ripples on the dune.

© Dick Hilton

● **FIGURE 9.12** Snow drifts accumulating behind snow fences in Wyoming.

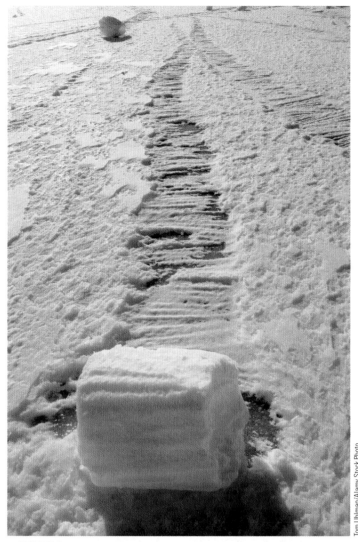

● **FIGURE 9.11** Snow rollers—natural cylindrical rolls of snow—grow larger as the wind blows them down a hillside.

allowing the snow to settle to the ground. Added snow cover is important for open areas because it acts like an insulating blanket that protects the ground from the bitter cold air, which often follows in the wake of a major snowstorm. In regions of low rainfall, moisture from the spring snowmelt can be a critical factor during long, dry summers. Snow fences are also built to protect major highways in these areas. Hopefully, the snow will accumulate behind the fence rather than in huge drifts on the road.

Strong winds can whip the snow about, reducing visibility to practically zero, often closing side streets and even interstate highways. When sustained winds or frequent gusts reach 35 mi/hr, the blowing or falling snow can produce *blizzard conditions*.

WIND AND VEGETATION   Strong winds can have an effect on vegetation, too. Armed with sand, winds can damage or destroy tender new vegetation, decreasing crop productivity. Most plants increase their rate of transpiration as wind speed increases.* This

---

*This effect actually drops above a certain wind speed and varies greatly among plant species.

leads to rapid water loss, especially in warmer areas having low humidities, and may actually dry out plants. If sustained, this drying-out effect may stunt plant growth, and, in some windy, dry regions, mature trees that might otherwise be many meters tall grow only to the height of a small shrub.

Wind-dried vegetation can result in an area of high fire danger. If a fire should begin here, any additional wind helps it along, directing its movement, adding oxygen for combustion, and carrying burning embers elsewhere to start new fires. On the open plains, where the wind blows practically unimpeded, wind-whipped prairie fires can imperil homes and livestock as the fires burn out of control over large areas.

Wind erosion is greatly reduced by a continuous cover of vegetation. The vegetation screens the surface from the direct force of the wind and anchors the soil. Soil moisture also helps to resist wind erosion by holding particles together. From this fact, we can see that land where natural vegetation has been removed for farming purposes—followed by several years of drought—is ripe for wind erosion. This situation happened in parts of the Great Plains in the middle 1930s, when winds carried millions of tons of dust into the air, creating vast dust storms that buried whole farmhouses, reduced millions of acres to unproductive wasteland, and financially ruined thousands of families. Because of these disastrous effects of the wind, portions of the western plains became known as the *Dust Bowl*.

To protect crops and soil, *windbreaks*—commonly called **shelterbelts**—are planted. Shelterbelts usually consist of a series of mixed conifer and deciduous trees or shrubs planted in rows perpendicular to the prevailing wind flow. They greatly reduce the wind speed behind them (see ● Fig. 9.13). As air filters through the belt, the flow is broken into small eddies, which have little mixing effect on the air near the surface. However, if trees are planted too close together, several unwanted effects may result. For one thing, the air moving past the belt may be broken into larger, more turbulent eddies, which swirl soil about. Furthermore, in high winds, strong downdrafts may damage the crops.

The use of properly designed shelterbelts has benefited agriculture. In some parts of the Central Plains, these belts have stabilized the soil and increased wheat yield. Despite their advantages,

● **FIGURE 9.14** Wind blowing over a wave creates a small eddy of air that helps to reinforce the up-and-down motion of the water.

many of the shelterbelts planted during the drought years of the mid-1930s have been removed. Some are economically unfeasible because they occupy valuable crop land. Others interfere with the large center pivot sprinkler systems now in use. At any rate, one wonders how the absence of these shelterbelts would affect this region if it were struck by a drought similar to that experienced in the 1930s.

WIND AND WATER   The impact of the wind on Earth's surface is not limited to land; wind also influences water—it makes waves. Waves forming by wind blowing over the surface of the water are known as **wind waves**. Just as air blowing over the top of a water-filled pan creates tiny ripples, so waves are created as the frictional drag of the wind transfers energy to the water. In general, the greater the wind speed, the greater the amount of energy added, and the higher will be the waves. Actually, the amount of energy transferred to the water (and thus the height to which a wave can build) depends on three factors:

1. the wind speed
2. the length of time that the wind blows over the water
3. the *fetch*, or distance, of deep water over which the wind blows

A sustained 50-knot wind blowing steadily for nearly three days over a minimum distance of 2600 km (1600 mi) can generate waves with an average height of 15 m (49 ft). Thus, a stationary storm system centered somewhere over the open sea is capable of creating large waves with wave heights occasionally measuring over 31 m (100 ft).

Microscale winds actually help waves grow taller. Consider, for example, the wind blowing over the small wave depicted in ● Fig. 9.14. Observe that both the wind and the wave are moving in the same direction, and that the wave crest deflects the wind upward, producing an undulation in the airflow just above the water. This looping air motion establishes a small eddy of air between the two crests. The upward and downward motions of the eddy reinforce the upward and downward motions of the water. Consequently, the eddy helps the wave to build in height.

Traveling in the open ocean, waves represent a form of energy. As they move into a region of weaker winds, they gradually change: Their crests become lower and more rounded, forming what are

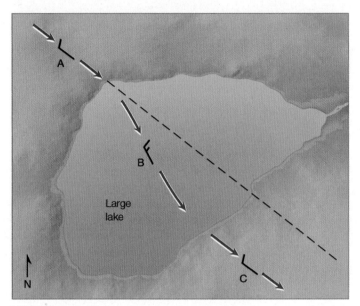

● **FIGURE 9.15** Wind can change in both speed and direction when crossing a large lake.

commonly called *swells*. When waves reach a shoreline they transfer their energy—sometimes catastrophically—to the coast and structures along it. High, storm-induced waves can hurl thousands of tons of water against the shore. If this happens during an unusually high tide, resort homes overlooking the ocean can be pounded into a twisted mass of board and nails by the surf. Bear in mind that the storms creating these waves may be thousands of kilometers away and, in fact, may never reach the shore. Some of the largest, most damaging waves ever to strike the beach communities of Southern California arrived on what was described as "one of the clearest days imaginable." On the more positive side, in the Hawaiian Islands these high waves are excellent for surfboarding.

Frequently, as winds cross a large body of water, they will change their speed and direction. ● Figure 9.15 shows the wind speed and direction as air flows over a large lake. At position A, on the upwind side, the wind is blowing at 10 knots from the northwest; at position B, the wind speed is 15 knots and has shifted to a more northerly direction; at position C, the wind speed is again blowing at 10 knots from the northwest. Why does the wind blow faster and from a slightly different direction in the center of the lake? As the air moves from the rough land over the relatively smooth lake, friction with the surface lessens, and the wind speed increases. The increase in wind speed, however, increases the Coriolis force, which turns the wind flow slightly to the right of its intended path as shown by the wind report at position B. When the air reaches the opposite side of the lake, it again encounters rough land, and its speed slows. This process reduces the Coriolis force, and the wind responds by shifting to a more westerly direction, as shown by the report at position C.

Changes in wind speed along the shore of a large lake can inhibit cloud formation on one side and enhance it on the other. Suppose warm, moist air flows over a lake, as illustrated in ● Fig. 9.16. Observe that clouds are forming on the downwind side, but not on the upwind side. The lake is slightly cooler than the air. Consequently, by the time the air reaches the downwind side of the lake, it will be cooler, denser, and less likely to rise. Why, then, are clouds forming on this side of the lake? As air moves from the land over the water, it travels from a region of greater friction into a region of less friction, so it increases in speed, which causes the surface air to diverge—to spread apart. Such spreading of air forces air from above to slowly sink, which, of course, inhibits the formation of clouds. Hence, there are no clouds on the upwind side of the lake. Out over the lake, the separation between air temperature and dew point lessens. As this nearly saturated air moves onshore, friction with the rougher ground slows it down, causing it to "bunch up" or converge (which forces the air upward). This slight upward motion coupled with surface heating is often sufficient to initiate the formation of clouds along the downwind side of the lake. When temperature and moisture contrasts are particularly strong over a large lake, heavy rain or snow squalls can develop in a narrow band along the downwind shore. *Lake-effect* precipitation often occurs along the south and east shores of the Great Lakes. Snow totals of up to several feet in a single day have been observed.

Strong winds blowing over an open body of water, such as a lake, can cause the water to slosh back and forth rhythmically. This sloshing causes the water level to periodically rise and fall, much like water does at both ends of a bathtub when the water is disturbed. Such water waves that oscillate back and forth are called **seiches** (pronounced "sayshes"). In addition to strong winds, seiches may also be generated by sudden changes in atmospheric pressure or by earthquakes.* Around the Great Lakes,

---

*Earthquakes and other disturbances on a lake floor can cause the water to slosh back and forth, producing a seiche. Earthquakes on the ocean basin floor can cause a tsunami, a Japanese word meaning "harbor waves" because these waves build in height as they enter a bay or harbor.

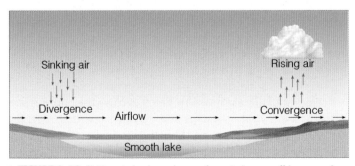

● **FIGURE 9.16** Sinking air develops where surface winds move offshore, speed up, and diverge. Rising air develops as surface winds move onshore, slow down, and converge.

seiche applies to any sudden rise in water level whether or not it oscillates. In November 2003, strong westerly winds gusting to more than 50 knots created a seiche on Lake Erie that caused a 4 m (12 ft) difference in lake level between Toledo, Ohio (on its western shore) and Buffalo, New York (on its eastern shore). Severe thunderstorms along Lake Erie's northern shore in May 2012 generated a seiche that pushed a 2 m (7 ft) wave over a seawall northeast of Cleveland, Ohio, on the lake's south shore. Several children were rescued after having been washed into the lake by the unexpected wave.

In summary, we've discussed how the wind blowing over Earth can produce a variety of features, from snow rollers to ocean waves. We also saw how the wind can influence a moving auto. To see how the wind can influence someone riding a bicycle, read Focus section 9.2.

BRIEF REVIEW

Up to this point we've been examining microscale winds and how they affect our environment. Before we turn our attention to winds on a larger scale, here is a brief review of some of the main points presented so far:

● Viscosity is the friction of fluid flow. The small-scale fluid friction due to the random motion of the molecules is called molecular viscosity. The larger-scale internal friction produced by turbulent flow is called eddy viscosity.

● Mechanical turbulence is created by twisting eddies that form as the wind blows past obstructions. Thermal turbulence results as rising and sinking air forms when Earth's surface is heated unevenly by the sun.

● The planetary boundary layer (or friction layer) is usually given as the first 1000 m (3300 ft) above the surface.

● Wind shear is a sudden change in wind speed or wind direction (or both).

● The wind can shape a landscape, influence crop production, transport material from one area to another, and generate waves.

Local Wind Systems

Every summer, millions of people flock to the New Jersey shore, hoping to escape the oppressive heat and humidity of the inland region. On hot, humid afternoons, these travelers often encounter thunderstorms about 30 km or so from the ocean, thunderstorms that invariably last for only a few minutes. In fact, by the time the vacationers arrive at the beach, skies are generally clear and air temperatures are much lower, as cool ocean breezes greet them. If the travelers return home in the afternoon a few days later, these "mysterious" showers often occur at just about the same location as before.

The showers are not really mysterious, of course. They are caused by a local wind system, the *sea breeze*. As cooler ocean air pours inland, it forces the warmer, conditionally unstable humid air to rise and condense, producing majestic clouds and rain showers along a line that separates the contrasting temperatures.

Pedaling into the Wind

Anyone who rides a bicycle knows that it is much easier to pedal with the wind than against it. (See ● Fig. 3.) The reason is obvious: As we saw earlier in this chapter, when wind blows against an object, it exerts a force upon it. The amount of force exerted by the wind over an area increases as the square of the wind velocity. This relationship is shown by

$$F \sim V^2$$

where F is the wind force and V is the wind velocity. From this we can see that, if the wind velocity doubles, the force goes up by a factor of 2^2, or 4, which means that pedaling into a 40-knot wind requires four times as much effort as pedaling into a 20-knot wind.

Wind striking an object exerts a pressure on it. The amount of pressure depends upon the object's shape and size, as well as on the amount of reduced pressure that exists on the object's downwind side. Without concern for all the complications, we can approximate the wind pressure on an object with a simple formula. For example, if the wind velocity (V) is in miles per hour, and the wind force (F) is in pounds, and the object's surface area (A) is measured in square feet, the wind pressure (P), in pounds per square feet, is

$$\frac{F}{A} = P = 0.004V^2$$

We can look at a practical example of this expression if we consider a bicycle rider going

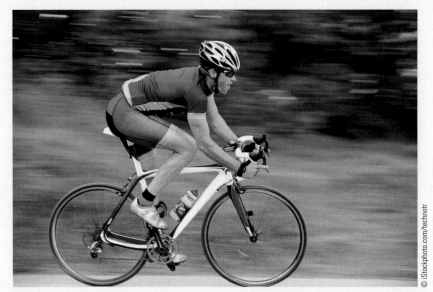

● **FIGURE 3** Pedaling into a 15-knot wind requires nine times as much effort as pedaling into a 5-knot wind.

10 mi/hr into a head wind of 40 mi/hr. With the total velocity of the wind against the rider (wind speed plus bicycle speed) being 50 mi/hr, the pressure of the wind is

$$P = 0.004 V^2$$
$$P = 0.004(50^2)$$
$$P = 10\,\text{lb}/\text{ft}^2$$

If the rider has a surface body area of $5\,\text{ft}^2$, the total force exerted by the wind becomes

$$F = P \times A$$
$$F = 10\,\text{lb}/\text{ft}^2 \times 5\,\text{ft}^2$$
$$F = 50\,\text{lb}$$

This force is enough to make pedaling into the wind extremely difficult. To remedy this adverse effect, cyclists—especially racers—bend forward as low as possible to expose a minimum surface area to the wind.

Runners also experience wind impacts. At competitions, wind affects records set during track events to the extent that when runners race for 200 meters or less (i.e., in one direction) with a tail wind of more than 2 meters per second (about 4.5 mi/hr), their results are asterisked with the qualifier "wind-aided."

The sea breeze forms as part of a thermally driven circulation. Consequently, we will begin our study of local winds by examining the formation of thermal circulations.

THERMAL CIRCULATIONS Consider the vertical distribution of pressure shown in ● Fig. 9.17a. The isobaric surfaces all lie parallel to Earth's surface; thus, there is no horizontal variation in pressure (or temperature), and there is no pressure gradient and no wind. Suppose in Fig. 9.17b the atmosphere is cooled to the north and warmed to the south. In the cold, dense air above the surface, the isobars bunch closer together, while in the warm, less-dense air, they spread farther apart. This dipping of the isobars produces a horizontal pressure gradient force aloft that causes the air to move from higher pressure (warm air) toward lower pressure (cold air).

At the surface, the air pressure changes as the air aloft begins to move. As the air aloft moves from south to north, air leaves the southern area and "piles up" above the northern area. This redistribution of air reduces the surface air pressure to the south and raises it to the north. Consequently, a pressure gradient force is established at Earth's surface from north to south and, hence, surface winds begin to blow from north to south.

We now have a distribution of pressure and temperature and a circulation of air, as shown in Fig. 9.17c. As the cool surface air flows southward, it warms and becomes less dense. In the region of surface low pressure, the warm air slowly rises, expands, cools, and flows out the top at an elevation of about 1 km (about 3000 ft) above the surface. At this level, the air flows horizontally northward toward lower pressure, where it completes the circulation by slowly sinking and flowing out the bottom of the surface

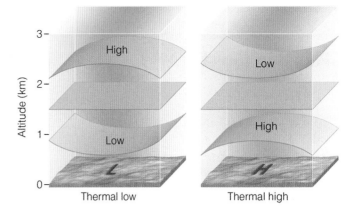

● **FIGURE 9.18** The vertical distribution of pressure with thermal highs and thermal lows.

● **FIGURE 9.17** A thermal circulation produced by the heating and cooling of the atmosphere near the ground. The H's and L's refer to atmospheric pressure. The lines represent surfaces of constant pressure (isobaric surfaces), where 1000 is 1000 millibars. For more information on isobaric surfaces, see Chapter 8, p. 205.

high. Circulations brought on by changes in air temperature, in which warmer air rises and colder air sinks, are termed **thermal circulations**.

The regions of surface high and low atmospheric pressure created as the atmosphere either cools or warms are called **thermal (cold-core) highs** and **thermal (warm-core) lows**. In general, they are shallow systems, usually extending no more than a few kilometers above the ground. These systems weaken with height. For example, at the surface, atmospheric pressure is lowest in the center of the warm thermal low in ● Fig. 9.18. In the warm air above the low, the isobars spread apart, and, at some intermediate level, the thermal low disappears and actually changes into a high. A similar phenomenon happens above the cold thermal high. The surface pressure is greatest in its center, but because the isobars aloft are crowded together due to the cold dense air, the surface thermal high becomes a low a kilometer or so above the ground.

We can summarize the typical characteristics of thermal pressure systems as being shallow, weakening with height, and being maintained, for the most part, by local surface heating and cooling.

SEA AND LAND BREEZES The sea breeze is a type of thermal circulation. The uneven heating rates of land and water (described in Chapter 3) cause these mesoscale coastal winds.

During the day, the land heats more quickly than the adjacent water, and the intensive heating of the air above produces a shallow thermal low. The air over the water remains cooler than the air over the land; hence, a shallow thermal high exists above the water. The overall effect of this pressure distribution is a **sea breeze** *that blows at the surface from the sea toward the land* (see ● Fig. 9.19a). Since the strongest gradients of temperature and pressure occur near the land-water boundary, the strongest winds typically occur right near the beach and diminish inland. Further, as the greatest contrast in temperature between land and water usually occurs in the afternoon, sea breezes are strongest at this time. (The same type of breeze that develops along the shore of a large lake is called a **lake breeze**.)

At night, the land cools more quickly than the water. The air above the land becomes cooler than the air over the water, producing a distribution of pressure such as the one shown in Fig. 9.19b. With higher surface pressure now over the land, the surface wind reverses itself and becomes a **land breeze**—*a surface breeze that flows from the land toward the water*. Temperature contrasts between land and water are generally much smaller at night; hence, land breezes are usually weaker than their daytime counterpart, the sea breeze. In regions where greater nighttime temperature contrasts exist, stronger land breezes occur over the water, off the coast. They are not usually noticed much onshore but are frequently observed by ships in coastal waters.

Look at Fig. 9.19 again and observe that the rising air is over the land during the day and over the water during the night. Therefore, along the humid east coast of the United States, daytime clouds tend to form over land and nighttime clouds over water. This explains why, at night, distant lightning flashes are sometimes seen over the ocean.

Sea breezes are best developed where large temperature differences exist between very warm land and less-warm water. Such conditions prevail year-round in many tropical regions. In middle latitudes, however, sea breezes are invariably spring and summer phenomena.

During the summer, a sea breeze usually sets in about mid-morning after the land has been warmed. By early afternoon, the breeze has increased in strength and depth. By late afternoon, the cool ocean air may reach a depth of more than 300 m (1000 ft) and extend inland for more than 20 km (12 mi).

● **FIGURE 9.19** Development of a sea breeze and a land breeze. (a) At the surface, a sea breeze blows from the water onto the land, whereas (b) the land breeze blows from the land out over the water. Notice that the pressure at the surface changes more rapidly with the sea breeze. This situation indicates a stronger pressure gradient force and higher winds with a sea breeze.

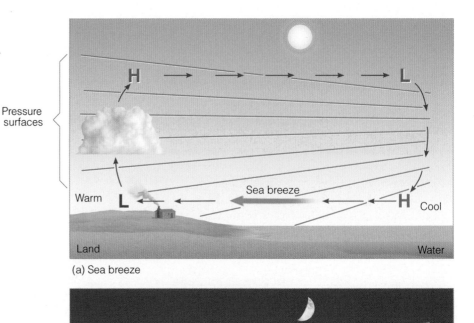

(a) Sea breeze

(b) Land breeze

The leading edge of the sea breeze is called the **sea-breeze front**. As the front moves inland, a rapid drop in temperature usually occurs just behind it. In some locations, this temperature change can be 5°C (9°F) or more during the first hours—a refreshing experience on a hot, sultry day. In regions where the water temperature is warm, the cooling effect of the sea breeze is hardly evident. Since cities near the ocean usually experience the sea breeze by noon, their highest temperature usually occurs much earlier than in inland cities. Along the east coast of North America, the passage of the sea-breeze front is marked by a wind shift, usually from west to east. In the cool ocean air, the relative humidity rises as the temperature drops. If the relative humidity increases beyond about 70 percent, water vapor begins to condense upon particles of sea salt or industrial smoke, producing haze. When the ocean air is highly concentrated with pollutants, the sea-breeze front may meet relatively clear air and thus appear as a *smoke front*, or a *smog front*. If the ocean air becomes saturated, a mass of low clouds and fog will mark the leading edge of the marine air.

When there is a sharp contrast in air temperature across the frontal boundary, the warmer, lighter air will converge and rise. In many regions, this makes for good sea-breeze glider soaring. If this rising air is sufficiently moist, a line of cumulus clouds will form along the sea-breeze front, and, if the air is also conditionally unstable, thunderstorms may form. As previously mentioned, on a hot, humid day one can drive toward the shore, encounter heavy showers several kilometers from the ocean, and arrive at the beach to find it sunny with a steady onshore breeze.

A sea breeze moving over a forest fire can be dangerous. First of all, gusty surface winds often make the fire difficult to control. Another problem is the return flow aloft. Along the sea-breeze frontal boundary, air can rise to elevations where it becomes part of the return flow. Should burning embers drift seaward with this flow and drop to the ground behind the fire, they could start additional fires. Flames from these fires pushed on by surface winds can trap firefighters between two or more blazes.

When cool, dense, stable marine air encounters an obstacle, such as a row of hills, the heavy air tends to flow around them rather than over them. When the opposing breezes meet on the opposite side of the obstruction, they form what is called a *sea-breeze convergence zone*. Such conditions are common along the mountainous Pacific coast of North America.

Sea breezes in Florida help produce that state's abundant summertime rainfall. On the Atlantic side of the state, the sea breeze blows in from the east; on the Gulf shore, it moves in from the west (see ● Fig. 9.20). The convergence of these two moist wind systems, coupled with daytime convection, produces cloudy conditions and showery weather over the land (see ● Fig. 9.21). Over the water (where cooler, more stable air lies close to the surface), skies often remain cloud-free. On many days during June and July of 1998, however, Florida's converging wind system did not materialize. The lack of converging surface air and its accompanying showers left much of the state parched. Huge fires broke out over northern and central Florida, which left hundreds of people homeless and burned many thousands of acres of grass and woodlands. A weakened sea breeze and dry conditions have produced wildfires on numerous occasions, including the spring of 2006.

Convergence of coastal breezes is not restricted to ocean areas, as large lakes are capable of producing well-defined lake breezes. For example, both Lake Superior and Lake Michigan can

produce strong lake breezes. In upper Michigan, these large bodies of water are separated by a narrow strip of land about 80 km (50 mi) wide. As can be seen from ● Fig. 9.22, the two breezes push inland and converge near the center of the peninsula, creating afternoon clouds and showers, while the lakeshore areas remain sunny, pleasantly cool, and dry.

MOUNTAIN AND VALLEY BREEZES Mountain and valley breezes develop along mountain slopes. Observe in ● Fig. 9.23 that, during the day, sunlight warms the valley walls, which in turn warm the air in contact with them. The heated air, being less dense than the air of the same altitude above the valley, rises as a gentle upslope wind known as a **valley breeze**. At night, the flow reverses. The mountain slopes cool quickly, chilling the air in contact with them. The cooler, more-dense air glides downslope into the valley, providing a **mountain breeze**. (Because gravity is the force that directs these winds downhill, they are also referred to as *gravity winds*, or *nocturnal drainage winds*.) This daily cycle of wind flow is best developed in clear summer weather when prevailing winds are light.

In many areas, the upslope winds begin early in the morning, reach a peak speed of about 6 knots by midday, and reverse direction by late evening. The downslope mountain breeze increases in intensity, reaching its peak in the early morning hours, usually

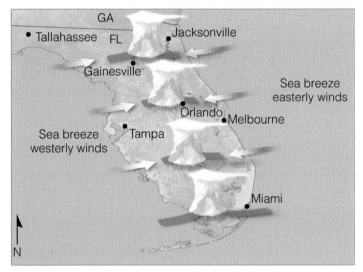

● **FIGURE 9.20** Typically, during the summer over Florida, converging sea breezes in the afternoon produce uplift that enhances thunderstorm development and rainfall, as shown above. However, when westerly surface winds dominate and a ridge of high pressure forms over the area, thunderstorm activity diminishes and dry conditions prevail.

● **FIGURE 9.21** Surface heating and lifting of air along a converging sea breeze combine to form thunderstorms almost daily during the summer in southern Florida.

© T. Ansel Toney

just before sunrise. In the Northern Hemisphere, valley breezes are particularly well developed on south-facing slopes, where sunlight is most intense. On partially shaded north-facing slopes, the upslope breeze may be weak or absent. Because upslope winds begin soon after the sun's rays strike a hill, valley breezes typically begin first on the hill's east-facing side. In the late afternoon, this side of the mountain goes into shade first, producing the onset of downslope winds at an earlier time than experienced on west-facing slopes. Hence, it is possible for campfire smoke to drift downslope on one side of a mountain and upslope on the other side.

When the upslope winds are well developed and have sufficient moisture, they can reveal themselves through cumulus clouds that build above mountain summits (see ● Fig. 9.24). Given that valley breezes usually reach their maximum strength

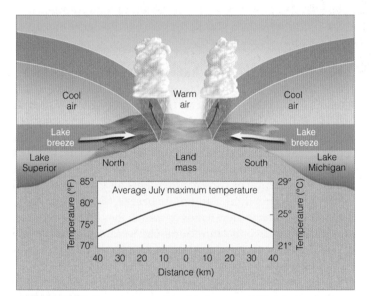

● **FIGURE 9.22** The convergence of two lake breezes and their influence on the maximum temperature during July in upper Michigan.

in the early afternoon, cloudiness, showers, and even thunderstorms are common over mountains during the warmest part of the day—a fact well known to seasoned hikers and climbers.

KATABATIC WINDS Although any downslope wind is technically a *katabatic wind*, the name is usually reserved for downslope winds that are much stronger than mountain breezes. **Katabatic (or *fall*) winds** can rush down elevated slopes at hurricane speeds, but most are not that intense and many are on the order of 10 knots or less.

The ideal setting for a katabatic wind is an elevated plateau surrounded by mountains, with an opening that slopes rapidly downhill. When winter snows accumulate on the plateau, the overlying air grows extremely cold and a shallow dome of high pressure forms near the surface (see ● Fig. 9.25). Along the edge of the plateau, the horizontal pressure gradient force is usually strong enough to cause the cold air to flow across the isobars through gaps and saddles in the hills. Along the slopes of the plateau, the wind continues downhill as a gentle or moderate cold breeze. If the horizontal pressure gradient increases substantially, such as when a storm approaches, or if the wind is confined to a narrow canyon or channel, the flow of air can increase, often destructively, as cold air rushes downslope like water flowing over a fall.

Katabatic winds are observed in various regions of the world. For example, along the northern Adriatic coast in the former Yugoslavia, a polar invasion of cold air from Russia descends the slopes from a high plateau and reaches the lowlands as the *bora*—a cold, gusty, northeasterly wind with speeds sometimes in excess of 100 knots. A similar, but often less violent, cold wind known as the *mistral* descends the western mountains into the Rhone Valley of France, and then out over the Mediterranean Sea. It frequently causes frost damage to exposed vineyards and makes people bundle up in the otherwise mild climate along the Riviera. Strong, cold katabatic winds also blow downslope off the ice sheets in Greenland and Antarctica, occasionally with speeds greater than 100 knots.

● **FIGURE 9.23** Valley breezes blow uphill during the day; mountain breezes blow downhill at night. (The L's and H's represent pressure, whereas the purple lines represent surfaces of constant pressure.)

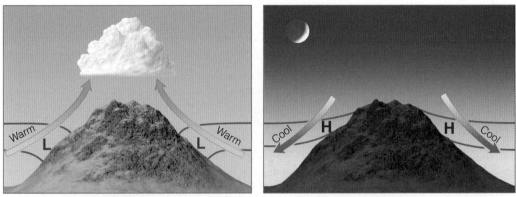

Valley breeze

Mountain breeze

CRITICAL THINKING QUESTION Suppose you are camping along the eastern side of a small mountain. Smoke from your campfire is blowing uphill. Meanwhile, a friend camping on the western side of the same hill finds her campfire smoke drifting downhill. Are you and your friend cooking breakfast or dinner? How did the direction of the smoke from your campfire and Fig. 9.23 help you come up with your answer?

● **FIGURE 9.24** As mountain slopes warm during the day, air rises and often condenses into cumuliform clouds, such as the ones shown here.

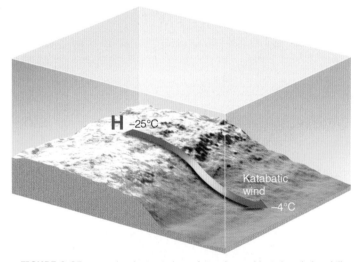

● **FIGURE 9.25** Strong katabatic winds can form where cold winds rush downhill from an elevated plateau covered with snow.

In North America, when cold air accumulates over the Columbia Plateau of Idaho, Oregon, and Washington, it may flow westward through the Columbia River Gorge as a strong, gusty, and sometimes violent wind. Even though the sinking air warms by compression, it is so cold to begin with that it reaches the ocean side of the Cascade Mountains much colder than the marine air it replaces. The *Columbia Gorge wind* (called the *coho*) is often the harbinger of a prolonged cold spell.

Strong downslope katabatic-type winds funneled through a mountain canyon can do extensive damage. In January 1984, a ferocious downslope wind blew through Yosemite National Park in California at speeds estimated at 100 knots. The wind toppled many trees and, unfortunately, caused a fatality when a tree fell on a park employee sleeping in a tent.

CHINOOK (FOEHN) WINDS   The **chinook wind** is a warm, dry, downslope wind that descends the eastern slope of the Rocky Mountains. The region of the chinook is rather narrow (only several hundred kilometers wide) and extends from northeastern New Mexico northward into Canada. Similar winds occur along the leeward slopes of mountains in other regions of the world. The general term for such a wind is the **foehn** (a name that originated in the European Alps), but there are many local names, such as the *zonda* in Argentina. When foehn winds move through an area, the temperature rises sharply, sometimes 20°C (36°F) or more in less than an hour, and a corresponding sharp drop in the relative humidity occurs, occasionally to less than 5 percent. (More information on temperature changes associated with U.S. chinooks is given in Focus section 9.3.)

In North America, chinooks occur when strong westerly winds aloft flow over a north-south–trending mountain range, such as the Rockies and Cascades. Such conditions (described in Chapter 12) can produce a trough of low pressure on the mountain's eastern side, a trough that tends to force the air downslope. As the air descends, it is compressed and warms at the dry adiabatic rate (10°C/km). So the main source of warmth for a chinook is *compressional heating*, as potentially warmer (and drier) air is brought down from aloft.

Clouds and precipitation on the mountain's windward side can enhance the chinook. For example, as the cloud forms on the upwind side of the mountain in ● Fig. 9.26a, the release of latent heat inside the cloud supplements the compressional heating on the downwind side. This phenomenon makes the descending air at the base of the mountain on the downwind side warmer than it was before it started its upward journey on the windward side. The air is also drier, since much of its moisture was removed as precipitation on the windward side (see Fig. 9.26b).

Snow Eaters and Rapid Temperature Changes

Chinooks are thirsty winds. As they move over a heavy snow cover, they can melt and evaporate a foot of snow in less than a day. This situation has led to some tall tales about these so-called snow eaters. Canadian folklore has it that a sled-driving traveler once tried to outrun a chinook. During the entire ordeal, as the story has it, his front runners were in snow while his back runners were on bare soil.

Actually, the chinook is important economically. It not only brings relief from the winter cold, but it uncovers prairie grass, so that livestock can graze on the open range. Also, these warm winds can help keep railroad tracks clear of snow, so that trains can keep running. On the other hand, the drying effect of a chinook can create an extreme fire hazard. And when a chinook follows spring planting, the seeds may die in the parched soil. Along with the dry air comes a buildup of static electricity, making a simple handshake a shocking experience. These warm, dry winds have sometimes adversely affected human behavior. During periods of chinook winds some people feel irritable and depressed and others become ill. The exact reason for this phenomenon is not clearly understood.

Chinook winds have been associated with rapid temperature changes. On January 11, 1980, the air temperature in Great Falls, Montana, rose from −32°F to 17°F (a 49°F rise in temperature) in just seven minutes. The nation's largest temperature swing within a 24-hour period occurred in Loma, Montana, on January 15, 1972, when the temperature soared from −54°F to 48°F. How such rapid changes in temperature can occur is illustrated in ● Fig. 4. Notice that a shallow layer of extremely cold air has moved out of Canada and is now resting against the Rocky Mountains.

The cold air behaves just as any fluid, and, in some cases, atmospheric conditions may cause the air to move up and down much like water does when a bowl is rocked back and forth. This rocking motion can cause extreme temperature variations for cities located at the base of the hills along the periphery of the cold air–warm air boundary, as they are alternately in and then out of the cold air. Such a situation is probably responsible for the extremely rapid two-minute temperature change of 49°F recorded at Spearfish, South Dakota, during the morning of January 22, 1943. On the same morning, in nearby Rapid City, the temperature fluctuated from −4°F at 5:30 a.m. to 54°F at 9:40 a.m., then down to 11°F at 10:30 a.m. and up to 55°F just 15 minutes later. At nearby cities, the undulating cold air produced similar temperature variations that recurred over several hours.

● **FIGURE 4** Cities near the warm air–cold air boundary can experience sharp temperature changes if cold air should slosh back and forth like water in a bowl.

(a)

(b)

● **FIGURE 9.26** (a) A chinook wind can be enhanced when clouds form on the mountain's windward side. Heat added and moisture lost on the upwind side produce warmer and drier air on the downwind sides. (b) A graphic representation of the rising and sinking air as it moves over the mountain.

Along the Front Range of the Rockies, a bank of clouds forming over the mountains is a telltale sign of an impending chinook. This cloud feature, called a *foehn wall* (which looks like a wall of clouds), usually remains stationary as air rises, condenses, and then rapidly descends the leeward slopes, often causing strong winds in foothill communities. ● Figure 9.27 shows how a foehn

● **FIGURE 9.27** A foehn wall forming over the Colorado Rockies (viewed from the plains).

wall appears as one looks west toward the Rockies from the Colorado plains. The photograph was taken on a winter afternoon with an air temperature of about −7°C (20°F). That evening, the chinook moved downslope at high speeds through foothill valleys, picking up sand and pebbles (which dented cars and cracked windshields). The chinook spread out over the plains like a warm blanket, raising the air temperature the following day to a mild 15°C (59°F). The chinook and its wall of clouds remained for several days, bringing with it a welcomed break from the cold grasp of winter.

As mentioned in the previous paragraph, chinook winds can be quite destructive. Strong chinook winds are especially notorious in winter in Boulder, Colorado. These *Boulder winds* have been recorded gusting to more than 100 knots, damaging roofs, uprooting trees, overturning mobile homes and trucks, and sandblasting car windows. Although the forces involved in these high winds are not completely understood, some evidence indicates that the highest gusts may be associated with large vertically oriented whirls of air that are sometimes call *mountainadoes*.

SANTA ANA WINDS   A warm, dry wind that blows downhill from the east or northeast into southern California is the **Santa Ana wind**. As the air descends from the elevated desert plateau, it funnels through mountain canyons in the San Gabriel and San Bernardino Mountains, finally spreading over the Los Angeles Basin and San Fernando Valley and out over the Pacific Ocean (see ● Fig. 9.28). The wind often blows with exceptional speed—occasionally over 90 knots—in the Santa Ana Canyon (the canyon from which it derives its name).

These warm, dry winds develop as a region of high pressure builds over the Great Basin. The clockwise circulation around the anticyclone forces air downslope from the high plateau. Thus, *compressional heating* provides the primary source of warming. The air is dry, since it originated in the desert, and it dries out even more as it is heated. ● Figure 9.29 shows a typical wintertime Santa Ana situation.

As the wind rushes through canyon passes, it lifts dust and sand and dries out vegetation, which sets the stage for serious brush fires, especially in autumn, when chaparral-covered hills are already

● **FIGURE 9.28** Warm, dry Santa Ana winds sweep downhill through mountain canyons into Southern California. The large H represents higher air pressure over the elevated desert.

● **FIGURE 9.29** Surface weather map showing Santa Ana conditions in January. Maximum temperatures for this particular day are given in °F. Observe that the downslope winds blowing into Southern California raised temperatures into the upper 80s, while elsewhere temperature readings were much lower.

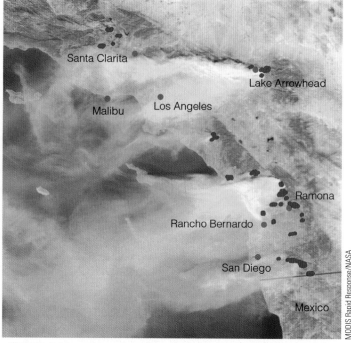

● **FIGURE 9.30** Satellite view showing strong northeasterly Santa Ana winds on October 23, 2007, blowing smoke from massive wild fires (red dots) across Southern California and out over the Pacific Ocean.

parched from the dry summer.* One such fire in November 1961—the infamous *Bel Air fire*—burned for three days, destroying 484 homes and causing over $25 million in damage in 1961 dollars (close to $200 million in today's dollars). During October 2003, massive wildfires driven by strong Santa Ana winds swept through Southern California. The fires charred more than 750,000 acres, destroyed over 2800 homes, took 20 lives, and caused over $2 billion in property damage. Only four years later (and after one of the driest years on record) in October 2007, wildfires broke out again in Southern California. Pushed on by hellacious Santa Ana winds that gusted to over 80 knots, the fires raced through dry vegetation, scorching virtually everything in their paths. The fires, which extended from north of Los Angeles to the Mexican border (see ● Fig. 9.30), burned over 500,000 acres, destroyed more than 1800 homes, and took 9 lives. The total costs of the fires exceeded $1.5 billion.

Four hundred miles to the north of Los Angeles in Oakland, California, a ferocious Santa Ana–type wind (often called a *diablo*, or devil, wind) was responsible for the disastrous Oakland hills fire during October 1991. The fire started in the parched Oakland hills, just east of San Francisco, where a firestorm driven by strong northeast winds blackened almost 2000 acres, damaged or destroyed over 3000 dwellings, caused more than $1.5 billion in damage, and took 25 lives.

When a Santa Ana fire removes the protective cover of vegetation, the land is ripe for erosion, as winter rains may wash away topsoil and, in some areas, create serious mudslides such as those that occurred in Southern California during May 2005. The adverse effects of a wind-driven Santa Ana fire may be felt long after the fire itself has been put out.

_____
*Chaparral denotes a shrubby environment in which many of the plant species contain highly flammable oils.

● **FIGURE 9.31** Blown by ferocious winds exceeding 50 mph, multiple fires swept across parts of northern California in October 2017. The fires rampaged across the state's famous wine country, destroying this house and more than 6000 other structures.

The deadliest and most destructive wildfire event in California history took place in early October 2017 just north of San Francisco, including parts of the city of Santa Rosa and nearby counties. More than 6000 structures were lost and more than 40 people were killed, many of them during a fierce firestorm that developed as powerful diablo winds struck late on October 8. As the strong winds blew burning embers well ahead of the existing fire, the situation deteriorated so quickly that many residents had only a few minutes to evacuate. The fires developed after California's second-wettest winter on record fostered the growth of lush vegetation, which then dried out during the state's hottest summer on record (see ● Fig. 9.31).

A similar downslope-type wind called a *California norther* can produce unbearably high temperatures in the northern half of California's Central Valley. On August 8, 1978, for example, a ridge of high pressure formed to the north of this region, while a thermal low was well entrenched to the south. This pressure pattern produced a north wind in the area. A summertime north wind in most parts of the country means cooler weather and a welcome relief from a hot spell, but not in Red Bluff, California,* where the winds moved downslope off the mountains. Heated by compression, these winds increased the air temperature in Red Bluff to an astounding 119°F for two consecutive days—amazing when you realize that Red Bluff is located at about the same latitude as Philadelphia, Pennsylvania.

CRITICAL THINKING QUESTION In the autumn, Santa Ana winds blowing into Southern California often cause air temperatures in cities around Los Angeles to rise above 90°F. Look back at Fig. 9.29 and come up with two possible reasons why on August 8, 1978, a Santa Ana–type wind blowing downhill into Red Bluff, California, was able to raise the air temperature there to 119°F—an air temperature typically much higher than those experienced during Santa Ana conditions in Southern California.

DESERT WINDS Winds of all sizes develop over the deserts. Huge *dust storms* form in dry regions, where strong winds are able to lift and fill the air with particles of fine dust. In February 2001, an exceptionally large dust storm (about the size of Spain) formed over the African Sahara and swept westward off the African coast, then northeastward for thousands of miles (see ● Fig. 9.32). During the drought years of the 1930s, large dust

*The location of Red Bluff is shown on the map in Fig. 9.29.

storms formed over the Great Plains of the United States (see ● Fig. 9.33). Some individual storms lasted for three days and spread dust for hundreds of miles to the east, to the Atlantic Ocean and beyond. In desert areas where loose sand is more prevalent, *sandstorms* develop, as high winds enhanced by surface heating rapidly carry sand particles close to the ground.

A spectacular example of a storm composed of dust or sand is the **haboob** (from Arabic *habb*: "to blow"). The haboob forms as cold downdrafts along the leading edge of a thunderstorm lift dust or sand into a huge, tumbling dark cloud that may extend horizontally for more than 100 kilometers and rise vertically to the base of the thunderstorm. Haboobs are most common in the African Sudan (where about twenty-four occur each year) and in the Desert Southwest of the United States, especially in southern Arizona. A particularly strong haboob swept into Phoenix, Arizona, in July 2011 (see ● Fig. 9.34), moving vast amounts of dust from a landscape parched by drought.

On a smaller scale, in dry areas, the wind can also produce rising, spinning columns of air that pick up dust or sand from the ground. Called **dust devils** or *whirlwinds*,* these rotating vortices generally form on clear, hot days over a dry surface where most of the sunlight goes into heating the surface, rather than into evaporating water from vegetation. The atmosphere directly above the hot surface becomes unstable, convection sets in, and the heated air rises, often lifting dust, sand, and dirt high into the air. Wind, often deflected by small topographic barriers, flows into this region, rotating the rising air as depicted in ● Fig. 9.35. Depending on the nature of the topographic feature, the spin of a dust devil around its central core may be cyclonic or anticyclonic, and both directions occur with about equal frequency. (Dust devils are too small and fleeting to be substantially influenced by the Coriolis force.)

Having diameters of only a few meters and heights that are usually less than 100 m (300 ft), most dust devils last only a short time (see ● Fig. 9.36). However, some dust devils reach sizable dimension, extending upward from the surface for several hundred meters. Such whirlwinds are capable of considerable damage; winds exceeding 75 knots can overturn mobile homes and tear the roofs off buildings. Fortunately, the majority of dust devils

*In Australia, the Aboriginal word *willy-willy* refers to a dust devil.

AFRICA

Sahara Desert

NASA

● **FIGURE 9.32** A large dust storm over the African Sahara Desert during February 2001 sweeps westward off the coast, then northward into a mid-latitude cyclonic storm west of Spain, as indicated by red arrows.

© Library of Congress

● **FIGURE 9.33** A truck tries to outrun one of the many devastating dust storms that roared over the Central Plains in the 1930s.

● **FIGURE 9.34** A large haboob (dust storm) moves through Phoenix, Arizona, on July 5, 2011.

Daniel J Bryant/Getty Images

are small. Also keep in mind that dust devils are *not* tornadoes. The circulations of most tornadoes (as we will see in Chapter 15) develop near the base of a thunderstorm, whereas the circulation of a dust devil begins at the surface, normally in sunny weather (although some dust devils form beneath convective-type clouds).

Desert winds are not confined to planet Earth; they form on the planet Mars as well. Most of the Martian dust storms are small and only cover a relatively small portion of that planet. However, dust storms can actually grow large enough to encircle Mars with a dusty haze, as happened in 2001. Dust devils also form on Mars when high winds sweep over uneven terrain.

Some of the hottest winds in the world blow over deserts. For example, extremely hot, dry winds originating over the Sahara Desert are given local names as they move into different regions. Over North Africa the general wind flow is from the northeast. However, when a cyclonic storm system is located west of Africa and southern Spain (position 1, ● Fig. 9.37), a hot, dry, and dusty easterly or southeasterly wind—the *leste*—blows over Morocco and out over the Atlantic. If the wind crosses the Mediterranean and enters southern Spain, it becomes the hot, dry *leveche*. When a low-pressure area is centered at position 2, a hot, dust-laden south or southeast wind—the *sirocco*—originates over the Sahara Desert and blows across North Africa.

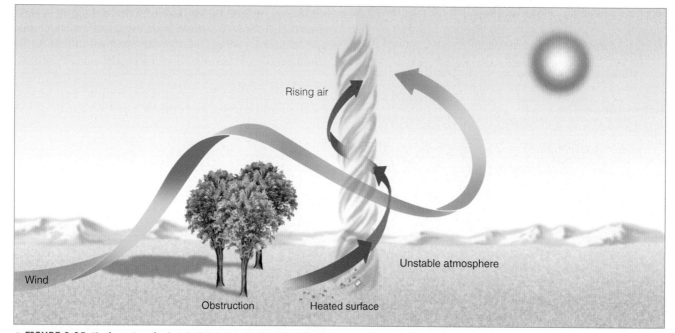

● **FIGURE 9.35** The formation of a dust devil. On a hot, dry day, the atmosphere next to the ground becomes unstable. As the heated air rises, wind blowing past an obstruction twists the rising air, forming a rotating air column or *dust devil*. Air from the sides rushes into the rising column, lifting sand, dust, leaves, or any other loose material from the surface.

● **FIGURE 9.36** A well-developed dust devil moves over a hot desert landscape on a clear summer day.

A storm located still farther to the east (position 3) may produce a dry, hot southerly wind, called the *khamsin*, which blows over Egypt, the Red Sea, and Saudi Arabia. This wind can raise air temperatures well above 100°F while lowering the relative humidity to less than 10 percent. If this wind moves into Israel, it is called the *Sharav*. As hot as these winds are, they don't hold a candle to one of the hottest winds on earth, the **simoom**. Called the "poison wind," this strong, dry, and dusty wind blows over Africa and the Arabian Desert. It can reportedly push air temperatures above 120°F.

SEASONALLY CHANGING WINDS—THE MONSOON On our planet there are thermal circulations that are much larger than those of the more local sea and land breezes described earlier. A good example of such a circulation is the **monsoon**, which derives from the Arabic *mausim* ("seasons"). A **monsoon wind system** is one that *changes direction seasonally*, blowing from one direction in summer and from the opposite direction in winter. This seasonal reversal of winds is especially well developed in eastern and southern Asia.

In some ways, the monsoon is similar to a large-scale sea breeze. During the winter, the air over northern Asia becomes much colder than the air over the adjacent ocean. A large, shallow high-pressure area develops over continental Siberia, producing a *clockwise* circulation of air that flows out over the Indian Ocean and South China Sea (see ● Fig. 9.38a). Subsiding air of the anticyclone and the downslope movement of northeasterly winds from the inland plateau provide eastern and southern Asia with generally fair weather. Hence, the *winter monsoon*, which lasts from about December through February, means clear skies and generally dry weather (*dry season*), with surface winds that blow from land to sea.

In summer, the wind-flow pattern reverses itself as air over the continents becomes much warmer than air above the water. A shallow thermal low develops over the continental interior. The heated air within the low rises, and the surrounding air responds by flowing *counterclockwise* into the low center. This condition results in moisture-bearing winds advancing from the ocean into the continent during the late spring. The humid air converges with a drier westerly flow of continental air, causing the humid air to rise; further lifting is provided by hills and mountains. Lifting cools the air to its saturation point, resulting in heavy showers and thunderstorms. Thus, the *summer monsoon* of southeastern Asia, which lasts from about June through September, means wet, rainy weather (the *wet season*) with surface winds blowing from sea to land (see Fig. 9.38b). Although the majority of rain falls during the wet season, it does not rain all the time. In fact, rainy periods lasting between 15 and 40 days are often followed by 3 to 15 days of hot, sunny weather, known as a *monsoon break*.

Many factors help create the monsoon wind system. The latent heat given off during condensation aids in the warming of the air over the continent and strengthens the summer

● **FIGURE 9.37** Exceptionally hot, dry local winds that form over North Africa and the Sahara Desert.

● **FIGURE 9.38** Changing annual surface wind flow patterns associated with the (a) winter and (b) summer Asian monsoon.

monsoon circulation. Rainfall is enhanced by weak, westward-moving low-pressure areas called *monsoon depressions*. The formation of these depressions is aided by an upper-level jet stream. Where winds in the jet diverge, surface pressures drop, the monsoon depressions intensify, and surface winds increase. The greater inflow of moist air supplies larger quantities of latent heat, which, in turn, intensifies the summer monsoon circulation.

The strength of the Indian monsoon appears to be related to the reversal of surface air pressure that occurs at irregular intervals about every two to seven years at opposite sides of the tropical South Pacific Ocean. As we will see in Chapter 10, this reversal of pressure (which is known as the *Southern Oscillation*) is linked to an ocean-warming phenomenon known as *El Niño*. During an El Niño event, surface water near the equator becomes warmer than average over the central and eastern Pacific. Over the region of warm water we find rising air, huge convective clouds, and heavy rain. Meanwhile, to the west of the warm water (over the region influenced by the summer monsoon), sinking air inhibits cloud formation and convection. Hence, during El Niño years, monsoon rainfall is likely to be deficient.

Summer monsoon rains over southern Asia can be truly extreme. The town of Cherrapunji (also known as Sohra), located about 300 km inland on the southern slopes of the Khasi Hills in northeastern India, receives an average of 1176 cm (463 in.) of rainfall each year, most of it between April and October (see ● Fig. 9.39). The town also holds world records for the heaviest rainfall measured anywhere in a 12-month period—2647 cm (1042 in) from August 1860 to July 1861—and the heaviest 48-hour rainfall, 249.3 cm (98.15 in.) on June 15–16, 1995.

The summer monsoon rains are essential to the agriculture of southern and eastern Asia. More than 2 billion people rely on the summer rains so that crops will grow and drinking water will be available. Unfortunately, the monsoon can be unreliable in both duration and intensity, and these are difficult to predict. Because the monsoon is vital to the survival of so many people, it is no wonder that meteorologists have investigated it extensively. They have tried to develop methods of accurately forecasting the intensity and duration of the monsoon. With the aid of current

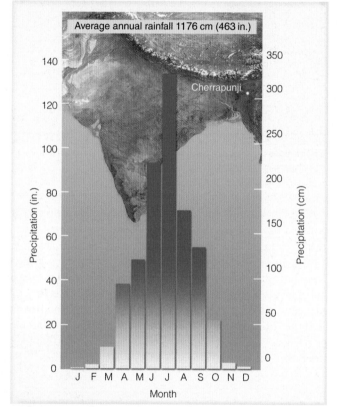

● **FIGURE 9.39** Average annual precipitation for Cherrapunji, India. Note the abundant rainfall during the summer monsoon (April through October) and the lack of rainfall during the winter monsoon (November through March).

research projects and the latest climate models, there is hope that monsoon forecasts will improve in accuracy.

Monsoon wind systems exist in other regions of the world, such as Australia, Africa, and North and South America, where large contrasts in temperature develop between oceans and continents. Usually, however, these systems are not as pronounced as in southeast Asia. For example, the North American monsoon affects northwest Mexico and the southwestern United States, especially Arizona, New Mexico, Nevada, and the southern part of California. In this region, spring and early summer

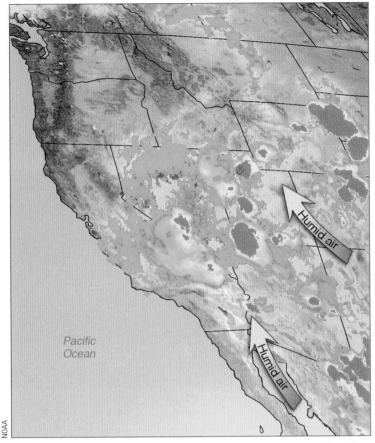

● **FIGURE 9.40** Enhanced infrared satellite image showing strong monsoonal circulation. Moist southerly winds from the *Gulf of California* (bottom arrow) and southeasterly winds from the *Gulf of Mexico* (top arrow) are feeding showers and thunderstorms (yellow and red areas) over the southwestern section of the United States during July 2001.

BRIEF REVIEW

Before moving on to the next section, here is a brief review of some of the main points about winds.

- Thermal pressure systems are shallow pressure systems that are driven by the unequal heating and cooling of Earth's surface.

- The sea breeze and the land breeze are types of thermal circulations that are due to uneven heating and cooling rates of land and water.

- At the surface, a sea breeze blows from water to land, whereas a land breeze blows from land to water.

- A valley breeze blows uphill during the day and a mountain breeze blows downhill at night.

- Chinook (foehn) winds are warm, dry winds that blow downhill along the eastern side of the Rocky Mountains.

- The main source of warmth for the chinook is compressional heating.

- Santa Ana winds are warm, dry downslope winds that warm by compressional heating and blow from the east or northeast into Southern California.

- Dust devils tend to form over dry terrain on clear, hot days. They are not tornadoes, although the winds of a large dust devil may cause minor damage to structures.

- Monsoon winds are winds that change direction seasonally. In southern Asia, the winter monsoon, which blows from land to water, is dry; the summer monsoon, which blows from water to land, is wet.

● **FIGURE 9.41** Clouds and thunderstorms forming over Arizona, as humid monsoonal air flows northward over the region during July 2007.

are normally dry, as warm westerly winds predominate. By mid-July, however, humid southerly or southeasterly winds are more common, and so are afternoon showers and thunderstorms (see ● Fig. 9.40 and ● Fig. 9.41).

Up to now, we've examined a variety of wind systems that are recognized more than just locally. Many other wind systems have gained local notoriety in different parts of the world, a few of which are listed in ▼ Table 9.1.

NAME	DESCRIPTION
Cold Winds	
athos	A strong northeasterly fall wind that descends from Mount Athos over the Aegean Sea
buran	A strong, cold wind that blows over Russia and central Asia
purga	A buran accompanied by strong winds and blowing snow
pampero	A cold wind blowing from the south over Argentina, Uruguay, and into the Amazon Basin
burga	A cold northeasterly wind in Alaska usually accompanied by snow; similar to the buran and purga of Russia
bise	Generally a cold north or northeast wind that blows over southern France; often brings damaging spring frosts
Papagayo wind	A cold northeasterly wind along the Pacific coast of Nicaragua and Guatemala; occurs when a cold air mass overrides the mountains of Central America
Tehuantepecer	A strong wind from the north or northwest funneled through the gap between the Mexican and Guatemalan mountains and out into the Gulf of Tehuantepec
blue norther	Cold northerly winds behind an intense middle-latitude cyclone crossing the U.S. Great Plains, which may penetrate into Central America, where the wind is called a *norte*
Mild Winds	
levant	A mild, humid, and often rainy east or northeast wind that blows across southern Spain
harmattan	A dry, dusty, and mild wind from the northeast or east that originates over the cool Sahara in winter and blows over the west coast of Africa; brings relief from the hot, humid weather along the coastal region

Determining Wind Direction and Speed

Wind—the horizontal movement of air—is characterized by its direction, speed, and gustiness. If we imagine air molecules as being a swarm of bees, the wind may be seen as the movement of the entire swarm. This analogy can be carried a little further: On a calm day, the swarm will remain in one spot with each bee randomly darting about; while on a windy day the entire swarm will move quickly from one place to another. The swarm's speed would be the rate at which it moves past you. Likewise, wind speed is the rate at which air moves by a stationary observer. This movement can be expressed as the distance in nautical miles traveled in one hour (knots) or as the number of meters traveled in one second (m/sec), as well as in miles per hour (mph).

Unlike a swarm of bees, air is invisible; we cannot really see it. Rather, we see things being moved by it. Therefore, we can estimate wind direction by watching the movement of objects as air passes them. For example, the rustling of small leaves, smoke drifting near the ground, and flags waving on a pole all indicate wind direction. In a light breeze, a tried-and-true method of determining wind direction is to raise a wet finger into the air. The dampness quickly evaporates on the wind-facing side, cooling the skin. Traffic sounds carried from nearby railroads or airports can be used to help figure out the direction of the wind. Even your nose can alert you to the wind direction, as the smell of fried chicken or broiled hamburgers drifts with the wind from a local restaurant.

We already know that *wind direction* is given as the direction *from which* it is blowing—a north wind blows from the north toward the south. However, near large bodies of water and in hilly regions, wind direction may be expressed differently. For example, wind blowing from the water onto the land is referred to as an **onshore wind** (see ● Fig. 9.42a). Hence, a sea breeze and a lake breeze are both onshore winds. Conversely, wind that blows from the land onto the water is called an **offshore wind** (see Fig. 9.42b). So, a land breeze is an offshore wind.

Air moving uphill is an *upslope wind;* air moving downhill is a *downslope wind.* Hence, valley breezes are upslope winds, and mountain breezes are downslope winds. The wind direction may also be given as degrees about a 360° circle. These directions are expressed by the numbers shown in ● Fig. 9.43. For example: A wind direction of 360° is a north wind; an east wind is 90°; a south wind is 180°; and calm is expressed as zero. It is also common practice to express the wind direction in terms of compass points, such as N, NW, NE, and so on.

THE INFLUENCE OF PREVAILING WINDS At many locations, the wind blows more frequently from one direction than from any other. The **prevailing wind** is the name given to the wind direction most often observed during a given time period. Prevailing winds can greatly affect the climate of a region. For example, where the prevailing winds are upslope, the rising, cooling

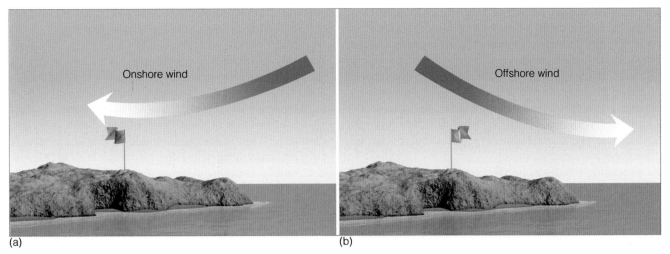

● **FIGURE 9.42** (a) An onshore wind blows from water to land; (b) An offshore wind blows from land to water.

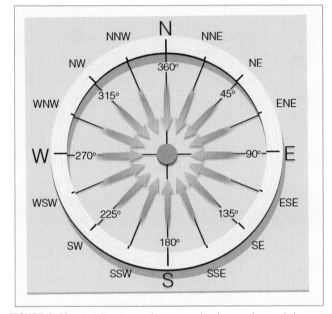

● **FIGURE 9.43** Wind direction can be expressed in degrees about a circle or as compass points.

● **FIGURE 9.44** In the high country, trees standing unprotected from the wind are often sculpted into "flag trees," such as these trees in Wyoming.

air makes clouds, fog, and precipitation more likely than where the winds are downslope. Prevailing onshore winds in summer carry moisture, cool air, and fog into coastal regions, whereas prevailing offshore breezes carry warmer and drier air into the same locations.

In city planning, the prevailing wind can help decide where industrial centers, factories, and city dumps should be built. All of these, of course, must be located so that the wind will not carry pollutants into populated areas. Sewage disposal plants must be situated downwind from large housing developments, and major runways at airports must be aligned with the prevailing wind to assist aircraft in taking off or landing. In the high country, strong prevailing winds can bend and twist tree branches toward the downwind side, producing wind-sculpted "*flag trees*" (see ● Fig. 9.44).

The prevailing wind can even be a significant factor in building an individual home. In the northeastern half of the United States, the prevailing wind in winter is northwest and in summer it is southwest. Thus, to maximize comfort and energy efficiency, houses built in the northeastern United States should have windows facing southwest to provide summertime ventilation and few, if any, windows facing the cold winter winds from the northwest. The northwest side of the house should be thoroughly insulated and even protected by a windbreak.

From the prevailing wind, biologists can predict the direction disease-carrying insects and plant spores will move and, hence, how a disease may spread. Geologists use the prevailing wind to predict where ejected debris from potentially active volcanoes will land.

Many local ground and landscape features show the effect of a prevailing wind. For example, smoke particles from an industrial stack settle to the ground on its downwind side. From the air, the prevailing wind direction can be seen as a discolored landscape on the downwind side of the stack. Wind blowing over surfaces of snow and sand produces ripples with a more gentle slope facing into the wind. As previously mentioned, sand dunes have similar shapes and, thus, show the prevailing wind direction. Look at ● Fig. 9.45 and see if you can determine the prevailing wind when this cinder cone in Iceland erupted.

The prevailing wind can be represented by a **wind rose**, which indicates the percentage of time the wind blows from different directions. Extensions from the center of a circle point to the wind direction, and the length of each extension indicates the percentage of time the wind blew from that direction. Wind speed does not affect the length of each extension. However, the strength of the wind blowing from different directions is very helpful to know, so wind roses may include this information, such as the wind rose in ● Figure 9.46. In this case, the prevailing wind is from the south, but winds from the west and north are almost as common. While this wind rose covers all four seasons and all times of day, a wind rose can be made for any particular time of the day, and it can represent the wind direction for any month or season of the year.

The prevailing wind in a town does not always represent the prevailing wind of an entire region. In mountainous regions, the wind is usually guided by topography and is often deflected by obstructions that cause its direction to change abruptly. Within this region, the wind may be blowing from one direction on one side of a valley and from an entirely different direction on the other side.

In an attempt to harness some of the prevailing wind's energy and turn it into electricity, large wind turbines and wind farms are becoming increasingly common around the world. More information on this topic is given in Focus section 9.4.

WIND MEASUREMENTS A very old, yet reliable, weather instrument for determining wind direction is the **wind vane**. Most wind vanes consist of a long arrow with a tail, which is allowed to move freely about a vertical post (see ● Fig. 9.47). The arrow always points into the wind and, hence, always gives the wind direction. Wind vanes can be made of almost any material. At airports, a cone-shaped bag opened at both ends so that it extends horizontally as the wind blows through it sits near the runway. This form of wind vane, called a *wind sock,* enables pilots to tell the surface wind direction when landing.

The instrument that measures wind speed is the **anemometer**. The oldest type of anemometer is the *pressure plate anemometer* developed by Robert Hooke in 1667. It consists of a rectangular metal plate, which hangs downward from a hinge. As the speed of the wind increases, the force of the wind on the plate pushes it outward at a greater angle. The wind speed is read from a scale mounted adjacent to the arm of the swinging plate. Many anemometers today consist of three (or more) hemispherical cups *(cup anemometer)* mounted on a vertical shaft as shown in Fig. 9.47. The difference in wind pressure from one side of a cup to the other causes the cups to spin about the shaft. The rate at which they rotate is directly proportional to the speed of the wind. The spinning of the cups is usually translated into wind

● **FIGURE 9.45** A cinder cone in Iceland that has a more gentle slope on its right side. Which way was the wind blowing when it erupted? Answer: During eruption, the prevailing wind was blowing from left to right because ejected particles blown by the wind accumulated on the volcano's right side, producing a more gentle slope.

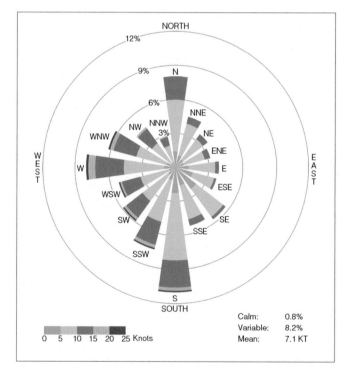

● **FIGURE 9.46** This wind rose represents the percentage of time the wind blew from various directions at Louisville Kentucky International Airport between 1977 and 2006. The most common directions are south (S) and north (N); the least common is north-northwest (NNW). Each bar includes color categories showing how often the wind blew at a particular speed from a given direction. The wind was variable or calm about 9 percent of the time (not shown in the wind rose) (Midwest Regional Climate Center).

Wind Energy

For centuries, thousands of small windmills—their arms spinning in a stiff breeze—pumped water, sawed wood, and even supplemented the electrical needs of homes and farms around the world. It was not until the energy crisis of the early 1970s, however, that we seriously considered wind-driven turbines, called *wind turbines*, to run generators that produce electricity. Recently, the amount of wind energy in the United States has doubled every few years: The nation's wind-generation capacity was more than twice as much in 2016 as it was in 2010. Only China had more capacity in its wind energy installations (about twice the U.S. total).

In several ways, wind is an effective way to produce energy—it is nonpolluting and, unlike the sun, is not restricted to daytime use. A single commercial wind turbine, which may cost several million dollars to build and install, can generate power for hundreds of homes or businesses at a cost that is comparable in some areas to electricity generated by fossil fuels. The sight of large wind turbines is not to everyone's liking, although other types of energy-related infrastructure, such as oil pumps and electrical transmission towers, can be eyesores as well. Unfortunately, each year the blades of spinning turbines kill many birds—perhaps hundreds of thousands in the United States alone. To help remedy this problem, many wind turbine companies hire avian specialists to study bird behavior, and some turbines are actually shut down during nesting time. And the blades of modern high-capacity turbines turn more slowly, thereby helping birds to avoid them.

● **FIGURE 5** Cows graze beneath the turbines of a wind farm in southern Wyoming.

Jim Havey/Alamy Stock Photo

If a wind turbine is to produce electricity, there must be wind—and not just any wind, but a flow of air neither too weak nor too strong. A slight breeze will not turn the blades, and a powerful wind gust could severely damage the machine. Thus, regions with the greatest potential for wind-generated power have moderate, relatively steady winds. So much of the Great Plains of the United States is well-suited to wind power that the region has been dubbed "the Saudi Arabia of wind energy."

Sophisticated technology allows many modern turbines to sense meteorological data from their surroundings. Some turbines can produce energy in winds as low as 5 knots and as high as 45 knots.

As of 2017, more than 48,000 wind turbines in the United States had the capacity of generating more than 82,000 megawatts of electricity, enough to supply the annual needs of more than 24 million homes. In Texas alone (the leading wind-energy state), there are thousands of wind turbines, many of which are on *wind farms*—clusters of 50 or more wind turbines (see ● Fig. 5). Wind provided close to 5 percent of the electricity needs in the United States in 2015, more than double the percentage of 2010. The U.S. Department of Energy has estimated that wind energy using current technology could provide 20 percent of the nation's electricity by 2030.

speed through a system of gears and may be read from a dial or transmitted to a recorder.

The **aerovane** *(skyvane)* is an instrument that indicates both wind speed and direction. It consists of a bladed propeller that rotates at a rate proportional to the wind speed. Its streamlined shape and a vertical fin keep the blades facing into the wind (see ● Fig. 9.48). When attached to a recorder, a continuous record of both wind speed and direction is obtained.

Over the last few years, the national ASOS network of automated weather stations has been replacing the original cup anemometers with *sonic anemometers* (see ● Fig. 9.49). These anemometers measure changes in ultrasonic signals that are produced when the wind blows across three sets of transmitters and receivers. The changes in signal speed are then converted into wind speeds in each of the three directions.

The wind-measuring instruments described thus far are "ground-based" and only give wind speed or direction at a particular fixed location. But the wind is influenced by local conditions, such as buildings, trees, and so on. Also, wind speed normally increases rapidly with height above the ground. Thus, wind instruments should be exposed to freely flowing air at a height of at least 10 m (30 ft) above the surface and well above

● **FIGURE 9.47** A wind vane and a cup anemometer.

● **FIGURE 9.49** The sonic anemometer shown here measures wind speed as part of the ASOS system.

● **FIGURE 9.48** The aerovane (skyvane).

the roofs of buildings. In practice, unfortunately, anemometers are placed at various levels; the result, then, is often erratic wind observations. Thus, placement of anemometers is critical to avoid erratic and unrepresentative wind readings.

A simple way to obtain wind data above the surface is with a **pilot balloon**. A small balloon filled with helium is released from the surface. The balloon rises at a known rate, but drifts freely with the wind. It is manually tracked with a small telescope called a *theodolite*. Every minute (or half minute), the balloon's vertical angle (height) and horizontal angle (direction) are measured. The data from the observations are fed into a computer or plotted on a special board, and the wind speed and direction are computed at specific intervals—usually every 300 m (1000 ft)—above the surface.

The pilot balloon principle can be used to obtain wind information during a radiosonde observation. During this type of observation, a balloon rises from the surface carrying a *radiosonde* (an instrument package designed to measure the vertical profile of temperature, pressure, and humidity—see Chapter 1, Focus section 1.3). Equipment located on the ground constantly tracks the balloon, measuring its vertical and horizontal angles, as well as its height above the ground. From this information, a computer determines and prints the vertical profile of wind from the surface up to where the balloon normally pops, typically in the stratosphere near 30 km (19 mi). The observation of winds using a radiosonde balloon is called a *rawinsonde observation*.

Radiosonde data, taken from balloons launched around the world every 12 hours, has been an important tool in tracking upper-level winds for more than 60 years. On special occasions, such as during field studies or when hurricanes are being monitored, an airplane may deploy an instrument package called a *dropsonde*, which falls to the ground via parachute. The resulting data are *dropwindsonde observations*.

Above about 30 km, rockets and radar provide information about the wind flow. One type of rocket ejects an instrument attached to a parachute that drifts with the wind as it slowly falls to earth. While descending, the instrument is tracked by a ground-based radar unit that determines wind information for that region of the atmosphere. Other rockets eject metal strips at some desired level. Again, radar tracks these drifting pieces of chaff, which provide valuable wind speed and direction data for elevations outside the normal radiosonde range.

A device similar to radar called **lidar** (*light detection and ranging*) uses infrared or visible light in the form of a laser beam to determine wind information. Basically, it sends out a narrow beam of light that is reflected from particles, such as smoke or dust, and measures wind velocity by measuring the movement of these particles.

With the aid of Doppler radar, a vertical profile of wind speed and direction up to an altitude of 16 kilometers or so above the ground can be obtained. Such a profile is called a *wind sounding*, and the radar, a **wind profiler** (or simply a *profiler*). Doppler radar, like conventional radar, emits pulses of microwave radiation that are returned (backscattered) from a target, in this case the irregularities in moisture and temperature created by turbulent, twisting eddies that move with the wind. Doppler radar works on the principle that, as these eddies move toward or away from the receiving antenna, the returning radar pulse will change in frequency. The Doppler radar wind profilers are so sensitive that they can translate the backscattered energy from these eddies into a vertical picture of wind speed and direction in a column of air 16 km (10 mi) thick. Wind profilers have been used at NASA's Kennedy Space Center and across the Great Plains.

In remote regions of the world where upper-air observations are lacking, wind speed and direction can be obtained from satellites. As we saw in Chapter 5, geostationary satellites positioned above a particular location show the movement of clouds. The direction of cloud movement indicates wind direction, and the

● **FIGURE 9.50** A satellite image of wind speed and cloud-top temperature associated with Cyclone Ita on April 10, 2014. This intense tropical cyclone struck far northeast Australia (left side of image). The colored background shows the temperature of the cyclone's cloud tops, which is correlated with the intensity of convection. Each wind barb represents 10 m/s (19 knots), with the satellite-observed winds reaching 50 m/s (95 knots) near the center of Ita. Wind speeds were gathered by the European Advanced Scatterometer (ASCAT) aboard the MetOp-A satellite.

horizontal distance the cloud moves during a given time period indicates the wind speed.

In addition, a specialized satellite-borne instrument called a **scatterometer** (a type of radar) can measure surface winds above the open ocean during all kinds of weather by observing the roughness of the sea. From the satellite, the scatterometer sends out a microwave pulse of energy that travels through the clouds, down to the sea surface. A portion of this energy is scattered (bounced) back to the satellite. The amount of energy returning to the scatterometer (called the *echo*) depends on the roughness of the sea—rougher seas have a stronger echo because they scatter back more incoming energy. Since the sea's roughness depends upon the strength of the wind blowing over it, the echo's intensity can be translated into surface wind speed and direction (see ● Fig. 9.50). Surface wind information of this nature can be extremely valuable to the shipping industry, as well as to coastal communities. Hurricanes and other storms over the open ocean can be carefully monitored to see how their winds are changing. And incorporating sea surface wind information into computer forecast models may have the benefit of improving weather forecasts.

SUMMARY

In this chapter, we concentrated on microscale and mesoscale winds. In the beginning of the chapter, we considered both how our environment influences the wind and how the wind influences our environment. We saw that the friction of airflow—viscosity—can be brought about by the random motion of air molecules (molecular viscosity) or by turbulent whirling eddies of air (eddy viscosity). The depth of the atmospheric layer near the surface that is influenced by surface friction (the boundary layer) depends upon atmospheric stability, the wind speed, and the roughness of the terrain. Although it may vary, the top of the boundary layer is usually near 1000 meters or about 3300 feet.

Air blowing past obstructions can produce a number of effects, from gusty winds at a sports stadium to howling winds on a blustery night. Aloft, winds blowing over a mountain range may generate hazardous rotors downwind of the range. And the eddies that form in a region of strong wind shear, especially in the vicinity of a jet stream, can produce extreme turbulence, even in clear air.

Wind blowing over Earth's surface can create a variety of features. In deserts, we see sand dunes and *desert pavement* (a solid surface that forms over time). Over a snow surface, the wind produces snow ripples, snow rollers, and huge drifts. Where high winds blow over a ridge, trees may be sculpted into "flag" trees. In unprotected areas, shelterbelts are planted to protect crops and soil from damaging winds.

We also examined winds on a slightly larger scale. Land and sea breezes are true mesoscale winds that blow in response to local pressure differences created by the uneven heating and cooling rates of land and water. A sea breeze is an onshore wind because it blows from sea to land, whereas a land breeze is an offshore wind because it blows from land to sea. When winds move across a large body of water, they often change in speed and direction. Where the winds change direction seasonally, they are termed *monsoon winds*. Monsoon winds exist in many parts of the world, including North America, Asia, Australia, and Africa.

Local winds that blow uphill during the day are called valley breezes and those that blow downhill at night, mountain breezes. A strong, cold downslope wind is the katabatic (or fall) wind.

A warm, dry wind that descends the eastern side of the Rocky Mountains is the chinook. The same type of wind in the Alps is called a foehn. A warm, dry, usually strong downslope wind that blows into Southern California from the east or northeast is the Santa Ana wind.

Local intense heating of the surface can produce small, rotating circulations, such as the dust devil, while downdrafts from a thunderstorm are responsible for the desert haboob. Some winds, such as those in a blizzard, are snow-bearing, whereas others, such as the sirocco, are dust-bearing.

KEY TERMS

The following terms are listed (with corresponding page numbers) in the order they appear in the text. Define each. Doing so will aid you in reviewing the material covered in this chapter.

scales of motion, 228
microscale, 228
mesoscale, 228
synoptic scale, 228
planetary (global) scale, 228
macroscale, 228
turbulence, 229
viscosity, 229
mechanical turbulence, 229
planetary boundary layer, 230
thermal turbulence, 230
rotors, 231
wind shear, 232
clear air turbulence (CAT), 232
snow rollers, 234
shelterbelts (windbreaks), 235
wind waves, 236
seiches, 237
thermal circulations, 239
thermal highs, 239
thermal lows, 239
sea breeze, 239
lake breeze, 239

land breeze, 239
sea-breeze front, 240
valley breeze, 241
mountain breeze, 241
katabatic (or fall) wind, 242
chinook wind, 243
foehn wind, 243
Santa Ana wind, 245
haboob, 247
dust devil (whirlwind), 247
simoom, 249
monsoon, 249
monsoon wind system, 249
onshore wind, 252
offshore wind, 252
prevailing wind, 252
wind rose, 254
wind vane, 254
anemometer, 254
aerovane (skyvane), 255
pilot balloon, 256
lidar, 257
wind profiler, 257
scatterometer, 257

QUESTIONS FOR REVIEW

1. Describe the various scales of motion, and give an example of each.
2. How does Earth's surface influence the flow of air above it?
3. What causes wind gusts?
4. How does mechanical turbulence differ from thermal turbulence?
5. Why are winds near the surface typically stronger and more gusty in the afternoon?
6. Describe several ways in which an eddy might form.
7. A friend has just returned from a transatlantic flight and reported that the plane dropped about 1000 m when it entered an "air pocket." Explain to your friend what apparently happened to cause this drop.

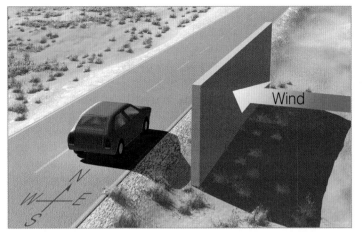

● **FIGURE 9.51**

8. Explain why the car in the diagram in ● Fig. 9.51 may experience a west wind as it travels past the wall.

9. What is wind shear and how does it relate to clear air turbulence?

10. Explain how shelterbelts protect sensitive crops from wind damage.

11. With the same wind speed, explain why a camper is more easily moved by the wind than a car.

12. What are the necessary conditions for the development of large wind waves?

13. How can a coastal area have large waves on a clear, non-stormy day?

14. Why do winds usually change direction and speed when moving over a large body of water?

15. Using a diagram, explain how a thermal circulation develops.

16. Which wind will most likely produce clouds: a valley breeze or a mountain breeze? Why?

17. Explain why chinook winds are warm and dry.

18. Name some of the benefits of a chinook wind.

19. What atmospheric conditions contribute to the development of a strong Santa Ana condition? Why is a Santa Ana wind warm?

20. How do strong katabatic winds form?

21. Why are haboobs more prevalent in Arizona than in Oklahoma?

22. Describe how dust devils usually form.

23. Discuss the factors that contribute to the formation of the summer monsoon and the winter monsoon in India.

24. An upper wind direction is reported as 315°. From what compass direction is the wind blowing?

25. List as many ways as you can of determining the wind direction and the wind speed.

26. Name and describe three instruments used to measure wind speed and direction.

27. If you are standing directly south of a smokestack and the wind from the stack is blowing over your head, what would be the wind direction?

28. How does a wind profiler obtain winds?

29. Explain how satellites are able to estimate surface winds over the ocean.

30. In what part of the world would you expect to encounter each of the following winds, and *what type of weather would each wind bring*?
    (a) foehn
    (b) California norther
    (c) Santa Ana
    (d) zonda
    (e) chinook
    (f) Columbia Gorge wind
    (g) sirocco
    (h) mistral

QUESTIONS FOR THOUGHT

1. A pilot enters the weather service office and wants to know what time of the day she can expect to encounter the least turbulent winds at 760 m (2500 ft) above central Kansas a month from now. If you were the meteorologist on hand, what would you tell her?

2. Why is it dangerous during hang gliding to enter the downwind side of the hill when the wind speed is strong?

3. After a winter snowstorm, an observer in downtown Cheyenne, Wyoming, reports a total snow accumulation of 48 cm (19 in.), while the maximum depth in the surrounding countryside is only 28 cm (11 in.). If the storm's intensity and duration were practically the same for a radius of 50 km around Cheyenne, explain why Cheyenne received so much more snow.

4. Why is the difference in surface wind speed between morning and afternoon typically greater on a clear, sunny day than on a cloudy, overcast day?

5. Might it be possible to have a city/suburb breeze? If so, would you expect it to be more prominent during the day or night? Describe how it would form. Use a diagram to help you.

6. Average annual wind speed information in knots is given at the bottom of this page for two cities located on the Great Plains. Which city would probably be the best site for a wind turbine? Why?

7. Which of the sites in ● Fig. 9.52 would probably be the best place to construct a wind turbine? A, B, or C? Which would be the worst? Explain.

● FIGURE 9.52

8. Explain why cities near large bodies of cold water in summer experience well-developed sea breezes but only poorly developed land breezes.
9. Why do clouds tend to form over land with a sea breeze and over water with a land breeze?

10. The convergence of two sea breezes in Florida frequently produces rain showers; the convergence of two sea breezes in California does not. Explain.
11. Why don't chinook winds form on the east side of the Appalachians?
12. Show, with the aid of a diagram, what atmospheric and topographic conditions are necessary for an area in the Northern Hemisphere to experience *hot* summer breezes from the *north*.
13. The prevailing winds in southern Florida are northeasterly. Knowing this, would you expect the strongest sea breezes to be along the east or west coast of southern Florida? What about the strongest land breezes?

PROBLEMS AND EXERCISES

1. A model city is to be constructed in the middle of an uninhabited region. The wind rose seen here (● Fig. 9.53) shows the annual frequency of wind directions for this

TIME

	MIDNIGHT	3	6	9	NOON	3	6	9	AVERAGE ANNUAL WIND SPEED (KNOTS)
City A:	12	7	8	13	15	18	14	13	12.5
City B:	8	6	6	13	20	22	15	10	12.5

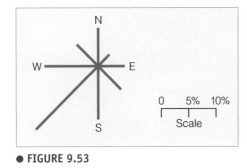
● **FIGURE 9.53**

region. With the aid of the wind rose, on a square piece of paper determine where the following should be located:
(a) industry
(b) parks
(c) schools
(d) shopping centers
(e) sewage disposal plants
(f) housing development
(g) an airport with two runways

2. What would be the total force exerted on a camper 15 ft long and 8 ft high, if a wind of 40 mi/hr blows perpendicular to one of its sides?
3. On a map of the United States show where each of the winds listed below would be observed. Then determine the direction of prevailing wind flow for each wind.
(a) Santa Ana wind
(b) chinook wind
(c) California norther
(d) monsoonal winds in summer over the Desert Southwest
(e) Columbia Gorge wind (downslope)
(f) sea breeze along the New Jersey shore
(g) breeze in Los Angeles, California

ONLINE RESOURCES

Visit www.cengagebrain.com to view additional resources, including video exercises, practice quizzes, an interactive eBook, and more.

Vast parts of the western United States, including this landscape in Utah, are dominated by circulation patterns that lead to dry conditions.

ONE SUNNY AUGUST MORNING, THREE VESSELS SET SAIL from Palos, Spain, on an established course. After halting at the Canary Islands (about 30°N) for a few days, the vessels sailed due west. For most of their Atlantic journey, steady winds from the northeast filled their sails and blew them along at a hundred miles a day. They were quite fortunate. The steady northeast winds had edged unusually far north this year, and only for about 10 days or so did the vessels encounter the light, variable winds more typical of this region—a notorious region where, years later, ships were frequently becalmed under a blistering hot sun. And on October 12, 1492, the flotilla landed on a small island in the Bahamas. Christopher Columbus had not only found a route to the New World, he had discovered the trade winds—the steady northeast winds that are an integral part of the world's global wind system.

In Chapter 9, we learned that local winds vary considerably from day to day and from season to season. As you may suspect, these winds are part of a much larger circulation—the little whirls within larger whirls that we spoke of before. Indeed, if the rotating high- and low-pressure areas in our atmosphere we see on a weather map are like spinning eddies in a huge river, then the flow of air around the globe is like the meandering river itself. When winds throughout the world are averaged over a long period of time, the local wind patterns vanish, and what we see is a picture of the winds on a global scale—what is commonly called the **general circulation of the atmosphere**. Just as the eddies in a river are carried along by the overall flow of water, so the highs and lows in the atmosphere are swept along by the general circulation. We will examine this large-scale circulation of air, its effects and its features, in this chapter.

General Circulation of the Atmosphere

Before we study the general circulation, we must remember that it only represents the *average* airflow around the world. Actual winds at any one place and at any given time may vary considerably from this average. Nevertheless, the average can answer why and how the winds blow around the world the way they do— why, for example, prevailing surface winds are northeasterly in Honolulu, Hawaii, and westerly in New York City. The average can also give a picture of the mechanism driving these winds, as well as a model of how heat and momentum are transported from equatorial regions poleward, keeping the climate in middle latitudes tolerable.

The underlying cause of the general circulation is the unequal heating of Earth's surface. We learned in Chapter 2 that, averaged over the entire Earth, incoming solar radiation is roughly equal to outgoing Earth radiation. However, we also know that this energy balance is not maintained for each latitude, since the tropics experience a net gain in energy, while polar regions suffer a net loss. To balance these inequities, the atmosphere transports warm air poleward and cool air equatorward. Although seemingly simple, the actual flow of air is complex; certainly not everything is known about it. To better understand it, we will first look at some models (that is, artificially constructed simulations) that eliminate some of the complexities of the general circulation.

SINGLE-CELL MODEL   The first model is the single-cell model, in which we assume that:

1. Earth's surface is uniformly covered with water (so that differential heating between land and water does not come into play).
2. The sun is always directly over the equator (so that the winds will not shift seasonally).
3. Earth does not rotate (so that the only force we need to deal with is the pressure-gradient force).

With these assumptions, the general circulation of the atmosphere on the side of Earth facing the sun would look much like the representation in ● Fig. 10.1a, a huge thermally driven convection cell in each hemisphere. (For reference, the names of the different regions of the world and their approximate latitudes are given in Fig. 10.1b.)

The circulation of air described in Fig. 10.1a is the **Hadley cell** (named after the eighteenth-century English meteorologist

● **FIGURE 10.1** Diagram (a) shows the general circulation of air on the side of Earth facing the sun on a nonrotating Earth uniformly covered with water and with the sun directly above the equator. (Vertical air motions are highly exaggerated in the vertical.) Diagram (b) shows the names that apply to the different regions of the world and their approximate latitudes.

George Hadley, who first proposed the idea). It is referred to as a *thermally direct cell* because it is driven by energy from the sun as warm air rises and cold air sinks. Excessive heating of the equatorial area produces a broad region of surface low pressure, while at the poles excessive cooling creates a region of surface high pressure. In response to the horizontal pressure gradient, cold surface polar air flows equatorward, while at higher levels air flows toward the poles. The entire circulation consists of a closed loop with rising air near the equator, sinking air over the poles, an equatorward flow of air near the surface, and a return flow aloft. In this manner, some of the excess energy of the tropics is transported as sensible and latent heat to the regions of energy deficit at the poles.*

Such a simple cellular circulation as this does not actually exist on Earth. For one thing, Earth rotates, so the Coriolis force would deflect the southward-moving surface air in the Northern Hemisphere to the right, producing easterly surface winds at practically all latitudes north of the equator. We know that this does not happen and that prevailing winds in middle latitudes actually blow from the west. Therefore, observations alone tell us that a closed circulation of air between the equator and the poles is not an accurate model for a rotating Earth. But this model does show us how a nonrotating planet would balance an excess of energy at the equator and a deficit at the poles. How, then, does the wind blow on a rotating planet? To answer, we will keep our model simple by retaining our first two assumptions—that is, that Earth is covered with water and that the sun is always directly above the equator.

*Additional information on thermal circulations is found on p. 239 in Chapter 9.

THREE-CELL MODEL If we allow Earth to spin, the simple convection system breaks into a series of cells as shown in ● Fig. 10.2. Although this model is considerably more complex than the single-cell model, there are some similarities. The tropical regions still receive an excess of heat and the poles, a deficit. In each hemisphere, three cells instead of one have the task of energy redistribution. A surface high-pressure area is located at the poles, and a broad trough of surface low pressure still exists at the equator. From the equator to latitude 30°, the circulation is the *Hadley cell*. Let's look at this model more closely by examining what happens to the air above the equator. (Refer to Fig. 10.2 as you read the following section.)

Over equatorial waters, the air is warm, horizontal pressure gradients are weak, and winds are light. This region is referred to as the **doldrums**. (The monotony of the weather in this area has given rise to the expression "down in the doldrums.") Here, warm, humid air rises, often condensing into huge cumulus clouds and thunderstorms called *convective "hot" towers* because of the enormous amount of latent heat they liberate. This heat makes the air more buoyant and provides energy to drive the Hadley cell. The rising air reaches the tropopause, which acts like a barrier, causing the air to move laterally toward the poles. The Coriolis force deflects this poleward flow toward the right in the Northern Hemisphere and to the left in the Southern Hemisphere, providing westerly winds aloft in both hemispheres. (We will see later that these westerly winds reach maximum velocity and produce jet streams near latitudes 30° and 60°.)

As air moves poleward from the tropics, it constantly cools by giving up infrared radiation, and at the same time it also begins to

● **FIGURE 10.2** The idealized wind and surface-pressure distribution over a uniformly water-covered rotating Earth.

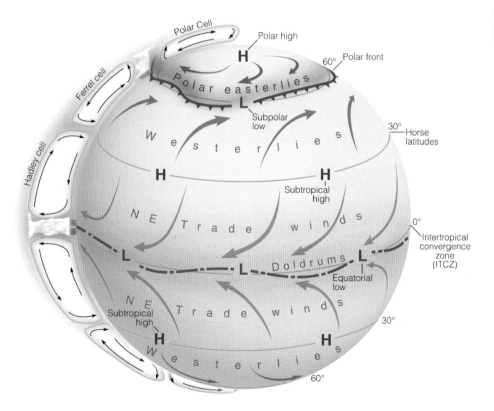

converge, especially as it approaches the middle latitudes.* This convergence (piling up) of air aloft increases the mass of air above the surface, which in turn causes the air pressure at the surface to increase. Hence, at latitudes near 30°, the convergence of air aloft produces belts of high pressure called **subtropical highs** (or anticyclones). As the converging, relatively dry air above the highs slowly descends, it warms by compression. This subsiding air produces generally clear skies and warm surface temperatures; hence, it is here that we find the major deserts of the world, such as the Sahara of Africa and the Sonoran of North America (see ● Fig. 10.3).

Over the ocean, the weak pressure gradients in the center of the high produce only weak winds. According to legend, sailing ships traveling to the New World were frequently becalmed in this region, and, as food and supplies dwindled, horses were either thrown overboard or eaten. Because of this, this region is sometimes called the **horse latitudes**.

From the horse latitudes, near latitude 30°, some of the surface air moves back toward the equator. It does not flow straight back, however, because the Coriolis force deflects the air, causing it to blow from the northeast in the Northern Hemisphere and from the southeast in the Southern Hemisphere. These steady winds provided sailing ships with an ocean route to the New World; hence, these winds are called the **trade winds**. Near the equator, the *northeast trades* converge with the *southeast trades* along a boundary called the **intertropical convergence zone (ITCZ)**. In this region of surface convergence, air rises and continues its cellular journey. Along the ITCZ, it is usually very wet as the rising air develops into huge thunderstorms that drop copious amounts of rain in the form of heavy showers (see ● Fig. 10.4).

---

*You can see why the air converges if you have a globe of the world. Put your fingers on meridian lines at the equator and then follow the meridians poleward. Notice how the lines and your fingers bunch together in the middle latitudes.

Meanwhile, at latitude 30°, not all of the surface air moves equatorward. Some air moves toward the poles and deflects toward the east, resulting in a more or less westerly airflow—called the *prevailing westerlies*, or, simply, **westerlies**—in both hemispheres. Consequently, from Texas northward into Canada, it is much more common to experience winds blowing out of the west than from the east. The westerly flow in the real world is not constant, as migrating areas of high and low pressure break up the surface flow pattern from time to time. In the middle latitudes of the Southern Hemisphere, where the surface is mostly water, winds blow more steadily from the west.

As this mild surface air travels poleward from latitude 30°, it encounters cold air moving down from the poles. These two air masses of contrasting temperature do not readily mix. They are separated by a boundary called the **polar front**, a zone of low pressure—the **subpolar low**—where surface air converges and rises, and storms and clouds develop. In our model in Fig. 10.2, some of the rising air returns at high levels to the horse latitudes, where it sinks back to the surface in the vicinity of the subtropical high. The middle cell (a *thermally indirect cell*, in which cool air rises and warm air sinks, called the *Ferrel cell*, after the American meteorologist William Ferrel) is completed when surface air from the horse latitudes flows poleward toward the polar front.

Notice in Fig. 10.2 that, in the Northern Hemisphere, behind the polar front, the cold air from the poles is deflected by the Coriolis force, so that the general flow of air is from the northeast. Hence, this is the region of the **polar easterlies**. In winter, the polar front with its cold air can move into middle and subtropical latitudes, producing a cold polar outbreak. Along the front, a portion of the rising air moves poleward, and the Coriolis force deflects the air into a westerly wind at high levels. Air aloft eventually reaches the poles, slowly sinks to the surface, and flows back toward the polar front, completing the weak *polar cell*.

● **FIGURE 10.3** Subtropical deserts, such as the Sonoran Desert shown here, are mainly the result of sinking air associated with subtropical high-pressure areas.

Hank Shiffman/Shutterstock.com

● **FIGURE 10.4** The solid red line in this visible satellite image marks the position of the ITCZ in the eastern Pacific. The bright white clouds are huge thunderstorms forming along the ITCZ.

We can summarize all of this by referring back to Fig. 10.2, p. 265, and noting that, at the surface, there are two major areas of high pressure and two major areas of low pressure. Areas of high pressure exist near latitude 30° and the poles; areas of low pressure exist over the equator and near 60° latitude in the vicinity of the polar front. Knowing the way surface winds blow around these pressure systems on the three-cell model gives us a generalized picture of how surface winds blow throughout the world. The trade winds extend from the subtropical high to the equator, the westerlies from the subtropical high to the polar front, and the polar easterlies from the poles to the polar front.

How does this three-cell model compare with actual observations of winds and pressure in the real world? We know, for example, that upper-level winds at middle latitudes generally blow from the west. The middle Ferrel cell in our model, however, suggests an east wind aloft as air flows equatorward, so we know that discrepancies exist between this model and atmospheric observations. The model does, however, agree closely with the winds and pressure distribution at the *surface*, and so we will examine this next.

AVERAGE SURFACE WINDS AND PRESSURE: THE REAL WORLD When we examine the real world with its continents and oceans, mountains and ice fields, we can obtain an average distribution of sea-level pressure and winds for January and July, as shown in ● Figs. 10.5a and 10.5b. Look closely at both maps and observe that there are regions where pressure systems appear to persist throughout the year. These systems are referred to as **semipermanent highs and lows** because they move only slightly during the course of a year.

In Fig. 10.5a, there are four semipermanent pressure systems in the Northern Hemisphere during January. In the eastern Atlantic, between latitudes 25° and 35°N is the *Bermuda–Azores high*, often called the **Bermuda high**, and, in the Pacific Ocean, its counterpart, the **Pacific high**. These are the subtropical

anticyclones that develop in response to the convergence of air aloft near an upper-level jet stream. Given that surface winds blow clockwise around these systems, we find the trade winds to the south and the prevailing westerlies to the north. In the Southern Hemisphere, where there is relatively less land area, there is less contrast between land and water, and the subtropical highs show up as well-developed systems with a clearly defined circulation.

Where we would expect to observe the polar front (between latitudes 40° and 65°), there are two semipermanent subpolar lows. In the North Atlantic, there is the *Greenland-Icelandic low*, or simply **Icelandic low**, which covers Iceland and southern Greenland, while the **Aleutian low** sits over the Gulf of Alaska and Bering Sea near the Aleutian Islands in the North Pacific. These zones of cyclonic activity actually represent regions where numerous storms, having traveled eastward, tend to converge, especially in winter. In the Southern Hemisphere, where there is very little land to disrupt the flow, the subpolar low forms a continuous trough that completely encircles the globe.

The January map (Fig. 10.5a) shows other pressure systems that are not semipermanent in nature but are still observed often. Over Asia, for example, there is a huge (but shallow) thermal anticyclone called the **Siberian high**, which forms because of the intense cooling of the land. South of this system, the winter monsoon shows up clearly, as air flows away from the high across Asia and out over the ocean. A similar (but less intense) anticyclone (called the *Canadian high*) is evident over North America.

As summer approaches, the land warms and the cold shallow highs disappear. In some regions, areas of surface low pressure replace areas of high pressure. The lows that form over the warm land are shallow *thermal lows*. On the July map (Fig. 10.5b), warm thermal lows are found over the Desert Southwest of the United States and over the plateau of Iran. Notice that these systems are located at the same latitudes as the subtropical highs. We can

(a) January

(b) July

● **FIGURE 10.5** Average sea-level pressure distribution and surface wind flow patterns for January (a) and for July (b). The solid red line represents the position of the ITCZ.

understand why they form when we realize that, during the summer, the subtropical high-pressure belt girdles the world *aloft* near 30° latitude.* Within this system, the air sinks and warms, producing clear skies (which allow intense surface heating by the sun). This air near the ground warms rapidly, rises only slightly, then flows laterally several hundred meters above the surface. The outflow lowers the surface pressure and, as we saw in Chapter 9, a shallow, thermal low forms. The thermal low over India, also called the *monsoon low*, develops when the continent of Asia warms. As the low intensifies, warm, moist air from the ocean is drawn into it, producing the wet summer monsoon so characteristic of India and Southeast Asia. Where these surface winds converge with the general westerly flow, rather weak monsoon depressions form. These enhance the position of the monsoon low on the July map.

When we compare the January and July maps (Figs. 10.5a and 10.5b), we can see several changes in the semipermanent pressure systems. The strong subpolar lows so well developed in January over the Northern Hemisphere are hardly discernible on the July map. The subtropical highs, however, remain dominant in both seasons. Because the sun is overhead in the Northern Hemisphere in July and overhead in the Southern Hemisphere in January, the zone of maximum surface heating shifts seasonally. In response to this shift, the major pressure systems, wind belts, and ITCZ (heavy red line in Fig. 10.5) *shift toward the north in July and toward the south in January.***

● Figure 10.6 illustrates a winter weather map where the main features of the general circulation have been displaced southward.

THE GENERAL CIRCULATION AND PRECIPITATION PATTERNS

The position of the major features of the general circulation and their latitudinal displacement (which annually averages about 10° to 15°) strongly influence the precipitation of many areas. For example, on the global scale, we would expect abundant rainfall where the air rises and very little where the air sinks. Thus, areas of high rainfall exist in the tropics, where humid air rises in conjunction with the ITCZ, and between about 40° and 55° latitude, where middle-latitude cyclonic storms and the polar front force air upward. Areas of low precipitation are found near 30° latitude in the vicinity of the subtropical highs and in polar regions where the air is cold and dry (see ● Fig. 10.7).

Poleward of the equator, between the doldrums and the horse latitudes, the area is influenced by both the ITCZ and the subtropical high. In summer (high sun period), the subtropical high moves poleward and the ITCZ invades this area, bringing with it ample rainfall. In winter (low sun period), the subtropical high moves equatorward, bringing with it clear, dry weather.

During the summer, the Pacific high drifts northward to a position off the California coast (see ● Fig. 10.8). Sinking air on its eastern side produces a strong upper-level subsidence inversion,

which tends to keep summer weather along the West Coast relatively dry. The rainy season typically occurs in winter when the high moves south and the polar front and storms are able to penetrate the region. Observe in Fig. 10.8 that along the East Coast, the clockwise circulation of winds around the Bermuda high brings warm tropical air northward into the United States and southern Canada from the Gulf of Mexico and the Atlantic Ocean. Because sinking air is not as well developed on this side of the high, the humid air can rise and condense into towering cumulus clouds and thunderstorms. In part, then, it is the air motions associated

● **FIGURE 10.6** A winter weather map depicting the main features of the general circulation over North America. Notice that the Canadian high, polar front, and subpolar lows have all moved southward into the United States, and that the prevailing westerlies exist south of the polar front. The arrows on the map illustrate wind direction.

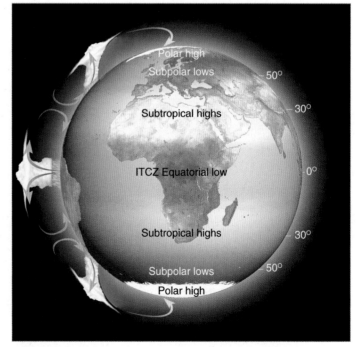

● **FIGURE 10.7** Rising and sinking air associated with the major pressure systems of Earth's general circulation. Where the air rises, precipitation tends to be abundant (blue shade); where the air sinks, drier regions prevail (tan shade). Note that the sinking air of the subtropical highs produces the major desert regions of the world.

---

*This belt of high pressure aloft shows up well in Fig. 10.10b, p. 271, the average 500-mb map for July.

**An easy way to remember the seasonal shift of pressure systems is to think of birds—in the Northern Hemisphere, they migrate south in the winter and north in the summer.

with the subtropical highs that keep summer weather dry in California and moist in Georgia. (Compare the patterns for Los Angeles, California, and Atlanta, Georgia, in ● Fig. 10.9.)

AVERAGE WIND FLOW AND PRESSURE PATTERNS ALOFT

● Figures 10.10a and 10.10b are average global 500-mb charts for the months of January and July, respectively. Look at both charts carefully and observe that some of the surface features of the general circulation are reflected on these upper-air charts. On the January map, for example, both the Icelandic low and Aleutian low are located to the west of their surface counterparts. On the July map, the subtropical high-pressure areas of the Northern Hemisphere appear as belts of high height (high pressure) that tend to circle the globe south of 30°N. In both hemispheres, the air is warmer over low latitudes and colder over high latitudes. This horizontal temperature gradient establishes a horizontal pressure (contour) gradient that causes the winds to blow from the west, especially in middle and high latitudes.*

---

*Remember that, at this level (about 5600 m or 18,000 ft above sea level), the winds are approximately geostrophic, and tend to blow more or less parallel to the contour lines.

Notice that the temperature gradients and the contour gradients across the Northern Hemisphere are steeper in January than in July. Consequently, the winds aloft are stronger in winter than in summer. The westerly winds, however, do not extend all the way to the equator, as easterly winds appear on the equatorward side of the upper-level subtropical highs.

In middle and high latitudes, the westerly winds continue to increase in speed above the 500-mb level. We already know that the wind speed increases up through the friction layer, but why should it continue to increase at higher levels? You may remember from Chapter 8 that the geostrophic wind at any latitude is directly related to the pressure gradient and inversely related to the air density. Therefore, a greater pressure gradient will result in stronger

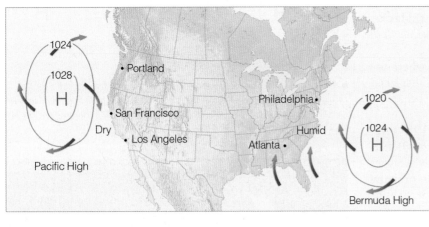

● **FIGURE 10.8** During the summer, the Pacific high moves northward. Sinking air along its eastern margin (over California) produces a strong subsidence inversion, which causes relatively dry weather to prevail. Along the western margin of the Bermuda high, southerly winds bring in humid air, which rises, condenses, and produces abundant rainfall.

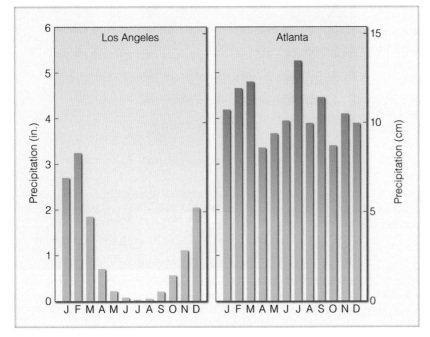

● **FIGURE 10.9** Average annual precipitation for Los Angeles, California (Los Angeles International Airport), and Atlanta, Georgia (Hartsfield-Jackson International Airport), for the climatological period 1981–2010.

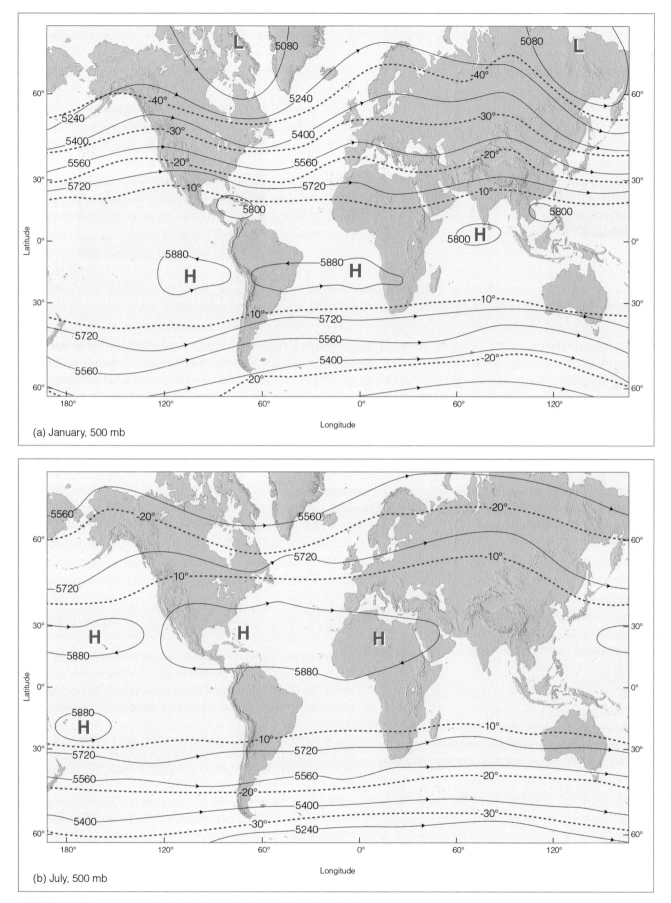

(a) January, 500 mb

(b) July, 500 mb

● **FIGURE 10.10** Average 500-mb chart for the month of January (a) and for July (b). Solid lines are contour lines in meters above sea level. Dashed red lines are isotherms in °C. Arrowheads illustrate wind direction.

The "Dishpan" Experiment

● **FIGURE 1** (a) A "dishpan" with a hot "equator" and a cold "pole" rotating at a speed corresponding to that of Earth (b) produces troughs, ridges, and eddies, which appear (when viewed from above) very similar to the patterns we see on an upper-level chart.

● **FIGURE 2** The circulation of the air aloft is in the form of waves—troughs and ridges—that encircle the globe.

We know that the primary cause of the atmosphere's general circulation is the unequal heating that occurs between tropical and polar regions. A laboratory demonstration that tries to replicate this situation is the *"dishpan"* experiment.

The dishpan experiment involves a flat, circular pan, filled with water several centimeters deep. The pan is positioned on a rotating table (see ● Fig. 1a). Around the edge of the pan, a heating coil supplies heat to the pan's "equator." In the center of the pan, a cooling cylinder represents the "pole," and ice water is continually supplied here. When the pan is rotated counterclockwise, the temperature difference between "equator" and "pole" produces a thermally driven circulation that transports heat poleward.

Aluminum powder (or dye) is added to the water so that the motions of the fluid can be seen. If the pan rotates at a speed that corresponds to the rotation of Earth, the flow develops into a series of waves and rotating eddies similar to those shown in Fig. 1b. The atmospheric counterpart of these eddies are the cyclones and anticyclones of the middle latitudes. At Earth's surface, they occur as winds circulating around centers of low and high pressure. Aloft, the waves appear as a series of troughs and ridges that encircle the globe and slowly migrate from west to east (see ● Fig. 2). The waves represent a fundamental feature of the atmosphere's circulation. Later in this chapter we will see how they transfer momentum, allowing the atmosphere to maintain its circulation. In Chapter 12, we will see that these waves are instrumental in the development of surface mid-latitude cyclonic storms. The center of the dishpan corresponds to the *polar vortex*, the zone of upper-level low pressure found in the polar regions that is sometimes disrupted or displaced during winter.

winds, and so will a decrease in air density. Owing to the fact that air density decreases with height, the same pressure gradient will produce stronger winds at higher levels. In addition, the north-to-south temperature gradient causes the horizontal pressure (contour) gradient to increase with height up to the tropopause. As a result, the winds increase in speed up to the tropopause (about 11 km or 33,000 ft above sea level). Above the tropopause, the temperature gradients reverse. This changes the pressure gradients and reduces the strength of the westerly winds. Where strong winds tend to concentrate into narrow bands at the tropopause, we find rivers of fast-flowing air—*jet streams*. (In the following section, you will read about a wavy jet stream. Focus section 10.1 describes an experiment that illustrates how these waves form.)

Jet Streams

Atmospheric **jet streams** are swiftly flowing air currents thousands of kilometers long, a few hundred kilometers wide, and only a few kilometers thick. Wind speeds in the central core of a jet stream often exceed 100 knots and occasionally exceed 200 knots. Jet streams are usually found at the tropopause at elevations between 10 and 15 km (6 and 9 mi), although they may occur at both higher and lower altitudes.

Jet streams were first encountered by high-flying military aircraft during World War II, but their existence was suspected before 1940. Ground-based observations of fast-moving cirrus clouds had revealed that westerly winds aloft must be moving rapidly.

● Figure 10.11 illustrates the average position of two jet streams, the tropopause, and the general circulation of air for the Northern Hemisphere in winter. Both jet streams are located at tropopause gaps, where mixing between tropospheric and stratospheric air takes place. The jet stream situated near 30° latitude at about 13 km (43,000 ft) above the subtropical high is the **subtropical jet stream**.* To the north, the jet stream situated at a lower altitude of about 10 km (33,000 ft), near the polar front, is known as the **polar front jet stream** or, simply, the *polar jet stream*. Since both are

---

*The subtropical jet stream is normally found between 20° and 30° latitude.

found at the tropopause, they are referred to as *tropopause jets*.

In Fig. 10.11, the wind in the center of the jet stream would be flowing as a westerly wind away from the viewer. This direction, of course, is only an average, as jet streams often flow in a wavy west-to-east pattern. When the polar jet stream flows in broad loops that sweep north and south, it may even merge with the subtropical jet. Occasionally, the polar jet splits into two jet streams. The jet stream to the north is often called the *northern branch* of the polar jet, whereas the one to the south is called the *southern branch*. ● Figure 10.12 illustrates how the polar jet stream and the subtropical jet stream might appear as they sweep around Earth in winter.

We can better see the looping pattern of the jet by studying ● Fig. 10.13a, which shows the position of the polar jet stream

and the subtropical jet stream at the 300-mb level (near 9 km or 30,000 ft) on March 9, 2005. The fastest flowing air, or *jet core*, is represented by the heavy dark arrows. The map shows a strong polar jet stream sweeping south over the Great Plains with an equally strong subtropical jet over the Gulf states. Notice that the polar jet has a number of loops, with one off the west coast of North America and another over eastern Canada. Observe in the satellite image (Fig. 10.13b) that the polar jet stream (blue arrows) is directing cold, polar air into the Plains states, while the subtropical jet stream (orange arrow) is sweeping subtropical moisture, in the form of a dense cloud cover, over the southeastern states.

The looping (meridional) pattern of the polar jet stream has an important function. In the Northern Hemisphere, where the air flows southward, swiftly moving air directs cold air

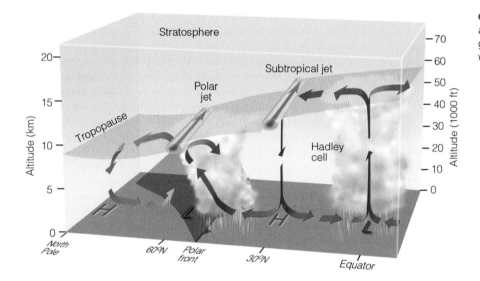

● **FIGURE 10.11** Average position of the polar jet stream and the subtropical jet stream, with respect to a model of the general circulation in winter. Both jet streams are flowing from west to east.

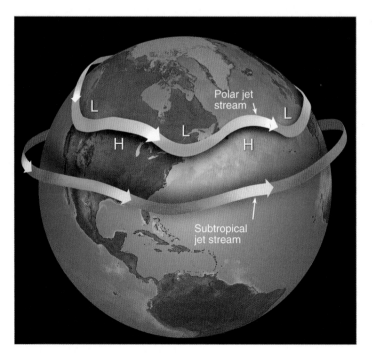

● **FIGURE 10.12** A jet stream is a swiftly flowing current of air that moves in a wavy west-to-east direction. The figure shows the position of the polar jet stream and subtropical jet stream in winter. Although jet streams are shown as one continuous river of air, in reality they are discontinuous, with their position varying from one day to the next.

(a)   (b)

● **FIGURE 10.13** (a) Position of the polar jet stream (blue arrows) and the subtropical jet stream (orange arrows) at the 300-mb level (about 9 km or 30,000 ft above sea level) on March 9, 2005. Solid lines are lines of equal wind speed (isotachs) in knots. (b) Satellite image showing clouds and positions of the jet streams for the same day.

equatorward; where the air flows northward, warm air is carried toward the poles. Jet streams, therefore, play a major role in the global transfer of heat. Moreover, since jet streams tend to meander around the world, we can easily understand how pollutants or volcanic ash injected into the atmosphere in one part of the globe could eventually settle to the ground many thousands of kilometers downwind. And, as we will see in Chapter 12, the looping nature of the polar jet stream has an important role in the development of mid-latitude cyclonic storms.

THE FORMATION OF JET STREAMS   The ultimate cause of jet streams is the energy imbalance that exists between high and low latitudes. How, then, do jet streams actually form?

The Polar Front Jet   Horizontal variations in temperature and pressure offer clues to the existence of the polar jet stream. ● Figure 10.14a is a 3-D model that shows a side view of the atmosphere in the region of the polar front. Because the polar front is a boundary separating cold polar air to the north from warm subtropical air to the south, the greatest contrast in air temperature occurs along the frontal zone. We can see this contrast as the −20°C isotherm dips sharply crossing the front. This rapid change in temperature produces a rapid change in pressure (as shown by the sharp bending of the constant pressure [isobaric] 500-mb surface as it passes through the front). *The sudden change in pressure along the front sets up a steep pressure (contour) gradient that intensifies the wind speed and causes the jet stream.*

Observe in Fig. 10.14a and on the 500-mb chart (Fig. 10.14b) that the wind is blowing along the front (from the west), parallel to the contour lines, with cold air on its left side.* The north-south temperature contrast along the polar front is strongest in

(a) 3-D view

(b) 500-mb chart

● **FIGURE 10.14** Diagram (a) is a model that shows a vertical 3-D view of the polar front in association with a sharply dipping 500-mb pressure surface, an isotherm (dashed line), and the position of the polar front jet stream in winter. The diagram is highly exaggerated in the vertical. Diagram (b) represents a 500-mb chart that cuts through the polar front as illustrated by the dipping 500-mb surface in (a). Sharp temperature contrasts along the front produce tightly packed contour lines and strong winds (contour lines are in meters above sea level).

CRITICAL THINKING QUESTION   If the air just to the south of the front warms, and the air just behind the front becomes much colder, how would these temperature changes influence the 500-mb surface and the winds of the jet stream?

---

*Recall from Chapter 8 that any horizontal change in temperature causes the isobaric surfaces to dip or slant. The greater the temperature difference, the greater the slanting, and the stronger the winds. The changing wind speed with height due to horizontal temperature variations is referred to as the *thermal wind*. The thermal wind always blows with cold air on its left side in the Northern Hemisphere and on its right side in the Southern Hemisphere.

winter and weakest in summer. This situation explains why the polar-front jet shows seasonal variations. In winter, the winds are stronger and the jet moves farther south as the leading edge of the cold air can extend into subtropical regions, as far south as Florida and Mexico. In summer, the jet is weaker and is usually found over more northerly latitudes.

The Subtropical Jet The subtropical jet stream, which is usually strongest slightly above the 200-mb level (above 12 km), tends to form along the poleward side of the Hadley cell as shown in Fig. 10.11, p. 273. Here, warm air carried poleward by the Hadley cell produces sharp temperature contrasts along a boundary sometimes called the *subtropical front*. In the vicinity of the subtropical front (which does not have a frontal structure extending to the surface), sharp contrasts in temperature produce sharp contrasts in pressure and strong winds.

When we examine jet streams carefully, we see that another mechanism (other than a steep temperature gradient) causes a strong westerly flow aloft. The cause appears to be the same as that which makes an ice skater spin faster when the arms are pulled in close to the body—the *conservation of angular momentum*.

Jet Streams and Momentum At the equator, Earth rotates toward the east at a speed close to 1000 knots. On a windless day, the air above moves eastward at the same speed. If somehow Earth should suddenly stop rotating, the air above would continue to move eastward until friction with the surface brought it to a halt; the air keeps moving because it has momentum.

Straight-line momentum—called *linear momentum*—is the product of the mass of the object times its velocity. An increase in either the mass or the velocity (or both) produces an increase in momentum. Air on a spinning planet moves about an axis in a circular path and has angular momentum. Along with the mass and the speed, angular momentum depends on the distance *(r)* between the mass of air and the axis about which it rotates. *Angular momentum* is defined as the product of the mass *(m)* times the velocity *(v)* times the radial distance *(r)*. Thus

$$\text{angular momentum} = mvr$$

As long as there are no external twisting forces (torques) acting on the rotating system, the angular momentum of the system does not change. We say that angular momentum is *conserved*; that is, the product of the quantity *mvr* at one time will equal the numerical quantity *mvr* at some later time. Hence, a decrease in radius must produce an increase in speed and vice versa. An ice skater, for instance, with arms fully extended rotates quite slowly. As the arms are drawn in close to the body, the radius of the circular path *(r)* decreases, which causes an increase in rotational velocity *(v)*, and the skater spins faster. As arms become fully extended again, the skater's speed decreases. The conservation of angular momentum, when applied to moving air, will help us to understand the formation of fast-flowing air aloft.

Consider heated air parcels rising from the equatorial surface on a calm day. As the parcels approach the tropopause, they spread laterally and begin to move poleward. If we follow the air moving northward (● Fig. 10.15), we see that, because of the curvature of Earth, air constantly moves closer to its axis of rotation

(r decreases). Because angular momentum is conserved (and since the mass of air is unchanged), the decrease in radius must be compensated for by an increase in speed. The air must, therefore, move faster to the east than a point on Earth's surface does. To an observer, this is a west wind. Hence, the conservation of angular momentum of northward-flowing air leads to the generation of strong westerly winds and the formation of a jet stream.

OTHER JET STREAMS There is another jet stream that forms in summer near the tropopause above Southeast Asia, India, and Africa. Here, the altitude of the summer tropopause and the jet stream is near 15 km (49,000 ft). Because the jet forms on the equatorward side of the upper-level subtropical high, its winds are easterly and, hence, it is known as the **tropical easterly jet stream**. Although the exact causes of this jet have yet to be completely resolved, its formation appears to be, at least in part, related to the warming of the air over large elevated landmasses, such as Tibet. During the summer, the air above this region (even at high elevations) is warmer than the air above the ocean to the south. This contrast in temperature produces a north-to-south pressure gradient and strong easterly winds that usually reach a maximum speed near 15°N latitude.

Not all jet streams form at the tropopause. For example, there is a jet stream that forms near the top of the stratosphere over polar latitudes. Because little, if any, sunlight reaches the polar region during the winter, air in the upper stratosphere is able to cool to low temperatures. By comparison, in equatorial regions, sunlight prevails all year long, allowing stratospheric ozone to absorb solar energy and warm the air. The horizontal temperature gradients between the cold poles and the warm tropics

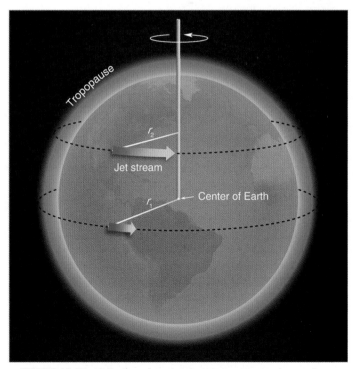

● **FIGURE 10.15** Air flowing poleward at the tropopause moves closer to the rotational axis of Earth (r_2 is less than r_1). This decrease in radius is compensated for by an increase in velocity and the formation of a jet stream.

create steep horizontal pressure gradients, and a strong westerly jet forms in polar regions at an elevation near 50 km (30 mi). Because this wind maximum occurs in the stratosphere during the dark polar winter, it is known as the *stratospheric polar night jet stream.*

In summer, the polar regions experience more hours of sunlight than do tropical areas. Stratospheric temperatures over the poles increase more than at the same altitude above the equator, which causes the horizontal temperature gradient to reverse itself. The jet stream disappears, and in its place there are weaker easterly winds.

Jet streams also form in the upper mesosphere and in the thermosphere. Not much is known about the winds at these high levels, but they are probably related to the onslaught of charged particles that constantly bombard this region of the atmosphere.

Jet streams form near Earth's surface as well. One such jet stream develops just above the central plains of the United States, where it occasionally attains speeds exceeding 60 knots. This wind speed maximum, which usually flows from the south or southwest, is known as a **low-level jet**. It typically forms at night above a temperature inversion, and so it is sometimes called a *nocturnal jet stream.* Apparently, the stable air reduces the interaction between the air within the inversion and the air directly above. Consequently, the air in the vicinity of the jet is able to flow faster because it is not being slowed by the lighter winds below. Also, the north-south-trending Rocky Mountains tend to funnel the air northward. Another important element contributing to the formation of the low-level jet is the downward sloping of the land from the Rockies to the Mississippi Valley, which causes nighttime air above regions to the west to be cooler than air at the same elevation to the east. This horizontal contrast in temperature causes pressure surfaces to dip toward the west. The dipping of pressure surfaces produces strong pressure-gradient forces directed from east to west which, in turn, cause strong southerly winds.

During the summer, these strong southerly winds carry moist air from the Gulf of Mexico into the Central Plains. This moisture, coupled with converging, rising air of the low-level jet, enhances thunderstorm formation. Therefore, on warm, moist, summer nights, when the low-level jet is present, it is common to have nighttime thunderstorms over the plains.

BRIEF REVIEW

Before going on to the next section, which describes the many interactions between the atmosphere and the ocean, here is a review of some of the important concepts presented so far:

- The two major semipermanent subtropical highs that influence the weather of North America are the Pacific high situated off the West Coast and the Bermuda high situated off the Southeast Coast.

- The polar front is a zone of low pressure where cyclonic storms often form. It separates the mild westerlies of the middle latitudes from the cold, polar easterlies of the high latitudes.

- In equatorial regions, the intertropical convergence zone (ITCZ) is a boundary where air rises in response to the convergence of the northeast trades and the southeast trades. Along the ITCZ huge thunderstorms produce heavy rain showers.

- In the Northern Hemisphere, the major global pressure systems and wind belts shift northward in summer and southward in winter.

- The northward movement of the Pacific high in summer tends to keep summer weather along the west coast of North America relatively dry.

- Jet streams exist where strong winds become concentrated in narrow bands. The polar-front jet stream is associated with the polar front. The polar jet stream meanders in a wavy west-to-east pattern, becoming strongest in winter when the contrast in temperature along the front is greatest.

- The subtropical jet stream is found on the poleward side of the Hadley cell, between 20° and 30° latitude. It is normally observed at a higher altitude than the polar jet stream.

- The conservation of angular momentum plays a role in producing strong, westerly winds aloft. As air aloft moves from lower latitudes toward higher latitudes, its axis of rotation decreases, which results in an increase in its speed.

Atmosphere-Ocean Interactions

The atmosphere and oceans are both dynamic fluid systems that interact with each other in many complex ways. For example, evaporation of ocean water provides the atmosphere with surplus water that falls as precipitation. The latent heat that is taken up by the water vapor during evaporation goes into the atmosphere during condensation to fuel storms. The storms, in turn, produce winds that blow over the ocean, which causes waves and currents. The currents, in turn, can modify the weather and climate of a region by bringing in vast quantities of warm or cold water.

The complexity of the interaction between the atmosphere and ocean makes our scientific understanding of how one influences the other on a global scale far from complete. What we will focus on in the remainder of this chapter is what we do know, beginning with ocean currents. Later, we will concentrate on some of the most important weather and climate oscillations that result from atmosphere-ocean interactions.

GLOBAL WIND PATTERNS AND SURFACE OCEAN CURRENTS

As the wind blows over the oceans, it causes the surface water to drift along with it. The moving water gradually piles up, creating pressure differences within the water itself. This leads to further motion several hundreds of meters down into the water. In this manner, the general wind flow around the globe starts the major surface ocean currents moving. The relationship between the general circulation and ocean currents can be seen by comparing ● Fig. 10.16 and ● Fig. 10.17.

Because of the larger frictional drag in water, ocean currents move more slowly than the prevailing winds above. Typically, these currents range in speed from several kilometers per day to several kilometers per hour. However, comparing Figs. 10.16 and 10.17, we can see that ocean currents do not follow the wind pattern exactly; rather, they spiral in semiclosed circular whirls called *gyres*. In the North Atlantic and North Pacific, the prevailing winds blow clockwise and outward from the subtropical highs. As the water moves beneath the wind, the Coriolis force deflects the water to the right in the Northern Hemisphere (to the left in the Southern Hemisphere). This deflection causes the surface water to move at an angle between 20° and 45° to the direction of the wind. Hence, surface water tends to move in a circular pattern as winds blow outward, away from the center of the subtropical highs.

Important interactions between the atmosphere and the ocean can be seen by examining the huge gyre in the North Atlantic. Flowing northward along the east coast of the United States is a tremendous warm water current called the **Gulf Stream**. The Gulf Stream carries vast quantities of warm tropical water into higher latitudes. Off the coast of North Carolina, the Gulf Stream provides warmth and moisture for the development of mid-latitude cyclonic storms.

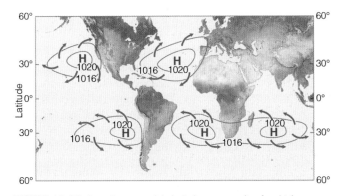

● **FIGURE 10.16** Annual average global wind patterns and surface high-pressure areas over the oceans.

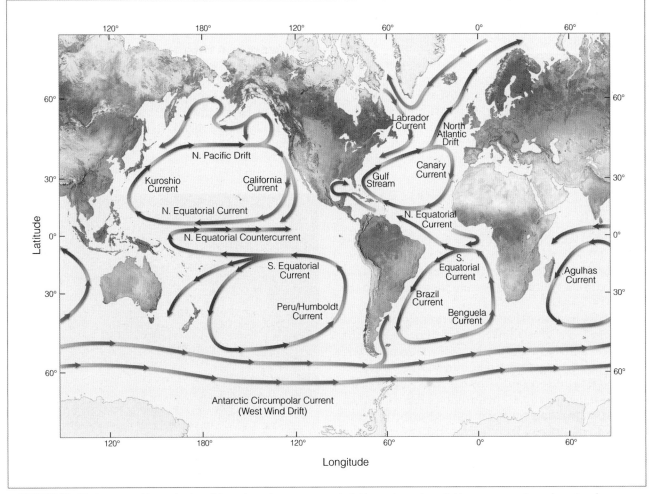

● **FIGURE 10.17** Average position and extent of the major surface ocean currents. Cold currents are shown in blue; warm currents are shown in red.

To the north, on the western side of the smaller subpolar gyre, cold water moves southward along the Atlantic coast of North America. This *Labrador Current* brings cold water as far south as Massachusetts in summer and North Carolina in winter. In the vicinity of the Grand Banks of Newfoundland, where the two opposing currents flow side by side, there is a sharp temperature gradient. When warm Gulf Stream air blows over the cold Labrador Current water, the stage is set for the formation of the fog so common to this region.

Meanwhile, steered by the prevailing westerlies, the Gulf Stream swings away from the coast of North America and moves eastward toward Europe. Gradually, it widens and slows as it merges into the broader *North Atlantic Drift*. As this current approaches Europe, it divides into two currents. A portion flows northward along the coasts of Great Britain and Norway, bringing with it warm water (which helps keep winter temperatures much warmer than one would expect this far north). The other part of the North Atlantic Drift flows southward as the *Canary Current*, which transports cool northern water equatorward. Eventually, the Atlantic gyre is completed as the Canary Current merges with the westward-moving *North Equatorial Current*, which derives its energy from the northeast trades.

The ocean circulation in the North Pacific is similar to that in the North Atlantic. On the western side of the ocean is the Gulf Stream's counterpart, the warm, northward-flowing *Kuroshio Current*, which gradually merges into the slower-moving *North Pacific Drift*. A portion of this current flows southward along the coastline of the western United States as the cool *California Current*. In the Southern Hemisphere, surface ocean circulations are much the same except that the gyres move counterclockwise in response to the winds around the subtropical highs. Notice that the ocean currents at higher latitudes tend to move in a more west-to-east pattern than do the currents of the Northern Hemisphere. This zonal pattern limits, to some extent, the poleward transfer of warm tropical water. Hence, there is a much smaller temperature difference between the ocean's surface and the atmosphere than exists over Northern Hemisphere oceans. This situation tends to limit the development of vigorous convective activity over the oceans of the Southern Hemisphere. In the Indian Ocean, monsoon circulations tend to complicate the general pattern of ocean currents.

To sum up: On the eastern edge of continents there usually is a warm current that carries huge quantities of warm water from the equator toward the pole; whereas on the western side of continents a cool current typically flows from the pole toward the equator.

Up to now, we have seen that atmospheric circulations and ocean circulations are closely linked; wind blowing over the oceans produces surface ocean currents. The currents, along with the wind, transfer heat from tropical areas, where there is a surplus of energy, to polar regions, where there is a deficit. This helps to equalize the latitudinal energy imbalance, with about 40 percent of the total heat transport in the Northern Hemisphere coming from surface ocean currents. The environmental implications of this heat transfer are tremendous. If the energy imbalance were to go unchecked, yearly temperature differences between low and high latitudes would increase greatly, and, as we will see in Chapter 17, the climate would gradually change.

Satellite pictures reveal that distinct temperature gradients exist along the boundaries of surface ocean currents. For example, off the east coast of the United States, where the warm Gulf Stream meets cold waters to the north, sharp temperature contrasts are often present. The boundary separating the two masses of water with contrasting temperatures and densities is called an **oceanic front**. Along this frontal boundary, a portion of the meandering Gulf Stream occasionally breaks away and develops into a closed circulation of either cold or warm water—a whirling eddy (see ● Fig. 10.18). Because these eddies transport heat and momentum from one region to another, they may have a far-reaching effect upon the climate and a more immediate impact upon coastal waters. Scientists are investigating the effects of these eddies.

UPWELLING   Earlier, we saw that the cool California Current flows roughly parallel to the west coast of North America. From this observation, we might conclude that summer surface water temperatures will be cool along the coast of Washington and gradually warmer as we move south. A quick glance at the water temperatures along the west coast of the United States during August (see ● Fig. 10.19) quickly alters that notion. The coldest water is observed along the northern California coast near Cape Mendocino. Why there? To answer this, we need to examine how the wind influences the movement of surface water.

As the wind blows over an open stretch of ocean, the surface water beneath it is set in motion. The Coriolis force bends the moving water to the right in the Northern Hemisphere. Thus, if we look at a shallow surface layer of water, we see in ● Fig. 10.20 that it moves at an average angle of about 45° to the direction of the wind. If we imagine the top layer of ocean water to be broken into a series of layers, then each layer will exert a frictional drag on the layer below. Not only will each successive layer move a little slower than the one above, but (because of the Coriolis effect) each layer will also rotate slightly to the right of the layer above. (The rotation of each layer is to the right in the Northern Hemisphere

● **FIGURE 10.18** The Gulf Stream (dark red band) and its eddies are revealed in this satellite mosaic of sea surface temperatures of the western North Atlantic during May 2001. Bright red shows the warmest water, followed by orange and yellow. Green, blue, and purple represent the coldest water.

and to the left in the Southern Hemisphere.) Consequently, descending from the surface, we would find water slowing and turning to the right until, at some depth (usually about 100 m), the water actually moves in a direction opposite to the flow of water at the surface. This turning of water with depth is known as the **Ekman spiral**.* The Ekman spiral in Fig. 10.20 shows us that the average movement of surface water down to a depth of about 100 m (330 ft) is at right angles (90°) to the surface wind direction. The Ekman spiral helps to explain why, in summer, surface water is cold along the west coast of North America.

---

*The Ekman spiral is also present in the atmospheric boundary layer, from the surface up to the top of the friction layer, which is usually about 1000 m above the surface.

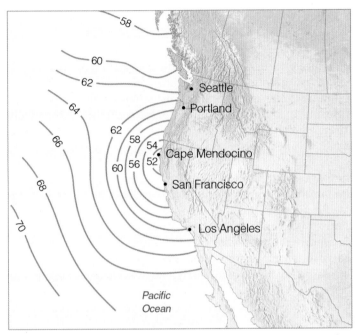

● **FIGURE 10.19** Average sea surface temperatures (°F) along the west coast of North America during August.

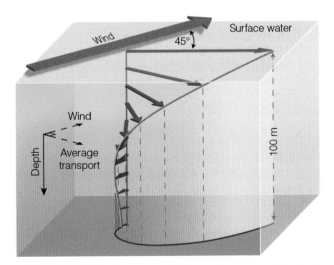

● **FIGURE 10.20** The Ekman spiral. Winds move the water, and the Coriolis force deflects the water to the right (Northern Hemisphere). Below the surface each successive layer of water moves more slowly and is deflected to the right of the layer above. The average transport of surface water in the Ekman layer is at right angles to the prevailing winds.

CRITICAL THINKING QUESTION Suppose Fig. 10.20 is in the Southern Hemisphere. If the surface wind is still blowing from the same direction, how would the Ekman spiral appear beneath the surface?

The summertime position of the Pacific high and the low coastal mountains cause winds to blow parallel to the California coastline (see ● Fig. 10.21). The net transport of surface water (called the **Ekman transport**) is at right angles to the wind, in this case, out to sea. As surface water drifts away from the coast, cold, nutrient-rich water from below rises to replace it. The rising of cold water is known as **upwelling**. Upwelling is strongest and surface water is coolest in this area because here the wind parallels the coast.

Summertime weather along the West Coast often consists of low clouds and fog, as the air over the water is chilled to its saturation point. Upwelling produces good fishing conditions, however, as higher concentrations of nutrients are brought to the surface. But swimming along the coast of Northern California is only for the hardiest of souls, as the average surface water temperature in summer is nearly 10°C (18°F) colder than the average coastal water temperature found at the same latitude along the Atlantic coast.

Between the ocean surface and the atmosphere, there is an exchange of heat, moisture, and momentum that depends, in part, on temperature differences between water and air. In winter, when air-water temperature contrasts are greatest, there is a substantial transfer of sensible and latent heat from the ocean surface into the atmosphere. This energy helps to maintain the global airflow. Because of the difference in heat capacity between water and air, even a relatively small change in surface ocean temperatures can modify atmospheric circulations. Such changes can have far-reaching effects on global weather and climate patterns. One ocean-atmospheric phenomenon that is linked to worldwide weather events is a periodic warming of the eastern tropical Pacific Ocean known as **El Niño**.

EL NIÑO, LA NIÑA, AND THE SOUTHERN OSCILLATION Along the west coast of South America, where the cool Peru Current sweeps northward, southerly winds promote upwelling of cold, nutrient-rich water that gives rise to large fish populations, especially anchovies. The abundance of fish supports a large population of sea birds whose droppings (called *guano*) produce huge phosphate-rich deposits, a valuable source of fertilizer. Every two to five years or so, a warm current of nutrient-poor tropical water moves southward, replacing the cold, nutrient-rich surface water. Because this condition frequently occurs around Christmas, local fishermen labeled this warm current *Corriente del Niño* more than a century ago, which translated means "current of the Christ Child"; hence, the warm current's name—*El Niño*.

It was once thought that El Niño was a local event that occurred only along the west coast of Peru and Ecuador. It is now known that the ocean warming can cover an area of the tropical Pacific much larger than the continental United States. Although in recent decades the term *El Niño* has gained global prominence, the large, prolonged warming that develops at irregular intervals every few years is often referred to as an *El Niño event*. During such an event, the surface water temperature

● **FIGURE 10.21** As winds blow parallel to the west coast of North America, surface water is transported to the right (out to sea). Cold water moves up from below (upwells) to replace the surface water. The large H represents the position of the Pacific High in summer. Blue arrows show the movement of water.

across much of the eastern tropical Pacific rises by at least 0.5°C (0.9°F) for periods of a few months to a year or more. Warmer-than-average water may be so widespread that it raises the average surface air temperature of the entire planet by several tenths of a degree Celsius.

Why does the ocean become so warm over such a large area during an El Niño event? Normally in the tropical Pacific Ocean, trade winds are persistently blowing westward from a region of higher pressure over the eastern Pacific toward a region of lower pressure in the western Pacific, centered near Indonesia (see ● Fig. 10.22a). As they push water away from the west coast of South America, the trades create upwelling that brings cold water to the surface. As this water moves westward, it is heated by sunlight and the atmosphere. Consequently, surface water along the equator usually is cooler in the eastern Pacific and warmer in the western Pacific. In addition, the dragging of surface water by the trades raises sea level in the western Pacific and lowers it in the eastern Pacific, which produces a thick layer of warm water over the tropical western Pacific Ocean. The higher level of water in the western Pacific causes warm water to flow slowly eastward toward South America as a weak, narrow ocean current called the *countercurrent*.

Every few years, the surface atmospheric pressure patterns break down. Air pressure rises over the western Pacific and falls over the eastern Pacific (see Fig. 10.22b). This change in pressure weakens the trades, and over a period of a few months, east winds may be replaced by west winds that strengthen the countercurrent. Warm water gradually extends farther east across the tropical Pacific, and the location of showers and thunderstorms may shift east as well, reinforcing the shift in atmospheric circulation. If the ocean warming reaches a certain threshold and remains at that level long enough, an El Niño is declared to be in place.

Toward the end of the warming period, which typically lasts about a year but may recur for another year or more, atmospheric pressure over the eastern Pacific reverses and begins to rise, whereas over the western Pacific, it falls. This seesaw pattern of reversing surface air pressure at opposite ends of the Pacific Ocean is called the **Southern Oscillation**. Because the reversals in air pressure and the ocean warming are more or less simultaneous, scientists call this phenomenon the *El Niño–Southern Oscillation* (**ENSO**). Although most El Niño episodes follow a similar evolution, each event has its own personality, differing in both strength and behavior. During especially strong El Niño events (such as in 1982–1983, 1997–1998, and 2014–2016) the easterly trades may actually become westerly winds, as shown in Fig. 10.22b. As these winds push eastward, they drag surface water with them. This dragging raises sea level in the eastern Pacific and lowers sea level in the western Pacific. The eastward-moving water gradually warms under the tropical sun, becoming as much as 6°C (11°F) warmer than normal in the eastern equatorial Pacific. The unusually warm water may extend from South America's coastal region for many thousands of kilometers westward along the equator (see ● Fig. 10.23).

Following an El Niño event, the trade winds often return to normal. However, if the trades become exceptionally strong, then unusually cold surface water extends over the eastern and central Pacific, and warm water and rainy weather are confined mainly to the western tropical Pacific (see Fig. 10.23b). This cold-water episode is termed **La Niña** (the girl child). In some ways La Niña can be thought of as the opposite of El Niño, since water temperatures in the eastern tropical Pacific are colder than average during La Niña while they are warmer than average during El Niño. However, keep in mind that while an El Niño event

WEATHER WATCH

The very strong El Niño of 2014–2016 disturbed the weather worldwide. Both Texas and Oklahoma had their wettest months on record in May 2015, as did South Carolina in October 2015. Massive flooding in all three states killed dozens of people and caused billions of dollars in damage. Further west, an onslaught of hurricanes and typhoons spun across the tropical Pacific, with a total of 48 named storms recorded in 2015 and 2016 in the Northeast Pacific. Two of the world's strongest landfalling storms on record occurred in 2016: Super Typhoon Meranti came ashore at Itbayat, Philippines, with sustained winds of 190 mph, and Cyclone Winston struck Fiji with 180-mph sustained winds.

(a) Non-El Niño conditions

(b) El Niño conditions

● **FIGURE 10.22** In diagram (a), under non-El Niño conditions higher pressure over the southeastern Pacific and lower pressure near Indonesia produce easterly trade winds along the equator. These winds promote upwelling and cooler ocean water in the eastern Pacific, while warmer water prevails in the western Pacific. The trades are part of a circulation that typically finds rising air and heavy rain over the western Pacific and sinking air and generally dry weather over the eastern Pacific. If the trades are exceptionally strong (not shown here), water along the equator in the eastern Pacific becomes quite cool. This cool event is called La Niña. During El Niño conditions—diagram (b)—atmospheric pressure decreases over the eastern Pacific and rises over the western Pacific. This change in pressure causes the trades to weaken or reverse direction. This situation enhances the countercurrent that carries warm water from the west over a vast region of the eastern tropical Pacific.

features a *reversal* of the typical east-to-west wind flow and ocean currents across the tropical Pacific, La Niña features a *strengthening* of the same east-to-west wind flow and ocean currents. La Niña events sometimes occur immediately after major El Niño events, but they can also develop independently.

● Figure 10.24a shows warm events (El Niño years) in red and cold events (La Niña years) in blue. Notice in Fig. 10.24a that it is possible to have two or more El Niño or La Niña periods in a row, separated by neutral conditions. El Niño events typically develop in the northern autumn, peak in the winter, and weaken in the spring and summer. Although El Niño events rarely last

more than a year, La Niña conditions may recur or persist for two or three years at a time. Over a long time frame (several decades), the tropical Pacific is divided roughly equally among El Niño, La Niña, and neutral conditions.

CRITICAL THINKING QUESTION Looking at Fig. 10.24a, what might you conclude about the typical frequency of El Niño and La Niña events if you looked only at the 1970s? The 1990s? How long a period of time would you need to analyze in order for the number of El Niño and La Niña events to be roughly equal?

(a) El Niño conditions, December 1998

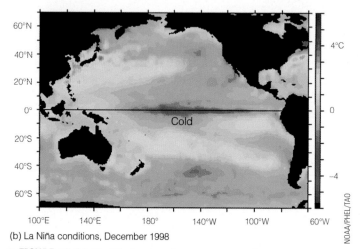

NOAA/PHEL/TAO

(b) La Niña conditions, December 1998

● **FIGURE 10.23** Monthly departures from average sea surface temperature as measured by satellite. (a) During El Niño conditions, upwelling is greatly diminished, and warmer-than-normal water (deep red color) extends from the coast of South America westward, across the Pacific. (b) During La Niña conditions, strong trade winds promote upwelling, and cooler-than-normal water (dark blue color) extends over the eastern and central Pacific.

Effects of El Niño and La Niña A major El Niño or La Niña event can have impacts over a large part of the globe. These effects were first recognized in coastal areas of Peru and Ecuador, where a thick layer of warm water brought by El Niño can choke off the upwelling that supplies cold, nutrient-rich water to South America's coastal region. As a result, large numbers of fish, birds, and marine plants can die, littering the water and beaches of Peru. Their decomposing carcasses deplete the water's oxygen supply, which leads to the bacterial production of huge amounts of smelly hydrogen sulfide. The El Niño event of 1972–1973, one of the first to be studied in depth, reduced the annual Peruvian anchovy catch by more than 50 percent. Since much of the harvest of this fish was normally converted into fishmeal and exported for use in feeding livestock and poultry, the world's fishmeal production in 1972 was greatly reduced. Countries such as the United States that relied on fishmeal for animal feed had to use soybeans as an alternative, which raised poultry prices in the United States by more than 40 percent.

The large areas of abnormally warm water associated with El Niño and abnormally cold water associated with La Niña can have effects far beyond South America. During El Niño, the usual region of warm water in the western tropical Pacific is moved thousands of kilometers to the east. There, the water fuels the atmosphere with additional warmth and moisture, which feeds into storminess and rainfall. Along with the added warmth from the oceans, latent heat is released during condensation. All of these factors help change the regional atmospheric circulation. These effects can propagate for thousands of miles, rearranging the locations where rising and sinking motions tend to predominate. As a result, some regions of the world experience too much rainfall during El Niño, whereas others have too little. For example, as we saw in Chapter 9, there is a tendency during El Niño events for monsoon conditions over India to weaken. Drought is typically felt in Indonesia, southern Africa, and Australia, while heavy rains and flooding often occur in Ecuador and Peru. In the Northern Hemisphere, a strong subtropical westerly jet stream often directs mid-latitude cyclonic storms into California and heavy rain into the Gulf Coast states. The total global impact of a major El Niño event due to flooding, winds, and drought may exceed many billions of dollars, although some areas can experience financial benefits. For example, milder-than-usual conditions are frequent during El Niño across Canada and the northern United States, and this can result in savings on winter heat costs.

Notice in Fig. 10.24a that the three strongest El Niño events since 1950 were in 1982–1983, 1997–1998, and 2014–2016. Global air temperature typically rises during El Niño, as the expanded area of warm surface water in the tropical Pacific Ocean allows heat to move from ocean to atmosphere. The short-term atmospheric warming produced by El Niño, on top of longer-term global warming, resulted in record-high global temperatures during each of the three strongest El Niño events: in 1983, in 1998, and in all three years from 2014 to 2016.

The 1997–1998 event had remarkable impacts worldwide. In parts of South America, rainfall was up to 10 times higher than normal, causing flooding, erosion, mudslides, and deaths. Parts of California and the southern United States also had near-record rainfall. Worldwide, more than 20,000 deaths were attributed to droughts, flooding, famine, and disease associated with this severe El Niño. Only a few months later, in mid-1998, one of the strongest La Niña events in decades arrived, lasting for almost three years. The 2014–2016 El Niño was another very strong event with a variety of global impacts, including heavy rain in Texas (● Fig. 10.25) and drought in Indonesia (● Fig. 10.26).

One of the most devastating ecological effects of a strong El Niño is the damage to coral reefs due to unusually warm waters. If the warming is especially strong and persistent, the algae that support the coral can die off, leaving the coral with a pale, bleached appearance and sometimes leading to their death. Warm waters toward the end of the 1997–1998 El Niño killed an estimated 16 percent of all the corals on Earth, and another mass bleaching event was associated with the 2014–2016 El Niño.

Hurricanes are one of the most prominent weather features affected by El Niño and La Niña. Because the tropical Northeast Pacific is warmer than normal during an El Niño event, the

NOAA and Jan Null

(a)

● **FIGURE 10.24** (a) The Oceanic Niño Index (ONI). The numbers on the left side of the diagram represent a running 3-month mean for sea surface temperature variations (from normal) over the tropical Pacific Ocean from latitude 5°N to 5°S and from longitude 120°W to 170°W (area outlined in Fig. 10.24(b)). Warm El Niño episodes are in red; cold La Niña episodes are in blue. Warm and cold events occur when the deviation from the normal is 0.5 or greater. An index value between 0.5 and 0.9 is considered weak; an index value between 1.0 and 1.4 is considered moderate, and an index value of 1.5 or greater is considered strong.

(b)

● **FIGURE 10.25** Vehicles left stranded on a flooded Interstate 45 in Houston, Texas, on May 26, 2015. Heavy rains put the city under massive amounts of water, closing roadways and trapping residents in their cars and buildings. An El Niño event was intensifying in the spring of 2015.

Aaron M. Sprecher/AFP/Getty Images

Ulet Ifansasti/Getty Images News/Getty Images

● **FIGURE 10.26** A firefighter holds a water pipe while working to extinguish a fire in South Sumatra, Indonesia, on October 2, 2015. Intense drought triggered by very strong El Niño conditions in fall 2015 left the landscape vulnerable to forest and peat-land fires, many set illegally to clear land for agricultural production. More than 100,000 Indonesians were treated for respiratory ailments caused by the widespread smoke and haze.

frequency of hurricanes usually increases in that region. However, over the tropical North Atlantic, El Niño tends to reduce the strength of easterly winds at high levels, which disrupts the organization of thunderstorms necessary for hurricane development. Hence, there are fewer hurricanes than usual in this region during El Niño events, and this contributes to a reduced risk in the United States for hurricane landfalls. The opposite occurs during La Niña events, with more frequent hurricanes observed in the North Atlantic and fewer hurricanes in the Northeast Pacific. On average, tropical storms and hurricanes are almost 50 percent more likely to strike the United States during La Niña years as opposed to El Niño years.

● Figure 10.27a illustrates typical winter weather patterns over North America during El Niño conditions. Notice that a persistent trough of low pressure forms over the North Pacific and, to the south of the low, the jet stream (from off the Pacific) steers wet weather and storms into California and the southern part of the United States. A weak polar jet stream forms over eastern Canada, allowing warmer-than-normal weather to prevail over a large part of North America.

Figure 10.27b shows typical North American winter weather patterns with a La Niña. Notice that a persistent high-pressure area (called a *blocking high*) forms south of Alaska, forcing the polar jet stream into Alaska, then southward into Canada and the western United States. The southern branch of the polar jet stream, which forms south of the high, directs moist air from the ocean into the Pacific Northwest, producing a wet winter for that region. Meanwhile, winter in the southern part of the United States tends to be warmer and drier than normal. In between these regions, the storm track across the central and eastern United States can be intensified during La Niña. Especially strong outbreaks of severe thunderstorms and tornadoes are somewhat more likely in these areas during La Niña.

As we have seen, El Niño, La Niña, and the Southern Oscillation are part of a large-scale ocean-atmosphere interaction that can take several years to run its course. Using data from previous ENSO episodes, scientists at the National Oceanic and Atmospheric Administration's Climate Prediction Center have obtained a global picture of where climatic abnormalities related to ENSO are most likely (see ● Fig. 10.28 for global effects of El Niño). Such ocean-atmosphere interactions, where warm or cold surface ocean temperatures can influence precipitation patterns in a distant part of the world, are called **teleconnections**. Notice in Figs. 10.27a and 10.27b that in some areas, the teleconnections related to La Niña are opposite to those of El Niño. However, this is not the case in every location, because the circulations in El Niño and La Niña are not exact opposites. Also, it is important to keep in mind that these preferred patterns are not *guaranteed* to occur with every El Niño or La Niña event, since each one is different. For example, Los Angeles, California, had a drier-than-average winter during the strong El Niño event of 2015–2016, even though strong El Niño winters tend to be wetter than average in Los Angeles. Over the long term, though, the effects shown are the most likely ones to expect. So while we cannot say with certainty that an El Niño winter will bring more rainfall than average to southern California, we can say that the odds are higher during El Niño than they might otherwise be, especially if the El Niño is a strong one. (For a look at some of the challenges facing scientists in predicting El Niño and La Niña, read the Focus section 10.2.)

Up to this point, we have looked at El Niño, La Niña, and the Southern Oscillation and how the reversal of surface ocean temperatures and atmospheric pressure combine to influence regional and global weather and climate patterns. There are other atmosphere-ocean interactions that can have an effect on large-scale weather patterns. Some of these are described in the following sections.

(a) El Niño winter conditions

(b) La Niña winter conditions

● **FIGURE 10.27** Typical winter weather patterns across North America during an El Niño warm event (a) and during a La Niña cold event (b). During El Niño conditions, a persistent trough of low pressure forms over the North Pacific and, to the south of the low, the jet stream (from off the Pacific) steers wet weather and storms into California and the southern part of the United States. During La Niña conditions, a persistent high-pressure area forms south of Alaska, forcing the polar jet stream and accompanying cold air over much of western North America. The southern branch of the polar jet stream directs moist air from the ocean into the Pacific Northwest, producing a wet winter for that region.

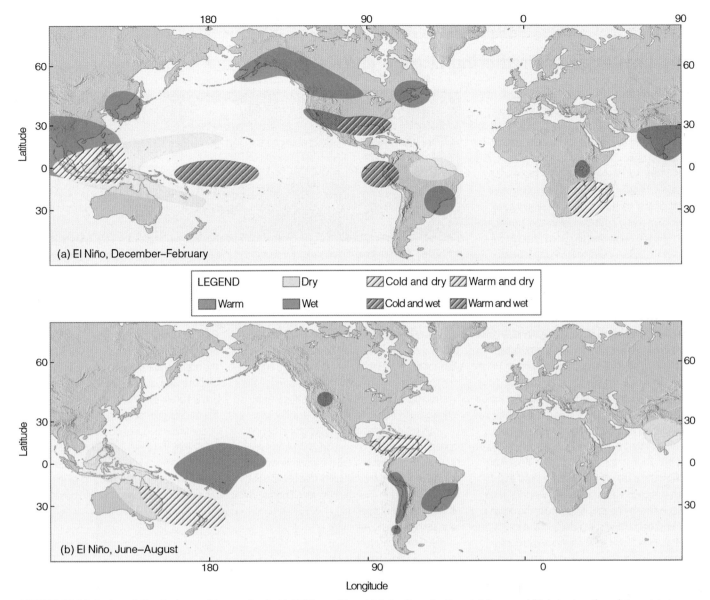

LEGEND ▢ Dry ▨ Cold and dry ▨ Warm and dry ▨ Warm ▨ Wet ▨ Cold and wet ▨ Warm and wet

(a) El Niño, December–February

(b) El Niño, June–August

Longitude

● **FIGURE 10.28** Regions of climatic abnormalities associated with El Niño conditions (a) during December through February and (b) during June through August. A strong El Niño event may trigger a response in nearly all indicated areas, whereas a weak event will likely play a role in only some areas. (After NOAA Climate Prediction Center.)

PACIFIC DECADAL OSCILLATION Over the North Pacific Ocean, periodic changes in surface water temperature can influence weather along the west coast of North America for much longer periods of time than El Niño and La Niña. The **Pacific Decadal Oscillation (PDO)** is like ENSO in that it has a warm phase and a cool phase, and the temperature patterns produced by the PDO are similar in some locations to those produced by ENSO. However, the PDO has more of an influence in the midlatitudes of the North Pacific rather than the tropical Pacific, and it operates on a much longer time scale than ENSO. Each PDO phase tends to predominate for 20 to 30 years before switching to the other phase.

During the warm (or positive) phase of the PDO, unusually warm surface water exists along the west coast of North America, whereas over the central North Pacific, cooler-than-normal surface water prevails (see ● Fig. 10.29a). At the same time, the Aleutian low in the Gulf of Alaska strengthens, which causes more Pacific storms to move into Alaska and California. This situation causes winters, as a whole, to be warmer and drier over northwestern North America. Elsewhere, winters tend to be drier over the Great Lakes and cooler and wetter in the southern United States. Salmon populations often increase in Alaska and diminish along the Pacific Northwest coast.

Cool (or negative) PDO phases have cooler-than-average surface water along the west coast of North America and an area of warmer-than-normal surface water extending from Japan into the central North Pacific (see Fig. 10.29b). Winters in the cool phase tend to be cooler and wetter than average over northwestern North America, wetter over the Great Lakes, and warmer and drier in the southern United States. Salmon fishing diminishes in Alaska and increases along the Pacific Northwest coast.

These climate patterns only represent average conditions, as individual years within either PDO phase may vary considerably, sometimes reverting to the opposite phase for short periods.

The Challenge of Predicting El Niño and La Niña

For several decades, atmospheric scientists have explored techniques for predicting the future state of the El Niño–Southern Oscillation (ENSO) and the onset of El Niño and La Niña events. This is an especially difficult challenge because ENSO events are not like typical day-to-day weather features. Instead, they unfold in a much more gradual way through a complex sequence of events involving both the ocean and atmosphere.

El Niño events are triggered through processes that allow warmer-than-usual surface water to build up over the central and eastern parts of the tropical Pacific. One important process is the *Madden-Julian Oscillation* (MJO), a recurring event across the tropics that influences weather across much of the globe. An MJO consists of a large area of showers and thunderstorms, together with a westerly wind burst that helps keep the MJO propagating eastward. MJO pulses can sometimes be tracked from the Indian Ocean all the way across the Pacific, the Americas, the Atlantic, and into Africa over a period of weeks to months. In the tropical Pacific, the westerly wind bursts associated with the MJO push against the trade winds and the normal east-to-west surface ocean currents. If a westerly wind burst is strong enough, it can cause the ocean currents below it to shift direction and thus move warm water from the western tropical Pacific toward the east. This effect is typically short-lived, however, disappearing after the MJO passes.

Another process that helps warm water to expand eastward into these regions is a surge known as a *Kelvin wave*, which can be triggered by westerly wind bursts. This is a slow-moving oceanic feature that brings warm water across the tropical Pacific. It takes about two months for a Kelvin wave to travel from Indonesia to South America. A Kelvin wave can be enormous, extending hundreds of kilometers north and south of the equator (see ● Fig. 3). Vertically, it may extend only 10 or 15 centimeters above the ocean surface but more than 100 meters below the surface. When it reaches the coast of South America, a strong Kelvin wave can help suppress the normal upwelling of cooler water and push warmer water into its place. However, this effect only lasts while the Kelvin wave is passing, and upwelling may resume afterward.

An El Niño becomes increasingly likely when several strong MJO pulses, westerly wind bursts, and Kelvin waves occur within a few weeks of each other. In this case, the repeated flows of warm water may persist across the surface of the eastern tropical Pacific long enough for the atmospheric circulation to reinforce the warming. At this point, if the water is warm enough, an El Niño event may be under way.

Using long-range models of global atmospheric and oceanic circulation, NOAA and several other agencies around the world now issue outlooks giving the probability that El Niño or La Niña will develop over the next few months. (See ● Fig. 4 for an example of one such forecast.) Although computer forecast models are gaining skill, it is still difficult for them to capture the features such as MJO pulses and westerly wind bursts that can help kick off an El Niño or La Niña event. The models that have greater skill at simulating these features appear to be more skillful at depicting ENSO.

When it becomes clear in the summer or autumn that an El Niño or La Niña event is actually developing, then forecasters can provide several months' notice of the type of conditions that are most likely to occur in the winter, the time of year when the events usually reach their full strength.

In-depth studies of the tropical Pacific and Indian Ocean are providing scientists with valuable information about the interactions that occur between the ocean and the atmosphere. One ultimate aim of these field studies is to pave the way for better prediction of ENSO and other climatic fluctuations that unfold over periods of months and years. A better understanding of the processes behind ENSO may increase our ability to reliably predict the formation of El Niño or La Niña months before it happens, thus leading to improved long-range forecasts of weather and climate.

NASA

● **FIGURE 3** These three images depict the evolution of a warm water Kelvin wave moving eastward in the equatorial Pacific Ocean during March and April 1997. The white areas near the equator represent ocean levels about 20 cm (8 in.) higher than average, while the red areas represent ocean levels about 10 cm (4 in.) higher than average. Notice how the wave (high region) moves eastward across the tropical Pacific Ocean. These data were collected by the altimeter on board the joint United States/French TOPEX/Poseidon satellite.

● **FIGURE 4** A depiction of the consensus among a team of forecasters at NOAA and the International Research Institute for Climate and Society on how ENSO conditions would evolve during 2014. The forecast was issued in February and predicted the likelihood of neutral conditions (green), El Niño (red), and La Niña (blue), looking ahead to the rest of the year. The time periods shown are overlapping three-month periods; for example, JFM 2014 is January-February-March 2014 and FMA is February-March-April 2014.

(a) Positive (warm) phase (b) Negative (cool) phase

−0.6 −0.2 0 0.2 0.4 0.8
Temperature anomaly (°C)

● FIGURE 10.29 Typical winter sea-surface temperature departure from normal in °C during the Pacific Decadal Oscillation's warm phase (a) and cool phase (b). *Source:* Retrieved from http://research.jisao.washington.edu/pdo/. Used with permission of Nate Mantua.

Obtained via the www.http://tao.atmos.washington.edu/pdo. Used with permission of N. Mantua

These variations, which can last several months to a year or more, make it more difficult to decipher exactly when the PDO has changed from one long-term phase to the other.

Knowing the long-term phase of the PDO can help improve seasonal climate prediction, because the warm phase of PDO tends to reinforce the climate patterns of El Niño, which helps strengthen the impact of an El Niño event. Likewise, La Niña events are generally reinforced when the PDO is in its cool phase.

The PDO was generally positive from 1922 to 1947, negative from 1947 to 1977, positive from 1977 to 1998, and negative from 1998 into the early 2010s (see ● Fig. 10.30). From 2014 into 2017, every month's PDO value was positive. This streak was the longest on record, and it suggested that the PDO had entered a new long-term positive phase. Scientists are researching the factors that cause the PDO to shift from one long-term phase to the other.

NORTH ATLANTIC OSCILLATION AND ARCTIC OSCILLATION

Over the Atlantic, a periodic reversal of pressure called the **North Atlantic Oscillation (NAO)** has a substantial effect on the weather in Europe and along the east coast of North America, especially during the winter. The NAO is measured by the difference in atmospheric pressure between the vicinity of the Icelandic low and the region of the Bermuda-Azores high. (Note that the NAO is defined by atmospheric change, whereas ENSO and the PDO are defined by oceanic change.) When the air pressure in the North Atlantic is unusually high near the Azores and unusually low near Iceland, the increased pressure gradient leads to stronger westerlies. These, in turn, drive frequent, powerful cyclonic storms into northern Europe, where winters tend to be wet and mild. During this *positive phase* of the NAO, winters in the eastern United States also tend to be wet and relatively mild, while northern Canada and eastern Europe are typically cold and dry (see ● Fig. 10.31a).

The *negative phase* of the NAO occurs when the atmospheric pressure in the vicinity of the Icelandic low rises, while the pressure drops in the region of the Bermuda high (see Fig. 10.31b). This results in a reduced pressure gradient and weaker westerlies, which steer fewer and weaker winter storms across the Atlantic.

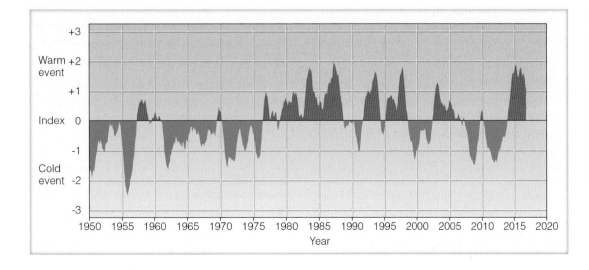

● FIGURE 10.30 The Pacific Decadal Oscillation (PDO) Index shown here is based on 12-month averages of sea surface temperatures (SST) of the Pacific Ocean north of 20°N from 1950 to 2013. Positive values (in red) indicate the warm phase, whereas negative values (in blue) represent the cool phase. *Source:* JISAO, University of Washington. Retrieved from http://jisao.washington.edu/pdo/PDO.latest. Courtesy Nate Mantua.

● **FIGURE 10.31** Change in surface atmospheric pressure and typical winter weather patterns associated with the (a) positive phase and (b) negative phase of the North Atlantic Oscillation (NAO).

The weaker jet stream also allows for storms to move or develop unusually far south, bringing wet weather to southern Europe and other areas near the Mediterranean Sea. The weaker, more variable jet stream during a negative NAO phase also allows cold air masses to move southward more readily into northern Europe and the eastern United States, which tends to make these regions generally colder and drier than usual (although the presence of cold air can sometimes lead to intense winter storms across the mid-Atlantic and northeastern states). Far eastern Canada and eastern Europe often experience milder-than-average winters during a negative NAO.

Although the NAO often varies from month to month, it can also tend to favor one phase or the other for several years. From the late 1960s into the early 2000s, the NAO was positive more often than negative, but that trend had faded by the late 2000s with the appearance of several periods of intensely negative NAO readings. During the severe winter of 2009–2010 in Europe and the eastern United States, the NAO reached its lowest average level in more than 100 years of records.

Closely related to the North Atlantic Oscillation, but analyzed at a more northerly latitude, is the **Arctic Oscillation (AO)**. The AO is defined by variations in atmospheric pressure between the Arctic and the North Pacific and Atlantic. During the *positive (warm) phase* of the AO (see ● Fig. 10.32a), higher pressures to the south and lower pressures across the Arctic produce strong westerly winds aloft. These winds wrap around the semipermanent zone of upper-level low pressure over the North Pole that is sometimes referred to as the **polar vortex**. When a positive AO is in place, the polar vortex is typically stronger than usual, and cold arctic air generally remains bottled up in and near the polar regions. Thus, winters in the United States tend to be warmer than normal, while winters over Newfoundland and Greenland tend to be very cold. Meanwhile, strong winds over the Atlantic direct storms into northern Europe, bringing with them wet, mild weather.

During the *negative (cold) phase* of the AO (Fig. 10.32b), pressure differences are smaller between the Arctic and regions to the south, leading to weaker and more variable westerly

● **FIGURE 10.32** Change in surface atmospheric pressure in polar regions and typical winter weather patterns associated with the (a) positive (warm) phase and the (b) negative (cold) phase of the Arctic Oscillation (AO). The arrows at the top of each globe show the relative strength of the winds encircling the polar vortex.

winds aloft. The weaker westerlies mean that the polar vortex can more easily shift to lower latitudes, and cold arctic air is now able to penetrate farther south, often producing colder-than-normal winters over much of the United States, as was the case during the winter of 2009–2010. Typically, cold air also invades northern Europe and Asia, while Newfoundland and Greenland experience warmer-than-normal winters.

Much like the NAO, the AO switches from one phase to another on an irregular basis, and one phase may predominate for several years in a row, bringing with it a succession of either cold or mild winters.*

*During the winter of 2009–2010, which produced severe cold over much of the United States and Europe, the NAO and AO were both strongly negative. However, these oscillation phases were not strongly negative during the winter of 2013–2014, which was also very cold over the eastern United States. This example shows us that the NAO and AO are not the only factors that can produce cold winters.

Unlike El Niño and La Niña, the NAO and AO can switch modes over just a few weeks' time, and these changes cannot be predicted more than a couple of weeks in advance. This influence makes it more challenging to produce seasonal outlooks over eastern North America and Europe than over western North America, where the more slowly varying ENSO and the PDO play a larger role. As our knowledge of the interactions between the ocean and atmosphere improves, we can expect scientists to gain skill at predicting shifts in all of these phenomena, as well as the resulting influences they have on regional weather and climate.

SUMMARY

In this chapter, we described the large-scale patterns of wind and pressure that persist around the world. We found that, at the surface in both hemispheres, the trade winds blow equatorward from the semipermanent high-pressure areas centered near 30° latitude. Near the equator, the trade winds converge along a boundary known as the intertropical convergence zone (ITCZ). On the poleward side of the subtropical highs are the prevailing westerly winds. The westerlies meet cold polar easterly winds along a boundary called the polar front, a zone of low pressure where middle-latitude cyclonic storms often form. The annual shifting of the major pressure areas and wind belts—northward in July and southward in January—strongly influences the annual precipitation of many regions.

Warm air aloft (high pressure) over low latitudes and cold air aloft (low pressure) over high latitudes produce westerly winds aloft in both hemispheres, especially at middle and high latitudes. Near the equator, easterly winds exist. The jet streams are located where strong winds concentrate into narrow bands. The polar jet stream forms in response to temperature contrasts along the polar front, while the subtropical jet stream forms at higher elevations above the subtropics, along an upper-level boundary called the subtropical front.

Near the surface, we examined the interaction between the atmosphere and oceans. We found the interaction to be an ongoing process where everything, in one way or another, seems to influence everything else. On a large scale, winds blowing over the surface of the water drive the major ocean currents; the oceans, in turn, release energy to the atmosphere, which helps to maintain the general circulation of winds. Where winds and the Ekman spiral move surface water away from a coastline, cold, nutrient-rich water upwells to replace it, creating good fishing conditions and cooler surface water.

When atmospheric circulation patterns change over the tropical Pacific, and the trade winds weaken or reverse direction, warm tropical water is able to flow eastward toward South America, where it chokes off upwelling and leads to potentially devastating effects on local economies. When the warm water extends over a vast area of the tropical Pacific and persists for several months to a year or more, the warming is called an El Niño event. La Niña is the name given to the situation where the surface water of the central and eastern tropical Pacific turns cooler than normal.

The reversals of atmospheric pressure over the Pacific Ocean associated with El Niño and La Niña events are called the Southern Oscillation. The large-scale interaction between the atmosphere and ocean during this process, which is sometimes referred to as the El Niño–Southern Oscillation (ENSO), affects global atmospheric circulation patterns. The changes in where rising and sinking air predominate lead to too much rain in some areas and not enough in others. Over the north-central Pacific and along the west coast of North America, the Pacific Decadal Oscillation affects surface water temperature, with its phase tending to change every 20 to 30 years. Over the Atlantic Ocean, a periodic reversal of pressure called the North Atlantic Oscillation influences weather in neighboring regions. A similar variation in atmospheric pressure at higher latitudes called the Arctic Oscillation also influences weather over parts of the Northern Hemisphere.

KEY TERMS

The following terms are listed (with corresponding page numbers) in the order they appear in the text. Define each. Doing so will aid you in reviewing the material covered in this chapter.

general circulation of the atmosphere, 264
Hadley cell, 264
doldrums, 265
subtropical highs, 266
horse latitudes, 266
trade winds, 266
intertropical convergence zone (ITCZ), 266
westerlies, 266
polar front, 266
subpolar low, 266
polar easterlies, 266
semipermanent highs and lows, 267
Bermuda high, 267
Pacific high, 267
Icelandic low, 267
Aleutian low, 267
Siberian high, 267
jet streams, 272

subtropical jet stream, 272
polar front jet stream, 273
tropical easterly jet stream, 275
low-level jet, 276
Gulf Stream, 277
oceanic front, 278
Ekman spiral, 279
Ekman transport, 279
upwelling, 279
El Niño, 279
Southern Oscillation, 280
ENSO, 280
La Niña, 280
teleconnections, 284
Pacific Decadal Oscillation (PDO), 285
North Atlantic Oscillation (NAO), 287
Arctic Oscillation (AO), 288
polar vortex, 288

QUESTIONS FOR REVIEW

1. Draw a large circle. Now, place the major surface pressure and wind belts of the world at their appropriate latitudes.
2. Explain how and why the average surface pressure features shift from summer to winter.
3. Why is it impossible on Earth for a Hadley cell to extend from the pole to the equator?
4. Along a meridian line running from the equator to the poles, how does the general circulation help to explain zones of abundant and sparse precipitation?
5. Most of the United States is in what wind belt?
6. Explain why summers in the United States tend to be dry along the West Coast but wet along the East Coast.
7. Explain why the winds in the middle and upper troposphere tend to blow from west to east in both the Northern and the Southern Hemispheres.
8. How does the polar front influence the development of the polar front jet stream?

9. Describe how the conservation of angular momentum plays a role in the formation of a jet stream.

10. Why is the polar front jet stream stronger in winter than in summer?

11. Explain the relationship between the general circulation of air and the circulation of ocean currents.

12. List at least four important interactions that exist between the ocean and the atmosphere.

13. Describe how the Ekman spiral forms.

14. What conditions are necessary for upwelling to occur along the west coast of North America? The east coast of North America?

15. What is an El Niño event?
    (a) What happens to the surface pressure at opposite ends of the Pacific Ocean during the Southern Oscillation?
    (b) Describe how the Southern Oscillation influences an El Niño event.

16. What are the conditions over the tropical eastern and central Pacific Ocean during the phenomenon known as La Niña?

17. What type of weather (cold/warm, wet/dry) would you expect over various parts of North America during a strong El Niño? During a strong La Niña?

18. Describe the ocean surface temperatures associated with the Pacific Decadal Oscillation. What climate patterns (cool/warm, wet/dry) tend to exist during the warm phase and the cool phase?

19. How does the positive phase of the North Atlantic Oscillation differ from the negative phase?

20. During the negative (cold) phase of the Arctic Oscillation, when Greenland is experiencing mild winters, what type of winters (cold or mild) is northern Europe usually experiencing?

QUESTIONS FOR THOUGHT

1. What effect would continents have on the circulation of air in the single-cell model?

2. How would the general circulation of air appear in summer and winter if Earth were tilted on its axis at an angle of 45° instead of $23\frac{1}{2}°$?

3. Summer weather in the southwestern section of the United States is influenced by a subtropical high-pressure cell, yet Fig. 10.5b (p. 268) shows an area of low pressure at the surface. Explain.

4. Explain why icebergs tend to move at right angles to the direction of the wind.

5. Give *two* reasons why pilots would prefer to fly in the core of a jet stream rather than just above or below it.

6. Why do the major ocean currents in the North Indian Ocean reverse direction between summer and winter?

7. Why are surface water temperatures along the northern California coast warmer in winter than in summer?

8. You are given an upper-level map that shows the position of two jet streams. If one is the polar front jet and the other the subtropical jet, how would you be able to tell which is which?

9. The Coriolis force deflects moving water to the right of its intended path in the Northern Hemisphere and to the left of its intended path in the Southern Hemisphere. Why, then, does upwelling tend to occur along the western margin of continents in both hemispheres?

PROBLEMS AND EXERCISES

1. Locate the following cities on a world map. Then, based on the general circulation of surface winds, predict the prevailing wind for each one during July and January.
    (a) Nashville, Tennessee
    (b) Oklahoma City, Oklahoma
    (c) Melbourne, Australia
    (d) London, England
    (e) Paris, France
    (f) Reykjavik, Iceland
    (g) Fairbanks, Alaska
    (h) Seattle, Washington

2. In the column below is a list of average weather conditions that prevail during the month of July at San Francisco, California, and Atlantic City, New Jersey. Both cities lie adjacent to an ocean at nearly the same latitude; however, the average weather conditions vary greatly. In terms of the average surface winds and pressure systems (see Fig. 10.5b, p. 268) and the interaction between the atmosphere and the ocean, explain what accounts for the variation between the two cities of each weather element.

Atlantic City, New Jersey (latitude 39°N)
Average weather, July

Temperature maximum/maximum	85°F / 67°F
Dew point	66°F
Monthly precipitation	3.72 in.
Prevailing wind	S
Water temperature	70°F

San Francisco, California (latitude 37°N)
Average weather, July

Temperature maximum/maximum	66°F / 54°F
Dew point	53°F
Monthly precipitation	0.00 in.
Prevailing wind	NW
Water temperature	59°F

ONLINE RESOURCES

Visit www.cengagebrain.com to view additional resources, including video exercises, practice quizzes, an interactive eBook, and more.

Cold fronts that push from the Andes across the Pampas (plains) of Argentina often produce intense thunderstorms.

Air Masses and Fronts

ABOUT TWO O'CLOCK IN THE AFTERNOON IT BEGAN to grow dark from a heavy, black cloud that was seen in the northwest. Almost instantly the strong wind, traveling at the rate of 70 miles an hour, accompanied by a deep bellowing sound, with its icy blast, swept over the land, and everything was frozen hard. The water in the little ponds in the roads froze in waves, sharp-edged and pointed, as the gale had blown it. The chickens, pigs, and other small animals were frozen in their tracks. Wagon wheels ceased to roll, froze to the ground. Men, going from their barns or fields a short distance from their homes, in slush and water, returned a few minutes later walking on the ice. Those caught out on horseback were frozen to their saddles and had to be lifted off and carried to the fire to be thawed apart. Two young men were frozen to death near Rushville. One of them was found with his back against a tree, with his horse's bridle over his arm and his horse frozen in front of him. The other was partly in a kneeling position, with a tinder box in one hand and a flint in the other, with both eyes wide open as if intent on trying to strike a light. Many other casualties were reported. As to the exact temperature, however, no instrument has left any record; but the ice was frozen in the stream, as variously reported, from six inches to a foot in thickness in a few hours.

John Moses, *Illinois: Historical and Statistical*

The opening quotation of this chapter details the passage of a spectacular cold front as it moved through Illinois on December 21, 1836. While some of the incidents described are likely exaggerated, at least one observing station, in Augusta, Illinois, reported air temperatures dropping from 40°F at dawn to 0°F by 2 p.m., and many people and animals are believed to have perished in the sudden cold. Fortunately, temperature changes of this magnitude with cold fronts are uncommon.

In this chapter, we will examine the more typical weather associated with cold fronts and warm fronts. We will address questions such as: Why are cold fronts usually associated with showery weather? How can warm fronts during the winter cause freezing rain and sleet to form over a vast area? And how can one read the story of an approaching warm front by observing its clouds? But, first, so that we may better understand fronts, we will examine air masses. We will look at where and how they form and the type of weather usually associated with them.

Air Masses

An **air mass** is an extremely large body of air whose properties of temperature and humidity are fairly similar in any horizontal direction at any given altitude. A single air mass may cover more than a million square kilometers. In ● Fig. 11.1, a large winter air mass, associated with a high-pressure area, covers over half of the United States. Note that, although the surface air temperature and dew point vary somewhat, everywhere the air is cold and dry, with the exception of the zone of snow showers on the eastern shores of the Great Lakes. This cold, shallow anticyclone will drift

eastward, carrying with it the temperature and moisture characteristic of the region where the air mass formed; hence, in a day or two, cold air will be located over the central Atlantic Ocean. Part of weather forecasting is, then, a matter of determining air mass characteristics, predicting how and why they change, and in what direction the systems will move.

SOURCE REGIONS Regions where air masses originate are known as **source regions**. For a huge mass of air to develop uniform characteristics, its source region should be generally flat and of uniform composition with light surface winds. The longer the air remains stagnant over its source region, or the longer the path over which the air moves, the more likely it will acquire properties of the surface below. Ideal source regions are usually those areas dominated by surface high pressure, which include the ice- and snow-covered arctic plains in winter and subtropical oceans in summer. The middle latitudes, where surface temperatures and moisture characteristics vary considerably, are not good source regions. Instead, this region is a transition zone where air masses with different physical properties move in, clash, and produce an exciting array of weather activity.

CLASSIFICATION Air masses are usually classified according to their temperature and humidity, both of which usually remain fairly uniform in any horizontal direction.* There are cold and warm air masses, humid and dry air masses. Air masses are grouped into five general categories according to their source

---

*In classifying air masses, it is common to use the *potential temperature* of the air. The potential temperature is the temperature that unsaturated (dry) air would have if moved from its original level to a pressure of 1000 millibars at the dry adiabatic rate (10°C/1000 m).

● **FIGURE 11.1** Here, a large, extremely cold winter air mass is dominating the weather over much of the United States. At almost all cities, the air is cold and dry. Upper number is air temperature (°F); bottom number is dew point (°F).

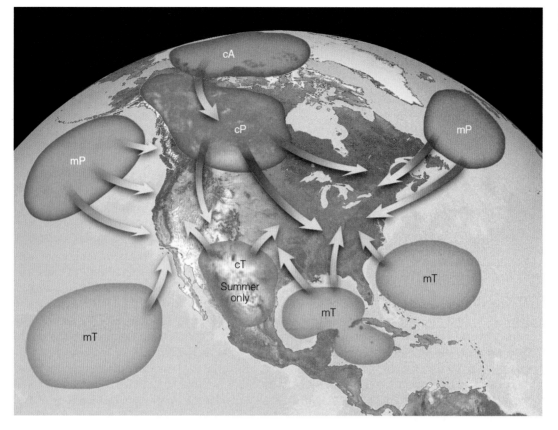

region. Air masses that originate in polar latitudes are designated by the capital letter P (for *polar*); those that form in warm tropical regions are designated by the capital letter T (for *tropical*). If the source region is land, the air mass will be dry and the lowercase letter c (for *continental*) precedes the P or T. If the air mass originates over water, it will be moist—at least in the lower layers—and the lowercase letter m (for *maritime*) precedes the P or T. We can now see that polar air originating over land will be classified cP on a surface weather map, whereas tropical air originating over water will be marked as mT. In winter, an extremely cold air mass that forms over the Arctic is designated as cA, *continental arctic*. Sometimes, however, it is difficult to distinguish between arctic and polar air masses, especially when the arctic air mass has traveled over warmer terrain. ▼ Table 11.1 lists the five basic air masses.

After the air mass spends some time over its source region, it may begin to move in response to strengthening winds aloft. As it moves away from its source region, the air mass encounters surfaces that may be warmer or colder than itself. When the air mass is colder than the underlying surface, it is warmed from below, which produces instability at low levels. In this case, increased

convection and turbulent mixing near the surface usually produce good visibility, cumuliform clouds, and showers of rain or snow. On the other hand, when the air mass is warmer than the surface below, the lower layers are chilled by contact with the cold Earth. Warm air above cooler air produces stable air with little vertical mixing. This situation causes the accumulation of dust, smoke, and pollutants, which restricts surface visibilities. In moist air, stratiform clouds accompanied by drizzle or fog may form.

AIR MASSES OF NORTH AMERICA   The principal air masses (with their source regions) that enter the United States are shown in ● Fig. 11.2. We are now in a position to study the formation and modification of each of these air masses and the variety of weather that accompanies them.

Continental Polar (cP) and Continental Arctic (cA) Air Masses

The bitterly cold weather that invades southern Canada and the United States in winter is associated with **continental polar** and **continental arctic air masses**. These air masses originate over the ice- and snow-covered regions of the Arctic, northern

▼ **TABLE 11.1   Air Mass Classification and Characteristics**

SOURCE REGION	ARCTIC REGION (A)	POLAR (P)	TROPICAL (T)
Land	cA	cP	cT
Continental (c)	Extremely cold, dry, stable; ice-and snow-covered surface	Cold, dry, stable	Hot, dry, stable air aloft; unstable surface air
Water		mP	mT
Maritime (m)		Cool, moist, unstable	Warm, moist; usually unstable

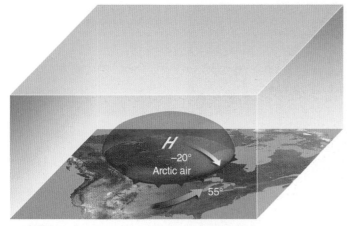

● **FIGURE 11.3** A shallow but large dome of extremely cold air—a continental arctic air mass—moves slowly southeastward across the upper plains. The leading edge of the air mass is marked by a cold front. (Numbers represent air temperature, °F.)

Canada, and Alaska where long, clear nights allow for strong radiational cooling of the surface. Air in contact with the surface becomes quite cold and stable. Since little moisture is added to the air, it is also quite dry, and dew-point temperatures are often less than −30°C (−22°F). Eventually a portion of this cold air breaks away and, under the influence of the air flow aloft, moves southward as an enormous, shallow high-pressure area, as illustrated in ● Fig. 11.3. (As we will see later in this chapter, air masses can be more than 100 times wider than they are tall.)

As the cold air moves into the interior plains, there are no topographic barriers to restrain it, so it continues southward, bringing with it frigid temperatures. The infamous *blue norther* is associated with continental arctic (and continental polar) air. As the air mass moves over warmer land to the south, the air temperature moderates slightly as it is heated from below. Even during the afternoon, when the surface air is most unstable, cumulus clouds are rare because of the extreme dryness of the air. At night, when the winds die down, rapid radiational surface cooling and clear skies can combine to produce very low minimum temperatures. If the cold air moves as far south as central or southern Florida, or south Texas, fruit and vegetable crops may be severely damaged. When the cold, dry air mass moves over a relatively warm body of water, such as the Great Lakes, heavy snow showers—called **lake-effect snows**—often form on downwind shores. Dense patches of *steam fog* can also develop on the upwind shores as frigid air passes over the water. (More information on lake-effect snows is provided in Focus section 11.1.)

In winter, the generally fair weather accompanying polar continental and arctic air masses is due to the stable nature of the atmosphere aloft. Sinking air develops above the large dome of high pressure. The subsiding air warms by compression and creates warmer air, which lies above colder surface air, often causing a strong upper-level subsidence inversion to form. Should the anticyclone stagnate over a region for several days, the visibility gradually drops as pollutants become trapped in the cold air near the ground. Usually, however, winds aloft move the cold air mass either eastward or southeastward.

The Rockies, Sierra Nevada, and Cascades normally protect the Pacific Northwest from the onslaught of arctic air, but,

occasionally, very cold air masses do invade these regions. When the upper-level winds over Washington and Oregon blow from the north or northeast on a trajectory beginning over northern Canada or Alaska, cold air can slip over the mountains and extend its icy fingers all the way to the Pacific Ocean. As the air moves off the high plateau, over the mountains, and on into the lower valleys, compressional heating of the sinking air causes its temperature to rise, so that by the time it reaches the lowlands, it is considerably warmer than it was originally. However, in no way would this air be considered warm. In some cases, the subfreezing temperatures slip over the Cascades and extend southward into the coastal areas of southern California.

A similar but less dramatic warming of continental polar and arctic air masses occurs along the eastern coast of the United States. Air rides up and over the lower Appalachian Mountains. Turbulent mixing and compressional heating increase the air temperatures on the downwind side. As a result, cities located to the east of the Appalachian Mountains usually do not experience temperatures as low as those on the west side. In Fig. 11.1, p. 294, notice that for the same time of day—in this case 7 a.m. EST—Philadelphia, on the eastern side of the mountains, with an air temperature of 14°F, is 16°F warmer than Pittsburgh, with an air temperature of −2°F, on the western side of the mountains.

● Figure 11.4 shows two upper-air wind patterns that led to extremely cold outbreaks of arctic air during December 1989 and 1990. Upper-level winds typically blow from west to east, but, in both of these cases, the flow, as indicated by the heavy, dark arrows, had a strong north-south (meridional) trajectory. The H represents the positions of the cold surface anticyclones. Numbers on the map represent minimum temperatures (°F) recorded during the cold spells. East of the Rocky Mountains, more than 350 record low temperatures were set between December 21 and 24, 1989, with the arctic outbreak causing an estimated $480 million in damage to the fruit and vegetable crops in Texas and Florida. Along the West Coast, the frigid air during December 1990, caused over $300 million in damage to the vegetable and citrus crops, as temperatures over parts of California plummeted to their lowest readings in more than 50 years. Notice in both cases how the upper-level wind directs the paths of the air masses.

Continental polar air that moves into the United States in summer has properties much different from its winter counterpart. The source region remains the same, but the air is now accompanied by long summer days that melt snow and warm the land. The air is only moderately cool, and surface evaporation adds water vapor to the air. A summertime continental polar air mass usually brings relief from the oppressive heat often occurring

Lake-Effect (Enhanced) Snows

During the winter, when the weather in the Midwest is dominated by clear and cold polar or arctic air, people living on the eastern or southern shores of the Great Lakes brace themselves for heavy snow showers. Snowstorms that form on the downwind side of one of these lakes are known as *lake-effect snows*. Because the lakes are responsible for enhancing the amount of snow that falls on its downwind side, these snowstorms are also called *lake-enhanced snows*, especially when the snow accompanies a cold front or mid-latitude cyclone. These storms are highly localized, extending from just a few kilometers to more than 100 km inland. The snow usually falls as a heavy shower or squall in a concentrated zone. So centralized is the region of snowfall that one part of a city may accumulate many inches of snow while in another part, the ground is bare. The amount of snow that falls can be enormous; for example, 65 inches fell near Buffalo, New York (on the eastern side of Lake Ontario), in less than 48 hours during November 2014.

Lake-effect snows are most numerous from November to January. During these months, cold air moves over the lakes when they are relatively warm and often not frozen. The contrast in temperature between water and air can be as much as 25°C(45°F). Studies show that the greater the contrast in temperature, the greater the potential for snow showers. In ● Fig. 1 we can see that, as the cold air moves over the warmer water, the air mass is quickly warmed from below, making it more buoyant and less stable. Rapidly, the air sweeps up moisture, soon becoming saturated. Out over the water, the vapor condenses into steam fog. As the air continues to warm, it rises and forms billowing cumuliform clouds, which continue to grow as the air becomes more unstable. Eventually, these clouds produce heavy showers of snow, which make the lake seem like a snow factory. Once the air and clouds reach the downwind side of the lake, additional lifting is provided by low hills and the convergence of air as it slows down over the rougher terrain. In late

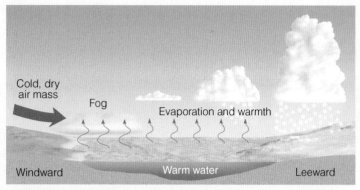

● **FIGURE 1** The formation of lake-effect snows. Cold, dry air crossing the lake gains moisture and warmth from the water. The more buoyant air now rises, forming clouds that deposit large quantities of snow on the lake's leeward (downwind) shores.

winter, the frequency and intensity of lake-effect snows often taper off as the temperature contrast between water and air diminishes and larger portions of the lakes freeze.

Generally, the longer the stretch of water over which the air mass travels (the longer the fetch), the greater the amount of warmth and moisture derived from the lake, and the greater the potential for heavy snow showers. In fact, studies show that, for significant snowfall to occur, the air must move across 80 km (50 mi) of open water. Consequently, forecasting lake-effect snowfalls depends to a large degree on determining the trajectory of the air as it flows over the lake. Regions that experience heavy lake-effect snowfalls are shown in ● Fig. 2.*

As the cold air moves farther east, the heavy snow showers usually taper off; however, the western slope of the Appalachian Mountains produces further lifting, enhancing the possibility of more and heavier showers. The heat given off during condensation warms the air and, as the air descends the eastern

---

*Buffalo, New York, is a city that experiences heavy lake-effect snows. Visit the National Weather Service website in Buffalo at www.weather.gov/buf/lakeeffect and read about lake-effect snowstorms measured in feet, as well as interesting weather stories related to lake-effect snow.

● **FIGURE 2** Areas shaded white show regions that experience heavy lake-effect snows.

slope, compressional heating warms it even more. Snowfall ceases, and by the time the air arrives in Philadelphia, New York, or Boston, the only remaining trace of the snow showers occurring on the other side of the mountains are the puffy cumulus clouds drifting overhead.

Lake-effect snows are not confined to the Great Lakes. In fact, any large unfrozen lake (such as the Great Salt Lake) can enhance snowfall when cold, relatively dry air sweeps over it. Moreover, a type of lake-effect snow occurs when cold air moves over a relatively warm ocean, then lifts slightly as it moves over a landmass. Such *ocean-effect snows* are common over Cape Cod, Massachusetts, in winter.

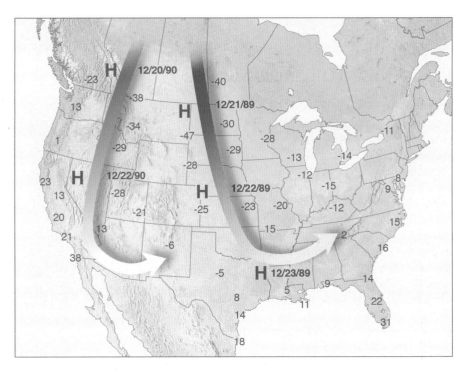

● **FIGURE 11.4** Average upper-level wind flow (heavy arrows) and surface position of anticyclones (H) associated with two extremely cold outbreaks of arctic air during late December 1989 and 1990. Numbers on the map represent minimum temperatures (°F) measured during each cold snap.

in the central and eastern states, as cooler air lowers the air temperature to more comfortable levels. Daytime heating warms the lower layers, producing surface instability. With its added moisture, the water vapor in the rising air may condense and create a sky dotted with fair-weather cumulus clouds (cumulus humilis). A typical profile of temperatures for a summer and a winter continental polar air mass is given in ● Fig. 11.5. Notice that the strong inversion so prevalent in winter is absent in summer.

When an air mass moves over a large body of water, its original properties may change considerably. For instance, cold, dry, continental polar air moving over the Gulf of Mexico warms rapidly and gains moisture. The air quickly assumes the qualities of a maritime air mass. Notice in ● Fig. 11.6 that rows of cumulus clouds (*cloud streets*) are forming over the Gulf of Mexico parallel to northerly surface winds as polar air is being warmed by the water beneath it, causing the air mass to destabilize. As the air continues its journey southward into Mexico and Central America, strong, moist northerly winds build into heavy clouds and showers along the northern coast. In this way, an air mass that was once cold, dry, and stable can be modified to such an extent that its original characteristics are no longer discernible. When this happens, the air mass is given a new designation.

Notice also in ● Fig. 11.7 that a similar modification of continental polar air is occurring along the Atlantic Coast, as northwesterly winds are blowing over the mild Atlantic. When this air encounters the much warmer Gulf Stream water, it warms rapidly and becomes conditionally unstable. Vertical mixing brings down faster-flowing cold air from aloft. This mixing creates strong, gusty surface winds and choppy seas, which can be hazardous to shipping.

In summary, polar and arctic air masses are responsible for the bitterly cold winter weather that can cover wide sections of North America. When the air mass originates over the Canadian Northwest Territories, frigid air can bring record-breaking low temperatures. Such was the case on Christmas Eve 1983, when arctic air covered most of North America. (A detailed look at this

air mass and its accompanying record-setting low temperatures is given in Focus section 11.2.)

Maritime Polar (mP) Air Masses   During the winter, cold polar and arctic air originating over Asia and frozen polar regions is carried eastward and southward over the Pacific Ocean by the circulation around the prevailing Aleutian low. The ocean water modifies these cold air masses by adding warmth and moisture to them. Since this air travels across many hundreds or even thousands of kilometers of water, it gradually changes into *maritime polar air*.

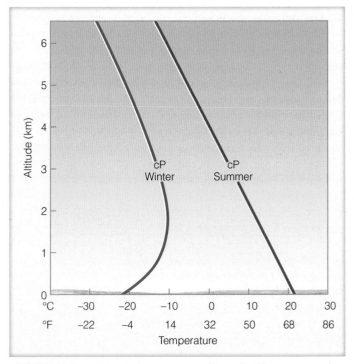

● **FIGURE 11.5** Typical vertical temperature profile over land for a summer and a winter cP air mass.

Cold, dry
(cP)

Gulf stream

Heavy
showers

● **FIGURE 11.6** Visible satellite image showing the modification of cold continental polar air as it moves over the warmer Gulf of Mexico and the Atlantic Ocean.

By the time this **maritime polar air mass** reaches the Pacific Coast, it is cool, moist, and conditionally unstable. The ocean's effect is to keep air near the surface warmer than the air aloft. Temperature readings in the 40s and 50s (°F) are common near the surface, while air at an altitude of about a kilometer or so above the surface may be at the freezing point. Within this colder air, characteristics of the original cold, dry air mass may still prevail. As the air moves inland, coastal mountains force it to rise, and much of its water vapor condenses into rain-producing clouds. In the colder air aloft, the rain changes to snow, with heavy amounts accumulating in mountain regions. Over the relatively warm open ocean, the cool moist air mass produces cumulus clouds that show up as tiny white splotches on a visible satellite image (see Fig. 11.7).

When the maritime polar air moves inland, it loses much of its moisture as it crosses a series of mountain ranges. Beyond these mountains, it travels over a cold, elevated plateau that chills the surface air and slowly transforms the lower level into dry, stable continental polar air. East of the Rockies this air mass is referred to as **Pacific air** (see ● Fig. 11.8). Here, it often brings fair weather and temperatures that are cool but not nearly as cold as the continental polar and arctic air that invades this region from northern Canada. In fact, when Pacific air from the west replaces retreating cold air from the north, chinook winds often develop. Furthermore, when the modified maritime polar air replaces moist subtropical air, thunderstorms can form along the boundary separating the two air masses.

Maritime polar air mass

L

● **FIGURE 11.7** Clouds and airflow aloft (large blue arrow) associated with maritime polar (mP) air moving into California. The large L shows the position of an upper-level low. Regions experiencing precipitation are also shown. The small, white clouds over the open ocean are cumulus clouds forming in the conditionally unstable air mass. (Precipitation symbols are given in Appendix B.)

● **FIGURE 11.8** After crossing several mountain ranges, cool, moist maritime polar (mP) air from off the Pacific Ocean descends the eastern side of the Rockies as modified, relatively dry Pacific air.

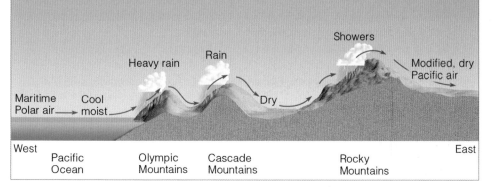

Showers

Heavy rain

Rain

Dry

Modified, dry
Pacific air

Maritime
Polar air

Cool
moist

West

Pacific
Ocean

Olympic
Mountains

Cascade
Mountains

Rocky
Mountains

East

The Return of the Siberian Express

The winter of 1983–1984 was one of the coldest on record across North America. Unusually frigid weather arrived in December, which ended up as the coldest December for the contiguous 48 states since records began in 1895 (a record it still holds). During the first part of the month, continental polar air covered most of the northern and central plains. As the cold air moderated slightly, far to the north a huge mass of bitterly cold arctic air was forming over the frozen reaches of the Canadian Northwest Territories.

By midmonth, the frigid air, associated with a massive high-pressure area, covered all of northwest Canada. Meanwhile, an upper-level ridge was forming over Alaska. On the eastern side of the ridge, strong northerly winds associated with the polar jet stream directed the frigid air southward over the prairie provinces of Canada. A portion of the extraordinarily cold air broke away, and, like a large swirling bubble, moved as a cold, shallow anticyclone southward into the United States. The frigid air was accompanied in some regions by winds gusting to 45 knots. At least one journalist labeled the onslaught "the Siberian Express." (The term has since become common in news coverage of cold waves in the United States, although it does not necessarily mean a particular air mass originated in Siberia.)

In many locations across the United States, the Siberian Express dropped temperatures to the lowest readings ever recorded during the month of December. On December 22, Elk Park, Montana, recorded an unofficial low of −64°F, only 6°F higher than the all-time low of −70°F for the United States (excluding Alaska), which was recorded at Rogers Pass, Montana, on January 20, 1954.

The center of the massive anticyclone gradually pushed southward out of Canada. By December 24, its center was over eastern Montana (see ● Fig. 3), where the sea-level pressure at Miles City reached an incredible 1064 mb (31.40 in.), a record for the 48 contiguous states. An enormous ridge of high pressure stretched from the Canadian arctic coast to the Gulf of Mexico. On the east side of the ridge, cold westerly winds brought lake-effect snows to the eastern shores of the Great Lakes. To the south of the high-pressure center, cold easterly winds, rising along the elevated plains, brought light amounts

● **FIGURE 3** Surface weather map for 7 a.m., EST, December 24, 1983. Solid lines are isobars. Areas shaded white represent snow. An extremely cold arctic air mass covers nearly 90 percent of the United States. (Weather symbols for the surface map are given in Appendix B.)

of *upslope snow** to sections of the Rocky Mountain states. Notice in Fig. 3 that, on Christmas Eve, arctic air covered almost 90 percent of the United States. As the cold air swept eastward and southward, a hard freeze caused hundreds of millions of dollars in damage to the fruit and vegetable crops in Texas, Louisiana, and Florida. On Christmas Day, 125 record-low temperature readings were set in 24 states. That afternoon, at 1:00 p.m., it was actually colder in Atlanta, Georgia, at 9°F than it was in Fairbanks, Alaska (10°F). One of the worst cold waves to occur in December during the twentieth century continued through the week, as many new record lows were established in the Deep South from Texas to Louisiana.

By January 1, the extreme cold had moderated, as the upper-level winds became more westerly. These winds brought milder Pacific air eastward into the Great Plains. The warmer pattern continued until about January 10, when the Siberian Express decided to make a return visit. Driven by strong upper-level northerly winds, impulse after impulse of arctic air from Canada

swept across the United States. On January 18, a low of −65°F was recorded at Middle Sinks, Utah. On January 19, temperatures plummeted to −7°F for the airports in Philadelphia and Baltimore, tying the coldest readings observed on any date at those locations. Toward the end of the month, the upper-level winds once again became more westerly. Over much of the nation, the cold air moderated. However, the express was to return at least one more time.

The beginning of February saw relatively warm air covering much of the nation from California to the Atlantic coast. However, on February 4, an arctic outbreak spread southward and eastward across the United States. Although freezing air extended southward into central Florida, the express ran out of steam, and a February heat wave soon engulfed most of the states east of the Rocky Mountains as warm, humid air from the Gulf of Mexico spread northward.

Even though February 1984 was a warmer-than-normal month over much of the United States, the winter of 1983–1984 (December, January, and February) will go down in the record books as one of the coldest winters for the United States as a whole since reliable record keeping began.

*Upslope snow forms as cold air moving from east to west over the Great Plains gradually rises (and cools even more) as it approaches the Rocky Mountains.

● **FIGURE 11.9** Winter and early spring surface weather pattern that usually prevails during the invasion of cold, moist maritime polar (mP) air into the mid-Atlantic and New England states. (Green-shaded area represents light rain and drizzle; pink-shaded region represents freezing rain and sleet; white-shaded area is experiencing snow.)

Along the East Coast, maritime polar air originates in the North Atlantic as continental polar air moves southward some distance off the Atlantic Coast. (Look back at Fig. 11.2, p. 295.) Because the water of the North Atlantic is very cold and the air mass travels only a short distance, wintertime Atlantic maritime polar air masses are usually much colder than their Pacific

counterparts. Because the prevailing winds aloft are westerly, Atlantic maritime polar air masses are also much less likely to move into the United States, although they often affect Europe.

● Figure 11.9 illustrates a typical late winter or early spring surface weather pattern that carries maritime polar air from the Atlantic into the New England and middle Atlantic states. A slow-moving, cold anticyclone drifting to the east (north of New England) causes a northeasterly onshore flow of cold, moist polar air to the south. The boundary separating this invading colder air from warmer air even farther south is marked by a stationary front. North of this front, northeasterly winds provide generally undesirable weather, consisting of cold, damp air and low, thick clouds from which light precipitation falls in the form of rain, drizzle, or snow. When upper atmospheric conditions are right, mid-latitude cyclonic storms may develop along the stationary front, move eastward, and intensify near the shores of Cape Hatteras. Such storms, called *Hatteras lows*, sometimes swing northeastward along the coast, where they become *northeasters* (commonly called *nor'easters*) bringing with them strong northeasterly winds, heavy rain or snow, and coastal flooding. (Such developing storms will be treated in detail in Chapter 12.)

Maritime Tropical (mT) Air Masses   The wintertime source region for Pacific **maritime tropical** air masses is the subtropical central and eastern Pacific Ocean. (Look back at Fig. 11.2, p. 295.) Air from this region must travel over 1600 km of water before it reaches the southern California coast. Consequently, these air masses are often very warm and moist by the time they arrive along the West Coast. In winter, the warm air produces heavy precipitation, usually in the form of rain, even at high elevations. Melting snow and rain quickly fill rivers, which overflow into the low-lying valleys. The rapid snowmelt can leave local ski slopes barren, and the heavy rain can cause disastrous mudslides in the steep canyons.

● Figure 11.10 shows maritime tropical air (usually referred to as *subtropical air*) streaming into northern California on

● **FIGURE 11.10** An infrared satellite image that shows maritime tropical air (heavy yellow arrow) moving into northern California on January 1, 1997. The warm, humid airflow (sometimes called "the Pineapple Express") produced heavy rain and extensive flooding in northern and central California. This event is one example of an atmospheric river, the periodic weather feature that provides the West Coast with much of its annual precipitation.

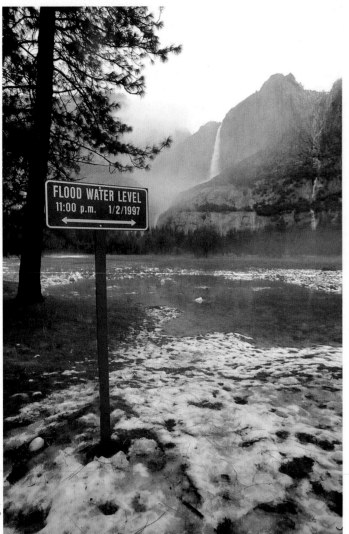

● **FIGURE 11.11** A sign atop flooding in Yosemite National Park in January 2017 indicates the height of the water level from a major flood at the same location in 1997. Both floods were related to the flow of moist, subtropical air masses into California.

The warm, humid subtropical air that influences much of the weather east of the Rockies originates over the Gulf of Mexico and Caribbean Sea. In winter, cold polar and arctic air tends to dominate the continental weather scene, so maritime tropical air is usually confined to the Gulf and extreme southern states. Occasionally, a slow-moving cyclonic storm system over the Central Plains draws warm, humid air northward. South or southwesterly winds carry this air into the central and eastern parts of the United States in advance of the storm. Because the land is still extremely cold, air near the surface is chilled to its dew point. Fog and low clouds form in the early morning, dissipate by midday, and reform in the evening. This mild winter weather in the Mississippi and Ohio valleys lasts, at best, only a few days. Soon cold polar or arctic air will move down from the north behind the eastward-moving storm system. Along the boundary between the two air masses, the warm, humid air is lifted above the more dense, cold polar air—a situation that often leads to heavy and widespread precipitation and storminess.

When a large mid-latitude cyclonic storm system stalls over the Central Plains, a constant supply of warm, humid air from the Gulf of Mexico can bring record-breaking maximum temperatures to the eastern half of the country. Sometimes the air temperatures are higher in the mid-Atlantic states than they are in the Deep South, as compressional heating warms the air even more as it moves downslope after crossing the Appalachian Mountains.

● Fig. 11.12 (p. 304) shows a surface weather map and the associated upper-level airflow (heavy arrow) that brought unseasonably warm maritime tropical air into the central and eastern states during March 2012. A large surface high-pressure area centered off the East Coast coupled with a strong southwesterly flow aloft carried warm, moist air into the Midwest and East, causing a record-breaking March heat wave. The flow aloft prevented the surface low and the cooler air behind it from making much eastward progress, so that the warm spell lasted a week or more in some locations. Michigan saw its first-ever 90°F reading to occur in March, and Burlington, Vermont, reached 80°F. More than 7000 daily record highs were set or tied in the United States during March 2012. Note that, on the west side of the surface low, the winds aloft funneled cool Pacific air into the western states. Although this cold air mass was not nearly as unusual as the warm air mass to its east, it still brought chilly conditions from California to the Rockies. Hence, while people in the Northwest were huddled around heaters, others several thousand miles away in the Midwest and Northeast were soaking up sunshine and warmth. We can see that it is the upper-level meridional flow, directing cooler air southward and warm subtropical air northward, that makes these contrasts in temperature possible.

In summer, the circulation of air around the Bermuda High (which sits off the southeast coast of North America—see Fig. 10.5b, p. 268) pumps warm, humid (maritime tropical) air northward from off the Gulf of Mexico and from off the Atlantic Ocean into the eastern half of the United States. As this humid air moves inland, it warms even more, rises, and

January 1, 1997. The flow of humid, subtropical air, which originated near the Hawaiian Islands, was termed by at least one forecaster "*the Pineapple Express*," a term that has since become common in media coverage. After battering the Pacific Northwest with heavy rain, the Pineapple Express roared into northern and central California, causing catastrophic floods that sent more than 100,000 people fleeing from their homes, mudslides that closed roads, property damage (including crop losses) that amounted to more than $1.5 billion, and eight fatalities. Yosemite National Park, which sustained over $170 million in damages due mainly to flooding, was forced to close for more than two months (see ● Fig. 11.11).

The Pineapple Express is one example of the feature known as an **atmospheric river (AR)**, a relatively narrow channel of very moist air. To learn more about atmospheric rivers, see Focus section 11.3.

Rivers in the Atmosphere

The Pineapple Express, which we discussed on p. 302, is one example of the feature known as an *atmospheric river* (AR). Behaving like a stream of moisture, but surrounded by air instead of soil, an atmospheric river transports huge amounts of water vapor from one place to another, typically from the warm, humid tropics and subtropics into the midlatitudes. Some of the water vapor in an atmospheric river also comes from the mid-latitude ocean itself, where it can evaporate directly into the surrounding air and become concentrated in the AR channel by converging winds. The strongest atmospheric rivers can carry up to 15 times more moisture in the form of water vapor than the amount of water flowing into the Gulf of Mexico at the mouth of the Mississippi River.

The bulk of the moisture in an atmospheric river is concentrated in the lowest 3 km (10,000 feet) of the atmosphere (see ● Fig. 4). Atmospheric rivers usually gather and concentrate moisture ahead of a mid-latitude storm system. A typical atmospheric river is roughly 400 to 640 km wide (250 to 400 miles) along a path that can span more than 1600 km (1000 miles). Sometimes an atmospheric river can extend in a nearly straight line all the way from the region near Hawaii to the U.S. West Coast, as in the Pineapple Express (see Fig. 11.10, p. 301).

The amount of time that atmospheric rivers persist at a given location is a critical factor in how much rain or snow occurs. Often an AR will develop one or more kinks in its path, as portions of the AR get pulled toward the center of a mid-latitude storm. This situation often produces a one- or two-day sequence of heavier precipitation alternating with lighter amounts. Periods of very heavy rain can occur while the AR is located across a given point, with lighter rain when the

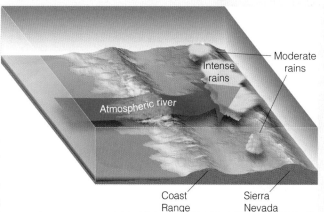

● **FIGURE 4** A three-dimensional portrayal of an atmospheric river heading into California. Precipitation amounts are heaviest near the core of the atmospheric river, with lighter amounts toward the north and south.

AR has shifted north or south of that point. Rainfall totals can be especially heavy if an AR remains stationary over a given point for a long period.

Along the west coast of the United States, a few atmospheric river events each year can provide as much as half of all annual precipitation. The rain and snow can be especially heavy when an AR slams directly (in perpendicular fashion) into the higher elevations from California to British Columbia, where its moisture can be forced upward and squeezed out. The winter of 2016–2017 brought a series of atmospheric rivers to California, producing one of the wettest years on record—a striking turnabout following five years of intense drought. Statewide, the total precipitation was the heaviest for any October-to-February period in more than a century of record keeping. A set of five benchmark snow observing points in the northern Sierra piled up more moisture in snowpack during 2016–2017 than for any other winter on record.

The central and eastern United States can also be affected by atmospheric rivers extending north from the Atlantic Ocean, the Gulf of Mexico, the Caribbean Sea, and the far eastern tropical Pacific. One such AR contributed to the catastrophic flooding that affected Nashville, Tennessee, and other parts of the mid-South in May 2010. Nashville received more than 13 inches of rain in two days, a total that would be expected only about every 500 to 1000 years on average.

Sometimes an AR affecting the southeast United States is called the *Mayan Express*, an analogy to the Pineapple Express but in this case referring to the source region of tropical and subtropical moisture in the waters surrounding eastern Mexico and Central America. One such Mayan Express in March 2016 brought 10 to 20 inches of rain and devastating floods to parts of Texas and Louisiana. Many nor'easters that bring heavy rain and snow from the mid-Atlantic to New England are injected with moisture from atmospheric rivers.

---

frequently condenses into cumuliform clouds, which produce afternoon showers and thunderstorms. You can count on thunderstorms developing along the Gulf Coast on almost every summer afternoon. As evening approaches, thunderstorm activity typically dies off. Nighttime cooling lowers the temperature of this hot, muggy air only slightly. Should the air become saturated, fog or low clouds usually form, and these normally dissipate by late morning as surface heating warms the air again.

A weak, but often persistent, flow around an upper-level anticyclone in summer will spread warm, humid air from the Gulf of Mexico and from the Gulf of California into the southern and central Rockies, where it causes afternoon thunderstorms. Occasionally, this easterly flow can work its way even farther west, producing shower activity in the otherwise dry southwestern desert.

Humid subtropical air that originates over the southeastern Pacific and Gulf of California remains south of the United States during much of the year. However, during the summer monsoon

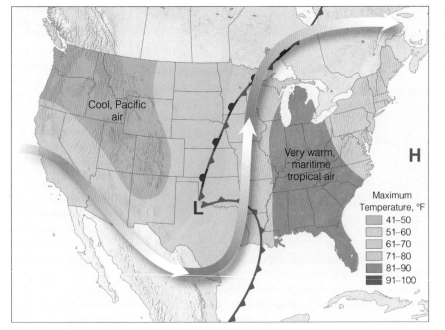

● **FIGURE 11.12** Weather conditions during an unseasonably warm spell in the eastern portion of the United States that occurred in the second half of March 2012. The surface low-pressure area, fronts, and maximum temperatures are shown for March 21. The heavy arrow is the average upper-level flow during the warm period.

period, a weak upper-level southerly flow can spread this humid air northward into the southwestern United States, most often in Arizona, Nevada, and the southern part of California. In many places, the moist, conditionally unstable air aloft only shows up as middle and high cloudiness, especially altocumulus and cirrocumulus castellanus. However, where the moist flow meets a mountain barrier, it tends to rise and condense into towering, shower-producing clouds. If the air is sufficiently moist, it can cause heavy showers over a broad area. (For an illustration of exceptionally strong flow of subtropical air into this region, look at Fig. 9.40 on p. 252.)

Continental Tropical (cT) Air Masses The only real source region for hot, dry **continental tropical air masses** in North America is found during the summer in northern Mexico and the adjacent arid southwestern United States. Here, the air mass is hot, dry, and conditionally unstable at low levels, with frequent dust devils forming during the day. Because of the low relative humidity (typically less than 10 percent during the afternoon), air must rise more than 3000 m (10,000 ft) before condensation begins. Furthermore, an upper-level ridge usually produces sinking air over the region, tending to make the air aloft rather stable and the surface air even warmer. Consequently, skies are generally clear, the weather is hot, and rainfall is practically nonexistent where continental tropical air masses prevail. If this air

mass moves outside its source region and into the Great Plains and stagnates over that region for any length of time, a severe drought may result. ● Figure 11.13 shows an instance where

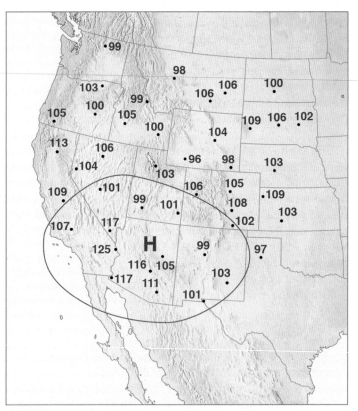

● **FIGURE 11.13** From July 14 through July 22, 2005, continental tropical air covered a large area of the southwestern United States. Numbers on the map represent maximum temperatures (°F) during this period. The large H with the isobar shows the upper-level position of the subtropical high. Sinking air associated with the high contributed to the hot, dry weather. Winds aloft were weak, with the main flow over central Canada.

WEATHER WATCH

A continental tropical air mass, stretching from California to Texas, brought widespread record warmth during the second half of June 2017. Temperatures soared to 127°F at Death Valley, California, and 119°F at Phoenix, Arizona. Dozens of flights were canceled at the Phoenix Sky Harbor Airport because the hot air was not dense enough to support take-offs and landings for certain types of aircraft.

continental tropical air covered a large portion of the southwestern United States and produced hot, dry weather during July 2005.

So far, we have examined the various air masses that enter North America annually. The characteristics of each depend on the air mass source region and the type of surface over which the air mass moves. The winds aloft determine the trajectories of these air masses. Occasionally, an air mass will control the weather in a region for some time. These persistent weather conditions are sometimes referred to as *air mass weather.*

Air mass weather is especially common in the southeastern United States during summer as, day after day, humid subtropical air from the Gulf brings sultry conditions and afternoon thunderstorms. It is also common in the Pacific Northwest in winter when conditionally unstable, cool moist air accompanied by widely scattered showers dominates the weather for several days or more. The real weather action, however, usually occurs not within air masses but at their margins, where air masses with sharply contrasting properties meet—in the zone marked by weather fronts.*

---

*The word *front* is used to denote the clashing or meeting of two air masses, probably because it resembles the fighting in Western Europe during World War I, when the term originated.

BRIEF REVIEW

Before we examine fronts, here is a review of some of the important facts about air masses:

- An air mass is a large body of air whose properties of temperature and humidity are fairly similar in any horizontal direction.

- Source regions for air masses tend to be generally flat, of uniform composition, and in an area of light winds, dominated by surface high pressure.

- Continental air masses form over land. Maritime air masses form over water. Polar air masses originate in cold, polar regions, and extremely cold arctic air masses form over arctic regions. Tropical air masses originate in warm, tropical regions.

- Continental polar (cP) air masses are cold and dry; continental arctic (cA) air masses are extremely cold and dry. It is the continental arctic air masses that produce the extreme cold of winter as they move across North America.

- Continental tropical (cT) air masses are hot and dry, and are responsible for the heat waves of summer in the western half of the United States.

- Maritime polar (mP) air masses are cold and moist, and are responsible for the cold, damp, and often wet weather along the northeast coast of North America, as well as for the cold, rainy winter weather along the west coast of North America.

- Maritime tropical (mT) air masses are warm and humid, and are responsible for the hot, muggy weather that frequently plagues the eastern half of the United States in summer.

Fronts

Although we briefly looked at fronts in Chapter 1, we are now in a position to study them in depth, which will aid us in forecasting the weather. We will now learn about the general nature of fronts—how they move and what weather patterns are associated with them.

A **front** is the transition zone between two air masses of different densities. As density differences are most often caused by temperature differences, fronts usually separate air masses with contrasting temperatures. Often, they separate air masses with different humidities as well. Remember that air masses have both horizontal and vertical extent; consequently, the upward extension of a front is referred to as a *frontal surface,* or a *frontal zone.*

● Figure 11.14 illustrates the vertical extent of two large-scale frontal zones—the *polar front* and the *arctic front.* The *polar front* boundary, which extends upward more than 5 km (3 mi) from the surface, separates warm, humid air to the south from cold polar air to the north. The *arctic front,* which separates cold air from extremely cold arctic air, is much shallower than the polar front and only extends upward to an altitude of about 1 or 2 kilometers. In the next several sections, as we examine individual fronts on a flat surface weather map, keep in mind that all fronts have horizontal and vertical extent.

● Figure 11.15 shows a surface weather map illustrating four different fronts. Notice that the fronts are associated with lower pressure and that the fronts separate differing air masses. As we move from west to east across the map, the fronts appear in the following order: a stationary front between points A and B; a cold front between points B and C; a warm front between points C and D; and an occluded front between points C and L. Let's examine the properties of each of these fronts.

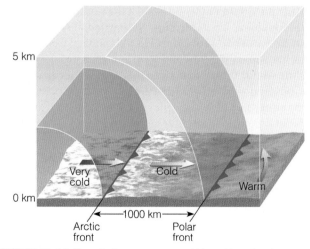

● **FIGURE 11.14** The polar front represents a cold frontal boundary that separates colder air from warmer air at the surface and aloft. The more shallow arctic front separates cold air from extremely cold air.

● **FIGURE 11.15** A surface weather map showing surface-pressure systems, air masses, fronts, and isobars (in millibars) as solid gray lines. Large arrows in color show airflow. (Green-shaded area represents rain; pink-shaded area represents freezing rain and sleet; white-shaded area represents snow.)

STATIONARY FRONTS   A **stationary front** has essentially no movement.* On a colored weather map, it is drawn as an alternating red and blue line. Red semicircles face toward colder air on the red line and blue triangles point toward warmer air on the blue line. The stationary front between points A and B in Fig. 11.15 marks the boundary where cold, dense continental polar (cP) air from Canada butts up against the north-south trending Rocky Mountains. Unable to cross the barrier, the cold air shows little or no westward movement. The stationary front is drawn along a line separating the continental polar air from the milder, more humid maritime polar air to the west. Notice that the surface winds tend to blow parallel to the front, but in opposite directions on either side of it. Upper-level winds often blow parallel to a stationary front. The weather along the front is clear to partly cloudy, with much colder air lying on its eastern side. Because both air masses are relatively dry, there is no precipitation. This is not, however, always the case. When warm moist air rides up and over the cold air, widespread cloudiness with light precipitation can cover a vast area. These are the conditions that prevail north of the east-west running stationary front depicted in Fig. 11.9 (p. 301). In some cases where a stationary front butts up against a mountain range, as shown in Fig. 11.15, winds blowing upslope

---

*They are usually called *quasi-stationary fronts* because they can show some movement.

can generate light rain or snow (called *upslope precipitation*) if there is enough moisture in the air.

If the colder air to the east begins to retreat and is replaced by the warmer air to the west, the front in Fig. 11.15 will no longer remain stationary; it will become a warm front. If, on the other hand, the colder air slides up over the mountain and replaces the warmer air on the other side, the front will become a cold front. If either a cold front or a warm front stops moving, it becomes a stationary front.

COLD FRONTS   The **cold front** between points B and C on the surface weather map in Fig. 11.15 represents a zone where cold, dry stable polar air is replacing warm, moist, conditionally unstable subtropical air. The front is drawn as a solid blue line with the triangles along the front showing its direction of movement. How did the meteorologist know to draw the front at that location? A closer look at the front will give us the answer.

The weather in the immediate vicinity of this cold front in the southern United States is shown in ● Fig. 11.16. The data plotted on the map represent the current weather at selected cities. The station model used to represent the data at each reporting station is a simplified one that shows temperature, dew point, present weather, cloud cover, sea-level pressure, and wind direction and speed. The little line in the lower right-hand corner of each station shows the *pressure tendency*—the pressure change, whether rising (/) or falling (\) — during the last three hours. With all of

● **FIGURE 11.16** A closer look at the surface weather associated with the cold front situated in the southern United States in Fig. 11.15. (Solid gray lines are isobars. Green-shaded area represents rain; white-shaded area represents snow.)

● **FIGURE 11.17** A Doppler radar image showing precipitation patterns along a cold front similar to the cold front in Fig. 11.16. Green represents light-to-moderate precipitation; yellow represents heavier precipitation; and red the most likely areas for thunderstorms. (The cold front is superimposed on the radar image.)

CRITICAL THINKING QUESTION   How would you expect the precipitation pattern to change along the cold front if the front were fast-moving and the air behind the front was continental arctic air?

this information, the front can be properly located.* (Appendix B explains the weather symbols and the station model more completely.)

We can use the following criteria to locate a front on a surface weather map:

1. Sharp temperature changes over a relatively short distance
2. Changes in the air's moisture content (as shown by marked changes in the dew point)
3. Shifts in wind direction
4. Pressure and pressure changes
5. Clouds and precipitation patterns

In Fig. 11.16, we can see a large contrast in air temperature and dew point on either side of the front. There is also a wind shift from southwesterly ahead of the front, to northwesterly behind it. Notice that each isobar kinks as it crosses the front, forming an elongated area of low pressure—a *trough*—which accounts for the wind shift. Because surface winds normally blow across the isobars toward lower pressure, we find winds with a southerly component ahead of the front and winds with a westerly component behind it.

Since the cold front is a trough of low pressure, sharp changes in pressure can be significant in locating the front's position. One important fact to remember is that the lowest pressure usually occurs just as the front passes a station. Notice that,

*Locating any front on a weather map is not always a clear-cut process. Even meteorologists can disagree on an exact position.

as you move toward the front, the pressure drops, and, as you move away from it, the pressure rises. This is clearly shown by the pressure tendencies for each station on the map. Just before the front passes, the pressure tendency shows the atmospheric pressure is falling (\), while just behind the front, the pressure is now beginning to rise (), and farther behind the front, the pressure is rising steadily (/).

The precipitation pattern along the cold front in Fig. 11.16 might appear similar to the Doppler radar image shown in ● Fig. 11.17. The region in color extending from northeast to southwest represents precipitation along a cold front. Notice that light-to-moderate rain (colored in green) occurs over a wide area along the front, while the heavier precipitation (yellow) is found in a narrow band along the front itself. Strong thunderstorms (red) are found only in certain areas along this front.

The cloud and precipitation patterns in Fig. 11.16 along the line X-X' are shown in a side view of the front in ● Fig. 11.18. We can see from Fig. 11.18 that, at the front, the cold, dense air wedges under the warm air, forcing the warm air upward, much like a snow shovel forces snow upward as the shovel glides through the snow. As the moist, conditionally unstable air rises, it condenses into a series of cumuliform clouds. Strong, upper-level westerly winds blow the delicate ice crystals (which form near the top of the cumulonimbus) into cirrostratus (Cs) and cirrus (Ci). These clouds usually appear far in advance of the approaching front. At the front itself, a relatively narrow band of thunderstorms (Cb) produces heavy showers with gusty winds. Behind the front, the air cools quickly. (Notice how the freezing level dips as it crosses the front.) The winds shift from southwesterly to northwesterly, pressure rises, and precipitation ends. As the air dries out, the skies clear, except for a few lingering cumulus clouds.

Observe that the leading edge of the front is steep. The steepness is due to friction, which slows the airflow near the ground. The air aloft pushes forward, blunting the frontal surface. Even so, the leading edge of the front does not extend very high into the atmosphere. If we could walk from where the front touches the surface back into the cold air, a distance of 50 km, the front would be about 1 km above us. Thus, the slope of the front—the ratio of vertical rise to horizontal distance—is 1:50. This is typical for a fast-moving cold front—those that move at about 25 knots or more. In a slower-moving cold front—one that moves at about 15 knots—the slope is much more gentle.

With slow-moving cold fronts, clouds and precipitation usually cover a broad area behind the front. When the ascending warm air is stable, stratiform clouds, such as nimbostratus, become the predominant cloud type and even fog may develop in the rainy area. Occasionally, out ahead of a fast-moving front, a line of active showers and thunderstorms, called a *squall line*, develops parallel to and often ahead of the advancing front, producing heavy precipitation and strong gusty winds. Scattered thunderstorms may also occur well ahead of a cold front.

As the temperature contrast across a front lessens, the front will often weaken and dissipate. Such a condition is known as **frontolysis**. On the other hand, an increase in the temperature contrast across a front can cause it to strengthen and regenerate into a more vigorous frontal system, a condition called **frontogenesis**.

An example of a regenerated front is shown in the infrared satellite images in ● Fig. 11.19. The cold front in Fig. 11.19a is weak, as indicated by the low clouds (gray tones) along the front. As the front moves offshore, over the warm Gulf Stream (Fig. 11.19b), it intensifies into a more vigorous frontal system as surface air becomes conditionally unstable and convective activity develops. Notice that the area of cloudiness is more extensive and thunderstorms are now forming along the frontal zone.

So far, we have considered the general weather patterns of "typical" cold fronts. There are, of course, many exceptions. In fact, no two fronts are exactly alike. In some, the cold air is very

shallow; in others, it is much deeper. If the rising warm air is dry and stable, scattered clouds are all that form, and there is no precipitation. In extremely dry weather, a marked change in the dew point, accompanied by a slight wind shift, may be the only clue to a passing cold front.

During the winter, a series of cold polar or arctic outbreaks may travel across the United States so quickly that warm air is unable to develop ahead of the front. In this case, frigid arctic air associated with an arctic front usually replaces cold polar air, and a drop in temperature is the only indication that a front has moved through your area. Along the West Coast, the Pacific Ocean modifies the air so much that typical cold fronts, such as those described in the previous section, are not seen. In fact, as a cold front moves inland from the Pacific Ocean, the surface temperature contrast across the front may be quite small. Topographic features usually distort the wind pattern so much that locating the position of the front and the time of its passage is exceedingly difficult. In this case, the pressure tendency is the most reliable indication of a frontal passage.

In some instances along the West Coast, an approaching cold front (or upper-level trough) will cause cool marine air at the surface to *surge* into coastal and inland valleys. The cool air (which is often accompanied by a wind shift) may produce a sharp drop in air temperature. This may give the impression that a rather strong cold front has moved through, when in reality, the front may be many kilometers offshore.

Cold fronts usually move toward the south, southeast, or east. But sometimes they will move southwestward. In New England,

NOAA

● **FIGURE 11.19** The infrared satellite image (a) shows a weakening cold front over land on Tuesday morning, November 21, intensifying into (b) a vigorous front over warm Gulf Stream water on Wednesday morning, November 22.

this movement occurs when northeasterly surface winds, blowing clockwise around an anticyclone centered to the north over Canada, push a cold front southwestward, often as far south as Boston. Because the cold front moves in from the east, or northeast, it is known as a **"back-door" cold front**. As the front passes, westerly surface winds usually shift to easterly or northeasterly and temperatures drop as moist maritime polar air flows in off the Atlantic Ocean.

An example of a "back-door" cold front is shown in ● Fig. 11.20. This is a springtime situation where, behind the front, the weather is cold and damp with drizzle, as northeasterly winds sweep into the region from off the chilly Atlantic. To the south of the front, the weather is much warmer. Should the front move through this area, the more summerlike weather would change, in a matter of hours, to more winterlike. The cold, dense air behind the front is rather shallow. Consequently, the Appalachian Mountains act as a dam to the front's forward progress, halting its westward movement. This situation, where the cold, damp air is confined to the eastern side of the mountains, is called **cold air damming**. The stalled cold front now becomes a stationary front. The cool air behind the front may linger for some time as warmer, less-dense air to the south rides up and over it. Forecasting how far south the "back-door" cold front will move and when the entrenched cold air will leave can be a bit tricky.

Even though cold-front weather patterns have many exceptions, learning these patterns can be to your advantage if you live in an area that experiences well-defined cold fronts. Knowing them improves your own ability to make short-range weather forecasts. For your reference, ▼ Table 11.2 summarizes idealized cold-front weather in winter in the Northern Hemisphere.

WARM FRONTS   In Fig. 11.15, p. 306, a **warm front** is drawn along the solid red line running from points C to D. Here, the leading edge of advancing warm, moist, subtropical (maritime

● **FIGURE 11.20** A "back-door" cold front moving into New England during the spring. Notice that, behind the front, the weather is cold and damp with drizzle, while to the south, ahead of the front, the weather is partly cloudy and warm.

tropical) air from the Gulf of Mexico replaces the retreating cold maritime polar air from the North Atlantic. The direction of frontal movement is given by the half circles, which point into the cold air; this front is heading toward the northeast. As the cold air recedes, the warm front slowly advances. The average speed of a warm front is about 10 knots, or about half that of an average cold front. During the day, as mixing occurs on both sides of the front, its movement may be much faster. Warm fronts often move in a series of rapid jumps, which show

WEATHER ELEMENT	BEFORE PASSING	WHILE PASSING	AFTER PASSING
Winds	South or southwest	Gusty, shifting	West or northwest
Temperature	Warm	Sudden drop	Steadily dropping
Pressure	Falling steadily	Minimum, then sharp rise	Rising steadily
Clouds	Increasing Ci, Cs, then either Tcu* or Cb*	Tcu or Cb	Often Cu, Sc* when ground is warm
Precipitation	Short period of showers	Heavy showers of rain or snow, sometimes with hail, thunder, and lightning	Decreasing intensity of showers, then clearing
Visibility	Fair to poor in haze	Poor, followed by improving	Good, except in showers
Dew point	High; remains steady	Sharp drop	Lowering

*Tcu stands for towering cumulus, such as cumulus congestus, whereas Cb stands for cumulonimbus. Sc stands for stratocumulus.

up on successive weather maps. At night, however, radiational cooling creates cool, dense surface air behind the front. This inhibits both lifting and the front's forward progress. When the forward surface edge of the warm front passes a station, the wind shifts, the temperature rises, and the overall weather conditions improve. To see why, we will examine the weather commonly associated with the warm front both at the surface and aloft.

● Figure 11.21 is a surface weather map showing the position of a wintertime warm front and its associated weather. ● Figure 11.22 is a vertical view of a model of the warm front in Fig. 11.21. Look at these two figures and observe that, in the model, the warmer, less-dense air rides up and over the colder, more-dense surface air. This rising of warm air over cold (called **overrunning** in this model), produces clouds and precipitation well in advance of the front's surface boundary. The warm front that separates the two air masses has an average slope of about 1:300*—a much more gentle slope than that of a typical cold front. Warm air overriding the cold air creates a stable atmosphere (see the vertical temperature profile in Fig. 11.22b). Notice that a temperature inversion—called a **frontal inversion**—exists in the region of the upper-level front at the boundary where the warm air overrides the cold air. Another fact to notice in Fig. 11.22b is that the wind *veers* (shifts clockwise) with altitude, so that the southeasterly (SE) surface winds become southwesterly (SW) and westerly (W) aloft.

Suppose that we are standing at the position marked P' in Figs. 11.21 and 11.22. Note that we are over 1200 km (750 mi) ahead of where the warm front is touching the surface. Here, the surface winds are light and variable, the air is cold, and about the only indication of an approaching warm front is the high cirrus clouds overhead. We know the front is moving slowly toward us and that within a day or so it will pass our area. Suppose that, instead of waiting for the front to pass us, we drive toward it, observing the weather as we go.

Heading toward the warm front, we notice that the cirrus (Ci) clouds gradually thicken into a thin, white veil of cirrostratus (Cs) whose ice crystals cast a halo around the sun.** Almost imperceptibly, the clouds thicken and lower, becoming altocumulus (Ac) and altostratus (As) through which the sun shows only as a faint spot against an overcast gray sky. Snowflakes begin to fall, and we are still over 600 km (370 mi) from the surface front. The snow increases, and the clouds thicken into a sheet-like covering of nimbostratus (Ns). The winds become brisk out of the southeast, while the atmospheric pressure slowly falls. Within 400 km (250 mi) of the front, the cold surface air mass is now quite shallow. The surface air temperature moderates and, as we approach the front, the light snow changes first into sleet. It then becomes freezing rain and finally rain and drizzle as the air temperature climbs above freezing. Overall, the precipitation remains light or moderate but covers a broad area. Moving still closer to the front, warm, moist air mixes with cold, moist air, producing ragged windblown stratus (St) and fog. (As you might deduce, flying in the vicinity of a warm front is quite hazardous.)

Finally, after a trip of over 1200 km, we reach the warm front's surface boundary. As we cross the front, the weather changes are noticeable, but much less pronounced than those experienced with the cold front; they show up more as a gradual transition rather than a sharp change. On the warm side of the front, the air temperature and dew point rise, the wind shifts from southeast to south or southwest, and the air pressure stops falling. The light rain ends and, except for a few stratocumulus (Sc), the fog and low clouds vanish.

This scenario of an approaching warm front represents average (if not idealized) warm-front weather in winter. In some instances, the weather can differ from this dramatically. For example, if the overrunning warm air is relatively dry and stable, only high and middle clouds will form, and no precipitation will occur. On the other hand, if the warm air is relatively moist and conditionally unstable (as is often the case during the summer), heavy showers can develop as thunderstorms become embedded in the cloud mass. Some of these thunderstorms may have bases at a relatively high level above the surface, and thus are called *elevated storms*.

---

*The slope of 1:300 is a much more gentle slope than that of most warm fronts. Typically, the slope of a warm front is on the order of 1:150 to 1:200.

**If the warm air is relatively unstable, ripples or waves of cirrocumulus clouds will appear as a "mackerel sky."

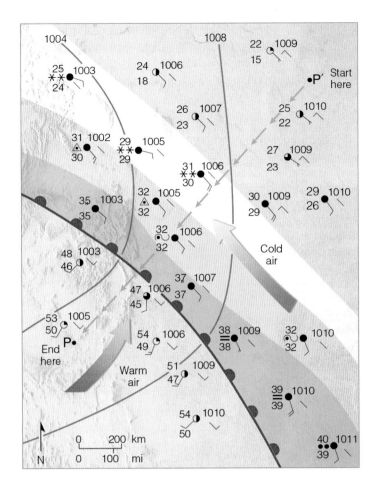

● **FIGURE 11.21** Surface weather associated with a typical warm front in winter. A vertical view along the dashed line P-P′ is shown in Fig. 11.22. (Green-shaded area represents rain; pink-shaded area represents freezing rain and sleet; white-shaded area represents snow.)

CRITICAL THINKING QUESTION If the warm front in Fig. 11.21 were located in the Southern Hemisphere, in which general direction would the front probably be moving? How would the wind direction change across the front? On which side of the front would you expect to find precipitation?

Along the West Coast, the Pacific Ocean significantly modifies the surface air so that warm fronts are difficult to locate on a surface weather map. Also, not all warm fronts move northward or northeastward. On rare occasions, a warm front will move into the eastern seaboard from the Atlantic Ocean as it spins all the way around a deep cyclonic storm positioned off the coast. Cold northeasterly winds ahead of the front usually become warm northeasterly winds behind it. Even with these exceptions, knowing the normal sequence of wintertime warm-front weather can be useful, especially if you live where warm fronts become well developed. You can look for certain cloud and weather patterns and make reasonably accurate short-range forecasts of your own. ▼ Table 11.3 summarizes typical winter warm-front weather. (Before going on to the next section, you may wish to read Focus section 11.4, which gives additional information about warm fronts.)

● **FIGURE 11.22** Vertical view of a model illustrating clouds, precipitation, and winds across the warm front in Fig. 11.21 along the line P-P′.

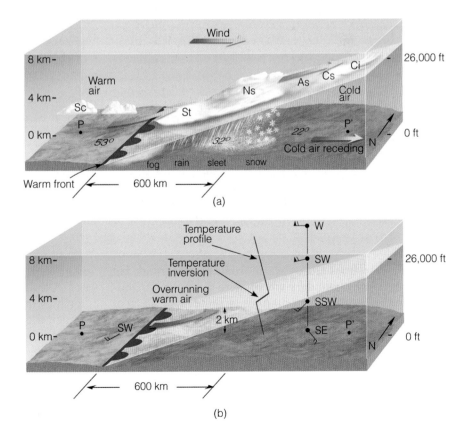

The Wavy Warm Front

Up to this point, we've examined idealized warm fronts on a surface weather map—like the one shown in Fig. 11.15, p. 306. Some warm fronts do look like this example; others, however, have an entirely different appearance. For instance, look at the warm front in ● Fig. 5. Notice that it has a wavelike shape as it approaches North Carolina from three different directions. So what causes the warm front to bend in this manner?

Look carefully at Fig. 5 and notice that at the surface cold air is flowing southwestward around a high-pressure area centered over southern Canada. As cold, dense, surface air pushes south into the southern states, it flows up against the Appalachian Mountains, which impede its westward progress. Since the shallow layer of cold air is unable to ride up and over the mountains, it becomes wedged along the mountains' eastern foothills. Recall that this trapping of cold air is called *cold air damming*.

Warm air pushing northward from the Gulf of Mexico rides up and over the cold, dry, surface air. Clouds and precipitation often form in this rising warm air. When rain falls into the shallow, cold air, it may evaporate, chilling the air even more. Sometimes the rain freezes before reaching the ground, producing sleet; other times, the rain freezes on impact, producing freezing rain. If the frozen precipitation falls for many hours, severe ice storms may result, with heavy accumulations of ice causing treacherous driving conditions and downed power lines.

The shallow layer of cold air usually becomes entrenched in low-lying areas and therefore retreats northward very slowly. As the cold air slowly recedes northward, warmer air pushes in from different directions, and the leading edge of the warm air (the warm front) no longer has a nice curved shape, but begins to take on a more wavy shape, such as the warm front in Fig. 5.

● **FIGURE 5** Surface weather map for 11:00 p.m. (EST), February 13, 2007.

▼ **TABLE 11.3** **Typical Weather Conditions Associated with a Warm Front in Winter in the Northern Hemisphere**

WEATHER ELEMENT	BEFORE PASSING	WHILE PASSING	AFTER PASSING
Winds	South or southeast	Variable	South or southwest
Temperature	Cool to cold, slowly warming	Steady rise	Warmer, then steady
Pressure	Usually falling	Leveling off	Slight rise, followed by fall
Clouds	Ci, Cs, As, Ns, St, and fog; occasionally Cb in summer	Stratus type	Clearing with scattered Sc, especially in summer; occasionally Cb in summer
Precipitation	Light-to-moderate rain, snow, sleet, or drizzle; showers in summer	Drizzle or none	Usually none; sometimes light rain or showers
Visibility	Poor	Poor, but improving	Fair in haze
Dew point	Steady rise	Steady	Rise, then steady

DRYLINES **Drylines** are not warm fronts or cold fronts, but represent a narrow boundary where there is a steep horizontal change in moisture. Thus, drylines separate moist air from dry air. Because dew-point temperatures may drop along this boundary by as much as 9°C(16°F) per km, drylines have been referred to as *dew-point fronts*.* Sometimes drylines are associated with mid-latitude cyclones, and sometimes they are not. Although drylines can occur in the United States as far north as the Dakotas, and as far east as the Texas-Louisiana border, they are most frequently observed in the western half of Texas, Oklahoma, and Kansas, especially during spring and early summer. In these locations, drylines tend to move eastward during the day, then westward toward evening. Drylines are observed in other regions of the world, too. They occur, for example, in Central West Africa and in India before the onset of the summer monsoon.

---

*Recall from Chapter 4 that the dew-point temperature is a measure of the amount of water vapor in the air.

● **FIGURE 11.23** A dryline represents a narrow boundary where there is a steep horizontal change in moisture as indicated by a rapid change in dew-point temperature. Here, a dryline moving across Texas and Oklahoma separates warm, moist air from warm, dry air during an afternoon in May.

● Figure 11.23 shows a dryline moving across Texas during May 2001. Notice that the dryline is represented as a line with hollow brown half circles, packed more closely together than they are on a warm front. Notice also that, to the west of the dryline, warm, dry continental tropical air is moving in from the southwest. Consequently, on this side, the weather is usually hot and dry with gusty southwesterly winds. To the east of the dryline, warm, very humid maritime tropical air is sweeping northward from the Gulf of Mexico. Here we typically find air temperatures to be slightly lower and the humidity (as indicated by the higher dew points) considerably higher than on the western side. The semicircles of the dryline point toward this humid air.

Even though the dryline represents a moisture boundary, its actual position on a weather map is plotted according to a shift in surface winds. When insects and insect-eating birds congregate along the dryline, Doppler radar may be able to locate it. On the radar screen, the echo from insects and birds shows up as a thin line, called a *fine line*.

Cumulus clouds and thunderstorms often form along or to the east of the dryline. This cloud development is caused in part by daytime convection and a sloping terrain. The Central Plains area of North America is higher to the west and lower to the east. Convection over the elevated western plains carries dry air high above the surface. Westerly winds sweep this dry air eastward over the lower plains where it overrides the slightly cooler but more humid air at the surface. This situation sets up a potentially unstable atmosphere that finds warm, dry air above warm, moist air. In regions where the air rises, cumulus clouds and organized bands of thunderstorms can form (see ● Fig. 11.24). We will examine in more detail the development of these storms in Chapter 14.

OCCLUDED FRONTS   If a cold front catches up to and overtakes a warm front, the frontal boundary created between the two air masses is called an **occluded front**, or, simply, an **occlusion** (meaning "closed off"). There are two main types of occlusions, *cold-type occlusions* and *warm-type occlusions*. The warm-type occlusion is by far the most commonly observed.

On the surface weather map, an occluded front is represented as a purple line with alternating cold-front triangles and warm-front half circles; both symbols point in the direction toward which the front is moving. Look back at Fig. 11.15, p. 306, and notice that the air behind the occluded front is colder than the air ahead of it. This is known as a *cold-type occluded front*, or **cold-type occlusion**. Let's see how this front develops.

● **FIGURE 11.24** Cumulus clouds and thunderstorms developing along a dryline in Kansas.

The classic model of a developing cold-type occluded front is shown in ● Fig. 11.25. Along line A-A', the cold front is rapidly approaching the slower-moving warm front. Along line B-B', the cold front overtakes the warm front, and, as we can see in the vertical view across C-C', underrides and lifts off the ground both the warm front and the warm air mass. This situation allows colder air to replace cool air at the surface. As a cold-type occluded front approaches, the weather sequence is similar to that of a warm front, with high clouds lowering and thickening into middle and low clouds and with precipitation forming well in advance of the surface front. Given that the front represents a trough of low pressure, southeasterly winds and falling atmospheric pressure occur ahead of it. The frontal passage, however, can bring weather similar to that of a cold front: heavy, often showery precipitation with winds shifting to west or northwest. After a period of wet weather, the sky begins to clear, atmospheric pressure rises, and the air turns colder. The most active weather usually occurs where the cold front is just overtaking the warm front, at the point of occlusion, where the greatest contrast in temperature occurs.

The classic model of a developing warm-type occluded front is shown in ● Fig. 11.26. Notice that continental polar air over eastern Washington and Oregon may be much colder than milder maritime polar air moving inland from the Pacific Ocean. Observe also that the air ahead of the warm front is colder than the air behind the cold front. Consequently, when the cold front catches up to and overtakes the warm front, the milder, lighter air behind the cold front is unable to lift the colder, heavier air off the ground. As a result, the cold front rides "piggyback" along the sloping warm front. This produces a *warm-type occluded front*, or a **warm-type occlusion**. The surface weather associated with a warm occlusion is similar to that of a warm front. In addition, the relatively mild winter air that moves into Europe from the North Atlantic causes many of the occlusions that move into this region in winter to be of the warm variety.

Contrast Fig. 11.25 and Fig. 11.26. Note that the primary difference between the warm- and cold-type occluded fronts is the location of the upper-level front. In a cold-type occlusion, there is an upper-level warm front that *follows* the surface occluded front, whereas in a warm occlusion, there is an upper-level cold front that *precedes* the upper-level cold front *precedes* the surface occluded front.

So far, the ideas presented here on the formation of occluded fronts are based on the surface air temperature on either side of the front. In recent years, the classic model of cold-type and warm-type occluded fronts forming due to temperature differences ahead of the advancing warm front and behind the cold front has been challenged. Studies show that the *stability* of the air (rather than the surface air temperature) is the most important factor in determining the type of occluded front that forms. The air within the warm frontal zone is typically more stable than the air within the cold frontal zone, so the air behind the cold front is more likely to ride up and over the warm front than to flow beneath it. For this reason, warm-type occluded fronts are far more common than cold-type occluded fronts.

● **FIGURE 11.25** A model that shows the formation of a cold-type occluded front. The faster-moving cold front (a) catches up to the slower-moving warm front (b) and forces it to rise off the ground (c). The top illustration represents a surface map of a cold occlusion. (Green-shaded area on the map represents precipitation.)

(c)

(b)

E   Warm occluded front   E′

(a)

D   D′

● **FIGURE 11.26** A model illustrating the formation of a warm-type occluded front. The faster-moving cold front in (a) overtakes the slower-moving warm front in (b). The lighter air behind the cold front rises up and over the denser air ahead of the warm front. The top illustration represents a surface map of a warm occlusion.

In the world of weather fronts, occluded fronts are the mavericks. In fact, they may show up on a surface chart as a trough of low pressure separating two cold air masses. Because of this, locating and defining occluded fronts at the surface is often difficult for the meteorologist.* Similarly, you too may find it hard to recognize an occlusion. In spite of this, we will assume that the weather associated with occluded fronts behaves in a similar way to that shown in ▼ Table 11.4.

The cold, warm, and occluded frontal systems described in this chapter are actually part of a much larger storm system—the middle-latitude cyclone. ● Figure 11.27 shows the cold front, warm front, and occluded front in association with a mid-latitudecyclonic storm. Notice that, as we would expect, clouds and precipitation form in a rather narrow band along the cold front, and in a much wider band along the warm front and the occluded front. In Chapter 12, we will look more closely at middle-latitude cyclonic storms, examining where, why, and how they form. Before we move on, however, we need to look at fronts that form in the upper troposphere—fronts that may, or may not, show up at the surface.

UPPER-AIR FRONTS   An **upper-air front** (which is also known as an upper front, or *upper-tropospheric front*) is a front that is present aloft. It may or may not extend down to the surface. ● Figure 11.28 shows a north-to-south side view of an idealized upper-air front. Notice that the front forms when the tropopause—the boundary separating the troposphere from the stratosphere—dips downward and folds under the polar jet stream. In the fold, the isotherms are tightly packed, marking the position of the upper front. Although the upper front may not connect with a surface front, the position of the surface front is shown in the diagram.

---

*In fact, in some countries, such as Canada, the surface occlusion is seldom analyzed on a surface weather map. Instead, the location of the occluded front aloft—where the cold air lifts the warm air above the surface—is marked by a *TROWAL* (which stands for *TRO*ugh of *W*arm air *AL*oft). So the position of the *TROWAL* on the weather map (as indicated by a "hook") marks the location where the cold and warm fronts intersect aloft.

▼ **TABLE 11.4**   **Typical Weather Most Often Associated with Occluded Fronts in Winter in North America**

WEATHER ELEMENT	BEFORE PASSING	WHILE PASSING	AFTER PASSING
Winds	East, southeast, or south	Variable	West or northwest

Temperature
  (a) Cold-type occluded
  (b) Warm-type occluded
	
	Cold or cool
	Cold
		
		Dropping
		Rising
			
			Colder
			Milder

Pressure	Usually falling	Low point	Usually rising
Clouds	In this order: Ci, Cs, As, Ns	Ns, sometimes Tcu and Cb	Ns, As, or scattered Cu
Precipitation	Light, moderate, or heavy precipitation	Light, moderate, or heavy continuous precipitation or showers	Light to moderate precipitation followed by general clearing
Visibility	Poor in precipitation	Poor in precipitation	Improving
Dew point	Steady	Usually slight drop, especially if cold-occluded	Slight drop, although may rise a bit if warm-occluded

● **FIGURE 11.27** A visible satellite image showing a mid-latitude cyclonic storm with its weather fronts over the Atlantic Ocean during March 2005. Superimposed on the photo is the position of the surface cold front, warm front, and occluded front. Precipitation symbols indicate where precipitation is reaching the surface.

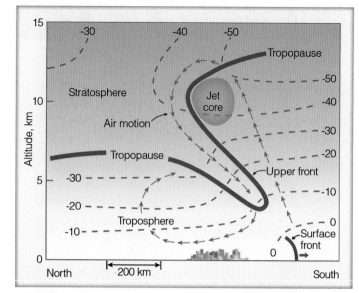

● **FIGURE 11.28** An idealized vertical view of an upper-air front showing the tropopause (heavy red line), isotherms in °C (dashed gray lines), and vertical air motions. The polar jet stream core (maximum winds) is flowing into the page (from west to east).

The small arrows in Fig. 11.28 show air motion associated with the upper front. On the north side of the front (and the north side of the jet stream), the air is slowly sinking. Here, in the folded troposphere, air from the stratosphere descends into the troposphere. To the south of the front (and south of the jet stream), the air slowly rises. These rising and descending air motions can aid in the development of middle-latitude cyclonic storms, as described in the next chapter.

SUMMARY

In this chapter, we considered the different types of air masses and the various weather each brings to a particular region. Continental arctic air masses are responsible for the extremely cold (arctic) outbreaks of winter, whereas continental polar air masses are responsible for cold, dry weather in winter and cool, pleasant weather in summer. Maritime polar air, having traveled over an ocean for a considerable distance, brings cool, moist weather to an area. The hot, dry weather of summer is associated with continental tropical air masses, while warm, humid conditions are due to maritime tropical air masses. Where air masses with sharply contrasting properties meet, we find weather fronts.

A front is a boundary between two air masses of different densities. Stationary fronts have essentially no movement, with cold air on one side and warm air on the other. Winds tend to blow parallel to the front, but in opposite directions on either side of it. Along the leading edge of a cold front, where colder air replaces warmer air, showers are prevalent, especially if the warmer air is moist and conditionally unstable. Along a warm front, warmer air rides up and over colder surface air, producing widespread cloudiness and light-to-moderate precipitation that can cover thousands of square kilometers. When the rising air is conditionally unstable (such as it often is in summer), showers and thunderstorms may form ahead of the advancing warm front. Cold fronts typically move faster and are more steeply sloped than warm fronts. Occluded fronts, which are often difficult to locate and define on a surface weather map, may have characteristics of both cold and warm fronts. A dryline is a type of front that represents a narrow boundary between hot, moist air and hot, dry air. Fronts that form in the upper troposphere, in the vicinity of the polar-front jet stream, are called upper-air fronts.

KEY TERMS

The following terms are listed (with corresponding page numbers) in the order they appear in the text. Define each. Doing so will aid you in reviewing the material covered in this chapter.

QUESTIONS FOR REVIEW

1. If an area is described as a "good air mass source region," what information can you give about it?
2. It is summer. What type of afternoon weather would you expect from an air mass designated as maritime tropical? Explain.
3. Why is continental polar air not welcome in the Central Plains in winter and yet very welcome in summer?
4. Explain why the central United States is not a good air mass source region.
5. Why do air temperatures tend to be a little higher on the eastern side of the Appalachian Mountains than on the western side at the same latitude, even though the same winter cP or cA air mass dominates both areas?
6. Explain how the airflow aloft regulates the movement of air masses.
7. List the temperature and moisture characteristics of each of the major air mass types.
8. What are lake-effect snows and how do they form? On which side of a lake do they typically occur?
9. Why are maritime polar air masses along the East Coast of the United States usually colder than those along the nation's West Coast? Why are they also *less* prevalent?
10. The boundaries between neighboring air masses tend to be more distinct during the winter than during the summer. Explain why.
11. What type of air mass would be responsible for the weather conditions listed as follows?
    (a) heavy snow showers and low temperatures at Buffalo, New York
    (b) hot, muggy summer weather in the Midwest and the East
    (c) daily afternoon thunderstorms along the Gulf Coast
    (d) heavy snow showers along the western slope of the Rockies
    (e) refreshing, cool, dry breezes after a long summer hot spell in the Midwest
    (f) heavy summer rain showers in southern Arizona
    (g) drought with high temperatures over the Great Plains
    (h) persistent cold, damp weather with drizzle along the East Coast
    (i) summer afternoon thunderstorms forming along the eastern slopes of the Sierra Nevada
    (j) record low winter temperatures in South Dakota
12. On a surface weather map, what do you know about a region where the word *frontogenesis* is marked?
13. Explain why barometric pressure usually falls with the approach of a cold front or occluded front.
14. How does the weather usually change along a dryline?

15. Based on the following weather forecasts, what type of front will most likely pass the area?
    (a) Light rain and cold today, with temperatures just above freezing. Southeasterly winds shifting to westerly tonight. Turning colder with rain becoming heavy and possibly changing to snow.
    (b) Cool today with rain becoming heavy at times by this afternoon. Warmer tomorrow. Winds southeasterly becoming westerly by tomorrow morning.
    (c) Increasing cloudiness and warm today, with the possibility of showers by evening. Turning much colder tonight. Winds southwesterly, becoming gusty and shifting to northwesterly by tonight.
    (d) Increasing high cloudiness and cold this morning. Clouds increasing and lowering this afternoon, with a chance of snow or rain tonight. Precipitation ending tomorrow morning. Turning much warmer. Winds light easterly today, becoming southeasterly tonight and southwesterly tomorrow.

16. Sketch side views of a typical cold front, warm front, and cold-occluded front. Include in each diagram cloud types and patterns, areas of precipitation, surface winds, and relative temperature on each side of the front.

17. During the spring, on a warm, sunny day in Boston, Massachusetts, the wind shifts from southwesterly to northeasterly and the weather turns cold, damp, and overcast. What type of front moved through the Boston area? From what direction did the front apparently approach Boston?

18. How does the tropopause show where an upper-level front is located?

QUESTIONS FOR THOUGHT

1. Suppose that a maritime polar air mass moving eastward from the Pacific Ocean travels across the United States. Describe all of the modifications that could take place as this air mass moves eastward in winter. In summer.

2. Explain how an anticyclone during autumn can bring record-breaking low temperatures and continental polar air to the southeastern states, and only a day or so later very high temperatures and maritime tropical air to the same region.

3. In Fig. 11.5 (p. 298), there is a temperature inversion. How does this inversion differ from the frontal inversion illustrated in Fig. 11.22b (p. 311)?

4. For Chicago, Illinois, to experience heavy lake-effect snows, from what direction would the wind have to be blowing?

5. When a very cold air mass covers half of the United States, a very warm air mass often covers the other half. Explain how this happens.

6. Explain why freezing rain more commonly occurs with warm fronts than with cold fronts.

7. In winter, cold-front weather is typically more violent than warm-front weather. Why? Explain why this is not necessarily true in summer.

8. Why does the same cold front typically produce more rain over Kentucky than over western Kansas?

9. You are in upstate New York and observe the wind shift from easterly to southerly. This shift in wind is accompanied by a sudden rise in both the air temperature and dew-point temperature. What type of front passed?

10. If Lake Erie freezes over in January, is it still possible to have lake-effect snows off Lake Erie in February? Why or why not?

11. Why are ocean-effect snow storms (described on p. 297) fairly common when a persistent cold northeasterly wind blows over Cape Cod, Massachusetts, but are not common when a cold northeasterly wind blows over Long Island, New York?

PROBLEMS AND EXERCISES

1. Make a sketch of North America and show the upper-air wind-flow pattern that would produce:
   (a) very cold continental arctic air moving into the far western states in winter
   (b) cold polar or arctic air over the Central Plains in winter
   (c) warm mT air over the Midwest in winter
   (d) warm, moist maritime tropical air over southern California and Arizona during the summer

2. You are presently taking a weather observation. The sky is full of wispy cirrus clouds estimated to be about 6 km (20,000 ft) overhead. If a warm front is approaching from the south, about how far away is it (assuming a slope of 1:200)? If it is moving toward you at an average warm-front speed of about 10 knots, how long will it take before it passes your area?

ONLINE RESOURCES

Visit www.cengagebrain.com to view additional resources, including video exercises, practice quizzes, an interactive eBook, and more.

A fierce middle-latitude cyclone over the eastern Atlantic pushes huge waves into the coast of Portugal.

Middle-Latitude Cyclones

CONTENTS

IT NEVER FOR A MOMENT OCCURRED TO ME to regard this storm as a thing to prevent me from getting to New York. The Jackson Avenue station of the New Jersey Central was only five or six blocks from the house, and I anticipated not the slightest trouble in getting there. It was still snowing and blowing . . . and not a print had been made in the snow about the house. I got out to the gate and into the "street," and was then able to discover the real snow.

It was everywhere. Great piles of it rose up like gigantic arctic graves . . . in all directions. Every way that I turned I was confronted with these awful mounds. I took my bearings and steered for the Jackson Avenue station. Every step I took I went into my knees in snow and every other step I fell over on my face and tried to see how much of the stuff I could swallow. The wind was at my back and its accompanying snowflakes cut the back of my head and ears like a million icy lashes. I . . . plowed my way, jumping, falling, and crawling over the drifts, some of which were nine or ten feet high . . . and after an hour and ten minutes I got to the end of my six blocks. There were trains there, two of them, but they were stuck.

I gave up the idea of going to New York. My trip back to the house was simply awful. The wind was straight in my face and beat so in my eyes that I couldn't see a rod* before me. My mustache was frozen stiff, and over my eyebrows were cakes of frozen snow. I stumbled along, falling down at almost every step, burying myself in the snow. Then, I began to feel like a crazy man. Every time I fell down, I shouted and cursed and beat the snow with my fists. Then it got dark, the wind howled and tore along, hurling the ice flakes in my face, and the very snow on the ground seemed to rise up and fling itself upon me.

In one of my crazy efforts to force ahead, I caught just a glimpse of the welcome gate posts, and then I laid down on my back and hollered. Somebody heard my cries, and just as I was going off comfortably to sleep my friend came plowing out through the snow, and he and this man dragged me into the house.

U.S. Department of Commerce, *American Weather Stories*

*A rod is a unit of measurement equal to 16½ feet.

The storm described in our opening is now referred to as the "Blizzard of '88." This legendary storm of March 11–14, 1888, was accompanied by high winds, measured at 60 miles per hour at Atlantic City, New Jersey; a severe cold wave; and unprecedented snowfall—up to 50 inches over portions of southeastern New York and southern New England, with drifts 30 to 40 feet high. In New York City, people died in the street, trapped in snowdrifts up to their hips. What atmospheric conditions are needed for such a monstrous storm to develop?

Early weather forecasters were aware that precipitation generally accompanied falling barometers and areas of low pressure. However, it was not until the early part of the twentieth century that scientists began to piece together the information that yielded the ideas of modern meteorology and storm development.

Working largely from surface observations, a group of scientists in Bergen, Norway, developed a model explaining the life cycle of an *extratropical*, or *middle-latitude cyclonic storm*; that is, a storm that forms at middle and high latitudes outside of the tropics. This extraordinary group of meteorologists included Vilhelm Bjerknes, his son Jakob, Halvor Solberg, and Tor Bergeron. They published their *Norwegian cyclone model* shortly after World War I. It was widely acclaimed and became known as the "polar front theory of a developing wave cyclone" or, simply, the **polar front theory.** What these meteorologists gave to the world was a working model of how a mid-latitude cyclone progresses through the stages of birth, growth, and decay. An important part of the model involved the development of weather along the polar front. As new information became available, the original work was modified, so that, today, it serves as a convenient way to describe the structure and weather associated with a migratory middle-latitude cyclonic storm system.

In the following sections, we will first examine, from a surface perspective, how a mid-latitude cyclone develops along the polar front. Then we will examine how the winds aloft influence the developing surface storm. Later on, we will obtain a three-dimensional view of a mid-latitude cyclone by observing how ribbons of air glide through the storm system.

Polar Front Theory

The development of a mid-latitude cyclone, according to the Norwegian model, begins along the polar front. Remember from the discussion of the general circulation in Chapter 10 that the polar front is a semicontinuous global boundary separating cold polar air from warm subtropical air. Because the mid-latitude cyclonic storm forms and moves along the polar front in a wave-like manner, the developing storm is called a **wave cyclone.** The stages of a developing wave cyclone are illustrated in the sequence of surface weather maps shown in ● Fig. 12.1.

Figure 12.1a shows a segment of the polar front as a stationary front. It represents a trough of lower pressure with higher pressure on both sides. Cold air to the north and warm air to the south flow parallel to the front, but in opposite directions. This type of flow sets up a cyclonic wind shear. You can conceptualize the shear more clearly if you place a pen between the palms of

your hands and slide your left palm toward your body; the pen turns counterclockwise, cyclonically.

Under the right conditions (described later in this chapter), a wavelike kink forms on the front, as shown in Fig. 12.1b. The wave that forms is known as a **frontal wave** or an *incipient cyclone*. Watching the formation of a frontal wave on a weather map is like watching a water wave from its side as it approaches a beach: It first builds, then breaks, and finally dissipates, which is why a mid-latitude cyclonic storm system is known as a *wave cyclone*. Figure 12.1b shows the newly formed wave with a cold front pushing southward and a warm front moving northward. The region of lowest pressure (called the *central pressure*) is at the junction of the two fronts. As the cold air displaces the warm air upward along the cold front, and as *overrunning* occurs ahead of the warm front, a narrow band of precipitation forms (shaded green area). Steered by the winds aloft, the system typically moves east or northeastward and gradually becomes a fully developed *open wave* in 12 to 24 hours (see Fig. 12.1c). The central pressure of the wave cyclone is now much lower, and several isobars encircle the wave's apex. These more tightly packed isobars create a stronger cyclonic flow, as the winds swirl counterclockwise and inward toward the low's center. Precipitation forms in a wide band *ahead* of the warm front and in *a narrow band* along the cold front. The region of warm air between the cold and warm fronts is known as the **warm sector.** Here, the weather tends to be partly cloudy, although scattered showers and thunderstorms may develop if the air is conditionally unstable.

Energy for the storm is derived from several sources. As the air masses try to attain equilibrium, warm air rises and cold air sinks, transforming potential energy into kinetic energy—energy of motion. Condensation supplies energy to the system in the form of latent heat. And, as the surface air converges toward the low's center, wind speeds may increase, producing an increase in kinetic energy.

As the open wave moves eastward, its central pressure continues to decrease, and the winds blow more vigorously as the wave quickly develops into a *mature cyclone*. The faster-moving cold front constantly inches closer to the warm front, squeezing the warm sector into a smaller area, as shown in Fig. 12.1d. In this model, the cold front eventually overtakes the warm front and the system becomes occluded. At this point, the storm is usually most intense, with clouds and precipitation covering a large area. The area of most intense weather is normally found to the northwest of the storm's center. Here, strong winds and blowing and drifting snow can create blizzard conditions in winter.

The point of occlusion where the cold front, warm front, and occluded front all come together in Fig. 12.1e is referred to as the *triple point*. Notice that in this region the cold and warm fronts appear similar to the open-wave cyclone in Fig. 12.1c. It is here where a new wave (called a **secondary low**) will occasionally form, move eastward or northeastward, and intensify into a cyclonic storm. The center of the intense storm system shown in Fig. 12.1e gradually dissipates, because cold air now lies on both sides of the occluded front. The warm sector is still present, but is far removed from the center of the storm. Without the supply of energy provided by the rising warm, moist air, the old storm system dies out and gradually disappears (see Fig. 12.1f). We can

● **FIGURE 12.1** The idealized life cycle of a mid-latitude cyclone (a through f) in the Northern Hemisphere based on the polar front theory. As the life cycle progresses, the system moves northeastward in a dynamic fashion. The small arrow next to each L shows the direction of storm movement.

think of the sequence of a developing wave cyclone as a whirling eddy in a stream of water that forms behind an obstacle, moves with the flow, and gradually vanishes downstream. The entire life cycle of a wave cyclone can last from a few days to more than a week.

● Figure 12.2 shows a series of wave cyclones in various stages of development along the polar front in winter. Such a succession of storms is known as a *"family" of cyclones.* Observe that to the north of the front are cold anticyclones; to the south over the Atlantic Ocean is the warm, semipermanent Bermuda high. The polar front itself has developed into a series of loops, and at the apex of each loop is a cyclonic storm system. The cyclone over the northern plains (Low 1) is just forming; the one along the East Coast (Low 2) is an open wave; and the occluded system near Iceland (Low 3) is dying out. If the average rate of movement of a wave cyclone from birth to decay is 25 knots, then it is entirely possible for a storm to develop over the central part of the United States, intensify into a large storm over New England, become occluded over the ocean, and reach the coast of England in its dissipating stage less than a week after it formed.

● Figure 12.3 is a visible satellite image of clouds and two mid-latitude cyclones in different stages of development along the polar front. Superimposed on the image are the weather fronts. Look again at Fig. 12.1 and determine what stages of development the two cyclones are in.

Up to now, we have considered the polar front model of a developing wave cyclone, which represents a rather simplified version of the stages that an extratropical cyclonic storm system must go through. Even though few (if any) storms adhere to the model exactly, it can still serve as a good foundation for understanding the structure of storms. So keep the model in mind as you read the following sections.

Where Do Mid-latitude Cyclones Tend to Form?

Any development or strengthening of a mid-latitude cyclone is called **cyclogenesis.** Certain regions of North America show a propensity for cyclogenesis, including the Gulf of Mexico, the Atlantic Ocean east of the Carolinas, and the eastern slopes of

● **FIGURE 12.2** A series of wave cyclones (a "family" of cyclones) forming along the polar front.

NOAA

● **FIGURE 12.3** Visible satellite image of the North Pacific with two mid-latitude cyclones in different stages of development during February 2000.

CRITICAL THINKING QUESTION Based on your knowledge of the polar front theory, in which general direction would you expect the low (that sits off the west coast of North America) to move? If the low moves in the direction you chose, what type of weather would you expect for the state of Oregon?

high mountain ranges, such as the Rockies and the Sierra Nevada. As westerly winds blow over a mountain range, the air expands vertically on the downwind (lee) side, which can help intensify any preexisting areas of low pressure. Troughs and developing cyclonic storms that form in this manner are called **lee-side lows** (see ● Fig. 12.4) and their development, *lee cyclogenesis.*

Another region of cyclogenesis lies near Cape Hatteras, North Carolina, where warm Gulf Stream water can supply moisture and warmth to the region south of a stationary front, thus increasing the contrast between air masses to a point where storms can suddenly spring up along the front. These cyclones, called **northeasters** or *nor'easters*, normally move northeastward

Nor'easters

Northeasters (commonly called *nor'easters*) are mid-latitude cyclonic storms that develop or intensify off the eastern seaboard of North America during the fall, winter, and spring. They usually move northeastward along the coast, often bringing strong northeasterly winds to coastal areas, hence the name, nor'easter. In addition to strong winds, these storms can bring heavy rain, snow, and sleet. Most often they deepen and become most intense off the coast of New England. Nor'easters are fueled by the large temperature gradient between the warm ocean and much colder air over the continental landmass. They gain additional energy from a strong upper-level jet stream and from moisture evaporating from the ocean, especially the warm Gulf Stream that flows northeastward parallel to the east coast of the United States.

The ferocious northeaster of January 22–24, 2016 (shown in ● Fig. 1), produced the heaviest snow totals on record at several locations, including Baltimore/Washington International Airport (29.2 in.) and New York's Central Park (27.5 in.). Much of the snow fell in less than 24 hours. High winds and 20-foot waves pounded the coastlines of Delaware and New Jersey, and high storm tides put many coastal areas and highways under water.

Studies suggest that some of the northeasters that batter the coastline in winter may actually possess some of the characteristics of a tropical hurricane. For example, a strong nor'easter that dumped between 10 and 40 inches of snow over parts of the

● **FIGURE 1** The surface weather map for 7 a.m. (EST) January 23, 2016, shows an intense low-pressure area (central pressure 987 mb, or 29.15 in.), which is generating strong northeasterly winds and heavy precipitation (areas shaded green for rain, white for snow, and salmon for sleet or freezing rain) from the mid-Atlantic states into New England. More than 30 million people were placed under blizzard warnings as a result of this storm, which dropped more than 2 feet of snow from Washington, D.C., to New York. Damage estimates were as high as $3 billion.

Northeast and New England in February 2013 developed an eyelike feature.

One of the most dramatic combinations of tropical and extratropical cyclone processes ever observed was Hurricane Sandy in late October 2012. Sandy briefly attained Category 3 strength before striking Cuba, weakened, then restrengthened as it approached the northeast United States. The storm grew to an immense size, with some circulation features comparable to a strong nor'easter, while it also maintained hurricane characteristics until just several hours before it struck New Jersey, when it was reclassified as a post-tropical cyclone. (We will examine hurricanes and their characteristics, including Sandy, in more detail in Chapter 16.)

---

along the Atlantic Coast, bringing high winds and heavy snow or rain to coastal areas. Before the age of modern satellite imagery and weather prediction, such coastal storms would often go undetected during their formative stages, and sometimes an evening weather forecast of "fair and colder" along the eastern seaboard would have to be changed to "heavy snowfall" by morning. Fortunately, with today's weather information-gathering and forecasting techniques, these storms rarely strike by surprise. (Additional information on nor'easters is given in Focus section 12.1.)

● Figure 12.5 shows the typical paths taken in winter by mid-latitude cyclones and anticyclones. Notice in Fig. 12.5a that some of the lows are named after the region where they form,

such as the *Hatteras low,* which develops off the coast near Cape Hatteras, North Carolina. The *Alberta Clipper* forms (or redevelops) on the eastern side of the Canadian Rockies in Alberta, then rapidly skirts across the northern tier of U.S. states. Similarly, the *Colorado low* forms (or redevelops) on the eastern side of the U.S. Central Rockies. Notice that the lows generally move eastward or northeastward, whereas the highs typically move southeastward, then eastward.

When mid-latitude cyclones deepen rapidly (in excess of 24 mb in 24 hours), the term *explosive cyclogenesis,* or *"bomb,"* is sometimes used to describe them. As an example, explosive cyclogenesis occurred in a storm that developed over the warm

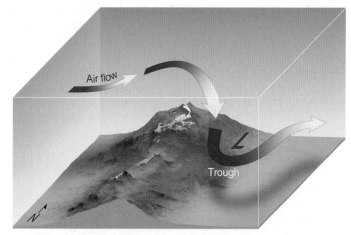

● **FIGURE 12.4** As westerly winds blow over a mountain range, the air can expand vertically on the downwind (lee) side, enhancing the development of a trough or cyclonic storm, called a *lee-side low*.

● **FIGURE 12.5** (a) Typical paths of winter mid-latitude cyclones. The lows are named after the region where they form. (b) Typical paths of winter anticyclones.

Atlantic just east of New Jersey on September 10, 1978. As the central pressure of the storm dropped nearly 60 mb (1.8 in.) in 24 hours, hurricane force winds battered the ocean liner *Queen Elizabeth II* and sank the fishing vessel *Captain Cosmo*. Another case of intense "bombogenesis" occurred on March 26, 2014, when a cyclone southeast of New England deepened by 24 mb in just six hours.

Some frontal waves form suddenly, grow in size, and develop into huge cyclonic storms, then slowly dissipate, with the entire process taking several days to a week to complete. Other frontal waves remain small and never grow into giant weather producers. Why is it that some frontal waves develop into huge cyclonic storms, whereas others simply dissipate in a day or so?

This question poses one of the real challenges in weather forecasting. The answer is complex. Indeed, there are many surface conditions that influence the formation of a mid-latitude cyclone, including mountain ranges and land-ocean temperature contrasts. However, the real key to the development of a wave cyclone is found in the *upper-wind flow,* in the region of the high-level westerlies. Therefore, before we can arrive at a reasonable answer to our question, we need to see how the winds aloft influence surface pressure systems.

Vertical Structure of Deep Dynamic Lows

In Chapter 9, we learned that thermal pressure systems are shallow systems that are typically weaker with increasing height above the surface. (Look back at Fig. 9.18, p. 239.) On the other hand, developing surface middle-latitude cyclones are deep *dynamic lows* that are usually stronger with height. Hence, they appear on an upper-level chart as either a closed low or a trough.

Suppose that the upper-level low is directly above the surface low as illustrated in ● Fig. 12.6. Notice that only at the surface (because of friction) do the winds blow inward toward the low's center. As these winds converge (flow together), the air "piles up." This piling up of air, called **convergence,** causes air density to increase directly above the surface low. This increase in mass causes surface pressures to rise; gradually, the low fills and the

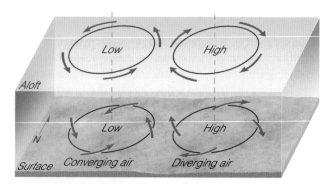

● **FIGURE 12.6** If lows and highs aloft were always directly above lows and highs at the surface, the surface systems would quickly dissipate.

surface low dissipates. The same reasoning can be applied to surface anticyclones. Winds blow outward, away from the center of a surface high. If a closed high or ridge lies directly over the surface anticyclone, **divergence** (the spreading out of air) at the surface will remove air from the column directly above the high. The decrease in mass causes the surface pressure to fall and the surface high-pressure area to weaken. Consequently, it appears that, if upper-level pressure systems were always located directly above those at the surface (such as shown in Fig. 12.6), cyclones and anticyclones would die out soon after they form (if they could form at all). What, then, is it that allows these systems to develop and intensify?

THE ROLES OF CONVERGING AND DIVERGING AIR ● Figure 12.7 is an idealized model of the vertical structure of a middle-latitude cyclone and anticyclone in the Northern Hemisphere. Note that behind the cold front there is cold air both at the surface and aloft. This cold, dense air is helping to maintain the surface anticyclone. But remember from Chapter 8 that aloft, in a region of cold air, constant pressure surfaces are packed closer together. This is because in the cold, dense air the atmospheric pressure decreased

rapidly with height, causing cold air aloft to be associated with low pressure. Consequently, in the cold air aloft we find the upper low; and it is located *behind,* or to the *west* of, the surface low.

Observe also in Fig. 12.7 how the surface low tilts toward the northwest as we move up from the surface, showing up as a closed system on the 500-mb chart and as a trough on the 300-mb chart. Directly above the surface low, at 300 mb, the air spreads out and diverges (as indicated by the wind flow). This allows the converging surface air to rise and flow out of the top of the air column just below the tropopause, which acts as a constraint to vertical motions. We now have a mechanism for developing mid-latitude cyclonic storms. *When upper-level divergence is stronger than surface convergence (more air is taken out at the top than is brought in at the bottom), surface pressures drop, and the low intensifies (deepens). By the same token, when upper-level divergence is less than surface convergence (more air flows in at the bottom than is removed at the top), surface pressures rise, and the system weakens (fills).*

We can also use Fig. 12.7 to explain the structure of the anticyclone. Notice that at the surface and aloft, warm air lies to the southwest of the surface high. Again, in Chapter 8 we saw that warm air aloft causes the isobaric surfaces to spread farther apart, which results in warm air aloft being associated with higher pressure. This situation causes the surface anticyclone to tilt toward the southwest—toward the warmer air—at higher altitudes. As we move upward from the surface, we observe that the closed area of high pressure at 500 mb becomes a ridge at the 300-mb level. Also notice that directly above the surface high at 300 mb there is convergence of air (as indicated by the wind flow lines). Convergence causes an accumulation of air above the surface high, which allows the air to sink slowly and replace the diverging surface air. Hence, *when upper-level convergence of air exceeds low-level divergence (inflow at top is greater than outflow near the surface), surface pressures rise, and the anticyclone builds.* On the other hand, *when upper-level convergence of air is less than low-level divergence, the anticyclone weakens as surface pressures fall.* (Additional information on the subject of convergence and divergence is provided in Focus section 12.2.)

Look at the wind direction at the 500-mb level in Fig. 12.7. Winds at this altitude tend to steer surface systems in the same direction that the winds are moving. Thus, the surface mid-latitude cyclone will move toward the northeast, while the surface anticyclone will move toward the southeast. As we can see in Fig. 12.5, these paths indicate the average movement of surface pressure systems in the eastern two-thirds of the United States. On average, surface storms travel across the United States at about 16 knots in summer and about 27 knots in winter. The faster winter velocity reflects the stronger upper-level flow during this time of year.[*]

So far, we have seen that deep pressure systems exist at the surface and aloft throughout much of the troposphere. When the upper-level trough lies to the west of the surface mid-latitude cyclone, the atmosphere is able to redistribute its mass. Regions of low-level converging air are compensated for by regions of upper-level diverging air and vice versa. Cyclones and anticyclones can intensify and, steered by the winds aloft, move away from their region of formation.

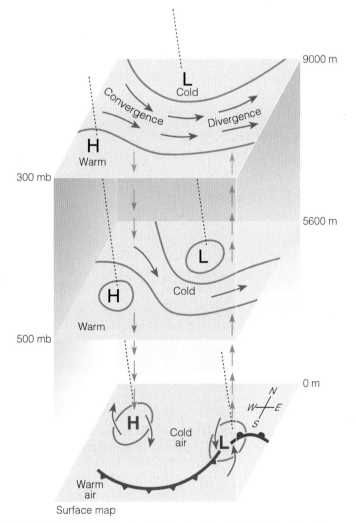

● **FIGURE 12.7** An idealized vertical structure of a middle-latitude cyclone and anticyclone.

---

[*]As a forecasting rule of thumb, surface pressure systems tend to move in the same direction as the wind at the 500-mb level. The speed at which the surface systems move is about half the speed of the 500-mb winds.

A Closer Look at Convergence and Divergence

We know that *convergence* is the piling up of air above a region, while *divergence* is the spreading out of air above some region. Convergence and divergence of air may result from changes in wind direction and wind speed. For example, convergence occurs when moving air is funneled into an area, much in the way cars converge when they enter a crowded freeway. Divergence occurs when moving air spreads apart, much as cars spread out when a congested two-lane freeway becomes three lanes. On an upper-level chart, this type of convergence (also called *confluence*) occurs when contour lines move closer together, as a steady wind flows parallel to them (as shown on the upper-level chart in ● Fig. 2). On the same chart, this type of divergence (also called *diffluence*) occurs when the contour lines move apart as a steady wind flows parallel to them. Notice in Fig. 2 that below the area of convergence lies the surface anticyclone (H), whereas below the area of divergence lies the surface middle-latitude cyclonic storm (L).

Convergence and divergence may also result from changes in wind speed. *Speed convergence* occurs when the wind slows down as it moves along, whereas *speed divergence* occurs when the wind speeds up. We can grasp these relationships more clearly if we imagine air molecules to be marching in a band. When the marchers in front slow down, the rest of the band members squeeze together, causing convergence; when the marchers in front start to run, the band members spread apart, or diverge.

● Figure 3 illustrates how this type of convergence and divergence can occur in the upper troposphere. The upper-air chart shows a trough and two ridges with evenly spaced

● **FIGURE 2** The formation of convergence (CON) and divergence (DIV) of air with a constant wind speed (indicated by flags) in the upper troposphere. Circles represent air parcels that are moving parallel to the contour lines on a constant pressure chart. Below the area of convergence the air is sinking, and we find the surface high (H). Below the area of divergence the air is rising, and we find the surface low (L).

● **FIGURE 3** As the faster-flowing air in the ridge moves toward the slower-flowing air in the trough, the air piles up and converges. As the slower-moving air in the trough moves toward the faster-flowing air in the ridge, the air spreads apart and diverges.

contour lines. Notice that even though the contours are evenly spaced, the winds blow faster in the ridge than they do in the trough (a concept discussed in Chapter 8 on p. 216).

As the faster-flowing air moves away from the ridge and approaches the slower-moving air in the trough, the air piles up, producing convergence. Where the slower-moving air in the trough approaches the faster-moving air in the ridge, the air spreads out, producing divergence in the air flow.

As you continue to read about mid-latitude cyclonic storms, remember from Figs. 2 and 3 that the most likely location for upper-level converging air is to the *left (west)* of the upper trough. The most likely location for diverging air is to the *right (east)* of the upper trough.

Since regions of strong upper-level divergence and convergence typically occur when deep troughs and ridges—waves—exist in the flow aloft, the next section examines these waves and their influence on a developing mid-latitude cyclonic storm.

Upper-Level Waves and Mid-latitude Cyclones

You may remember from the "dishpan" experiment (found in Focus section 10.1, p. 272) that, aloft, waves are a fundamental feature of an unevenly heated, rotating sphere, such as Earth. If we examine an upper-level chart that shows almost the entire Northern (or Southern) Hemisphere, such as ● Fig. 12.8, the waves appear as a series of troughs and ridges with significant amplitude that encircle the globe. The distance from trough to trough (or ridge to ridge), is known as the *wavelength*. When the wavelength is on the order of many thousands of kilometers, the wave is called a **longwave**. At any one time there are usually between three and six longwaves looping around the Northern Hemisphere. The fewer the number of waves, the longer their wavelengths. Because mountain ranges tend to disturb the upper-level wind flow, these waves are often found to the east of such topographic barriers as the Rockies and Tibetan Plateau. Sometimes longwaves exhibit a relatively small amplitude (or north-to-south extent) and the flow is mainly zonal, or west to

● **FIGURE 12.8** A 500-mb map of the Northern Hemisphere from a polar perspective shows five longwaves encircling the globe. Note that the wavelength at position 1 is as great as the width of the United States. Solid lines are contours. Dashed lines show the position (axis) of longwave troughs.

east. On other occasions, the waves exhibit considerable amplitude and the flow has a strong north-to-south (meridional) component to it.

Longwaves are also known as *planetary waves* and as **Rossby waves,** after C. G. Rossby, a famous meteorologist who carefully studied their motion. Embedded in longwaves are **shortwaves,** which are small disturbances or ripples that move with the wind flow (see ● Fig. 12.9a). Rossby found that the shorter the wavelength of a particular wave, the faster it moved downstream. Shortwaves tend to move eastward at a speed proportional to the average wind flow near the 700-mb level, about 3 km above

sea level. Longwaves, on the other hand, often remain stationary, move eastward very slowly at less than 4° of longitude per day (about eight knots), or even move westward *(retrograde).** We can obtain a better idea of this wave movement if we think of longwaves as being huge meanders (loops) in a swiftly flowing stream of water. Water moves through the loops quickly, while the loops themselves move eastward very slowly, as the fast-flowing water cuts away at one bank and deposits material on the other. Suppose debris tumbles into the stream, disturbing the flow. The disturbed flow appears as a small wrinkle that travels downstream through the loops at a speed near the average stream flow. This wrinkle in the flow is analogous to a shortwave in the atmosphere.

Notice in Fig. 12.9b that, while the longwaves move eastward very slowly, the shortwaves move fairly quickly around the longwaves. Notice also that the shortwaves tend to deepen (that is, increase in size) when they approach a longwave trough and weaken (become smaller) when they approach a ridge. Moreover, when a shortwave moves into a longwave trough, the trough tends to deepen. (Look at shortwave 3 in Fig. 12.9b.)

Where the contour lines in Fig. 12.9b are roughly parallel to the isotherms (dashed red lines), the atmosphere is said to be **barotropic.** Since the winds at this level more or less parallel the contour lines, in a barotropic atmosphere the winds blow parallel to the isotherms. By comparison, where the isotherms *cross* the contour lines, temperature advection occurs and the atmosphere is said to be **baroclinic.*** Notice in Fig. 12.9b that the baroclinic region tends to be in a narrow zone in the vicinity of shortwaves 1 and 3. The shortwaves actually disturb the flow and accentuate the region of baroclinicity.

In the region of baroclinicity, winds cross the isotherms and produce *temperature advection.* **Cold advection** (or *cold air advection*) is the transport of cold air by the wind from a region of lower (colder) temperatures to a region of higher (warmer)

---

*Retrograde wave motion means that the wave is actually moving in the opposite direction of the wind flow.

**Actually, on a constant pressure surface, baroclinic conditions exist where the air density varies, and barotropic conditions exist where the air density does not vary.

(a) DAY 1 (b) DAY 2 (24 hours later)

● **FIGURE 12.9** (a) Upper-air chart showing a longwave with three shortwaves (heavy dashed lines) embedded in the flow. (b) Twenty-four hours later the shortwaves have moved rapidly around the longwave. Notice that the shortwaves labeled 1 and 3 tend to deepen the longwave troughs, while shortwave 2 has weakened as it moves into a ridge. Dashed red lines are isotherms (lines of constant temperature) in °C. Solid gray lines are contours. Blue arrows indicate cold advection and red arrows, warm advection.

temperatures. In the region of cold advection, the air temperature normally decreases. On the other hand, **warm advection** (or *warm air advection*) is the transport of warm air by the wind from a region of higher (warmer) temperatures to a region of lower (colder) temperatures. In the region of warm advection, the air temperature normally increases. For cold advection to occur, the wind must blow across the isotherms from colder to warmer regions, whereas for warm advection, the wind must blow across the isotherms from warmer to colder regions.

In the baroclinic region in Fig. 12.9b, observe that strong winds cross the isotherms, producing cold advection (blue arrows) on the trough's west side and warm advection (red arrows) on its east side. Below the baroclinic zone lies the polar front; above it flows the polar-front jet stream. The disturbed flow created by the shortwaves is now capable of aiding in the development or intensification of a surface mid-latitude cyclonic storm. The theory explaining how this phenomenon occurs is known as the *baroclinic wave theory of developing cyclones.*

The Necessary Ingredients for a Developing Mid-latitude Cyclone

To better understand how a wave cyclone may develop and intensify into a huge mid-latitude cyclonic storm, we need to examine atmospheric conditions at the surface and aloft. Suppose that a portion of a longwave trough at the 500-mb level lies directly above a surface stationary front, as illustrated in ● Fig. 12.10a. On the 500-mb chart, contour lines (solid gray lines) and isotherms (dashed red lines) parallel each other and are crowded close together. Colder air is located in the northern half of the map, while warmer air is located to the south. Winds are blowing at fairly high velocities, which produce a sharp change in wind speed—a strong *wind speed shear*—from the surface up to this level. Suppose that a shortwave moves through this region,

disturbing the flow as shown in Fig. 12.10b. This sets up a kind of instability in the flow (as warmer air rises and colder air sinks) known as **baroclinic instability**.

UPPER-AIR SUPPORT With the onset of baroclinic instability, horizontal and vertical air motions begin to enhance the formation of a cyclonic storm. For example, as the flow aloft becomes disturbed, it begins to lend support for the intensification of surface pressure systems, as a region of converging air forms above position 1 in Fig. 12.10b and a region of diverging air forms above position 2.* The converging air aloft causes the surface air pressure to rise in the region marked *H* in Fig. 12.10b. Surface winds begin to blow out away from the region of higher pressure, and the air aloft gradually sinks to replace it. Meanwhile, diverging air aloft causes the surface air pressure to decrease beneath position 2, in the region marked *L* on the surface map. This initiates rising air, as the surface winds blow in toward the region of lower pressure. As the converging surface air develops cyclonic spin, cold air flows southward and warm air northward. We can see in Fig. 12.10b that the western half of the stationary front is now a cold front and the eastern half a warm front. Cold air moves in behind the cold front, while warm air slides up along the warm front. These regions of cold and warm advection occur all the way up to the 500-mb level.

On the 500-mb chart in Fig. 12.10b, cold advection is occurring at position 1 (blue arrow) as the wind crosses the isotherms, bringing cold air into the trough. The cold advection makes the air more dense and lowers the height of the air column from the surface up to the 500-mb level. (Recall that, on a 500-mb chart, lower heights mean the same as lower pressures.) Consequently, the pressure in the trough lowers and the trough deepens. The deepening of the upper trough causes the contour lines to crowd

_____
*Look back at Fig. 12.7, p. 327, and the upper-air chart in Fig. 2 and Fig. 3 on p. 328, and note the regions of converging air and diverging air on these maps.

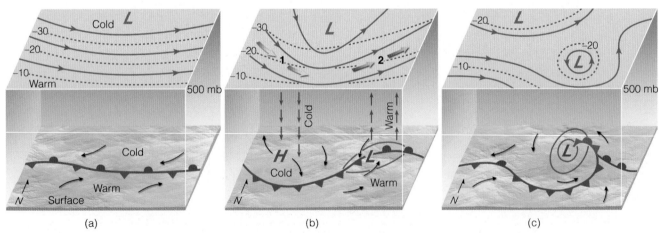

● **FIGURE 12.10** An idealized 3-D view of the formation of a mid-latitude cyclone during baroclinic instability. (a) A longwave trough at 500 mb lies parallel to and directly above the surface stationary front. (b) A shortwave (not shown) disturbs the flow aloft, initiating temperature advection (blue arrow, cold advection; red arrow, warm advection). The upper trough intensifies and provides the necessary vertical motions (as shown by vertical arrows) for the development of the surface wave cyclone. (c) As the surface storm moves northeastward, it occludes, and without upper-level diverging air to compensate for surface converging air, the cyclonic storm system dissipates.

closer together and the winds aloft to increase. Meanwhile, at position 2 (red arrow) warm advection is taking place, which has the effect of raising the height of a column of air; here, the 500-mb heights increase and a ridge builds (strengthens). Therefore, *the overall effect of differential temperature advection is to amplify the upper-level wave.* As the trough aloft deepens, its curvature increases, which in turn increases the region of divergence above the developing surface storm. At this point, the surface mid-latitude cyclone rapidly develops as surface pressures fall.

Regions of cold and warm advection are associated with vertical motions. Where there is cold advection, some of the cold, heavy air sinks; where there is warm advection, some of the warm, light air rises. Hence, due to advection, air must be sinking in the vicinity of position 1 and rising in the vicinity of position 2.

The sinking of cold air and the rising of warm air provide energy for a developing cyclone, as potential energy is transformed into kinetic energy. Further, if clouds form, condensation in the ascending air releases latent heat, which warms the air. The warmer air lowers the surface pressure, which strengthens the surface low even more. So, we now have a full-fledged middle-latitude cyclone with all of the necessary ingredients for its development.

Eventually, the warm air curls around the north side of the low, and the storm system occludes (see Fig. 12.10c).* Some storms may continue to deepen, but most do not as they move out from under the region of upper-level divergence. Additionally, at the surface the storm may weaken as the supply of warm air is cut off and cold, dry air behind the cold front (called a *dry slot*) is drawn in toward the surface low.

Sometimes, an upper-level pool of cold air (which has broken away from the main flow) lies almost directly above the surface low. Occasionally the upper low will break away entirely from the main flow, producing a **cut-off low,** which often appears as a

---

*If the occluded front should extend west of the low's center, it is sometimes referred to as a *bent-back occlusion.*

single contour line on an upper-level chart. When the upper low lies directly above the surface low (as in Fig. 12.10c), the storm system is said to be *vertically stacked.* Usually the isotherms around the upper low parallel the contour lines, which indicates that no significant temperature advection is occurring. Without the necessary energy transformations, the surface system gradually dissipates. As its winds slacken and its central pressure gradually rises, the low is said to be *filling.* The upper-level low, however, may remain stationary for many days. If air is forced to ascend into this cold pocket, widespread clouds and precipitation may persist for some time, even though the surface storm system itself has moved east out of the picture.

THE ROLE OF THE JET STREAM As we have seen, for middle latitude cyclones to develop and intensify there must be upperlevel diverging air above the surface storm. The polar jet stream can provide such areas of divergence. In Chapter 10, we learned that the axis of the polar-front jet stream pretty much coincides with the position of the polar front. The region of strongest winds in the jet stream is known as a *jet stream core,* or **jet streak.** When the polar jet stream flows in a wavy west-to-east pattern, a jet streak tends to form in the trough of the jet, where pressure gradients are tight. The curving of the jet stream coupled with the changing wind speeds around the jet streak produces regions of strong convergence and divergence of air along the flanks of the jet. (If you are curious as to why these regions of converging and diverging air develop around the jet streak, read Focus section 12.3.)

Notice that the region of diverging air (marked *D* in ● Fig. 12.11a) draws warm surface air upward to the jet, which quickly sweeps the air downstream. Since the air above the mid-latitude cyclone is being removed more quickly than converging surface winds can supply air to the storm's center, the central pressure of the storm drops rapidly. As surface pressure gradients increase, the wind speed increases. Above the high-pressure area, a region of converging air (marked *C* in Fig. 12.11a) feeds air downward into the

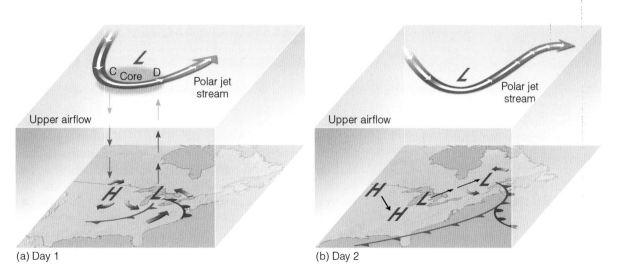

(a) Day 1   (b) Day 2

● **FIGURE 12.11** (a) As the polar jet stream and its area of maximum winds (the jet streak, or core) swings over a developing mid-latitude cyclone, an area of divergence *(D)* draws warm surface air upward, and an area of convergence *(C)* allows cold air to sink. The jet stream removes air above the surface storm, which causes surface pressures to drop and the storm to intensify. (b) When the surface storm moves northeastward and occludes, it no longer has the upper-level support of diverging air, and the surface storm gradually dies out.

Jet Streaks and Storms

● Figure 4 shows an area of maximum winds, a *jet streak*, on a 300-mb chart. Jet streaks have winds of at least 50 knots, and represent small segments (ranging in length from a few hundred kilometers to over 3000 kilometers) within the meandering jet stream flow.

Jet streaks are important in the development of surface mid-latitude cyclones because areas of convergence and divergence form at specific regions around them. To understand why, consider air moving through a straight jet streak (shaded area) in ● Fig. 5. As the air enters the back of the streak (known as the *entrance region*), it increases in speed; as it leaves the front of the streak (known as the *exit region*), it decreases in speed. At this elevation in the atmosphere (about 10,000 m or 33,000 ft above the surface), the wind flow is nearly in

● **FIGURE 5** Changing air motions within a straight jet streak (shaded area) cause strong convergence of air at point 1 (left entrance region) and strong divergence at point 3 (left exit region).

geostrophic balance with the pressure gradient force (directed north) and the Coriolis force (directed south). As the air enters the jet streak, it increases in speed because the contour lines are closer together, causing an increase in the pressure gradient force. The greater force temporarily exceeds the Coriolis force, and the air swings slightly to the north across the contour lines, which causes a piling up of air (called a *bottleneck effect*) and *strong convergence* at point 1; *weak divergence* occurs at point 2.

Toward the middle of the jet streak, the increase in wind speed causes the Coriolis force to increase and the wind to become nearly geostrophic again. However, as the air exits the jet streak, the pressure gradient force is reduced as the contour lines spread farther apart. Hence, the Coriolis force temporarily exceeds the pressure gradient force, causing the air to cross the

contour lines and swing slightly to the south. This process produces *strong divergence at point 3* and weak convergence at point 4.

The conditions described so far exist in a straight jet streak that shows no curvature. When the jet stream becomes wavy, and the jet streak exhibits cyclonic curvature (as it does in Fig. 4), the areas of weak divergence at point 2 and weak convergence at point 4 all but disappear. What we are left with is a curving jet streak that exhibits strong divergence at point 3 (in the left exit region) and strong convergence at point 1 (in the left entrance region). Notice in ● Fig. 6 that below the area of strong divergence the air rises, cools and, if sufficiently moist, condenses into clouds. Moreover, the removal of air in the region of strong divergence causes surface pressures to fall, which results in the development of an area of surface low pressure.

● **FIGURE 6** An area of strong divergence (DIV) can form with a curving jet streak. Below the area of divergence are rising air, clouds, and the developing mid-latitude cyclonic storm.

● **FIGURE 4** A portion of a 300-mb chart (about 33,000 ft above sea level) that shows the core of the jet—the region of maximum winds (MAX)—called a *jet streak*. Dashed lines are equal lines of wind speed (isotachs) in knots.

anticyclone. Hence, we find *the polar jet stream removing air above the surface cyclone and supplying air to the surface anticyclone.*

As the jet stream steers the mid-latitude cyclonic storm along—toward the northeast in this case—the surface cyclone occludes, and cold air surrounds the surface low (see Fig. 12.11b). Since the surface low has moved out from under the pocket of diverging air aloft, the occluded storm gradually fills as surface air flows into the system.

Because the Northern Hemisphere's polar jet stream is strongest and moves farther south in winter, we can see why mid-latitude cyclonic storms are better developed and move more quickly during the colder months. During the summer, when the polar jet stream shifts northward, developing mid-latitude storm

activity shifts northward as well, occurring principally in Canada over the province of Alberta and the Northwest Territories.

In general, we now have a fairly good picture as to why some surface lows intensify into huge mid-latitude cyclones while others do not. For a surface cyclonic storm to intensify, there must be an upper-level counterpart—a trough of low pressure—that lies to the *west* of the surface low. As shortwaves disturb the flow aloft, they cause regions of differential temperature advection to appear, leading to an intensification of the upper-level trough. At the same time, the polar jet stream forms into waves and swings slightly south of the developing storm. When these conditions exist, zones of converging and diverging air, along with rising and sinking air, provide energy conversions for the storm's growth.

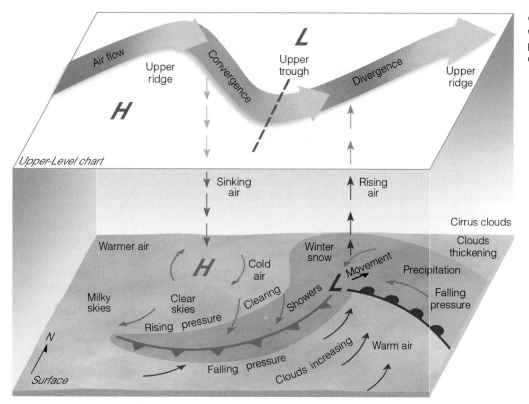

● **FIGURE 12.12** Summary of clouds, weather, vertical motions, and upper-air support associated with a developing mid-latitude cyclone. Dark green area represents precipitation.

With this atmospheric situation, storms may form even where there are no preexisting fronts.* In regions where the upper-level flow is not disturbed by shortwaves or where no upper trough or jet stream exists, the necessary vertical and horizontal motions are insufficient to enhance cyclonic storm development and we say that the surface storm does not have the necessary *upper-air support.* The horizontal and vertical motions, cloud patterns, and weather that typically occur with a developing open-wave cyclone are summarized in ● Fig. 12.12.

CONVEYOR BELT MODEL OF MID-LATITUDE CYCLONES A three-dimensional model of a developing mid-latitude cyclone is illustrated in ● Fig. 12.13. The model describes rising and sinking air as traveling along three main "conveyor belts." Just as people ride escalators to higher levels in a department store, so air glides along through a constantly evolving mid-latitude cyclone. According to the **conveyor belt model,** a warm air stream (known as the *warm conveyor belt*—orange arrow in Fig. 12.13) originates at the surface in the warm sector, ahead of the cold front. As the warm air stream moves northward, it slowly rises along the sloping warm front, up and over the cold air below. As the rising air cools, water vapor condenses, and clouds form well out ahead of the surface low and its surface warm front. From these clouds, steady precipitation usually falls in the form of rain or snow. Aloft, the warm air flow gradually turns toward the northeast, parallel to the upper-level winds.

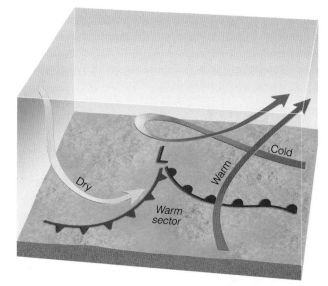

● **FIGURE 12.13** The conveyor belt model of a developing mid-latitude cyclone. The warm conveyor belt (in orange) rises along the warm front, causing clouds and precipitation to cover a vast area. The cold conveyor belt (in blue) slowly rises as it carries cold, moist air westward ahead of the warm front but under the rising warm air. The cold conveyor belt lifts rapidly and wraps counterclockwise around the center of the surface low. The dry conveyor belt (in yellow) brings very dry, cold air downward from the upper troposphere.

Notice in Fig. 12.13 that directly below the warm conveyor belt, a cold airstream—the *cold conveyor belt*—moves slowly westward. As the air moves west ahead of the warm front, precipitation and surface moisture evaporates into the cold air, making it moist. As the cold, moist airstream moves into the vicinity of the surface low, rising air gradually forces the cold conveyor

---

*The beginning stage of a wave cyclone almost always takes place when an area of upper-level divergence passes over a surface front. Even if initially there are no fronts on the surface map, they may begin to form where air masses having contrasting properties are brought together in the region where the surface air rises and the surrounding air flows inward.

belt upward. As the cold, moist air sweeps northwest of the surface low, it often brings heavy winter snowfalls to this region of the cyclone. The rising airstream usually turns counterclockwise, around the surface low, first heading south, then northeastward, when it gets caught in the upper air flow. It is the counterclockwise turning of the cold conveyor belt that produces the comma-shaped cloud similar to the one shown in ● Fig. 12.14.

The last conveyor belt is a dry one that forms in the cold, very dry region of the upper troposphere. Called the *dry conveyor belt*, and shaded yellow in Fig. 12.13, this airstream slowly descends from the northwest behind the surface cold front, where it brings generally clear, dry weather and, occasionally, blustery winds. If a branch of the dry air sweeps into the storm, it produces a clear area called a **dry slot,** which appears to pinch off the comma cloud's head from its tail. This phenomenon tends to show up on satellite images as the mid-latitude storm becomes more fully developed (see Fig. 12.14).

We are now in a position to tie together many of the concepts we have learned about developing mid-latitude cyclones by examining a monstrous storm that formed during March 1993—one of the strongest mid-latitude cyclones ever recorded in the United States.

A DEVELOPING MID-LATITUDE CYCLONE—THE MARCH STORM OF 1993

A strengthening mid-latitude cyclone on the morning of March 13, 1993, can be seen in a color-enhanced infrared satellite image in ● Fig. 12.15. Notice that its cloud band is in the shape of a comma that covers the entire eastern seaboard. Such **comma clouds** indicate that the storm is still developing and intensifying. However, this storm is not an ordinary wave cyclone—this storm intensified into a superstorm that some forecasters and journalists dubbed the "Storm of the Century."

The surface weather map for the morning of March 13 (see ● Fig. 12.16) shows that the center of the open wave is over northern Florida. Observe in Fig. 12.15 that this position is in the head of the comma cloud. The central pressure of the storm is 975 mb (28.79 in.), which indicates an incredibly deep system, considering that a typical open wave would have a central pressure closer to 996 mb (29.41 in.). A strong cold front stretches from the storm's center through western Florida. Behind the front, cold arctic air pours into the Deep South. Well ahead of the advancing front, an elongated band of heavy thunderstorms called a *squall line* has just pounded Florida with heavy rain, high winds, and tornadoes. (A smaller line of thunderstorms can be seen in Fig. 12.15 moving across southwest Florida along the cold front.) Warm humid air in the warm sector is streaming northward, overrunning cold surface air ahead of the warm front, which is causing precipitation in the form of rain, snow, and sleet to fall over a vast area extending from Florida to New York.

The 500-mb chart for the morning of March 13 (see ● Fig. 12.17) shows that a deep trough extending southward out of Canada lies to the west of the surface low. Around the trough where strong winds cross the isotherms, there is temperature advection.

Notice that warm advection (indicated by red barbs) is occurring on the trough's eastern side, ahead of the surface warm front

● **FIGURE 12.14** Visible satellite image of a mature mid-latitude cyclone with the three conveyor belts superimposed on the storm. As in Fig. 12.13, the warm conveyor belt is in orange, the cold conveyor belt is in blue, and the dry conveyor belt (forming the *dry slot*) is in yellow.

● **FIGURE 12.15** An enhanced infrared satellite image that shows a developing mid-latitude cyclone at 7 a.m. (EST) on March 13, 1993. The dark red color represents clouds with the coldest and highest tops. The red cloud band just east of Florida represents a line of severe thunderstorms associated with a strong cold front. Notice that a dry slot exists behind the front and that the cloud pattern is in the shape of a comma.

Cooperative Institute for Meteorological Satellite Studies (CIMSS), University of Wisconsin-Madison with data from (c)EUMETSAT Meteosat-3.

● **FIGURE 12.16** Surface weather map for 4 a.m. (EST) on March 13, 1993. Lines on the map are isobars. A reading of 96 is 996 mb and a reading of 00 is 1000 mb. (To obtain the proper pressure in millibars, place a 9 before those readings between 80 and 96, and place a 10 before those readings of 00 or higher.) Green shaded areas are receiving precipitation. Heavy arrows represent surface winds. The orange arrow represents warm, humid air; the light blue arrow, cold, moist air; and the dark blue arrow, cold, arctic air.

● **FIGURE 12.17** The 500-mb chart for 7 a.m. (EST) March 13, 1993. Solid lines are contours where 564 equals 5640 meters. Dashed lines are isotherms in °C. Wind entries in red show warm advection. Those in blue show cold advection. Those in black indicate no appreciable temperature advection is occurring.

shown in Fig. 12.16. Cold advection (indicated by blue barbs) is occurring on the trough's western side, behind the position of the surface cold front. As temperature advection deepens the trough, rising and sinking air provide energy for the developing surface storm. Higher in the upper troposphere, a strong jet stream and jet streak exist over northern Florida (see ● Fig. 12.18). On the eastern side of the jet streak, diverging air forms over the surface low, causing surface air to rapidly rise. The rising air is then swept northeastward by the jet stream and the surface low deepens into an intense area of low pressure.

● **FIGURE 12.18** Air flow aloft at an altitude above 10,000 m (33,000 ft) on March 13, 1993. Notice that a jet streak (orange shading) swings over northern Florida. The letters *DIV* represent an area of strong divergence that formed above the surface low.

● **FIGURE 12.19** The development of the ferocious mid-latitude cyclonic storm of March 1993. A small wave in the western Gulf of Mexico intensifies into a deep open-wave cyclone over Florida. It moves northeastward and becomes occluded over Virginia where its central pressure drops to 960 mb (28.35 in.). As the occluded storm continues its northeastward movement, it gradually fills and dissipates. The number next to the storm is its central pressure in millibars. Arrows show direction of movement. Time is Eastern Standard Time. Gray shaded represents total snowfall during storm.

We now have a pretty good idea as to why this wave cyclone developed into such a deep low-pressure area. The storm began on March 12 as a frontal wave off the Texas coast (see ● Fig. 12.19). In the upper air, a shortwave, moving rapidly around a longwave,

disturbed the flow, setting up the necessary ingredients for the surface storm's development. By the morning of March 13, the storm had intensified into a deep open-wave cyclone centered over Florida. In the upper air, a region of diverging air positioned above the storm caused the storm's surface pressures to drop rapidly. Upperlevel southwesterly winds (Fig. 12.18) directed the surface low northeastward, where it became occluded over Virginia during the afternoon of March 13. At this point, the storm's central pressure dropped to an incredibly low 960 mb (28.35 in.). Although the surface winds associated with this low pressure were quite strong and gusty, they were not as strong as those in a major hurricane because the isobars around the storm were spread farther apart than those in a hurricane and because surface friction slowed the winds. Higher up, away from the influence of the surface, the winds were much stronger, as a wind of 125 knots (144 mi/hr) was reported at the top of 1900-meter-high Mount Washington, New Hampshire.

The upper trough remained to the west of the surface low, and the storm continued its northeastward movement. In Fig. 12.19, we can see that by the morning of March 14 the storm (which was now a deep, bent-back occluded system) had weakened slightly and was centered along the coast of Maine. Moving out from under its area of upper-level divergence, the storm weakened even more as it continued its northeastward journey, out over the North Atlantic.

In all, the March 1993 storm was one of the strongest and most widespread in U.S. weather history. It blanketed a vast area from Alabama to Canada with deep snow. Fierce winds piled the snow into huge drifts that closed roads, leaving vehicles stranded. The storm shut down every major airport along the East Coast, and more than 3 million people lost electric power at some point during the storm. The 1993 Storm of the Century damaged or destroyed hundreds of homes, produced 27 tornadoes, stranded hikers in North Carolina and Tennessee, caused an estimated $800 million in damage, and claimed the lives of at least 270 people.

BRIEF REVIEW

Up to this point, we have looked at the structure and development of mid-latitude cyclones. Before going on, here is a summary of a few of the important ideas presented so far:

● The polar front (or Norwegian) model of a developing mid-latitude cyclonic storm represents a simplified but useful model of how an ideal storm progresses through the stages of birth, maturity, and dissipation.

● For a surface mid-latitude cyclone to develop or intensify (deepen), the upper-level low must be located to the west of (behind) the surface low.

● For a surface mid-latitude cyclonic storm to form, there must be an area of upper-level diverging air above the surface low. For the surface storm to intensify, the region of upper-level diverging air must be greater than surface converging air (that is, more air must be removed above the storm than is brought in at the surface).

● When the upper-air flow develops into waves, winds often cross the isotherms, producing regions of cold advection and warm advection, which tend to amplify the wave. At the same time, vertical air motions begin to enhance the formation of the surface storm as the rising of warm air and the sinking of cold air provide the proper energy conversion for the storm's growth.

- When the polar-front jet stream develops into a looping wave, it provides an area of upper-level diverging air for the development of surface mid-latitude cyclonic storms.

- The curving nature of the polar-front jet stream tends to direct surface mid-latitude cyclonic storms northeastward and surface anticyclones southeastward.

Vorticity, Divergence, and Developing Mid-latitude Cyclones

We know that for a surface mid-latitude cyclone to develop into a deep low-pressure area there must be an area of strong upper-level divergence situated above the developing storm. Meteorologists, then, are interested in locating regions of divergence on upper-level charts so that developing storms can be accurately predicted. We know from an earlier discussion that divergence and convergence of air are due to changes in either wind speed or wind direction. The problem is that it is a difficult task to measure divergence (or convergence) with any degree of accuracy using upper-level wind information. Therefore, meteorologists must look for something else that can be measured and, at the same time, can be related to regions of diverging (and converging) air. That something is called *vorticity*.

When something spins, it has vorticity. The faster it spins, the greater its vorticity. In meteorology, **vorticity** is a measure of the spin of small air parcels. Although the spin can be in any direction, our concern will be with the spin of horizontally flowing air about a vertical axis, much like an ice skater spins about an imaginary vertical axis. Because our goal is to see how vorticity can be used to identify regions of divergence and convergence, we must give vorticity some quantitative value. When viewed from above, air that spins cyclonically (counterclockwise) has *positive vorticity* and air that spins anticyclonically (clockwise) has *negative vorticity*.

We can see in ● Fig. 12.20 how divergence aloft and the vorticity of surface air are related. Suppose that the air column in Fig. 12.20 represents an area of low pressure with weak cyclonic (positive) spin. Further, suppose that an invisible barrier separates the column (and the ice skater inside) from the surrounding air. Now suppose that an area of divergence aloft (at the 300-mb level, about 30,000 ft above the surface) moves directly over the

● **FIGURE 12.20** When upper-level divergence moves over an area of weak cyclonic circulation, the cyclonic circulation increases (that is, it becomes more positive), and air is forced upward.

air column. Divergence aloft means that within this region more air is leaving than is entering. This removal of air above the column lowers the atmospheric pressure at the surface. As the surface pressure lowers, the air surrounding the column converges on it, pushing inward on its sides. If we assume that the total mass of air in the column does not change, then squeezing the column causes it to shrink horizontally and stretch vertically, forcing the air in the column upward. Meanwhile, as the column stretches, the ice skater's arms are pulled in close, and the skater spins much faster. Hence, the rate at which air flows around the center of the column must also increase. Consequently, the vorticity of the column increases, becoming more positive. So we can see that divergence aloft causes an increase in the cyclonic (positive) vorticity of surface cyclones, which usually results in cyclogenesis and upward air motions.

Before we consider how vorticity on an upper-level chart ties in with developing mid-latitude cyclones, we need to examine two important types of this phenomenon: Earth's vorticity and relative vorticity.

VORTICITY ON A SPINNING PLANET Because Earth spins, it has vorticity. In the Northern Hemisphere, **Earth's vorticity** (also called *planetary vorticity*) is always positive because the Earth spins counterclockwise about its vertical North Pole axis. The amount of Earth vorticity imparted to any object—even to those that are not moving relative to Earth's surface—depends upon the latitude. In ● Fig. 12.21, an observer standing on the equator would not spin about his or her own *vertical axis*; farther north, the observer would spin very slowly, while at the North Pole the observer would spin at a maximum rate of one revolution per day. It is now apparent that any object on Earth has vorticity simply because Earth is spinning, and the amount of this Earth vorticity increases from zero at the equator to a maximum at the poles.*

---

*Earth's vorticity at any latitude is equal to the product of twice Earth's angular rate of spin (2Ω) and the sin of the latitude (ϕ); that is, $2\Omega \sin (\phi)$. This expression is referred to as the *Coriolis parameter*, and it is usually expressed by the letter f.

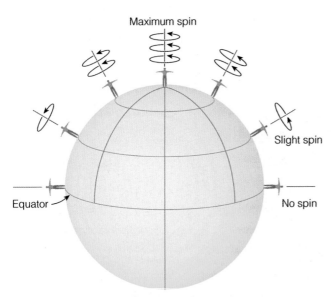

FIGURE 12.21 Due to the rotation of Earth, the rate of spin of observers about their vertical axes increases from zero at the equator to a maximum at the poles.

Moving air will generally have additional vorticity relative to Earth's surface. This type of vorticity, called **relative vorticity,** is the sum of two effects: the curving of the air flow *(curvature)* and the changing of the wind speed over a horizontal distance *(shear).* ● Figure 12.22 illustrates vorticity due to curvature. Air moving through a trough tends to spin cyclonically (counterclockwise), increasing its relative vorticity. In the ridge, the spin tends to be anticyclonic (clockwise), and the relative vorticity of the air increases, but in a negative direction. Whenever the wind blows faster on one side of an air parcel than on the other, a shear force is imparted on the parcel and it will spin and gain (or lose) relative vorticity. ● Figure 12.23 illustrates relative vorticity due to horizontal wind shear.

The sum of *Earth's vorticity* and the *relative vorticity* is called the **absolute vorticity.** To further illustrate this concept, suppose that you are watching an ice skater spin counterclockwise on a frozen lake. The ice skater possesses positive relative vorticity. If you could suddenly leave Earth and watch the same spinning skater from space, you would see the ice skater spinning on a rotating platform—Earth. The combination of the skater spinning about her vertical axis (her relative vorticity) plus the small spin imparted to her from the spinning Earth (Earth's vorticity) yields the absolute vorticity of the skater. We are now in a position to see how vorticity aloft ties in with divergence and the development of middle-latitude cyclones. (The concept of absolute vorticity can explain why the westerly flow aloft tends to form into waves. This topic is presented in Focus section 12.4.)

● Figure 12.24 shows an air parcel—a blob of air in this case—moving through a ridge and a trough in the upper troposphere. (For simplicity, we will assume that the parcel has a constant speed and that there is no shear acting on it.) Notice that at every position the parcel has cyclonic spin. This situation occurs because the parcel's total, or absolute, vorticity is a combination of Earth's vorticity plus its relative vorticity. In middle and high latitudes, Earth's vorticity is great enough to make the parcel's spin cyclonic everywhere on Earth.

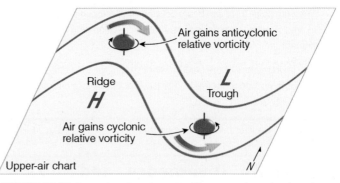

● FIGURE 12.22 In a region where the contour lines curve, air moving through a ridge spins clockwise and gains anticyclonic relative vorticity. In the trough, the air spins counterclockwise and gains cyclonic relative vorticity.

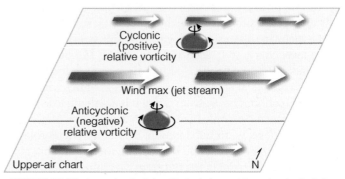

● FIGURE 12.23 Areas of cyclonic (positive) relative vorticity and anticyclonic (negative) relative vorticity can form in a region of strong horizontal wind-speed shear that can occur near a jet stream. Notice that parcels of air on either side of the jet stream have opposite directions of spin.

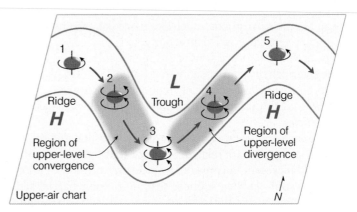

● FIGURE 12.24 The vorticity of an air parcel changes as we follow it through a wave. From position 1 to position 3, the parcel's absolute vorticity increases with time. In this region (shaded blue), we normally experience an area of upper-level converging air. As the air parcel moves from position 3 to position 5, its absolute vorticity decreases with time. In this region (shaded green), we normally experience an area of upper-level diverging air.

At position 1 (in the ridge), the air parcel has only a slight cyclonic spin because the relative vorticity, due to curvature, is anticyclonic and subtracts from Earth's vorticity. At position 2, the relative vorticity due to curvature is zero, which allows Earth's vorticity to spin the parcel faster. At position 3, the parcel spins even faster as the curvature is cyclonic, which adds to Earth's vorticity. At position 4, the parcel spins more slowly as the curvature is once again zero.

Vorticity and Longwaves

We know that longwaves develop aloft in the atmosphere and that longwaves are important, because they transfer heat and momentum from one latitude to another. But how do these waves form in the first place? The concept of vorticity can help explain longwave formation. For example, let's assume that in the atmosphere there is no divergence or convergence of air so that air columns cannot stretch or contract. Where these conditions prevail, the absolute vorticity of air will be conserved, meaning that the numerical value of the sum of Earth's vorticity and the relative vorticity will not change with time. Thus

$$\text{Absolute vorticity} =$$
$$\text{Earth's vorticity} + \text{relative vorticity} =$$
$$\text{constant}$$

If ζ_a* is the absolute vorticity, ζ_r the relative vorticity, and f Earth's vorticity, then the expression becomes

$$\zeta_a = \zeta_r + f = \text{constant}$$

Hence, any decrease in Earth's vorticity must be compensated for by an increase in the relative vorticity and vice versa.

Consider, for example, that air in ● Fig. 7 is moving horizontally at a constant speed at an altitude near 5.5 km (about 18,000 ft) above

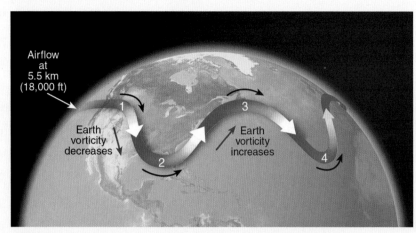

● **FIGURE 7** The wavy path of air aloft due to the conservation of absolute vorticity.

sea level.** At this level, divergence is usually near zero, so our initial assumption approaches a real situation. Because there is no wind speed shear, any change in the relative vorticity will be due to curvature.

Suppose that the air flow is disturbed by a mountain range such that the air at position 1 is flowing southeastward. Heading equatorward, the air moves into a region of decreasing Earth vorticity. To keep the absolute vorticity of the air constant, there must be a corresponding increase in the relative vorticity. Given that increasing relative vorticity implies cyclonic

curvature, the air turns counterclockwise at position 2 and heads northeastward. But now the air is moving into a region where Earth's vorticity steadily increases. To offset this increase, the relative vorticity must decrease. The relative vorticity will decrease if the curvature becomes anticyclonic, so at position 3 air turns clockwise and heads toward the equator once again. This again brings the air into a region where Earth's vorticity decreases. To compensate, the air must now turn cyclonically at position 4, and so on. In this manner, a series of upper-level longwaves may develop, encircling the entire Earth.

---

*The symbol ζ is the Greek letter zeta. Meteorologists often use ζ to represent vorticity.

**You may recall that the atmospheric pressure at this level is about 500 millibars.

We can see in Fig. 12.24 that as the parcel moves from position 1 in the ridge to position 3 in the trough, its absolute vorticity *increases* as it moves along. Within this region is typically found an area of upper-level convergence. As the parcel moves from position 3 in the trough to position 5 in the ridge, its absolute vorticity *decreases* as it moves along. Within this region is typically found an area of upper-level divergence. We can now summarize the information in Fig. 12.24 by stating that as a parcel of air moves with the upper-level flow, an *increase in its absolute vorticity with respect to time is related to upper-level converging air, and a decrease in its absolute vorticity with respect to time is related to upper-level diverging air.**

On upper-air charts that show absolute vorticity (such as on a 500-mb chart), we find that even though the clockwise flow around a high-pressure area produces negative relative vorticity, because of Earth's (positive) vorticity, the absolute vorticity is normally positive everywhere on the map. However, there are regions on the map (about the size of the state of Iowa) where the absolute vorticity is considerably greater. An area of high absolute vorticity is referred to as a *vorticity maximum* or *vort max*. A region of low absolute vorticity is known as a *vorticity minimum*.

● Figure 12.25 illustrates how vorticity, divergence, vertical air motions, and surface mid-latitude cyclonic storm development are linked together. The 500-mb chart (middle chart) in Fig. 12.25 shows a vorticity maximum. If you move from the max eastward toward point 1, notice that, as you move along, vorticity decreases with time. Associated with this region we find upper-level divergence (at 300 mb) and low-level convergence (at the surface), which implies rising air and the possibility of clouds, precipitation,

---

*Viewed another way, in the upper troposphere, an area of converging air increases the total spin—the absolute vorticity—of air parcels, whereas an area of diverging air decreases the absolute vorticity of air parcels.

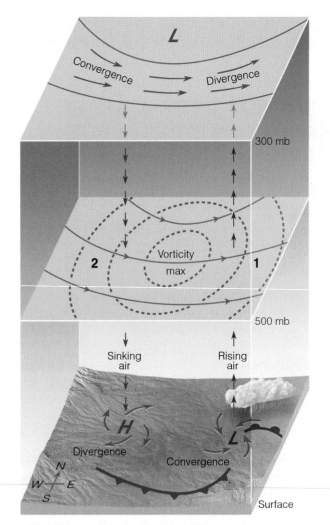

● **FIGURE 12.25** A region of high absolute vorticity—a vorticity maximum—on its downwind (eastern) side has diverging air aloft, converging surface air, and ascending air motions. On its upwind (western) side, there is converging air aloft, diverging surface air, and descending air motions.

and cyclonic storm development. It follows that when a vorticity maximum aloft moves toward a stationary front, a wave will form along the front, and a mid-latitude cyclonic storm will have a good chance of developing. Even in the absence of fronts, a zone of organized clouds with precipitation may form in conjunction with a vorticity maximum. On the other hand, if you move toward the vorticity maximum from position 2, we find that vorticity increases with time. Here, to the west of the vorticity maximum in Fig. 12.25, exists upper-level convergence (at 300 mb), low-level divergence (at the surface), slowly sinking air, and the generally fair weather we associate with surface high-pressure areas.

Cloud patterns in visible and infrared satellite images can be very helpful in identifying vorticity maxima in regions where clouds are present. In cloud-free areas, a type of infrared image called a *water vapor image* measures wavelengths of radiation emitted by water vapor. The swirling patterns of moisture clearly identify the position of vorticity centers. In ● Fig. 12.26, observe the cyclonic swirl of water vapor associated with regions of maximum vorticity over the north Pacific Ocean.

VORTICITY ADVECTION AND SHORTWAVES Recall from an earlier discussion that shortwave troughs can be an important component in the development of mid-latitude cyclonic storms (see p. 329–330 and especially Figs. 12.9 and 12.10). Sometimes it is difficult to pinpoint the location of a shortwave trough on a 500-mb chart simply by looking at the contour lines. However, with the aid of vorticity, we can locate a shortwave on a map and determine how it will move.

● Figure 12.27 is a 500-mb chart that shows a vorticity maximum and a shortwave trough. Notice that the shortwave aligns itself with the region of maximum vorticity. So where we find a

NOAA, U of WI (CIMSS)

● **FIGURE 12.26** This infrared image of water vapor shows regions of maximum vorticity as cyclonic swirls of moisture over the North Pacific Ocean on April 2, 2011. (Also look back at Fig. 1.15 on p. 18.)

● **FIGURE 12.27** A 500-mb chart that shows a region of maximum vorticity, a shortwave trough (heavy dashed line), and wind flow (heavy purple arrows). The solid gray lines are contour lines. Dashed green lines are lines of constant absolute vorticity.

vorticity maximum on the chart, we also find a shortwave trough. To understand how the vorticity maximum moves, we need to examine the concept of **vorticity advection**.

You may recall from Chapter 2 (p. 38) that *advection* is the horizontal transfer of any atmospheric property by the wind. *Vorticity advection* therefore is the transfer of vorticity (the rate of horizontal spin of air parcels) by the wind. Look at Fig. 12.27 and notice that as the wind crosses position 2, air is moving from a region of lower vorticity toward a region of higher vorticity. Since lower values of vorticity are moving into position 2, this region represents an area of *negative vorticity advection (NVA)*. On the other hand, note that at position 1 air is moving from a region of higher vorticity toward a region of lower vorticity. So, position 1 represents a region of *positive vorticity advection (PVA)*. Over time, vorticity values will decrease at position 2 and increase at position 1. The result of these changes in vorticity is the eastward movement of the vorticity maximum and its associated shortwave trough. Generally, the vorticity maximum and the shortwave move along at about half the speed of the wind at this level.

PUTTING IT ALL TOGETHER—A MASSIVE SNOWSTORM
We are now in a position to place much of what we have learned into a real-life weather situation. ● Figure 12.28 illustrates the atmospheric conditions that turned an open-wave cyclone into a ferocious storm. The lower map shows a developing mid-latitude cyclone off the North Carolina coast on February 11, 1983. The counterclockwise circulation around the low is causing warm, moist air from subtropical waters (heavy red arrow) to ride up and over very cold surface air that is entrenched over the eastern seaboard. Snow is falling in the cold air in advance of a warm front that extends from the low northeastward over the Atlantic Ocean.

Above the surface, the 500-mb chart (middle chart, Fig. 12.28) shows a broad longwave trough with a shortwave (heavy dashed line) moving through it. Notice that the shortwave is to the west of the surface low. Apparently, the shortwave has disturbed the flow, as a strong area of baroclinicity exists to the west of the shortwave. Here, isotherms (dashed lines) are intersecting contour lines (solid lines), and cold advection is occurring.

At the 200-mb level (upper chart, Fig. 12.28), the polar front jet stream (blue arrow) swings just to the south of the surface low. The region in orange represents the zone of strongest winds, the jet streak. The area of strong divergence (marked by the letters *DIV* on the 200-mb chart) is almost over the surface low.

● Figure 12.29 is a 500-mb chart that shows absolute vorticity for the same date and time as Fig. 12.28. Notice that the vorticity maximum of 14 (which is actually $14 \times 10^{-5}/\text{sec}$) is located in the same position as the shortwave in Fig. 12.28.* Notice also that aloft, to the east of the vorticity maximum, lies the region of strong upper-level divergence associated with the jet stream. Directly below this region is the open-wave cyclone. Hence, the stage is set, and the necessary ingredients are in place for the surface low to develop into a major cyclonic storm system.

---

*Vorticity values on an upper-level chart are derived using mathematical equations and high-speed computers.

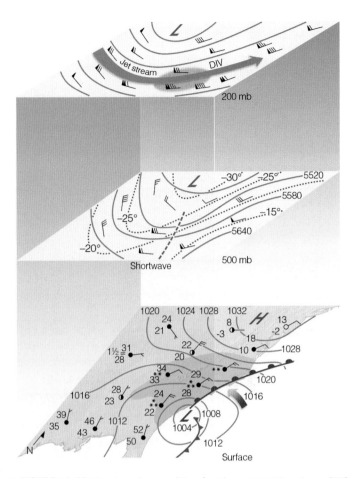

● **FIGURE 12.28** The atmospheric conditions for February 11, 1983, at 7 a.m., (EST). The bottom chart is the surface weather map. The middle chart is the 500-mb chart that shows contour lines (solid lines) in meters above sea level, isotherms (dashed lines) in °C, and the position of a shortwave trough (heavy dashed line). The upper chart is the 200-mb chart that illustrates contours, winds, and the position of the polar jet stream (dark blue arrow). The letters *DIV* represent an area of strong divergence. The region shaded orange represents the jet stream core—the jet streak.

The winds aloft steered the surface low northeastward, but the strong blocking high to the north over southern Canada (surface map, Fig. 12.28) slowed the storm's forward pace, allowing the region of upper-level divergence to intensify the system. With the deep low just off the coast and the strong high to the north, pressure gradients increased, and howling winds in excess of 30 knots battered New Jersey, eastern Pennsylvania, and southeastern New York. Meanwhile, huge quantities of moist ocean air produced copious snowfalls along the Atlantic coastal states. High above the surface, as cold air from the west moved over the region, the air became conditionally unstable; strong convective cells developed, and heavy "downpours" of snow were accompanied by lightning and thunder, producing "thundersnow." At one point in the storm, Allentown, Pennsylvania, reported snow falling at the rate of 5 inches per hour.

The storm left a buried landscape from northern Virginia to Connecticut, where snow drifts in some areas rose to the house-tops. The storm shut down entire cities. It crippled travel on the ground and in the air; closed businesses, government agencies, and schools; knocked out power to many thousands of homes; and caused several deaths. In addition, many towns and cities reported their greatest 24-hour snowfall ever recorded up to that

● **FIGURE 12.29** The 500-mb chart for February 11, 1983, at 7 a.m. (EST). Solid lines are height contours in meters above sea level. Dashed lines are lines of constant absolute vorticity × 10⁻⁵ / sec.

CRITICAL THINKING QUESTION Look at the map in Fig. 12.29 carefully. In which direction will the vorticity max move? How were you able to come up with your answer?

point. Because the heavy snow was accompanied by high winds and low temperatures, the storm has come to be known as "the Blizzard of '83," or the "Megapolitan snowstorm," referring to its impact across the East Coast's biggest cities. It represents one of the 10 most intense snowstorms to hit the east coast of the United States during the twentieth century.

POLAR LOWS Up to now, we have concentrated on middle-latitude cyclones, especially those storms that form along the polar front. There are storms, however, that develop over polar water behind (or poleward of) the main polar front. Such storms are called **polar lows**. Although polar lows develop in both hemispheres, our discussion will center on those storms that form in the Northern Hemisphere in the cold polar air of the North Pacific, North Sea, and North Atlantic, especially in the region south of Iceland.

With diameters of 1000 to 500 km (600 to 300 mi) or less, polar lows are generally smaller in size than their mid-latitude cousins, the wave cyclones that tend to form along the polar front. Some polar lows have a comma-shaped cloud band. Other polar lows have a tight spiral of convective clouds that swirl counterclockwise around a clear area, or "eye," that resembles the eye of a tropical hurricane (see ● Fig. 12.30). In fact, like hurricanes, these smaller intense storms normally have a warmer central core, strong winds (often gale force or higher), and heavy showery precipitation that, unlike a hurricane, is in the form of snow.*

---

*Tropical storms such as hurricanes are covered in Chapter 16. As we will see, the input of heat from the ocean surface into the hurricane increases as the wind speed increases, causing a positive feedback. Moreover, in hurricane environments, the ocean and air temperatures are about the same, whereas in the Arctic, the transfer of sensible heat from the surface is large because the air–sea temperature difference is large.

● **FIGURE 12.30** An enhanced infrared satellite image of an intense polar low situated over Hudson Bay on November 23, 2005. Notice that convective clouds swirl counterclockwise about a clear area, or eye. Surprising similarities exist between polar lows and tropical hurricanes, described in Chapter 16.

Polar lows form most often during the winter, from November through March. During this time, the sun is low on the horizon and absent for extended periods. This situation allows the air next to snow-and-ice-covered surfaces to cool rapidly and become incredibly cold, forming a *continental arctic* air mass. As this frigid air sweeps off the winter ice that covers much of the Arctic Ocean, it may come in contact with warmer *maritime arctic* air that is resting above a relatively warm ocean current. Where these two masses of contrasting air meet, the boundary separating them is called an **arctic front.**

Along the arctic front, the warmer (less-dense) air rises, while the much colder (more-dense) air slowly sinks beneath it. Recall from our discussion of developing middle-latitude cyclones (p. 330) that the rising of warm air and the sinking of cold air establishes a condition known *as baroclinic instability.* As the warm air rises, some of its water vapor condenses, resulting in the formation of clouds and the release of latent heat, which warms the atmosphere. The warmer air has the effect of lowering the surface air pressure. Meanwhile, at the ocean surface there is a transfer of sensible heat from the relatively warm water to the cold air above. This transfer drives convective updrafts directly from the surface. In addition, it tends to destabilize the atmosphere, as heat is gained at the surface and lost to space at the top of the clouds as they radiate infrared energy upward.

The storm's development is enhanced if an upper trough lies to the west of the surface system and a shortwave disturbs the flow aloft. Similarly, the storm may intensify if a band of maximum winds—a jet streak—moves over the surface storm and a region of upper-level divergence draws the surface air upward. The developing cyclonic storm may attain a central pressure of 980 mb (28.94 in.) or lower. Generally, polar lows dissipate rapidly when they move over land.

SUMMARY

In this chapter, we discussed where, why, and how a mid-latitude cyclone forms. We began by examining the early polar front theory proposed by Norwegian scientists after World War I. We saw that the wave cyclone goes through a series of stages from birth to maturity to, finally, decay as an occluded storm.

We looked at the important effect that the upper air flow has on the intensification and movement of surface mid-latitude cyclones. We saw that when an upper-level trough lies to the west of a surface low-pressure area and when a shortwave disturbs the flow aloft, horizontal and vertical air motions begin to enhance the formation of the surface storm. Aloft, regions of divergence remove air from above surface mid-latitude cyclones, and regions of convergence supply air to surface anticyclones. The rising of warm air and the sinking of cold air provide the proper energy conversions for the storm's growth, as potential energy is transformed into kinetic energy. A region of maximum winds—a jet streak—associated with the polar jet stream provides additional support as an area of divergence removes air above the surface mid-latitude cyclone, allowing it to develop into a deep low-pressure area. The curving nature of the jet stream tends to direct mid-latitude cyclones northeastward and anticyclones southeastward.

We looked at the concept of vorticity and how it relates to developing mid-latitude cyclones. We found that on the downwind (eastern) side of a vorticity maximum, there is normally diverging air aloft, converging air at the surface, and cyclonic storm development. Finally, we examined polar lows—those storms that form over water in the cold air of the polar regions.

KEY TERMS

The following terms are listed (with corresponding page numbers) in the order they appear in the text. Define each. Doing so will aid you in reviewing the material covered in this chapter.

polar front theory, 322	cold advection, 329
wave cyclone, 322	warm advection, 330
frontal wave, 322	baroclinic instability, 330
warm sector, 322	cut-off low, 331
secondary low, 322	jet streak, 331
cyclogenesis, 323	conveyor belt model, 333
lee-side low, 324	dry slot, 334
northeaster, 324	comma clouds, 334
convergence, 326	vorticity, 337
divergence, 327	Earth's vorticity, 337
longwave, 328	relative vorticity, 338
Rossby waves, 329	absolute vorticity, 338
shortwave, 329	vorticity advection, 341
barotropic (atmosphere), 329	polar low, 342
baroclinic (atmosphere), 329	arctic front, 342

QUESTIONS FOR REVIEW

1. On a piece of paper, draw the different stages of a mid-latitude cyclonic storm as it goes through birth to decay according to the polar front (Norwegian) model.
2. Why do mid-latitude cyclones usually "die out" after they become occluded?
3. List four regions in North America where mid-latitude cyclones tend to develop or redevelop.
4. Explain this fact: Without upper-level divergence, a surface open wave would probably persist for less than a day.
5. Why do middle-latitude surface low-pressure areas tilt westward with increasing height?
6. If upper-level diverging air above a surface area of low pressure exceeds converging air around the surface low, will the surface low-pressure area weaken or intensify? Explain.
7. What is an Alberta Clipper? Where does it form and how does it move?
8. What are nor'easters? Why are they given that name?
9. How are longwaves in the upper-level westerlies different from shortwaves?
10. What are the necessary ingredients for a mid-latitude cyclonic storm to develop into a huge storm system?
11. Why do surface storms tend to dissipate or "fill" when the upper-level low and the surface low become vertically stacked?
12. How does the polar-front jet stream influence the formation of a mid-latitude cyclone?
13. Explain why, even though the polar-front jet stream coincides with the polar front, some surface regions are more favorable for the development of mid-latitude cyclones than others.
14. Using a diagram, explain why a surface high-pressure area over North Dakota will typically move southeastward while, at the same time, a deep mid-latitude cyclone over the Great Lakes will generally move northeastward.
15. What are the sources of energy for a developing mid-latitude cyclone?
16. How do warm and cold advection aid in the development of a surface mid-latitude cyclone?
17. What are the roles of warm, cold, and dry conveyor belts in the development of a mid-latitude cyclonic storm system?
18. Explain why surface lows tend to deepen when a vorticity maximum aloft moves in from the west.
19. What are polar lows? How and where do they form? What do some of them have in common with tropical hurricanes?

QUESTIONS FOR THOUGHT

1. A British friend of yours says that last night's rain over Northern England was caused by a storm that originally formed east of the Colorado Rockies. Explain how this could happen.

2. Would a mid-latitude cyclone intensify or dissipate if the upper trough were located to the *east* of the surface disturbance? Explain your answer with the aid of a diagram.

3. Explain why, at 500 millibars, when cold advection is occurring, the air temperature does not drop as fast as it should. (Hint: What type of vertical air motions are also occurring?)

4. Over Earth as a whole, would you expect the atmosphere to be mainly barotropic or baroclinic? Explain.

5. Baroclinic waves seldom form in the tropics. Why not?

6. Suppose that Earth stops rotating. How would this affect Earth's vorticity? What would happen to the absolute vorticity of a moving air parcel? If the parcel were initially moving southwestward, how would its direction change, if at all?

7. Why do Pacific storms often redevelop on the eastern side of the Sierra Nevada?

8. If you only had isotherms on an upper-level chart, how would a cut-off low appear?

9. If polar lows form in frigid polar air over water, how is the atmosphere made conditionally unstable so that towering convective cumulus clouds can form?

10. The 500-mb level (at about 5600 m above the surface) is referred to as the "level of nondivergence." Give an explanation as to what this statement means with reference to what is taking place above and below this level.

PROBLEMS AND EXERCISES

1. On the 300-mb chart (● Fig. 12.31), suppose that the winds are blowing parallel to the contour lines.
   (a) On the chart, mark where regions of convergence and divergence are occurring.
   (b) Put an *L* on the chart where you might expect to observe a developing mid-latitude cyclone at the surface.
   (c) Put an *H* on the chart where you might expect to observe a surface anticyclone.
   (d) In which directions would the surface cyclonic storm and anticyclone most likely move?

(e) In terms of convergence and divergence, what are the necessary conditions for the intensification of the surface mid-latitude storm? For the building of the anticyclone?

2. Suppose that there is a region of maximum absolute vorticity at position M on the chart in ● Fig. 12.31.

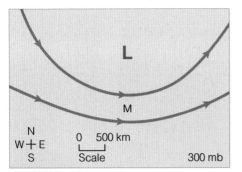

● FIGURE 12.31

(a) How would the absolute vorticity be changing with time as an air parcel moves downstream (from position M) with the flow?

(b) Based on absolute vorticity, circle on the map where you would expect to find upper-level divergence, surface convergence, and a developing mid-latitude cyclone.

(c) Draw a dashed line where you would expect to observe a shortwave trough.

ONLINE RESOURCES

Visit www.cengagebrain.com to view additional resources, including video exercises, practice quizzes, an interactive eBook, and more.

Specialized weather forecasts can help
utilities maximize the power generated from
wind turbines.

Weather Forecasting

CONTENTS

SOMETIMES THERE IS NO JOB SECURITY IN weather forecasting. In fact, at least one weather forecaster actually lost his job for not altering his prediction. On April 15, 2001, a function honoring a well-known radio talk show host was scheduled outdoors at the Madera, California, fairgrounds. The story goes that a local forecaster at the radio station that sponsored the event had called for a "chance of rain" on April 15. Upset that such a forecast might discourage people from attending the function, the station manager told the forecaster to alter his forecast and predict a greater possibility of sunshine. The forecaster refused and was promptly fired. Apparently, retribution reigned supreme—it poured on the event.

Weather forecasts are issued to save lives, to save property and crops, and to tell us what to expect in our atmospheric environment. In addition, knowing what the weather will be like in the future is vital to many human activities. For example, a summer forecast of extended heavy rain and cool weather would have construction supervisors planning work under protective cover, department stores advertising umbrellas instead of bathing suits, and ice cream vendors vacationing as their business declines. The forecast would alert farmers to harvest their crops before their fields became too soggy to support the heavy machinery needed for the job. And the commuter? Well, the commuter knows that prolonged rain could mean clogged gutters, flooded highways, stalled traffic, blocked railway lines, and late dinners.

On the other side of the coin, a forecast calling for extended high temperatures with low humidity has an entirely different effect. As ice cream vendors prepare for record sales, the dairy farmer anticipates a decrease in milk and egg production. The forest ranger prepares warnings of fire danger in parched timber and grasslands. The construction worker is on the job outside once again, but the workday begins in the early morning and ends by early afternoon to avoid the oppressive heat. And the commuter prepares for increased traffic stalls due to overheated car engines. Put yourself in the shoes of a weather forecaster: It is your responsibility to predict the weather accurately so that thousands (possibly millions) of people in your area will know whether to carry an umbrella, wear an overcoat, or prepare for a winter storm. Since weather forecasting is not an exact science, your predictions will occasionally be incorrect. If your erroneous forecast misleads many people, you may become the target of jokes, insults, and even anger. There are people who expect you to be able to predict the unpredictable. For example, on Monday you may be asked whether two Mondays from now will be a nice day for a picnic. And, of course, what about next winter? Will it be bitterly cold?

Unfortunately, accurate answers to such questions are beyond meteorology's present technical capabilities, but "useful" answers may be possible by applying different techniques to current forecast methods. Will forecasters ever be able to answer such questions confidently? If so, what steps are being taken to improve the forecasting art? How are forecasts made, and why do they sometimes go awry? These are just a few of the questions we will address in this chapter.

Weather Observations

Weather forecasting basically entails predicting how the present state of the atmosphere will change. Consequently, if we wish to make a weather forecast, we must know the present weather conditions over a large area. A network of observing stations located across the world provides the forecaster with this information. Forecasters have access to many maps and charts showing the present conditions at various atmospheric heights, as well as vertical profiles (called *soundings*) of temperature, dew point, and winds. Also available are visible and infrared satellite images, as well as Doppler radar information that can detect and monitor

the severity of precipitation and thunderstorms. All of these sources are used by forecasters to monitor current weather and anticipate future conditions. Many of these observations are also brought into computer-based **atmospheric models** that project the weather forward, as we will see later in this chapter.

One of the biggest challenges for modern forecasters is to deal with an avalanche of observational and forecast data and zero in on the elements that are most important on any given day. The **forecast funnel** describes the steps used by forecasters to steer their attention from large scales to smaller scales and from short time frames to longer periods. More information about the forecast funnel can be found in Focus section 13.1.

SURFACE AND UPPER-AIR DATA

A network of surface-observing stations extends across the world, reporting the weather close to ground level. More than 10,000 land-based stations and hundreds of ships and buoys provide surface weather information at least four times a day.* Most airports observe conditions hourly, and hundreds of Automated Surface Observing Systems (ASOS) send data even more often from selected airports and other sites throughout the United States. The ASOS system is designed to provide nearly continuous information about wind, temperature, pressure, cloud-base height, and runway visibility at various airports.**

To sample the atmosphere above ground level, radiosondes are launched at more than 800 locations around the world, including almost 100 U.S sites. (Radiosondes are described in more detail in Focus section 1.3 on p. 15.) Radiosonde data are regularly collected twice a day, at 0000 and 1200 UTC. Special launches may occur in association with research projects or to help provide more data when a major weather threat is looming, such as a high-impact winter storm or an outbreak of tornadic thunderstorms. Data on upper-air conditions may also be gathered and provided by some aircraft as they travel their usual routes.

SATELLITE PRODUCTS

Many types of observations collected by satellites are available to forecasters, providing a clearer representation of the atmosphere.† Visible, enhanced infrared, and water vapor images provide a wealth of information, some of which comes from inaccessible regions, that can be examined to analyze fast-changing conditions. Special instruments aboard satellites can detect a wide variety of important phenomena, including lightning, sea-surface temperature, and smoke from forest fires. Many kinds of satellite observations can be incorporated into forecast models (see ● Fig. 13.1).

DOPPLER RADAR

A network of more than 100 Doppler radar units covers nearly all of the 48 contiguous United States. These radars provide round-the-clock information on the evolution

---

*Observations are usually taken at 0000, 0600, 1200, and 1800 Coordinated Universal Time (UTC), which is also called *Greenwich Mean Time* (GMT)—local time at the Greenwich Observatory in England. To convert from UTC to your local time, see Appendix F, "Changing GMT and UTC to Local Time."

**Additional information on ASOS is given in Chapter 3 on p. 85.

†Information provided by satellites is located in various sections of this book. For example, see Chapter 5, pp. 133–139.

The Forecast Funnel

Every day on the job, a working forecaster must quickly assess the state of the atmosphere, then create forecasts that may be used by many thousands of people. Even for professionals well trained in meteorology, the pressure of forecasting can sometimes be quite intense. To help produce the best possible predictions in a limited amount of time, forecasters commonly follow a sequence of steps described as the *forecast funnel* (see ● Fig. 1). This flowchart, sometimes called the Snellman Funnel, was developed by meteorologist Leonard Snellman in the 1970s and 1980s. It has since become a standard approach in daily forecasting for many meteorologists at the National Weather Service and in the private sector.

Based on the forecast funnel, the first step in a forecaster's day is to assess the current state of the atmosphere on the largest relevant scale—the *global scale*. In most cases, the forecaster will examine the flow of winds and the progression of longwaves at the jet-stream level across the hemisphere where the forecaster is located (either Northern or Southern). At this stage, forecasters typically look at the current 500-millibar chart and recent satellite imagery, as well as computer model projections of how the 500-millibar flow might evolve. Where are the most prominent features? Are they sweeping quickly eastward in zonal flow, or are they moving more slowly in meridional flow? If it is winter, is the polar jet stream consolidated, or is it split into branches that might encourage a meandering jet-stream flow and weather features that could stall for days? Which of these features are most likely to affect the local area over the next few days? Because computer models are now quite skilled at mapping out the large-scale jet-stream patterns, the forecaster typically spends the least amount of time here at the top of the funnel.

The forecaster then zooms in, focusing on the *synoptic scale* (roughly a state to a continent in size). Often the forecaster looks for the "problem of the day"—the most uncertain aspect of a given day's weather picture—and focuses his or her attention on that issue. Sometimes the problem of the day is the approach of a strong shortwave that is likely to trigger heavy rain or snow, or it might be the passage of a cold front expected to produce frigid wind chills. At this point, the forecaster may refer to the 850-millibar chart to analyze areas of high moisture content and

converging winds. Satellite water vapor images can also help determine where upper-level moisture is present. The problem of the day becomes more difficult to diagnose when computer models disagree strongly on the future of important weather features. In this case, the forecaster may check the model's initial conditions against current observations and satellite images, making sure that the initial conditions contain no major errors and that they are not missing important pieces of data. Looking toward the future, the forecaster may use one or more of the analysis tools outlined on p. 350 to better understand how weather features will unfold at various atmospheric levels.

Next, the forecaster moves to the *mesoscale* (roughly a county to a state in size). Here, the emphasis is usually on precipitation: how much and exactly where? The important variables include where the air is likely to be rising or sinking and how much moisture is in the atmosphere. The forecaster is likely to turn to high-resolution satellite images, as well as Doppler radar, local networks of surface stations, and other sources of local and regional data. He or she will look for evolving features such as small-scale boundaries on which thunderstorms or bands of snow might develop. The forecaster may try to determine where heavy rains are most likely to fall on soaked ground, raising the risk of flash floods, or where tornadic thunderstorms are most likely to develop later that day. The effect of nearby coastlines or mountains must also be taken into account when focusing on the mesoscale. The forecaster will often spend considerably more time in this stage, near the bottom of the funnel, than at the top, as it takes more time to evaluate mesoscale features and local data sources that are not necessarily reflected in the output of larger-scale computer models.

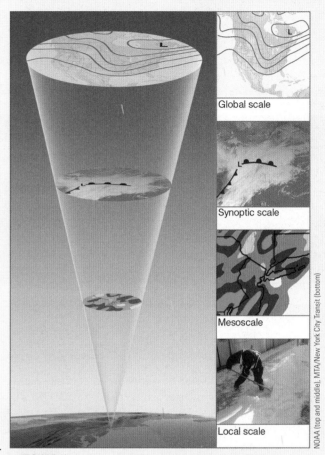

Global scale

Synoptic scale

Mesoscale

Local scale

NOAA (top and middle), MTA/New York City Transit (bottom)

● **FIGURE 1** The forecast funnel applied to a region of the northeast coast of the United States.

After completing a trip through the forecast funnel, our meteorologist is now well prepared to begin crafting *local-scale* forecasts. Usually he or she will start with the very short range and move on to longer time frames, including outlooks that may extend out 8 to 10 days. Sometimes forecasters at the National Weather Service will split the forecasting duties on a given shift, with one person dedicated to the shorter-term outlook and another focusing on the longer-range part of the forecast.

The forecast funnel helps ensure that the ever-increasing wealth of information now available to forecasters is used in the most helpful and efficient way. An example of how the forecast funnel might be applied is given in the section "Using Forecasting Tools to Predict the Weather," which appears at the end of this chapter.

(a) Visible image

(b) Enhanced infrared imaage

(c) Water vapor image

NASA

● **FIGURE 13.1** These three satellite images were each collected by the *GOES-East* satellite at 11:15 a.m. on June 16, 2014. (a) Visible imagery that details solar radiation being reflected is useful for detecting areas of snow cover as well as low-level cloud features, such as cumulus development. (b) Infrared imagery analyzes the temperature at the top of clouds, which is related to the amount of infrared energy being emitted; the bright colors (yellow and orange) over the northern Plains indicate the high, cold tops of severe thunderstorms. (c) Water vapor imagery detects the amount of energy absorbed by water vapor at a particular wavelength and thereby reveals moisture present in the middle and upper troposphere. Lighter colors indicate more moisture, while darker colors (such as in western Kansas and eastern Colorado) show where the upper troposphere is drier.

of rain, snow, sleet, and hail. Because Doppler radar can track winds as well as precipitation, the network is also a valuable tool in providing warnings of destructive windstorms and tornadoes, as we will explore in the following chapters. Some of the most advanced computer forecast models are now being designed to incorporate information from radars, which could help improve forecasting significantly.

Acquisition of Weather Information

Collecting weather data is only the beginning of the process that leads to a forecast. Meteorologists at government weather services and private firms across the world rely on an accurate supply of weather data to make predictions for their areas. A United Nations agency—the *World Meteorological Organization* (WMO), which includes more than 175 nations—is responsible for the international exchange of weather data and certifies that the observation procedures do not vary among nations. This is an extremely important task, as the observations must be comparable.

Weather information from all over the world is transmitted electronically to government meteorological centers worldwide. This includes the National Centers for Environmental Prediction (NCEP), a branch of the National Weather Service (NWS) located at the University of Maryland in College Park, just outside Washington, D.C. Here, the massive job of analyzing the data, running models, preparing weather maps and charts, and predicting the weather on a global and national basis begins.* From NCEP, observations and computer model output are transmitted to private forecasting firms and public agencies in the United States. Many of NCEP's products are also posted on the web. Across the nation, dozens of NWS *Weather Forecast Offices* (WFOs) use the information to issue local and regional weather forecasts. Standard forecasts are prepared every 12 hours and updated as needed in between these intervals.

The public gets weather forecasts through a variety of media, including radio, television, computers, and smartphones. Many broadcasting stations hire private meteorological companies or professional meteorologists to modify a National Weather Service forecast or to make their own forecasts, assisted by data and products from government agencies. Other stations may use announcers without meteorological training who paraphrase NWS forecasts, or read them word for word. On the web or on smartphones, the public can access local forecasts from the NWS or from private firms, often presented with eye-catching graphics. (For more on how forecast information is presented to the public, see Focus section 13.2.)

Weather Forecasting Tools

During the course of a single workday, a typical forecaster may examine and compare dozens or even hundreds of individual weather maps. To help forecasters handle all the available charts and maps, the NWS employs the high-speed *Advanced Weather Interactive Processing System* (**AWIPS**). A second-generation version, called *AWIPS II*, was adopted by the NWS starting in 2013 (see ● Fig. 13.2).

---

*By international agreement, data are plotted using symbols illustrated in Appendix B.

The Forecast in Words and Pictures

From the earliest days of science-based weather forecasting, predictions have been issued graphically as well as verbally. The "weather map," produced by compiling and mapping observations gathered by teletype, became a popular feature in U.S. and British newspapers in the late 1800s. By the middle of the twentieth century, most newspapers included a weather map showing the expected locations of high- and low-pressure centers later that day or on the next day. Some of these maps also showed the locations of warm, cold, and occluded fronts, but it was the advent of television weathercasting in the late 1940s and 1950s that made weather fronts truly familiar to millions of Americans.

With the growth of the Internet since the 1990s, and especially the huge increase in the use of smartphones in the 2000s, it is now more important than ever to convey the weather in a compact graphic form that displays well on a small screen. On its public website, the National Weather Service presents its forecasts in both iconic and verbal forms (see ● Fig. 2).

Many smartphone "apps" are now available that display today's weather and the forecast looking days ahead. These forecasts are typically generated from software created by private-sector meteorologists, drawing on the observations and computer-model predictions that are gathered or produced by the National Weather Service. These forecasters may also use proprietary models and software that generate forecasts customized to their particular needs. A typical app-based forecast will include an iconic image capturing the expected weather on a given day (a bright sun, for example, or a

Three-day forecast

Tonight	Sunday	Sunday Night	Monday	Monday Night	Tuesday
Chance of Thunderstorms	Partly Cloudy	Mostly Cloudy	Partly Cloudy	Thunderstorms Likely	Thunderstorms Likely
Low: 62° F	High: 79° F	Low: 62° F	High:78° F	Low: 62° F	High: 74° F

Detailed Forecast

Overnight   20 percent chance of showers and thunderstorms before 4 a.m. Mostly cloudy, with a low of 62°. North-northwest wind of 11 mi/hr, gusting to 17 mi/hr.

Sunday   Partly sunny, with a high of 79°. North wind of 8 to 11 mi/hr, with gusts up to 15 mi/hr.

Sunday Night   Mostly cloudy, with a low of 62°. East-northeast wind of 7 mi/hr.

Monday   Partly sunny, with a high of 78°. East wind of 8 mi/hr.

Monday Night   Showers and thunderstorms likely. Cloudy with a low of 62°. East winds of 9 mi/hr, with gusts to 15 mi/hr. 70 percent chance of precipitation.

Tuesday   Showers and thunderstorms likely. Cloudy, with a high of 74°. 60 percent chance of precipitation.

● **FIGURE 2** The first three days of the National Weather Service forecast for the Springfield, Illinois, area, issued on Saturday evening, June 7, 2014.

thunderstorm with a lightning bolt and sheets of rain below it). Many apps include a single word or two describing the weather (such as "sunny" or "T-storms"), along with the expected high and low temperatures. Some apps attempt to provide the timing of rain or snow down to the hour, or even the minute, although you should keep in mind that the forecast with the most fine-grained timing is not necessarily the most accurate one.

Sophisticated software allows the forecasts on your smartphone to be personalized

to your location—sometimes even down to a single small community. Here again, it is important to keep in mind that the prediction from a computer model may not be as precise as the impression given by your smartphone app. On most days, outside of mountainous locations or coastlines, there is likely to be little major difference in the weather experienced in your town or city and the next one over. However, there can be large variations across small distances during thunderstorms or other forms of extreme weather.

The AWIPS II system has data communications, storage, processing, and display capabilities (including graphical overlays) to better help the individual forecaster extract and assimilate information from the mass of available data. In addition, AWIPS is able to integrate information received from satellites and surface stations as well as from the national Doppler radar network (the WSR-88D), which now includes dual-polarization technology. (Dual-polarization technology is covered in Chapter 15, p. 434.)

Much of the information from ASOS and Doppler radar is processed by software according to predetermined formulas, or *algorithms*, before it goes to the forecaster. Certain criteria or combinations of measurements can alert the forecaster to an impending weather situation.

A software component of AWIPS II called the *Graphical Forecast Editor* allows forecasters to look at the daily prediction of weather elements, such as temperature and dew point, in a grid format with spacing as small as 2.5 km (1.6 mi). Presenting the

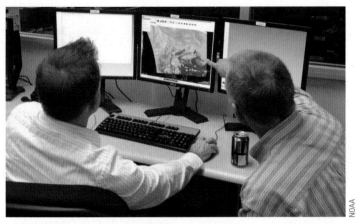

● **FIGURE 13.2** Two NWS forecasters testing the AWIPS II system, adopted by the NWS in 2013.

data in this format allows the forecaster to predict the weather more precisely over a relatively small area.

With so much information at the forecaster's disposal, it is essential that the data be easily accessible and in a format that allows several weather variables to be viewed at one time. The **meteogram** is a chart that shows how one or more weather variables has changed at a station over a given period of time. As an example, the chart may represent how air temperature, dew point, and sea-level pressure have changed over the past five days, or it may illustrate how these same variables are projected to change over the next five days (see ● Fig. 13.3).

Another aid in weather forecasting is the *sounding*—a two-dimensional vertical profile of temperature, dew point, and winds (see ● Fig. 13.4).* The analysis of a sounding can be especially helpful when making a short-range forecast that covers a relatively small area, such as the mesoscale. The forecaster examines the sounding of the immediate area (or closest proximity), as well as the soundings of those sites upwind, to see how the atmosphere might be changing. Figure 13.4 is a simplified version of the more complex-sounding plot typically used by forecasters; a larger depiction of a sounding can be found in Appendix H, "Adiabatic Chart."

Computer programs automatically calculate from the sounding a number of meteorological *indexes* that can aid the forecaster in determining the likelihood of smaller-scale weather phenomena, such as thunderstorms, tornadoes, and

*A sounding is obtained from a radiosonde or from satellite data.

● **FIGURE 13.3** Meteogram illustrating predicted weather at the surface and aloft at St. Louis, Missouri, from 6 a.m., November 19, 2007, to noon on November 21, 2007. The forecast is derived from the Global Forecast System (GFS) model. (NOAA)

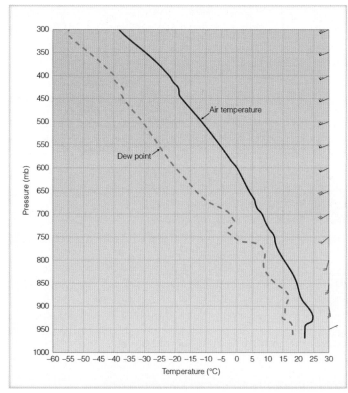

● **FIGURE 13.4** A sounding of air temperature, dew point, and winds near Oklahoma City, Oklahoma, during the evening of May 20, 2013, on the same day a violent tornado (pictured in Fig. 13.14) tore through the town of Moore, Oklahoma, which lies just outside of Oklahoma City.

hail. Soundings also provide information that can aid in the prediction of fog, air pollution alerts, and the downwind mixing of strong winds.

Satellite information is also a valuable tool for the forecaster. Visible, enhanced infrared, and water vapor images provide a wealth of information, some of which comes from inaccessible regions, that can be examined to analyze fast-changing conditions. Special instruments aboard satellites can detect a wide variety of important phenomena, including lightning, sea surface temperature, and smoke from forest fires. Many kinds of satellite observations are incorporated into forecast models.

Forecasters also use a chart called the *thickness chart* to scrutinize the atmosphere. This is especially helpful when analyzing temperatures at different altitudes and how they are changing. If you are interested in learning how this chart can be used to predict whether falling precipitation will be in the form of rain or snow, read Focus section 13.3.

Weather Forecasting Methods

Up to this point, we have examined some of the weather data and tools a forecaster might use in making a weather prediction. With all of this information available to the forecaster, including countless charts and maps, just *how* does a meteorologist make a weather forecast, and what types of forecasts are there?

THE COMPUTER AND WEATHER FORECASTING: NUMERICAL WEATHER PREDICTION Each day the many thousands of observations transmitted to NCEP are mapped by computers and depicted in surface and upper-air charts. Meteorologists interpret the weather patterns and then correct any errors that may be present. The final chart is referred to as an **analysis**.

Computers not only plot and analyze data, but they also predict the weather—a much more challenging task. Today's supercomputers can analyze large quantities of data extremely quickly, carrying out trillions of calculations per second. Because the atmosphere is so complex, some of the most powerful supercomputers on Earth are devoted to weather and climate prediction. The routine daily forecasting of weather by computers using mathematical equations is known as **numerical weather prediction**.

Because weather variables are constantly changing, meteorologists have devised atmospheric models that describe the present state of the atmosphere. These are not physical models that paint a picture of a developing storm; they are, rather, mathematical models consisting of many equations that describe how atmospheric temperature, pressure, winds, and moisture will change with time. The models do not fully represent the real atmosphere, since the processes occurring around the world at every instant are too complex to represent completely. Instead, the models are very useful approximations, formulated to retain the most important aspects of the atmosphere's behavior.

How do these models actually work? The equations are translated into complex software, and surface and upper-air observations of temperature, pressure, moisture, winds, and air density are fed into the equations at regular intervals. The process of integrating these data into numerical models is called **data assimilation**. As more and more types of data are assimilated into the models, the quality of the model forecasts often improves.

To determine how each of these key meteorological variables will change, each equation is solved for a small increment of future time—say, five minutes—for a large number of locations called *grid points*, each situated a given distance from the next.* In addition, each equation is solved for as many as 50 levels in the atmosphere. The results of these computations are then fed back into the original equations. The computer again solves the equations with the new "data," thus predicting weather over the following five minutes. This procedure is done repeatedly until it reaches some desired time in the future. For example, one standard NWS model produces weather depictions every hour out to 18 hours. Another covers a longer period, but with larger steps in between; it produces snapshots of weather conditions every 3 hours out to 84 hours (3.5 days). And one model even forecasts the state of the atmosphere 384 hours (16 days) into the future. Once the calculations of future weather are completed, the computer then analyzes the data and draws the projected positions of pressure systems with their isobars or contour lines. The final forecast chart representing the atmosphere at a specified future time is called a **prognostic chart**, or, simply, a **prog**. Computer-drawn progs have come to be known as "machine-made" forecasts.

*Some models have a grid spacing as small as 0.5 km, whereas the spacing in others exceeds 100 km. There are models that actually describe the atmosphere using a set of mathematical equations with wavelike characteristics rather than a set of discrete numbers associated with grid points.

The Thickness Chart—A Forecasting Tool

The thickness chart can be a valuable forecasting tool for the meteorologist. It can help identify air masses and locate fronts, and a prognostic thickness chart can help predict the daily max and min temperature. It can also help predict whether falling precipitation will be in the form of rain or snow. What, then, is a thickness chart?

A thickness chart shows the difference in height between two constant pressure surfaces. The vertical depth or thickness between any two pressure surfaces is related to the average air temperature between the two surfaces. Recall that air pressure decreases more rapidly with height in cold air than in warm air. This fact is illustrated in ● Fig. 3, which shows a 1000-mb pressure surface and a 500-mb pressure surface. The difference in height between these two pressure surfaces is called the *1000-mb to 500-mb* thickness.* Notice that the vertical distance (thickness) between these two pressure surfaces is greater in warm air than in cold air. Consequently, warm air produces high thickness, and cold air low thickness. In fact, thickness is directly related to the layer's average temperature.

*Because the 1000-mb pressure surface is often close to the surface of Earth, the 1000-mb to 500-mb thickness is sometimes referred to as the *surface to 500-mb thickness.*

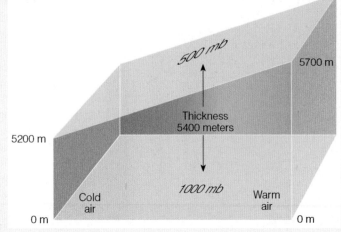

● **FIGURE 3** The vertical separation (thickness) between the 1000-mb pressure surface and the 500-mb pressure surface is greater in warm air than in cold air.

▼ **TABLE 1** A general rule of thumb relating 1000-mb to 500-mb thickness values for snow levels for mountainous areas west of the Rocky Mountains

1000-MB TO 500-MB THICKNESS VALUES (METERS)	APPROXIMATE SNOW LEVEL	
	Meters	Feet
5220	near sea level	near sea level
5280	500	1500
5340	1000	3300
5400	1500	5000
5460	2000	6500

Each of the major meteorological centers around the world has its own computer models, including some that are tailored to a particular continent or region. The centers may also share the forecasts issued by their various large-scale models, which often predict weather going out a week or more across the entire Northern (or Southern) Hemisphere. The NOAA National Centers for Environmental Prediction runs several models in *operational mode,* meaning that they have been thoroughly tested and are reliable enough for everyday use. Other models are run in *experimental mode;* these models show promise and may be consulted by forecasters, but they may not yet be robust enough for operational use. Currently, the NCEP operational models include the *North American Mesoscale* (NAM) model, which is run every six hours and predicts conditions at three-hour intervals out to 84 hours, and the *Global Forecast System* (GFS) model, which is run every 12 hours and issues predictions extending out to 384 hours, or 16 days. An experimental model recently converted to operational use at NCEP is the *High-Resolution Rapid Refresh* (HRRR) model, which will be run every hour and is designed to predict weather in sharp detail at hourly time steps out to 15 hours.

Supercomputers can solve the equations of atmospheric motion far more quickly and efficiently than could possibly be done by hand. For example, just to produce a 24-hour forecast chart for the Northern Hemisphere requires many hundreds of millions of mathematical calculations. It would, therefore, take a group of meteorologists working full time with hand calculators years to produce a single chart; by the time the forecast was available, the weather for that day would already be ancient history. Today, computer models are taken for granted in weather forecasting. They have become so important that our current level of skill in weather forecasting would be impossible without them. In fact, in some cases a computer model may do just as well as a human being in predicting high and low temperatures during tranquil weather. However, forecasters must be careful not to take the prediction of any model as the gospel truth. If forecasters rely too much on models, without bringing their own knowledge and experience into the forecasting process, they may become victims of what has been termed "meteorological cancer." During the most unusual and threatening weather situations, the best forecasts occur when the output of computer models is carefully adjusted as needed by an experienced, knowledgeable meteorologist.

An ever-increasing variety of models (and, hence, progs) is now available to forecasters, each producing a slightly different interpretation of the weather for the same projected time and

When 1000-mb to 500-mb thickness lines are drawn on a chart, they may appear similar to those shown in ● Fig. 4. On the chart, regions of low thickness correspond to cold air, and regions of high thickness to warm air. There are a number of forecasting rules for predicting air temperature and precipitation using this chart. One forecasting rule is that the 5400-meter thickness line often represents the dividing line between rain and snow, especially for cities receiving precipitation east of the Rocky Mountains. If precipitation is falling, cities with a thickness greater than 5400 m should be receiving rain, whereas cities with a thickness less than 5400 m should be receiving snow. Look at Fig. 4. St. Louis is receiving precipitation. Would you expect it to be in the form of rain or snow? What about Detroit? If you live east of the Rockies, find your city on the chart and see whether precipitation would be in the form of rain or snow on this day.

During the winter in mountainous regions, the elevation where rain changes to snow is called the *snow level*. For locations west of the Rockies, ▼ Table 1 gives the snow level based on the 1000-mb to 500-mb thickness. The information in Table 1 works well when the atmosphere is well mixed, but not when a cold, stable layer rests near the surface.

● **FIGURE 4** A 1000-mb to 500-mb thickness chart for a January morning. The lines on the chart represent the vertical depth in meters between the 1000-mb and 500-mb pressure surfaces. Low thickness lines correspond to cold air, and high thickness lines to warm air. For reference, the 5400-meter thickness line represents a vertical layer of air 5400 meters thick with an average temperature of -7°C (19°F). The 5200 thickness line is roughly the boundary for arctic air.

atmospheric level. ● Fig. 13.5 shows four progs for different levels in the atmosphere 24 hours into the future. How the forecaster might use each prog in making a prediction is given in ▼ Table 13.1.

The differences between progs can result from the way the models use the equations, or from the distance between grid points, called *resolution*. Some models predict certain features better than others: One model may work best in predicting the position of troughs on upper-level charts, whereas another may forecast the position of surface lows quite well. Although a model with a higher resolution can provide more detail, such enhanced detail does not necessarily mean the model is more accurate.

A good forecaster knows the idiosyncrasies of each model and carefully scrutinizes all the progs. The forecaster then makes a prediction based on the *guidance* from the computer, his or her interpretation of the weather situation, and any local geographic features that influence the weather within the specific forecast area.

Currently, forecast models predict the weather reasonably well four to six days into the future, with decreasing accuracy for longer intervals. The models tend to do a better job at predicting temperature and jet-stream patterns than precipitation. However, even with all of the modern advances in weather forecasting provided by ever more powerful computers, forecasts by the National Weather Service and by private firms are sometimes wrong. (An example of making a 24-hour forecast using different progs is given toward the end of this chapter on p. 373.)

WHY COMPUTER-BASED FORECASTS CAN GO AWRY AND STEPS TO IMPROVE THEM Why do forecasts sometimes go wrong? There are a number of reasons for this unfortunate situation. For one, computer models have inherent flaws that limit the accuracy of weather forecasts. For example, computer-forecast models idealize the real atmosphere, meaning that each model makes certain

WEATHER WATCH

When a weather forecast calls for "fair weather," does the "fair" mean that the weather is better than "poor" but not up to being "good"? According to the National Weather Service, the subjective term "fair" implies a rather pleasant weather situation where there is no precipitation, no extremes in temperature occur, visibility is good, and less than 40 percent of the sky is covered by opaque clouds such as stratus.

● **FIGURE 13.5** Computer-drawn 24-hour progs using the GFS (Global Forecast System) model for 850 mb, 700 mb, 500 mb, and 300 mb. The progs were drawn on March 5, 2013, and became valid on March 6 at 7 a.m. Solid lines represent height contours. Orange shade shows the wind speed in knots. Green shade shows where the predicted relative humidity is 70 percent or greater.

▼ **TABLE 13.1 The Use of Various Charts as a Forecasting Tool**

FORECAST CHART	APPROXIMATE ALTITUDE ABOVE SEA LEVEL	ELEMENTS THAT MAY BE SPOTTED AND TRACKED
Surface map		• Location and motion of frontal systems, centers of high and low pressure • Areas of cloudiness, precipitation, high wind, and fog • Cross-isobar winds that indicate strengthening or weakening low pressure • Warm, moist air that can foster shower and thunderstorm development if conditional instability is present
850 mb	1500 m (4900 ft)	• High moisture values that can contribute to heavy precipitation • Convergent winds associated with strengthening low pressure areas • A low-level jet stream that can help intensify thunderstorm development • Temperatures that determine whether precipitation will fall as snow, rain, sleet, or a mixture
700 mb	3000 m (9800 ft)	• Moisture to feed precipitation and mid-latitude storm systems • Temperature advection that could strengthen or weaken fronts • A dry, warm layer that can inhibit thunderstorm development • Temperatures that help determine ice crystal and snowfall type
500 mb	5600 m (18,400 ft)	• General steering flow for mid-latitude cyclonic storm systems, hurricanes, and tropical cyclones • Location and motion of ridges, troughs, and short waves that generate and strengthen surface features • Areas of cold advection that can help increase conditional instability and support thunderstorm development • Large areas of high or low heights that correspond to unusually warm or cold conditions at the surface, depending on region and time of year
300 mb	9180 m (30,100 ft)	• Location of core of jet stream • Jet streaks and areas of divergence within jet stream that may correspond to intensifying low pressure at the surface • Areas of high pressure and light, divergent wind in the tropics and subtropics that can support hurricane development

assumptions about the atmosphere. These assumptions may be on target for some weather situations and be way off for others. Consequently, the computer may produce a prog that comes quite close to describing the actual state of the atmosphere on one day and not so close on another. A forecaster who bases a prediction on an "off day" computer prog may find a forecast of "rain and windy" turning out to be a day of "clear and colder."

Another forecasting problem arises because the majority of models are not global in their coverage, and errors can creep in along the model's boundaries. For example, a model that predicts the weather for North America may not accurately treat weather systems that move in along its boundary from the western Pacific. Obviously, a global model would usually be preferred. However, a global model of similar sophistication with a high resolution requires an incredible number of computations.

Even though many thousands of weather observations are taken worldwide each day, there are still regions where observations are sparse, particularly over the oceans and at higher latitudes. To help alleviate this problem, newer satellites are providing a more accurate profile of temperature and humidity for the computer models. Wind information now comes from a variety of sources, such as Doppler radar, commercial aircraft, buoys, and satellites that translate ocean surface roughness into surface wind speed (see Chapter 9, p. 258).

Grid Spacing   Earlier, we saw that the computer solves the equations that represent the atmosphere at many locations called grid points, each spaced from 100 km to as low as 0.5 km apart. On computer models with large spacing between grid points (say 60 km), larger weather systems, such as extensive mid-latitude cyclones and anticyclones, show up on progs whereas much smaller systems, such as thunderstorms, do not. Because such models are too coarse to generate showers and thunderstorms directly, these features are instead *parameterized*, meaning that they are approximated for broad areas instead of being predicted by the model for specific points. The computer models that forecast for a large area such as North America are, therefore, better at predicting the widespread precipitation associated with a large cyclonic storm than predicting where localized showers and thunderstorms will occur. In summer, when much of the precipitation falls as local showers, a computer prog may have indicated fair weather, while outside it is pouring rain.

To capture smaller-scale weather features as well as the terrain of the region, the distance between grid points on some models is being reduced. For example, the High-Resolution Rapid Refresh model has a grid spacing of 3 km, and some experimental models have spacing as low as 0.5 km. Instead of parameterizing showers and thunderstorms, a model such as HRRR with its small grid spacing (high resolution) can actually incorporate radar information and simulate how showers and thunderstorms might evolve (see ● Fig. 13.6). The downside of high-resolution models is that, as the horizontal spacing between grid points decreases, the number of computations increases. When the distance is halved, there are eight times as many computations to perform, and the time (and computational expense) required to run the model goes up by a factor of 16.

NOAA/ESRL

● **FIGURE 13.6** A forecast of shower and thunderstorm activity across the Great Plains for 7 p.m. CDT May 17, 2017, issued 13 hours in advance. The forecast precipitation (left) closely matches the actual precipitation that was mapped with a blend of radar and other sensor data (right). The forecast was produced by the NOAA High-Resolution Rapid Refresh model (HRRR) with a resolution of 3 kilometers. The radar observations at right are on a higher-resolution 1-kilometer grid, so they contain more detail than the forecast at left.

Another forecasting problem is that many computer models cannot adequately interpret many of the factors that influence surface weather, such as the interactions of water, ice, surface friction, and local terrain on weather systems. Many large-scale models now take mountain regions and oceans into account. Some models (such as HRRR) take even smaller factors into account—features that large-scale computers miss due to their larger grid spacing. Given the effect of local terrain, as well as the impact of some of the other problems previously mentioned, computer models that forecast the weather over a vast area do an inadequate job of predicting local weather conditions, such as surface temperatures, winds, and precipitation.

Even with better observing techniques and high-quality computer models, there are countless small, unpredictable atmospheric fluctuations, referred to as **chaos**, that limit model accuracy. For example, tiny eddies are much smaller than the grid spacing on the computer model and, therefore, go unaccounted for in the model. These small disturbances, as well as small errors (uncertainties) in the data, generally amplify with time as the computer tries to project the weather further and further into the future. After a number of days, these initial imperfections tend to dominate, and the forecast shows little or

no accuracy in predicting the behavior of the real atmosphere. In essence, what happens is that the small uncertainty in the initial atmospheric conditions eventually leads to a huge uncertainty in the model's forecast. There is, therefore, a limit as to how far into the future we will ever be able to accurately forecast the weather at a specific place and time. However, it is still possible to make climatological projections that give the *likelihood* of particular types of weather far into the future. We will discuss the difference between these kinds of climate projections and weather forecasts in Chapter 18.

● **FIGURE 13.7** Ensemble 500-mb forecast chart for March 1, 2013, issued on the morning of February 25. The blue lines represent the 5460-meter contour; the red lines, the 5640-meter contour; and the green lines, a 16-year average, represents the average locations of the two contours. The yellow lines show where the main or "operational" member of the ensemble placed the contours.

━━ 5460 m   ━━ 5640 m   ━━ Center line   ━━ Climatology

NOAA

Ensemble Forecasts Because of the atmosphere's chaotic nature, meteorologists have turned to a technique called **ensemble forecasting** to improve short- and medium-range forecasts. The ensemble approach is based on running several forecast models—or different versions (simulations) of a single model—each beginning with slightly different weather information to reflect the errors inherent in the measurements. Suppose, for example, a forecast model predicts the state of the atmosphere 24 hours into the future. For the ensemble forecast, the entire model simulation is repeated, but only after the initial conditions are "tweaked" just a little. The "tweaking," of course, represents the degree of uncertainty in the observations. Repeating this process several times creates an ensemble of forecasts for a range of small initial changes.

● Figure 13.7 shows an ensemble 500-mb forecast chart for March 1, 2013 (96 hours into the future) using the global atmospheric circulation model. The chart is constructed by running the model 17 different times, each time starting with slightly different initial conditions. Notice that the ensemble numbers are in strong agreement in some locations, such as the eastern Pacific, while there is large uncertainty in the other locations, such as the midwestern United States. As the forecast goes further and further into the future, the lines look more and more like scrambled spaghetti, which is why an ensemble forecast chart such as this one is often referred to as a *spaghetti plot*.

If, at the end of a specific time, the progs, or model runs, match each other fairly well, the forecast is considered *robust*. This situation allows the forecaster to issue a prediction with a high degree of confidence. If the progs disagree, the forecaster has less faith in the computer model prediction and will issue a forecast with more limited confidence. In essence, *the less agreement among the progs, or model runs, the less predictable the weather.* Consequently, it would not be wise to make outdoor plans for Saturday when on Monday the weekend forecast calls for "sunny and warm" with a low degree of confidence. Most everyday weather forecasts do not indicate the degree of forecaster confidence, but it is sometimes mentioned in TV weather reports or special NWS advisories, especially when threatening weather is a possibility.

The example in ● Fig. 13.8 shows how widely the members of an ensemble can diverge for a critical weather situation, such as the blizzard that struck New England in early February 2013. The ensemble includes three different models, each run six times with slightly different initial conditions, plus two other runs. The result is 20 predictions for total snowfall in Boston, Massachusetts. Notice that the forecasts vary tremendously, from only 3.5 inches to an eye-opening 51 inches. The solid black line indicates the ensemble mean, which is 24 inches. Since the actual snow total in Boston in this storm was 24.9 inches, a forecast that relied on the ensemble mean would have been an excellent one. The ensemble mean is not a perfect predictor in every case, but on average it tends to perform as well or better than any single model.

In summary, imperfect numerical weather predictions may result from flaws in the computer models, from errors that creep in along the models' boundaries, from the sparseness of data, and/or from inadequate representation of many pertinent processes, interactions, and inherently chaotic behavior that occurs within the atmosphere. However, by watching carefully for potential model errors and by using ensembles made up of a number of different model runs, a forecaster can increase the likelihood of making an accurate prediction.

OTHER FORECASTING TECHNIQUES Because the weather affects every aspect of our daily lives, attempts to predict it have been made for centuries. One of the earliest attempts was undertaken by Theophrastus, a pupil of Aristotle, who in 300 B.C. compiled all sorts of weather indicators in his *Book of Signs*. A dominant influence in the field of weather forecasting for 2000 years, this work consists of ways to foretell the weather by examining natural signs, such as the color and shape of clouds, and the intensity at which a fly bites. Some of these signs have validity and are a part of our own weather folklore; "a halo around the moon portends rain" is one of these. Today, we realize that the halo is caused by the bending of light as it passes through ice crystals and that ice crystal-type clouds (cirrostratus) are often the forerunners of an approaching cyclonic storm. (See ● Fig. 13.9.)

Although numerical weather prediction has proven to be much more powerful a tool than these ancient methods, some types of weather predictions can still be made by observing the sky and using a little weather wisdom. If you keep your eyes

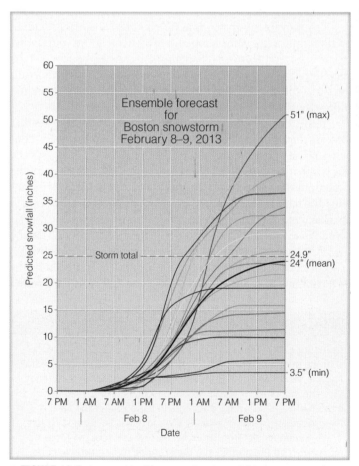

● **FIGURE 13.8** An ensemble of forecasts of total snowfall in Boston, Massachusetts, for February 8 to 9, 2013, issued early on February 6. Each trace represents the forecast from a different model run, all carried out at the same time but with slightly different initial conditions.

● **FIGURE 13.9** A halo around the sun (or moon) means that rain is on the way, a weather forecast made by simply observing the sky.

open and your senses keenly tuned to your environment, you should, with a little practice, be able to make fairly good short-range local weather forecasts by interpreting the messages written in the weather elements. ▼ Table 13.2 is designed to help you with this endeavor.

The movement of clouds at different levels can assist you in predicting changes in the temperature of the air above you. Knowing how the temperature is changing aloft can help you predict the stability of the air (described in Chapter 6), as well as whether falling snow will change to rain, or vice versa. (This topic is explored more extensively in Focus section 13.4.)

To further help you forecast the weather, an instant weather forecast chart (see Online Appendices) has been prepared by considering the relationship that the pressure and wind have on various weather systems. While the chart is applicable to much of the United States and southern Canada, local influences, such as mountains and large bodies of water, can affect the local weather to such an extent that the large-scale weather patterns on which the chart is based do not always show up clearly. (The chart works best during the fall, winter, and spring when the weather systems are active.)

Official weather forecasting activities were launched by the governments of many nations in the late 1800s and expanded

as observational techniques improved in the 1900s. During the years before the advent of computer-based weather models, many forecasting methods were based largely on the experience of the forecaster. Many of these techniques were of value, but typically they gave more of a general overview of what the weather should be like, rather than a specific forecast. As late as the mid-1950s, all weather maps and charts, including those depicting current as well as future conditions, were plotted by hand and analyzed by individuals. Meteorologists predicted the weather using certain rules that related to the particular weather system in question. For short-range forecasts of six hours or less, surface weather systems were moved along at a steady rate. Upper-air charts were used to predict where surface storms would develop and where pressure systems aloft would intensify or weaken. The predicted positions of these systems were extrapolated into the future using linear graphical techniques and current maps. Experience played a major role in making the forecast. In many cases, these forecasts turned out to be amazingly accurate. However, with the advent of modern supercomputers, along with our present observing techniques, today's forecasts are more consistently better.

Probably the easiest weather forecast to make is a **persistence forecast**, which is simply a prediction that future weather will be the same as present weather. If it is snowing today, a persistence forecast would call for snow through tomorrow. Such forecasts are most accurate for time periods of several hours and become less and less accurate after that. Persistence forecasts are also more useful at those times and places where the weather tends to change less dramatically.

Another method of forecasting is the **steady-state**, or **trend forecast**. The principle involved here is that surface weather systems tend to move in the same direction and at approximately the same speed as they have been moving, providing no evidence exists to indicate otherwise. Suppose, for example, that a cold front is moving eastward at an average speed of 30 km per hour and it is 90 km west of your home. Using the steady-state method, we might extrapolate and predict that the front should pass through your area in about three hours.

The **analog method** is yet another form of weather forecasting. Basically, this method relies on the fact that existing features on a weather chart (or a series of charts) may strongly resemble features that produced certain weather conditions sometime in the past. To the forecaster, the weather map "looks familiar," and for this reason the analog method is often referred to as **pattern recognition**. A forecaster might look at a prog and say, "I've seen this weather situation before, and this happened." Previous weather events can then be used as a guide to the future. The problem here is that, even though weather situations may appear similar, they are never *exactly* the same. There are always sufficient differences in the variables to make applying this method a challenge.

Even so, the analog method can be used to predict a number of weather elements, such as maximum temperature. Suppose that in New York City the average maximum temperature on a particular date for the past 30 years is $10°C$ ($50°F$). By statistically relating the maximum temperatures on this date to other weather elements—such as the wind, cloud cover, and humidity—a relationship between these variables and maximum temperature can be drawn.

LISTED BELOW ARE A FEW FORECASTING RULES THAT MAY BE APPLIED WHEN MAKING A SHORT-RANGE LOCAL WEATHER FORECAST		
OBSERVATION	**INDICATION**	**LOCAL WEATHER FORECAST**
Surface winds from the S or from the SW; clouds building to the west; warm (hot) and humid (pressure falling)	Possible cool front and thunderstorms approaching from the west	Possible showers; possibly turning cooler; windy
Surface winds from the E or from the SE, cool or cold; high clouds thickening and lowering; halo (ring of light) around the sun or moon (pressure falling)	Possible approach of a warm front	Possibility of precipitation within 12–24 hours; windy (rain with possible thunderstorms during the summer; snow changing to sleet or rain in winter)
Strong surface winds from the NW or W; cumulus clouds moving overhead (pressure rising)	A low-pressure area may be moving to the east, away from you; and an area of high pressure is moving toward you from the west	Continued clear to partly cloudy, cold nights in winter; cool nights with low humidity in summer
WINTER NIGHT		
(a) If clear, relatively calm with low humidity (low dew-point temperature)	(a) Rapid radiational cooling will occur	(a) A very cold night
(b) If clear, relatively calm with low humidity and snow covering the ground	(b) Rapid radiational cooling will occur	(b) A very cold night with minimum temperatures lower than in (a)
(c) If cloudy, relatively calm with low humidity	(c) Clouds will absorb and radiate infrared (IR) energy to surface	(c) Minimum temperature will not be as low as in (a) or (b)
SUMMER NIGHT		
(a) Clear, hot, humid (high dew points)	(a) Strong absorption and emission of IR energy to surface by water vapor	(a) High minimum temperatures
(b) Clear and relatively dry	(b) More rapid radiational cooling	(b) Lower minimum temperatures
SUMMER AFTERNOON		
(a) Scattered cumulus clouds that show extensive vertical growth by mid-morning	(a) Atmosphere is relatively unstable	(a) Possible showers or thunderstorms by afternoon with gusty winds
(b) Afternoon cumulus clouds with limited vertical growth and with tops at just about the same level	(b) Stable layer above clouds (region dominated by high pressure)	(b) Continued partly cloudy with no precipitation; probably clearing by nightfall

By comparing these relationships with current weather information, the forecaster can predict the maximum temperature for the day.

Statistical forecasts of weather elements are based on the past performance of computer models. Known as *Model Output Statistics* (MOS), these predictions, in effect, are statistically weighted analog forecast corrections incorporated into the computer model output. For example, a forecast of tomorrow's maximum temperature for a city might be derived from a statistical equation that uses a numerical model's forecast of relative humidity, cloud cover, wind direction, and air temperature.

When the NWS issues a forecast calling for rain, it is usually followed by a probability. For example: "The chance of rain is 60 percent." Does this mean (a) that it will rain on 60 percent of the forecast area or (b) that there is a 60 percent chance that it will rain within the forecast area? Neither one! The expression means there is a 60 percent chance that any random place in the forecast area, such as your home, will receive measurable rainfall.[*]

---

[*] The 60 percent chance of rain does not apply to a situation that involves rain showers. In the case of showers, the percentage refers to the expected area over which the showers will fall.

Forecasting Temperature Advection by Watching the Clouds

We know from Chapter 12 that when cold air is being brought into a region by the wind, we call this *cold advection*. When warm air is brought into a region, we call this *warm advection*.

A knowledge of temperature advection aloft is a valuable tool in forecasting the weather. In summer, when the surface is warm, cold advection aloft sets up instability and increases the likelihood of towering cumulus clouds and showers. On the other hand, warm advection aloft usually increases the temperature of the air, thus making it more stable. During the winter, this often leads to smoke and haze accumulating in the colder air near the surface.

By watching the movement of clouds, we get a good indication as to the wind direction at cloud level and also the type of advection. For instance, a cloud moving from the west indicates a west wind, a cloud from the south a south wind, and so on. Clouds at different levels frequently move in different directions, meaning that the wind direction is changing with height. Wind that changes direction in a clockwise sense (north to northeast to east, etc.) is a *veering wind*. Wind that changes direction in a counterclockwise sense (north to northwest to west, etc.) is a *backing wind*. There are two general rules that will help us determine whether cold or warm advection is occurring in a layer of air above us:

1. Winds that *back* with height (change counterclockwise) indicate *cold advection*.
2. Winds that *veer* with height (change clockwise) indicate *warm advection*.

As an example, suppose we observe lower clouds moving from a southerly direction (south wind) and higher clouds moving from a westerly direction (west wind) (see ● Fig. 5a). The wind direction is veering with height; warm advection is occurring between the cloud layers, and the air should be getting warmer. If, on another day, we see lower clouds moving from a southerly direction and higher clouds moving from an easterly direction (Fig. 5b), the wind is

● **FIGURE 5** The wind veers with height, suggesting warm advection is occurring between the cloud layers. (b) The wind is backing with height, and cold advection is occurring between the cloud layers.

backing with height and the atmosphere between the cloud layers is probably becoming colder.*

An example of the relationship between winds and advection is seen in the vertically shifting winds that accompany weather fronts. Look at ● Fig. 6 and observe that Fig. 6b (middle diagram) is a 3-D model of a typical open-wave cyclone with its accompanying warm and cold front. Behind the cold front, swiftly moving cumulus clouds indicate a northwesterly wind exists about a kilometer or so above the surface. Ahead of the advancing warm front, stratiform clouds indicate that here southeasterly winds prevail about a kilometer above the ground. We know from Chapter 12 that warm advection takes place *ahead* of the warm front and cold advection *behind* the cold front. In both cases, the advection usually occurs in a layer from the surface up to at least the 500-mb level, about 5500 m or 18,000 ft above sea level.

On the upper-air 500-mb chart (Fig. 6a), the position of the upper trough and the region of coldest air is to the west of the surface low

The direction of the wind and also the cloud movement is shown by arrows. Because of the upper trough's position, the winds aloft are westerly behind the cold front and southwesterly ahead of the warm front. Figure 6c (bottom diagram) shows how the wind direction changes from the surface to the 500-mb level. Behind the cold front, the winds back from northwesterly to westerly as we move upward. Cold advection is taking place as chilling air moves in from the west. Just ahead of

the approaching warm front, the wind veers with height from southeasterly at the surface to southwesterly aloft as warm air glides up and over the cool surface air.

We can use this information to improve upon a weather forecast. For instance, if you happen to be located ahead of an advancing warm front and the winds above you are veering with height, the chances are that even if precipitation begins as snow it may change to rain as warm air moves in overhead. Behind a cold front where winds are backing with height, the influx of cold air may lower the temperature sufficiently so that rain first becomes mixed with snow, and then changes to snow before the storm moves eastward.

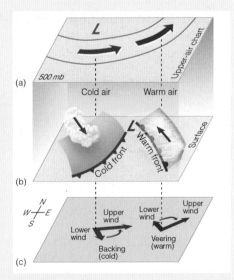

● **FIGURE 6** Clouds, winds, and advection associated with a cold and a warm front.

---

*In both of these examples, we are assuming horizontal air motion only.

Looking at the forecast in another way, if the forecast for 10 days calls for a 60 percent chance of rain, it should rain where you live on six of those days. The verification of the forecast (as to whether it actually rained or not) is usually made at the Weather Service office, but remember that the computer models produce forecasts for a given region, not for an individual location. When the National Weather Service issues a forecast calling for a "slight chance of rain," what is the probability (percentage) that it will rain? ▼ Table 13.3 provides this information.

An example of a **probability forecast** using climatological data is given in ● Fig. 13.10. The map shows the probability of a "White Christmas"—an inch or more of snow on the ground—across the United States. The map is based on the average of 30 years of data and gives the likelihood of snow in terms of a probability. For instance, the chances are greater than 90 percent (9 Christmases out of 10) that portions of northern Minnesota, Michigan, and Maine will experience a white Christmas. In Chicago, it is close to 50 percent; and in Washington, D.C., about 20 percent. Many places in the far west and south have probabilities less than 5 percent, but nowhere is the probability exactly zero, for there is always some chance (no matter how small) that a mantle of white will cover the ground on Christmas Day. For example, on December 24–25, 2004 (see ● Fig. 13.11), Corpus Christi, Texas, reported 4.4 inches of snowfall, and Brownsville, at the very southern part of the state, had 1.5 inches of snow, making it the first measurable snow in Brownsville since 1899. For both cities, it was the heaviest snowfall on record for any date.

Predicting the weather using **weather type forecasting** employs the analog method. In general, weather patterns are categorized into similar groups or "types," using such criteria as the position of the subtropical highs, the upper-level flow, and the prevailing storm track. As an example, when the Pacific high is weak or depressed southward and the flow aloft is zonal (west-to-east), surface storms tend to travel rapidly eastward across the Pacific Ocean and into the United States without developing into deep systems. But when the Pacific high is to the north of its normal position and the upper airflow is meridional (north-south), looping waves form in the flow with surface lows usually

▼ **TABLE 13.3** Forecast wording used by the National Weather Service to describe the percentage probability of measurable precipitation (0.01 inch or greater) for steady precipitation and for convective, showery precipitation

PERCENT PROBABILITY OF PRECIPITATION	FORECAST WORDING FOR STEADY PRECIPITATION	FORECAST WORDING FOR SHOWERY PRECIPITATION
10 to 20 percent	*Slight chance* of precipitation	*Isolated* showers
30 to 50 percent	*Chance* of precipitation	*Scattered* showers
60 to 70 percent	Precipitation *likely*	*Numerous* showers
≥80 percent	Precipitation,* *rain, snow*	*Showers**

*A forecast that calls for an 80 percent chance of rain in the afternoon might read like this: "cloudy today with rain this afternoon. . . ." For an 80 percent chance of rain showers, the forecast might read "cloudy today with rain showers this afternoon. . . ."

developing into huge cyclonic storms. As we saw in Chapter 12, these upper-level longwaves move slowly, usually remaining almost stationary for perhaps a few days to a week or more. Consequently, the particular surface weather at different positions around the wave is likely to persist for some time. ● Figure 13.12 presents an example of weather conditions most likely to prevail with a winter meridional weather type.

A forecast based on the climate* of a particular region is known as a **climatological forecast**. Anyone who has lived in Los Angeles for a while knows that July and August are practically rain-free. In fact, rainfall data for the summer months taken over many years reveal that rainfall amounts of more than a trace occur in Los Angeles about 1 day in every 90, or only about

---

*The climate of a region represents the total accumulation of daily and seasonal weather events for a specific interval of time, most often 30 years.

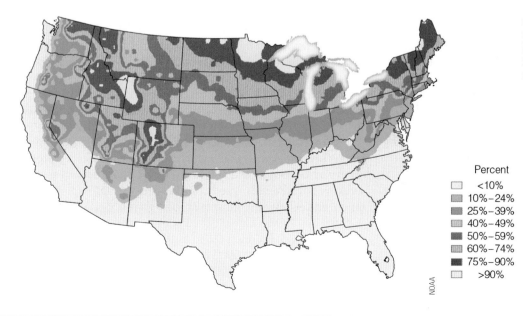

● **FIGURE 13.10** Probability of a "White Christmas"—1 inch or more of snow on the ground—based on a 30-year average, from 1981 to 2010, inclusive. The probabilities do not include all of the mountainous areas in the western United States.

Percent
- <10%
- 10%–24%
- 25%–39%
- 40%–49%
- 50%–59%
- 60%–74%
- 75%–90%
- >90%

NOAA

● **FIGURE 13.11** Satellite view of south Texas along the Gulf Coast on Christmas Day 2004. The white area covering Corpus Christi and Brownsville is snow. The probability of measurable snow on the ground in either of these two cities on Christmas Day is less than 1 percent (see Fig. 13.8). Yet, Corpus Christi received 4.4 inches of snow and Brownsville 1.5 inches. Just days later, the temperature climbed into the 80s (°F).

● **FIGURE 13.12** Winter weather type showing upper-airflow (heavy arrow), surface position of Pacific high, and general weather conditions that should prevail.

1 percent of the time. Therefore, if we predict that it will not rain on a particular date next year during July or August in Los Angeles, our chances are nearly 99 percent that the forecast will be correct based on past records. As it is unlikely that this pattern will significantly change in the near future, we can confidently make the same forecast for the year 2020.

Time Range of Forecasts

Weather forecasts are normally grouped according to how far into the future the forecast extends. For example, a weather forecast for up to a few hours (usually not more than six hours) is called a **very short-range forecast**, or *nowcast*. The techniques used in making such a forecast normally involve both subjective and definitive interpretations of surface observations, satellite imagery, and Doppler radar information. Often the forecaster moves weather systems along by the steady-state or trend method of forecasting, with human experience and pattern recognition coming into play.

When severe or hazardous weather is likely or is occurring, the National Weather Service issues short-range alerts in the form of weather watches, warnings, and advisories. A **weather watch** indicates that atmospheric conditions favor hazardous weather occurring over a particular region during a specified time period. These hazards may or may not actually develop, and their timing and location are uncertain, so a watch simply means to be on "watch" for that threat and to be prepared to act if necessary. When hazardous weather has developed, or is about to develop, two types of alerts may be issued. A **weather warning** indicates that the hazard now occurring or imminent is considered to be a threat to life and/or property (such as tornadoes, flash floods, severe thunderstorms, and winter storms). An *advisory* is similar to a warning, except that it is used to indicate hazards that are usually less severe (such as light snow, light freezing rain, or dense fog). Note that sometimes even an advisory-level hazard can produce major problems, such as when light snow falls, melts, and quickly freezes on road surfaces.

Weather forecasts that range from about 12 hours to a few days (generally three days or 72 hours) are called **short-range forecasts**. The forecaster may incorporate a variety of techniques in making a short-range forecast, such as satellite imagery, Doppler radar, surface weather maps, upper-air wind data, and pattern recognition. As the forecast period extends beyond about 12 hours, the forecaster tends to weight the forecast heavily on computer-drawn forecast maps and statistical information.

A **medium-range forecast** is one that extends from about 3 days up to 8 days (192 hours) into the future. Medium-range forecasts are almost entirely based on computer-derived products, such as forecast charts and statistical forecasts. A forecast that extends beyond 3 days is often called an *extended forecast*.

A forecast that extends beyond about 8 days is called a **long-range forecast**. Although computer forecast charts (called progs) are available for up to 16 days into the future, those beyond about a week are not accurate in predicting local temperature and precipitation, and at best only show the broad-scale upper-level weather features. The NOAA Climate Prediction Center summarizes these general trends in products called **outlooks** that cover 6- to 10-day and 8- to 14-day periods. These are not forecasts in the strict sense, but rather an overview of how the expected precipitation and temperature patterns may compare with average conditions.

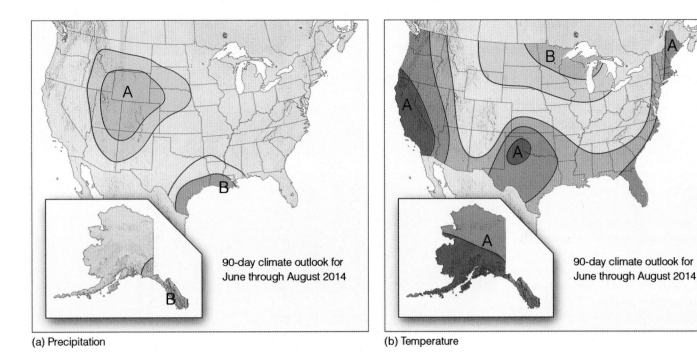

(a) Precipitation   (b) Temperature

● **FIGURE 13.13** The 90-day outlook for (a) precipitation and (b) temperature for June, July, and August 2014, issued in mid-May 2014. For precipitation (a), the darker the green color the greater the probability of precipitation being above normal, whereas the deeper the brown color the greater the probability of precipitation being below normal. For temperature (b), the darker the orange/red colors the greater the probability of temperatures being above normal, whereas the darker the blue color, the greater the probability of temperatures being below normal. On both maps, the letter A stands for *above normal* and the letter B for *below normal*.

NOAA also issues *seasonal outlooks* every month. These cover three-month periods that overlap and extend out to roughly a year. Again, rather than depicting specific weather features, these outlooks show the odds that a given area might experience temperatures or precipitation that are above or below average. ● Figure 13.13 gives a typical 90-day outlook. Initially, these outlooks were based mainly on the relationship between the projected average upper-air flow and the surface weather conditions that the type of flow will create. Today, long-range forecasts call on models that link the atmosphere with sea-surface temperature, such as the *Climate Forecast System version 2* (CFSv2). Many of the outlooks also take into account persistence statistics that carry over the general weather pattern from immediately preceding months, seasons, and years.

As we saw in Chapter 10, a vast warming (El Niño) or cooling (La Niña) of the tropical Pacific can affect the weather in different regions of the world. These interactions, such as where a warmer tropical Pacific can influence rainfall in California, are called **teleconnections**.* These types of interactions between widely separated regions are identified through statistical correlations. For example, look back at Fig. 10.27, p. 284, and observe where seasonally averaged temperature and precipitation patterns over North America tend to depart from normal during El Niño and La Niña events. Using this type of information, the Climate Prediction Center can issue—months in advance—a seasonal outlook of an impending wetter or drier

winter, months in advance. Seasonal outlooks using teleconnections have become increasingly useful.

Up to now we have looked at how weather forecasts are made and how forecasts can influence our daily lives. (For a look at how weather forecasts can influence the marketplace, read Focus section 13.5.)

In most locations throughout North America, the weather is fair more often than rainy. Consequently, there is a forecasting bias toward fair weather, which means that, if you made a forecast of "no rain" where you live for each day of the year, your forecast would be correct more than 50 percent of the time. But did you show any *skill* in making your correct forecast? What constitutes skill and accuracy, anyway?

Accuracy and Skill in Weather Forecasting

In spite of the complexity and ever-changing nature of the atmosphere, forecasts made by the National Weather Service out to between 12 and 24 hours are usually quite accurate. Those made for between two and five days are fairly good. Beyond about seven days, due to the chaotic nature of the atmosphere, computer prog forecast accuracy falls off rapidly. Although weather predictions made for up to three days are by no means perfect, they are far better than simply flipping a coin. But just how accurate are they?

One problem with determining forecast accuracy is deciding what constitutes a right or wrong forecast. Suppose tomorrow's forecast calls for a minimum temperature of 35°F. If the

---

*Teleconnections include not only El Niño and La Niña but other indices, such as the Pacific Decadal Oscillation, the North Atlantic Oscillation, and the Arctic Oscillation. (See Chapter 10, pp. 279–289.)

Weather Prediction and The Marketplace

A good forecast cannot only make or break plans for a picnic, but it can spell the difference between profit and loss for an entire business. Weather predictions are a critical tool for many parts of the economy. Short-term forecasts can help an orange grower deal with the threat of a hard freeze or tip off a construction company to the risk of work delays. On a broader scale, the prices of stocks and commodities* can swing up or down based on the approach of a major storm, the forecasts of its behavior, and the damage left behind. For example, the price of frozen concentrated orange juice rose more than 40 percent in the month after Hurricane Charley struck many of Florida's citrus groves in August 2004.

For many companies, seasonal outlooks are even more important than day-to-day forecasts. A corporation that makes bread or pasta might pay close attention to long-term outlooks for temperature and precipitation across the wheat-growing areas of North America (see ● Figure 7) in order to anticipate potential drops in supply. For energy companies, even a small seasonal shift can play a huge role in the demand for summer cooling or winter heating. An unusually mild winter might provide a boost to airlines and trucking companies, which would suffer fewer delays from snow and ice, but it could also cut into the sales of cold-weather clothing. Long-term outlooks for El Niño and La Niña can provide months of valuable lead time on where winter temperatures in the United States are likely to run warmer or cooler than average.

*Commodities represent a vast array of goods bought and sold in large quantities, from oranges to oil.

● **FIGURE 7** A baking company might arrange to buy this wheat in advance at a guaranteed price if long-range weather forecasts point toward a poor crop.

© Dmitry Pichugin/Shutterstock.com

The most direct protection against the risk of weather-related financial downturns comes from *insurance* for hail, flooding, drought, and the like. Weather insurance typically covers only the most dire meteorological threats, much like a catastrophic health-care plan that covers heart attacks but not chronic illness.

Several other tools can help a company use weather predictions to smooth out the potential ups and downs in profit linked to the atmosphere. Many commodities can be traded through contracts called *futures* (a type of *derivative**). Futures contracts are agreements to buy or sell a commodity at a fixed price at some later date. For example, a bread-baking company might buy wheat futures based on a projected precipitation outlook. This forecast would help the company plan with more confidence, knowing that the cost it will pay for wheat won't change even if a drought should strike and the price of wheat should go up dramatically.

*Derivatives are contracts that derive their value from some other quantity, such as the price of a commodity.

It is also possible to trade futures contracts based on indices of the weather itself. Rather than specifying the future cost of a commodity, a weather derivative contract puts a price tag on a particular weather outcome, such as a record-hot summer that boosts demand for air-conditioning. Many such contracts are based on heating or cooling degree days, described in Chapter 3 on p. 80.

Many investors and speculators try to make a profit on the twists and turns of the atmosphere, often by looking at weather predictions and by buying and selling futures or weather derivatives. Traders keep a close eye on both seasonal weather projections and short-term forecasts, such as the track of a hurricane that could knock out oil and gas production. For example, as Tropical Storm Rita gathered strength on September 20, 2005, and forecasts called for Rita to approach the Gulf Coast as a major hurricane, the price of oil rose by the largest single-day amount on record—$4.39, or about 7 percent.

official minimum turns out to be 37°F, is the forecast incorrect? Is it as incorrect as one 10 degrees off? By the same token, what if snow is predicted over a large city, and the snow line cuts the city in half with the southern portion receiving heavy amounts and the northern portion none? Is the forecast right or wrong? Or what if the snow started just 3 hours earlier than predicted, falling during the morning rush hour instead of at mid-morning? At present, there is no clear-cut answer to the question of determining forecast accuracy, so meteorologists use a variety of mathematical techniques to measure the quality of their predictions. These might take into account how much the weather naturally varies at a given location and time of year, or how much data are actually available to determine whether a forecast was correct.

How does forecast accuracy compare with forecast skill? Suppose you are forecasting the daily summertime weather in Los Angeles. It is not raining today and your forecast for tomorrow calls for "no rain." Suppose that tomorrow it doesn't rain. You made an accurate forecast, but did you show any skill in so doing? Earlier, we saw that the chance of measurable rain in Los Angeles on any summer day is very small indeed; chances are good that day after day it will not rain. For a forecast to show skill, it should be better than one based solely on the current weather (*persistence*) or on the "normal" weather (*climatology*) for a given region. Therefore, during the summer in Los Angeles, a forecaster will have many accurate forecasts calling for "no measurable rain," but will need skill to predict correctly on which summer days it will rain. If on a sunny July day in Los Angeles you forecast rain for tomorrow and it rains, you not only made an accurate forecast, but you have also showed skill in making your forecast because your forecast was better than both persistence and climatology.

A meteorological forecast, then, shows skill when it is more accurate than a forecast utilizing only persistence or climatology. Persistence forecasts are usually difficult to improve upon for a period of time of several hours or less. Weather forecasts ranging from 12 hours to a few days generally show much more skill than persistence forecasts. However, as the range of the forecast period increases, the skill drops quickly because of the effects of chaos discussed earlier in this chapter. The six- to fourteen-day mean outlooks show some skill (which has been increasing over the last several decades) in predicting both temperature and precipitation, although the accuracy of precipitation forecasts is less than that for temperature. Today, seven-day forecasts of major weather features are roughly as skillful as three- to four-day forecasts were in the 1990s. Beyond 15 days, specific forecasts are only slightly better than climatology. However, skill in making forecasts of average monthly temperature and precipitation approximately doubled from 1995 to 2006.

Forecasting large-scale weather events several days in advance (such as the disruptive Groundhog Day blizzard of 2011 across the Midwest) is far more accurate than forecasting the precise evolution and movement of small-scale, short-lived weather systems, such as tornadoes and severe thunderstorms. In fact, three-day forecasts of the development and movement of a major low-pressure system show more skill today than 36-hour forecasts did in the 1990s. Even though determining the *precise* location where a tornado will form is presently beyond modern forecasting techniques, the regions where tornadic storms are *likely* to form can often be predicted several days in advance.

With improved observing systems, such as Doppler radar and advanced satellite imagery, the lead time* of watches and warnings for severe storms has increased. In fact, the lead time for tornado warnings has more than doubled since the 1980s, with the average lead time today being close to 15 minutes, and the lead time for the most deadly and destructive tornadoes often being more than 30 minutes (see ● Fig. 13.14).

Although scientists may never be able to skillfully predict the weather beyond about 15 days using available observations, the prediction of *climatic trends* is more promising. Whereas individual weather systems vary greatly and are difficult to forecast very far in advance, global-scale patterns of winds and pressure frequently show a high degree of persistence and predictable change over periods of a few weeks to a month or more. With the latest generation of high-speed supercomputers, general circulation models (GCMs) are doing a far better job at predicting large-scale atmospheric behavior than did the earlier models. (The GCMs are numerical computer models that simulate global patterns of wind, pressure, and temperature, and how these phenomena change over time.) In fact, the new GCMs are able to

---

Lead time is the interval of time between the issuance of the warning and actual observance of the event, in this case, the tornado.

● **FIGURE 13.14** (left) This violent tornado ripped through the town of Moore, Oklahoma, on March 20, 2013. Even with a lead time of more than 15 minutes, the tornado took 24 lives, including 10 children, and injured 377 others. (right) A Doppler radar observation taken at 2:38 p.m. CDT on May 20, 2013, shows the supercell thunderstorm at about the time the tornado was developing west of Newcastle, Oklahoma. The red box shows the tornado warning issued by the National Weather Service. The track that was actually taken by the tornado is outlined in dark brown.

CRITICAL THINKING QUESTION Suppose that a forecaster had tracked the first few miles of the tornado's path up to point number 1, then made a very specific prediction, using persistence, that the tornado would continue on that track and pass just north of Moore. Would that forecast have been accurate? Would you consider the warning box outlined in red to be accurate?

simulate a number of global patterns quite well, such as *blocking highs** that can cause precipitation and temperature patterns to deviate considerably from average conditions. As new knowledge and methods of modeling are fed into the GCMs, they are expected to become an increasingly reliable tool in the forecasting of weather and climate. (In Chapter 18, we will examine in more detail the climatic predictions based on numerical models.)

Through their broadcasts as well as social media, television weathercasters play an important role in communicating forecasts extending out a few days, discussing local and national climatology, and advising viewers of immediate threats, such as watches and warnings for severe weather. Focus section 13.6 describes how TV weather forecasters present weather visuals.

---

*Blocking highs are high-pressure areas that tend to remain nearly stationary for some time, thus "blocking" the west-to-east movement of mid-latitude cyclonic storms.

BRIEF REVIEW

Up to this point, we have looked at the various methods of weather forecasting. Before going on, here is a review of some of the important ideas presented so far:

- Forecasters have access to a wide array of observations and a number of tools that can be used when making a forecast, including surface and upper-air observations, satellite imagery, Doppler radar, meteograms, soundings, and computer progs.

- The forecasting of weather by high-speed computers is known as numerical weather prediction. Mathematical models that describe how atmospheric temperature, pressure, winds, and moisture will change with time are programmed into the computer. The computer then draws surface and upper-air charts and produces a variety of forecast charts called progs.

- Imperfections of the computer models—atmospheric chaos and small errors in the data—greatly limit the accuracy of weather forecasts for periods beyond a few days.

- Ensemble forecasting is a technique based on running several forecast models (or different versions of a single model), each beginning with slightly different weather information to serve as an approximation of errors in the measurements.

- A persistence forecast is a prediction that future weather will be the same as the present weather, whereas a climatological forecast is based on the climatology of a particular region.

- For a forecast to show skill, it must be better than a persistence forecast or a climatological forecast.

- Weather forecasts that range from about 12 hours to about three days are called *short-range forecasts*. Those that extend from about three days to eight days are called *medium-range forecasts*, and forecasts that extend beyond about eight days are called *long-range forecasts*.

- Seasonal outlooks provide an overview of how temperature and precipitation patterns may compare with normal conditions.

Weather Forecasting Using Surface Charts

The best forecasts incorporate multiple layers of the atmosphere using numerical modeling, as we saw earlier in this chapter. However, even when computer models are skilled at advancing large-scale weather features forward in time, a capable forecaster also needs a strong sense of how surface features typically evolve, which can help him or her when the progs disagree. Let's look at how a forecaster might combine this knowledge with limited or conflicting data.

Suppose that we wish to make a *short-range weather prediction* and the only information available is a surface weather map. Can we make a forecast from such a chart? Most definitely. And our chances of that forecast being correct improve markedly if we have maps available from several days back. We can use these past maps to locate the previous position of surface features and predict their movement.

A simplified surface weather map is shown in ● Fig. 13.15. The map portrays early winter weather conditions on Tuesday morning at 6:00 a.m. A single isobar is drawn around the pressure centers to show their positions without cluttering the map. Note that an open-wave cyclone is developing over the Central Plains. The weather conforms to the cyclone model (see Fig. 12.12, p. 333), with showers forming along the cold front and light rain, snow, and sleet ahead of the warm front. The dashed lines on the map represent the position of the weather systems six hours ago. Our first question is: How will these systems move?

DETERMINING THE MOVEMENT OF WEATHER SYSTEMS We can use several methods to forecast the movement of surface pressure systems and fronts. The following are a few of these forecasting rules of thumb:

1. For short-time intervals, mid-latitude cyclonic storms and fronts tend to move in the same direction and at approximately the same speed as they did during the previous six hours (providing, of course, there is no evidence to indicate otherwise).
2. Low-pressure areas tend to move in a direction that parallels the isobars in the warm air (the warm air sector) ahead of the cold front.
3. Lows tend to move toward the region of greatest pressure drop, while highs tend to move toward the region of greatest rise.
4. Surface pressure systems tend to move in the same direction as the wind at 5500 m (18,000 ft)—the 500-mb level. The speed at which surface systems move is about half the speed of the winds at this level.

When the surface map (Fig. 13.15) is examined carefully and when rules of thumb 1 and 2 are applied, it appears that—based on present trends—the low pressure centered over the Central Plains should move northeast. If *pressure tendencies** are

---

*The *pressure tendency* is the rate at which the pressure is changing during a given time, usually the past three hours.

TV Weathercasters—How Do They Do It?

As you watch the TV weathercaster, you typically see a person describing and pointing to specific weather information, such as satellite and radar images, and weather maps, as shown in ● Fig. 8. What you may not know is that in many instances the weathercaster is actually pointing to a blank screen (typically green) on which there is nothing (● Fig. 9).* This process of electronically superimposing weather information in the TV camera against a blank wall is called color-separation overlay, or *chroma key.*

The chroma key process works because the studio camera is constructed to pick up all colors except (in this case) green. The various maps, charts, satellite photos, and other graphics are electronically inserted from a computer to this green area of the color spectrum. The person in the TV studio should not wear green clothes because such clothing would not be picked up by the camera—what you would see on your home screen would be a head and hands moving about the weather graphics!

How, then, does a TV weathercaster know where to point on the blank wall? Positioned on each side of the green screen are TV monitors (look carefully at Fig. 9) that weathercasters watch so that they know where to point.

*In some stations, forecasters point to weather information on a large TV screen.

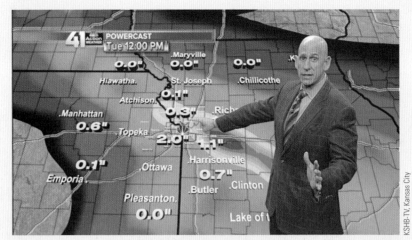

● **FIGURE 8** On your home television, weather forecaster Gary Lezak appears to be pointing to weather information directly behind him.

● **FIGURE 9** In the studio, however, he is actually standing in front of a blank green screen.

plotted on our map, we can draw lines connecting points of equal pressure change. These lines, called **isallobars**, help us to visualize the regions of falling and rising pressure. The distribution of pressure change for our map might look like the one in ● Fig. 13.16. Drawn at 2-mb intervals, the isallobars show a broad region of falling pressure ahead of the warm front, with the largest drop occurring to the northeast of the low-pressure area. This pattern fits with the previous observations and strengthens the prediction that the center of the low will move toward the northeast. The area of rising pressure immediately behind the cold front suggests that the high-pressure area over Montana will continue to move southeastward.

Pressure tendencies not only help predict the movement of highs and lows, they also indicate how the pressure systems are changing with time. The rapid fall in pressure in advance of the low indicates that the storm center is deepening as it moves. A deepening low means more closely spaced isobars, a greater pressure gradient, and stronger winds—something to take into account when we make our weather forecast. A drop in pressure, on the other hand, in the vicinity of an anticyclone suggests that it is weakening, while a rise in pressure means that its central pressure is increasing. Hence, the high-pressure area over Montana is strong (1034 mb) and will remain so, whereas the high centered off the South Carolina coast is either moving eastward or weakening rapidly as indicated by the falling pressure in that area.

Before we complete our prediction about the movement of the pressure centers in Fig. 13.15, we need to look closely at the high-pressure area off the South Carolina coast. Strong highs, especially slow-moving ones, often retard the eastward progress of lows, deflecting them either north or south. From all indications—falling pressures and past movement—this anticyclone is weakening and

● **FIGURE 13.15** Surface weather map for 6:00 a.m. Tuesday. Dashed lines indicate positions of weather features six hours ago. Areas shaded green are receiving rain, while areas shaded white are receiving snow, and those shaded pink, freezing rain or sleet.

● **FIGURE 13.16** Isallobars—lines of equal three-hour pressure change—for 6:00 a.m. Tuesday. The "F" represents the region of greatest pressure fall, while the "R" shows the region of greatest pressure rise. A +2 indicates a rise of 2 millibars.

drifting slowly eastward. It should, therefore, pose no immediate problem to the northeastward movement of the storm center.

Even if we do not have access to pressure tendencies or previous weather maps, we can make an initial approximation of how pressure systems will move by using Fig. 12.5, p. 326, which shows the average tracks of lows and highs during the winter months. From this diagram, it appears that the cyclones and anticyclones in ● Fig. 13.15 are following rather typical trajectories.

If a 500-mb chart is available (such as ● Fig. 13.17), it would also indicate how the surface pressure systems should move, since the winds at this level tend to steer these systems along. From Fig. 13.17, it appears that, indeed, the surface low should move northeastward.

A FORECAST FOR SIX CITIES Imagine that you are now tasked with making weather forecasts for six cities scattered across the United States, and the only information you have is the surface map in Fig. 13.15. To make these forecasts, we will project the surface pressure systems, fronts, and current weather into the future by assuming steady-state conditions exist. ● Figure 13.18 gives the 12- and 24-hour projected positions of these features.

A word of caution before we make our forecasts. We are assuming that the pressure systems and fronts are moving at a constant rate. This may or may not occur. Low-pressure areas, for example, tend to accelerate until they occlude, after which their rate of movement may slow. Furthermore, the direction of moving systems can change due to blocking highs and lows that exist in their path or because of shifting upper-level wind patterns. We will assume a constant rate of movement and forecast accordingly, always keeping in mind that the longer our forecasts extend into the future, the more susceptible they are to error. Keep in mind that the surface map in Fig. 13.15 is the primary data source we will use in this exercise, until it comes time to see how well our forecasts turned out.

If we move the low- and high-pressure areas eastward, as illustrated in Fig. 13.18, we can make a basic weather forecast for various cities. For example, the cold front moving into north Texas on Tuesday morning is projected to pass Dallas by that evening, so a forecast for the Dallas area would be "warm with showers, then turning colder." But we can do much better than this. Knowing the weather conditions that accompany advancing pressure areas and fronts, we can make more detailed weather

● **FIGURE 13.17** A 500-mb chart for 6:00 a.m. Tuesday, showing wind flow. The light orange L represents the position of the surface low. The winds aloft tend to steer surface pressure systems along and, therefore, indicate that the surface low should move northeastward at about half the speed of the winds at this level, or 25 knots. Solid lines are contours in meters above sea level.

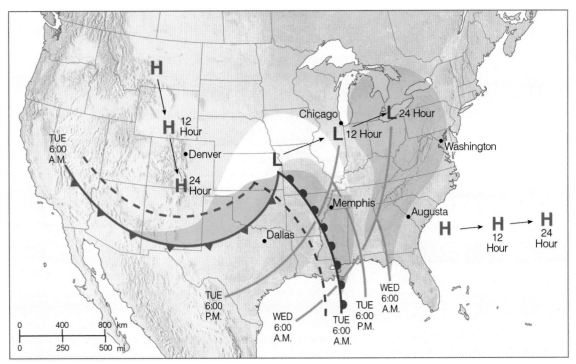

● **FIGURE 13.18** Projected 12- and 24-hour movement of fronts, pressure systems, and precipitation from 6:00 a.m. Tuesday until 6:00 a.m. Wednesday. (The dashed lines represent frontal positions six hours ago.)

forecasts that will take into account changes in temperature, pressure, humidity, cloud cover, precipitation, and winds. Our forecast will include the 24-hour period from Tuesday morning to Wednesday morning for the cities of Augusta, Georgia; Washington, D.C.; Chicago, Illinois; Memphis, Tennessee; Dallas, Texas; and Denver, Colorado. We will begin with Augusta.

Weather Forecast for Augusta, Georgia   On Tuesday morning, cold, dry polar air associated with a high-pressure area brought freezing temperatures and fair weather to the Augusta area (see Fig. 13.15, p. 370). Clear skies, light winds, and low humidities allowed rapid nighttime cooling so that, by morning, temperatures were in the low 30s (°F). Now look closely at Fig. 13.18 and observe that the anticyclone is moving slowly eastward, away from Augusta. Southerly winds on the western side of this system will bring warmer and more humid air to the region. Therefore, afternoon temperatures will be warmer than those of the day before. As the warm front approaches from the west, clouds will increase, appearing first as cirrus, then thickening and lowering into the normal sequence of warm-front clouds. Barometric pressure should fall. Clouds and high humidity should keep minimum temperatures well above freezing on Tuesday night. Note that the projected area of precipitation (green-shaded region) does not quite reach Augusta. With all of this in mind, our forecast might sound something like this:

> Clear and cold this morning with moderating temperatures by afternoon. Increasing high clouds with skies becoming overcast by evening. Cloudy and not nearly as cold tonight and tomorrow morning. Winds will be light and out of the south or southeast. Barometric pressure will fall slowly.

Wednesday morning we discover that the weather in Augusta is foggy with temperatures in the upper 40s (°F). However, fog was not in the forecast. What went wrong? We forgot to consider that the ground was still cold from the recent cold snap. The warm moist air moving over the cold surface was chilled below its dew point, resulting in fog. Above the fog were the low clouds we predicted. The minimum temperatures remained higher than anticipated because of the release of latent heat during fog formation and the absorption of infrared energy by the fog droplets. Not bad for a start. Now we will forecast the weather for Washington, D.C.

Rain or Snow for Washington, D.C.?

Look at Fig. 13.18 and observe that the low-pressure area over the Central Plains is slowly approaching Washington, D.C., from the west. Hence, the clear weather, light southwesterly winds, and low temperatures on Tuesday morning (Fig. 13.15) will gradually give way to increasing cloudiness, winds becoming southeasterly, and slightly higher temperatures. By Wednesday morning, the projected band of precipitation will be over the city. Will it be in the form of rain or snow? Without a vertical profile of temperature (a sounding) or a thickness chart, this question is difficult to answer. We can see in Fig. 13.18, however, that on Tuesday morning cities south of Washington, D.C.'s latitude are receiving snow. So a reasonable forecast would call for snow, possibly changing to rain as warm air moves in aloft in advance of the approaching fronts. A 24-hour forecast for Washington, D.C., might sound like this:

> Increasing clouds today and continued cold. Snow beginning by early Wednesday morning, possibly changing to rain. Winds will be out of the southeast. Pressures will fall.

Wednesday morning a friend in Washington, D.C., calls to tell us that the sleet began to fall but has since changed to rain. Sleet? Another fractured forecast! Well, almost. What we forgot to account for this time was the intensification of the storm. As the low-pressure area moved eastward, it deepened; central pressure lowered, pressure gradients tightened, and southeasterly winds blew stronger than anticipated. As air moved inland off the warmer Atlantic, it rode up and over the colder surface air. Snow falling into this warm layer at least partially melted; it then refroze as it entered the colder air near ground level. The advection of warmer air from the ocean slowly raised the surface temperatures,

and the sleet soon became rain. Although we did not see this possibility when we made our forecast, a forecaster more familiar with local surroundings would have. Let's move on to Chicago.

Big Snowstorm for Chicago

From Figs. 13.15 and 13.18, it appears that Chicago is in for a major snowstorm. Overrunning of warm air has produced a wide area of snow, which, from all indications, is heading directly for the Chicago area. Since cold air north of the low center will be over Chicago, precipitation reaching the ground should be frozen. On Tuesday morning (Fig. 13.18), the leading edge of precipitation is less than 6 hours away from Chicago. Based on the projected path of the storm in Fig. 13.18 light snow should begin to fall around noon on Tuesday.

By evening, as the storm intensifies, snowfall should become heavy. It should taper off and finally end around midnight as the storm moves east. If it snows for a total of 12 hours—6 hours as light snow (around 1 inch every 3 hours) and 6 hours as heavy snow (around 1 inch per hour)—then the total expected accumulation will be between 6 and 10 inches. As the center of the low moves eastward, passing south of Chicago, winds on Tuesday will gradually shift from southeasterly to easterly, then northeasterly by evening. Since the storm system is intensifying, it should produce strong winds that will swirl the snow into huge drifts, which may bring traffic to a crawl.

The winds will continue to shift as they become northerly and finally northwesterly by Wednesday morning. By then the storm center will probably be far enough east so that skies should begin to clear. Cold air advected from the northwest behind the storm will cause temperatures to drop further. Barometer readings during the storm will fall as the low center approaches and reach a low value sometime Tuesday night, after which they will begin to rise. A weather forecast for Chicago might be:

> Cloudy and cold with light snow beginning by noon, becoming heavy by evening and ending by Wednesday morning. Total accumulations will range between 6 and 10 inches. Winds will be strong and gusty out of the east or northeast today becoming northerly tonight and northwesterly by Wednesday morning. Barometric pressure will fall sharply today and rise tomorrow.

A text message Wednesday morning from a friend in Chicago reveals that our forecast was largely correct—except that the total snow accumulation so far is 13 inches. We were off in our forecast because the storm system slowed as it became occluded. We did not consider this because we moved the system by the steady-state forecast method. At this time of year (early winter), Lake Michigan is not quite frozen over, and the added moisture picked up from the lake by the strong easterly and northeasterly winds enhanced the snowfall. Again, a knowledge of the local surroundings would have helped make a more accurate forecast. The weather 500 miles south of Chicago should be much different from this.

Mixed Bag of Weather for Memphis

Observe in Fig. 13.18 that, within 24 hours, both a warm and a cold front should move past Memphis, Tennessee. The light rain that began Tuesday morning (Fig. 13.15) should saturate the cool air, creating a blanket of low clouds and fog by midday. The warm front, as it moves through sometime Tuesday afternoon, should cause temperatures to rise slightly as winds shift to the south or southwest. At night, clear to partly cloudy skies should allow the ground and air above to cool, offsetting any tendency for a rapid rise in temperature. Falling pressures should level off in the warm air, then fall once again as the cold front approaches. According to the projection in Fig. 13.18, the cold front should arrive sometime before midnight on Tuesday, bringing with it gusty northwesterly winds, showers, the possibility of thunderstorms, rising pressures, and colder air. Taking all of this into account, our weather forecast for Memphis will be:

> Cloudy and cool with light rain, low clouds, and fog early today, becoming partly cloudy and warmer by late this afternoon. Clouds increasing with possible showers and thunderstorms later tonight and turning colder. Winds southeasterly this morning, becoming southerly or southwesterly this evening and shifting to northwesterly tonight. Pressures falling this morning, leveling off this afternoon, then falling again, but rising after midnight.

A friend who lives near Memphis calls Wednesday to inform us that our forecast was correct except that the thunderstorms did not materialize and that dense fog formed in low-lying valleys on Tuesday night, but by Wednesday morning it had dissipated. Apparently, in the warm sector of the storm (the region ahead of the advancing cold front), winds were not strong enough to mix the cold, moist air that had settled in the valleys with the warm air above. It's on to Dallas.

Cold Wave for Dallas   From Fig. 13.18, it appears that our weather forecast for Dallas should be straightforward, since a cold front is expected to pass the area around noon on Tuesday. Weather along the front (Fig. 13.15) is showery with a few thunderstorms developing; behind the front the air is clear but cold. By Wednesday morning (Fig. 13.18) it looks as if the cold front will be far to the east and south of Dallas and an area of high pressure will be centered over southern Colorado. North or northwesterly winds on the east side of the high will bring cold arctic air into Texas, dropping temperatures as much as 40°F within a 24-hour period. With minimum temperatures well below freezing, Dallas will be in the grip of a cold wave. Our weather forecast should therefore read something like this:

> Increasing cloudiness and mild this morning with the possibility of showers and thunderstorms this afternoon. Clearing and turning much colder tonight and tomorrow. Winds will be southwesterly today, becoming gusty north or northwesterly this afternoon and tonight. Pressures falling this morning, then rising later today.

How did our forecast turn out? A quick check of Dallas weather on your smartphone on Wednesday morning reveals that the weather there is cold but not as cold as expected, and the sky is overcast. Cloudy weather? How can this be?

The cold front moved through on schedule Tuesday afternoon, bringing showers, gusty winds, and cold weather with it. Moving southward, the front gradually slowed and became stationary along a line stretching from the Gulf of Mexico westward through southern Texas and northern Mexico. (From the surface map alone we had no way of knowing this would happen.) Along the stationary front a wave of low pressure formed. This disturbance caused warm, moist Gulf air to slide northward up and over the cold surface air. Clouds formed, minimum temperatures did not go as low as expected, and we are left with a fractured forecast. Let's give Denver a try.

Clear but Cold for Denver   In Fig. 13.18, we can see that, based on our projections, the cold high-pressure area will be centered slightly to the south of Denver by Wednesday morning. Sinking air aloft associated with this high-pressure area should keep the sky relatively free of clouds. Weak pressure gradients will produce only weak winds and this, coupled with dry air, will allow for intense radiational cooling. Minimum temperatures will probably drop to well below 0°F. Our forecast should therefore read:

> Clear and cold through tomorrow. Northerly winds today becoming light and variable by tonight. Low temperatures tomorrow morning will be below zero. Barometric pressure will continue to rise.

Almost reluctantly, we look up the weather conditions at Denver on Wednesday morning and find that it is clear and very cold. A successful forecast at last! We find out, however, that the minimum temperature did not go below zero; in fact, 13°F was as cold as it got. A downslope wind coming off the mountains to the west of Denver kept the air mixed and the minimum temperature higher than expected. Again, a forecaster familiar with the local topography of the Denver area would have foreseen the conditions that lead to such downslope winds and would have taken this into account when making the forecast.

A complete picture of the surface weather systems for 6:00 a.m. Wednesday morning is given in ● Fig. 13.19. By comparing this chart with Fig. 13.18, we can summarize why our forecasts did not turn out exactly as we had predicted. For one thing, the storm center near the Great Lakes moved slower than expected. This slow movement allowed a southeasterly flow of mild Atlantic air to overrun cooler surface air ahead of the storm while, behind the low, cities remained in the snow area for a longer time. The weak wave that developed along the trailing cold front over south Texas brought cloudiness and precipitation to Texas and prevented the really cold air from penetrating deep into the south. Farther west, the high-pressure area originally over Montana moved more southerly than southeasterly, which set up a pressure gradient that brought westerly downslope winds to eastern Colorado.

The subjective forecasting techniques demonstrated for these six cities are those one might use in making a short-range weather prediction with very limited resources. The following section describes how a meteorologist predicts the weather in a region where, to the west, surface weather features are greatly modified by a vast body of water and only scanty surface and upper-air data are available. Here, the forecaster must rely heavily on experience as well as more sophisticated tools, which include satellite data, upper-air charts, and computer progs.

Using Forecasting Tools to Predict the Weather

It is late afternoon, and outside the weather forecast office near San Francisco the meteorologist mulls over what is going on in the sky. Overhead is a thin covering of cirrostratus; to the west, draped over the foothills, is the ever-present stratus and fog. The air is cool and the winds are westerly. It is Sunday, March 25, and the forecaster's task is to make a 24-hour weather forecast for the coastal area of central California.

CRITICAL THINKING QUESTION

Look closely at the eastern half of the map in Fig. 13.19, especially the middle Atlantic states. If you were analyzing this map, would you place a warm front in this region? What evidence is present that suggests a warm front could be drawn? Why do you feel the person who analyzed the map did not draw a warm front in this region?

What will tomorrow's weather be like? Will it be similar to today's or will it change markedly? A slowly falling barometer of 1016 mb (30 in.) and the high clouds moving in from the west point to an approaching storm system. A forecast of persistence might be good for the next several hours, but what about tomorrow morning or tomorrow afternoon? One of the biggest challenges for modern forecasters is to zero in on the elements that are most important on any given day.

Based on the forecast funnel approach described in Focus section 13.1, a forecaster would first analyze the major features that could affect the area on a large scale, then zero in on smaller-scale conditions. Before looking at the large-scale upper-level features, however, let's take a quick look at current surface conditions.

The surface map for 4:00 p.m. (PST) Sunday, ● Fig. 13.20, shows there are no weather fronts approaching the West Coast. In fact, the nearest front is a stationary one that has stalled over the Rockies. There is, however, a region of low pressure centered about 1100 km (700 mi) west of San Francisco, which (according to previous maps) has been there for several days. With a central pressure of only about 1012 mb (29.88 in.), the system is fairly weak. Could this weak storm system be causing the increase in high cloudiness and the falling barometer? And will this pattern lead to rain tomorrow? A look at the 500-mb chart may help with these questions.

HELP FROM THE 500-MB CHART ● Figure 13.21 shows the 500-mb analysis for 4:00 p.m. Sunday afternoon. While examining the chart, the meteorologist recognizes certain clues that will aid in making the forecast. For one thing, the 5640-m height line is over northern California. The forecaster knows that when this contour line is situated here or farther south, the statistical probability of receiving measurable rainfall over central California increases greatly.

West of San Francisco the flow is meridional with a cut-off warm, upper high situated just south of Alaska. To the south both east and west of the high are troughs. Because the shape of this flow around the high resembles the Greek letter omega (Ω), the high and its accompanying ridge is known as an **omega high**. The forecaster recognizes the omega high as a *blocking high*, one that tends to persist in the same geographic location for many days. This blocking pattern also tends to keep the troughs in their respective positions, which has been the case for several days

● **FIGURE 13.20** Surface weather map for 4:00 p.m. Sunday, March 25.

● **FIGURE 13.21** The 500-mb chart for 4:00 p.m. Sunday, March 25. Arrows indicate wind flow. Red arrows indicate warm advection and blue arrows, cold advection. Solid lines are height contours where 564 equals 5640 meters above sea level. Dashed lines are isotherms in °C. Heavy dashed purple line shows the position of a shortwave trough and the region where we would find a vorticity maximum.

now. However, the chart indicates that the cold upper trough located west of San Francisco may be changing somewhat.

Observe the spacing of the contour lines around this trough. Even with a limited number of actual wind observations, the close spacing of the contours to the west and northwest of the trough, and the more widely spaced contours to the east of the trough, hint that stronger winds exist to the west of the trough. The forecaster knows from past experience that this usually means the trough will deepen. Also note that on the west side of the trough, cold air is flowing southward (blue arrows), indicating that cold advection is occurring here. The heavy dashed purple line on the west side of the trough represents the position of a shortwave trough and a vorticity maximum,* which is moving rapidly southward. The injection of cold air and the shortwave into the main trough should cause it to intensify. To the east of the main trough, warm air is moving northeastward (red arrows), indicating that warm advection is occurring. It is the lifting and condensing of this moist air that is producing the high clouds over San Francisco. All of these conditions—high wind speeds, cold air moving southward, warm air moving northward, and a shortwave moving into a longwave trough—manifest themselves as a deepening of the longwave trough. As the upper trough deepens, it should be capable of providing the necessary conditions favorable for the development of the surface low into a major mid-latitude cyclonic storm. (You may remember from Chapter 12 that the generation of this type of *dynamic instability* is called baroclinic instability.) The forecaster will likely consult maps for other levels of the atmosphere as well, such as 700 and 850 mb (see Fig. 13.5, p. 356). These maps can reveal, for example, where moisture and warm air are flowing into the developing low-pressure center at lower levels of the atmosphere. Such knowledge gives the forecaster information as to how quickly and strongly the surface storm will develop.

*The position of the shortwave trough pretty much coincides with the position of a vorticity maximum discussed in Chapter 12. For quick reference, look at Fig. 12.27 on p. 340.

One of the main ingredients necessary for the development and intensification of the surface low is divergence of the airflow aloft. The forecaster knows that divergence aloft is associated with a decrease in surface pressure. This decrease, in turn, causes surface air to converge and rise, and its moisture to potentially condense into widespread cloudiness. But where will regions of divergence, convergence, and rising air be found on tomorrow's map? And how will tomorrow's map be different from today's? This is where the computer and the forecaster work together to come up with a prediction.

THE MODELS PROVIDE ASSISTANCE   The computer progs predict the future positions of weather systems. Some of the progs also predict where shortwave troughs and vorticity maximums will be located. It is important to know where the shortwaves and vorticity maximums will be found, because to the east of them there is usually upper-level divergence, lower-level convergence, rising air, clouds, and precipitation. Hence, predicting the position of a shortwave (and a vorticity max) means predicting regions of inclement weather.

Three 24-hour forecast models that predict the positions of the shortwaves, upper-level pressure systems, and the flow aloft at the 500-mb level for Monday, March 26, at 4 p.m. (PST) are shown in ● Fig. 13.22.* (Each prediction is made on Sunday afternoon.) Observe that there is good agreement among the models in that each model moves the upper trough slowly eastward and positions it off the coast. However, the actual positioning of the trough and the shortwaves (heavy dashed lines) differ somewhat for each model. For example, model A and model C move the upper trough eastward more quickly than does model B.

After examining each prog carefully, the forecaster must decide which model most accurately describes the future state of the atmosphere. Over the years, the forecaster knows that model A has performed well in predicting the positions of upper troughs that develop off the coast. Likewise, model C, because it uses more closely spaced grid points and a greater number of data points, and thus better resolution, has done an admirable job of forecasting the positions of upper troughs, shortwaves, and vorticity maximums. On the other hand, although model B has its strengths, it tends to move the shortwaves along too slowly. Consequently, the forecaster puts more confidence into model A and model C for this particular situation.

Using experience and the progs, the meteorologist sets out to predict the weather for the next 24 hours. The 24-hour progs for both model A and model C show a shortwave (labeled 1) approaching the California coast. As the shortwave (vorticity max) approaches the California coast, clouds will increase and thicken, and the likelihood of rain will increase. Therefore, a forecast for the next 24 hours might sound like this:

> Increasing cloudiness Sunday night, with rain beginning Monday morning and continuing through Monday afternoon.

*Explaining the differences among the three models is beyond the scope of this book. Each model treats the atmosphere in a slightly different way. Some models have closer grid points. Some models have better resolution in the lower part of the atmosphere, whereas others have better resolution in the higher regions of the troposphere. The idea in this forecasting example is *not* to illustrate the different models in use, but rather to show how a forecaster might use *any* numerical computer model as a forecasting tool.

● **FIGURE 13.22** Three computer-drawn progs (model A, model B, and model C) show the 24-hour projected 500-mb chart. Solid lines are contours where 564 represents a height of 564 decameters (5640 meters) above sea level. Dashed lines represent projected positions of shortwaves, which are also the positions of vorticity maximums. (Predictions were made on Sunday, March 25, at 4:00 p.m., PST.)

A VALID FORECAST By early Monday morning, the maps begin to show the changes that the computer progs predicted. The surface map for 4:00 a.m. (PST) Monday morning (● Fig. 13.23) shows that the surface low in the Pacific has moved eastward and developed into a broad trough west of California. (Compare its position with Fig. 13.20.) The surface low has deepened considerably, as indicated by its central pressure of 1004 mb (29.65 in.). The approach of the storm is evidenced in San Francisco by thick middle clouds, southerly winds, and a falling barometer, nearly 4 mb lower than 12 hours ago. All these signs suggest that rain is on the way.

On the 500-mb chart for 4:00 a.m. Monday morning, March 26 (● Fig. 13.24), we can see that the movement of cold air southward around the upper trough, along with the swift movement of the shortwave, has caused the upper trough to deepen. Note that the height contours are now displaced farther south and that the contour in the middle of the trough is lower than on the previous 500-mb map (Fig. 13.21). Compare Fig. 13.24 with the 24-hour progs in Fig. 13.22 and notice that both model A and model C are projecting the movement of the shortwave (number 1) quite well. The forecaster made a wise choice in showing confidence in these two computer models as they did a good job predicting the position of the upper-level low and shortwave. Since the shortwave (vorticity maximum) is moving with the flow toward San Francisco, it should rain today. But at what time will the rain begin? The forecaster must now go deeper into the forecast funnel and examine local conditions in greater detail. Here is where satellite and radar information come in.

SATELLITE AND UPPER-AIR ASSISTANCE The infrared satellite image taken at 6:45 a.m. Monday (see ● Fig. 13.25) shows that the middle clouds presently over California will soon give way to an organized band of cumuliform clouds in the shape of a comma. Such comma clouds tell the forecaster that the low-pressure area off the coast is developing into a mature mid-latitude cyclone. Observe that this comma-shaped cloud band lies slightly to the

● **FIGURE 13.23** Surface weather map for 4:00 a.m. (PST) Monday, March 26.

east of the shortwave approaching the coast, shown in Fig. 13.24. Also note that to the west of the comma cloud, a relatively unorganized mass of cumulus congestus clouds is beginning to form near the second shortwave in Fig. 13.24. The 300-mb chart for 4:00 a.m. Monday morning (● Fig. 13.26) shows strong jet stream winds over Northern California and off the coast, suggesting that southwesterly winds aloft will carry the large comma-shaped cloud and its weather directly into California. And an area of strong divergence aloft associated with the jet stream (orange color on the map) will aid in deepening the cyclonic storm.

By examining the movement of the cloud mass on successive satellite images, the forecaster can predict its arrival time and, hence, when rainfall will begin. According to satellite images, the leading edge of the comma cloud should be just offshore by Monday afternoon. Also, Doppler radar indicates that, just off the

● **FIGURE 13.24** The 500-mb analysis for 4:00 a.m. (PST) Monday, March 26. Heavy dashed lines show position of shortwaves. Solid lines are height contours where 564 equals 5640 meters above sea level. (Compare with Fig. 13.21, the 24-hour progs for model A and model C.)

● **FIGURE 13.26** The 300-mb chart for 4:00 a.m. (PST) Monday, March 26. Solid lines are height contours, where 900 equals 9000 meters above sea level. The area shaded orange represents a region of strong divergence of upper-level air.

● **FIGURE 13.25** Infrared satellite image taken at 6:45 a.m. (PST) Monday, March 26. The cloud in the shape of a comma indicates that the mid-latitude cyclonic storm is deepening. (The heavy dashed line shows the tail of the comma cloud.)

coast, light rain is now falling from the middle cloud layer. The position of the upper shortwave trough and the jet stream above the area of surface low pressure should provide enough support for the low to continue to intensify. Pressure gradients around the low will likely increase, creating strong, gusty winds from the south as the storm approaches. An amended forecast for San Francisco might read:

> Rain beginning this morning becoming heavy by this afternoon. Strong and gusty southerly winds.

A DAY OF RAIN AND WIND The first raindrops falling from altostratus clouds dampen city streets near the end of the morning rush hour. Quickly, the rain spreads inland, and by late Monday afternoon, Doppler radar shows that precipitation is falling throughout northern and central California (see ● Fig. 13.27), as gusty southerly winds and moderate rain greet commuters on their way home.

The barometer has fallen sharply all day at San Francisco and by 4 p.m. the barometer reading is 1004 mb, a drop of 7 mb in just 6 hours. We can see the reason for this on the surface map for 4 p.m. Monday afternoon (● Fig. 13.28). The mid-latitude cyclone has not only moved closer to the coast, it has intensified considerably, as indicated by the drop of 11 mb in central pressure in just 12 hours. Spiraling around the low, a cold front marks the position of the comma-shaped cloud. At first, this may seem surprising, since no fronts were drawn on the previous map. However, strong diverging air aloft caused surface air with contrasting temperatures and moisture properties to spiral counterclockwise into the surface low. With the coldest air on the western side of the surface low, the meteorologist saw fit to draw a cold front on the weather map.

As the surface low intensifies, it and the spiraling band of clouds move eastward more slowly. The front will, therefore, move through later than anticipated, sometime late Monday night or early Tuesday morning. The forecaster expects that the winds will remain strong and precipitation will be heavy as the front passes.

Early Tuesday morning, the front moves onshore, bringing with it heavy rain and winds with gusts exceeding 45 knots.

● **FIGURE 13.27** Doppler radar image showing precipitation falling across northern and central California at 5:00 p.m. (PST) on Monday, March 26. Areas in blue and green indicate light to moderate precipitation; yellow indicates heavier precipitation; orange and red heaviest precipitation.

● **FIGURE 13.28** Surface weather map for 4:00 p.m. (PST) Monday, March 26.

● **FIGURE 13.29** Visible satellite image for 9:00 a.m. (PST) Tuesday, March 27. Included in the picture are the positions of surface fronts, the upper-level flow (heavy arrows), and precipitation patterns.

Billowing cumulus clouds, and in some areas thunderstorms, drench the entire Pacific Coast with rain, and with snow at higher elevations. The storm center, constantly being drained of air by upper-level divergence associated with a strong polar jet stream, has deepened into a furious system with a central pressure of 988 mb (29.17 in.).

The satellite image for 10 a.m. PDT Tuesday morning, March 27, (see ● Fig. 13.29) provides us with a visual interpretation of the storm. We see superimposed on the image the positions of the surface low, fronts, and the winds aloft (heavy arrow). Note that a front with its heavy band of clouds stretches from Idaho southward into Nevada and southern California, while the surface low is still positioned off the northern California coast. Moving through central California is a band of clouds and showers associated with a shortwave (vorticity maximum) spinning counterclockwise around the low.

As the surface low moves onshore, will it intensify or weaken? Will the unstable cold air behind the front develop into towering clouds and thunderstorms? It's back to the drawing board—to the computer progs, the forecast charts, Doppler radar, and the satellite images. The challenge and anticipation of making another forecast are at hand.

SUMMARY

Forecasting tomorrow's weather entails a variety of techniques and methods. Persistence, surface maps, satellite imagery, and Doppler radar are all useful when making a very short-range (0–6 hour) prediction. For short- and medium-range forecasts, the current analysis, satellite data, pattern recognition, meteorologist intuition, and experience, along with statistical information and guidance from the many computer progs supplied by the National Weather Service, all go into making a prediction. For monthly and seasonal long-range forecasts, meteorologists incorporate changes in sea surface temperature in the Pacific and Atlantic Oceans into seasonal outlooks of temperature and precipitation in North America.

Different computer progs are based upon different atmospheric models that describe the state of the atmosphere and how it will change with time. The atmosphere's chaotic behavior, along with tiny errors (uncertainties) in the data, generally amplify as the computer tries to project weather further and further into the future. At present, computer progs that predict the weather over a vast region are better at forecasting the position of mid-latitude highs and lows and their development than at forecasting local showers and thunderstorms. To skillfully forecast smaller features, the grid spacing on some models is reduced to as low as 0.5 kilometers.

As new information from atmospheric research programs is fed into the latest generation of computers, progs are expected to show increasing skill in predicting the weather up to 10 days in the future. The simulation of large-scale climatic trends by the most recent general circulation models also shows promise.

In the latter part of this chapter, we learned how one can make short-range weather predictions even with limited information, such as several days' worth of surface weather maps. We also saw some of the problems facing anyone who attempts to predict the behavior of the churning mass of air we call our atmosphere.

Most of the forecasting methods in this chapter apply mainly to skill in predicting events associated with large-scale weather systems, such as fronts and mid-latitude cyclones. The next three chapters deal with the formation and forecasting of smaller-scale (mesoscale) systems, such as thunderstorms, squall lines, tornadoes, and hurricanes.

KEY TERMS

The following terms are listed (with corresponding page numbers) in the order they appear in the text. Define each. Doing so will aid you in reviewing the material covered in this chapter.

atmospheric models, 348
forecast funnel, 348
AWIPS, 350
meteogram, 352
analysis, 353
numerical weather prediction, 353
data assimilation, 353
prognostic chart (prog), 353
chaos, 358
ensemble forecasting, 359
persistence forecast, 360
steady-state (trend) forecast, 360
analog forecasting method, 360
pattern recognition 360
statistical forecast, 361
probability forecast, 363
weather type forecasting, 363
climatological forecast, 363
very short-range forecast, 364
weather watch, 364
weather warning, 364
short-range forecast, 364
medium-range forecast, 364
long-range forecast, 364
outlooks, 364
teleconnections, 365
isallobars, 369
omega high, 374

QUESTIONS FOR REVIEW

1. What is the function of the National Centers for Environmental Prediction?
2. List at least four tools a weather forecaster might use when making a short-range forecast.
3. How does a *weather watch* differ from a *weather warning*?
4. Do monthly and seasonal forecasts make specific predictions of rain or snow? Explain.
5. Describe four methods to forecast the weather and give an example for each one.
6. Explain how teleconnections are used in making a long-range seasonal forecast?
7. Make a persistence forecast for your area for this same time tomorrow. Did you use any skill in making this prediction? Explain.
8. If today's weather forecast calls for a "chance of snow," what is the probability that it will snow today? (Hint: See Table 13.3, p. 363.)
9. Would a forecast calling for a 20 percent chance of rain be high enough for you to cancel your plans for a picnic? Explain.
10. In what ways has the computer assisted the meteorologist in making weather forecasts?
11. How does a prog differ from an analysis?
12. How are computer-generated weather forecasts prepared?
13. What are some of the problems associated with computer-model forecasts?
14. How does pattern recognition aid a forecaster in making a prediction?

15. How can ensemble forecasts improve medium-range weather forecasts?

16. Do all accurate forecasts show skill? Explain.

17. If low clouds at an elevation of 3000 feet above you are moving from the southeast, and clouds about 8000 feet higher are moving from the southwest, is cold or warm advection taking place between the cloud layers? Explain. (Hint: Read Focus section 13.4, found on p. 362.)

18. List four methods that you could use to predict the movement of a surface mid-latitude cyclone.

19. What is an omega high? What influence does it have on the movement of surface highs and lows?

20. Suppose that where you live, the middle of January is typically several degrees warmer than the rest of the month. If you forecast this "January thaw" for the middle of next January, what type of a weather forecast will you have made?

21. Given a map with isallobars, where will high- and low-pressure systems move?

QUESTIONS FOR THOUGHT

1. From Fig. 13.10, p. 363, determine the probability of a white Christmas for your area.

2. Suppose the chance for a white Christmas at your home is 10 percent. Last Christmas was a white one. If for next year you forecast a "nonwhite" Christmas, will you have shown any skill if your forecast turns out to be correct? Explain.

3. Suppose that it is presently warm and raining. A cold front will pass your area in 3 hours. Behind the front it is cold and snowing. Make a persistence forecast for weather that will occur in your area six hours from now. Would you expect this forecast to be correct? Explain. Now, make a forecast for your area using the steady-state, or trend, method.

4. Since computer models have difficulty in adequately considering the effects of small-scale geographic features on a weather map, why don't numerical weather forecasts simply reduce the grid spacing to about 1 kilometer?

5. Explain how the phrase "sensitive dependence on initial conditions" relates to the final outcome of a computer-based weather forecast.

6. You are in Calgary, Alberta, Canada, 100 km (62 mi) east of the Rockies, in January. The current wind is from the north. Looking at a prog for tomorrow, you see that the wind will be from the west. Will tomorrow's temperature be warmer or cooler than today? Explain.

PROBLEMS AND EXERCISES

1. When a persistent winter pattern at 500 mb appears similar to that shown in ● Fig. 13.30, show on the map where you would forecast the following: (a) good chance of precipitation; (b) above seasonal temperatures; (c) below seasonal temperatures; and (d) generally dry weather. (e) What type of forecasting method did you use to figure out (a) through (d)?

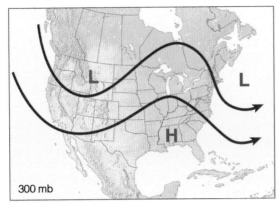

● **FIGURE 13.30** Diagram for Exercise 1.

2. In Fig. 13.15, p. 370, mark the position of the following cities: Cleveland, Ohio; Albuquerque, New Mexico; and New Orleans, Louisiana. Based on the projected movement of the surface weather systems in Fig. 13.18, p. 371, make a short-range 24-hour weather forecast for each of these cities. In your forecast, include temperature, pressure, cloud cover, humidity, winds, and precipitation (if any). Compare your forecasts with the actual weather at the end of the period in Fig. 13.19, p. 374.

3. Go outside and observe the weather. Make a weather forecast using the weather signs you observe. Explain the rationale for your forecast.

ONLINE RESOURCES

Visit www.cengagebrain.com to view additional resources, including video exercises, practice quizzes, an interactive eBook, and more.

The Great Plains of the United States are home to some of the world's most intense thunderstorm activity.

Thunderstorms

CONTENTS

WHEN I STARTED OUT ON FOOT THAT AUGUST afternoon, the thunderstorm was blowing in fast. On the face of the mountain, a mile ahead, hard westerly gusts and sudden updrafts collided, pulling black clouds apart. Yet the storm looked harmless. When a distant thunderclap scared the dogs, I called them to my side and rubbed their ears: "Don't worry, you're OK as long as you're with me."

I woke in a pool of blood, lying on my stomach some distance from where I should have been, flung at an odd angle to one side of the dirt path. The whole sky had grown dark. Was it evening, and if so, which one? How many minutes or hours had elapsed since I lost consciousness, and where were the dogs? I tried to call out to them but my voice didn't work. The muscles in my throat were paralyzed and I couldn't swallow. Were the dogs dead? Everything was terribly wrong: I had trouble seeing, talking, breathing, and I couldn't move my legs or right arm. Nothing remained in my memory—no sounds, flashes, smells, no warnings of any kind. Had I been shot in the back? Had I suffered a stroke or heart attack? . . .

The sky was black. Was this a storm in the middle of the day, or was it night with a storm traveling through? When thunder exploded over me, I knew I had been hit by lightning.

Gretel Ehrlich, *A Match to the Heart*

The excerpt on the preceding page shows the intense impact that a thunderstorm can have in a mere fraction of a second. Most of us have seen many thunderstorms in our life, even if we have never been struck by lightning. Although thunderstorms are very common, they still evoke awe with their towering clouds, spectacular lightning, pelting hailstones, torrential rains, powerful winds, and—occasionally—tornadoes. In this chapter, we will examine the different types of thunderstorms, how they develop, and the dangerous phenomena they can produce. In the following chapter, we will focus on tornadoes, examining how and where they form, and why they are so destructive.

Thunderstorm Types

It probably comes as no surprise that a *thunderstorm* is merely a storm containing lightning and thunder. Sometimes a thunderstorm produces gusty surface winds with heavy rain and hail. The storm itself may be a single cumulonimbus cloud, or several thunderstorms may form into a cluster. In some cases, a line of thunderstorms will form that may extend for hundreds of kilometers.

Thunderstorms are *convective storms* that form with rising air. So the birth of a thunderstorm typically involves warm, moist air rising in a conditionally unstable environment.* The rising air may be a parcel of air ranging in size from a large balloon to a city block. Or an entire layer, or slab of air, may be lifted. As long as the rising air is warmer (less dense) than the air surrounding it, there is an upward-directed *buoyant force* acting on it. The warmer the parcel is compared to its surroundings, the greater the buoyant force and the stronger the convection. The trigger (or "forcing mechanism") needed to start air moving upward may be

1. random, turbulent eddies that lift small bubbles of air
2. unequal heating at the surface
3. the effect of terrain (such as small hills) or the lifting of air along shallow boundaries of converging surface winds

4. large-scale uplift along mountain barriers and rising terrain
5. diverging upper-level winds, coupled with converging surface winds and rising air
6. warm air rising along a frontal zone

Usually, several of these mechanisms work together with vertical wind shear to generate severe thunderstorms. The vertical wind shear may be produced by winds increasing with height (*speed shear*), changing direction with height (*directional shear*), or both.

Although we often see thunderstorms forming where the surface air is quite warm and humid, they can also form when the surface air temperature is 10°C (50°F) or even lower. This latter situation often occurs in winter along the west coast of North America, when cold air aloft moves over the region. The cold air aloft destabilizes the atmosphere to the point where air parcels, given an initial push upward, are able to continue their upward journey because they remain warmer (less dense) than the colder air surrounding them. The cold air aloft may even cause sufficient instability to generate *thundersnow*—thunder and lightning observed in wintertime snowstorms.

Most thunderstorms that form over North America are short-lived, producing rain showers, gusty surface winds, thunder and lightning, and sometimes small hail. Many have an appearance similar to the mature thunderstorm shown in ● Fig. 14.1. The majority of these storms do not reach severe status. *Severe thunderstorms* are defined by the National Weather Service as thunderstorms that produce at least one of the following: large hail with a diameter of at least 1 inch, surface wind gusts of at least 50 knots (58 mi/hr), or a tornado.

Scattered thunderstorms (sometimes called "pop-up" or "popcorn" storms) that typically form on warm, humid days are often referred to as *ordinary cell thunderstorms** or air-mass thunderstorms because they tend to form in warm, humid air masses away from significant weather fronts. These thunderstorms can be considered "simple storms" because they rarely become severe, typically are less than a kilometer wide, and they go through a rather predictable life cycle from birth to maturity to decay that

---

*A conditionally unstable atmosphere exists when cold, dry air aloft overlies warm, moist surface air. Additional information on atmospheric instability is given in Chapter 6, beginning on p. 147.

*In convection, the cell may be a single updraft or a single downdraft, or a combination of the two.

● **FIGURE 14.1** An ordinary thunderstorm in its mature stage. Note the distinctive anvil top.

© C. Donald Ahrens

usually takes less than an hour to complete. However, under the right atmospheric conditions (described later in this chapter), more intense "complex thunderstorms" may form, such as the *multicell thunderstorm* and the *supercell thunderstorm*—an intense, rotating storm that can last for hours and produce severe weather such as strong surface winds, large damaging hail, flash floods, and violent tornadoes.

We will examine the development of ordinary cell (air mass) thunderstorms first, before we turn our attention to the more complex multicell and supercell storms.

ORDINARY CELL THUNDERSTORMS Ordinary cell (air mass) **thunderstorms** or, simply, *ordinary thunderstorms,* tend to form in a region where there is limited vertical wind shear—that is, where the wind speed and wind direction *do not* abruptly change with increasing height above the surface.* Many ordinary thunderstorms appear to form as parcels of air are lifted from the

---

*As we will see later in this chapter, vertical wind shear is different from the horizontal wind shear (the changing of wind direction and/or speed in the horizontal) that pilots are concerned about.

surface by turbulent overturning in the presence of wind. Moreover, ordinary storms often form along shallow zones where surface winds converge. Such zones can be due to any number of things, such as topographic irregularities, sea-breeze fronts, or the cold outflow of air from inside a thunderstorm that reaches the ground and spreads horizontally. These converging wind boundaries are normally zones of contrasting air temperature and humidity and, hence, air density.

Ordinary thunderstorms go through a fairly predictable cycle of development from birth to maturity to decay. The first stage is known as the **cumulus stage**, or *growth stage*. As a parcel of warm, humid air rises, it cools and condenses into a single cumulus cloud or a cluster of clouds (see ● Fig. 14.2). If you have ever watched a thunderstorm develop, you may have noticed that at first the cumulus cloud grows upward only a short distance before the top of the cloud becomes less distinct. This dissipation occurs because the cloud droplets evaporate as the drier air surrounding the cloud mixes with it. However, after the water drops evaporate, the air is more moist than before. So, the rising air is now able to condense at successively higher levels, and the cumulus cloud grows taller, often appearing as a rising dome or tower.

As the cloud builds, the transformation of water vapor into liquid or solid cloud particles releases large quantities of latent heat, a process that keeps the rising air inside the cloud warmer (less dense) than the air surrounding it. The cloud continues to grow in the unstable atmosphere as long as it is constantly fed by rising air from below. In this manner, a cumulus cloud can show extensive vertical development and grow into a towering cumulus cloud (cumulus congestus) in just a few minutes. The cumulus stage does not usually last long enough for precipitation to form,

(a) Cumulus (b) Mature (c) Dissipating

● **FIGURE 14.2** Simplified model depicting the life cycle of an ordinary cell thunderstorm that is nearly stationary as it forms in a region of low wind shear. (Arrows show vertical air currents. Dashed line represents freezing level, 0°C isotherm.)

and the updrafts keep water droplets and ice crystals suspended within the cloud. Also, there is no lightning or thunder during this stage.

As the cloud builds well above the freezing level, the cloud particles collide and join with one another and thus grow larger and heavier. These liquid and solid (ice) particles continue to rise until the size of the growing particles exceeds the ability of the updraft to keep them suspended. Meanwhile, drier air from around the cloud is being drawn into it in a process called *entrainment*. The entrainment of drier air causes some of the raindrops to evaporate, which chills the air. This air, now colder and heavier than the air around it, begins to descend as a *downdraft*. As the air descends, the cold ice particles begin to melt, which chills the air and enhances the downdraft. The downdraft may be further enhanced as falling precipitation drags some of the air along with it.

The appearance of the downdraft marks the beginning of the **mature stage**. The downdraft and updraft within the mature thunderstorm now constitute the cell. In some storms, there are several cells, each of which may last for less than 30 minutes.

During its mature stage, the thunderstorm is most intense. The top of the cloud, having reached a stable region of the atmosphere (which may be as high as the stratosphere), begins to take on the familiar anvil shape, as upper-level winds spread the cloud's ice crystals horizontally (see Fig. 14.2b). The cloud itself may extend upward to an altitude of over 12 km (40,000 ft) and be several kilometers in diameter near its base. Updrafts and downdrafts are strongest in the middle of the cloud, creating severe turbulence.

WEATHER WATCH

On July 13, 1999, in Sattley, California, a strong downdraft from a mature thunderstorm dropped the air temperature from 97°F at 4:00 p.m. to a chilly 57°F one hour later.

Lightning and thunder are also present in the mature stage. Heavy rain (and occasionally small hail) often falls from the cloud. And, at the surface, there is often a downrush of cooler air with the onset of precipitation.

Where the cold downdraft reaches the surface, the air spreads out horizontally in all directions. The surface boundary that separates the advancing cooler air from the surrounding warmer air is called a *gust front*. Along the gust front, winds rapidly change both direction and speed. Look at Fig. 14.2b and notice that the gust front forces warm, humid air up into the storm, which enhances the cloud's updraft. In the region of the downdraft, rainfall may or may not reach the surface, depending on the relative humidity beneath the storm. In the dry air of the Desert Southwest, for example, a mature thunderstorm may look ominous and contain all of the ingredients of any other storm, except that the raindrops evaporate before reaching the ground. However, intense downdrafts from the storm may reach the surface, producing strong, gusty winds and a gust front.

After the storm enters the mature stage, it begins to dissipate in about 15 to 30 minutes. The **dissipating stage** occurs when the updrafts weaken as the gust front moves away from the storm and no longer enhances the updrafts. At this stage, as illustrated in Fig. 14.2c, downdrafts tend to dominate throughout much of the cloud. The reason an ordinary cell thunderstorm does not normally last very long is that the downdrafts inside the cloud tend to cut off the storm's fuel supply by destroying the humid updrafts. Deprived of the rich supply of warm, humid air, cloud droplets no longer form. Light precipitation now falls from the cloud, accompanied by only weak downdrafts. As the storm dies, the lower-level cloud particles evaporate rapidly, sometimes leaving only the cirrus anvil as the reminder of the once-mighty presence (see ● Fig. 14.3). A single ordinary cell thunderstorm may go through its three stages in one hour or less.

Not only do these thunderstorms produce summer rainfall for a large portion of the United States, but they can also bring

● **FIGURE 14.3** A dissipating thunderstorm near Naples, Florida. Most of the cloud particles in the lower half of the storm have evaporated.

© Howard B. Bluestein

with them momentary cooling after an oppressively hot day. The cooling comes during the mature stage, as the downdraft reaches the surface in the form of a blast of welcome relief. Sometimes, the air temperature may drop by more than $10°C(18°F)$ in just a few minutes. Unfortunately, the cooling effect often is short-lived, as the downdraft diminishes or the thunderstorm moves on. In fact, after the storm has ended, the air temperature usually rises, and as the moisture from the rainfall evaporates into the air, the humidity increases—sometimes to a level that actually feels more oppressive after the storm than it did before.

Up to this point, we've looked at ordinary cell thunderstorms that are short-lived, rarely become severe, and form in a region with weak vertical wind shear. As these storms develop, the updraft eventually gives way to the downdraft, and the storm ultimately collapses on itself. However, in a region where strong vertical wind shear exists, thunderstorms often take on a more complex structure. Strong, vertical wind shear can cause the storm to tilt in such a way that it becomes a multicell thunderstorm—a thunderstorm with more than one cell.

MULTICELL THUNDERSTORMS   Thunderstorms that contain a number of cells, each in a different stage of development, are called **multicell thunderstorms** (see ● Fig. 14.4). Such storms tend to form in a region of moderate-to-strong vertical wind speed shear. Look at ● Fig. 14.5 and notice that on the left side of the illustration the wind speed increases rapidly with height, producing strong wind speed shear. This type of shearing causes the cell inside the storm to tilt in such a way that the updraft actually rides up and over the downdraft. Note that the rising updraft is capable of generating new cells that go on to become mature thunderstorms.

● **FIGURE 14.4** This multicell storm complex is composed of a series of cells in successive stages of growth. The thunderstorm in the middle is in its mature stage, with a well-defined anvil. Heavy rain is falling from its base. To the right of this cell, a thunderstorm is in its cumulus stage. To the left, a well-developed cumulus congestus cloud is about ready to become a mature thunderstorm. With new cells constantly forming, the multicell storm complex can exist for hours.

© C. Donald Ahrens

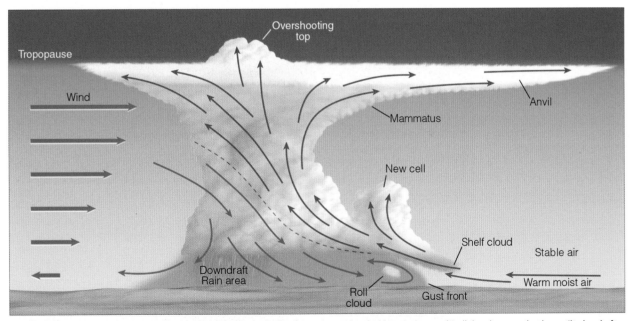

● **FIGURE 14.5** A simplified model describing air motions and other features associated with an intense multicell thunderstorm that has a tilted updraft. The severity depends on the intensity of the storm's circulation pattern.

Notice also that precipitation inside the storm does not fall into the updraft (as it does in the ordinary cell thunderstorm), so the storm's fuel supply is maintained and the storm complex can survive for a long time. Long-lasting multicell storms can become intense and produce severe weather for brief periods.

When convection is strong and the updraft intense (as it is in Fig. 14.5), the rising air may actually intrude well into the stable stratosphere, producing an **overshooting top**. As the air spreads laterally into the anvil, sinking air in this region of the storm can produce beautiful mammatus clouds. At the surface, below the thunderstorm's cold downdraft, the cold, dense air may cause the surface air pressure to rise—sometimes several millibars. The relatively small, shallow area of high pressure is called a *mesohigh* (meaning "mesoscale high"). The mesohigh increases the pressure gradient between the storm-cooled air and the warmer, unstable air that lies beyond the storm, a situation that raises the risk of high winds.

The Gust Front   When the cold downdraft reaches Earth's surface, it pushes outward in all directions, producing a strong **gust front** that represents the leading edge of the cold outflowing air (see ● Fig. 14.6). To an observer on the ground, the passage of the gust front resembles that of a cold front. During its passage, the temperature drops sharply and the wind shifts and becomes strong and gusty, with speeds occasionally exceeding 55 knots. These high winds behind a strong gust front are called **straight-line winds** to distinguish them from the rotating winds of a tornado. As we will see later in this chapter, straight-line winds can inflict a great deal of damage, such as by blowing down trees and overturning mobile homes.

Along the leading edge of the gust front, the air is quite turbulent. Here, strong winds can pick up loose dust and soil and lift them into a huge tumbling cloud (see ● Fig. 14.7).* The cold surface air behind the gust front may linger close to the ground for hours, well after thunderstorm activity has ceased.

---

*In dry, dusty areas or desert regions, the leading edge of the gust front is the haboob described in Chapter 9, p. 247.

As warm, moist air rises along the forward edge of the gust front, a **shelf cloud** (also called an *arcus cloud*) may form, such as the one shown in ● Fig. 14.8. These clouds are especially prevalent when the atmosphere is very stable near the base of the thunderstorm. Look again at Fig. 14.5 and notice that the shelf cloud is attached to the base of the thunderstorm. On rare occasions, a narrow, elongated cloud will form along the gust front, but completely detached from the thunderstorm base. These clouds, which appear to slowly spin about a horizontal axis, are called **roll clouds** (see ● Fig. 14.9).

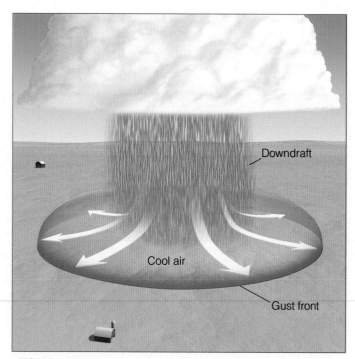

● **FIGURE 14.6** When a thunderstorm's downdraft reaches the ground, the air spreads out, forming a gust front.

● **FIGURE 14.7** A swirling mass of dust forms along the leading edge of a gust front as it moves across western Nebraska.

When the atmosphere is conditionally unstable, the leading edge of the gust front may force the warm, moist air upward, producing a complex of multicell storms, each with new gust fronts.

These gust fronts may then merge into a huge gust front called an **outflow boundary**. Along the outflow boundary, air is forced upward, often generating new thunderstorms (see ● Fig. 14.10).

Microbursts   Beneath an intense thunderstorm, the downdraft can become localized so that it hits the ground and spreads horizontally in a radial burst of wind, much like water pouring from a tap and striking the sink below. (Look at the downdraft in Fig. 14.6.) Such downdrafts are called **downbursts**. A downburst with winds spanning less than 4 km is termed a **microburst**. In spite of its small size, an intense microburst can induce damaging straight-line winds well over 100 knots (115 mi/hr). (A larger downburst with winds spanning 4 km or more is termed a

● **FIGURE 14.8** A dramatic example of a shelf cloud (or arcus cloud) associated with an intense thunderstorm. The photograph was taken in central Oklahoma as the thunderstorm approached from the northwest.

● **FIGURE 14.9** A roll cloud moves over Calgary, Alberta, on the morning of June 18, 2013.

● **FIGURE 14.10** Radar image of an outflow boundary. As cool (more-dense) air from inside the severe thunderstorms (red and orange colors) spreads outward, away from the storms, it comes in contact with the surrounding warm, humid (less-dense) air, forming a density boundary (blue line) called an *outflow boundary* between cool air and warm air. Along the outflow boundary, new thunderstorms often form.

macroburst.) *Air Force One* narrowly avoided disaster on August 1, 1983, when the aircraft landed with President Ronald Reagan at Andrews Air Force Base only a few minutes before a microburst produced winds exceeding 130 knots (150 mi/hr). ● Figure 14.11 shows the dust clouds generated from a microburst north of Denver, Colorado. Since a microburst is an intense downdraft, its leading edge can evolve into a gust front. The leading edge of a microburst can contain an intense horizontally rotating vortex that, in a relatively dry region, is often filled with dust.

Microbursts can blow down trees and inflict heavy damage upon poorly built structures as well as upon sailing vessels that encounter microbursts over open water. In fact, microbursts may be responsible for some damage once attributed to tornadoes. They pose an especially serious hazard to aircraft, largely because of the accompanying horizontal wind shear. (Recall that wind shear is caused by rapid changes in wind speed and/or wind direction across a short distance.) When an aircraft flies through a microburst at a relatively low altitude, say 300 m (1000 ft) above the ground, it first encounters a headwind that generates extra lift. This is position (a) in ● Fig. 14.12. At this point, the aircraft tends to climb (it gains lift), and if the pilot noses the aircraft downward there could be grave consequences. In a matter of seconds the aircraft encounters the powerful downdraft (position b), and the headwind is replaced by a tail wind (position c). This situation causes a sudden loss of lift and a subsequent decrease in the performance of the aircraft, which is now accelerating toward the ground.

In the United States, hundreds of air passengers died in microburst-related accidents in the 1970s and 1980s. Recognizing the danger that microbursts posed to aviation, scientists carried out intensive research that led to a warning system installed at airports in the 1990s throughout the United States. This system includes automated weather stations, Doppler radars, and computer algorithms designed to detect microbursts and low-level wind shear. The system has virtually eliminated microburst-related accidents in U.S. commercial flights.

Microbursts can be associated with severe thunderstorms, producing strong, damaging winds. But studies show that they can also occur with ordinary cell thunderstorms and with clouds that produce only isolated showers—clouds that may or may not contain thunder and lightning. In the western United States, many microbursts emanate from virga—rain falling from a cloud but evaporating before reaching the ground. In these "dry" microbursts (Fig. 14.11), evaporating rain cools the air. The cooler, heavier air then plunges downward through the warmer, lighter air below. In humid regions, many microbursts are "wet" in that they are accompanied by blinding rain.

Up to this point, you might think that thunderstorm-related downdrafts are always cool. Most are cool, but occasionally

● **FIGURE 14.11** Dust clouds rising in response to the outburst winds of a microburst north of Denver, Colorado.

they can be extremely hot. For example, just after midnight on June 9, 2011, a blast of hot, dry air from a dissipating thunderstorm raised the surface air temperature at Wichita (Kansas) Dwight D. Eisenhower National Airport from 85°F to 102°F in only 20 minutes, with winds gusting to more than 40 knots. Such sudden warm downbursts are called **heat bursts**. Studies indicate that the heat burst originates high in the thunderstorm, toward the back edge of the anvil, and warms by compressional heating as it plunges toward the surface. Some research suggests that the source of heat bursts may be air from outside a thunderstorm that is forced downward as it encounters the storm, with light precipitation falling into it. However, the exact cause is not yet known.

Squall-Line Thunderstorms Multicell thunderstorms may form as a line of thunderstorms, called a **squall line**. The line of storms may form directly along a cold front and extend for hundreds of kilometers, or the storms may form in the warm air 100 to 300 km out ahead of the cold front. These *prefrontal squall-line thunderstorms* of the middle latitudes represent the largest and most severe type of squall line, with huge thunderstorms causing severe weather over much of its length (see ● Fig. 14.13).*

There is still debate as to exactly how prefrontal squall lines form. Models that simulate their formation suggest that, initially, convection begins along the cold front, then redevelops farther away. Moreover, the surging nature of the main cold front itself, or developing cumulus clouds along the front, may cause the air aloft to develop into waves (called *gravity waves*), much like the waves that form downwind of a mountain chain (see ● Fig. 14.14). Out ahead of the cold front, the rising motion of the wave may be the trigger that initiates the development of cumulus clouds and a prefrontal squall line. In some instances, low-level converging air is better established out ahead of the advancing cold front.

Rising air along the frontal boundary (and along the gust front), coupled with the tilted nature of the updraft, promotes the development of new cells as the storm moves along. Hence, as old cells decay and die out, new ones constantly form, and the squall line can maintain itself for hours on end. Occasionally, a new squall line will actually form out ahead of the front as the gust front pushes forward, beyond the main line of storms.

Squall lines that exhibit weaker updrafts and downdrafts tend to be more shallow than prefrontal squall lines, and usually they have shorter life spans. These storms are referred to as *ordinary squall lines*. Severe weather may occur with them, but more typically they form as a line of thunderstorms that exhibit characteristics of ordinary cell thunderstorms. Ordinary squall lines may form along a gust front, with a stationary front, with a weak wave cyclone, or where no large-scale cyclonic storms are present. Many of the ordinary squall lines that form in the middle latitudes exhibit a structure similar to squall lines that form in the tropics.

---

*Within a squall line there may be multicell thunderstorms, as well as supercell storms—violent thunderstorms that contain a single rapidly rotating updraft. We will look more closely at supercells in the next section.

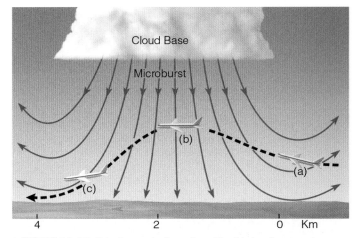

● **FIGURE 14.12** Flying into a microburst. At position (a), the pilot encounters a headwind; at position (b), a strong downdraft; and at position (c), a tailwind that reduces lift and causes the aircraft to lose altitude. (This horizontal wind shear is different from the vertical wind shear that acts to increase storm severity.)

CRITICAL THINKING QUESTION Moving from right to left in the figure above, would you expect the greatest risk of the aircraft crashing near position (c), if the aircraft at position (a) is taking off or just beginning its descent to land?

In some squall lines, the leading area of thunderstorms and heavy precipitation is followed by a region of extensive stratified clouds and light precipitation (see ● Fig. 14.15). The stratiform clouds represent a region where the anvil cloud trails behind the

● **FIGURE 14.13** Doppler radar display superimposed on a map shows a prefrontal squall line extending from Texas into Oklahoma and Arkansas during February 2011. Some of the thunderstorms embedded within the squall line (dark red and orange color) produced high winds, heavy rain, and large hail.

● **FIGURE 14.14** Prefrontal squall-line thunderstorms may form ahead of an advancing cold front as the upper-air flow develops waves downwind from the cold front.

● **FIGURE 14.15** A model describing air motions and precipitation associated with a squall line that has a trailing stratiform cloud layer.

thunderstorm. Slowly rising air within the region keeps the air saturated. Beneath the rising air, the air slowly descends in association with the falling rain.

Strong downdrafts often form to the rear of the squall line, as some of the falling precipitation evaporates and chills the air. The heavy cooler air then descends, dragging some of the surrounding air with it. If the cool air rapidly descends, it may concentrate into a rather narrow band of fast-flowing air called the *rear-inflow jet*, because it enters the storm from the west as shown in Fig. 14.15. Sometimes the rear-inflow jet will bring with it the strong upper-level winds from aloft. Should these winds reach the surface, they rush outward producing damaging *straight-line winds* that may exceed 90 knots (104 mi/hr). (See ● Fig. 14.16.)

As the strong winds rush forward along the ground, they sometimes push the squall line outward so that it appears as a *bow* (or a series of bows) on a radar screen. Such a bow-shaped squall line is called a **bow echo** (see ● Fig. 14.17). Sometimes the rush of strong winds will produce relatively small bows only about 8 to 15 km long (*mini-bows*). If the wind shear ahead of the advancing squall line is strong, much larger bows (over 150 km long) may form, similar to the bow echo in Fig. 14.17. The strongest straight-line winds tend to form near the center of the bow, where the sharpest bending occurs. Tornadoes can form, especially near the left (northern) side of the bow where cyclonic rotation often develops, but they are usually small and short-lived.

● **FIGURE 14.16** A side view of the lower half of a squall-line thunderstorm with the rear-inflow jet carrying strong winds from high altitudes down to the surface. These strong winds push forward along the surface, causing damaging straight-line winds that may reach 100 knots or more.

● **FIGURE 14.17** A Doppler radar image showing an intense squall line in the shape of a bow—called a *bow echo*—moving eastward across Missouri on the morning of May 8, 2009. The strong thunderstorms (red and orange in the image) are producing damaging straight-line winds over a wide area. A region of straight-line winds that extends for a good distance along a squall line is called a *derecho*.

Look at Fig. 14.17 closely and observe that at the northern end of the bow echo the thunderstorms are organizing into a region of cyclonic rotation. In this region, the release of large quantities of latent heat during cloud formation leads to the development of a small area of low pressure that spins counterclockwise. Because the size of the spinning low (called a *vortex*) is relatively small— usually less than 250 km (155 mi) in diameter—the spinning low and its band of clouds is called a **mesoscale convective vortex**. Notice in ● Fig. 14.18 that the vortex becomes more well defined several hours later. Even after the thunderstorms gradually dissipate, the vortex may continue for one or more days, serving as a focus for additional storms.

If straight-line winds gusting to more than 50 knots (58 mi/hr) persist along a path at least 400 km (250 mi) long, the windstorm is called a **derecho** (day-ray-sho), after the Spanish word for "straight ahead." In an average year, about 20 derechos occur in the United States. Typically, derechos form in the early evening and last throughout the night. An especially powerful derecho roared through New York State during the early morning of July 15, 1995, where it blew down an estimated 1 million trees in Adirondack State Park. Damage from Ontario to New England totaled more than $500 million. Another extremely strong derecho swept from the Midwest into the Washington, D.C., area on June 29, 2012. With winds gusting to more than 70 knots (80 mi/hr), the derecho killed 22 people and left more than 4 million people without power, in some places for days.

It is common for the damaging effects of a derecho to be attributed to a tornado. The degree of damage from a derecho can be extensive, just as with a tornado.* However, with a derecho, most debris is blown in one direction, generally over a wide area. Debris with a tornado is usually thrown in many directions along a narrow swath with embedded circular features.

Squall lines are one type of convective phenomenon called a *mesoscale convective system* (MCS). Squall lines come under this heading because they are driven by convective processes and because they are mesoscale (middle scale) in size. Mesoscale convective systems are organized thunderstorms that can take on a variety of configurations, from the elongated squall line, to the circular mesoscale convective vortex, to the much larger *mesoscale convective complex* described in the next section.

---

*To refresh your memory on the different sizes of atmospheric wind systems, look at Fig. 9.2 on p. 229.

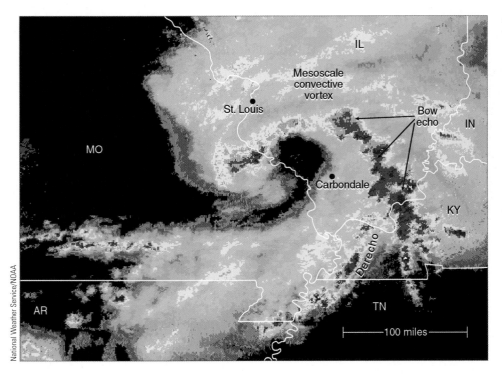

● **FIGURE 14.18** Doppler radar image showing that the bow echo in Fig. 14.17 has developed into a *mesoscale convective vortex* by noon on May 8, 2009. Strong, straight-line winds are still occurring with severe thunderstorms, as Carbondale, Illinois, reported an unofficial wind gust of 106 mi/hr.

Mesoscale Convective Complexes Where conditions are favorable for convection, a number of individual multicell thunderstorms may occasionally grow in size and organize into a large circular convective weather system. These convectively driven systems, called **mesoscale convective complexes (MCCs)**, are quite large, sometimes as much as 1000 times larger than an individual ordinary cell thunderstorm. In fact, they often span an area in excess of 100,000 square kilometers, large enough to cover some entire states (see ● Fig. 14.19).

Within the MCCs, the individual thunderstorms apparently work together to generate a large weather system that appears on satellite to be relatively circular. MCCs typically last at least 6 hours, and sometimes more than 12 hours. Thunderstorms within MCCs support the growth of new thunderstorms as well as a region of widespread precipitation. MCCs are beneficial in that they provide a significant portion of the growing season rainfall over much of the corn and wheat belts of the United States. However, MCCs can also produce a wide variety of severe weather, including hail, high winds, destructive flash floods, and tornadoes.

Mesoscale convective complexes tend to form during the summer in regions where the upper-level winds are weak, which is often beneath a ridge of high pressure. If a weak cold front should stall beneath the ridge, surface heating and moisture may be sufficient to generate thunderstorms on the cool side of the front. Moisture from the south is often brought into the system by a low-level jet stream typically found within about 1500 m (5000 ft) of the surface. In addition, the low-level jet can provide shearing so that multicell storms can form within the complex.

Most MCCs reach their maximum intensity in the early morning hours, which is partly due to the fact that the low-level jet reaches its maximum strength late at night or in the early morning. Moreover, at night, the cloud tops cool rapidly by emitting infrared energy to space. Gradually, the atmosphere destabilizes as a vast amount of latent heat is released in the lower and middle part of the clouds. Within the multicell storm complex new thunderstorms form as older ones dissipate. With only weak upper-level winds, most MCCs move very slowly toward the east or southeast while dumping torrential rain.

SUPERCELL THUNDERSTORMS In a region where there is strong vertical wind shear (speed and/or directional shear), a thunderstorm may form in such a way that the outflow of cold air from the downdraft never undercuts the updraft. In such a storm, the wind shear may be so strong as to create horizontal spin, which, when tilted into the updraft, causes it to rotate. An intense, long-lasting thunderstorm with a single violently rotating updraft is called a **supercell**.* As we will see later in this chapter, it is the rotating aspect of the supercell that can lead to the formation of tornadoes.

● Figure 14.20 shows a supercell with a tornado. The internal structure of a supercell is organized in such a way that the storm may maintain itself as a single entity for hours. Storms of this type are capable of generating an updraft that may exceed 90 knots (104 mi/hr) and may produce damaging surface winds and strong tornadoes. In some cases, the top of the storm may extend to as high as 18 km (60,000 ft) above the surface, and its width may exceed 40 km (25 mi). (See ● Fig. 14.21.)

The largest hail observed on Earth forms within supercells because of updrafts that are both wide and intense. In rare cases, such hailstones can grow to the size of grapefruits or even larger. One hailstone that fell in Vivian, South Dakota, on July 23, 2010, weighed nearly 2 pounds and measured 8 inches across, which made it the heaviest and widest hailstone on record. The broad and powerful updrafts within supercells can keep hailstones airborne for relatively long periods, which allows many water droplets to accumulate and freeze on them. Once the hailstones are large enough, they may fall out of the bottom of the cloud with the downdraft, or the violent spinning updraft may whirl them out the side of the cloud or even from the base of the anvil. Aircraft have actually encountered hail in clear air several kilometers from a storm. ● Fig. 14.22 shows a cross section through a supercell thunderstorm producing large hail.

Although no two supercells are exactly alike, for convenience they are often divided into three types. *Classic supercells* produce heavy rain, large hail, high surface winds, and the majority of tornadoes. The classic supercell serves as an excellent model for all supercells and is the one normally shown in diagrams. When a supercell becomes dominated by heavy precipitation, strong downdrafts (downbursts), and large hail, it is referred to as an *HP (high precipitation)* supercell. Tornadoes in an HP supercell may be wrapped in heavy rain and difficult to see. On the other hand, tornadoes and cloud features are often quite visible in an *LP (low precipitation)* supercell (see Fig. 14.20). Although LP supercells tend to produce

● **FIGURE 14.19** An enhanced infrared satellite image showing the cold cloud tops (dark red and orange colors) of a mesoscale convective complex extending from central Kansas across western Missouri. This organized mass of multicell thunderstorms brought hail, heavy rain, and flooding to this area.

---

*Smaller thunderstorms that occur with rotating updrafts are referred to as *mini supercells*.

● **FIGURE 14.20** A tornado descends from beneath a low-precipitation (LP) supercell thunderstorm in eastern Colorado on June 10, 2010.

● **FIGURE 14.21** A high-precipitation (HP) supercell bearing heavy rain, large hail, strong winds, and lightning plows across eastern Nebraska.

● **FIGURE 14.22** A vertical cross section of reflectivities observed by Doppler radar in a supercell thunderstorm that has a top exceeding 40,000 ft and hail (shown as red and purple inside the storm) as large as golf balls in the El Paso, Texas, area on September 16, 2009. With damage of $150 million, it was the most destructive hailstorm in the city's history.

little rain, they are still capable of producing large hail as well as tornadoes. The rotation of an LP supercell is often visible in a bell-shaped central tower, with a corkscrew-like pattern along its sides.

A model of a classic supercell with many of its features is given in ● Fig. 14.23. In the diagram, we are viewing the storm from the southeast, and the storm is moving from southwest to northeast. The rotating air column on the south side of the

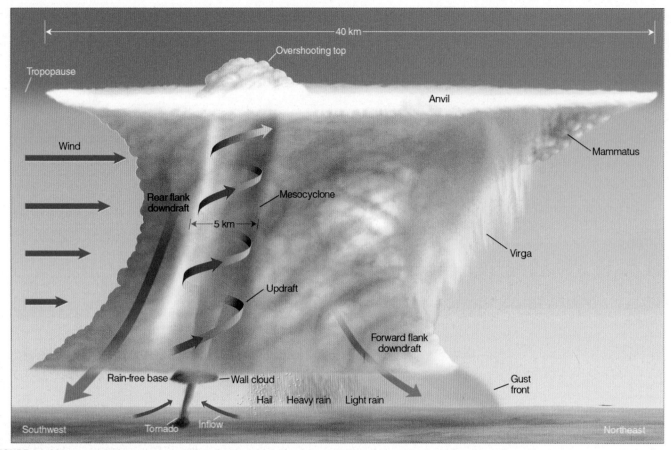

● **FIGURE 14.23** Some of the features associated with a classic tornado-breeding supercell thunderstorm as viewed from the southeast. The storm is moving to the northeast.

storm, usually 5 to 10 kilometers across, is called a **mesocyclone** (meaning "mesoscale cyclone"). The rotating updraft associated with the mesocyclone is so strong that precipitation cannot fall through it. This situation produces a rain-free area (called a *rain-free base*) beneath the updraft. Strong southwesterly winds aloft typically blow the precipitation northeastward. Notice that large hail, having remained in the cloud for some time, usually falls just north of the updraft. The heaviest rain occurs just north of the falling hail, with the lighter rain falling in the northeast quadrant of the storm. If humid low-level air is drawn into the updraft, a rotating cloud, called a **wall cloud**, may descend from the base of the storm as shown in Fig. 14.23 and ● Fig. 14.24.

We can obtain a better picture of how vertical wind shear plays a role in the development of supercell thunderstorms by observing ● Fig. 14.25. The illustration represents atmospheric conditions often observed during the spring over the Central Plains. At the surface, we find an open-wave middle-latitude cyclone with cold, dry air moving in behind a cold front, and warm humid air pushing northward from the Gulf of Mexico behind a warm front. Above the warm surface air, a wedge or "tongue" of warm, moist air is streaming northward. It is in this region we find a relatively narrow band of strong winds, sometimes exceeding 50 knots, called the low-level jet. Directly above the moist layer is a wedge of cooler, drier air moving in from the southwest. Higher up, at the 500-mb level, a trough of low pressure exists to the west of the surface low. At the 300-mb level, the

polar front jet stream swings over the region, often with an area of maximum wind (a jet streak) above the surface low. At this level, the jet stream provides an area of divergence that enhances surface convergence and rising air. The stage is now set for the development of supercell thunderstorms.

The light yellow area on the surface map (Fig. 14.25) shows where supercells are likely to form. They tend to form in this region because (1) the position of cold air above warm air produces a conditionally unstable atmosphere and because (2) strong vertical wind shear induces rotation.

Rapidly increasing wind speed from the surface up to the low-level jet provides strong vertical wind speed shear. Within this region, wind shear causes the air to spin about a horizontal axis. You can obtain a better idea of this spinning by placing a pen (or pencil) in your left hand, parallel to the edge of the table in front of you. Now take your right hand and push it over the pen away from you. The pen rotates much like the air rotates. If you tilt the spinning pen into the vertical by lifting its left side, the pen rotates counterclockwise from the perspective of looking down on it. A similar situation occurs with the rotating air. As the spinning air rotates counterclockwise about a horizontal axis, an updraft from a developing thunderstorm can draw the spinning air into the cloud, causing the updraft to rotate. It is this rotating updraft that is characteristic of all supercells. The increasing wind speed with height up to the 300-mb level, coupled with the changing wind direction with height from

● **FIGURE 14.24** A wall cloud develops above the Kansas prairie on April 29, 2010.

more southerly at low levels to more westerly at high levels, further induces storm rotation.*

Ahead of the advancing cold front, we might expect to observe many supercells forming as warm, conditionally unstable air rises from the surface. Often, however, this is not the case, as the atmospheric conditions that promote the formation of large supercell thunderstorms tend to prevent many smaller ones from forming. To see why, we need to examine the vertical profile of temperature and moisture—a sounding—in the warm air ahead of the advancing cold front.

● Figure 14.26 shows a typical sounding of temperature and dew point in the warm air before supercells form. From the surface up to 800 mb—in a layer of air perhaps 2000 m (6000 ft) thick—the air is warm, very humid, and conditionally unstable. At 800 mb, a shallow inversion (or simply, a very stable layer) acts like a *cap* (or a *lid*) on the moist air below. Above the inversion, the air is cold and much drier. This air is also conditionally unstable, as the temperature drops at just about the dry adiabatic rate (10°C/1000 m). The cooling of this upper layer is due, mainly, to cold air moving in from the west. Cold, dry, unstable air sitting above a warm, humid layer produces a type of atmospheric instability called *convective instability*, which means that the atmosphere will destabilize even more if a layer of air is somehow forced to rise. (See Chapter 6, p. 152, for more information on this

---

*As we will see in the next chapter, it is this rotation that sets the stage for tornado development.

● **FIGURE 14.25** Conditions leading to the formation of severe thunderstorms, and especially supercells. The area in yellow shows where severe thunderstorms are likely to form.

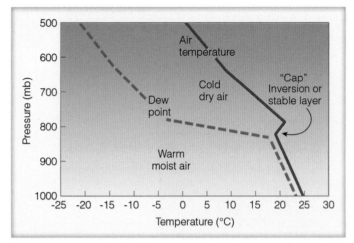

● **FIGURE 14.26** A typical sounding of air temperature and dew point that frequently precedes the development of supercell thunderstorms. The thickness of the warm, moist air from the surface up to the cap at 800 mb is usually on the order of about 2000 m or 6000 ft. (An actual sounding on the day that a severe thunderstorm tore through Moore, Oklahoma, May 20, 2013, is shown in Fig. 13.4 on p. 353.)

topic.) In addition, the warm, humid air beneath the capping stable layer represents potential energy for the thunderstorm. This potential energy is converted into kinetic energy when parcels of air are lifted and form into clouds.

The lifting of warm surface air can occur at the frontal zones, but the air may also begin to rise anywhere in the region of warm air when the surface air heats up during the day. However, there may be a persistent inversion above the layer of surface mixing. This layer, called a **cap**, serves as a lid on rising thermals, often allowing only small cumulus clouds to form. Sometimes the cap is strong enough to prevent any showers or thunderstorms from developing through the day. At other times, the cap may be absent or weak, allowing many thunderstorms to develop but reducing the chance of an isolated supercell. The most violent supercells tend to occur when a cap is just strong enough to prevent a large number of thunderstorms from developing early in the day. As the day progresses (and the surface air heats even more), rising air may break through the cap at isolated places. If this happens, clouds build rapidly—sometimes explosively—as the moist air is vented upward through the opening. When the surface air is finally able to puncture the inversion, one or more supercells may quickly develop to great height. When a jet streak associated with the upper-level jet stream (at the 300-mb level) is present, it can enhance the lifting and strengthen the possibility that a supercell will form.

Violent thunderstorms have very strong updrafts. A measure of how much energy is available to produce these updrafts is the *convective available potential energy (CAPE)*. We know that thunderstorms may begin as parcels of air rise from the surface, eventually becoming saturated, and their moisture condenses into clouds. CAPE is a measure of how rapidly an air parcel will rise inside a cloud (its positive buoyancy) when the parcel becomes warmer than the air surrounding it. The higher the value of CAPE, the more likely a supercell will form. For example,

a value of CAPE ranging from 0 to 500 joules*/kg means that there is only a marginal chance for the development of strong thunderstorms, whereas values greater than 3500 indicate that the environment is ripe for strong convection (strong updrafts) and the formation of supercell thunderstorms. (We will look at CAPE again in the next chapter when we examine the forecasting of severe thunderstorms and tornadoes.)

Most thunderstorms move roughly in the direction of the winds in the middle troposphere. However, most supercell storms in the Northern Hemisphere are *right-movers*; that is, they move to the right of the steering winds aloft. These right-movers tend to move about 30 degrees to the right of the mean wind in the middle troposphere. This motion occurs because the rapidly rising air of the storm's updraft interacts with increasing horizontal winds that change direction with height (from more southerly to more westerly) in such a way that vertical pressure gradients are able to generate new updrafts on the right side of the storm. Hence, as the supercell moves along, its development is focused toward the right of the winds aloft.** This rightward motion can actually increase the amount of wind shear relative to the storm, which can aid its intensification.

THUNDERSTORMS AND THE DRYLINE Thunderstorms may form along or just east of a boundary called a *dryline*. Recall from Chapter 11 that the dryline represents a narrow zone where there is a sharp horizontal change in moisture. In the United States, drylines are most frequently observed in the western half of Texas, Oklahoma, and Kansas. In this region, drylines occur most often during spring and early summer, where they are observed about 40 percent of the time.

● Figure 14.27 shows springtime weather conditions that can lead to the development of a dryline and intense thunderstorms. The map shows a developing mid-latitude cyclone with a cold front, a warm front, and three distinct air masses. Behind the cold front, cold, dry continental polar (cP) air or modified cool, dry Pacific air pushes in from the northwest. In the warm air, ahead of the cold front, hot, dry continental tropical (cT) air moves in from the southwest. Farther east, warm, humid maritime tropical (mT) air sweeps northward from the Gulf of Mexico. The dryline is the north–south oriented boundary that separates the warm, dry air and the warm, humid air.

Along the cold front—where cold, dry air replaces warm, dry air—there is insufficient moisture for thunderstorm development. The surface moisture boundary lies along the dryline. Because the Central Plains of North America are elevated to the west, some of the hot, dry air from the southwest is able to ride over the slightly cooler, more humid air from the Gulf. This condition sets up a potentially unstable atmosphere just east of the dryline. Converging surface winds in the vicinity of the dryline, coupled with

---

*Recall from Chapter 2 that a joule is a unit of energy where 1 joule equals 0.239 calories.

**Some thunderstorms move to the left of the steering winds aloft. This movement may happen as a thunderstorm splits into two storms, with the northern half of the storm often being a left-mover and the southern half a right-mover. Typically the right-mover intensifies and the left-mover weakens.

upper-level outflow, may result in rising air and the development of thunderstorms. As thunderstorms form, the cold downdraft from inside the storm may produce a blast of cool air that moves along the ground as a gust front and initiates the uplift necessary for generating new (possibly more severe) thunderstorms.

BRIEF REVIEW

In the last several sections, we examined different types of thunderstorms. Listed below for your review are important concepts we considered:

- All thunderstorms need three basic ingredients: (1) moist surface air; (2) a conditionally unstable atmosphere; and (3) a mechanism "trigger" that forces the air to rise.

- Ordinary cell (air mass) thunderstorms tend to form where warm, humid air rises in a conditionally unstable atmosphere and where vertical wind shear is weak. They are usually short-lived and go through their life cycle of growth (cumulus stage), maturity (mature stage), and decay (dissipating stage) in less than an hour. They rarely produce severe weather.

- An ordinary cell thunderstorm dies because its downdraft falls into the updraft, which cuts off the storm's fuel supply.

- As wind shear increases (and the winds aloft become stronger), multicell thunderstorms are more likely to form as the storm's updraft rides up and over the downdraft. The tilted nature of the storm allows new cells to form as old ones die out.

- Multicell storms often form as a complex of storms, such as the squall line (a long line of thunderstorms that form along or out ahead of a frontal boundary) and the mesoscale convective complex (a large circular cluster of thunderstorms).

- Supercell thunderstorms are long-lasting violent thunderstorms, with a single rotating updraft that forms in a region of strong vertical wind shear. A rotating supercell is more likely to develop when (a) the winds aloft are strong and change direction from southerly at the surface to more westerly aloft and (b) a low-level jet exists just above Earth's surface.

- Although supercells are likely to produce severe weather, such as strong surface winds, large hail, heavy rain, and tornadoes, not all do.

- A gust front, or outflow boundary, represents the leading edge of cool air that originates inside a thunderstorm, reaches the surface as a downdraft, and moves outward away from the thunderstorm.

- Strong downdrafts from a thunderstorm, called downbursts (or microbursts if the downdrafts are smaller than 4 km), have been responsible for several airline crashes, because upon striking the surface, these winds produce extreme horizontal wind shear.

- A derecho is a long-lived straight-line wind produced by strong downbursts from intense thunderstorms that often appear as a bow (bow echo) on a radar screen.

- Intense thunderstorms often form along a dryline, a narrow zone that separates warm, dry air from warm, humid air.

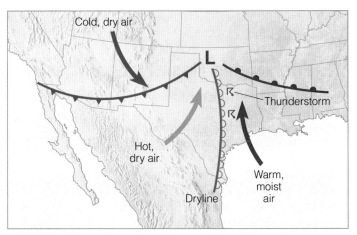

● **FIGURE 14.27** Surface conditions that can produce a dryline with intense thunderstorms.

Thunderstorms and Flooding

Intense thunderstorms can be associated with floods and **flash floods**—floods that develop rapidly with little or no advance warning. Such flooding often results when thunderstorms stall or move very slowly, causing heavy rainfall over a relatively small area. Such flooding occurred over parts of New England and the mid-Atlantic states during June 2006, when a stationary front stalled over the region and moist tropical air, lifted by the front, produced thunderstorms and heavy rainfall that caused extensive flooding and damage to thousands of homes. Flooding may also occur when thunderstorms move quickly, but keep passing over the same area, a phenomenon called *training* (similar to railroad cars, one after another, passing over the same tracks). In recent years, flooding and flash floods in the United States (apart from flooding directly related to hurricanes, such as Katrina in 2005) have claimed an average of more than 100 lives a year and have accounted for untold property and crop damage. An example of a terrible flash flood that took the lives of more than 140 people is given in Focus section 14.1.

In some areas, flooding occurs primarily in the spring when heavy rain and melting snow cause rivers to overflow their banks. For example, during March 1997, heavy downpours over the Ohio River Valley caused extensive flooding that forced thousands from their homes along rivers and smaller streams in Ohio, Kentucky, Tennessee, and West Virginia. One month later, heavy rain coupled with melting snow caused the Red River to overflow its banks, inundating 75 percent of the city of Grand Forks, North Dakota. These spring rainstorms may or may not feature extensive thunder and lightning. Flooding also occurs in the summer and autumn with tropical storms that deposit torrential rains over an extensive area.

If thunderstorms bring repeated heavy rain to a region for days or weeks, the result can be one or more *river floods*. A river flood occurs when a major river system rises slowly but ends up flooding a large area, whereas a flash flood may devastate a smaller area in minutes to hours. During the summer of 1993,

The Terrifying Flash Flood in the Big Thompson Canyon

Saturday, July 31, 1976, was a banner day in Colorado, as residents and visitors were celebrating the weekend of the state's centennial. Meteorologically speaking, though, July 31 was like any other summer day in the Colorado Rockies, as small cumulus clouds with flat bases and dome-shaped tops began to develop over the eastern slopes near the Big Thompson and Cache La Poudre rivers. At first glance, there was nothing unusual about these clouds, as almost every summer afternoon they form along the warm mountain slopes. Normally, strong upper-level winds push them over the plains, causing rainshowers of short duration. But the cumulus clouds on this day were different. For one thing, their bases were much lower than usual, indicating that the southeasterly surface winds were bringing in a great deal of moisture. Also, their tops were somewhat flattened, suggesting that an inversion (or "cap") aloft was stunting their growth. But these harmless-looking clouds gave no clue that later that evening in the Big Thompson Canyon more than 140 people would lose their lives in a terrible flash flood.

By late afternoon, a few of the cumulus clouds were able to break through the cap. Fed by moist southeasterly winds, these clouds soon developed into gigantic multicell thunderstorms with tops exceeding 18 km (60,000 ft). By early evening, these same clouds were producing amazing downpours in the mountains.

In the narrow canyon of the Big Thompson River, some places received as much as 30.5 cm (12 in.) of rain in the four hours between 6:30 p.m. and 10:30 p.m. local time. This is an incredible amount of precipitation, considering that the area normally receives about 40.5 cm (16 in.) for an entire year. The heavy downpours turned small creeks into raging torrents, and the Big Thompson River was quickly filled to capacity. Where the canyon narrowed, the river overflowed its banks and water covered the road. The relentless pounding of water caused the road to collapse.

Soon cars, tents, mobile homes, resort homes, and campgrounds were being claimed by the river (see ● Fig. 1), affecting many hundreds of people who had planned to spend the weekend celebrating the state centennial. Where the debris entered a narrow constriction, it became a dam. Water backed up behind it, then broke through, causing a wall of water to rush downstream.

● **FIGURE 1** This car is one of more than 400 destroyed by floodwaters in the Big Thompson Canyon on July 31, 1976.

● Figure 2 shows the weather conditions during the evening of July 31, 1976. A cool front moved through earlier in the day and remained south of Denver. The weak inversion layer associated with the front kept the cumulus clouds from building to great heights earlier in the afternoon. However, the strong southeasterly flow behind the cool front pushed unusually moist air upslope along the mountain range. Heated from below, the conditionally unstable air eventually punctured the inversion and developed into a huge multicell thunderstorm complex that remained nearly stationary for several hours due to the weak southerly winds aloft. The deluge may have deposited 19 cm (7.5 in.) of rain on the main fork of the Big Thompson River in about one hour. Of the approximately 2000 people in the canyon that evening, more than 140 lost their lives, and property damage exceeded $35.5 million. Nearly all of the deaths were from blunt trauma due to the raging waters and debris, rather than from drowning. As a result of this disaster, signs have been placed in major canyons encouraging people to leave their homes or vehicles and "climb to safety" if they encounter flood waters.

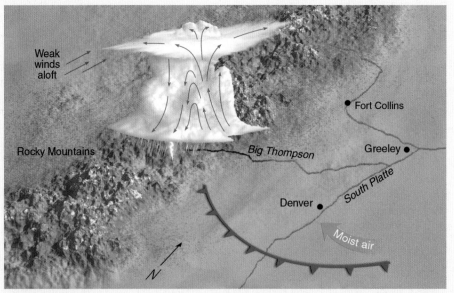

● **FIGURE 2** Weather conditions that led to the development of intense multicell thunderstorms that remained nearly stationary over the Big Thompson Canyon in the Colorado Rockies. The arrows within the thunderstorm represent air motions.

thunderstorm after thunderstorm rumbled across the upper Midwest, causing the worst flooding ever recorded in that part of the United States. What began as a wetter-than-normal winter and spring for most of the upper Midwest turned into "The Great Flood of 1993" by the end of July. In mid-June, thunderstorms began to occur almost daily along a persistent frontal boundary that stretched across the upper Midwest. The front (which remained nearly stationary for days on end) was positioned beneath the polar jet stream, which was situated much farther south than usual for this time of year (see ● Fig. 14.28).* The jet stream provided pockets of upper-level divergence for the development of weak surface waves that rippled along the front, while converging surface air provided uplift for thunderstorm growth.

Fed by warm, humid air from the Gulf of Mexico, thunderstorms almost daily rolled through an area that stretched eastward from Nebraska and South Dakota into Minnesota and Wisconsin, and southward into Iowa, Illinois, and Missouri. Torrential rains from these storms quickly saturated the soil, and soon runoff began to raise the water level in creeks and rivers. By the end of June, communities in the northern regions of the Mississippi River Valley were experiencing flooding.

As the thunderstorms continued into July, city after city was claimed by the rising waters (see ● Fig. 14.29). Between April and July, many areas received twice their normal rainfall, and rivers continued to crest well above flood stage through July. By the time the water began to recede in August, more than 60 percent of the levees along the Mississippi River had been destroyed, and an area larger than the state of Texas (the blue-shaded area in Fig. 14.28) saw significant amounts of land flooded. Estimates are that $6.5 billion in crops was lost as millions of acres of valuable farmland were inundated by flood waters. The worst flooding residents of this area had ever seen, much of it along

---

*As a note, the position of the jet stream caused the weather to be cooler than normal in the Pacific Northwest and warmer than normal in the East. While the Midwest was deluged with rain, the southeastern section of the United States was experiencing an extensive dry period.

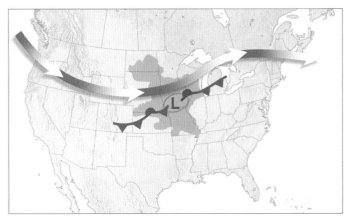

● **FIGURE 14.28** The heavy arrow represents the average position of the upper-level jet stream from mid-June through July 1993. The jet stream helped fuel thunderstorms that developed in association with a stationary front that seemed to oscillate back and forth over the region as an alternating cold front and warm front. The "L" marks the center of a frontal wave that is moving along the front. Many of the thunderstorms that formed in conjunction with this pattern were severe, and over a period of weeks produced "The Great Flood of 1993." Counties within the blue-shaded area were declared federal disaster areas due to flooding.

the Mississippi River, took 45 lives, damaged or destroyed 45,000 homes, and forced the evacuation of 74,000 people. Parts of the Mississippi Valley were again struck by catastrophic flooding in the spring of 2008, when intense thunderstorms fell atop ground that had received heavy winter snow. In much of eastern Iowa, water levels topped those seen in the Great Flood of 2003. Some of the worst flooding was in Cedar Rapids, where most of the downtown area was inundated.

Sometimes a region may experience both flash flooding and river flooding. This was the case across much of Texas and Oklahoma in May 2015. Many rivers overflowed for days, as persistent heavy rain led to the wettest month on record in both states. On the night of May 24, thunderstorms dumped more than 30 cm (12 in.) of rain in the Texas Hill Country, turning the

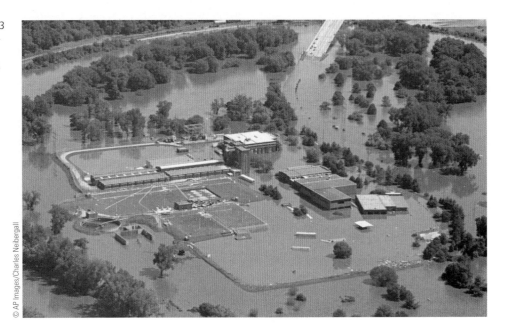

● **FIGURE 14.29** Flooding during the summer of 1993 covered a vast area of the upper Midwest. Here, floodwaters near downtown Des Moines, Iowa, during July 1993 inundate buildings of the Des Moines waterworks facility. Flood-contaminated water left 250,000 people without drinking water.

© AP Images/Charles Neibergall

normally calm Blanco River into a raging torrent that destroyed hundreds of homes. At least 31 people were killed in Texas and Oklahoma. Catastrophic flash floods and river floods also struck northeast Colorado in September 2013. Several days of rainfall included periods when showers and thunderstorms passed over the same areas repeatedly for hours, producing flash flooding and huge rainfall totals. In one day, Boulder, Colorado, reported 23 cm (9.08 in.), almost doubling its previous 24-hour record. The waters eventually merged to cause record flooding along the South Platte River from northeast Colorado into Nebraska. In all, the multiday event produced an estimated $2 billion in damage, caused at least 10 deaths, and destroyed more than 1800 homes and 750 businesses, as well as some 200 miles of state highway.

Distribution of Thunderstorms

It is estimated that more than 50,000 thunderstorms occur each day throughout the world. Hence, over 18 million occur annually. The combination of warmth and moisture make equatorial landmasses especially conducive to thunderstorm formation.

Thunderstorms occur here on about one out of every three days. Thunderstorms are also prevalent over water along the intertropical convergence zone, where the low-level convergence of air helps to initiate uplift. The heat energy liberated in these storms helps Earth maintain its heat balance by distributing heat poleward (see Chapter 10). Thunderstorms are much less prevalent in dry climates, such as the polar regions and the desert areas dominated by subtropical highs.

● Figure 14.30 shows the average annual number of days having thunderstorms in various parts of the United States and southern Canada. Notice that they occur most frequently in the southeastern United States along the Gulf Coast, with a maximum in Florida. The region with the fewest thunderstorms is the Pacific coastal and interior valleys.

In many areas, thunderstorms form primarily in summer during the warmest part of the day when the surface air is most unstable. There are some exceptions, however. During the summer in the valleys of central and southern California, dry, sinking air produces an inversion that inhibits the development of towering cumulus clouds. In these regions, thunderstorms are most frequent in winter and spring, particularly when cold, moist, conditionally unstable air aloft moves over moist, mild surface air. The surface air remains relatively warm because of its proximity to the ocean.

On many summer days, thunderstorms develop just east of the Rocky Mountains in the afternoon, then intensify into mesoscale convective systems in the evening as they move eastward across the central Great Plains. Often these thunderstorms congeal into large mesoscale convective systems and continue through the night. One of the most common times to experience thunder and lightning in parts of Iowa and Missouri is between

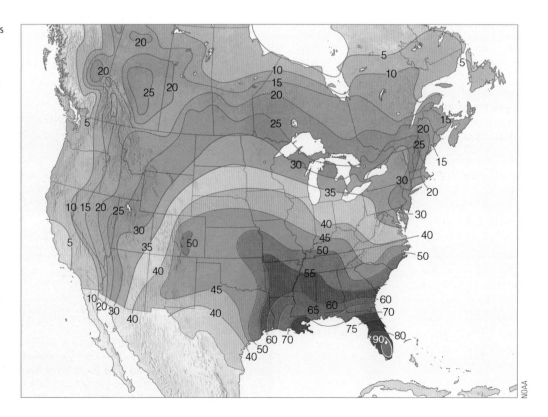

● **FIGURE 14.30** The average number of days each year on which thunderstorms are observed throughout the United States and southern Canada. (Due to the scarcity of data, the number of thunderstorms is underestimated in the mountainous far west.)

● **FIGURE 14.31** The average number of days each year on which hail is observed throughout the United States and southern Canada.

midnight and dawn. Such storms can be fueled by a southerly low-level jet that often strengthens after sunset, bringing humid air northward and helping trigger convergence and uplift of surface air. As the thunderstorms build, their tops cool by radiating infrared energy to space. This cooling process tends to destabilize the atmosphere around the storms, making it even more suitable for nighttime thunderstorm development.

At this point, it is interesting to compare Fig. 14.30 and ● Fig. 14.31. Notice that, even though the greatest frequency of thunderstorms is near the Gulf Coast, the greatest frequency of hailstorms is over the western Great Plains. One reason for this situation is that conditions over the Great Plains are more favorable for the development of severe thunderstorms, and especially supercells that have strong updrafts capable of keeping hailstones suspended within the cloud for a long time so that they can grow to an appreciable size before plunging to the ground. We also find that, in summer along the Gulf Coast, a thick layer of warm, moist air extends upward from the surface. Most hailstones falling into this layer will melt before reaching the ground.* In contrast, when a thunderstorm is located on the high plains east of the Rocky Mountains, a much larger portion of the storm will typically be located above the freezing level, so hail has a much better chance of occurring.

Now that we have looked at the development and distribution of thunderstorms, we are ready to examine an interesting, though yet not fully understood, aspect of all thunderstorms—lightning.

---

*The formation of hail is described in Chapter 7 on p. 186.

Lightning and Thunder

Lightning is simply a discharge of electricity, a giant spark, which usually occurs in mature thunderstorms. Lightning may take place within a cloud, from one cloud to another, from a cloud to the surrounding air, or from a cloud to the ground (see ● Fig. 14.32). (The majority of lightning strikes occur within the cloud, while only about 20 percent or so occur between cloud and ground.) The lightning stroke can heat the air through which it travels to an incredible 30,000°C (54,000°F), which is five times hotter than the surface of the sun. This extreme heating causes

● **FIGURE 14.32** The lightning stroke can travel in a number of directions. It can occur within a cloud, from one cloud to another cloud, from a cloud to the air, or from a cloud to the ground. Notice that the cloud-to-ground lightning can travel out away from the cloud, then turn downward, striking the ground many miles from the thunderstorm. When lightning behaves in this manner, it is often described as a *"bolt from the blue."*

the air to expand explosively, thus initiating a shock wave that becomes a booming sound wave—called **thunder**—that travels outward in all directions from the flash.

A sound occasionally mistaken for thunder is the **sonic boom**. Sonic booms are produced when an aircraft exceeds the speed of sound at the altitude at which it is flying. (The speed of sound at sea level is 661 knots or 761 mi/hr, but it decreases gradually with height.) The aircraft compresses the air, forming a shock wave that trails out as a cone behind the aircraft. Along the shock wave, the air pressure changes rapidly over a short distance. The rapid pressure change causes the distinct boom. (Exploding fireworks generate a similar shock wave and a loud bang.)

HOW FAR AWAY IS THE LIGHTNING? START COUNTING
When you see a flash of lightning, how can you tell how far away (or how close) it is? Light travels so fast that you see light instantly after a lightning flash. But the sound of thunder, traveling at only about 330 m/sec (1100 ft/sec), takes much longer to reach your ear. If you start counting seconds from the moment you see the lightning until you hear the thunder, you can determine how far away the stroke is. Because it takes sound about 5 seconds to travel 1 mile, if you see lightning and hear the thunder 15 seconds later, the lightning stroke (and the thunderstorm) is about 3 miles away.

When the lightning stroke is very close—on the order of 100 m (330 ft) or less—thunder sounds like a clap or a crack followed immediately by a loud bang. When it is farther away, it often rumbles. The rumbling can be due to the sound emanating from different areas of the stroke (see ● Fig. 14.33). Moreover, the rumbling is accentuated when the sound wave reaches an observer after having bounced off obstructions, such as hills and buildings.

In some instances, lightning is seen but no thunder is heard. Does this mean that thunder was not produced by the lightning? Actually, there is thunder, but the atmosphere has refracted (bent) and attenuated the sound waves, making the thunder inaudible. Sound travels faster in warm air than in cold air.* Because

---

*The speed of sound in calm air is equal to $20\sqrt{T}$, where T is the air temperature in Kelvins.

thunderstorms form in a conditionally unstable atmosphere, where the temperature normally drops rapidly with height, a sound wave moving outward away from a lightning stroke will often bend upward, away from an observer at the surface. Consequently, an observer closer than about 5 km (3 mi) to a lightning stroke usually will hear thunder, whereas an observer about 8 km (5 mi) away usually will not.

However, even when a viewer is as close as several kilometers to a lightning flash, thunder may not be heard. For one thing, the complex interaction of sound waves and air molecules tends to attenuate the thunder. In addition, turbulent eddies of air less than 50 meters in diameter scatter the sound waves. Hence, when thunder from a low-energy lightning flash travels several miles through turbulent air, it may become inaudible.

Earlier, we learned that lightning occurs with mature thunderstorms. But lightning may also occur in snowstorms, in dust storms, in the gas cloud of an erupting volcano, and on very rare occasions in nimbostratus clouds. Lightning may also shoot from the top of thunderstorms into the upper atmosphere. More on this topic is given in Focus section 14.2.

What causes lightning? The normal fair weather electric field of the atmosphere is characterized by a negatively charged surface and a positively charged upper atmosphere. For lightning to occur, separate regions containing opposite electrical charges must exist within a cumulonimbus cloud. Exactly how this charge separation comes about is not fully understood; however, many theories have tried to account for it.

ELECTRIFICATION OF CLOUDS
One theory proposes that clouds become electrified when graupel (small ice particles called *soft hail*) and hailstones fall through a region of supercooled liquid droplets and ice crystals. As liquid droplets collide with a hailstone, they freeze on contact and release latent heat. This process keeps the surface of the hailstone warmer than that of the surrounding ice crystals. When the warmer hailstone comes in contact with a colder ice crystal, an important phenomenon occurs: *There is a net transfer of positive ions (charged molecules) from the warmer object to the colder object.*

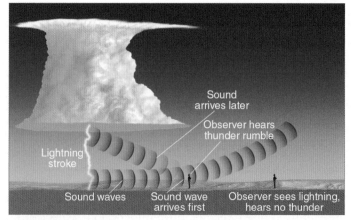

● **FIGURE 14.33** Thunder travels outward from the lightning stroke in the form of waves. If the sound waves from the lower part of the stroke reach an observer before the waves from the upper part of the stroke, the thunder appears to rumble. If the sound waves bend upward away from an observer, the lightning stroke may be seen, but the thunder will not be heard.

ELVES in the Atmosphere

For many years airline pilots reported seeing strange bolts of light shooting upward, high above the tops of intense thunderstorms. These faint, mysterious flashes did not receive much attention, however, until they were first photographed in 1989. Photographs from sensitive, low-light-level cameras on board jet aircraft revealed that the mysterious flashes were actually a colorful display called *red sprites* and *blue jets*, which seemed to dance above the clouds.

Sprites are massive, but dim, light flashes that appear directly above an intense thunderstorm system (see ● Fig. 3). Usually red, and lasting but a few thousandths of a second, sprites tend to form almost simultaneously with lightning in the cloud below and with severe thunderstorms that have positive cloud-to-ground lightning strokes. (Most cloud-to-ground lightning is negative.) Although it is not entirely clear how they form, the thinking now is that sprites form when positive lightning disrupts the atmosphere's electrical field in such a way that charged particles in the upper atmosphere are accelerated downward toward the thunderstorm and upward to higher levels in the atmosphere.

Blue jets usually dart upward in a conical shape from the tops of thunderstorms that are experiencing vigorous lightning activity, although they do not seem to be directly caused by lightning (Fig. 3). Although faint and

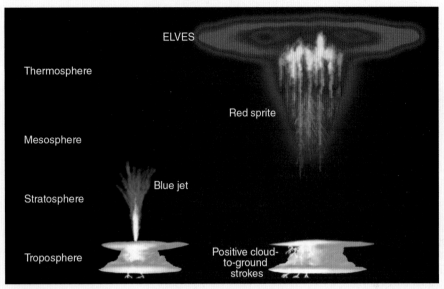

● **FIGURE 3** Various electrical phenomena observed in the upper atmosphere.

very brief, lasting only a fraction of a second, blue jets can occasionally be seen with the naked eye. The color of blue jets appears to be related to emissions produced by ionized atmospheric nitrogen. They are not well understood, but appear to transfer large amounts of electrical energy into the upper atmosphere.

The acronym *ELVES* is from *Emissions of Light and Very Low Frequency Perturbations due to Electromagnetic Pulse Sources*. ELVES, as illustrated in Fig. 3, appear as a faint halo—too faint to be seen with the naked eye, only with sensitive cameras. They occur in the ionized region of the upper atmosphere. ELVES

occur at night and are extremely short-lived, lasting less than a thousandth of a second. They appear to form when a lightning bolt from an intense thunderstorm gives off a strong electromagnetic pulse that causes electrons in the ionosphere to collide with molecules that become excited and give off light.

The roles that red sprites, blue jets, and ELVES play in the Earth's global electrical system have yet to be determined. However, scientists continue to find other types of electromagnetic phenomena related to thunderstorms. Three of the most recently discovered ones have been named *trolls, gnomes*, and *pixies*.

Hence, the hailstone (larger, warmer particle) becomes negatively charged and the ice crystal (smaller, cooler particle) positively charged, as the positive ions are incorporated into the ice crystal (see ● Fig. 14.34).

The same effect occurs when colder, supercooled liquid droplets freeze on contact with a warmer hailstone and tiny splinters of positively charged ice break off. These lighter, positively charged particles are then carried to the upper part of the cloud by updrafts. The larger hailstones (or graupel), left with a negative charge, either remain suspended in an updraft or fall toward the bottom of the cloud. By this mechanism, the cold upper part of the cloud becomes positively charged, while the middle of the cloud becomes negatively charged. The lower part of the cloud is generally of negative and mixed charge except for an occasional positive region located in the falling precipitation near the melting level (see ● Fig. 14.35).

Another school of thought proposes that during the formation of precipitation, regions of separate charge exist within tiny cloud droplets and larger precipitation particles. In the upper part of these particles we find negative charge, while in the lower part we find positive charge. When falling precipitation collides with smaller particles, the larger precipitation particles become negatively charged and the smaller particles, positively charged. Updrafts within the cloud then sweep the smaller positively charged particles into the upper reaches of the cloud, while the larger negatively charged particles either settle toward the lower part of the cloud or updrafts keep them suspended near the middle of the cloud. These two theories of cloud electrification do not rule each other out. It is possible that both processes are at work during the evolution of a thunderstorm.

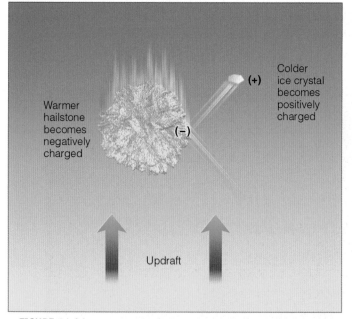

● **FIGURE 14.34** When the tiny colder ice crystals come in contact with the much larger and warmer hailstone (or graupel), the ice crystal becomes positively charged and the hailstone negatively charged. Updrafts carry the tiny positively charged ice crystal into the upper reaches of the cloud, while the heavier hailstone falls through the updraft toward the lower region of the cloud.

THE LIGHTNING STROKE Because unlike charges attract one another, the negative charge at the bottom of the cloud causes a region of the ground beneath it to become positively charged. As the thunderstorm moves along, this region of positive charge follows the cloud like a shadow. The positive charge is most dense on protruding objects, such as trees, poles, and buildings. The difference in charges causes an electric potential between the cloud and ground. In dry air, however, a flow of current does not occur because the air is a good electrical insulator. Gradually, the electric potential gradient builds, and when it becomes sufficiently large (on the order of one million volts per meter), the insulating properties of the air break down, a current flows, and lightning occurs.

Cloud-to-ground lightning begins within the cloud when the localized electric potential gradient exceeds 3 million volts per meter along a path perhaps 50 m long. This situation causes a discharge of electrons to rush toward the cloud base and then toward the ground in a series of steps (see ● Fig. 14.36a). Each discharge covers about 50 to 100 m, then stops for about 50-millionths of a second, then occurs again over another 50 m or so. This **stepped leader** is very faint and is usually invisible to the human eye. As the tip of the stepped leader approaches the ground, the potential gradient (the voltage per meter) increases, and a current of positive charge starts upward from the ground (usually along elevated objects) to meet it (see Fig. 14.36b). After they meet, large numbers of electrons flow to the ground and a much larger, more luminous **return stroke** several centimeters in diameter surges upward to the cloud along the path followed by the stepped leader (Fig. 14.36c). Hence, the downward flow of electrons establishes the bright channel of upward propagating current. Even though

the bright return stroke travels from the ground up to the cloud, it happens so quickly—in one ten-thousandth of a second—that our eyes cannot resolve the motion, and we see what appears to be a continuous bright flash of light (see ● Fig. 14.37).

Sometimes there is only one lightning stroke, but more often the leader-and-stroke process is repeated in the same ionized channel at intervals of about four-hundredths of a second. The subsequent leader, called a **dart leader**, proceeds from the cloud along the same channel as the original stepped leader; however, it proceeds downward more quickly because the electrical resistance of the path is now lower. As the leader approaches the ground, normally a less energetic return stroke than the first one travels from the ground to the cloud. Typically, a lightning flash will have three or four leaders, each followed by a return stroke. Even a lightning flash consisting of many strokes (one photographed flash had 26 strokes) usually lasts less than a second. During this short period of time, our eyes may barely be able to perceive the individual strokes, and the flash appears to flicker.

The lightning described so far (where the base of the cloud is negatively charged and the ground positively charged) is called *negative cloud-to-ground lightning*, because the stroke carries negative charges from the cloud to the ground. About 90 percent of all cloud-to-ground lightning is negative. However, when the base of the cloud is positively charged and the ground negatively charged, a *positive cloud-to-ground lightning* flash may result.

● **FIGURE 14.35** The generalized charge distribution in a mature thunderstorm.

● **FIGURE 14.36** The development of a lightning stroke. (a) When the negative charge near the bottom of the cloud becomes large enough to overcome the air's resistance, a flow of electrons—the stepped leader—rushes toward the earth. (b) As the electrons approach the ground, a region of positive charge moves up into the air through any conducting object, such as trees, buildings, and even humans. (c) When the downward flow of electrons meets the upward surge of positive charge, a strong electric current—a bright return stroke—carries positive charge upward into the cloud.

Positive lightning, most common with supercell thunderstorms, has the potential to cause more damage because it generates a much higher current level and its flash lasts for a longer duration than negative lightning.

Notice in ● Fig. 14.38 that lightning may take on a variety of shapes and forms. When a dart leader moving toward the ground deviates from the original path taken by the stepped leader, the lightning appears crooked or forked, as shown in Fig. 14.38a. Lightning that takes on this shape is called *forked lightning*. An interesting type of lightning is *ribbon lightning*, which forms when the wind moves the ionized channel between each return stroke, causing the lightning to appear as a ribbon hanging from the cloud (see Fig. 14.38b). If the lightning channel breaks up, or appears to break up, the lightning (called *bead lightning*) looks like a series of beads tied to a string (see Fig. 14.38c).

Ball lightning looks like a luminous sphere, often about the size of a football, that appears to float in the air or slowly dart about for several seconds, as illustrated in Fig. 14.38d. For centuries, scientists were unable to confirm the existence of ball lightning, despite many reports from observers of luminous spheres striking or entering buildings. Finally, the first detailed observation of a ball lightning event, including high-speed video, was made over the Qinghai

● **FIGURE 14.37** Time exposure of an evening thunderstorm with an intense lightning display near Denver, Colorado. The bright flashes are return strokes. The lighter forked flashes are probably stepped leaders that did not make it to the ground.

(a) Forked lightning (b) Ribbon lightning (c) Bead lightning

(d) Ball lightning (e) Sheet lightning (f) Heat lightning

● **FIGURE 14.38** Different forms of lightning.

Plateau in western China in 2012. Although many theories have been proposed, the actual cause of ball lightning remains an enigma.

Sheet lightning forms when either the lightning flash occurs inside a cloud or intervening clouds obscure the flash, such that a portion of the cloud (or clouds) appears as a luminous white sheet (see Fig. 14.38e). Distant lightning from thunderstorms that is seen but not heard is commonly called **heat lightning** because it frequently occurs on hot summer nights when the overhead sky is clear (see Fig. 14.38f). As the light from distant electrical storms is refracted through the atmosphere, air molecules and fine dust scatter the shorter wavelengths of visible light, often causing heat lightning to appear orange to a distant observer. When cloud-to-ground lightning occurs with thunderstorms that do not produce rain, the lightning is often called **dry lightning**. Such lightning often starts forest fires in regions of dry timber. Lightning may also shoot upward from the tops of thunderstorms into the upper atmosphere as a dim red flash called a *red sprite*, or as a narrow blue cone called a *blue jet*. These phenomena were seen for years by pilots, but they were not widely studied until they were photographed by sensitive low-light cameras, beginning in 1989. More information about these phenomena is found in Focus section 14.2 on p. 405.

As the electric potential near the ground increases during a thunderstorm, a current of positive charge moves up pointed objects, such as antennas and masts of ships. However, instead of a lightning stroke, a luminous greenish or bluish halo may appear above them, as a continuous supply of sparks—a *corona discharge*—is sent into the air. This electric discharge, which can cause the top of a ship's mast to glow, is known as **St. Elmo's Fire**, named after the patron saint of sailors (see ● Fig. 14.39). St. Elmo's Fire is also seen around power lines and the wings of aircraft. When St. Elmo's Fire is visible

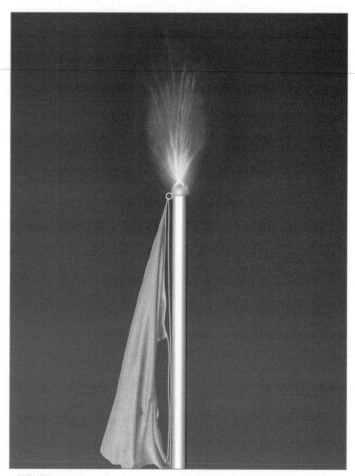

● **FIGURE 14.39** St. Elmo's Fire tends to form above objects, such as aircraft wings, ships' masts, and flagpoles.

and a thunderstorm is nearby, a lightning flash may occur in the near future, especially if the electric field of the atmosphere is increasing.

Lightning rods are placed on buildings to protect them from lightning damage. The rod is made of metal and has a pointed tip, which extends well above the structure. The positive charge concentration will be maximum on the tip of the rod, thus increasing the probability that the lightning will strike the tip and follow the metal rod harmlessly down into the ground, where the other end is deeply buried.

When lightning enters sandy soil, the extremely high temperature of the stroke may fuse sand particles together, producing a rootlike system of tubes called a **fulgurite**, after the Latin word for "lightning" (see ● Fig. 14.40). However, when lightning strikes an object such as a car, the lightning bolt typically leaves

● **FIGURE 14.40** A fulgurite that formed by lightning fusing sand particles.

the passengers unharmed because it usually takes the quickest path to the ground along the outside metal casing of the vehicle. The lightning then jumps to the road through the air, or it enters the roadway through the tires (see ● Fig. 14.41). The same type of protection is provided by the metal skin of a jet airliner, as hundreds of aircraft are struck by lightning each year with no harm to passengers.

If you should be caught in the open in a thunderstorm, what should you do? Of course, seek shelter immediately, but under a tree? If you are not sure, then you should read Focus section 14.3.

LIGHTNING DETECTION AND SUPPRESSION For many years, lightning strokes were detected primarily by visual observation. In recent decades, cloud-to-ground lightning has been located by means of an instrument called a *lightning direction-finder*, which works by detecting the radio waves produced by lightning. Such waves are called *sferics*, a contraction from their earlier designation, *atmospherics*. Networks of lightning direction-finders can be used to pinpoint the location of cloud-to-ground flashes throughout the United States and Canada. The National Lightning Detection Network, which includes more than 100 lightning direction-finders, has compiled data on more than 160 million flashes across the United States since 1989. Lightning detection devices allow scientists to examine in detail the lightning activity inside a storm as it intensifies and moves. Such investigations give forecasters a better idea where intense lightning strokes might be expected. Researchers have also installed lightning mapping arrays at several locations around the United States. These arrays use special VHF antennas to detect the very high frequency radiation produced along a lightning flash. The resulting data can

● **FIGURE 14.41** The four marks on the road surface represent areas where lightning, after striking a car traveling along Florida's turnpike, entered the roadway through the tires. Lightning flattened three of the car's tires and slightly damaged the radio antenna. The driver and a six-year-old passenger were taken to a nearby hospital, treated for shock, and released.

Don't Sit Under the Apple Tree

Because a single lightning stroke may involve a current as great as 100,000 amperes, animals and humans can be electrocuted when struck by lightning. Although the per-capita rate of lightning deaths in the United States has decreased by more than 90 percent in the last century, several dozen people are killed by lightning each year, with Florida accounting for the most fatalities. Many victims are struck in open places, operating farm equipment, playing golf, attending sports events, or sailing in a small boat. Some live to tell about it, as did the retired champion golfer Lee Trevino. Others are less fortunate, as about 10 percent of people struck by lightning are killed. Most die from cardiac arrest. Consequently, when you see someone struck by lightning, immediately give CPR (cardiopulmonary resuscitation), as lightning normally leaves its victims unconscious without heartbeat and without respiration. Those who do survive often suffer from long-term psychological disorders, such as personality changes, depression, and chronic fatigue.

Many lightning fatalities occur in the vicinity of relatively isolated trees (see ● Fig. 4). As a tragic example, during June 2004, three people were killed near Atlanta, Georgia, seeking shelter under a tree. Because a positive charge tends to concentrate in upward projecting objects, the upward return stroke that meets the stepped leader is most likely to originate from such objects. Clearly, seeking shelter under a tree during an electrical storm is not wise. What *should* you do?

When caught outside in an electrical storm, the best protection, of course, is to get inside a building. But stay away from electrical appliances and corded phones, and avoid taking a shower. Automobiles with metal frames and trucks (but not golf carts) may also

● **FIGURE 4** A cloud-to-ground lightning flash hitting a 65-foot sycamore tree. It should be apparent why one should *not* seek shelter under a tree during an electrical thunderstorm.

provide protection. If no such shelter exists, be sure to avoid elevated places and isolated trees. If you are on level ground, try to keep your head as low as possible, but do not lie down. Because lightning channels usually emanate outward through the ground at the point of a lightning strike, a surface current may travel through your body and injure or kill you. Therefore, crouch down as low as possible and minimize the contact area you have with the ground by touching it with only your toes or your heels.

There are some warning signs to alert you to a strike. If your hair begins to stand on end or your skin begins to tingle and you hear clicking sounds, beware—lightning may be about to strike. And if you are standing upright, you may be acting as a lightning rod (see ● Fig. 5).

What happens when lightning strikes a vehicle? As long as the windows are rolled up, the occupants may be unharmed by the flash, because the lightning will typically travel across the car's outer surface or through wiring. However, the complex electronic systems of a modern vehicle can easily be damaged by a lightning strike, and tires can be blown out. During an electrical storm, occupants should avoid touching any metallic interior object that may be connected to the exterior, such as a door handle or a gear shifter. If it is possible to do so safely, pull over and wait until the storm has passed.

● **FIGURE 5** Lightning can be both hair-raising and deadly. This photograph, taken by Mary McQuilken, shows her younger brother, Sean (on the left), and older brother Michael, standing during a thunderstorm atop Moro Rock in California's Sequoia National Park. Shortly after this photo was taken, Sean was struck by lightning and seriously injured, and a nearby hiker was killed by the same lightning strike.

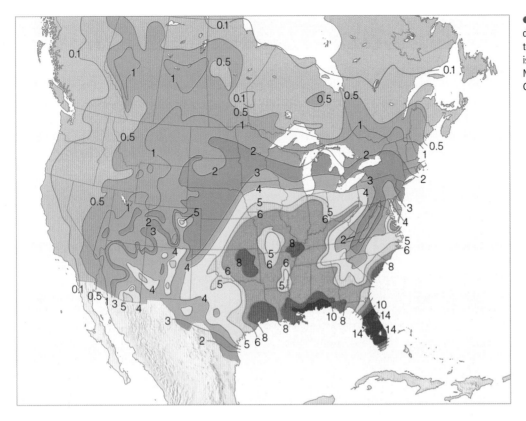

● **FIGURE 14.42** Average lightning flash density per square kilometer per year from 1997 to 2010. Notice that in the United States, Florida is the most lightning-prone state. (Data from the North American Lightning Detection Network. Courtesy of Vaisala.)

be used to map out each segment of a lightning channel in three-dimensional detail.

Satellites now have the capability of providing even more lightning information than ground-based sensors, because satellites can continuously detect all forms of lightning over land and over water (see ● Fig. 14.42). Lightning information correlated with satellite images provides a more complete and precise structure of a thunderstorm. The new *GOES-R* series of satellites includes a Geostationary Lightning Mapper that is beginning to provide continuous monitoring of lightning activity across the Americas.

WEATHER WATCH

Lightning strikes can be troublesome for travelers on roadways. A Canadian couple got quite the "shock" in May 2014 when their truck was hit by lightning as they drove along a highway in central Alberta. Surveillance video taken from a nearby facility shows a quick, fiery blast, after which smoke filled the vehicle. Unable to open windows or doors because of malfunctioning electronic systems, the couple was finally rescued by a passing police officer.

SUMMARY

In this chapter, we examined thunderstorms and the atmospheric conditions that produce them. Thunderstorms are convective storms that produce lightning and thunder. Lightning is a discharge of electricity that occurs in mature thunderstorms. The lightning stroke momentarily heats the air to an incredibly high temperature. The rapidly expanding air produces a sound called thunder.

The ingredients for the isolated ordinary cell thunderstorm are humid surface air, plenty of sunlight to heat the ground, a conditionally unstable atmosphere, a "trigger" to start the air rising, and weak vertical wind shear. When these conditions prevail, and the air begins to rise, small cumulus clouds may grow into towering clouds and thunderstorms within 30 minutes.

When conditions are ripe for thunderstorm development, and moderate or strong vertical wind shear exists, the updraft in the thunderstorm may tilt and ride up and over the downdraft. As the forward edge of the downdraft (the gust front) pushes outward along the ground, the air is lifted and new cells form, producing a multicell thunderstorm. Some multicell storms form as a complex of thunderstorms, such as the squall line (which forms as a line of thunderstorms), and the mesoscale convective complex (which forms as a cluster of storms). When convection in the multicell storm is strong, it may produce severe weather, such as strong damaging surface winds, hail, and flooding.

Supercell thunderstorms are intense thunderstorms with a single rotating updraft. The updraft and the downdraft in a supercell are nearly in balance, so that the storm may exist for many hours. Supercells are capable of producing severe weather, including strong, damaging tornadoes.

KEY TERMS

The following terms are listed (with corresponding page numbers) in the order they appear in the text. Define each. Doing so will aid you in reviewing the material covered in this chapter.

ordinary cell (air mass) thunderstorms, 385
cumulus stage, 385
mature stage, 386
dissipating stage, 386
multicell thunderstorm, 387
overshooting top, 388
gust front, 388
straight-line winds, 388
shelf cloud, 388
roll cloud, 388
outflow boundary, 389
downburst, 389
microburst, 389
heat burst, 391
squall line, 391
bow echo, 392
mesoscale convective vortex, 393
derecho, 393
mesoscale convective complexes (MCCs), 394
supercell, 394
mesocyclone, 396
wall cloud, 396
cap, 398
flash floods, 399
lightning, 403
thunder, 404
sonic boom, 404
stepped leader, 406
return stroke, 406
dart leader, 406
ball lightning, 407
sheet lightning, 408
heat lightning, 408
dry lightning, 408
St. Elmo's Fire, 408
fulgurite, 409

QUESTIONS FOR REVIEW

1. What is a thunderstorm?
2. What atmospheric conditions are necessary for the development of ordinary cell (air mass) thunderstorms?
3. Describe the stages of development of an ordinary cell thunderstorm.
4. How do downdrafts form in ordinary cell thunderstorms?
5. Why do ordinary cell thunderstorms most frequently form in the afternoon?
6. Explain why ordinary cell thunderstorms tend to dissipate much sooner than multicell storms.
7. How does the National Weather Service define a severe thunderstorm?
8. Give two examples of vertical wind shear.
9. What atmospheric conditions are necessary for a multicell thunderstorm to form?

10. (a) How do gust fronts form? (b) What type of weather does a gust front bring when it passes?

11. (a) Describe how a microburst forms. (b) Why is the term *horizontal wind shear* often used in conjunction with a microburst?

12. How do derechos form?

13. How does a squall line differ from a mesoscale convective complex (MCC)?

14. Give a possible explanation for the generation of a pre-frontal squall-line thunderstorm.

15. How do supercell thunderstorms differ from ordinary cell (air mass) thunderstorms?

16. Describe the atmospheric conditions at the surface and aloft that are necessary for the development of most supercell thunderstorms. (Include in your answer the role that the low-level jet plays in the rotating updraft.)

17. What is the difference between an HP supercell and an LP supercell?

18. When thunderstorms are *training*, what are they doing?

19. In what region in the United States do dryline thunderstorms most frequently form? Why there?

20. Where does the highest frequency of thunderstorms occur in the United States? Why there?

21. Why is large hail more common in Kansas than in Florida?

22. Describe one process by which thunderstorms become electrified.

23. How is thunder produced?

24. Explain how a cloud-to-ground lightning stroke develops.

25. Why is it unwise to seek shelter under an isolated tree during a thunderstorm? If caught out in the open, what should you do?

26. How does negative cloud-to-ground lightning differ from positive cloud-to-ground lightning?

QUESTIONS FOR THOUGHT

1. Why does the bottom half of a dissipating thunderstorm usually "disappear" before the top?

2. Sinking air warms, yet the downdrafts in a thunderstorm are usually cold. Why?

3. Explain why squall-line thunderstorms often form ahead of advancing cold fronts but seldom behind them.

4. A forecaster may say that he or she looks for "right-moving" thunderstorms when predicting severe weather. What does this mean?

5. Why is the old adage "lightning never strikes twice in the same place" wrong?

6. Why are left-moving supercell thunderstorms uncommon in the Northern Hemisphere, yet are very common in the Southern Hemisphere?

PROBLEMS AND EXERCISES

1. On a map of the United States, place the surface weather conditions as well as weather conditions aloft (jet stream and so on) that are necessary for the formation of most supercell thunderstorms.

2. If you see lightning and 10 seconds later you hear thunder, how far away is the lightning stroke?

ONLINE RESOURCES

Visit www.cengagebrain.com to view additional resources, including video exercises, practice quizzes, an interactive eBook, and more.

Three simultaneous tornadoes emerge from a severe thunderstorm near Dodge City, Kansas, on May 24, 2016.

Ryan McGinnis/Moment/Getty Images

WEDNESDAY, MARCH 18, 1925, WAS A DAY THAT BEGAN uneventfully, but within hours it turned into a day that changed the lives of thousands of people and made meteorological history. Shortly after 1 p.m., the sky turned a dark greenish-black and the wind began whipping around the small town of Murphysboro, Illinois. Arthur and Ella Flatt lived on the outskirts of town with their only son, Art, who would be four years old in two weeks. Arthur was working in the garage when he heard the roar of the wind and saw the threatening dark clouds whirling overhead.

Instantly concerned for the safety of his family, he ran toward the house as the tornado began its deadly pass over the area. With debris from the house flying in his path and the deafening sound of destruction all around him, Arthur reached the front door. As he struggled in vain to get to his family, whose screams he could hear inside, the porch and its massive support pillars caved in on him. Inside the house, Ella had scooped up young Art in her arms and was making a panicked dash down the front hallway towards the porch when the walls collapsed, knocking her to the floor, with Art cradled beneath her. Within seconds, the rest of the house fell down upon them. Both Arthur and Ella were killed instantly, but Art was spared, nestled safely under his mother's body.

As the dead and survivors were pulled from the devastation that remained, the death toll mounted. Few families escaped the grief of lost loved ones. The infamous Tri-State Tornado killed 234 people in Murphysboro and leveled 40 percent of the town.

The devastating tornado described in our opening cut a mile-wide path for a distance of more than 150 miles through the states of Missouri, Illinois, and Indiana. The tornado (which was most likely a series of tornadoes) totally obliterated four towns, killed an estimated 695 people, and left more than 2000 injured. Tornadoes such as these, as well as much smaller ones, are associated with severe thunderstorms. In the previous chapter, we examined the different types of thunderstorms. Here, we will focus on tornadoes, examining how and where they form, and why they are so destructive.

Tornadoes: A Few Facts

A **tornado** is a rapidly rotating column of air, extending down from a cumuliform cloud, that blows around a small area of intense low pressure with a circulation that reaches the ground. A tornado's circulation is present on the ground either as a funnel-shaped cloud or as a swirling cloud of dust and debris. Sometimes called *twisters* or *cyclones*, tornadoes can assume a variety of shapes and forms that range from twisting ropelike funnels, to cylinder-shaped funnels, to massive black wedge-shaped funnels, to funnels that resemble an elephant's trunk hanging from a large cumulonimbus cloud (see ● Fig. 15.1). Sometimes the term **funnel cloud** is used to refer to a visible funnel that does not touch the ground. However, a tornado can still be present even if the visible cloud does not extend to the surface. Perhaps only about 30 percent of funnel clouds become active tornadoes.

When viewed from above, the majority of North American tornadoes rotate counterclockwise about their central core of low pressure. Sometimes a tornado is observed rotating clockwise, but those are infrequent, accounting for as few as 1 percent of all tornadoes. A clockwise-rotating tornado may occur within the same thunderstorm as a larger, counterclockwise-rotating tornado.

The majority of tornadoes have wind speeds of less than 100 knots (115 mi/hr), although violent tornadoes may have winds exceeding 220 knots (253 mi/hr). The diameter of most tornadoes is between 100 and 600 m (about 300 to 2000 ft), although some are just a few meters wide and others have diameters exceeding 1600 m (1 mi). The largest tornado on record, spanning 4.2 km (2.6 mi), took the lives of three tornado researchers as it plowed across western Oklahoma near El Reno on May 31, 2013.

Tornadoes that form ahead of an advancing cold front are often steered by southwesterly winds and, therefore, tend to move from the southwest toward the northeast at speeds usually between 20 and 40 knots. However, some have been clocked at speeds greater than 70 knots. Most tornadoes last only a few minutes and have an average path length of about 7 km (4 mi). There are cases where they have reportedly traveled for hundreds of kilometers and have existed for many hours. The 1925 Tri-State Tornado mentioned at the start of this chapter featured the longest official damage track on record: 352 km (219 mi).*

TORNADO LIFE CYCLE   Major tornadoes usually evolve through a series of stages. The first stage is the *dust-whirl stage*, where dust swirling upward from the surface marks the tornado's circulation

*A recent analysis found that the damage path of the 1925 Tri-State Tornado was bracketed by destruction that may have been produced by two other tornadoes, which brings the total damage swath to as much as 378 km (235 mi).

● **FIGURE 15.1** A large wedge-shaped violent tornado moves northwestward directly toward Windsor, Colorado, on May 22, 2008. The photo (taken by a webcam) shows hail the size of golf balls falling from the thunderstorm and covering the ground.

© Tony Hake/Denver Weather Examiner

on the ground and a short funnel often extends downward from the thunderstorm's base. Damage during this stage is normally light. The next stage, called the *organizing stage*, finds the tornado increasing in intensity with an overall downward extent of the funnel. During the tornado's *mature stage*, damage normally is most severe as the funnel reaches its greatest width and is almost vertical (see ● Fig. 15.2). The *shrinking stage* is characterized by an overall decrease in the funnel's width, an increase in the funnel's tilt, and a narrowing of the damage swath at the surface, although the tornado may still be capable of intense and sometimes violent damage. The final stage, called the *decay stage*, usually finds the tornado stretched into the shape of a rope (see ● Fig. 15.3). Normally, the tornado becomes greatly contorted before it finally dissipates. Although these are the typical stages of a major tornado, minor tornadoes may evolve only through the organizing stage. Some even skip the mature stage and go directly into the decay stage. However, when a tornado reaches its mature stage, its circulation usually stays in contact with the ground until it dissipates. Sometimes a new tornado will emerge from the same supercell thunderstorm just before or shortly after an existing tornado dissipates.

TORNADO OCCURRENCE AND DISTRIBUTION Tornadoes occur in many parts of the world, but no country experiences more tornadoes than the United States, which, in recent years, has averaged more than 1000 annually. In 2004, a record was set with 1819 tornadoes observed. The number of total tornadoes reported each year has more than doubled since the 1950s (see ● Fig. 15.4), even though the number of strong tornadoes (those with winds exceeding 117 knots or 135 mi/hr) has shown no significant trend. The difference in tornado numbers is most likely because many weaker, more short-lived tornadoes are being reported (and

photographed) than was the case decades ago. Although tornadoes have occurred in every state, including Alaska and Hawaii, the greatest number occur in the tornado belt, or **tornado alley**, of the Central Plains, which stretches from central Texas to Nebraska* (see ● Fig. 15.5). The belt of tornadoes that occurs over Mississippi and Alabama is sometimes called *Dixie Alley*.

The Central and Southern Plains region is most susceptible to tornadoes because it often provides the proper atmospheric setting for the development of the severe thunderstorms that spawn tornadoes. Recall from Fig. 14.25 on p. 397 that over the Central and Southern Plains (especially in spring) warm, humid surface air is overlain by cooler, drier air aloft, producing a conditionally unstable atmosphere. When a strong vertical wind shear exists (usually provided by a low-level jet and by the polar jet stream) and the surface air is forced upward, large supercell thunderstorms capable of spawning tornadoes may form. Therefore, tornado frequency is highest during the spring and lowest during the winter when the warm surface air is normally absent.

The frequency of tornadic activity shows a seasonal shift. For example, during the winter, tornadoes are most likely to form over the southern Gulf states when the polar-front jet is above this region, and the contrast between warm and cold air masses is greatest. In spring, humid Gulf air surges northward; contrasting air masses and the jet stream also move northward and tornadoes become more prevalent from the southern Atlantic states westward into the southern Great Plains. In summer, the contrast between air masses lessens, and the jet stream is normally near the Canadian border; hence, tornado activity tends to be concentrated from the northern plains eastward to New York State.

---

*Many of the tornadoes that form along the Gulf Coast are generated by thunderstorms embedded within the circulation of hurricanes.

● **FIGURE 15.2** A tornado in its mature stage roars over the Great Plains.

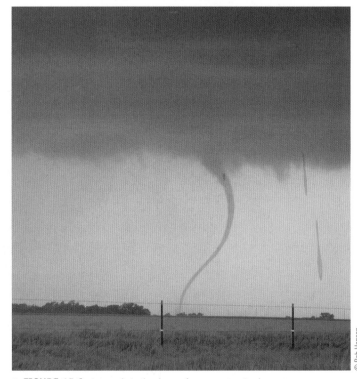

● **FIGURE 15.3** A tornado in the shape of a rope enters its decay stage near Cherokee, Oklahoma, on April 14, 2012

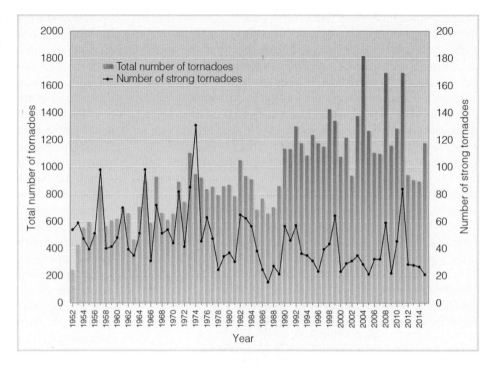

● **FIGURE 15.4** Total number of tornadoes reported in the United States for each year from 1952 to 2015 (blue bar); and the total number of strong tornadoes with winds exceeding 117 knots or 135 mi/hr (EF3 on the Enhanced Fujita Scale) reported during the same period (red line). (Note: Tornadoes that occurred before the introduction of the Enhanced Fujita Scale in 2007 have been converted to the new scale. (Data from NOAA)

In ● Fig. 15.6 we can see that about 70 percent of all tornadoes in the United States develop from March to July. The month of May normally has the greatest number of tornadoes (the average is about nine per day), while the most violent tornadoes tend to occur in April, when vertical wind shear tends to be present as well as when horizontal and vertical temperature and moisture contrasts are greatest.* Although tornadoes have occurred at all times of the day and night, they are most frequent in the late afternoon (between 3 p.m. and 7 p.m.), when the surface air is most unstable; they are least frequent in the early morning before sunrise, when the atmosphere is most stable.

Although large, destructive tornadoes are most common in the Central Plains, they can develop anywhere—and anytime—if the conditions are right. For example, a supercell thunderstorm roared across the Florida Peninsula during the predawn hours on February 2, 2007. The storm produced two strong tornadoes that damaged more than 1000 homes and killed 21 people.

On July 31, 1987, a violent tornado with winds exceeding 180 knots moved through the city of Edmonton, Alberta, Canada, well north of where such intense tornadoes are usually observed.

*During May 2003, 516 tornadoes touched down in the United States, an average of more than 16 per day. However, during April 2011, a record 748 tornadoes were reported, the most in any month since records began in 1950.

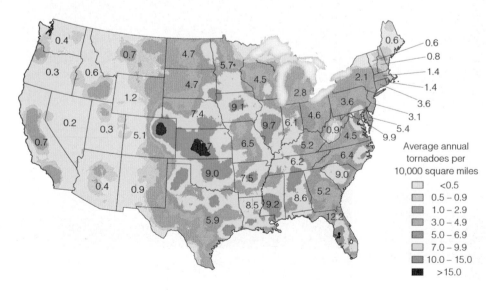

● **FIGURE 15.5** The average annual number of observed tornadoes per 10,000 square miles in each state from 1991 to 2012. (Data from NOAA)

Average annual tornadoes per 10,000 square miles

☐	<0.5
	0.5 – 0.9
	1.0 – 2.9
	3.0 – 4.9
	5.0 – 6.9
	7.0 – 9.9
	10.0 – 15.0
	>15.0

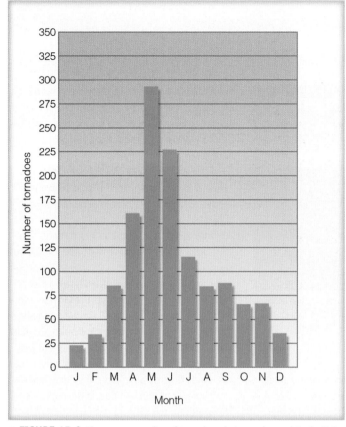

● **FIGURE 15.6** The average number of tornadoes during each month in the United States from 2000 to 2010.

There is a myth that larger cities such as Los Angeles and New York City are somehow protected from tornadoes. But on March 1, 1983, a rare tornado cut a 5-km swath of destruction through downtown Los Angeles, damaging more than 100 homes and businesses, and injuring 33 people. And during the summer of 2010, three tornadoes actually touched down in New York City. One of these tornadoes (on July 25) caused only minimal damage but injured seven people.

Even in the central part of the United States, the statistical chance that a tornado will strike a particular place this year is quite small. However, tornadoes can provide many exceptions to statistics. Oklahoma City, for example, has been struck by tornadoes at least 35 times in the past 100 years. The adjacent suburb of Moore experienced destructive tornadoes on May 3, 1999; May 8, 2003; and May 20, 2013. And the little town of Codell, Kansas, was hit by tornadoes in three consecutive years—1916, 1917, and 1918—and each time on the same date: May 20! Considering the many millions of tornadoes that must have formed during the geological past, it is very probable that at least one actually moved across the land where your home is located, especially if it is in the Central Plains.

TORNADO WINDS   At one point, our knowledge of the furious winds of a tornado came mainly from observations of the damage done and the analysis of motion pictures. Today more accurate wind measurements are made with Doppler radar. Because of the destructive nature of the tornado, it was once thought that it packed winds greater than 500 knots. However, mobile radar observations since the 1990s confirm that even the most powerful twisters seldom have winds exceeding 220 knots, and most tornadoes have winds of less than 125 knots. Nevertheless, even a small tornado can be terrifying and dangerous. (Focus section 15.1 includes more background on the quirky types of damage that tornadoes can inflict.)

(See ● Fig. 15.7.) This massive tornado cut a destruction path over 40 km (25 mi) long. It destroyed more than 3000 homes, caused more than $330 million in damages, injured more than 300 people, and took 27 lives.

● **FIGURE 15.7**   A violent tornado moves through Edmonton, Alberta, Canada, on July 31, 1987.

The Weird World of Tornado Damage

The strong winds of a tornado can destroy buildings, uproot trees, and hurl all sorts of lethal missiles into the air. People, animals, and home appliances all have been picked up, carried several kilometers, and then deposited. Even motor vehicles have been tossed a kilometer or more, and lighter objects such as checkbooks have been blown 100 km (62 mi) or more, sometimes from one state to another. One tornado lifted a railroad coach with its 117 passengers and dumped it into a ditch about 25 m (82 ft) away. In one instance, a schoolhouse was demolished, and the 85 students inside were carried more than 90 m (295 ft) without any fatalities. In another freakish example, a house in Michigan survived a tornado but was turned onto its side so that the front door couldn't be reached without a ladder.

Showers of toads and frogs have poured out of a cloud after tornadic winds sucked them up from a nearby pond. Other oddities include pieces of straw being driven into metal pipes, frozen hot dogs being driven into concrete walls, and chickens losing all of their feathers. (Actually, the more likely explanation for the defeathering is a process called *flight molt*, in which chickens involuntarily release feathers when threatened.)

● **FIGURE 1** This bicycle tire was wrapped around a utility pole by the fierce winds of an EF4 tornado that struck Picher, Oklahoma, on May 10, 2008.

The power delivered by the winds of a strong tornado is far beyond anything in our everyday experience. The pressure of the wind (the force per unit area) increases with the square of the wind speed, which means that if you double the wind speed, you get four times as much destructive potential. Imagine a very windy day, with gusts of 40 knots (45 mi/hr) flinging dust into your eyes and tossing loose objects down the street. In the most violent tornadoes on Earth, winds can exceed 174 knots

(200 mi/hr), which means they carry more than 16 times the power of the gusts already making your day miserable. When a long object such as a straw or a two-by-four board (or a frozen hot dog) is tossed lengthwise in a violent tornado, its force is packed into the tiny area at its tip. This concentration gives the wind-blown projectile a surprising ability to penetrate solid objects.

Another factor that makes tornado damage so quirky is the huge variability of tornado behavior. Tornadoes can strengthen or weaken dramatically, and shrink or enlarge, in just a few seconds. The suction vortices inside a tornado illustrated in Fig. 15.9 can be as small as 10 m (30 ft) in diameter, so it is quite possible for a suction vortex to hit one house but miss the next-door neighbor's house entirely. Even as winds destroy a building, they may leave a few objects untouched due to the complex flow around the distintegrating structure. (Such cases often make the news when the intact item is a cherished keepsake.) In a large tornado, pieces of debris can be spun a mile or more away from the main tornadic circulation. These objects can cause localized pockets of damage in areas that are otherwise unscathed.

When a tornado is approaching from the southwest, its strongest winds are on its southeast side. We can see why in ● Fig. 15.8. The tornado is heading northeast at 50 knots. If its rotational speed is 100 knots, then its forward speed will add 50 knots to its southeastern side (position D) and subtract 50 knots from its northwestern side (position B). Hence, the most destructive and extreme winds will be on the tornado's southeastern side.

Many violent tornadoes (with winds exceeding 180 knots) contain smaller whirls that rotate within them. Such tornadoes are called *multivortex tornadoes* and the smaller whirls are called **suction vortices** (see ● Fig. 15.9). Suction vortices can be as small as 10 m (30 ft) in diameter, but they rotate very fast and can do a great deal of damage. Suction vortices help explain how one building may experience major tornado damage while the next one is virtually untouched. It is also possible for a separate twister called a *satellite tornado* to develop just outside a larger, stronger tornado.

Seeking Shelter The pressure in the center of a tornado may be more than 100 millibars (3 in.) lower than that of its surroundings. It was once thought that opening windows and allowing the inside and outside pressures to equalize would minimize the chances of a building exploding. However, we now know that the wind itself is the main factor in tornado damage. Opening windows during a tornado actually increases the pressure on the opposite wall and ceiling and *increases* the chances that the building will collapse. (The windows are usually shattered by flying debris anyway.) As the winds push against walls and under eaves, they can easily unroof a building. Once the roof is gone, walls of buildings can quickly buckle and collapse when blasted by the extreme wind force and by debris carried by the wind. Damage from tornadoes may also be inflicted on people and structures by flying debris. Hence, the wisest course to take when confronted with an approaching tornado is to *seek shelter immediately*.

At home, take shelter in a basement, storm shelter, or a dedicated "safe room" (a hardened structure built to FEMA standards

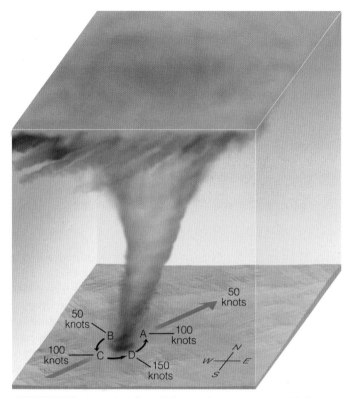

● **FIGURE 15.8** The total wind speed of a tornado is greater on one side than on the other. If a tornado is approaching, the strongest winds will be on your left side (unless it is one of those unusual tornadoes that rotates clockwise).

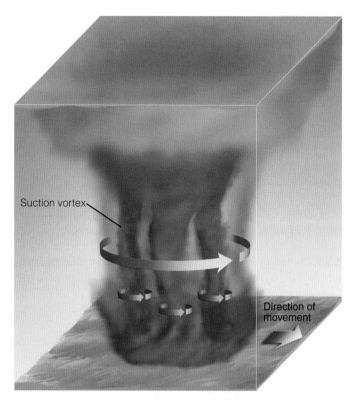

● **FIGURE 15.9** A powerful multivortex tornado with three suction vortices.

to survive high winds with as little damage as possible). In a large building without a basement, the safest place is usually in a small room, such as a bathroom, closet, or interior hallway, preferably on the lowest floor and near the middle of the structure. Pull a mattress or sleeping bag around you. Wear a bike or football helmet, if one is available, to protect your head from flying debris, and stay away from windows. At school, move to the hallway and lie flat with your head covered. In a mobile home, leave immediately and seek substantial shelter. If none exists, lie flat on the ground in a depression or ravine.

CRITICAL THINKING QUESTION Suppose the tornado in Fig. 15.9 is moving toward the west (instead of toward the east) as in the illustration. Further, suppose that the tornado and its suction vortices are all rotating clockwise, as viewed from above. Would the extreme winds of this tornado now be greater than, less than, or the same as the winds of the tornado in the illustration? On which side of the tornado (north or south) would you expect the strongest winds?

Don't try to outrun an oncoming tornado in a car or truck, as tornadoes often cover erratic paths with speeds sometimes exceeding 70 knots (80 mi/hr). Instead, if the tornado is still at some distance, drive at a right angle to the tornado's path, away from the core of the parent thunderstorm. (If the tornado appears to be stationary, but getting bigger, it is most likely moving directly toward you.) If there is any question about how close the tornado is, abandon your vehicle and seek shelter immediately—but not under a freeway overpass, as the tornado's winds may be funneled (strengthened) by the overpass structure. If caught outdoors in an open field, look for a ditch, streambed, or ravine, and lie flat with your head covered.

The National Weather Service provides several kinds of services related to tornadoes. A **tornado watch** is issued by the Storm Prediction Center to alert the public that tornadoes may develop within a specific area during a certain time period, usually a few hours long. Many communities have trained volunteer spotters, who look for tornadoes after the watch is issued. When a tornado is spotted by a reliable observer, or its telltale signs are evident on radar, a **tornado warning** is issued by the local NWS office, typically covering parts of one or several counties and lasting about 30 to 45 minutes.[*] A **tornado emergency** may also be issued on rare occasions when an especially strong tornado threatens a populated area. In many communities, sirens are sounded to alert people of the approaching storm. Radio and television stations (including NOAA Weather Radio) and many

---

[*]In 2007, the National Weather Service launched a new, more specific tornado warning system called *Storm-Based Warnings*. The new system provides more precise information on where a tornado is located and where it is heading.

cable TV systems will interrupt regular programming to broadcast the warning. Most newer cell phones will display the warning automatically through the Wireless Emergency Alert system, assuming the phone's owner has not blocked this feature.

Although not completely effective, the current warning system appears to be saving many lives. Despite the large increase in population in the tornado belt during the past 50 years, tornado-related deaths have generally held steady during this period, with 2011 being a major exception (see ▼Table 15.1). (For additional information on tornado watches and warnings, see Focus section 15.2.)

The Enhanced Fujita Scale In the 1960s, the late Dr. T. Theodore Fujita, a noted authority on tornadoes at the University of Chicago, proposed a scale (called the **Fujita scale**) for classifying tornadoes according to their rotational wind speed. The tornado winds are estimated based on the damage caused by the storm.

The original Fujita scale, implemented in 1971, was based mainly on the extent of tornado damage to frame houses. Because many types of structures are susceptible to tornado damage and not all tornadoes strike frame homes, a new scale came into effect in 2007. Called the **Enhanced Fujita Scale**, or simply the **EF Scale**, it attempts to provide a wide range of criteria in estimating a tornado's winds by using a set of 28 damage indicators, including small barns, mobile homes, schools, and trees. The quality of building construction is also taken into account. Each structure or object is examined for the degree of damage it sustained. The combination of the damage indicators along with the degree of damage provides a range of probable wind speeds and an EF rating for the tornado. The wind estimates for the Enhanced Fujita Scale are given in ▼Table 15.2.

● Figure 15.10 shows a house situated somewhere on the Great Plains of the United States or Canada, and ● Fig. 15.11 shows the damaging effect that tornadoes ranging in intensity from EF0 to EF5 can have on this structure and its surroundings.

▼ **TABLE 15.2 Enhanced Fujita Scale for Damaging Tornado Winds**

EF SCALE	CATEGORY	MI/HR*	KNOT
EF0	Weak	65–85	56–74
EF1		86–110	75–95
EF2	Strong	111–135	96–117
EF3		136–165	118–143
EF4	Violent	166–200	144–174
EF5		>200	>174

*The wind speed is a 3-second gust estimated at the point of damage, based on a judgment of damage indicators.

Notice that an EF0 tornado causes only minimal damage, whereas an EF5 completely demolishes the house and sweeps it off its foundation. In this example, we assume that the home is well constructed; otherwise, the damage for a given strength of tornado could be even greater than shown. A devastating tornado that struck central Arkansas on April 27, 2014, was rated EF4 rather than EF5, even though some homes were swept off their foundations, because damage surveyors found that these homes were generally fastened to their foundations with nails rather than anchor bolts.

Manufactured or mobile homes are especially vulnerable in tornadoes. During recent years, roughly 45 percent of all tornado fatalities in the United States have occurred in mobile homes. Note that damage can vary greatly along a tornado's path; for example, one section may exhibit EF2 or EF3 damage while another area may show EF4 damage. Such a tornado would earn an EF4 rating, corresponding to the most severe damage observed. Statistics reveal that the vast majority of tornadoes are relatively weak, with wind speeds less than about 100 knots (115 mi/hr). Only a few percent each year are classified as violent, with an average of one or two EF5 tornadoes reported annually

▼ **TABLE 15.1 Average Annual Number of Tornadoes and Tornado Deaths by Decade**

DECADE	TORNADOES/YEAR*	DEATHS/YEAR
1950–59	480	148
1960–69	681	94
1970–79	858	100
1980–89	819	52
1990–99	1220	56
2000–09	1277	56
2010–16	1123	121†

*More tornadoes are being reported as populations increase, more citizens watch for tornadoes, and tornado-spotting technology improves.
†This seven-year average includes the especially deadly year of 2011, when 553 deaths occurred. Otherwise, the average for 2010–2016 was around 50 deaths per year.

● **FIGURE 15.10** A house situated on the Great Plains. Observe in Fig. 15.11 how tornadoes of varying EF intensity can damage this house and its surroundings.

| EF0 | [65-85 mi/hr] | EF1 | [86-110 mi/hr] | EF2 | [111-135 mi/hr] |
| EF3 | [136-165 mi/hr] | EF4 | [166-200 mi/hr] | EF5 | [Over 200 mi/hr] |

● **FIGURE 15.11** Damage to the house in Fig. 15.10 and its surroundings caused by tornadoes of varying EF intensity.

(although several years may pass without the United States experiencing an EF5). It is the violent tornadoes that account for the majority of tornado-related deaths.

For example, an EF5 tornado roared through the town of Greensburg, Kansas, on the evening of May 4, 2007. The tornado, with winds estimated at 180 knots (205 mi/hr) and a width approaching 2 miles, completely destroyed over 95 percent of the town. The tornado took 11 lives, and probably more would have perished had it not been for the tornado warning issued by the National Weather Service and the sirens in the town, signaling "take cover," that sounded about 20 minutes before the tornado struck.

On May 22, 2011, a violent EF5 multivortex tornado struck Joplin, Missouri, completely demolishing part of the city.* The tornado injured almost 1000 people and took 159 lives—the greatest death toll attributed to a single tornado in the United States since the Woodward, Oklahoma, tornado on April 9, 1947. (See ▼ Table 15.3 for a summary of the ten deadliest tornadoes observed in the United States.) Even more people were killed just weeks before the Joplin tornado by a swarm of twisters—an event referred to as a **tornado outbreak**.

TORNADO OUTBREAKS As we have seen in the previous situations, tornadoes take the lives of many people each year. The annual average is less than 100, although more than 100 people can die in a single tornado, as tragically occurred in Joplin, Missouri. The deadliest tornadoes are those that occur in *families*—that is, a group or series of tornadoes spawned by the same thunderstorm. (Some thunderstorms produce a sequence of several tornadoes

over two or more hours and over distances of 100 km or more.) Tornado families are typically the result of a single, long-lived supercell thunderstorm. Sometimes a large, well-organized frontal system will spawn multiple supercells, each producing its own family of tornadoes. When a large number of tornadoes develop in association with a particular weather system (typically six tornadoes or more, although there is no strict definition), it is referred to as a tornado outbreak. An outbreak may extend over multiple days, as long as no more than a few hours (typically six) elapse without at least one tornado being observed in the outbreak region. The most severe tornado outbreaks are typically very well predicted because they involve large areas of conditionally unstable air and extremely strong vertical wind shear, along with an upper-level trough that helps trigger widespread supercell formation. These features can often be spotted by computer forecast models several days in advance. The NOAA Storm Prediction Center in Norman, Oklahoma, issues convective outlooks each day that depict where outbreaks of severe weather and tornadoes are possible over the next eight days. Researchers are now exploring techniques to provide several weeks' notice of when there may be an elevated risk of tornado outbreaks.

A particularly devastating outbreak occurred on May 3, 1999, when 78 tornadoes marched across parts of Texas, Kansas, and Oklahoma. One tornado, whose width at times reached 1 mile and whose wind speed was measured by a portable Doppler radar at 262 knots (301 mi/hr), moved through parts of Oklahoma City and the suburb of Moore. Within its 38-mile path, it damaged or destroyed thousands of homes, injured nearly 600 people, claimed 36 lives, and caused over $1 billion in property damage. On May 20, 2013, an EF5 tornado with maximum winds estimated at 183 knots (210 mi/hr) took a path similar to that of the deadly 1999 tornado, with the

*A photo of the Joplin tornado is shown in Fig. 1.22, on p. 24. The destruction caused by this tornado is shown in Figure 1.23.

The Evolution of Tornado Watches and Warnings

Few weather situations are as terrifying as the approach of a churning tornado. From the 1890s until the 1930s, U.S. government forecasters were forbidden from even mentioning tornadoes in public statements. Fortunately, the United States now has a well-established system for letting people know when twisters are possible and when one is imminent. The system of tornado watches and warnings in the United States was created in the 1950s, shortly after a remarkable coincidence. On March 20, 1948, a destructive tornado struck Tinker Air Force Base, just southeast of Oklahoma City. A total of 50 aircraft were destroyed, at a cost of $10 million (likely more than $100 million in current dollars). The next day, two Air Force meteorologists, E. J. Fawbush and Robert Miller, were asked to develop a technique for predicting when tornadoes were likely. Fawbush and Miller quickly created a scheme based on such factors as instability, wind shear, and the approach of a front. Incredibly, another tornado struck the base only five days after the first one. This time the brand-new Fawbush-Miller technique provided notice that tornadoes were very likely, so aircraft were safely sheltered and the damage toll was far less.

As we previously learned, a tornado watch is issued when conditions favor the formation of severe thunderstorms and tornadoes. The average tornado watch covers a period of six to eight hours and an area of roughly 65,000 square km, about the size of South Carolina. Tornado watches are issued by the National Weather Service's Storm Prediction Center in Norman, Oklahoma, in coordination with local NWS offices. When conditions are most threatening, a watch will receive the "PDS" tag, which denotes a *particularly dangerous situation*. The center also issues more general outlooks that highlight where severe weather is possible up to eight

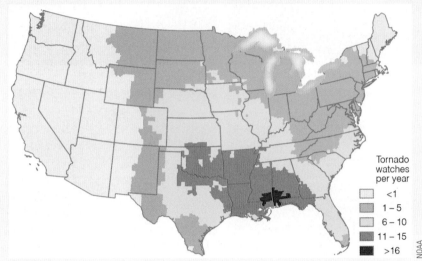

● **FIGURE 2** Average annual number of tornado watches per county issued by the Storm Prediction Center during 1999–2008.

days in advance. If you are in a region under a tornado watch, stay aware of changing weather conditions and watch or listen for possible warnings.

When a tornado is actually sighted, or when a tornadic circulation is evident on radar, the local NWS office will issue a *tornado warning*. In October 2007, the NWS launched a new, more specific tornado warning system called *Storm-Based Warnings*. This system provides more precise information on where a tornado is located and where it is heading. Warnings typically cover parts of one to several counties and are in effect for 30 to 60 minutes. If your smartphone is equipped to receive Wireless Emergency Alerts (WEA), you will normally receive a text message if a tornado warning is issued in your area. Should you find yourself in a tornado warning, you need to take cover immediately—ideally in an interior room on the lowest floor of a well-built structure, but not in a vehicle or a mobile home.

The tornado warning system is not perfect. As many as 30 percent of all tornadoes come and go before a warning can be issued, and more than half of all warnings are "false alarms," typically because a tornadic signature on radar has failed to produce a twister. Even when a tornado is accurately warned, it will usually strike only part of a warned area. Still, the risk of injury or death is very real in these situations, and many lives are saved by quick response to a tornado warning. The advent of the NEXRAD Doppler radar network in the 1990s helped improve the warning system substantially. The average lead time for a tornado warning rose from 3 minutes in the late 1970s to 13 minutes by the late 1990s, and the lead time is now often 20 minutes or more for the longest-lived and most violent tornadoes. A new NWS initiative is exploring whether high-resolution computer models that incorporate radar data could help increase the lead time of tornado warnings to as much as 1 to 2 hours.

paths actually crossing at one point (see Fig. 13.14 on p. 367). The 2013 tornado was on the ground for 27 km (17 mi) and cut a destructive swath through a highly populated section of Moore, killing 24 people, including 10 children, and causing an estimated $2 billion in damage.

One of the most violent outbreaks ever recorded occurred on April 3 and 4, 1974. During a 16-hour period, 148 tornadoes cut through parts of 13 states and one Canadian province, killing 319 people, injuring more than 5000, and causing an estimated $600 million in damage in the United States. Some of these

RANKING	DATE	LOCATION(S)	DEATHS
1	March 18, 1925	Tri-State (Missouri/Illinois/Indiana)	695
2	May 6, 1840	Natchez, Mississippi	317
3	May 27, 1896	St. Louis, Missouri	255
4	April 5, 1936	Tupelo, Mississippi	216
5	April 6, 1936	Gainesville, Georgia	203
6	April 9, 1947	Woodward, Oklahoma	181
7	May 22, 2011	Joplin, Missouri	158
8	April 24, 1908	Amite, Louisiana/Purvis, Mississippi	143
9	June 12, 1899	New Richmond, Wisconsin	117
10	June 8, 1953	Flint, Michigan	116

tornadoes were among the most powerful ever recorded, as seven tornadoes were rated at F5 intensity and 23 tornadoes at F4 intensity. The combined path of all the tornadoes during this *"Super Outbreak"* amounted to 4181 km (2598 mi), well over half of the total path for an average year.

The greatest one-day loss of life attributed to tornadoes in the United States occurred during the Tri-State Outbreak mentioned earlier, when an estimated 695 people died as at least seven tornadoes traveled a total of 703 km (437 mi) across portions of Missouri, Illinois, and Indiana.

The only recorded outbreak on par with the 1974 Super Outbreak occurred on April 25–28, 2011, when 357 tornadoes (four of which reached EF5 intensity) moved across portions of the eastern United States, plus one in southeastern Canada. The tornadoes claimed 316 lives, injured thousands of people, and caused more than $10 billion in damages. One particularly strong EF4 tornado, with winds estimated at 190 mi/hr, moved through the city of Tuscaloosa, Alabama, on April 27 (see ● Fig. 15.12). The tornado, which had a damage path width of about 1.5 miles, left 43 dead in Tuscaloosa and injured more than 1000 (see ● Fig. 15.13).

Tornado Formation

Although not everything is known about the formation of a tornado, we do know that tornadoes develop within thunderstorms and that a conditionally unstable atmosphere is essential for their development. Most often, tornadoes form with supercell thunderstorms in an environment with strong vertical wind shear.[*] The rotating air of the tornado may begin within a thunderstorm and work its way downward, or it may begin at the surface and work its way upward. First, we will examine tornadoes that form with supercells; then we will examine nonsupercell tornadoes.

SUPERCELL TORNADOES Tornadoes that form with supercell thunderstorms are called **supercell tornadoes**. In Chapter 14, we learned that a supercell is a thunderstorm that has a single rotating updraft that can exist for hours. ● Figure 15.14 illustrates

this updraft and the pattern of precipitation associated with the storm. Notice that as warm, humid air is drawn into the supercell, it spins counterclockwise as it rises. Near the top of the storm, strong winds push the rising air to the northeast. Heavy precipitation falling northeast of the updraft mixes with drier air. Evaporative cooling chills the air. The heavy rain-chilled air then descends as a strong downdraft called the *forward-flank downdraft*. The separation of the updraft from the downdraft means that the downdraft is unable to fall into the updraft and suppress it. This is why the storm is able to maintain itself as a single entity for hours.

Tornadoes are rapidly rotating columns of air, so what is it that starts the air rotating? We can see how rotation can develop by looking at ● Fig. 15.15a. Notice that there is directional shear in the vertical, as the surface winds are southeasterly and a kilometer or so above the surface they are westerly. There is also speed shear in the vertical, as the wind speed increases rapidly with height. This vertical wind shear causes the air near the surface to rotate about a horizontal axis much like a pencil would rotate around its long axis. Such horizontal tubes of spinning air are called *vortex tubes*. (These spinning vortex tubes also form when a southerly low-level jet exists just above weaker surface winds.) If the strong updraft of a developing thunderstorm should *tilt* the rotating tube upward and draw it into the storm, as illustrated in Fig. 15.15b, the tilted rotating tube then becomes a rotating air column inside the storm. The rising, spinning air is now part of the storm's structure called the **mesocyclone**—an area of lower pressure (a small cyclone) perhaps 5 to 10 kilometers across. The rotation of the updraft lowers the pressure in the mid-levels of the thunderstorm, which acts to increase the strength of the updraft.[**]

As we learned earlier in the chapter, the updraft is so strong in a supercell (sometimes exceeding 90 knots or 104 mi/hr) that

---

[*]Atmospheric conditions favorable for the formation of supercell thunderstorms are presented in Chapter 14, beginning on p. 394.

[**]You can obtain an idea of what might be taking place in the supercell by stirring a cup of coffee or tea with a spoon and watching the low pressure form in the middle of the beverage.

● **FIGURE 15.12** This huge EF4 multivortex tornado devastated sections of Tuscaloosa, Alabama, on April 27, 2011. (See also Fig. 15.13.)

AP Images/Dusty Compton

● **FIGURE 15.13** Damage in Tuscaloosa, Alabama, after a massive EF4 tornado (shown in Fig. 15.12) plowed through the city on April 27, 2011.

NOAA

● **FIGURE 15.14** A simplified view of a supercell thunderstorm with a strong updraft and downdraft, forming in a region of strong vertical wind speed shear. Regions beneath the supercell receiving precipitation are shown in color: green for light rain, yellow for heavier rain, and red for very heavy rain and hail.

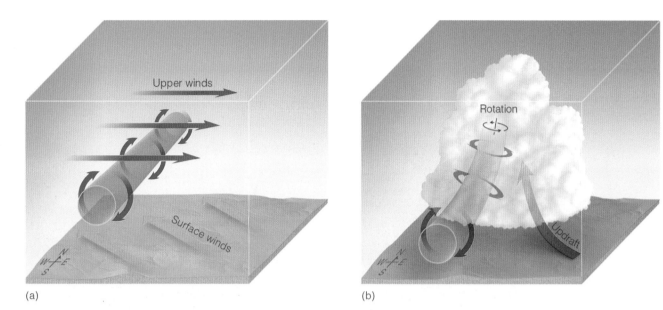

● **FIGURE 15.15** (a) A spinning vortex tube created by vertical wind shear. (b) The strong updraft in the developing thunderstorm carries the vortex tube into the thunderstorm producing a rotating air column that is oriented in the vertical plane.

● **FIGURE 15.16** A tornado-spawning supercell thunderstorm over west central Oklahoma during the afternoon May 20, 2013, shows a hook echo west of Moore, Oklahoma, in its rainfall reflectivity pattern on a Doppler radar screen. The colors red and orange represent the heaviest precipitation. Compare this precipitation pattern with the precipitation pattern illustrated in Fig. 15.14. (A photo of this tornado can be found in Chapter 13, Fig. 13.14a on p. 367.)

precipitation cannot fall through it. Southwesterly winds aloft usually blow the precipitation northeastward. If the mesocyclone persists, it can circulate some of the precipitation counterclockwise around the updraft. This swirling precipitation shows up on the radar screen, whereas the area inside the mesocyclone (nearly void of precipitation at lower levels) does not. The region inside the supercell where radar is unable to detect precipitation is known as the *bounded weak echo region* (BWER). Meanwhile, as the precipitation is drawn into a cyclonic spiral around the mesocyclone, the rotating precipitation may, on the Doppler radar screen, unveil itself in the shape of a hook, called a **hook echo**, as shown in ● Fig. 15.16.

At this point in the storm's development, the updraft, the counterclockwise swirling precipitation, and the surrounding air may all interact to produce the *rear-flank downdraft* (to the south of the updraft), as shown in Fig. 15.17. The strength of the downdraft is driven by the amount of precipitation-induced cooling in the upper levels of the storm. Downdrafts appear to be essential to the process of tornado formation. However, if a downdraft is too cold and strong, it may inhibit the lifting needed for tornado formation. In fact, relatively warm rear-flank downdrafts appear

to be more conducive to the formation of strong tornadoes. Researchers are still looking into many questions about how the rear-flank downdraft may evolve in a given storm.

When the rear-flank downdraft strikes the ground as illustrated in Fig. 15.17, it may (under favorable shear conditions) interact with the forward-flank downdraft beneath the mesocyclone to initiate **tornadogenesis**—the formation of a tornado. A preexisting boundary, such as an old gust front (outflow boundary), may also supply the surface air with horizontal spin that can be tilted and lifted into the storm by its updraft. At the surface of the mature supercell the strong winds of the rear-flank downdraft now wrap around the updraft at the center of the mesocyclone. This situation may initiate additional spin which can be lifted into the mesocyclone. At this point, the lower half of

● FIGURE 15.17 A classic mature tornadic supercell thunderstorm showing updrafts and downdrafts, along with surface air flowing counterclockwise and in toward the tornado. The flanking line is a line of cumulus clouds that form as surface air is lifted into the storm along the gust front.

the updraft begins to rise more slowly than the updraft aloft. The rising updraft, which we can imagine as a column of air, now shrinks horizontally and stretches vertically. This *vertical stretching* of the spinning column of air causes the rising, spinning air to spin faster.* If this stretching process continues, the rapidly rotating air column may shrink into a narrow column of rapidly rotating air—a *tornado vortex*.

As air rushes upward and spins around the low-pressure core of the vortex, the air expands, cools, and, if sufficiently moist, condenses into a visible cloud—the **funnel cloud**. As the air beneath the funnel cloud is drawn into its core, the air cools rapidly and its moisture condenses, and the funnel cloud descends toward the surface. Upon reaching the ground, the tornado's circulation usually picks up dirt and debris, making it appear both dark and ominous. This ground-based debris may become visible before the funnel cloud is connected to it. While the air along the outside of the funnel is spiraling upward, Doppler radar reveals that, within the core of large violent tornadoes, the air is descending toward the extreme low pressure at the ground (which may be 100 mb lower than that of the surrounding air). As the air descends, it warms, causing the cloud droplets to evaporate. This process leaves the core free of clouds.

Tornadoes within supercells usually develop near the right rear sector of the storm, on the southwestern side of a northeastward-moving storm, as shown in ● Fig. 15.17.

Many atmospheric situations can suppress tornado formation. For example, if the precipitation in the cloud is swept too far away from the updraft, or if too much precipitation wraps around the mesocyclone, the necessary interactions that produce the rear-flank downdraft are disrupted, and a tornado is unlikely to form. Moreover, tornadoes are not likely if the supercell is fed warm, moist air that is elevated above a deep layer of cooler surface air. And tornadoes usually will not form when the rain-chilled air of the rear-flank downdraft is too cold. Even among supercells that have mesocyclones evident on radar, only about 25 percent produce tornadoes. One of the main goals of tornado researchers is to better understand which supercells are most likely to generate tornadoes, especially the violent tornadoes that are the most destructive and deadly.

To an observer on the ground, the first sign that a supercell may give birth to a tornado is the sight of *rotating clouds* at the base of the storm.* If the area of rotating clouds lowers, it becomes the *wall cloud*. Notice in Fig. 15.17 that the tornado extends from within the wall cloud to Earth's surface. Sometimes the air is so dry that the swirling, rotating wind remains invisible until it reaches the ground and begins to pick up dust. Unfortunately, people have sometimes mistaken these "invisible tornadoes" for dust devils, only to find out (often too late) that they were not. Occasionally, the funnel is obscured by falling rain, clouds of dust, or darkness. When a tornado is not visible because it is surrounded by rain, it is referred to as being *"rainwrapped."* Even when not clearly visible, many tornadoes have a distinctive roar that can be heard as the tornado approaches. This sound, which has been described as "a roar like a thousand freight trains," appears to be loudest when the tornadic

---

*As the rotating air column stretches vertically into a narrow column, its rotational speed increases. You may recall from Chapter 10, on p. 275, that this situation is called the *conservation of angular momentum*.

*Occasionally, people will call a sky dotted with mammatus clouds "a tornado sky." Mammatus clouds may appear with both severe and nonsevere thunderstorms as well as with a variety of other cloud types (see Chapter 5). Mammatus clouds are not funnel clouds, they do not rotate, and their appearance has no relationship to tornadoes.

circulation is touching the surface. However, not all tornadoes make such sounds; when these twisters strike, they can become relatively silent killers.

Certainly, the likelihood of a thunderstorm producing a tornado increases when the storm becomes a supercell, but not all supercells produce tornadoes. And, as we will see in the next section, not all tornadoes come from rotating thunderstorms. (For additional information on predicting severe thunderstorms and tornadoes, read Focus section 15.3.)

NONSUPERCELL TORNADOES Tornadoes that do not occur in association with a mid-level mesocyclone (or a preexisting wall cloud) of a supercell are called **nonsupercell tornadoes**. These tornadoes may occur with intense multicell storms as well as with ordinary cell thunderstorms, even relatively weak ones. Some nonsupercell tornadoes extend from the base of a thunderstorm, whereas others may begin on the ground and build upwards in the absence of a condensation funnel.

Strong squall lines, and especially bow echoes (see Chapter 14, pp. 391–393), can produce nonsupercell tornadoes, particularly on the north side of a bow echo where cyclonic circulation is maximized. Tornadoes associated with squall lines and bow echoes are most frequent across the Mississippi and Ohio River valleys and over the Great Lakes states. These tornadoes are typically weaker than supercell tornadoes and tend not to last as long, but they can still be quite destructive and deadly.

Nonsupercell tornadoes may also form along a gust front where the cool downdraft of the thunderstorm forces warm, humid air upwards. Tornadoes that form along a gust front are commonly called **gustnadoes** (see ● Fig. 15.18). These relatively weak tornadoes normally are short-lived and rarely inflict significant damage. Gustnadoes are often seen as a rotating cloud of dust or debris rising above the surface.

Occasionally, rather weak, short-lived tornadoes will occur with rapidly building cumulus congestus clouds. Tornadoes such as these commonly form over northeastern Colorado and other parts of the High Plains. Because they look similar to waterspouts that form over water, they are sometimes called **landspouts*** (see ● Fig. 15.19). ● Figure 15.20 illustrates how a landspout can form. Suppose, for example, that the winds at the surface converge along a boundary, as illustrated in Fig. 15.20a.** Notice that along the boundary, the air is rising, condensing, and forming into a cumulus congestus cloud. Notice also that along the surface at the boundary there is horizontal rotation (spin) created by the wind blowing in opposite directions along the boundary. If the developing cloud should move over the region of rotating air (Fig. 15.20b), the spinning column may be drawn up into the cloud by the storm's updraft. In this case, the column of rising air typically narrows and its rotation intensifies as it is

---

*Landspouts occasionally form on the backside of a squall line where southerly winds ahead of a cold front and northwesterly winds behind it create swirling eddies that can be drawn into thunderstorms by their strong updrafts.

**The wind may converge due to topographic irregularities or any number of other factors, including temperature and moisture variations. In northeast Colorado, winds stirred by the Rocky Mountains and a smaller topographic feature (called the Palmer Divide) can converge and initiate uplift for landspout development.

● **FIGURE 15.18** A gustnado that formed along a gust front swirls across the plains of eastern Nebraska.

● **FIGURE 15.19** A well-developed landspout moves over eastern Colorado.

stretched upward. As the spinning, rising air shrinks in diameter, it produces a tornadic structure, a *landspout*, similar to the one shown in Fig. 15.19. Landspouts usually dissipate when rain falls through the cloud and destroys the updraft. Although they are

Forecasting Severe Thunderstorms and Tornadoes

We know that violent thunderstorms capable of producing severe weather form in a conditionally unstable environment with strong vertical wind shear (see Fig. 14.25, p. 397). We also know that, when atmospheric conditions are ripe for the development of tornadic thunderstorms, a *tornado watch* is issued by the Storm Prediction Center. But where exactly in the watch area will severe thunderstorms form, and which of those thunderstorms will produce tornadoes? These questions cannot yet be answered precisely on a given day. However, forecasters using a variety of indices can determine with increasing skill which areas are most likely to produce severe thunderstorms and tornadoes.

One important index for predicting the intensity of thunderstorms and the likelihood of tornado development is the *convective available potential energy* or *CAPE*, which we briefly looked at in Chapter 14 on p. 398.

Recall that CAPE represents the rising air's positive buoyancy, which is a measure of the amount of energy available to create updrafts in a thunderstorm. It is also proportional to the rate at which the air rises inside the storm. Thus, CAPE is an indication of the storm's intensity or strength. ▼ Table 1 shows how different values of CAPE relate to the likelihood of severe thunderstorm development.*

● Figure 3a is a 12-hour forecast map for values of CAPE for April 10, 2011. Notice that the highest forecast values extend over northeastern Iowa into southwestern Wisconsin, and

_____
*To compute the value of CAPE you need a temperature sounding similar to the one shown in Fig. 14.26 on p. 398. On the sounding, you lift a parcel of air from the surface and compare the rising parcel's temperature with that of the surrounding air, similar to what you did in Chapter 6 on p. 145. The area between the temperature of the rising parcel and that of the surrounding air is then converted to energy, or CAPE.

across Illinois into Indiana. Figure 3b shows the number of severe weather reports (damaging winds, large hail, and tornadoes) for April 10. It appears in Fig. 3b that a number of severe thunderstorms formed in the region of highest CAPE value (>2500 J/kg), then moved eastward and northeastward across Wisconsin, possibly producing a family of tornadoes along their paths, as Wisconsin reported a total of 19 tornadoes on this day.

Although a higher CAPE increases the potential for strong updrafts and significant tornadoes, CAPE is not the only factor involved in tornado formation. For example, we know that vertical wind shear and a rotating updraft (mesocyclone) are critical in the development of supercell storms. How then do forecasters estimate the chances that a thunderstorm will develop rotation? In an environment of sufficient vertical wind shear, a mesocyclone first appears in the mid-level of a thunderstorm (between about 10,000 and 18,000 ft). Forecasters estimate this region of favorable vertical wind shear (called the *deep layer shear*) by subtracting the surface wind direction and speed (the *surface wind vector*) from the wind vector at the 500-mb level (at about 18,000 ft). When the magnitude of this difference is on the order of 35 knots or greater, then mid-level mesocyclones and supercells become more likely.

Even in regions of favorable vertical deep layer shear, supercells may not develop rotating updrafts unless the surface-layer wind is at

▼ TABLE 1 Values of CAPE Related to Convection (Updrafts), and the Likelihood of Thunderstorms and Tornadoes

CAPE VALUE (J/KG)*	CONVECTION (UPDRAFTS)	FORECAST
0–1000	Weak	Isolated (ordinary cell/airmass) thunderstorms possible
1000–2500	Moderate	Intense thunderstorms possible
2500–3500	Strong	Severe thunderstorms possible
>3500	Very strong	Severe thunderstorms possibly with tornadoes

*The units of CAPE, J/kg, represent the amount of energy in joules (J) in 1 kilogram of air where 1 joule (J) equals 0.24 calories. Tornadoes are possible with relatively low CAPE if adequate wind shear is present.

(a) (b)

● **FIGURE 15.20** (a) Along the boundary of converging winds, the air rises and condenses into a cumulus congestus cloud. At the surface the converging winds along the boundary create a region of counterclockwise spin. (b) As the cloud moves over the area of rotation, the updraft draws the spinning air up into the cloud producing a nonsupercell tornado, or landspout. (Modified after Wakimoto and Wilson)

(a) CAPE forecast for April 10, 2011 (b) Storm reports for April 10, 2011

● **FIGURE 3** (a) Twelve-hour forecast for values of CAPE using the NAM model, valid April 10, 2011; (b) reports of wind damage, large hail, and tornadoes in the forecast area for April 10, 2011.

a substantial angle relative to the winds aloft. A measure of the degree to which the surface wind possesses this ability to rotate is called the *Storm Relative Helicity* (SRH), which is a measure of the low-level wind shear and how helical (corkscrew-like) the updraft in a growing thunderstorm will be. In general, the higher the value of SRH near the surface in layers about 3 km and 1 km thick, the greater the possibility for low-level rotation and tornado development, assuming a supercell is present. For helicity to be important, however, there must be some buoyancy. Therefore,

forecasters, when predicting supercells and tornado development, often take into account all three factors: CAPE, deep layer shear, and helicity. If the shear and helicity are strong enough, then even moderate amounts of CAPE can be enough to produce violent tornadoes.

The height of the thunderstorm cloud base—the lifted condensation level, or LCL— is another important factor. A relatively low LCL allows the rapidly rising air near the base of a supercell to coincide with strong low-level wind shear, which can make tornado production more likely.

There is one more factor to consider that can prevent supercells and tornadoes from developing even when all of the elements above are present: a low-level inversion, or "cap." If such a cap is strong enough, it can block parcels of warm, moist air from rising through it. Isolated tornadic supercells are most likely to develop when the cap is not so strong that it blocks all storms from developing, but strong enough to keep a large number of storms from forming (which would otherwise compete with each other).

typically brief and affect only a small area, landspouts are still capable of serious damage. Non-supercell tornadoes can form in this manner along many types of converging wind boundaries, including sea breezes and gust fronts. Nonsupercell tornadoes and funnel clouds may also form with thunderstorms when cold air aloft (associated with an upper-level trough) moves over a region. Common along the west coast of North America, these often short-lived tornadoes are sometimes called *cold-air funnels* (see ● Fig. 15.21).

WATERSPOUTS A **waterspout** is a rotating column of air that is connected to a cumuliform cloud over a large body of water. The waterspout may be a tornado that formed over

land and then traveled over water. In such a case, the waterspout is sometimes referred to as a *tornadic waterspout*. Such tornadoes can inflict major damage to ocean-going vessels, especially when the tornadoes are of the supercell variety. Strong waterspouts that form over water and then move over land can cause considerable damage. For example, on August 30, 2009, an intense waterspout formed over the warm Gulf of Mexico, then moved onshore into Galveston, Texas, where it caused EF1 damage over several blocks and injured three people.

Waterspouts that are not associated with supercells and that form over water, especially above warm, tropical coastal waters (such as in the vicinity of the Florida Keys, where almost 100 occur

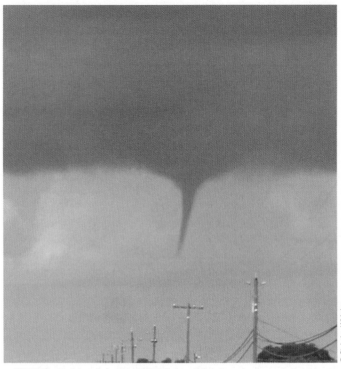

● **FIGURE 15.21** A funnel cloud—called a *cold-air funnel*—descends from a thunderstorm in California's Central Valley near Lodi.

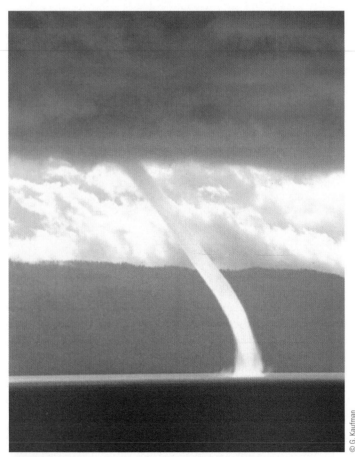

● **FIGURE 15.22** A powerful waterspout moves across Lake Tahoe, California. Compare this photo of a waterspout with the photo of a landspout in Fig. 15.19.

each month during the summer), are often referred to as "*fair weather*" *waterspouts*.* These waterspouts are generally much smaller than an average tornado, as they have diameters usually between 3 and 100 meters. Fair weather waterspouts are also less intense, as their rotating winds are typically less than 45 knots. In addition, they tend to move more slowly than tornadoes, and they typically only last for about 10 to 15 minutes, although some have existed for up to an hour.

Fair weather waterspouts tend to form in much the same way that landspouts do—when the air is conditionally unstable and cumulus clouds are developing. Some form with small thunderstorms, but most form with developing cumulus congestus clouds whose tops are frequently no higher than 3600 m (12,000 ft) and do not extend to the freezing level. Apparently, the warm, humid air near the water helps to create atmospheric instability, and the updraft beneath the resulting cloud helps initiate uplift of the surface air. Studies even suggest that gust fronts and converging sea breezes play a role in the formation of some of the waterspouts that form over the Florida Keys. As with a landspout, a waterspout becomes more likely when a preexisting boundary of converging air moves beneath a thunderstorm's updraft.

The waterspout funnel is similar to the tornado funnel in that both are clouds of condensed water vapor with converging winds that rise about a central core. Contrary to popular belief, the waterspout does not draw water up into its core; however, swirling spray can be lifted several meters when the waterspout funnel touches the water. A photograph of a particularly well-developed and intense waterspout is shown in ● Fig. 15.22.

Observing Tornadoes and Severe Weather

Starting as far back as the 1940s, organized networks of storm spotters kept an eye to the sky and reported the development of severe thunderstorms and tornadoes. Spotter networks are still important in notifying the National Weather Service of threatening conditions so that warnings can be promptly issued. In addition, laboratory models of tornadoes in chambers (called *vortex chambers*) as well as high-resolution computer models have provided insights into the three-dimensional processes at work in tornadic circulations.

In recent years, most of our knowledge about what goes on inside a tornado-generating thunderstorm has been gathered through the use of *Doppler radar*. Remember from Chapter 7 that a radar transmitter sends out microwave pulses and that, when this energy strikes an object, a small fraction is reflected and scattered back to the antenna. Precipitation particles are large enough to bounce microwaves back to the antenna. Consequently, as we saw earlier, the colorful area on the radar screen in Fig. 15.16, p. 427, represents the amount of reflected microwave energy translated into precipitation intensity inside a supercell thunderstorm.

Doppler radar can do more than measure rainfall intensity; it can actually measure the speed at which precipitation is moving

*Fair weather waterspouts can form over any large body of warm water. Hence, they occur frequently over the Great Lakes in summer.

NOAA/National Severe Storms Laboratory

● **FIGURE 15.23** Doppler radar display of winds within a supercell thunderstorm that moved through Tuscaloosa, Alabama, on April 27, 2011. The close packing of the horizontal winds blowing toward the radar (green shades) and those blowing away from the radar (red shades) indicate strong cyclonic rotation. A tornadic circulation (*tornado vortex signature*) exists where a small packet of red is adjacent to a small packet of green. (Doppler radar reflectivity for this tornado is shown in Figure 15.24.)

NOAA/NWS

● **FIGURE 15.24** Doppler radar display showing precipitation inside a large supercell that takes on the shape of a hook. This *hook echo* is associated with a violent multivortex tornado that is moving through Tuscaloosa, Alabama, just south of the University of Alabama, on April 27, 2011. The purple area in the center of the hook is debris that, having been picked up by the tornado, is now swirling counterclockwise around it. Debris on a radar screen such as this is referred to as a *debris ball*. (Damage caused by this tornado is shown in Fig. 15.13 on p. 426.)

horizontally toward or away from the radar antenna. Because precipitation particles are carried by the wind, Doppler radar can peer into a severe storm and unveil its winds.

Doppler radar works on the principle that, as precipitation moves toward or away from the antenna, the returning radar pulse will change in frequency when compared to the transmitted frequency. A similar change occurs when the high-pitched sound (high frequency) of an approaching noise source, such as a siren or train whistle, becomes lower in pitch (lower frequency) after it passes by the person hearing it. This change in frequency in sound waves or microwaves is called the *Doppler shift* and this, of course, is where the Doppler radar gets its name.

To help distinguish the storm's air motions, wind velocities can be displayed in color. Winds blowing toward the radar antenna are usually displayed in blue or green; those winds blowing away from the antenna are usually shown in shades of red. Color contouring the wind field gives a good picture of how winds are changing within a storm and the possibility of a tornado (see ● Fig. 15.23).

Doppler radar can uncover many of the features of a severe thunderstorm. For example, studies conducted in the 1970s revealed, for the first time, the existence of the swirling winds of the mesocyclone inside a supercell storm. Mesocyclones have a distinct image (signature) on the radar display. Tornadoes also have a distinct signature on the radar screen, known as the *tornado vortex signature* (TVS), which shows up as a region of rapidly (or abruptly) changing wind directions within the mesocyclone, as shown in Fig. 15.23.

When Doppler radar displays precipitation intensity (reflectivity) inside a supercell thunderstorm, a signature of a mesocyclone (or tornado) may appear on the radar screen as a hook-shaped appendage, or *hook echo*, as shown in ● Fig 15.24. The hook becomes visible as precipitation (and sometimes debris) swirls counterclockwise around the mesocyclone (or tornado). The doughnut-shaped dark red area at the end of the hook in Fig. 15.24 represents a massive multivortex tornado that is moving through Tuscaloosa, Alabama, on April 27, 2011, at about 5:10 p.m. CST.* The purple area in the center of the hook represents debris that, having been picked up by the tornado, is now swirling counterclockwise around it. Debris on a radar screen such as this is referred to as a *debris ball*. With the addition of dual-polarization technology to Doppler radar, debris can be observed even more clearly (see ● Fig. 15.25b). Although the hook echo in Fig. 15.25a has a tornado embedded in it, it should be noted that not all hook echoes are associated with tornadoes and not all tornadoes show a distinctive hook echo or a debris ball on the radar screen.

Unfortunately, the resolution of most Doppler radars is not high enough to measure actual wind speeds of most small tornadoes. However, a system called *Doppler lidar* uses a light beam (instead of microwaves) to measure the change in frequency of falling precipitation, cloud particles, and dust. Because it uses a shorter wavelength of radiation, it has a narrower beam and a higher resolution than does Doppler radar.

*See Fig. 15.12 on p. 426 for a photo of this massive tornado.

● FIGURE 15.25 (a) Doppler radar (reflectivity) shows precipitation intensity inside a large supercell thunderstorm on May 31, 2013, with heavy rain (red and yellow shades) falling west of Oklahoma City. A hook echo with a tornado exists to the south of El Reno, Oklahoma. (b) Doppler radar with dual-polarization technology more clearly depicts the different types of precipitation, as well as the tornado's debris, which shows up as a *tornado debris signature*.

(a) Doppler radar (reflectivity)

(b) Doppler radar with dual-polarization technology

The network of more than 150 Doppler radar units deployed at selected weather stations within the continental United States is referred to as **NEXRAD** (an acronym for *NEX*t Generation Weather *RAD*ar). The NEXRAD system consists of the WSR-88D* Doppler radar and a set of computers that perform a variety of functions.

The computers take in data, display them on a monitor, and run computer programs called *algorithms*, which, in conjunction with other meteorological data, detect severe weather phenomena, such as storm cells, hail, mesocyclones, and tornadoes. Algorithms provide a great deal of information to the forecasters that allows them to make better decisions as to which thunderstorms are most likely to produce severe weather and possible flash flooding. In addition, the algorithms give advanced and improved warning of an approaching tornado. More reliable warnings, of course, will cut down on the number of false alarms.

Because the Doppler radar shows horizontal air motion within a storm, it can help to identify the magnitude of other severe weather phenomena, such as gust fronts, derechos, microbursts, and horizontal wind shear that are dangerous to aircraft.

One of the latest advances in Doppler radar technology is the *polarimetric radar* (or *dual-polarization radar*) that transmits both a horizontal and a vertical radar pulse. This type of radar allows forecasters to better distinguish between very heavy rain and hail (see ● Fig. 15.26). The national NEXRAD radar network was upgraded in the early 2010s so that each Doppler radar now has dual-polarization capability. This should help improve the ability of forecasters to pinpoint areas where large hail, torrential rain, and flash flooding may occur. Researchers are also exploring the use of *phased array* radars, which allow a single radar unit to gather much more information than current Doppler radars by including many small transmitters and receivers on a flat plate.

Storm Chasing and Mobile Radar

Many people venture onto the Great Plains each spring to observe severe weather through "storm chasing," either informally or through organized tour groups. Most of this activity is unrelated to meteorological research. However, scientists have

---

*The name WSR-88D stands for *Weather Surveillance Radar, 1988 Doppler.*

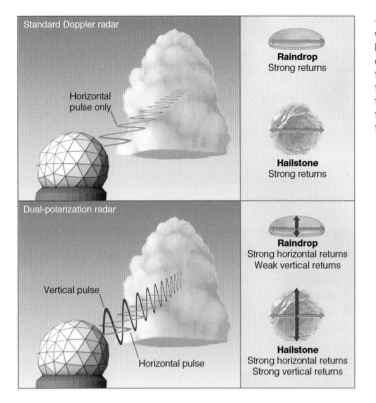

● **FIGURE 15.26** Signals returned to a Doppler radar from raindrops and hailstones within a thunderstorm. A standard Doppler radar (top) sends pulses that are oriented in a horizontal direction, while a dual-polarization Doppler radar (bottom) sends pulses that are oriented in both the horizontal and vertical. Raindrops will produce stronger radar returns for the horizontal pulses than for the vertical pulses, because the raindrops are fairly flat as they fall (see Focus section 7.3, p. 179). In contrast, because hailstones tumble as they fall, they tend to produce similar returns for both horizontal and vertical pulses. By comparing the returns for both types of pulses, a dual-polarization radar can distinguish heavy rain from hail.

been venturing into the field for more than 40 years in an attempt to directly observe tornadoes and collect observations as well as photos, films, and videos that might help unravel the secrets of tornado formation. By analyzing tornado films and videos, scientists can discern how fast cloud tags and pieces of debris are moving, which can then be compared to Doppler radar data. One of the earliest accomplishments of tornado field research occurred on May 24, 1973, with the filming of a tornado near Union City, Oklahoma. At the same time, the tornado was being sampled by experimental Doppler radar. For the first time, scientists observed a tornado vortex signature (TVS) on radar, and they noted that the TVS appeared more than 20 minutes before a tornado actually formed.

The two most comprehensive field studies on tornadoes to date were VORTEX (*V*erification of the *O*rigins of *R*otation in *T*ornadoes *Ex*periment), which took place in 1994 and 1995, and VORTEX2, which occurred in 2009 and 2010. In both studies, scientists used an armada of observational vehicles and state-of-the-art equipment. The original VORTEX included a turboprop aircraft and dozens of weather stations attached to the tops of cars, as well as the debut of a series of mobile Doppler radar units (Doppler on Wheels) mounted on trucks. VORTEX2 included a wider variety of mobile Doppler radar and lidar units, as well as compact surface weather stations and unmanned aircraft systems (see ● Fig. 15.27). To obtain as much information as possible, some instruments were placed directly in the paths of approaching storms, while others surrounded particular storms. The data obtained from these studies is being combined with high-resolution computer models of tornadic supercells to

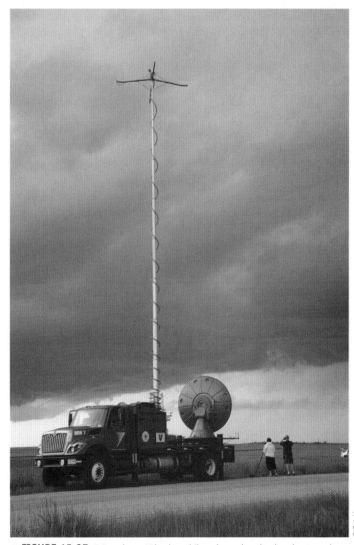

● **FIGURE 15.27** A Doppler on Wheels mobile radar probes the thunderstorm that produced a tornado near Lagrange, Wyoming, on June 5, 2009. The unit was one of several involved with the VORTEX2 field study.

provide new insights about the inner workings of supercells and tornadoes.

Storm chasing is a risky activity, even for veteran researchers. Three research-based storm chasers were killed in a violent tornado near El Reno, Oklahoma, on May 31, 2013—the same one shown on Doppler radar in Fig. 15.25a and 15.25b. With the research team positioned just ahead of it, the tornado suddenly changed direction, accelerated, and expanded to a record width of 4.2 km (2.6 mi), with multiple vortices spinning around the main circulation. One other person, a local resident, was killed while chasing the storm, and several other chase teams were struck by the tornado or by high winds nearby.

A mobile polarimetric Doppler radar operated by the University of Oklahoma detected near-surface winds estimated at 254 knots (291 mi/hr). However, because EF ratings of tornadoes are based on damage rather than radar observations, and because the strongest winds of the tornado apparently did not hit any structures, the tornado was classified as an EF3. In recent years, experts have been collaborating on a system that would expand the ways in which tornado wind speeds could be estimated and reported. The new system may incorporate mobile radar data as well as information about the damage patterns that emerge when tornadoes pass through a heavily treed area.

SUMMARY

Tornadoes are rapidly rotating columns of air with a circulation that reaches the ground. Tornadoes can form with supercells, as well as with less intense thunderstorms. Most tornadoes are less than a few hundred meters wide with wind speeds less than 100 knots, although violent tornadoes may have wind speeds that exceed 250 knots. A violent tornado may actually have smaller whirls (suction vortices) rotating within it. A normally small and less destructive cousin of the tornado is the "fair weather" waterspout that commonly forms above warm bodies of water. With the aid of Doppler radar and other tools, scientists are probing tornado-spawning thunderstorms, hoping to better predict tornadoes and to better understand where, when, and how they form.

KEY TERMS

The following terms are listed (with corresponding page numbers) in the order they appear in the text. Define each. Doing so will aid you in reviewing the material covered in this chapter.

tornado, 416	suction vortices, 420
funnel cloud, 416	tornado watch, 421
tornado alley, 417	tornado warning, 421

tornado emergency, 421	hook echo, 427
Fujita scale, 422	tornadogenesis, 427
Enhanced Fujita Scale (EF Scale), 422	nonsupercell tornadoes, 429
	gustnadoes, 429
tornado outbreak, 423	landspout, 429
supercell tornadoes, 425	waterspout, 431
mesocyclone, 425	NEXRAD, 434

QUESTIONS FOR REVIEW

1. What is the primary difference between a tornado and a funnel cloud?
2. Give some average statistics about tornado size, winds, and direction of movement.
3. At what point in its life cycle would a tornado resemble a rope?
4. Why should you *not* open windows when a tornado is approaching?
5. Why is the central part of the United States more susceptible to tornadoes than any other region of the world?
6. Explain both how and why there is a shift in tornado activity from early spring to midsummer.

7. How does a tornado *watch* differ from a tornado *warning*?

8. If you are in a single-story home (without a basement) during a tornado warning, what should you do?

9. Why was the original Fujita scale for tornado damage replaced by the Enhanced Fujita Scale?

10. Supercell thunderstorms that produce tornadoes form in a region of strong vertical wind shear. Explain how the wind changes in speed and direction to produce this shear.

11. What is the most common shape for a tornadic circulation when looking at precipitation on a Doppler radar? Why does that shape appear?

12. Explain how a nonsupercell tornado, such as a landspout, might form.

13. What atmospheric conditions lead to the formation of "fair weather" waterspouts?

14. Describe how Doppler radar measures the winds inside a severe thunderstorm.

15. How has Doppler radar helped in the prediction of severe weather?

QUESTIONS FOR THOUGHT

1. The number of strong tornadoes reported in the United States has held fairly steady over the last 60 years, while the number of weak tornadoes reported has doubled. What are at least two possible explanations for this trend? How would you investigate which of those explanations would be more likely?

2. Why might the average number of tornadoes in West Virginia be smaller than to the east and west of that state?

3. If you are confronted in an open field by a large tornado and there is no way that you could outrun it, probably the only thing that you could do would be to run and lie down in a depression. If given the choice, would you run toward your right or left as you face the approaching tornado? Explain your reasoning.

4. Tornadoes apparently form in the region of a strong updraft rather than in a downdraft, yet they descend from the base of a cloud. Why?

5. Why are left-moving supercell thunderstorms uncommon in the Northern Hemisphere, yet very common in the Southern Hemisphere?

PROBLEMS AND EXERCISES

1. A multivortex tornado with a rotational wind speed of 125 knots is moving from southwest to northeast at 30 knots. Assume the suction vortices within this tornado have rotational winds of 100 knots:

   a. What is the maximum wind speed of this multivortex tornado?

   b. If you are facing the approaching tornado, on which side (northeast, northwest, southwest, or southeast) would the strongest winds be found? the weakest winds? Explain both of your answers.

   c. According to Table 15.2, p. 422, how would this tornado be classified on the Enhanced Fujita Scale?

2. Suppose you and several of your friends were invited to join a storm-chasing research expedition in the central United States. Each night, you have access to a computer in your hotel room. Which current weather and forecast maps would you use to estimate the likelihood of severe weather in your study area on the next day? Explain why you choose those maps.

ONLINE RESOURCES

Visit www.cengagebrain.com to view additional resources, including video exercises, practice quizzes, an interactive eBook, and more.

Damage litters Canal Street in New Orleans after the passage of Hurricane Katrina in August 2005.

Hurricanes

CONTENTS

THE INDELIBLE MEMORY I HAVE OF COVERING the aftermath of Hurricane Katrina in New Orleans [August 2005] was how quickly and completely a modern, powered, connected, policed, orderly American city can descend into utter chaos. There was no 911 service; streets were impassable. No emergency rooms, no electricity, no stores, no communication system. The city had become preindustrial, medieval. And no one was coming. No one was in charge. No one knew what to do. Once people were rescued from rooftops and deposited on dry land, they were left completely on their own. Reporters became, in some cases, first responders. We were the first individuals that some evacuees encountered when they trudged out of their underwater neighborhoods. They wanted food, water, diapers, medicine—they didn't want an interview. All we could do was get their stories. I recall an overwhelming sense of powerlessness that we could not do more to help people, instead of just making deadline.

John Burnett, National Public Radio

The introduction on the previous page describes the terrible conditions in New Orleans after Hurricane Katrina made landfall in August 2005.* Born over warm tropical waters and nurtured by a rich supply of water vapor, *hurricanes* can grow into ferocious storms that generate enormous waves, heavy rains, severe flooding, and winds that can exceed 150 knots. What exactly are hurricanes? How do they form? And why do they strike the east coast of the United States far more frequently than the west coast? These are some of the questions we will consider in this chapter.

Tropical Weather

In the broad belt around the earth known as the tropics—the region between 23½° north and 23½° south of the equator—the weather is much different than it is in the middle latitudes. In the tropics, the noon sun is always high in the sky, and so diurnal and seasonal changes in temperature are small. The daily heating of the surface and high humidity favor the development of cumulus clouds and afternoon thunderstorms. Most of these are individual thunderstorms that are not severe. Sometimes, however, they group together into loosely organized systems called

*Landfall is the position along the coast where the center of a hurricane passes from ocean to land.

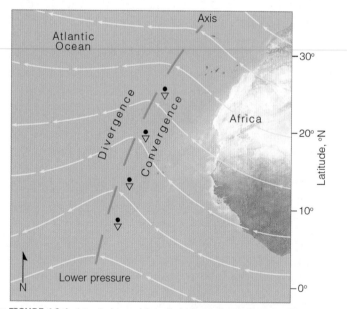

● **FIGURE 16.1** A tropical wave (also called an easterly wave) moving off the coast of Africa over the Atlantic. The wave is shown by the bending of streamlines—lines that show wind-flow patterns. (The dashed green line is the axis of the trough.) The wave moves slowly westward, bringing fair weather on its western side and rain showers on its eastern side.

CRITICAL THINKING QUESTION If the tropical wave in Fig. 16.1 were located over tropical water in the Southern Hemisphere, would you expect the wave to be moving from east to west? If yes, why? If no, why not? Would you expect the rain showers to still be located on the wave's eastern side?

non-squall clusters. On other occasions, the thunderstorms will align into a row of vigorous convective cells known as a *tropical squall cluster,* or *squall line.* The passage of a squall line is usually noted by a sudden wind gust followed immediately by a heavy downpour that may produce 3 cm (more than 1 in.) of rainfall in about 30 minutes. This deluge is then followed by several hours of relatively steady rainfall. Many of these tropical squall lines are similar to the middle-latitude ordinary squall lines described in Chapter 14 on p. 391.

Because it is warm all year long in the tropics, the weather is not characterized by four seasons, which, for the most part, are determined by temperature variations. Most of the tropics are marked instead by seasonal differences in precipitation. The greatest cloudiness and precipitation occur during the high-sun period, when the intertropical convergence zone moves into the region. Even during the dry season, precipitation can be irregular, as periods of heavy rain, lasting for several days, may follow an extreme dry spell.

The winds in the tropics generally blow from the east, northeast, or southeast. Because the variation in sea-level pressure is normally quite small, drawing isobars on a weather map provides little useful information. Instead of isobars, forecasters typically analyze **streamlines** that depict wind flow. Streamlines are useful because they show where surface air converges and diverges. Occasionally, the streamlines will be disturbed by a weak trough of low pressure called a **tropical wave**, or **easterly wave**, because it tends to move from east to west (see ● Fig. 16.1).

Tropical waves have wavelengths on the order of 2500 km (1550 mi) and travel from east to west at speeds of between 10 and 20 knots. Look at Fig. 16.1 and observe that, on the western side of the trough (dashed green line), where easterly and northeasterly surface winds diverge, sinking air produces generally fair weather. On its eastern side, southeasterly surface winds converge. The converging air rises, cools, and often condenses into showers and thunderstorms. Consequently, the main area of showers forms *behind* the trough. Occasionally, an easterly wave will intensify and grow into a hurricane.

Anatomy of a Hurricane

A **hurricane** is an intense storm of tropical origin, with sustained winds of at least 64 knots (74 mi/hr) and with considerably higher gusts, that forms over the warm northern Atlantic and eastern North Pacific Oceans. Hurricanes are composed of intense convective clouds (thunderstorms) that sometimes grow to over 15 km (50,000 ft) in height. This same type of storm is given different names in different regions of the world. In the

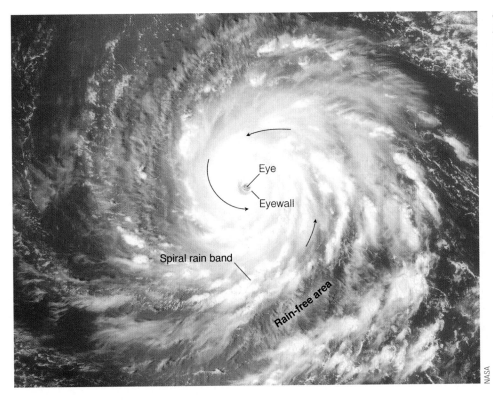

● **FIGURE 16.2** Hurricane Igor over the North Atlantic Ocean, about 1370 km (850 mi) east of the Leeward Islands, as photographed from NASA's *Aqua* satellite on September 13, 2010. Because this storm is situated north of the equator, surface winds are blowing counterclockwise about its center (eye). The central pressure of the storm is 933 mb, with sustained winds of 130 knots (150 mi/hr) near its eye.

Eye

Eyewall

Spiral rain band

Rain-free area

NASA

western North Pacific, it is called a **typhoon**, in India a *cyclone*, and in Australia a *tropical cyclone*. By international agreement, **tropical cyclone** is the general term for all hurricane-type storms that originate over tropical waters. For simplicity, we will refer to all of these storms as hurricanes.

● Figure 16.2 is a satellite image of Hurricane Igor, situated over the North Atlantic Ocean well east of the Caribbean Sea on September 13, 2010. The storm's thickest clouds cover an area approximately 500 km (310 mi) in diameter, which is about average for hurricanes. The relatively clear area at the center is the **eye**. Igor's eye is almost 40 km (25 mi) wide. Within the eye, winds are light and clouds are mainly broken. The surface air pressure is very low, around 933 mb (27.55 in.).* Notice that the clouds align themselves into spiraling bands (called *spiral rainbands*) that swirl in toward the storm's center, where they wrap themselves around the eye. Surface winds increase in speed as they blow counterclockwise and inward toward this center. (In the Southern Hemisphere, the winds blow clockwise around the center.) Adjacent to the eye is the **eyewall**, a ring of intense thunderstorms that whirl around the storm's center and may extend upward to almost 18 km (59,000 ft) above sea level. Within the eyewall, we find the heaviest precipitation and the strongest winds, which, in this storm, are 130 knots (150 mi/hr), with peak gusts of 160 knots (184 mi/hr).

If we were to venture from west to east (left to right) at the surface through the storm in Fig. 16.2, what might we experience? As we approach the hurricane, the sky becomes overcast with cirrostratus clouds; barometric pressure drops slowly at first, then more rapidly as we move closer to the center. Winds blow from the north and northwest with ever-increasing speed as we near the eye. The high winds, which generate huge waves over 10 m (33 ft) high, are accompanied by heavy rain showers. As we move into the eye, the winds slacken, rainfall ceases, and the sky brightens, as middle and high clouds appear overhead. The atmospheric pressure is now at its lowest point (933 mb), some 80 mb lower than the pressure measured on the outskirts of the storm. The brief respite ends as we enter the eastern region of the eyewall. Here, we are greeted by heavy rain and strong southerly winds. As we move away from the eyewall, the pressure rises, the winds diminish, the heavy rain lets up, and eventually the sky begins to clear.

This brief, imaginary venture raises many unanswered questions. Why, for example, is the surface pressure lowest at the center of the storm? And why is the weather clear almost immediately outside the storm area? To help us answer such questions, we need to look at a vertical view, a profile of the hurricane along a slice that runs through its center. A model that describes such a profile is given in ● Fig. 16.3.

The model shows that the hurricane is composed of an organized mass of thunderstorms** that are an integral part of the storm's circulation. Near the surface, moist tropical air flows in toward the hurricane's center. Adjacent to the eye, this air rises and the water vapor condenses into huge cumulonimbus clouds that produce heavy rainfall, as much as 15 cm (6 in.) per hour or more. Near the top of the clouds, the relatively dry air, having lost much of its moisture, begins to flow outward away from the center. This divergence aloft actually produces a clockwise flow

---

*An extreme low pressure of 870 mb (25.70 in.) was recorded in Typhoon Tip (while it was over the tropical Pacific Ocean) during October 1979, and Hurricane Patricia (while it was over the Northeast Pacific) had a pressure reading of 872 mb (25.75 in.) during October 2015.

**These huge convective cumulonimbus clouds have surprisingly little lightning (and, hence, thunder) associated with them. Even so, for simplicity we will refer to these clouds as thunderstorms throughout this chapter.

● **FIGURE 16.3** A model that shows a vertical view of air motions and clouds in a typical hurricane in the Northern Hemisphere. The diagram is exaggerated in the vertical.

of air (*anticyclonic* in the Northern Hemisphere) several hundred kilometers from the eye. As this outflow reaches the storm's periphery, it begins to sink and warm, inducing clear skies. In the vigorous convective clouds of the eyewall, the air warms due to the release of large quantities of latent heat. This produces slightly higher pressures aloft, which initiate downward air motion within the eye. As the air descends, it warms by compression. This process helps to account for the warm air and the absence of convective clouds in the eye of the storm (see ● Fig. 16.4).

● **FIGURE 16.4** A close-up of the eye and eyewall of Typhoon Maysak in the Northwest Pacific on April 2, 2015, taken from the International Space Station. Notice that the thunderstorms of the eyewall completely encircle the eye, while the eye itself is almost cloud-free. The eye is about 30 km (19 mi) wide.

● **FIGURE 16.5** A three-dimensional *TRMM* satellite view of Hurricane Katrina passing over the central Gulf of Mexico on August 28, 2005. The cutaway view shows concentric bands of heavy rain (red areas inside the clouds) encircling the eye. Notice that the heaviest rain (largest red area) occurs in the eyewall. The isolated tall cloud tower (in red) in the northern section of the eyewall indicates a cloud top of 16 km (52,000 ft) above the ocean surface. Such tall clouds in the eyewall often indicate that the storm is intensifying.

As surface air rushes in toward the region of much lower surface pressure, it should expand and cool, and we might expect to observe cooler air around the eye, with warmer air farther away. But, apparently, so much heat is added to the air from the warm ocean surface that the surface air temperature remains fairly uniform throughout the hurricane.

● Figure 16.5 is a three-dimensional radar composite of Hurricane Katrina as it passes over the central area of the Gulf of Mexico. Compare Katrina's features with those of typical hurricanes illustrated in Fig. 16.2 and Fig. 16.3. Notice that the strongest radar echoes (heaviest rain) near the surface are located in the eyewall, adjacent to the eye.

Hurricane Formation and Dissipation

We are now left with an important question: Where and how do hurricanes form? While there is no widespread agreement on how hurricanes actually form, it is known that certain necessary ingredients are required before a weak tropical disturbance will develop into a full-fledged hurricane.

THE RIGHT ENVIRONMENT   Hurricanes form over tropical waters where the winds are light, the humidity is high in a deep layer extending up through the troposphere, and the surface water temperature is warm, typically 26.5°C (80°F) or greater, over a vast area.* These conditions usually prevail over the tropical and subtropical North Atlantic and Northeast Pacific oceans

---

*It was once thought that for hurricane formation, the ocean must be sufficiently warm through a depth of about 200 meters. It is now known that hurricanes can form in the eastern North Pacific when the warm layer of ocean water is only about 20 m (65 ft) deep.

during the summer and early fall. These oceans usually reach their peak sea surface temperature in August or September because of the lag in seasonal temperature discussed in Chapter 3 (p. 65). The official hurricane season extends from May 15 to November 30 in the Northeast Pacific and from June 1 to November 30 in the North Atlantic, although a few tropical storms and hurricanes have formed outside these dates (as was the case for both regions in 2017). ● Figure 16.6 shows the number of tropical storms and hurricanes that one might expect per century over the tropical Atlantic in a 100-year period, based on data from 1870 to 2006. Notice that hurricane activity tends to pick up in August, peak in September, and then drop off rapidly.

For a mass of unorganized thunderstorms to develop into a hurricane, the surface winds must converge. In the Northern Hemisphere, converging air spins counterclockwise about an area of surface low pressure. Because this type of rotation will not develop on the equator where the Coriolis force is zero (see Chapter 8), hurricanes usually form at some distance from the equator, between about 5° and 20° latitude. (In fact, about two-thirds of all tropical cyclones form between 10° and 20° latitude.)

Hurricanes do not form spontaneously—they require a "trigger" to start the air converging. We know, for example, from Chapter 10 that surface winds converge along the intertropical convergence zone (ITCZ). Occasionally, when a wave forms along the ITCZ, an area of low pressure develops, convection becomes organized, and the system grows into a hurricane. Weak convergence also occurs on the eastern side of a tropical wave, where hurricanes sometimes form. In fact, many if not most Atlantic hurricanes can be traced to tropical waves that form over Africa. However, only a small fraction of all of the tropical disturbances that form over the course of a year ever grow into hurricanes.

Major Atlantic hurricanes are more numerous when the semiarid Sahel region of Africa is relatively wet. During the wet years, tropical waves are stronger, better organized, and more likely to develop into strong Atlantic hurricanes.

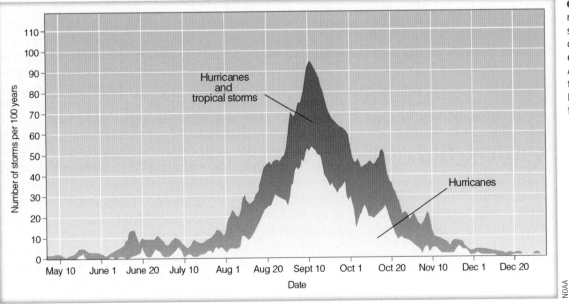

● **FIGURE 16.6** The total number of hurricanes and tropical storms (red shade) and hurricanes only (yellow shade) that one would expect every 100 years in the Atlantic Basin—the Atlantic Ocean, the Caribbean Sea, and the Gulf of Mexico—based on data extending from 1870 to 2006.

● **FIGURE 16.7** The top diagram shows an intensifying tropical cyclone. As latent heat is released inside the clouds, the warming of the air aloft creates an area of high pressure, which induces air to move outward, away from the high. The warming of the air lowers the air density, which in turn lowers the surface air pressure. As surface winds rush in toward the surface low, they extract sensible heat, latent heat, and moisture from the warm ocean. As the warm, moist air flows in toward the center of the storm, it is swept upward into the clouds of the eyewall. As warming continues, surface pressure lowers even more, the storm intensifies, and the winds blow even faster. This situation increases the transfer of heat and moisture from the ocean surface. The middle diagram illustrates how the air pressure drops rapidly as you approach the eye of the storm. The lower diagram shows how surface winds normally reach maximum strength in the region of the eyewall.

Convergence of surface winds may also occur along a pre-existing atmospheric disturbance, such as a front that has moved into the tropics from middle latitudes. Although the temperature contrast between the air on both sides of the front is gone, converging winds may still be present so that thunderstorms are able to organize.

Even when all of the surface conditions appear nearly perfect for the formation of a hurricane (for example, warm water, humid air, converging winds, and the necessary trigger), the storm may not develop if the weather conditions aloft are not just right. For instance, in the region of the trade winds, and especially near latitude 20°, the air is often sinking in association with the subtropical high-pressure area. The sinking air warms and creates an inversion above the surface, known as the **trade wind inversion**. When the inversion is strong, it can inhibit the formation of intense thunderstorms and hurricanes.

In addition, hurricanes do not form where the upper-level winds are strong. The resulting strong vertical wind shear tends to disrupt the organized pattern of convection and disperses heat and moisture, which are necessary for the growth of the storm. Yet another unfavorable factor for hurricane formation is the *Saharan air layer* (SAL), a mass of very dry air and blowing dust about 1.5 to 6 km (1 to 4 mi) above ground level that can cover an area as large as the contiguous United States. Pulses of the SAL move west from the Sahara Desert every few days. The dry air and strong winds associated with the SAL can weaken hurricanes and inhibit tropical cyclones from developing.

The warmer-than-normal water over the eastern tropical Pacific during El Niño conditions favors the development of a greater number of hurricanes than average over the Northeast Pacific.* These same El Niño conditions help to generate strong vertical wind shear over the tropical North Atlantic, so fewer hurricanes than average tend to form there. When the water over the eastern tropical Pacific turns cooler than normal (La Niña conditions), vertical wind shear tends to weaken over the tropical North Atlantic, producing more favorable conditions for Atlantic hurricane development.

THE DEVELOPING STORM The energy for a hurricane comes from the direct transfer of sensible heat and latent heat from the warm ocean surface. For a hurricane to form, a cluster of thunderstorms must become organized around a central area of surface low pressure to create a tropical cyclone. It appears that when a tropical wave moves at roughly the same speed as the surrounding upper-level flow, a protective zone of low wind shear and deep moisture may form that allows thunderstorms to gradually coalesce over several days. However, many questions remain about how such a cluster of thunderstorms goes on to evolve into a hurricane.

One theory proposes that a hurricane forms in the following manner: Suppose, for example, that the trade wind inversion is weak and that thunderstorms start to organize along the ITCZ, or along a tropical wave. In the deep, moist conditionally unstable environment, a huge amount of latent heat is released inside the clouds during condensation. The process warms the air aloft, causing the temperature near the cluster of thunderstorms to be much higher than the air temperature at the same level farther away. This warming of the air aloft causes a region of higher pressure to form in the upper troposphere (see ● Fig. 16.7), which, in turn, causes a horizontal pressure gradient aloft. The gradient induces the air aloft to move outward, away from the region of higher pressure in the anvils of the cumulonimbus clouds. This diverging air aloft, coupled with warming of the vertical air column, causes the surface pressure to drop and a small area of surface low pressure to form. The air now begins to spin counterclockwise (in the Northern Hemisphere) and in toward the region of surface low pressure. As the air moves inward, its speed increases, just as figure skaters spin faster when they bring their arms closer to their bodies (the conservation of angular momentum).

---

*Recall from Chapter 10, p. 279, that an El Niño event is a condition where extensive ocean warming occurs over the eastern tropical Pacific.

As the air moves over the warm water, small swirling eddies transfer heat energy from the ocean surface into the overlying air. The warmer the water and the greater the wind speed, the greater the transfer of sensible and latent heat into the air above. As the air sweeps in toward the center of lower pressure, the rate of heat transfer increases because the wind speed increases. Similarly, the higher wind speed causes greater evaporation rates, and the overlying air becomes nearly saturated. The turbulent eddies then transfer the warm, moist air upward, where the water vapor condenses to fuel new thunderstorms. As the surface air pressure lowers, wind speeds increase, more evaporation occurs at the ocean surface, and thunderstorms become more organized. The increasing winds also generate a large amount of sea spray that can partially evaporate, adding further heat and moisture to the lower atmosphere. At the top of the thunderstorms, heat is lost by the clouds radiating infrared energy into space.

The driving force behind a hurricane is similar to that of a heat engine. In a heat engine, heat is taken in at a high temperature, converted into work, then ejected at a low temperature. In a hurricane, heat is taken in near the warm ocean surface, converted to kinetic energy (energy of motion or wind), and lost at its top through radiational cooling.

In a heat engine, the amount of work done is proportional to the difference in temperature between its input and output region. The maximum strength a hurricane can achieve is proportional to the difference in air temperature between the tropopause and the surface, and to the potential for evaporation from the sea surface. As a consequence, the warmer the ocean surface, the lower the minimum pressure of the storm, and the higher its winds. Because there is a limit to how intense the storm can become, peak wind gusts seldom exceed 200 knots (230 mi/hr).

After a hurricane forms, it may go through an internal cycle of intensification. In strong hurricanes, for example, the eyewall may become encircled by a second eyewall, as another band of strong thunderstorms forms perhaps 5 to 24 km (3 to 15 mi) out from the original eyewall. The growing outer eyewall cuts off the moisture supply to the original eyewall, causing it to dissipate. The dissipation of the original eyewall and the formation of a new one farther out from the eye is called **eyewall replacement**. As the replacement of the eyewall is taking place, the central pressure of the storm may rise, and its maximum winds may lessen. Eventually, however, the newly formed eyewall will usually contract toward the center of the storm as the hurricane re-intensifies, provided that the surface water remains warm and other conditions remain favorable.

THE STORM DIES OUT   If a hurricane remains over warm water, it may survive for a long time. For example, Hurricane Tina (1992) traveled for thousands of kilometers over deep, warm, tropical waters and maintained hurricane force winds for 24 days, making it one of the longest-lasting North Pacific hurricanes on record. However, most tropical cyclones maintain hurricane strength for less than a week.

Hurricanes weaken rapidly when they travel over colder water and lose their heat source. Studies show that if the water beneath the eyewall of the storm (the region of thunderstorms adjacent to the eye) cools by 2.5°C (4.5°F), the storm's energy source is cut off, and the storm will dissipate. Even a small drop in water temperature beneath the eyewall will noticeably weaken the storm. A hurricane can also weaken if the layer of warm water beneath the storm is shallow. In this situation, the strong winds of the storm generate powerful waves that produce turbulence in the ocean water under the storm. Such turbulence creates currents that bring to the surface cooler water from below. If the storm is moving slowly, it is more likely to lose intensity, as the eyewall will remain over the cooler water for a longer period.

Hurricanes also dissipate rapidly when they move over a large landmass. Here, they not only lose their energy source but friction with the land surface causes surface winds to decrease and blow more directly into the storm, an effect that causes the hurricane's central pressure to rise. And a hurricane, or any tropical system for that matter, will rapidly dissipate should it move into a region of strong vertical wind shear.

Our understanding of hurricane behavior is far from complete. However, with the aid of computer model simulations, enhanced observations, and research projects, scientists are gaining new insight into how tropical cyclones form, intensify, and ultimately dissipate. For example, three major field campaigns in the hurricane season of 2010 used aircraft and specialized computer models to study the formation and intensification of tropical cyclones in the North Atlantic. These were followed from 2012 to 2014 by a mission called Hurricane and Severe Storm Sentinel, which used NASA's remotely piloted Global Hawk aircraft to probe hurricanes and the environmental factors that can affect their strength.

HURRICANE STAGES OF DEVELOPMENT   Hurricanes go through a set of stages from birth to death. Initially, a *tropical disturbance* shows up as a mass of thunderstorms with only slight wind circulation. The tropical disturbance becomes a **tropical depression** when the winds increase to between 20 and 34 knots (23 and 39 mi/hr) and several closed isobars appear about its center on a surface weather map. When the isobars are packed together and the winds are between 35 and 63 knots (40 and 74 mi/hr), the tropical depression becomes a **tropical storm**. (At this point, the storm gets a name.) If the sustained winds reach 64 knots (74 mi/hr), the tropical storm is classified as a *hurricane*. A tropical storm or hurricane will normally keep its name even after it weakens back to a tropical depression. If it moves to higher latitudes and begins to take on characteristics of a midlatitude cyclone, it may be classified as a post-tropical cyclone (still keeping its original name). Sometimes a cyclonic storm over the ocean will have characteristics of both tropical and midlatitude cyclones as it develops. If so, it would be classified as a subtropical storm and given a name as if it were a tropical storm.

● Figure 16.8 shows three tropical systems in various stages of development on July 10, 2012. Moving from east to west, we first see a small tropical disturbance (labeled 98E) south of Mexico. Over the next several days, this system will organize, strengthen, and eventually become Hurricane Fabio. Farther west, Hurricane Emilia is near its peak strength, with sustained winds of 120 knots. The more isolated system even farther to the west is Hurricane Daniel, now weakening over colder water. Its peak winds are down to 80 knots from a peak of 100 knots the previous day.

● **FIGURE 16.8** Infrared satellite image from 8 a.m. (EDT) on July 10, 2012, showing three tropical systems over the eastern tropical Pacific, each in a different stage of its life cycle. From left to right are weakening Tropical Storm Daniel, formerly a hurricane, with sustained winds of 55 knots (63 mi/hr); Hurricane Emilia, near its peak strength with sustained winds of 120 knots (138 mi/hr); and a small center of low pressure (a tropical disturbance called "98E") that developed several days later into Hurricane Fabio.

INVESTIGATING THE STORM   There are a variety of ways to obtain information about a developing hurricane and its environment. Visible, infrared, and enhanced infrared satellite images all provide a bird's-eye view of the storm, while sophisticated onboard radar instruments can actually peer into the storm and unveil its clouds to produce a three-dimensional image (see ● Fig. 16.9 and also Fig. 16.5 on p. 442). Some satellites are even equipped with onboard instruments capable of obtaining surface wind information in and around the storm (see ● Fig 16.10). A visible satellite image can be important in determining whether a developing hurricane will continue to strengthen. For example,

● **FIGURE 16.10** Arrows show surface winds spinning counterclockwise around Hurricane Dora situated over the eastern tropical Pacific during August 1999. Colors indicate surface wind speeds. Notice that winds of 80 knots (92 mi/hr) are encircling the eye (the dark dot in the center). Wind speed and direction obtained from NASA's *QuikSCAT* satellite, which gathered wind data from 1999 to 2009.

● **FIGURE 16.9** Three-dimensional satellite image of Hurricane Karl over the Bay of Campeche on September 16, 2010, with precipitation intensity from the *CloudSat* satellite. Karl made landfall as a major hurricane along the coast of Mexico, northeast of Veracruz.

the huge thunderstorms in the eyewall of the storm often produce a dense cirrus cloud shield that extends outward away from the eye, as illustrated in Fig. 16.3 on p. 442. If the storm in a visible satellite image has a well-defined eye and a dense cirrus cloud shield when it reaches hurricane strength, the storm will most likely continue to strengthen, as there appears to be insufficient wind shear to tear it apart.

Detailed information about a hurricane can also come from aircraft that fly directly into the storm. These so-called **hurricane hunters** carry instruments directly on the aircraft as well as instruments, such as the *dropsonde*, that are dropped from the aircraft into the storm. On its way down to the ocean surface, the dropsonde measures air temperature, humidity, and atmospheric pressure, which are transmitted back to the aircraft. Because the dropsonde is equipped with a Global Positioning System (GPS) that constantly monitors its changing position, it has the capability of providing wind information as well. Another temperature-measuring device dropped from the aircraft is the *bathythermograph*, which falls into the ocean where it measures water temperature as it slowly descends beneath the surface. Such probes can also measure the speed of ocean currents and the salinity (saltiness) of the water, an important factor in determining water density.

BRIEF REVIEW

Before reading the next several sections, here is a review of some of the important points about hurricanes.

- Hurricanes are tropical cyclones, comprised of an organized mass of thunderstorms.

- Hurricanes have sustained winds about a central core (eye) that reach at least 64 knots (74 mi/hr), with considerably higher gusts.

- The strongest winds and the heaviest rainfall normally occur in the eyewall—a ring of intense thunderstorms that surround the eye.

- Hurricanes form over warm tropical waters, where light surface winds converge, the humidity is high in a deep layer, and the winds aloft are weak.

- For a mass of thunderstorms to organize into a hurricane there must be some mechanism that triggers the formation, such as converging surface

winds along the ITCZ, a preexisting atmospheric disturbance, such as a weak front from the middle latitudes, or a tropical wave.

- Hurricanes derive their energy from the warm, tropical oceans and by evaporating water from the ocean's surface. Heat energy is converted to wind energy when the water vapor condenses and latent heat is released inside deep convective clouds.

- When hurricanes lose their source of warm water (either by moving over colder water or over a large landmass), they dissipate rapidly.

- The three primary stages in a developing hurricane are: tropical depression, tropical storm, and hurricane (tropical cyclone).

Up to this point, it is probably apparent that tropical cyclones called hurricanes are similar to middle-latitude cyclonic storms in that, at the surface, both have central cores of low pressure and winds that spiral counterclockwise (in the Northern Hemisphere) about their respective centers. However, there are many differences between the two systems, as described in Focus section 16.1.

Hurricane Movement

● Figure 16.11 shows where most hurricanes are born and the general direction in which they move, whereas ● Fig. 16.12 shows the actual paths taken by all hurricanes and tropical storms in the North Atlantic from 1980 to 2012. Notice that hurricanes that form over the warm, tropical North Pacific and North Atlantic generally move toward the west or northwest. Steered by easterly winds, they move at an average speed of about 10 knots, often for a week or so. Gradually, they swing poleward around the subtropical high, and when they move far enough north, they become caught in the westerly flow, which curves them to the north or northeast. In the middle latitudes, a hurricane's forward

● **FIGURE 16.11** Regions where tropical storms form (orange shading), the names given to storms, and the typical paths they take (red arrows).

● **FIGURE 16.12** Paths taken by hurricanes and tropical storms in the northern Atlantic from 1980 to 2012.

speed normally increases, sometimes to more than 50 knots. The actual path of a hurricane (which appears to be determined by the structure of the storm and the storm's interaction with the environment) may vary considerably. Some take erratic paths and make odd turns that have occasionally caught weather forecasters by surprise (see ● Fig. 16.13), although modern forecasting techniques are more likely to identify such twists and turns at least a day or two in advance. There have been many instances where a storm heading directly for land suddenly veered away and spared the region from almost-certain disaster. As a case in point, Hurricane Elena, with peak sustained winds of 90 knots, moved northwestward into the Gulf of Mexico on August 29, 1985. It then veered eastward toward the west coast of Florida. After stalling offshore, it headed northwest, weakened, and finally moved onshore near Biloxi, Mississippi, on the morning of September 2.

Look at Fig. 16.11 and notice that hurricanes apparently do not form over the South Atlantic and the eastern South Pacific—directly east and west of South America. Cooler water, vertical wind shear, and the unfavorable position of the ITCZ discourages hurricanes from developing in these regions. But guess what? For the first time since satellites began observing the south Atlantic, a hurricane formed off the coast of Brazil during March 2004. ● Figure 16.14 is a satellite image of the storm. So rare are tropical cyclones in this region that no government agency has an effective warning system for them, which is why the tropical cyclone was not given a name. It was informally dubbed *Catarina* because it struck the state of Santa Catarina. The storm caused more than $350 million in damages and resulted in seven deaths. Six years later, in March 2010, another tropical storm (informally named Anita), formed east of Brazil, prompting the government of Brazil to begin maintaining a list of names for tropical cyclones. It is quite possible other such storms developed before the era of satellite imagery.

Eastern Pacific Hurricanes　As we saw in an earlier section, many hurricanes form off the coast of Mexico over the Northeast Pacific. In fact, this area usually spawns about nine hurricanes each year, which is slightly more than the yearly average of six hurricanes born over the North Atlantic. We can see in Fig. 16.11 that eastern North Pacific hurricanes normally move westward, away from the coast. Because there is no landfall, little is heard about them. When one does move northwestward, it normally weakens rapidly over the cool water of the North Pacific. Occasionally, however, one will curve northward or even northeastward and slam into Mexico, causing destructive flooding. Hurricane Tico left 25,000 people homeless and caused an estimated $66 million in property damage after passing over Mazatlán, Mexico, in

● **FIGURE 16.13** Some erratic paths taken by hurricanes.

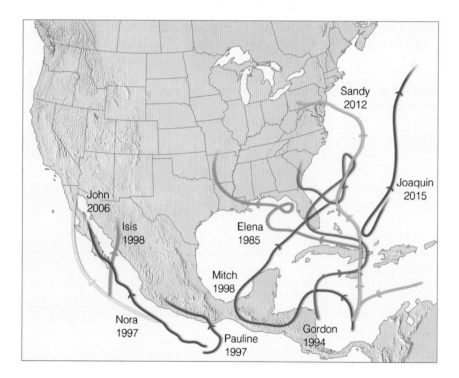

How Do Hurricanes Compare with Middle-Latitude Cyclones?

By now, it should be apparent that a hurricane is much different from the midlatitude cyclone that we discussed in Chapter 12. A hurricane derives its energy from the warm water and the latent heat of condensation, whereas the midlatitude storm derives its energy from horizontal temperature contrasts. The vertical structure of a hurricane is such that its central column of air is warm from the surface upward; this is why hurricanes are called *warm-core lows*. A hurricane weakens with height, and the area of low pressure at the surface may actually become an area of high pressure above 12 km (40,000 ft). Midlatitude cyclones, on the other hand, are *cold-core lows* that usually intensify with increasing height, with a cold upper-level low or trough often located above or to the west of the surface low.

A hurricane usually contains an eye where the air is sinking, while midlatitude cyclones are characterized by centers of rising air. Hurricane winds are strongest near the surface, whereas the strongest winds of the midlatitude cyclone are found aloft in the jet stream.

Further contrasts can be seen on a surface weather map. ● Figure 1 shows Hurricane Rita over the Gulf of Mexico and a midlatitude cyclonic storm north of New England. Around the hurricane, the isobars are more circular, the pressure gradient is much steeper, and the winds are stronger. The hurricane has no fronts and is smaller (although Rita happens to be a large Category 5 hurricane). There are similarities between the two systems: Both are areas of surface low pressure, with winds moving counterclockwise about their respective centers. ▼ Table 1 summarizes the similarities and differences between the two systems.

Some nor'easters (winter storms that move northeastward along the coastline of North America, bringing with them heavy precipitation, high surf, and strong winds) actually possess some of the characteristics of a hurricane (see Chapter 12, p. 325). For example, a particularly powerful nor'easter during January 1989—one of the strongest on record—was observed to have a cloud-free eye, with surface winds in excess of 85 knots (98 mi/hr) spinning about a warm inner core.

Occasionally, a hurricane will transition into a *post-tropical cyclone* such as a nor'easter. One particularly dramatic example is Superstorm Sandy in 2012 (see p. 462). Moreover, some *polar lows*—lows that develop over polar waters during winter—may exhibit many of the observed characteristics of a hurricane, such as a symmetric band of thunderstorms spiraling inward around a cloud-free eye, a warm-core area of low pressure, and strong winds near the storm's center. In fact, when surface winds within these polar storms reach 58 knots (67 mi/hr), they are sometimes referred to as *Arctic hurricanes*. (Observe the satellite image of the polar low in Fig. 12.30 on p. 342.)

Even though hurricanes weaken rapidly as they move inland, their circulation may draw in air with contrasting properties. If the hurricane links with an upper-level trough, it may actually become a midlatitude cyclone, sometimes bringing very heavy rain over a wide area. Swept eastward by upper-level winds, the remnants of an Atlantic hurricane can become an intense midlatitude autumn storm in Europe.

● **FIGURE 1** Surface weather map for the morning of September 23, 2005, showing Hurricane Rita over the Gulf of Mexico and a middle-latitude cyclonic storm system north of New England.

▼ **TABLE 1** **Comparison of Hurricanes with Midlatitude Cyclonic Storms**

	TYPE OF STORM	
CONDITIONS	Hurricane	Mid-Latitude Cyclone
Wind flow	Counterclockwise (NH) Clockwise (SH)	Counterclockwise (NH) Clockwise (SH)
Strongest winds	Near surface; around eye	Aloft, in jet stream
Surface pressure	Lowest at center	Lowest at center
Vertical structure	Weaken with height; high pressure aloft; warm-core low	Strengthens with height, low pressure aloft; cold-core low
Air in center	Sinking	Rising
Weather fronts	No	Yes
Energy source	Warm water; release of latent heat	Horizontal temperature contrasts

FIGURE 16.14 An extremely rare tropical cyclone (with no official name) near 28°S latitude spins clockwise over the south Atlantic off the coast of Brazil during March 2004. Due to cool water and vertical wind shear, storms rarely form in this region of the Atlantic Ocean. In fact, this was the first hurricane-strength tropical cyclone ever officially reported there.

NASA

October 1983. The remains of Tico even produced record rains and flooding in Texas and Oklahoma. The record for top wind speed for any tropical cyclone on Earth in recent decades is held by Hurricane Patricia, whose sustained winds reached an estimated 185 knots (215 mi/hr) southwest of Manzanillo, Mexico, in October 2015. Fortunately, Patricia was a very small hurricane, and it weakened dramatically before landfall.

Even less frequently, a hurricane in the Northeast Pacific will stray far enough north to bring summer rains to southern California and Arizona, as did the remains of Hurricane Nora during September 1997. (Nora's path is shown in Fig. 16.13.) The only hurricane on record to reach the west coast of the United States slammed into extreme southern California near San Diego in October 1858.

The Hawaiian Islands, which are situated in the central North Pacific between about 20° and 23°N, appear to be in the direct path of many eastern Pacific hurricanes and tropical storms. By the time most of these storms reach the islands, however, they have weakened considerably, and pass harmlessly to the south or northeast. Exceptions were Hurricane Iwa during November 1982 and Hurricane Iniki during September 1992. Iwa lashed part of Hawaii with 100-knot winds and huge surf, causing an estimated $312 million in damages. Iniki, the worst hurricane to hit Hawaii in the twentieth century, battered the island of Kauai with torrential rain, sustained winds of 114 knots that gusted to 140 knots, and 20-foot waves that crashed over coastal highways. Major damage was sustained by most of the hotels and about 50 percent of the homes on the island. Iniki, the costliest hurricane in Hawaiian history with damage estimates of $1.8 billion, flattened sugarcane fields, destroyed the macadamia nut crop, injured about 100 people, and caused at least 7 deaths. Recently, the Big Island of Hawaii experienced its first two named-storm landfalls on record. After weakening from Category 4 hurricane status in August 2014, Tropical Storm Iselle destroyed the bulk of the island's papaya crop when it made landfall as a tropical

storm. Similarly, Tropical Storm Darby weakened from Category 4 status before it made landfall in July 2016.

North Atlantic Hurricanes   Hurricanes that form over the tropical North Atlantic also move westward or northwestward on a collision course with Central or North America. Most hurricanes, however, swing away from land and move northward, parallel to the coastline of the United States (see Fig. 16.12).* On average, one or two hurricanes will strike the United States coast each year, bringing high winds, huge waves, and torrential rain that may last for days. There is a great deal of year-to-year variability in United States hurricane landfalls: Some years may see none, while other years bring four or more. In addition, several tropical storms can be expected to make landfall on the United States coast in a typical year. These bring less-severe winds and waves than hurricanes do, but the rainfall accompanying a tropical storm can be just as heavy as in a hurricane.

● Figure 16.15 shows the regions where Atlantic Basin hurricanes tend to form and the typical paths they take during the active hurricane months of August, September, and October. Observe that, during August, hurricanes are most likely to form over the western tropical Atlantic, where they then either track westward into the Gulf of Mexico toward Texas, move northwestward into Florida, or follow a path parallel to the coast of the United States. In September, notice that the region where hurricanes are most likely to form stretches westward into the Gulf of Mexico and northward along the Atlantic seaboard. Typical hurricane paths take them into the central Gulf of Mexico or northeastward out over the Atlantic. Should an Atlantic hurricane track close to the coastline, it could make landfall anywhere from Florida to the mid-Atlantic states. In October, hurricanes

---

*Sometimes hurricanes that remain over water and pose no threat to land are called "fish hurricanes" or "fish storms" because their greatest impact is on the fish in the open ocean.

(a) August

(b) September

(c) October

● **FIGURE 16.15** Regions where Atlantic Basin hurricanes tend to form, and the paths they are most likely to take during the months of (a) August, (b) September, and (c) October. (Data from NOAA)

are most likely to form in the western Caribbean and adjacent to the coast of North America, where they tend to take a more northerly trajectory. Note that Fig. 16.15 shows only the most common tracks observed over the long term. Any actual hurricane could behave much differently from these averages.

A hurricane moving northward over the Atlantic will normally survive as a hurricane for a much longer time than will its counterpart at the same latitude over the eastern Pacific. Why? An Atlantic hurricane moving northward usually stays over warmer water. In contrast, an eastern Pacific hurricane heading north will quickly move over much cooler water and, with its energy source cut off, will rapidly weaken.

Naming Hurricanes and Tropical Storms

In an earlier section, we learned that hurricanes are given a name when they reach tropical storm strength. Before hurricanes and tropical storms were assigned names, they were identified according to their latitude and longitude. This method was confusing, especially when two or more storms were present over the same ocean. When radio and television weathercasting became more important, the need to identify hurricanes for the public increased. To reduce the confusion, hurricanes were identified by letters of the alphabet. Starting in 1950, names such as Able and Baker were used. (These names correspond to the radio code words associated with each letter of the alphabet.) This method also led to confusion so, beginning in 1953, the National Weather Service began using female names to identify hurricanes. The list of names for each year was in alphabetical order, so that the names of the season's first storm began with the letter *A*, the second with *B*, and so on.

From 1953 to 1977, only female names were used. However, beginning in 1978, tropical storms in the eastern Pacific were alternately assigned female and male names—but not just English names, as Spanish and French ones were used, too. This practice was adopted for North Atlantic hurricanes in 1979. The World Meteorological Organization now coordinates the naming systems for each tropical cyclone region around the world. The Northeast Pacific and Central Pacific each has its own list of storm names. In the Northwest Pacific, for example, most typhoons are named after birds, flowers, or other items rather than people. If a storm causes great damage and/or loss of life, its name may be retired and a replacement name added to the list.

▼ Table 16.1 gives the current list of names for North Atlantic hurricanes. The lists are recycled every six years, so the list for 2019 will be used again in 2025. If the number of named storms in any year should exceed the names on the list, which occurred in 2005 for the first time, then tropical storms are assigned names from the Greek alphabet, such as Alpha, Beta, and Gamma. The last of the 27 named tropical systems in 2005 was Zeta, which formed on December 30. If it had developed just two days later, it would have been dubbed Alberto, the first name on the 2006 list.

Devastating Winds, the Storm Surge, and Flooding

When a hurricane is approaching from the south, its highest winds are usually on its eastern (right) side. The reason for this phenomenon is that the winds that push the storm along add to the winds on the east side and subtract from the winds on the west (left) side. The hurricane illustrated in Fig. 16.16 is moving northward along the east coast of the United States with winds of 100 knots swirling counterclockwise about its center. Because the storm is moving northward at about 25 knots, sustained winds on its eastern side are about 125 knots, while on its western side, winds are only 75 knots.

The stronger winds on the storm's eastern side will likely cause the highest storm surge and most damage just east of the eye as the storm moves onshore. If the hurricane in Fig. 16.16 should suddenly change direction and move toward the west, its strongest winds, highest storm surge, and greatest potential for damage would now be just north of the eye.

2018	2019	2020	2021	2022	2023
Alberto	Andrea	Arthur	Ana	Alex	Arlene
Beryl	Barry	Bertha	Bill	Bonnie	Bret
Chris	Chantal	Cristobal	Claudette	Colin	Cindy
Debby	Dorian	Dolly	Danny	Danielle	Don
Ernesto	Erin	Edouard	Elsa	Earl	Emily
Florence	Fernand	Fay	Fred	Fiona	Franklin
Gordon	Gabrielle	Gonzalo	Grace	Gaston	Gert
Helene	Humberto	Hanna	Henri	Hermine	Harvey*
Isaac	Imelda	Isaias	Ida	Ian	Irma*
Joyce	Jerry	Josephine	Julian	Julia	Jose
Kirk	Karen	Kyle	Kate	Karl	Katia
Leslie	Lorenzo	Laura	Larry	Lisa	Lee
Michael	Melissa	Marco	Mindy	Martin	Maria*
Nadine	Nestor	Nana	Nicholas	Nicole	Nate
Oscar	Olga	Omar	Odette	Owen	Ophelia
Patty	Pablo	Paulette	Peter	Paula	Philippe
Rafael	Rebekah	Rene	Rose	Richard	Rina
Sara	Sebastien	Sally	Sam	Shary	Sean
Tony	Tanya	Teddy	Teresa	Tobias	Tammy
Valerie	Van	Vicky	Victor	Virginie	Vince
William	Wendy	Wilfred	Wanda	Walter	Whitney

*The names Harvey, Irma, and Maria will likely be retired because of the devastation these hurricanes wrought in 2017 (see p. 468).

Even though the hurricane in ● Fig. 16.16 is moving northward, there is a net transport of water directed from the east toward the coast. To understand this behavior, recall from Chapter 10 that as the wind blows over open water, the water beneath is set in motion. If we imagine the top layer of water to be broken into a series of layers, then we find each layer moving to the *right* of the layer above (in the Northern Hemisphere). This type of movement (bending) of water with depth (called the *Ekman Spiral*) causes a net transport of water (known as **Ekman transport**) to the right of the surface wind in the Northern Hemisphere. Hence, the north wind on the hurricane's left (western) side causes a net transport of water toward the shore. Here, the water piles up and rapidly inundates the region.

The high winds of a hurricane also generate large waves, sometimes 10 to 15 m (33 to 49 ft) high. These waves move outward, away from the storm, in the form of *swells* that carry the storm's energy to distant beaches. Consequently, the effects of the storm can be felt days before the hurricane arrives.

Although the hurricane's high winds inflict a great deal of damage, it is the huge waves, high seas, and *flooding* that normally cause most of the destruction. The flooding is also responsible for the loss of many lives. In fact, the majority of hurricane-related deaths during the past century has been due to flooding. The flooding is due, in part, to winds pushing water onto the shore and to the heavy rains, which may exceed 63 cm (25 in.) in 24 hours. Flooding is also enhanced by the low pressure of the

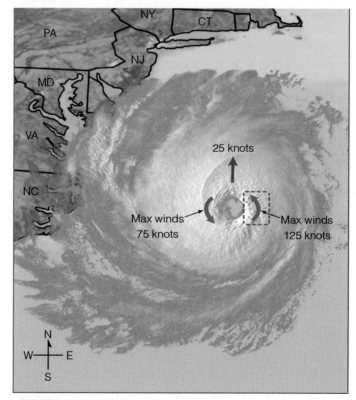

● FIGURE 16.16 A hurricane moving northward will have higher sustained winds on its eastern side than on its western side. The boxed area represents the region of strongest winds.

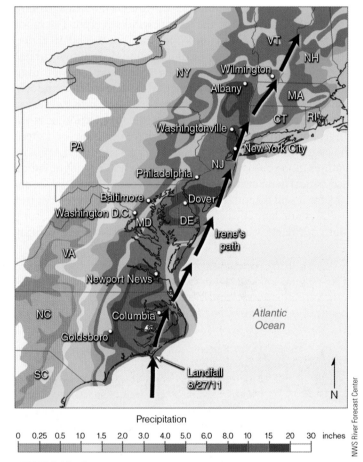

● **FIGURE 16.17** The high water observed in this hurricane is a combination of high tide (2 ft), the storm surge (15 ft), and waves (10 ft). The high-water level for the hurricane in this example is 27 ft. The high tide and waves add to the effect of the storm surge, allowing water to move inland for great distances and to destroy a wide swath of coastal structures.

storm. The region of low pressure allows the ocean level to rise (perhaps half a meter), much like liquid rising through a straw as air is withdrawn. (A drop of 1 millibar in air pressure produces a rise in ocean levels of 1 centimeter.) The combined effect of high water (which is usually well above the high-tide level), high winds, and the net Ekman transport toward the coast produces the **storm surge**—an abnormal rise of up to several meters in the ocean level—which can inundate low-lying areas and turn beachfront homes into piles of splinters (see ● Fig. 16.17). The storm surge is particularly damaging when it coincides with normal high tides.

The heavy rains produced by a hurricane can have a beneficial effect by providing much-needed rainfall to drought-stricken regions. However, extreme flooding can occur well inland with strong hurricanes as well as relatively weak storms, such as Hurricane Irene, a large Category 1 hurricane that made landfall along the coast of North Carolina on August 26, 2011. Irene produced heavy rain and record flooding from North Carolina to Vermont (see ● Fig. 16.18). Flooding is not just associated with hurricane-strength tropical cyclones, as destructive floods can occur with tropical storms that do not reach hurricane strength. More on this topic is presented in Focus section 16.2.

CLASSIFYING HURRICANE STRENGTH   In an effort to estimate the possible damage a hurricane's sustained winds and storm surge could do to a coastal area, the **Saffir-Simpson scale** was developed. The scale numbers (which range from 1 to 5) are based on actual conditions at some time during the life of the storm. As the hurricane intensifies or weakens, the category, or scale number, is reassessed accordingly. Major hurricanes are classified as Category

● **FIGURE 16.18** Hurricane Irene's path (dark arrows) and estimated rainfall totals over the eastern United States from August 26 through August 29, 2011. Irene, the only hurricane to make landfall in the United States in 2011, was a massive but relatively weak storm that produced heavy rainfall and record flooding over sections of the Northeast.

WEATHER WATCH

The largest tropical cyclone on record was Super Typhoon Tip, which formed in the western Pacific on October 5, 1979. At its peak it had a circulation of tropical storm–force winds that spanned 1350 miles, the distance from Key West, Florida, to Amarillo, Texas. On October 12, the storm's central pressure fell to 870 mb (25.69 in.), the lowest ever measured in any tropical system, and its winds reached an estimated 165 knots (190 mi/hr), a strength that ranks second only to that of Hurricane Patricia (2015). Fortunately, Typhoon Tip never made landfall.

3 and above. In the Northwest Pacific, a typhoon with sustained winds of at least 130 knots (150 mi/hr)—near the top end of Category 4 on the Saffir-Simpson scale—is designated a **super typhoon** by the United States Joint Typhoon Warning Center.*

---

*The United States Joint Typhoon Warning Center and the National Hurricane Center use top winds averaged over 1-minute intervals to gauge hurricane strength. In other parts of the world, a 10-minute interval is used, which yields a somewhat lower number for top wind speed. Some other countries, such as Australia, use rating systems that differ from the Saffir-Simpson scale.

▼ **TABLE 16.2** **Saffir-Simpson Hurricane Wind Scale**

| SCALE CATEGORY | WINDS (ONE-MINUTE SUSTAINED) | | SUMMARY* |
	mi/hr	knots	
1	74–95	64–82	Very dangerous winds will produce some damage
2	96–110	83–95	Extremely dangerous winds will cause extensive damage
3	111–129	96–112	Devastating damage will occur
4	130–156	113–136	Catastrophic damage will occur
5	>156	>136	Catastrophic damage will occur

*The scale provides extensive information for each category on the potential harm to people and pets and potential damage to structures such as mobile homes, houses, apartments, shopping centers, and so on.

The original Saffir-Simpson scale used a hurricane's central pressure as a measure of the storm's wind strength. The modified *Saffir-Simpson Hurricane Wind Scale* (see ▼ Table 16.2) does not use central pressure, as maximum winds today are accurately determined by modern observing techniques and are more directly relevant to a storm's impact. Also removed from the original scale is the storm surge, as it has become clear that other factors besides wind speed play a major role in determining storm surge. For example, if two hurricanes have the same wind speed, the larger one will normally produce a greater surge. Landscape features along the coast, as well as underwater topography, are also critical in determining how high the storm surge will be and how far it will extend inland. For example, Hurricane Andrew (1992) made landfall as a Category 5 storm, but its peak surge south of Miami was less than 17 ft high. In contrast, Hurricane Ike (2008) made landfall as a Category 2, but it brought a surge as high as 20 ft to parts of Texas.

Because a hurricane storm surge can occur on top of natural high or low tides, it can be difficult to convey the actual high-water levels that coastal residents in the path of a hurricane might experience. The National Hurricane Center has recently been revising its procedures for communicating the risk of storm surge during hurricane threats. These include storm surge watches and warnings, implemented in 2017, that are now issued in sync with hurricane watches and warnings. Another of the newly designed tools is a map that incorporates storm surge, tidal levels, and river flooding to show the maximum height of water above ground level that could occur during a given time span near the coast as a hurricane approaches. For an example of an inundation map, see ● Fig. 16.19.

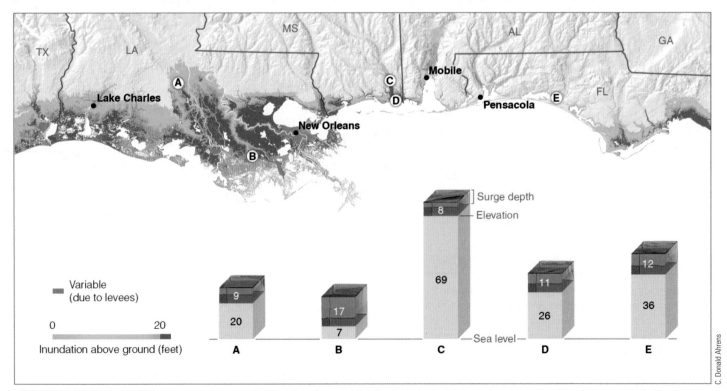

● **FIGURE 16.19** Illustration of near-worst-case storm surge flooding along the central Gulf Coast at high tide. This map is based on a composite of many hypothetical hurricanes; no one hurricane would produce all of the flooding shown above. The map colors show how high the near-worst-case storm surge would extend above local ground level. Columns show the local elevation and water depth at five points (A through E). During an actual hurricane, storm surge warnings and statements from local officials will provide more specific information.

Devastation from a Tropical Storm: The Case of Allison

Tropical storms that never become hurricanes can still produce catastrophic floods. Tropical Storm Agatha—the first named storm in the Eastern Pacific in 2010—brought torrential rains and flooding to Central America during May 2010. Flooding and mudslides resulted in $1.6 billion in damage, and more than 200 deaths, with 152 deaths in Guatemala alone. During August 2008, Tropical Storm Fay moved slowly over Florida where it dumped more than 25 inches of rain over portions of east-central Florida, which resulted in deadly flooding and significant damage. But probably the most infamous tropical storm in recent years is Allison, which was the first tropical storm in the North Atlantic to have its name retired without ever reaching hurricane strength.

In late May 2001, Allison began as a tropical wave that moved westward across the Atlantic. The wave continued its westward journey, and by the first of June it had moved across Central America and out over the Pacific Ocean. Here, it organized into a band of thunderstorms and a tropical depression. Upper-level winds guided the depression northward over the Gulf of Mexico, where the warm water fueled the circulation, and just east of Galveston, Texas, the depression became Tropical Storm Allison. Packing winds of 53 knots (61 mi/hr), Allison made landfall over the east end of Galveston Island on June 5. It drifted inland and weakened (see ● Fig. 2).

On the eastern side of the storm, heavy rain fell over parts of Texas and Louisiana. Some areas of southeast Texas received as much as 10 inches of rain in less than 5 hours. Homes, streets, and highways flooded as heavy rain continued to pound the area. But the worst was yet to come.

On June 7, as the upper-level winds began to change, the remnants of Allison drifted southwestward toward Houston. Heavy rain again fell over southeast Texas and Louisiana, where several tornadoes touched down. Over the Houston area, more than 20 inches of rain fell within a 12-hour period, submerging a vast part of the city. In six days, the Port of Houston received a staggering 37 inches of rain.

The center of circulation drifted southward, moving off the Texas coast and out over the Gulf of Mexico on the evening of June 9. The flow aloft then guided the storm northeastward, where the storm made landfall again, but this time in southeastern Louisiana. Heavy rain continued to pound Louisiana, creating one of the worst floods on record—a station in southern Louisiana reported a rainfall total of 30 inches. On June 11, a zone of maximum winds aloft (a jet streak) associated with the subtropical jet stream enhanced the outflow above the surface storm, and the remains of Tropical Storm Allison actually began to reintensify over land. As the storm entered Mississippi, its central pressure lowered, wind gusts reached 52 knots (60 mi/hr), and the center of circulation developed a weak-looking eye (see ● Fig. 3). As the system trekked eastward, it weakened and lost its eye, but continued to dump heavy rain over the southern Gulf States. Eventually, on June 14, the storm reached the Carolina coast.

Unfortunately, the storm slowed, then turned northward over North Carolina. Flooding became a major problem—Doppler radar estimated that up to 21 inches of rain had fallen over parts of the state. Severe weather broke out in Georgia and in the Carolinas, where some areas reported hail and downed trees due to gusty winds. The storm moved northeastward, parallel to the coast. A cold front moving in from the west eventually hooked up with the moisture from Allison. This situation caused heavy rain to fall over the mid-Atlantic states and southern New England. The storm finally accelerated to the northeast, away from the coast on June 18.

Allison, which never developed hurricane strength winds, claimed the lives of 43 people, whose deaths were mainly due to flooding. The total damage from the storm totaled in the billions of dollars, with the Houston area alone sustaining over $2 billion in damage. If all the rain that fell from Allison could be placed in Texas, it would cover two-thirds of the state with water a foot deep.

● **FIGURE 2** Visible satellite image showing the remains of Tropical Storm Allison centered over Texas on the morning of June 6, 2001. Heavy rain is falling from the thick clouds over Louisiana and eastern Texas.

● **FIGURE 3** Doppler radar display on June 11, 2001, showing bands of heavy rain swirling counterclockwise into the center of once–Tropical Storm Allison. The center of the storm, which is over Mississippi, has actually deepened and formed something of an eye.

● Figure 16.20 shows the number of hurricanes that have made landfall along the coastline of the United States from 1901 through 2016. Out of a total of 193 hurricanes striking the American coastline, 69 (36 percent) were major hurricanes—Category 3 or higher. Hence, along the Gulf and Atlantic coasts, on the average, about five hurricanes make landfall every three years, two of which are major hurricanes with winds in excess of 95 knots (110 mi/hr). However, the number of hurricanes in the North Atlantic varies greatly from year to year and even decade to decade. The activity appears to be enhanced for periods of roughly 25 to 35 years in connection with the warm phase of a variation in sea surface temperatures (called the *Atlantic Multidecadal Oscillation*, or *AMO*) which influences water temperatures in the tropical North Atlantic where hurricanes most often develop. Atlantic hurricanes have formed more often since the AMO shifted to a warm phase in the mid-1990s. However, a catastrophic hurricane can strike even when overall activity is low (such as Hurricane Andrew in 1992). Likewise, there are not necessarily more U.S. landfalls even when overall hurricane activity in the Atlantic is enhanced. Not a single major hurricane struck the United States from 2006 through 2016, the longest such interval on record.

HURRICANE-SPAWNED TORNADOES   Although the high winds of a hurricane can devastate a region, considerable damage may also occur from hurricane-spawned tornadoes. About one-fourth of the hurricanes that strike the United States produce tornadoes. In fact, in 2004 six tropical systems produced just over 300 tornadoes in the southern and eastern United States. Most of these tornadoes develop in small supercell thunderstorms, which tend

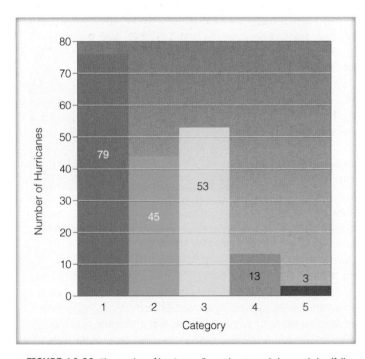

● **FIGURE 16.20** The number of hurricanes (by each category) that made landfall along the Gulf and Atlantic coasts of the United States from 1901 to 2016. Categories 3, 4, and 5 are considered major hurricanes.

to occur within rainbands or along preexisting boundaries that a hurricane may encounter as it moves inland. Tornadoes have been observed more than 500 km (310 mi) away from a hurricane's center. Moreover, tornadoes tend to form in the right front quadrant of an advancing hurricane, where vertical wind speed shear is greatest.

Researchers are still analyzing the factors that produce the strongest winds and damage within hurricane eyewalls, which are very difficult to observe directly. Studies of 1992's catastrophic Hurricane Andrew (see p. 457) revealed that the most severe damage occurred close to the eyewall, where wind gusts of up to 154 knots (170 mi/hr) were reported. It appears that the strongest winds lasted only a few seconds and struck in narrow bands less than 100 meters wide. Some research indicates that at least one tornado was involved. Other studies suggest the presence of tiny eddies, called *mini-whirls*, produced by very strong wind shear at low levels but perhaps not linked to a parent thunderstorm as a tornado would be. Some damage may be caused by strong downdrafts (microbursts) associated with the large, intense thunderstorms around the eyewall.

HURRICANE FATALITIES   Up until 2005, the annual death toll from hurricanes in the United States, over a span of about 30 years, averaged fewer than 50 persons.* Most of these fatalities occurred because of flooding. The relatively low total was achieved in part because of the advanced warning provided by the National Weather Service and in part because only a few really intense storms made landfall during this time. However, it became clear that massive loss of life was still possible with U.S. hurricanes when Hurricane Katrina slammed into Mississippi and Louisiana in 2005.

Even before it threatened the Gulf Coast, Hurricane Katrina inflicted more than $1 billion in damage and took 14 lives when it crossed South Florida. As Katrina moved into the Gulf of Mexico, intensified to Category 5 strength, and headed toward the upper Gulf coast, evacuation orders were given to residents living in low-lying areas, including the city of New Orleans. Hundreds of thousands of people moved to higher ground but, unfortunately, many thousands of others either refused to leave their homes or

---

*In other countries, the annual death toll was considerably higher during this period. For example, estimates are that more than 3000 people died in Haiti from flooding and mudslides when Hurricane Jeanne moved through the Caribbean during September 2004.

had no means of leaving and were forced to ride out the storm. Tragically, more than 1500 people died in southeast Louisiana, many as a result of catastrophic flooding across the New Orleans area that occurred when several levees broke and parts of the city were inundated with water more than 20 ft deep. In addition, more than 200 people were killed along the Mississippi coast, where a giant storm surge of up to 27 ft pushed well inland.

The aftermath of an intense hurricane can be devastating. The supply of fresh drinking water may be contaminated and food may become scarce, as grocery stores and markets are forced to close, sometimes for days or even weeks. Roads may be blocked by fallen trees and debris, or by sand that was deposited during the storm surge. Electrical and telephone service may be disrupted or completely lost. And many people may be displaced from their damaged or destroyed homes. Even the cleanup efforts can prove deadly, as, in certain areas, poisonous snakes often find their way into various nooks and crannies of the debris.

Some Notable Hurricanes

GALVESTON, 1900 Before the era of satellites, radar, and hurricane warnings on radio and television, catastrophic loss of life was all too common in the United States when hurricanes made landfall. An estimated 8000 people (perhaps as many as 12,000) lost their lives when a hurricane slammed into Galveston, Texas, in September 1900 with a huge storm surge estimated at 15 ft high. This remains the deadliest natural disaster in United States history. Galveston had no seawall at the time, and the flood waters quickly swept into homes and businesses as the water pushed inland.

The Galveston hurricane struck just a few years after two other storms that produced major loss of life. More than 1000 people were killed in South Carolina and Georgia by the Sea Islands Hurricane of August 1893, and in October of the same year, some 2000 people perished in far southeast Louisiana as a giant storm surge swept that region. Years later, another disastrous hurricane caused more than 2500 deaths in September 1928 as it passed across Florida's Lake Okeechobee. The resulting flood waters drowned many hundreds of people along the lakeshore and left thousands of others homeless.

NEW ENGLAND, 1938 A powerful September hurricane slammed into the south shore of Long Island in 1938 as a strong Category 3 storm with a central pressure of 946 mb (27.94 in.) and a storm surge exceeding 12 ft, combined with additional water from rising tides. Pulled northward by a strong, meridional jet stream, this hurricane was dubbed the "Long Island Express" because of its extremely fast forward motion, estimated at close to 61 knots (70 mi/hr)—the fastest hurricane motion on record. The rapid movement allowed less time for the storm to weaken over cooler waters, and it also led to especially high winds on the storm's eastern side. A record-high wind gust of 162 knots (186 mi/hr) was reported at Blue Hill Observatory near Boston.

Because forecasters had expected the storm to move out to sea, residents had very little warning of the impending disaster.

After blasting Long Island, the hurricane continued barreling northward, making a second landfall in Connecticut. Its intensity was little diminished, as a storm surge inundated downtown Providence, Rhode Island, and many nearby locations. In all, the hurricane damaged or destroyed more than 25,000 homes and took more than 600 lives, making it the deadliest hurricane in New England history.

CAMILLE, 1969 Hurricane Camille (1969) stands out as one of the most intense hurricanes to reach the coastline of the United States during the twentieth century (see ▼ Table 16.3). With a central pressure of 909 mb, winds reaching 160 knots (184 mi/hr), and a storm surge more than 7 m (23 ft) above the normal high-tide level, Camille, as a Category 5 storm, unleashed its fury on Mississippi, destroying thousands of buildings. During its rampage, it caused an estimated $1.5 billion in property damage and took more than 200 lives.

HUGO, 1989 With maximum winds estimated at about 120 knots (138 mi/hr), and a central pressure near 934 mb, Hugo made landfall as a Category 4 hurricane near Charleston, South Carolina, about midnight on September 21 (see ● Fig. 16.21). The high winds and storm surge, which ranged between 2.5 and 6 m (8 and 20 ft), hurled a thundering wall of water against the shore. This knocked out power, flooded streets, and caused widespread destruction to coastal communities. The total damage in the United States attributed to Hugo was over $7 billion, with a death toll of 21 in the United States and 49 overall.

ANDREW, 1992 On August 21, 1992, as Tropical Storm Andrew churned westward across the Atlantic, it began to weaken, prompting some forecasters to surmise that this tropical storm would never grow to hurricane strength. But Andrew moved into a region favorable for hurricane development. Even though it was outside the tropics near latitude 25°N, warm surface water and weak winds aloft allowed Andrew to intensify rapidly. And in just two days Andrew's winds increased from 45 knots to 122 knots, turning an average tropical storm into one of the most intense hurricanes to strike Florida in the last century (see Table 16.3).

With sustained winds estimated at 140 knots (167 mi/hr) and a powerful storm surge, Andrew made landfall south of Miami on the morning of August 24 (see ● Fig. 16.22). The eye of the storm moved over Homestead, Florida. Most of Andrew's destruction came not from the storm surge but from fierce

RANK	HURRICANE (MADE LANDFALL)	YEAR	CENTRAL PRESSURE (MILLIBARS/INCHES)	CATEGORY	DEATH TOLL
1	Florida (Keys)	1935	892/26.35	5	408
2	Camille (Mississippi)	1969	909/26.85	5	256
3	Andrew (South Florida)	1992	922/27.23	5	53
4	Katrina (Louisiana)	2005	920/27.17	3*	>1500
5	Florida (Keys)/South Texas	1919	927/27.37	4	>600**
6	Florida (Lake Okeechobee)	1928	929/27.43	4	>2000
7	Donna (Long Island, New York)	1960	930/27.46	4	50
8	Texas (Galveston)	1900	931/27.49	4	>8000
9	Louisiana (Grand Isle)	1909	931/27.49	4	350
10	Louisiana (New Orleans)	1915	931/27.49	4	275
11	Carla (South Texas)	1961	931/27.49	4	46
12	Hugo (South Carolina)	1989	934/27.58	4	49
13	Florida (Miami)	1926	935/27.61	4	243

*Although the central pressure in Katrina's eye was quite low, Katrina's maximum sustained winds of 110 knots at landfall made it a Category 3 storm.

**More than 500 of this total were lost at sea on ships. (The > symbol means "greater than.")

● FIGURE 16.21 Satellite image of Hurricane Hugo approaching Charleston, South Carolina, on September 21, 1989.

● FIGURE 16.22 Radar image of Hurricane Andrew as it moves onshore over south Florida on the morning of August 24, 1992. The dark red and purple show where the heaviest rain is falling. Miami Beach is just to the north of the eye and the National Hurricane Center (NHC) is about 20 miles to the northwest of the eye.

winds that ravaged the area (see ● Fig. 16.23), as 50,000 homes were destroyed, trees were leveled, and steel-reinforced tie beams weighing tons were torn free of townhouses and hurled as far as several blocks. A wind gust of 142 knots (164 mi/hr) blew down a radar dome and inactivated several satellite dishes on the roof of the National Hurricane Center in Coral Gables, Florida. The hurricane roared westward across southern Florida, weakened slightly, then regained strength over the warm Gulf of Mexico. Surging northwestward, Andrew slammed into Louisiana as a Category 3 with 100-knot winds after midnight on August 6.

All told, Hurricane Andrew was one of the costliest natural disasters ever to hit the United States. It destroyed or damaged more than 200,000 homes and businesses, left more than 160,000 people homeless, caused over $30 billion in damages, and took 53 lives, including 41 in Florida.

IVAN, 2004   Hurricane Ivan moved onshore just west of Gulf Shores, Alabama, on September 15, 2004 (see ● Fig. 16.24) as a strong Category 3 hurricane with winds of 105 knots

● **FIGURE 16.23** A community in Homestead, Florida, devastated by Hurricane Andrew during August 1992.

● **FIGURE 16.24** Visible satellite image of Hurricane Ivan as it makes landfall near Gulf Shores, Alabama, on September 15, 2004. Ivan is a major hurricane with winds of 105 knots (121 mi/hr) and a surface air pressure of 945 mb (27.91 in.).

(121 mi/hr) and a storm surge of about 5 m (16 ft). The strongest winds and greatest damage occurred over an area near the border between Alabama and Florida (see ● Fig. 16.25).

As Ivan moved inland, it weakened and eventually linked up with a midlatitude low. The remains of Ivan then split from the low and drifted southward, eventually ending up in the Gulf of Mexico, where it regained tropical storm strength. It made landfall for the second time along the Gulf Coast, but this time as a tropical depression. All told, Ivan took 26 lives in the United States, produced a record 117 tornadoes over the southern and eastern states, and caused an estimated $14 billion in damages. Ivan was one of five hurricanes to make landfall in the United States during 2004. Out of the five hurricanes that hit the United States, four affected the state of Florida. (More information on the record-setting Atlantic hurricane seasons of 2004 and 2005 is given in Focus section 16.3.)

KATRINA AND RITA, 2005   Hurricane Katrina was the most damaging hurricane to ever hit the United States (although it is estimated that the 1926 Miami hurricane could be more than twice as costly if it struck today). Katrina was also by far the deadliest U.S. hurricane in more than 70 years.

Forming over warm tropical water south of Nassau in the Bahamas, Katrina became a tropical storm on August 24, 2005, and a Category 1 hurricane just before making landfall in south Florida on August 25. (Katrina's path is shown in Fig. 4, in Focus section 16.3.) It moved southwestward across Florida and out over the eastern Gulf of Mexico. As Katrina then moved westward, it passed over a deep band of warm water called the *Loop Current* that allowed Katrina to rapidly intensify. Within 12 hours, the hurricane increased from a Category 3 to a Category 5 storm with winds of 152 knots (175 mi/hr) and a central pressure of 902 mb (see ● Fig. 16.26).

The Record-Setting Atlantic Hurricane Seasons of 2004 and 2005

Both 2004 and 2005 were active years for hurricane development over the tropical North Atlantic. During 2004, nine storms became full-fledged hurricanes. Out of the five hurricanes that made landfall in the United States, three (Charley, Frances, and Jeanne) plowed through Florida, and one (Ivan) came onshore just west of the Florida panhandle (see ● Fig. 4), making this the first time since record-keeping began in 1861 that four hurricanes have impacted the state of Florida in one year. Total damage in the United States from the five hurricanes exceeded $40 billion.

Then, in 2005, a record 27 named storms developed (the most in a single season), of which 15 (another record) reached hurricane strength. The 2005 Atlantic hurricane season also had four hurricanes (Emily, Katrina, Rita, and Wilma) reach Category 5 intensity for the first time since reliable record keeping began. And Hurricane Wilma had the lowest central pressure ever measured in an Atlantic hurricane up to that point—882 mb (26.04 in.). Out of five hurricanes that made landfall in the United States, three (Dennis, Katrina, and Wilma) made landfall in hurricane-weary Florida and one (Ophelia) skirted northward along Florida's east coast, giving Florida the dubious distinction of being the only state on record to experience eight hurricanes during the span of

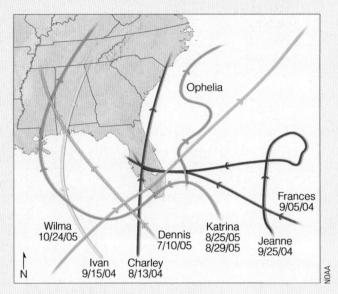

● **FIGURE 4** The paths of eight hurricanes that impacted Florida during 2004 and 2005. Notice that in 2004, hurricanes Frances and Jeanne made landfall at just about the same spot along Florida's southeast coast. The date under each hurricane's name indicates when the hurricane made landfall (except for Ophelia, which never made landfall). After a very large loop in its path, Ivan made a second landfall in southwest Louisiana as a tropical depression on September 23.

16 months (Fig. 4). To illustrate how unusual this event was, no major hurricanes (Category 3 or greater) made landfall in the United States between 2006 and 2015, and no hurricanes at all struck Florida during that period. Total damage in the United States from the five hurricanes that made landfall in 2005 exceeded $100 billion.

In both 2004 and 2005, very warm ocean water and weak vertical wind shear provided

favorable conditions for hurricane development. In previous years, winds associated with a persistent upper-level trough over the eastern United States steered many tropical systems away from the coast before they could make landfall. However, in 2004 and in 2005, an area of high pressure replaced the trough, and winds tended to steer tropical cyclones on a more westerly track, toward the coastline of North America.

(a)   (b)

● **FIGURE 16.25** Beach homes along the Gulf Coast at Orange Beach, Alabama (a) before, and (b) after Hurricane Ivan made landfall during September 2004. (Red arrows are for reference.)

● **FIGURE 16.26** Visible satellite image of Hurricane Katrina over the Gulf of Mexico. With sustained winds of 175 mi/hr and a central pressure near 902 mb (26.64 in.), this large and powerful Category 5 hurricane takes aim on Louisiana and Mississippi.

● **FIGURE 16.27** Hurricane Katrina just after making landfall along the Mississippi/ Louisiana coast on the morning of August 29, 2005. Shown here, the storm is moving north with its eye due east of New Orleans, marked X on the image. At landfall, Katrina had sustained winds of 110 knots (127 mi/hr), a central pressure of 920 mb (27.17 in.), and a storm surge of more than 20 ft.

CRITICAL THINKING QUESTION In the satellite image in Fig. 16.27, observe where New Orleans and the eye of Hurricane Katrina are located. Based on this observation, from what direction would the wind be blowing over New Orleans? Would this wind direction produce an Ekman transport toward the east or west? What effect, if any, might this transport of water have on the levee system around New Orleans?

Over the Gulf of Mexico, Katrina gradually turned northward toward Mississippi and Louisiana. As the powerful Category 5 hurricane moved slowly toward the coast, its rainbands near the center of the storm began to converge toward the storm's eye, thus cutting off moisture to the eyewall. As the old eyewall dissipated, a new one formed farther away from the center. This weakened the storm, and there was not enough time before landfall for the eyewall replacement process to be completed and for Katrina to restrengthen. Katrina made landfall near Buras, Louisiana, on August 29 (see ● Fig 16.27) as a strong Category 3 hurricane with sustained winds of 110 knots (127 mi/hr), a central pressure of 920 mb. Despite its weakened winds, Katrina still carried an extremely high storm surge of between 6 m and 9 m (20 and 30 ft).

The strong winds and high storm surge on Katrina's eastern side devastated southern Mississippi, with Biloxi, Gulfport, and Pass Christian being particularly hard hit (see ● Fig. 16.28). The winds demolished all but the strongest structures, and the huge storm surge scoured areas up to 6 km (10 mi) inland. The surge height of 27.8 ft measured at Pass Christian was the highest surge value ever recorded in the United States. The high winds and severe flooding from Katrina caused more than 200 deaths in Mississippi.

New Orleans and the surrounding parishes actually escaped the brunt of Katrina's winds, as the eye passed just to the east of the city (Fig. 16.27). In addition, the storm surge in the New Orleans area was considerably less than in Mississippi. However, the combination of high winds, large waves, and storm surge led to disastrous breaches in the levee system that protects New Orleans from the Mississippi River, the Gulf of Mexico, and Lake Pontchartrain. When the levees gave way, water up to 20 ft deep invaded a large part of the city, tragically before thousands of people could escape (see ● Fig. 16.29). The immense devastation was compounded by the inability of rescuers to provide immediate assistance to

● **FIGURE 16.28** High winds and huge waves crash against a boat washed onto Highway 90 in Gulfport, Mississippi, as Hurricane Katrina makes landfall on the morning of August 29, 2005.

many people, as flood waters persisted for days in parts of the New Orleans area that lie below sea level. The death toll due to Hurricane Katrina eventually reached more than 1800, and the devastation wrought by the storm totaled more than $75 billion.

● **FIGURE 16.29** Floodwaters inundate New Orleans, Louisiana, during August 2005, after the winds and storm surge from Hurricane Katrina caused several levee breaks.

Less than a month later, powerful Hurricane Rita, another Category 5 storm with sustained winds of 152 knots (175 mi/hr), moved across the Gulf of Mexico south of New Orleans. Strong, tropical storm–force easterly winds, along with another storm surge, caused some of the repaired levees in the New Orleans area to break again, flooding parts of the city that just days earlier had been pumped dry. Rita made landfall in southeast Texas, killing more than 100 people in the state. However, many of the deaths from Rita were caused by heat stress, as people evacuated from Rita's path in huge numbers (perhaps in part because of the Hurricane Katrina disaster a month earlier), only to be caught in massive traffic jams during a major heat wave.

SANDY, 2012 Although it lost its hurricane status about three hours before making landfall, Hurricane Sandy (often referred to as *Superstorm Sandy*) brought the most deadly and damaging storm surge in more than 70 years to coastal areas of New Jersey and New York, including New York City. Sandy formed as a tropical storm late in the 2012 hurricane season, on October 22, in the western Caribbean. It struck Jamaica as a Category 1 hurricane, then intensified to Category 3 strength with winds of 100 knots (115 mi/hr) before striking Cuba. Afterward, Sandy weakened to a tropical storm and followed a typical track east of the Bahamas, moving toward the north-northwest and then to the northeast, Normally, such a storm would be expected to continue moving out to sea, especially so late in the hurricane season. However, some computer models provided more than five days' notice that Sandy might curve to the northwest and strike the eastern coastline of the United States. No other hurricane on record had made such a sharp left turn so far north, and the various models disagreed for several days about Sandy's forecast path. Eventually, however, they converged to accurately predict a very dangerous track. Sandy's entire track is shown in Fig. 16.13, on p. 448.

Sandy regained Category 1 strength on October 27, and the storm struck the coast just northeast of Atlantic City, New Jersey, on the evening of October 29 as a powerful *post-tropical cyclone*,* with a central pressure of 945 mb (27.91 in.). Despite having been downgraded from hurricane status because it lacked a core of warm air, Sandy made landfall with sustained hurricane-force winds of 70 knots (80 mi/hr). Because Sandy was an extremely large storm, its winds were weaker than usual for a central pressure so low, but they were spread across a vast area (see ● Fig. 16.30). This situation helped generate an enormous storm surge that swept into New York Harbor and caused severe damage along much of the Long Island and New Jersey coast (see ● Fig. 16.31).

Sandy struck at close to high tide in the New York City area, which added roughly 5 ft of water on top of the 9-ft storm surge. Waters poured into the New York subway system and inundated much of lower Manhattan Island. More than 70 people died in the United States as a result of Sandy, and estimated damages totaled more than $50 billion, making it the second costliest storm of tropical origin in the United States going back to 1900.

Because Sandy was not expected to make landfall as a hurricane, coastal residents had been alerted to the storm's serious dangers but had not been placed under a hurricane warning. In the aftermath of Sandy, the National Hurricane Center changed its policy so that hurricane warnings can now be issued and kept in place up to landfall even if a hurricane is expected to become a post-tropical cyclone before it strikes.

---

*A post-tropical cyclone is a storm of tropical origin that, at some point during its life span, loses its tropical characteristics and sometimes takes on characteristics of a middle-latitude cyclonic storm.

0   20   40   60

Wind speed (miles per hour)

● **FIGURE 16.30** Hurricane Sandy over the Atlantic Ocean just after midnight (EDT) on October 29, 2012. Dark arrows show winds around the storm; color shading indicates wind speed (mi/hr). Heavy dashed lines show Sandy's path.

Destructive Tropical Cyclones Around the World

Though horrifying enough, the statistics from even the worst hurricanes to strike the United States pale when compared to the stunning tolls that have occurred in other nations, especially those that have large populations in low-lying river deltas. More than 300,000 lives were taken as a killer tropical cyclone and storm surge ravaged the coast of Bangladesh with floodwaters in November 1970. In April 1991, a similar cyclone devastated the area with reported winds of 127 knots (146 mi/hr) and a storm surge of 7 m (23 ft). In all, the storm destroyed 1.4 million houses and killed 140,000 people and 1 million cattle. And again in November 2007, Tropical Cyclone Sidr, a Category 4 storm with winds of 135 knots (155 mi/hr) moved into the region, killing more than 3000 people, damaging or destroying over one million houses, and flooding more than two million acres. Estimates are that Sidr adversely affected more than 8.5 million people. Unfortunately, the potential for a repeat of this type of disaster remains high in Bangladesh, because many people live along the relatively low, wide floodplain that slopes outward to the bay, and, historically, this region is in a path frequently taken by tropical cyclones.

In May 2008, Tropical Cyclone Nargis first took aim on Bangladesh, but then moved east, striking Myanmar (Burma), where strong tropical cyclones are much less common (see ● Fig. 16.32). Although the cyclone was accompanied by strong winds, it was the 16-foot storm surge and huge waves that caused much of the damage. Nargis pushed floodwaters inland for at least 50 km (31 mi) along the Irrawaddy Delta. No major tropical cyclone had been recorded in the delta in modern history, and millions of people were in flood-prone homes less than 10 ft above sea level. The cyclone killed at least 140,000 people as flooding washed away entire villages, in some places without leaving a single structure.

● **FIGURE 16.31** A home destroyed by Hurricane Sandy in Union Beach, New Jersey. Sandy devastated parts of coastal New Jersey and New York, causing more than $50 billion in damages.

Ramin Talaie/Corbis via Getty Images

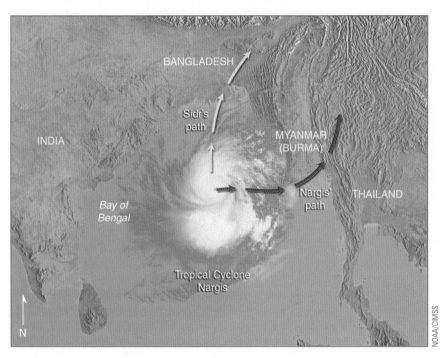

● **FIGURE 16.32** Visible satellite image of Tropical Cyclone Nargis on May 2, 2008, as it begins to move eastward over the Bay of Bengal toward Myanmar (Burma), where its storm surge and floodwaters killed more than 140,000 people. (The red arrows show the path of Nargis.) Also on the image is the path of Tropical Cyclone Sidr (yellow arrows), which caused widespread destruction in Bangladesh during November 2007.

WEATHER WATCH

The "Great Hurricane of 1780" is the deadliest Atlantic hurricane in history. During October 1780, it wreaked havoc on many Caribbean islands and completely devastated a large fleet of British ships. The storm took the lives of more than 27,000 sailors and island residents, including thousands of slaves.

One of the strongest tropical cyclones to make landfall in modern times was Super Typhoon Haiyan, which struck the Philippines in November 2013. (In the Philippines, Haiyan was referred to as Yolanda.) Haiyan intensified over waters that were unusually warm well below the surface, so that the strong mixing from high winds did not bring up as much cooler water as usual. The typhoon was at its peak strength when it came ashore, with sustained winds estimated at 169 knots (195 mi/hr), the highest value for any landfalling tropical cyclone in the world (see ● Fig. 16.33). The powerful winds drove a massive storm surge into the island of Leyte. The city of Tacloban experienced the worst damage, as a storm surge of 5 m (17 ft) completely submerged the first level of many buildings. More than 6000 people died in the Philippines, with additional damage occurring as a weakened Haiyan moved on to bring high winds and heavy rains to China and Vietnam.

The deadliest known hurricane to strike the Western Hemisphere was the Great Hurricane of 1780, which claimed approximately 27,000 lives in the eastern Caribbean. The next-deadliest was Hurricane Mitch in October 1998. Mitch's high winds, huge waves (estimated maximum height 44 ft), and torrential rains destroyed vast regions of coastal Central America (for Mitch's path, see Fig. 16.13, p. 448). In the mountainous regions of Honduras and Nicaragua, rainfall totals from the storm may have reached 190 cm (75 in.). The heavy rains produced floods and deep mudslides that swept away entire villages, including structures and inhabitants. Mitch caused over $5 billion in damages, destroyed hundreds of thousands of homes, and killed more than 11,000 people. More than 3 million others were left homeless or were otherwise severely affected by this deadly storm.

Are major hurricanes on the increase worldwide? Will the intensity of hurricanes increase as the world warms? These questions are addressed in Focus section 16.4.

● **FIGURE 16.33** Infrared satellite image of Super Typhoon Haiyan as it approaches the central Philippines on November 7, 2013. The large central ring indicates very intense convection, with the tops of the clouds so high and cold that their temperatures range from −80°C to −90°C (−112°F to −130°F).

Hurricanes in a Warmer World

In Focus section 16.3 (p. 460), we saw that 2005 was a record year for Atlantic hurricanes, with 27 named storms, 15 hurricanes, and 5 storms reaching Category 5 status on the Saffir-Simpson Hurricane Wind Scale. It is clear from ● Fig. 5 that the number of hurricanes reported in the Atlantic Basin has increased over the last several decades. Could the large numbers and high intensity of hurricanes in recent years be related to global warming?

We know that hurricanes are fueled by warm tropical water—the warmer the water, the more fuel available to drive the storm. A mere 0.6°C(1°F) increase in sea-surface temperature will increase the maximum winds of a hurricane by about 5 knots (almost 6 mi/hr), everything else being equal. Since the mid-1990s, sea-surface temperatures across the tropical North Atlantic have generally trended above the longer-term average. Some experts have attributed this phenomenon to a cyclic rise and fall in temperatures across this region, which apparently tend to peak about every 30 to 40 years. However, at least some of the warming appears to be related to an overall temperature rise throughout the world's oceans.

Between June and October 2005, sea-surface temperatures in the tropical Atlantic were about 0.9°C(1.6°F) warmer than the long-time (1901–1970) average for that region. One study concluded that about half of the warming (about 0.4°C) was due to climate change caused by increasing concentrations of greenhouse gases in the atmosphere. Other studies indicate a global rise in the average intensity of the strongest tropical cyclones over the last several decades. These results suggest that warmer sea-surface temperatures may indeed be helping to boost the strength of hurricanes. One difficulty in accepting this conclusion is that reliable and complete records of tropical cyclones have only been available since the 1970s, when observations from satellites became more extensive.

Climate models predict that, as the world continues to warm, sea-surface temperatures in the tropics will rise by about 0.6°C to 2.0°C(1.1°F to 3.6°F) by the end of this century. Should the upper end of these projections prove correct, a hurricane forming in today's atmosphere with maximum sustained winds of 130 knots (a strong Category 4 storm) could, in the warmer world, have maximum sustained winds of 140 knots (a Category 5 storm). Future hurricanes may also deposit more rain as they make landfall, as warmer temperatures globally are expected to increase evaporation from the oceans and increase the amount of water vapor available for precipitation.

As sea-surface temperatures rise, will hurricanes become more frequent? Over the last 40 years, there has been no significant change in the total number of tropical cyclones observed around the world, except for the increase in the number of Atlantic hurricanes noted earlier. Most climate models actually predict that the number of tropical cyclones will hold steady or decrease slightly in the long term, with one analysis suggesting the numbers could drop by anywhere from 6 percent to 34 percent by the end of this century. This decrease may be due in part to a projected weakening of the global tropical circulation that supports rising motion and hurricane formation. There could also be changes in the distribution of tropical cyclones, since it is possible that some oceans will warm more than others, which could lead to wind shear favoring the areas that are warming most rapidly and inhibiting hurricane formation in other areas. Overall, then, it appears we may see slightly fewer tropical cyclones over time, but the ones that do form are more likely to be stronger.

Sophisticated instruments today allow scientists to peer into hurricanes and examine their structure and winds with much greater clarity than in the past. Any trends in hurricane frequency or intensity will likely become clearer as these observing tools become more established, and when more reliable information on past tropical cyclone activity becomes available. One potential source of data is sea sediment cores, which hold clues to past tropical cyclone occurrences.

● **FIGURE 5** The total number of tropical storms and hurricanes (red bars), hurricanes only (yellow bars), and Category 3 hurricanes or greater (green bars) in the Atlantic Basin for the period 1851 through 2016.

Hurricane Watches and Warnings

With the aid of ship reports, satellites, radar, buoys, and reconnaissance aircraft, the location and intensity of tropical cyclones are pinpointed and their movements carefully monitored. Once a tropical depression forms, its motion and strength are predicted out to the following five days by the National Hurricane Center in Miami, Florida (for depressions in the North Atlantic), or by the Central Pacific Hurricane Center in Honolulu, Hawaii (for depressions in the Central and Northeast Pacific). These forecasts are then updated every six hours, or more often if a hurricane or tropical storm is nearing shore. When a hurricane poses a direct threat to an area, a **hurricane watch** is issued, typically 24 to 48 hours before the storm arrives. When it becomes more certain that the storm will strike an area, a **hurricane warning** is issued (see ● Fig. 16.34). A map of surface wind speed probabilities, updated every few hours, provides the likelihood that winds of a particular speed (such as tropical storm force or hurricane force) will occur in a particular area over various time periods (see ● Fig. 16.35).

Hurricane warnings are designed to give residents ample time to secure property and, if necessary, to evacuate the area. Even if a hurricane loses its tropical characteristics before striking land, the National Hurricane Center may opt to continue any existing hurricane warnings up until landfall. This new policy was implemented in 2013 as a result of Hurricane Sandy (see p. 462).

| 5% | 10% | 20% | 30% | 40% | 50% | 60% | 70% | 80% | 90% | 100% |

Probability of tropical storm force winds

NOAA

● **FIGURE 16.35** A map showing the likelihood that sustained winds from Hurricane Sandy will reach tropical storm force (34 knots or 39 mi/hr) at various locations, as predicted by NOAA's National Hurricane Center at 2 a.m. EDT on October 28, 2012, for the upcoming 120-hour period. The forecast was issued about 42 hours before Sandy made landfall on the evening of October 29 in New Jersey.

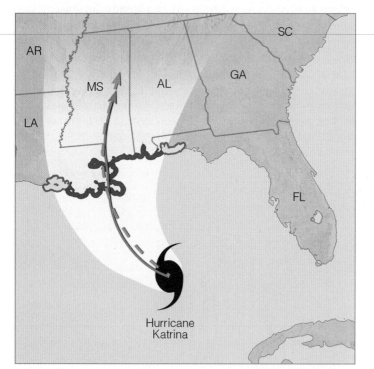

● **FIGURE 16.34** Hurricane Katrina over the Gulf of Mexico with sustained winds of 126 knots (145 mi/hr) on August 28, 2005, at 1 a.m. CDT. The current movement of the storm is west-northwest at 8 mi/hr. The dashed orange line shows the projected path of the hurricane's center; the solid blue line, the hurricane's actual path. Areas under a hurricane warning are in red. Those areas under a hurricane watch are in purple, while those areas under a tropical storm warning are in yellow. The light-shaded area denotes the track forecast cone, an estimate of the possible forecast error. On average, hurricanes will stay within the track forecast cone about 60 percent to 70 percent of the time.

Because hurricane-force winds can extend a considerable distance on either side of where the storm is expected to make landfall, and because of uncertainty in the landfall forecast, hurricane warnings are issued for rather large swaths of coastline, sometimes as wide as 500 km (310 mi) on average. Since the typical swath of damage normally spans only about one one-third of this area, much of the area is intentionally "over-warned," and people in a warning area may feel they are needlessly forced to evacuate. However, if the warning area were much narrower, it would increase the risk that hurricane damage might occur outside the warning area. Since hurricane impacts can cover a broad area, it is critical for people not to focus on the "skinny line" that denotes the best estimate of the track of the hurricane, but rather on the entire span of the hurricane warning. The *track forecast cone*, or "cone of uncertainty" (illustrated in Fig. 16.34) gives an estimate of the possible forecast error based on the five previous years of Atlantic tropical cyclone activity. Each hurricane will generally stay within the track forecast cone about 60 to 70 percent of the time.

Evacuation orders are given by local authorities, typically only for those low-lying coastal areas directly affected by a storm surge. The newly developed storm surge watches and warnings will greatly assist in this process. People at higher elevations or farther from the coast are not usually requested to leave, in part because of the added traffic problems this would create, as was the case in Hurricane Rita (see p. 462). The time it takes to complete an evacuation puts a special emphasis on the timing and accuracy of the warning.

Hurricane Forecasting Techniques

As a potentially devastating storm such as Hurricane Katrina (seen in Fig. 16.34) approaches land, will it intensify, maintain its strength, or weaken? Also, will it continue to move in the same direction and at the same speed? Such questions have challenged forecasters for decades. To forecast the intensity and movement of a hurricane, meteorologists use numerical weather prediction models, which are computer models that represent the hurricane and its environment in a greatly simplified manner. Information from satellites, buoys, and reconnaissance aircraft (that deploy dropsondes* into the eye of the storm) is fed into the models. The models then forecast the intensity and movement of the storm. There are a variety of forecast models, each one treating some aspect of the atmosphere (such as evaporation of water from the ocean's surface) in a slightly different manner. Sometimes the models do not agree on where the storm will move and on how

---

*Dropsondes are instruments (radiosondes) that are dropped from reconnaissance aircraft into a storm. As the instrument descends, it measures and relays data on temperature, pressure, and humidity back to the aircraft. Also obtained are data concerning wind speed and wind direction.

strong it will be. However, the best models in use today are more skillful than ever in their forecasting of hurricane movement.

The problem of different models forecasting different paths for the same hurricane has been addressed by using the method of *ensemble forecasting*. You may recall from Chapter 13, p. 359, that an ensemble forecast is based on running several forecast models (or different simulations of the same model), each beginning with slightly different weather information. If the forecast models (or different versions of the same model) all agree that a hurricane will move in a particular direction (as they do in ● Fig. 16.36a) the forecaster will have confidence in making a forecast of the storm's movement. If, on the other hand, the models do not agree (as in Fig. 16.36b), then the forecaster will have to decide which model (or models) is most likely correct in forecasting the hurricane's track. The use of ensemble forecasting along with better forecast models has helped raise the level of skill in forecasting hurricane paths. For example, in the early 1990s, the projected position of a hurricane three days into the future was off by an average of around 540 km (300 mi). Today, the average three-day error has dropped to around 160 km (100 mi).

Predicting hurricane intensity is a more challenging task. Forecasting of hurricane intensity showed little improvement for decades, but signs of progress began to appear by the 2010s, especially at longer time ranges of 3 to 5 days. To help predict hurricane intensity, forecasters traditionally used statistical models that compare the behavior of the present storm with that of similar tropical storms in the past. These models show little skill in intensity prediction, however. More recently, forecasters have relied more heavily on *dynamical models*, some of which take into

(a) Hurricane Irene (2011)   (b) Hurricane Sandy (2012)

● **FIGURE 16.36** The projected path of (a) Hurricane Irene, made by six numerical forecast models on August 24, 2011, and (b) Hurricane Sandy, made by six numerical forecast models on October 25, 2012. The models are in close agreement for Hurricane Irene, but vary greatly for Hurricane Sandy.

A Forecast Challenge: The Devastating Hurricanes of 2017

Forecasters had their hands full during late August and September 2017, when the Atlantic produced a virtually nonstop sequence of powerful hurricanes. From August 24 to September 27, a total of five hurricanes reached Category 4 or 5 strength. Three of these major hurricanes—Harvey, Irma, and Maria—produced catastrophic damage. Each member of this trio affected millions of people, and together they inflicted more than $100 billion in destruction. While several Caribbean islands were devastated, the most extensive damage from each of the three hurricanes occurred in the United States and its territories.

Harvey The wrath of Hurricane Harvey was focused on Texas, which experienced a one-two punch from the storm. Harvey entered the eastern Caribbean Sea as a weak tropical storm, then degenerated into a tropical wave in the central Caribbean. For more than three days, it was unclear whether Harvey would revive at all, but computer models and observations agreed that if Harvey did redevelop, it could quickly gain strength. Harvey ended up reorganizing into a tropical depression and then passed over a pool of unusually warm water in the western Gulf of Mexico, where it strengthened explosively. Harvey intensified from Category 1 to Category 4 strength in just 19 hours

on September 24–25 and made landfall several hours later just north of Corpus Christi, Texas.

Harvey's powerful winds and heavy rains at landfall produced extensive damage and several deaths. A much greater calamity unfolded over the next three days as Harvey moved into southeast Texas and then stalled as a tropical storm, moving very slowly offshore and onshore. Harvey's slowdown and the potential for extreme rains were well predicted by computer models. Short-range models on September 26 accurately called for mammoth amounts of rain that night centered on the Houston area.

Harvey ended up producing 64.58 in. at Nederland, Texas—the heaviest rainfall ever recorded from any tropical storm or hurricane in the United States. Tens of thousands of homes were flooded by Harvey across southeast Texas and southwest Louisiana (see ● Fig. 6), and as many as one million vehicles were ruined. At least 63 people died as a direct result of Harvey.

Irma Hurricane Irma was a vicious storm for more than a week before it struck the United States mainland. Irma reached Category 5 intensity on September 5 while approaching the northern Leeward Islands, which were ravaged by the storm's fierce winds. Especially hard hit were Barbuda, Anguilla, Saint Barthélemy, Saint Martin, and the Virgin Islands (see ● Fig. 7).

● **FIGURE 6** Residents of Port Arthur, Texas, carry a bucket to try to recover items from their flooded home in the aftermath of Hurricane Harvey.

● **FIGURE 7** Enhanced infrared satellite image of Hurricane Irma as it passed directly over the island of Barbuda (outlined within Irma's eye) at 1:35 am EDT on September 6, 2017. At this point, Irma was a Category 5 storm with 185 mi/hr winds.

---

account the depth of warm ocean water in front of the storm's path to predict the storm's intensity. Recall from an earlier discussion (p. 445) that if the reservoir of warm water ahead of the storm is relatively shallow, ocean waves generated by the hurricane's wind will turbulently bring deeper, cooler water to the surface. The cooler water will cut off the storm's energy source, and the hurricane will weaken. On the other hand, should a deep layer of warm water exist ahead of the hurricane, cooler water will not be brought to the surface, and the storm will either maintain its strength or intensify, as long as other factors remain the same. So, knowing the depth of warm surface water* ahead of the storm is important in predicting whether a hurricane will intensify or weaken.

The National Oceanic and Atmospheric Administration (NOAA) has been conducting a decade-long Hurricane Forecast Improvement Project through the 2010s, with the goal of reducing track and intensity errors by 50 percent for one- to five-day forecasts. As our understanding of hurricanes increases, as new hurricane-prediction models with greater resolution are implemented, and as enhanced observations are incorporated into those models, there is great potential for the forecasting of hurricane motion and intensity to improve (For a glimpse of how forecasters dealt with three devastating hurricanes in 2017, see Focus box 16.5.)

Modifying Hurricanes

Because of the potential destruction and loss of lives that hurricanes can inflict, scientists have long thought about how the power of hurricanes might be reduced. During the 1960s, an experiment called Project STORMFURY carried out seeding

---

*Sophisticated satellite instruments carefully measure ocean height, which is translated into ocean temperature beneath the sea surface. This information is then fed into the forecast models.

Computer models gave plenty of notice that Irma was headed west-northwest toward Florida, where they agreed it would make a sharp north turn. But there was disagreement for days among different models—and among the members of each model ensemble—on whether Irma would track along Florida's west coast, inland through the peninsula, or along the state's west coast. The National Hurricane Center ended up placing the entire Florida peninsula under hurricane warnings, because of the model uncertainty and because Irma was such a large hurricane that it could bring dangerous winds throughout the peninsula regardless of where the center tracked. Florida was spared the very worst from Irma, in part because the hurricane moved just far enough inland along Cuba's north coast to weaken its structure substantially. (Even though Cuba was on Irma's weaker left-hand side, the nation experienced more than $2 billion in damage.)

On September 10, Irma made landfall as a Category 4 hurricane in the Florida Keys and again as a Category 3 on the state's far southwest coast, then moved north just inland from the west coast. Because Irma's winds weakened during its passage over Cuba, structural damage across Florida was less than feared. However, more than half of all structures in the Florida Keys were significantly damaged by Irma. There was a destructive storm surge as far north as Jacksonville, Florida, where record flooding occurred, and power was knocked out to millions of Floridians for days on end. The power loss caused temperatures to soar inside a nursing home north of Miami, leading to 14 heat-related deaths. All told, Irma caused more than 130 deaths across 13 nations and territories.

Maria The third group of Americans who experienced major hurricane impacts in 2017 were the 3 million residents of Puerto Rico, which was hammered by Hurricane Maria on September 20 (see ● Fig. 8). Maria intensified from Category 1 to 5 strength in a mere 15 hours, just before it rampaged across the eastern Caribbean island of Dominica, inflicting colossal damage. Although Maria had strengthened even more quickly than forecasters had expected, its west-northwest track was well predicted. Maria remained a Category 5 hurricane as it skirted the U.S. territory of St. Croix and crashed into the southeast coast of Puerto Rico as a high-end Category 4, with sustained winds of 155 mi/hr, making it the strongest hurricane to hit the island in 89 years.

Puerto Rico experienced the gamut of major hurricane impacts from Maria, including powerful storm surge, torrential rains, flooding, mudslides, and wind damage. Thousands of homes were left unlivable. The island's public infrastructure was especially hard hit: many roads were blocked by debris, the power grid was destroyed, and running water was unavailable to most residents. Recovery was a painfully slow process, plagued by logistical and political difficulties and hampered by the region's geography and its distance from the U.S. mainland. A number of hurricane-related deaths in Puerto Rico occurred days after Maria's landfall. Although Hurricane Maria was well forecast, nobody quite foresaw how extensive and long-lasting its impacts on Puerto Rico would be.

● **FIGURE 8** Trees stripped of their vegetation surround a damaged road in Toa Alta, west of San Juan, Puerto Rico, on September 24, 2017, following the passage of Hurricane Maria.

on several hurricanes. The idea was to seed the clouds just outside the eyewall with just enough artificial ice nuclei so the latent heat given off would stimulate cloud growth in this area of the storm, causing a new eyewall to replace the old one. This new, larger eyewall would have a weaker pressure gradient and weaker winds. (The process of eyewall replacement is described on p. 445.) Several storms were seeded in Project STORMFURY, with some encouraging results. However, it became evident that eyewall replacement also occurs naturally in many hurricanes that are not seeded, so it was impossible to know whether the seeding was responsible. No such attempts to modify hurricanes have taken place since the 1970s.

More recent studies using computer models have shown that tiny particles from air pollution may have the same effect as cloud seeding, helping the outer bands of a hurricane to strengthen at the expense of the inner core. However, if these particles made it into the inner core, they might actually cause strengthening, so trying this in an actual hurricane would be risky.

Another idea for weakening a hurricane is to place some form of oil (monomolecular film) on the water to retard the rate of evaporation and hence cut down on the release of latent heat inside the clouds. Some sailors, even in ancient times, would dump oil into the sea during stormy weather, claiming it reduced the winds around the ship. However, it would be much more difficult to maintain a coating of oil over the much larger area of ocean over which a hurricane might travel. One group of researchers has speculated that ocean spray has an effect on the winds of a hurricane. Their computer modeling suggests that the tiny spray reduces the friction between the wind and the sea surface. Consequently, with the same pressure gradient, the more ocean spray, the higher the winds. If this idea were to prove correct, limiting ocean spray from entering the air above might reduce the storm's winds. This concept has not yet been proven, however, so it is too soon to tell whether the ancient sailors were on to an idea that could actually help combat the terrible effects of hurricanes.

SUMMARY

Hurricanes are tropical cyclones with sustained winds of at least 64 knots (74 mi/hr) blowing counterclockwise about their centers in the Northern Hemisphere. A hurricane consists of a mass of organized thunderstorms that spiral in toward the extreme low pressure of the storm's eye. The most intense thunderstorms, the heaviest rain, and the highest winds occur outside the eye, in the region known as the eyewall. In the eye itself, the air is warm, winds are light, and skies may be broken or overcast.

Hurricanes (and all tropical cyclones) are born over warm tropical waters where the air is humid, surface winds converge, and thunderstorms become organized in a region of weak upper-level winds and weak vertical wind shear. Surface convergence may occur along the ITCZ, on the eastern side of a tropical wave, or along a front that has moved into the tropics from higher latitudes. If the disturbance becomes more organized, it becomes a tropical depression. If central pressures drop and surface winds increase, the depression becomes a tropical storm. At this point, the storm is given a name. Some tropical storms continue to intensify into full-fledged hurricanes, as long as they remain over warm water and are not disrupted by either strong vertical wind shear or by a large landmass.

The energy source that drives the hurricane comes primarily from the warm tropical oceans and from the release of latent heat. A hurricane is like a heat engine in that energy for the storm's growth is taken in at the surface in the form of sensible and latent heat, converted to kinetic energy in the form of winds, then lost at the cloud tops through radiational cooling.

The easterly winds in the tropics usually steer hurricanes westward. In the Northern Hemisphere, most storms then gradually swing northwestward around the subtropical high. If the storm moves into middle latitudes, the prevailing westerlies steer it northeastward. Because hurricanes derive their energy from the warm surface water and from the latent heat of condensation, they tend to dissipate rapidly when they move over cold water or over a large mass of land, where surface friction causes their winds to decrease and flow into their centers.

Although the high winds of a hurricane can inflict a great deal of damage, it is usually the huge waves and the flooding associated with the storm surge that cause the most destruction and loss of life. The Saffir-Simpson Hurricane Wind Scale was developed to estimate the potential destruction that a hurricane can cause. Computer forecast models have led to gradual improvement in our ability to predict the behavior of a hurricane several days in advance.

KEY TERMS

The following terms are listed (with corresponding page numbers) in the order they appear in the text. Define each. Doing so will aid you in reviewing the material covered in this chapter.

streamlines, 440	hurricane, 440
tropical wave (easterly wave), 440	typhoon, 441
	tropical cyclone, 441
eye (of hurricane), 441	Ekman transport, 452
eyewall, 441	storm surge, 453
trade wind inversion, 444	Saffir-Simpson scale, 453
eyewall replacement, 445	super typhoon, 453
tropical depression, 445	hurricane watch, 466
tropical storm, 445	hurricane warning, 466
hurricane hunter, 447	

QUESTIONS FOR REVIEW

1. What is a tropical (easterly) wave? How do these waves generally move in the Northern Hemisphere? Are showers found on the eastern or western side of the wave?

2. Why are streamlines, rather than isobars, used on surface weather maps in the tropics?

3. What is the name given to a hurricane-like storm that forms over the tropical western North Pacific Ocean?

4. Describe the horizontal and vertical structure of a hurricane.

5. Why are skies often clear or partly cloudy in a hurricane's eye?

6. What conditions at the surface and aloft are necessary for hurricane development?

7. List three "triggers" that help in the initial stage of hurricane development.

8. (a) Hurricanes are sometimes described as a heat engine. What is the "fuel" that drives the hurricane?
   (b) What determines the maximum strength (the highest winds) that the storm can achieve?

9. Would it be possible for a hurricane to form over land? Explain.

10. If a hurricane is moving westward at 10 knots, will the strongest winds be on its northern or southern side? Explain. If the same hurricane turns northward, will the strongest winds be on its eastern or western side?

11. What factors tend to weaken hurricanes?

12. Distinguish among a tropical depression, a tropical storm, and a hurricane.

13. In what ways is a hurricane different from a midlatitude cyclone? In what ways are these two systems similar?

14. Why do most hurricanes move westward over tropical waters?

15. If the high winds of a hurricane are not responsible for inflicting the most damage, what is?

16. Most hurricane-related deaths are due to what?

17. Explain how a storm surge forms. How does it inflict damage in hurricane-prone areas?

18. Hurricanes are given names when the storm is in what stage of development?

19. When Hurricane Andrew moved over south Florida during August 1992, what was it that caused the relatively small areas of extreme damage?

20. As Hurricane Katrina moved toward the Louisiana coast, it underwent eyewall replacement. What actually happened to the eyewall during this process?

21. How do meteorologists forecast the intensity and paths of hurricanes?

22. How does a hurricane watch differ from a hurricane warning?

23. Why were some hurricanes once seeded with silver iodide, and why are they not seeded today?

24. Give two reasons why hurricanes are more likely to strike New Jersey than Oregon.

QUESTIONS FOR THOUGHT

1. Why are North Atlantic hurricanes more apt to form in October than in May?

2. A friend tells you that his grandparents experienced a terrible hurricane named David in 1975 while living on the Gulf Coast. How do you know there is something wrong with your friend's story even before you consult the climatological records?

3. If you were offered a year of study abroad in either Singapore (latitude 1°N), Hong Kong (latitude 22°N), or Tokyo (latitude 36°N), which city would give you the least chance of encountering a typhoon? Explain your reasoning. Hint: See Figure 16.11, p. 447.

4. Explain why the ocean surface water temperature is usually cooler after the passage of a hurricane. (Hint: The answer is not because the hurricane extracts heat from the water.)

5. Suppose, in the North Atlantic, an eastward-moving ocean vessel is directly in the path of a westward-moving hurricane. What would be the ship's wisest course—to veer to the north of the storm or to the south of the storm? Explain.

6. Suppose this year five tropical storms develop into full-fledged hurricanes over the North Atlantic Ocean. Would the name of the third hurricane begin with the letter C? Explain.

7. You are in Darwin, Australia (on the north shore), and a hurricane approaches from the north. Where would the highest storm surge be, to the east or west? Explain.

8. Occasionally when a hurricane moves inland, it will encounter a mountain range. Describe what will happen when this occurs. What will happen to the hurricane's intensity? What will cause the most damage (winds, storm surge, flooding)? Why?

9. Give several reasons how a hurricane that once began to weaken can strengthen.

10. Suppose a tropical storm in the Gulf of Mexico moves westward across Central America and out over the Pacific. If the storm maintains tropical storm strength the entire time, do you feel it should be given a new name over the Pacific? Explain your reasoning.

PROBLEMS AND EXERCISES

1. A hurricane just off the coast of northern Florida is moving northeastward, parallel to the eastern seaboard. Suppose that you live in North Carolina along the coast:
   (a) How will the surface winds in your area change direction as the hurricane's center passes due *east* of you? Illustrate your answer by making a sketch of the hurricane's movement and the wind flow around it.
   (b) If the hurricane passes east of you, the strongest winds would most likely be blowing from which direction? Explain your answer. (Assume that the storm does not weaken as it moves northeastward.)
   (c) The lowest sea-level pressure would most likely occur with which wind direction? Explain.

2. (a) Use the Saffir-Simpson Hurricane Wind Scale (see Table 16.2, p. 454) to determine the category of Hurricane Igor (Fig. 16.2, p. 441).
   (b) Would Hurricane Igor be classified as a major hurricane?
   (c) If Igor were over the western Pacific, would it be classified as a super typhoon?

3. You are on the coast of Long Island, facing south. Hurricane Jessica is moving north toward you and is expected to make landfall with a forward speed of 30 mi/hr. Which of the two scenarios below, (a) or (b), would you expect to bring the highest winds to your location? would you expect to bring higher winds to your location? Explain your answer with a diagram showing your location and the landfall trajectory of the hurricane. Hint: Refer to the Saffir-Simpson Hurricane Wind Scale, Table 16.2, p. 454.
   (a) Jessica making landfall as a minimal Category 2 hurricane just to your west
   (b) Jessica making landfall as a minimal Category 3 hurricane just to your east

ONLINE RESOURCES

Visit www. cengagebrain.com to view additional resources, including video exercises, practice quizzes, an interactive eBook, and more.NOAA

A brilliant sunset lights up the sky over the Kalahari Desert of southern Africa, which features a semi-arid climate.

Global Climate

THE CLIMATE IS UNBEARABLE . . . . AT NOON today the highest temperature measured was −33°C. We really feel that it is late in the season. The days are growing shorter, the sun is low and gives no warmth, katabatic winds blow continuously from the south with gales and drifting snow. The inner walls of the tent are like glazed parchment with several millimeters thick ice-armour . . . . Every night several centimeters of frost accumulate on the walls, and each time you inadvertently touch the tent cloth a shower of ice crystals falls down on your face and melts. In the night huge patches of frost from my breath spread around the opening of my sleeping bag and melt in the morning. The shoulder part of the sleeping bag facing the tent-side is permeated with frost and ice, and crackles when I roll up the bag . . . . For several weeks now my fingers have been permanently tender with numb fingertips and blistering at the nails after repeated frostbites. All food is frozen to ice and it takes ages to thaw out everything before being able to eat. At the depot we could not cut the ham, but had to chop it in pieces with a spade. Then we threw ourselves hungrily at the chunks and chewed with the ice crackling between our teeth. You have to be careful with what you put in your mouth. The other day I put a piece of chocolate from an outer pocket directly in my mouth and promptly got frostbite with blistering of the palate.

Ove Wilson (Quoted in David M. Gates, *Man and His Environment*)

Our opening comes from a report by Norwegian scientists on their encounter with one of nature's cruelest climates—that of Antarctica. Their experience illustrates the profound effect that climate can have on even ordinary events, such as eating a piece of chocolate. Though we may not always think about it, climate profoundly affects nearly everything in the middle latitudes, too. For instance, it influences our housing, clothing, the shape of landscapes, agriculture, how we feel and live, and even where we reside, as most people will choose to live on a sunny hillside rather than in a cold, dark, and foggy river basin. Entire civilizations have flourished in favorable climates and have moved away from, or perished in, unfavorable ones. We learned early in this text that *climate* is the average of the day-to-day weather over a long duration. But the concept of climate is much larger than this, for it encompasses, among other things, the daily and seasonal extremes of weather within specified areas.

When we speak of climate, then, we must be careful to specify the spatial location we are talking about. For example, the Chamber of Commerce of a rural town may boast that its community has mild winters with air temperatures seldom below freezing. This may be true several meters above the ground in an instrument shelter, but near the ground the temperature may drop below freezing on many winter nights. **Microclimate** typically refers to the climate very near the ground in a small zone (sometimes as small as a few square feet) where the surface properties are consistent, such as a city park, an airport tarmac, or a forest clearing. Because a much greater extreme in daily air temperatures exists near the ground than several feet above, the microclimate for small plants is far more harsh than the thermometer in an instrument shelter would indicate.

When we examine the climate of a small area of Earth's surface, we are looking at the **mesoclimate**, which includes the conditions across regions such as forests, valleys, beaches, and towns. The climate of an even bigger area, such as a large state or a small nation, is called **macroclimate**. The climate extending over the entire planet is often referred to as **global climate**.

In this chapter, we will concentrate on the larger scales of climate. We will begin with the factors that regulate global climate; then we will discuss how climates are classified. Finally, we will examine the different types of climate. (In Chapter 18, we will explore how both natural processes and human factors can result in climate change.)

A World with Many Climates

The world is rich in climatic types. From the teeming tropical jungles to the frigid polar "wastelands," there seems to be an almost endless variety of climatic regions. The factors that produce the climate in any given place—the **climatic controls**—are the same that produce our day-to-day weather. Briefly, the controls are:

1. intensity of sunshine and its variation with latitude
2. distribution of land and water
3. ocean currents
4. prevailing winds
5. positions of high- and low-pressure areas
6. mountain barriers
7. altitude

Human settlements can also serve as a climate control, such as when a large forest is replaced with an open pasture. In the following sections we will focus on large-scale controls that are unrelated to human alteration of the environment. We can ascertain the effect these controls have on climate by observing the global patterns of two weather elements—temperature and precipitation.

GLOBAL TEMPERATURES ● Figure 17.1 shows mean annual temperatures for the world. To eliminate the distorting effect of topography, the temperatures are corrected to sea level.* Notice that in both hemispheres the isotherms are oriented east-west, reflecting the fact that locations at the same latitude receive nearly the same amount of solar energy. In addition, the annual sunlight that each latitude receives decreases from low to high latitude; hence, annual temperatures tend to decrease from equatorial toward polar regions.**

The bending of the isotherms along the coastal margins is due in part to the unequal heating and cooling properties of land and water. Recall from Chapter 2 that oceans take longer than land areas to heat up in summer and cool down in winter, because water has a higher *specific heat capacity* than soil, meaning that it takes more energy to increase the temperature of water than it does to increase the temperature of soil by the same amount. The bending of the isotherms is also related to ocean currents and upwelling. For example, along the west coast of North and South America, ocean currents transport cool water equatorward. In addition to this, the wind in both regions blows toward the equator, parallel to the coast. This situation favors upwelling of cold water (see Chapter 10), which cools the coastal margins. In the area of the eastern North Atlantic Ocean (north of 40°N), the poleward bending of the isotherms is due to the Gulf Stream and the North Atlantic Drift, which carry warm water northward.

The fact that landmasses heat up and cool off more quickly than do large bodies of water means that variations in temperature between summer and winter will be far greater over continental interiors than along the west coastal margins of continents. By the same token, the climates of interior continental regions will be more extreme, as they have (on the average) higher summer temperatures and lower winter temperatures than their west-coast counterparts. In fact, west-coast climates are typically quite mild for their latitude.

When calculated across an entire year, the highest mean temperatures on Earth occur in the tropics, but the hottest individual readings are observed in the subtropical deserts of the Northern Hemisphere. Here, in summer, the subsiding air associated with the subtropical anticyclones produces generally clear skies and low humidity, and the high sun beating down upon a relatively barren landscape produces scorching heat.

The lowest annual mean temperatures occur over large landmasses at high latitudes. The coldest areas in the Northern Hemisphere are found in the interior of Siberia and Greenland, whereas the coldest area of the world is the Antarctic. During part of the

---

*This correction is made by adding to each station above sea level an amount of temperature that would correspond to the normal (standard) temperature lapse rate of 6.5°C per 1000 m (3.6°F per 1000 ft).

**Average global temperatures for January and July are given in Figs. 3.27 and 3.28, respectively, on p. 78.

● **FIGURE 17.1** Average annual sea-level temperatures throughout the world (°F).

year, the sun is below the horizon; when it is above the horizon, it is low in the sky and its rays do not effectively warm the surface. Consequently, most of Greenland and Antarctica remain snow- and ice-covered year-round. The snow and ice reflect perhaps 80 percent of the sunlight that reaches the surface. Much of the unreflected solar energy is used to transform the ice and snow into water vapor. The relatively dry air and the Antarctic's high elevation permit rapid radiational cooling during the dark winter months, producing extremely cold surface air.

The Antarctic continent covers the area around the South Pole and remains snow- and ice-covered all year long, whereas the North Pole is surrounded by ocean. In summer a vast amount of arctic sea ice melts, allowing sunlight to be absorbed and mixed into the Arctic Ocean. This added heat keeps the average temperature of the Arctic higher than that of the Antarctic.

The extremely cold Antarctic is one reason why, overall, the Southern Hemisphere is cooler than the Northern Hemisphere.

GLOBAL PRECIPITATION Appendix C shows the worldwide general pattern of annual precipitation, which varies from place to place. There are, however, certain regions that stand out as being wet or dry. For example, equatorial regions are typically wet, while the subtropics and the polar regions are relatively dry. The global distribution of precipitation is closely tied to the general circulation of winds in the atmosphere described in Chapter 10, and to the distribution of mountain ranges and high plateaus.

● Figure 17.2 shows in simplified form how the general circulation influences the north-to-south distribution of precipitation to be expected on a uniformly water-covered Earth. Precipitation is most abundant where the air rises; least abundant

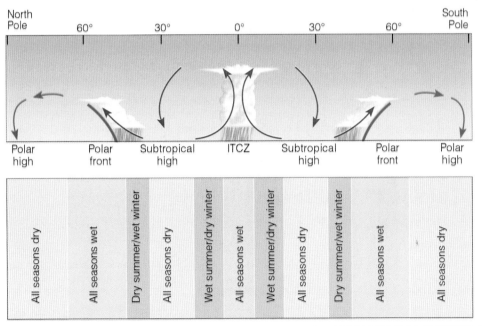

● **FIGURE 17.2** A vertical cross section along a line running north to south illustrates the main global regions of rising and sinking air and how each region influences precipitation.

where it sinks. Hence, one expects a great deal of precipitation in the tropics and along the polar front, and little near subtropical highs and at the poles. Let's look at this in more detail.

In tropical regions, the trade winds converge along the Intertropical Convergence Zone (ITCZ), producing rising air, towering clouds, and heavy precipitation all year long. Poleward of the equator, near latitude 30°, the sinking air of the subtropical highs produces a "dry belt" around the globe (although we will see that not all locations at this latitude have dry climates). The Sahara Desert of North Africa and the Mojave Desert of southwestern North America lie within this belt. Here, annual rainfall is exceedingly light and varies considerably from year to year. Because the major wind belts and pressure systems shift with the season—northward in July and southward in January—the area between the rainy tropics and the dry subtropics is influenced by both the ITCZ and the subtropical highs.

In the cold air of the polar regions there is little moisture, so there is little precipitation. Winter storms drop light, powdery snow that remains on the ground for a long time because of the low evaporation rates. In summer, a ridge of high pressure tends to block storm systems that would otherwise travel into the area; hence, precipitation in polar regions is meager in all seasons.

There are exceptions to this idealized pattern. For example, in middle latitudes the migrating position of the subtropical anticyclones also has an effect on the west-to-east distribution of precipitation. The sinking air associated with these systems is more strongly developed on their eastern side. Hence, the air along the eastern side of an anticyclone tends to be more stable; it is also drier, as cooler air moves equatorward because of the circulating winds around these systems. In addition, along coastlines, cold upwelling water cools the surface air even more, adding to the air's stability. Consequently, in summer, when the Pacific high moves to a position centered off the California coast, a strong, stable subsidence inversion forms above coastal regions. With the strong inversion and the fact that the anticyclone tends to steer storms to the north, central and southern California areas experience little, if any, rainfall during the summer months.

On the western side of subtropical highs, the air is less stable and more moist, as warmer air moves poleward. In summer, over the North Atlantic, the Bermuda high pumps moist tropical air northward from the Gulf of Mexico into the eastern two-thirds of the United States. The humid air is conditionally unstable to begin with, and by the time it moves over the heated ground, it becomes even more unstable. If conditions are right, the moist air will rise and condense into cumulus clouds, which may build into towering thunderstorms. Along the Gulf Coast, subtropical cities such as New Orleans, Louisiana, receive more than 127 cm (50 in.) of precipitation in a typical year (much more rain than in cities at the same latitude in some other locations).

In winter, the subtropical North Pacific high moves south, allowing storms traveling across the ocean to penetrate the western states, bringing much-needed rainfall to California after a long, dry summer. The Bermuda high also moves south in winter. Across much of the United States, intense winter storms develop and travel eastward, frequently dumping heavy precipitation as they go. Usually, however, the heaviest precipitation is concentrated in the eastern states, as moisture from the Gulf of Mexico moves northward ahead of these systems. Therefore, cities on the Plains typically receive more rainfall in summer and those on the West Coast have maximum precipitation in winter. Cities in the Midwest and East usually have abundant precipitation all year long. ● Figure 17.3 shows the average annual precipitation across the United States, as well as the contrasts in seasonal precipitation among a West Coast city (San Francisco), a Central Plains city (Kansas City), and an eastern city (Baltimore).

Mountain ranges disrupt the idealized pattern of global precipitation (1) by promoting convection (because their slopes

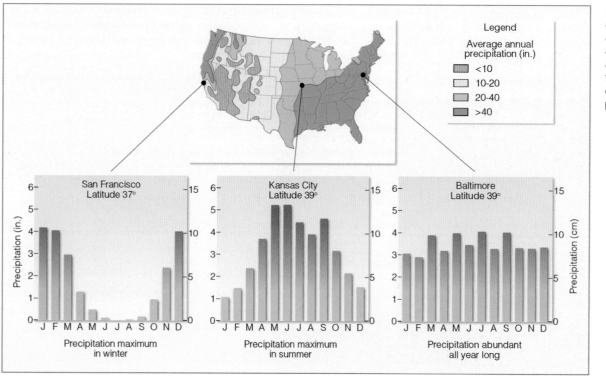

● **FIGURE 17.3** Average annual precipitation across the United States, along with variation in annual precipitation for three Northern Hemisphere cities (averaged for the period 1981–2010).

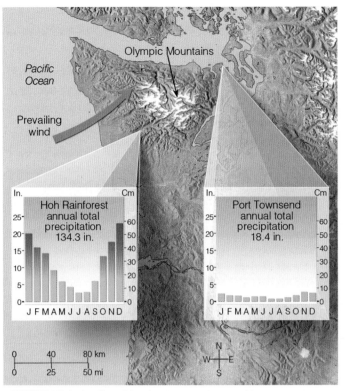

● **FIGURE 17.4** The effect of the Olympic Mountains in Washington State on average annual precipitation.

are warmer than the surrounding air) and (2) by forcing air to rise along their windward slopes *(orographic uplift)*. Consequently, the windward side of mountains tends to be "wet." As air descends and warms along the leeward side, there is less likelihood of clouds and precipitation. Thus, the leeward (downwind) side of mountains tends to be "dry." As Chapter 7 points out, a region on the leeward side of a mountain where precipitation is noticeably less is called a *rain shadow*.

A good example of the rain shadow effect occurs in the northwestern part of Washington State. Situated on the western side at the base of the Olympic Mountains, the Hoh Rainforest annually receives an average of 380 cm (150 in.) of precipitation (see ● Fig. 17.4). On the eastern (leeward) side of this range, only about 100 km (62 mi) from the Hoh Rainforest, the mean annual precipitation is less than 43 cm (17 in.), and irrigation is necessary to grow certain crops. ● Figure 17.5 shows a classic example of how topography produces several rain-shadow effects. (Additional information on precipitation extremes is given in Focus section 17.1.)

BRIEF REVIEW

Before going on to the section on climate classification, here is a brief review of some of the facts we have covered so far:

● The climate controls are the factors that govern the climate of any given region.

● The hottest temperatures on Earth tend to occur in the subtropical deserts of the Northern Hemisphere, where clear skies and sinking air, coupled with low humidity and a high summer sun beating down upon a relatively barren landscape, produce extreme heat. The warmest annual average temperatures are found in the tropics, where temperatures vary little through the year.

● The coldest temperatures on Earth tend to occur in the interior of high-latitude landmasses. The coldest areas of the Northern Hemisphere are found in the interior of Siberia and Greenland, whereas the coldest area of the world is the Antarctic.

● The wettest places in the world tend to be located on the windward side of mountains where warm, humid air rises upslope. On the downwind (leeward) side of a mountain there often exists a "dry" region, known as a rain shadow.

● **FIGURE 17.5** The effect of topography on average annual precipitation along a line running from the Pacific Ocean through central California into western Nevada.

Precipitation Extremes

Most of the "rainiest" places in the world are located on the windward side of mountains. For example, Mount Waialeale on the island of Kauai, Hawaii, has the greatest annual average rainfall in the United States: 1164 cm (458 in.). Mawsynram, on the crest of the southern slopes of the Khasi Hills in northeastern India, is often considered the wettest place in the world as it receives an average of 1187 cm (467 in.) of rainfall each year, the majority of which falls during the summer monsoon, between April and October. Cherrapunji, which is only about 10 miles from Mawsynram, holds the greatest 12-month rainfall total of 2647 cm (1042 in.), and once received 249 cm (98 in.) in just two days.

Record rainfall amounts are often associated with tropical storms. On the island of La Réunion (about 650 km east of Madagascar in the Indian Ocean), a tropical cyclone dumped 114 cm (45 in.) of rain on Foc-Foc in 12 hours. Heavy rains of short duration often occur with severe thunderstorms that move slowly or stall over a region. On July 4, 1956, 3 cm (1.2 in.) of rain fell from a thunderstorm on Unionville, Maryland, in one minute.

Snowfalls tend to be heavier where cool, moist air rises along the windward slopes of mountains. One of the snowiest places in North America is located at the Paradise Ranger Station in Mt. Rainier National Park, Washington. Situated at an elevation of 1646 m (5400 ft) above sea level, this station receives an average 1758 cm (692 in.) of snow annually, and holds the world's record 12-month snowfall total of 3109 cm (1224 in.), which fell between February 1971 and February 1972. A record seasonal snowfall total for North America of 2896 cm (1140 in.) fell on Mt. Baker Ski Area, Washington, during the winter of 1998–1999. The largest snow accumulations on Earth are believed to occur in the Japanese Alps, where a snow depth of 1181 cm (465 in.) was reported on Mt. Ibuki on February 14, 1927.

As we noted earlier, the driest regions of the world lie in the frigid polar region, the leeward side of mountains, and in the belt of subtropical high pressure, between 15° and 30° latitude. Arica in northern Chile holds the world record for lowest annual rainfall—0.08 cm (0.03 in.) per year, averaged over a period of 59 years—and for the longest period without measurable rainfall (more than 14 years, from October 1903 through January 1918). In the United States, the driest regions are found in the Desert Southwest, the southern San Joaquin Valley of California, and Death Valley in southern California, which averages only 5.9 cm (2.36 in.) of precipitation annually. ● Figure 1 gives additional information on world precipitation records.

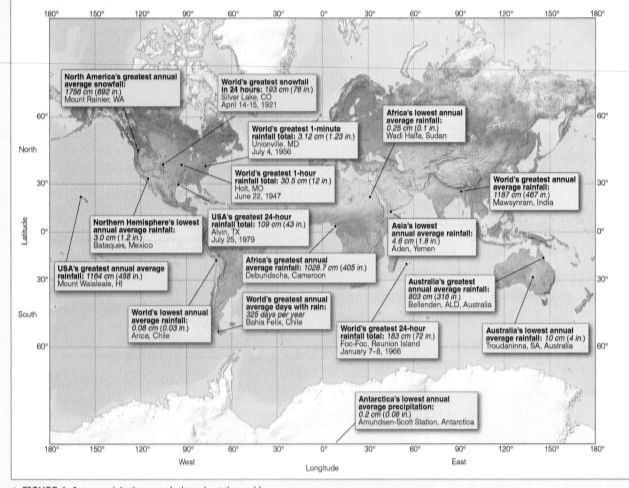

● **FIGURE 1** Some precipitation records throughout the world.

Climatic Classification

The climatic controls interact to produce such a wide array of different climates that no two places experience exactly the same climate. However, the similarity of climates within a given area allows us to divide Earth into climatic regions.

THE ANCIENT GREEKS By considering temperature and worldwide sunshine distribution, the ancient Greeks categorized the world into three climatic regions:

1. A low-latitude *tropical* (or *torrid*) *zone;* bounded by the northern and southern limit of the sun's vertical rays ($23\frac{1}{2}°$N and $23\frac{1}{2}°$S); here, the noon sun is always high, day and night are of nearly equal length, and it is warm year-round.

2. A high-latitude *polar* (or *frigid*) *zone;* bounded by the Arctic or Antarctic Circle; cold all year long due to long periods of winter darkness and a low summer sun.

3. A middle-latitude *temperate zone;* sandwiched between the other two zones; has distinct summer and winter, so exhibits characteristics of both extremes.

Such a sunlight-based, or temperature-based, climatic scheme is, of course, far too simplistic. It excludes precipitation, so there is no way to differentiate between wet and dry regions. The best classification of climates would take into account as many meteorological factors as can possibly be obtained.

THE KÖPPEN SYSTEM A widely used classification of world climates based on the annual and monthly averages of temperature and precipitation was devised by the German scientist Waldimir Köppen (1846–1940). In the absence of adequate observing stations throughout the world, Köppen related the distribution and type of native vegetation to the various climates. In this way, climatic boundaries could be approximated where no climatological data were available. Initially published in 1918, the **Köppen classification system** has since been modified and refined by a number of researchers. Köppen's scheme identifies five major climatic types, with each type designated by a capital letter:

A. *Tropical moist climates:* All months have an average temperature above $18°C$ ($64°F$). Because all months are warm, there is no real winter season.

B. *Dry climates:* Deficient precipitation most of the year. Potential evaporation and transpiration exceed precipitation.

C. *Moist midlatitude climates with mild winters:* Warm-to-hot summers with mild winters. The average temperature of the coldest month is below $18°C$ ($64°F$) and above $-3°C$ ($27°F$).

D. *Moist midlatitude climates with severe winters:* Warm summers and cold winters. The average temperature of the warmest month exceeds $10°C$ ($50°F$), and the coldest monthly average drops below $-3°C$ ($27°F$).

E. *Polar climates:* Extremely cold winters and summers. The average temperature of the warmest month is below $10°C$ ($50°F$). Given that all months are cold, there is no real summer season.

In mountainous country, where rapid changes in elevation bring about sharp changes in climatic type, delineating the climatic regions is impossible. These regions are designated by the letter *H*, for *highland climates.*

● Figure 17.6 gives a simplified overview of the major climate types throughout the world, according to Köppen's system. Superimposed on the map are some of the climatic controls. These include the average annual positions of the semipermanent high- and low-pressure areas, the average position of the Intertropical Convergence Zone in January and July, the major mountain ranges and deserts of the world, and some of the major ocean currents. Notice how the climate controls impact the climate in different regions of the world. As we would expect, because of changes in the intensity and amount of solar energy, polar climates are found at high latitudes and tropical climates at low latitudes. Dry climates tend to be located on the downwind side of major mountain chains and near 30° latitude, where the subtropical highs (with their sinking air) are found. Climates with more moderate winters (C climates) tend to be equatorward of those with severe winters (D climates). Along the west coast of North America and Europe, warm ocean currents and prevailing westerly winds modify the climate such that coastal regions experience much milder winters than do regions farther inland.

Keep in mind that within the Köppen system each major climatic group contains subgroups that describe special regional characteristics, such as seasonal changes in temperature and precipitation. The complete Köppen climatic classification system, including the criteria for the various subgroups, is given in ▼ Table 17.1.

Köppen's system has been criticized primarily because his boundaries (which relate vegetation to monthly temperature and precipitation values) do not correspond to the natural boundaries of each climatic zone. In addition, the Köppen system implies there is a sharp boundary between climatic zones, when in reality there is a gradual transition.

The Köppen system has been revised several times, most notably by the German climatologist Rudolf Geiger, who worked with Köppen on amending the climatic boundaries of certain regions. A popular modification of the Köppen system was developed by the American climatologist Glenn Trewartha, who redefined some of the climatic types and altered the climatic world map by putting more emphasis on the lengths of growing seasons and average summer temperatures.

THORNTHWAITE'S SYSTEM To correct some of the Köppen deficiencies, the American climatologist C. Warren Thornthwaite (1899–1963) devised a new classification system in the early 1930s. Both systems utilized temperature and precipitation measurements and both related natural vegetation to climate. However, to emphasize the importance of precipitation (P) and evaporation (E) on plant growth, Thornthwaite developed a *P/E ratio,* which is essentially

● **FIGURE 17.6** A simplified overview of the major climate types according to Köppen, along with some of the climatic controls. The large Hs and Ls on the map represent the average position of the semipermanent high- and low-pressure areas. The solid red lines show the average position of the ITCZ in January and July. The ocean currents in red are warm, whereas those in blue are cold. The major mountain ranges and deserts of the world also are included.

CRITICAL THINKING QUESTION Dry climates occupy more land area than any other major climatic type. If the world's average annual surface air temperature increases by 3°C over the next 100 years or so (and worldwide average precipitation does not change), would you expect the area occupied by dry climates to increase, decrease, or show no change?

monthly precipitation divided by monthly evaporation. The annual sum of the P/E ratios gives the **P/E index**. Using this index, the Thornthwaite system defines five major humidity provinces and their characteristic vegetations: rain forest, forest, grassland, steppe, and desert.

To better describe the moisture available for plant growth, Thornthwaite proposed a new classification system in 1948 and slightly revised it in 1955. His new scheme emphasized the concept of *potential evapotranspiration (PE)*, which is the amount of moisture that would be lost from the soil and vegetation if the moisture were available.*

Thornthwaite incorporated potential evapotranspiration into a moisture index that depends essentially on the differences between precipitation and PE. The index is high in moist climates and negative in arid climates. An index of 0 (zero) marks the boundary between wet and dry climates.

Evapotranspiration refers to evaporation from soil and transpiration from plants.

The Global Pattern of Climate

● Figure 17.7 gives a more detailed view of how the major climatic regions and subregions of the world are distributed based mainly on the work of Köppen. (The major climatic types along with their subdivisions are given in Table 17.1.) We will first examine humid tropical climates in low latitudes and then we'll look at middle latitude and polar climates. Bear in mind that each climatic region has many subregions of local climatic differences wrought by such factors as topography, elevation, and large bodies of water. Remember, too, that boundaries of climatic regions represent gradual transitions, not exact demarcations. Thus, the major climatic characteristics of a given region are best observed away from its periphery.

TROPICAL MOIST CLIMATES (GROUP A)

General characteristics: year-round warm temperatures (all months have a mean temperature above 18°C, or 64°F; abundant rainfall (typical annual average exceeds 150 cm, or 59 in.).

▼ TABLE 17.1 Köppen's Climatic Classification System

1st	2nd	3rd	Climatic Characteristics	Criteria
A			Tropical moist	All months have an average temperature of 18°C (64°F) or higher
	f		Tropical wet (rain forest)	Wet all seasons; all months have at least 6 cm (2.4 in.) of rainfall
	w		Tropical wet and dry (savanna)	Winter dry season; rainfall in driest month is less than 6 cm (2.4 in.) and less than 10 - $P/25$ (P is mean annual rainfall in cm)
	m		Tropical monsoon	Short dry season; rainfall in driest month is less than 6 cm (2.4 in.) but equal to or greater than 10 – $P/25$
B			Dry	Potential evaporation and transpiration exceed precipitation. The dry/humid boundary is defined by the following formulas: $p = 2t + 28$ when 70% or more of rain falls in warmer 6 months (dry winter) $p = 2t$ when 70% or more of rain falls in cooler 6 months (dry summer) $p = 2t + 14$ when neither half year has 70% or more of rain (p is the mean annual precipitation in cm and t is the mean annual temperature in °C*
	S		Semi-arid (steppe)	The BS/BW boundary is exactly one-half the dry/humid boundary
	W		Arid (desert)	
		h	Hot and dry	Mean annual temperature is 18°C (64°F) or higher
		k	Cool and dry	Mean annual temperature is below 18°C (64°F)
C			Moist with mild winters	Average temperature of coolest month is below 18°C (64°F) and above −3°C (27°F)
	w		Dry winters	Average rainfall of wettest summer month at least 10 times as much as in driest winter month
	s		Dry summers	Average rainfall of driest summer month less than 4 cm (1.6 in.); average rainfall of wettest winter month at least three times as much as in driest summer month
	f		Wet all seasons	Criteria for w and s cannot be met
		a	Summers long and hot	Average temperature of warmest month above 22°C (72°F); at least 4 months with average above 10°C (50°F)
		b	Summers long and cool	Average temperature of all months below 22°C (72°F); at least 4 months with average above 10°C (50°F)
		c	Summers short and cool	Average temperature of all months below 22°C (72°F); 1 to 3 months with average above 10°C (50°F)
D			Moist with cold winters	Average temperature of coldest month is −3°C (27°F) or below; average temperature of warmest month is greater than 10°C (50°F)
	w		Dry winters	Same as under Cw
	s		Dry summers	Same as under Cs
	f		Wet all seasons	Same as under Cf
		a	Summers long and hot	Same as under Cfa
		b	Summers long and cool	Same as under Cfb
		c	Summers short and cool	Same as under Cfc
		d	Summers short and cool; winters severe	Average temperature of coldest month is −38°C (−36°F) or below
E			Polar climates	Average temperature of warmest month is below 10°C (50°F)
	T		Tundra	Average temperature of warmest month is greater than 0°C (32°F) but less than 10°C (50°F)
	F		Ice cap	Average temperature of warmest month is 0°C (32°F) or below

*The dry/humid boundary is defined in English units as: $p = 0.44t − 3$ (dry winter); $p = 0.44t − 14$ (dry summer); and $p = 0.44t − 8.6$ (rainfall evenly distributed); where p is mean annual rainfall in inches and t is mean annual temperature in °C.

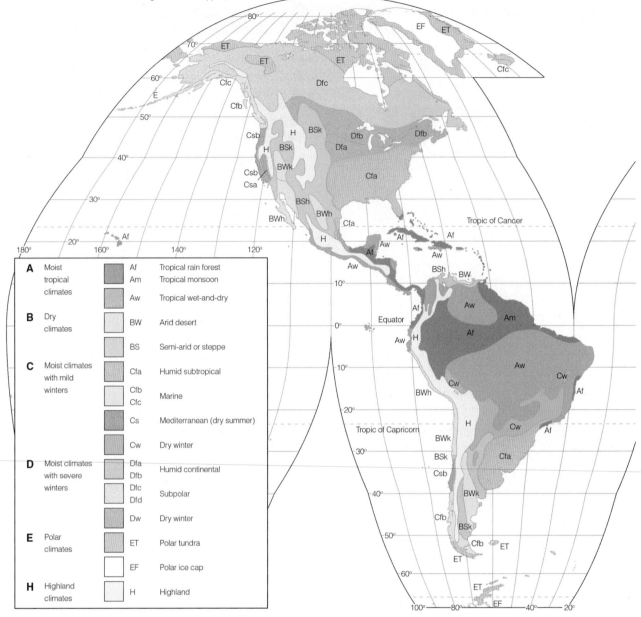

A	Moist tropical climates		Af	Tropical rain forest
			Am	Tropical monsoon
			Aw	Tropical wet-and-dry
B	Dry climates		BW	Arid desert
			BS	Semi-arid or steppe
C	Moist climates with mild winters		Cfa	Humid subtropical
			Cfb / Cfc	Marine
			Cs	Mediterranean (dry summer)
			Cw	Dry winter
D	Moist climates with severe winters		Dfa / Dfb	Humid continental
			Dfc / Dfd	Subpolar
			Dw	Dry winter
E	Polar climates		ET	Polar tundra
			EF	Polar ice cap
H	Highland climates		H	Highland

Extent: northward and southward from the equator to about latitude 15° to 25°.

Major types (based on seasonal distribution of rainfall): *tropical wet* (Af), *tropical monsoon* (Am), and *tropical wet and dry* (Aw).

At low elevations near the equator, in particular, the Amazon lowland of South America, the Congo River Basin of Africa, and the East Indies from Sumatra to New Guinea, high temperatures and abundant yearly rainfall combine to produce a dense, broadleaf, evergreen forest called a **tropical rain forest**. Here, many different plant species, each adapted to differing light intensity, present a crudely layered appearance of diverse vegetation. In the forest, little sunlight is able to penetrate through the thick crown cover to the ground. As a result, little plant growth is found on the forest floor. However, at the edge of the forest, however, or where a clearing has been made,

abundant sunlight allows for the growth of tangled shrubs and vines, producing an almost impenetrable *jungle* (see ● Fig. 17.8).

Within the **tropical wet climate** (Af), seasonal temperature variations are small (normally less than 3°C) because the noon sun is always high and the number of daylight hours is relatively constant.* However, there is a greater variation in temperature between day (average high about 32°C) and night (average low about 22°C) than there is between the warmest and coolest months. This is why people remark that winter comes to the tropics at night. The weather in a tropical wet climate is monotonous and sultry, with little change in temperature from one day to the next. Almost every day, towering cumulus clouds form and produce heavy, localized showers by early afternoon. As evening approaches, the showers usually end and skies clear. Typical

*The tropical wet climate is also known as the *tropical rainforest climate.*

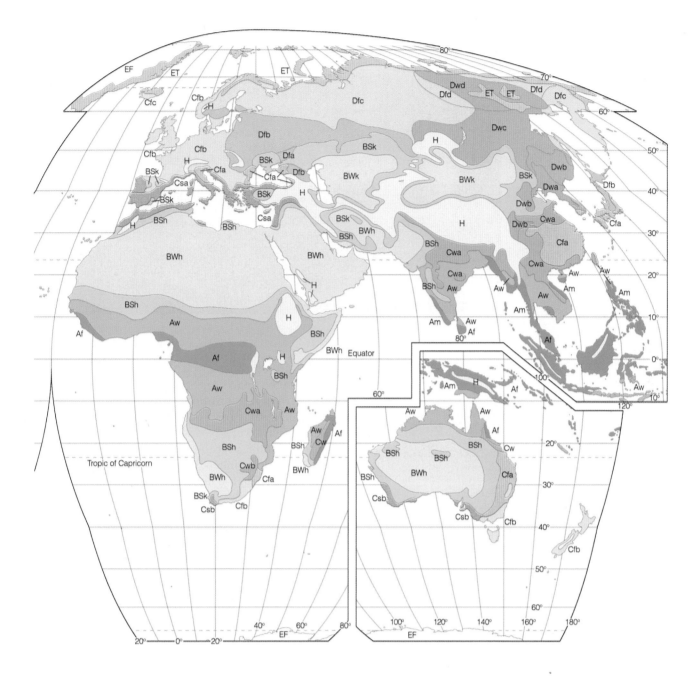

annual rainfall totals are greater than 150 cm (59 in.) and, in some cases, especially along the windward side of hills and mountains, the total may exceed 400 cm (157 in.)

The high humidity and cloud cover tend to keep maximum temperatures from reaching extremely high values. In fact, summer afternoon temperatures are normally higher in middle latitudes than here. Nighttime radiational cooling can produce saturation and, hence, a blanket of dew and—occasionally—fog covers the ground.

An example of a station with a tropical wet climate (Af) is Iquitos, Peru (see ● Fig. 17.9). Located near the equator (latitude 4°S), in the low basin of the upper Amazon River, Iquitos has an average annual temperature of 25°C (77°F), with an annual temperature range of only 2.2°C (4°F). Notice also that monthly rainfall totals are more variable than monthly temperatures. This is due primarily to the migrating position of the ITCZ and its

associated wind-flow patterns. Although monthly precipitation totals vary considerably, the average for each month exceeds 6 cm, and consequently no month is considered deficient of rainfall.

Take a minute and look again at Fig. 17.8. From the photo, one might think that the soil beneath the forest's canopy would be excellent for agriculture, but actually this is not true. As heavy rain falls on the soil, the water works its way downward, removing nutrients in a process called *leaching*. Strangely enough, many of the nutrients needed to sustain the lush forest actually come from dead trees that decompose. The roots of the living trees absorb this matter before the rains leach it away. When the forests are cleared for agricultural purposes, or for the timber, what is left is a thick red soil called **laterite**. When exposed to the intense sunlight of the tropics, the soil may harden into a bricklike consistency, making cultivation almost impossible.

● **FIGURE 17.9** Temperature and precipitation data for Iquitos, Peru, latitude 4°S. A station with a tropical wet climate (Af). (This type of diagram is called a *climograph*. It shows monthly mean temperatures with a solid red line and monthly mean precipitation with blue bar graphs.)

Köppen classified those tropical wet regions where the monthly precipitation totals drop below 6 cm for perhaps one or two months as **tropical monsoon climates** (Am). Here, yearly rainfall totals are similar to those of the tropical wet climate, usually exceeding 150 cm a year. Because the dry season is brief and copious rains fall throughout the rest of the year, there is sufficient soil moisture to maintain the tropical rain forest through the short dry period. Tropical monsoon climates can be seen in Fig. 17.7 along the coasts of Southeast Asia and southwestern India, and in northeastern South America.

Poleward of the tropical wet region, total annual rainfall diminishes, and there is a gradual transition from the tropical wet climate to the **tropical wet-and-dry climate** (Aw), where a distinct dry season prevails. Even though the annual precipitation usually exceeds 100 cm, the dry season, where the monthly rainfall is less than 6 cm (2.4 in.), lasts for more than two months. Because tropical rain forests cannot survive this "drought," the jungle gradually gives way to tall, coarse **savanna grass**, scattered with low, drought-resistant deciduous trees (see ● Fig. 17.10). The dry season occurs during the winter (the low-sun period), when the region is under the influence of the subtropical highs. In summer, the ITCZ moves poleward, bringing with it heavy precipitation, usually in the form of showers. Rainfall is enhanced by shallow lows moving slowly through the region.

Tropical wet-and-dry climates not only receive less total rainfall than the tropical wet climates, but the rain that does occur is much less reliable, as the total rainfall often fluctuates widely from one year to the next. In the course of a single year, for example, destructive floods may be followed by serious droughts. As with tropical wet regions, the daily range of temperature usually exceeds the annual range, but the climate here is much less monotonous. There is a cool season in winter when the maximum temperature averages 30°C to 32°C (86°F to 90°F). At night, the low humidity and clear skies allow for rapid radiational cooling and, by early morning, minimum temperatures typically drop to 20°C (68°F) or below.

● **FIGURE 17.10** Baobob and acacia trees illustrate typical trees of the East African grassland savanna, a region with a tropical wet-and-dry climate (Aw).

From Fig. 17.7, pp. 482–483, we can see that the principal areas having a tropical wet-and-dry climate (Aw) are those located in western Central America, in the region both north and south of the Amazon Basin (South America), in south-central and eastern Africa, in parts of India and Southeast Asia, and in northern Australia. In many areas (especially within India and Southeast Asia), the marked variation in precipitation is associated with the *monsoon*—the seasonal reversal of winds.

As we saw in Chapter 9, the Asian monsoon circulation stems in part from differential heating of landmasses and oceans. During winter in the Northern Hemisphere, winds blow outward, away from a cold, shallow high-pressure area centered over continental Siberia. These downslope, relatively dry northeasterly winds from the interior provide India and Southeast Asia with generally fair weather and the dry season. In summer, the wind-flow pattern reverses as air flows into a developing thermal low over the continental interior. The humid air from the water rises and condenses, resulting in heavy rain and the wet season. (A more detailed look at the winter and summer monsoon is shown in Fig. 9.38 on p. 250.)

An example of a station with a tropical wet-and-dry climate (Aw) is given in ● Fig. 17.11. Located at latitude 11°N in west Africa, Timbo, Guinea, receives an annual average 163 cm (64 in.) of rainfall. Notice that the rainy season is during the summer when the ITCZ has migrated to its most northern position. Note also that practically no rain falls during the months of December, January, and February, when the region comes under the domination of the subtropical high-pressure area and its sinking air.

The monthly temperature patterns at Timbo are characteristic of most tropical wet-and-dry climates. As spring approaches, the noon sun is slightly higher, and the more intense sunshine produces greater surface heating and higher afternoon temperatures—usually above 32°C (90°F) and occasionally above 38°C (100°F)—creating hot, dry, desertlike conditions. After this brief hot season, persistent cloud cover and the evaporation of rain tends to lower the temperature during the summer.

● **FIGURE 17.11** Climatic data for Timbo, Guinea, latitude 11°N, a station with a tropical wet-and-dry climate (Aw).

The warm, muggy weather of summer often resembles that of the tropical wet climate (Af). The rainy summer is followed by a warm, relatively dry period, with afternoon temperatures usually climbing above 30°C(86°F).

Poleward of the tropical wet-and-dry climate, the dry season becomes more severe. Clumps of trees are more isolated and the grasses dominate the landscape. When the potential annual water loss through evaporation and transpiration exceeds the annual water gain from precipitation, the climate is described as dry.

DRY CLIMATES (GROUP B)

General characteristics: deficient precipitation most of the year; potential evaporation and transpiration exceed precipitation.

Extent: the subtropical deserts extend from roughly 20° to 30° latitude in large continental regions of the middle latitudes, often surrounded by mountains.

Major types: arid (BW)—the "true desert"—and semi-arid (BS).

A glance at Fig. 17.7, pp. 482–483, reveals that, according to Köppen, the dry regions of the world occupy more land area (about 26 percent) than any other major climatic type. These dry regions are deficient in water, meaning that the potential annual loss of water through evaporation is greater than the annual water gained through precipitation. Thus, classifying a climate as dry depends not only on precipitation totals but also on temperature, which greatly influences evaporation. For example, 35 cm (14 in.) of precipitation in a hot climate will support only sparse vegetation, while the same amount of precipitation in much colder north-central Canada will support a conifer forest. A region with a low annual rainfall total is also more likely to be classified as dry if the majority of precipitation is concentrated during the warm summer months, when evaporation rates are greater.

Precipitation in a dry climate is both meager and irregular. Typically, the lower the average annual rainfall, the greater its variability. For example, a station that reports an annual rainfall of 5 cm (2 in.) may actually measure no rainfall for two years; then, in a single downpour, it may receive 10 cm (4 in.).

The major dry regions of the world can be divided into two primary categories. The first includes those sections of the subtropics (between latitude 15° and 30°) where the sinking air of the subtropical anticyclones produces generally clear skies. The second is found in the continental areas of the middle latitudes. Here, far removed from a source of moisture, areas are deprived of precipitation. Dryness in these areas is often accentuated by mountain ranges that produce a rain shadow effect.

Köppen divided dry climates into two types based on their degree of dryness: the *arid* (BW)* and the *semi-arid*, or *steppe* (BS). These two climatic types can be divided even further. For example, if the climate is hot and dry with a mean annual temperature above 18°C (64°F), it is either BWh or BSh (the *h* is for *heiss*, meaning "hot" in German). On the other hand, if the climate is cold (in winter, that is) and dry with a mean annual temperature below 18°C, then it is either BWk or BSk (where the *k* is for *kalt*, meaning "cold" in German).

*The letter *W* is for *Wüste*, the German word for "desert."

The **arid climates** (BW) occupy about 12 percent of the world's land area. From Fig. 17.7, pp. 482–483, we can see that this climatic type is found along the west coast of South America and Africa and over much of the interior of Australia. Notice, also, that a swath of arid climate extends from northwest Africa all the way into central Asia. In North America, the arid climate extends from northern Mexico into the southern interior of the United States and northward along the leeward slopes of the Sierra Nevada. This region includes both the Sonoran and Mojave deserts and the Great Basin.

The southern desert region of North America is dry because it is dominated by the subtropical high most of the year, and cyclonic winter storm systems tend to weaken before they move into the area. The northern region is in the rain shadow of the Sierra Nevada. These regions are deficient in precipitation all year long, with many stations receiving less than 13 cm (5 in.) annually. As noted earlier, the rain that does fall is spotty, often in the form of scattered summer afternoon showers. Some of these showers can be downpours that change a parched gully into a raging torrent. More often than not, however, the rain evaporates into the dry air before ever reaching the ground, and the result is rain streamers (virga) dangling beneath the clouds (see ● Fig. 17.12).

Contrary to popular belief, few deserts are completely without vegetation. Although sparse, the vegetation that does exist must depend on the infrequent rains. Thus, most of the native plants are **xerophytes**—those capable of surviving prolonged periods of drought (see ● Fig. 17.13). Such vegetation includes various forms of cacti and short-lived plants that spring up during the rainy periods.

In low-latitude deserts (BWh), intense sunlight produces scorching heat on the parched landscape. Here, air temperatures can reach values as high as anywhere in the world. Maximum daytime readings during the summer can exceed 50°C (122°F), although 40°C to 45°C (104°F to 113°F) are more common. In the middle of the day, the relative humidity is usually between

● **FIGURE 17.12** Rain streamers (virga) are common in dry climates, as falling rain evaporates into the drier air before ever reaching the ground.

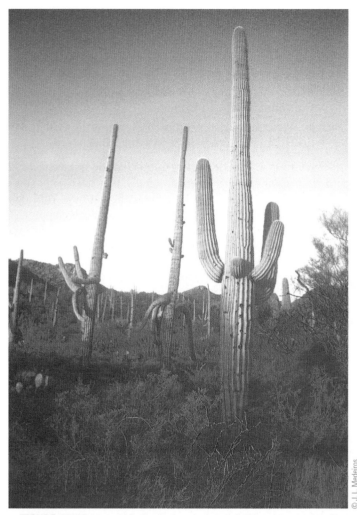

● **FIGURE 17.13** Creosote bushes and cacti are typical of the vegetation found in the arid southwestern American deserts (BWh).

5 and 25 percent. At night, the air's relatively low water-vapor content allows for rapid radiational cooling. Minimum temperatures often drop below 25°C (77°F). Thus, arid climates have large daily temperature ranges, often between 15°C and 25°C (27°F and 45°F) and occasionally higher.

During the winter, temperatures are more moderate, and minimums may, on occasion, drop below freezing. The variation in temperature from summer to winter produces large annual temperature ranges. We can see this in the climate record for Phoenix, Arizona (see ● Fig. 17.14), a city in the southwestern United States with a BWh climate. Notice that the average annual temperature in Phoenix is 21°C (70°F), and that the average temperature of the warmest month (July) reaches a sizzling 32°C (90°F). As we would expect, rainfall is scant in all months. There is, however, a slight maximum in July and August. This is due to the summer monsoon, when more humid, southerly winds are likely to sweep over the region and develop into afternoon showers and thunderstorms (see Fig. 9.40, p. 251).

In middle-latitude deserts (BWk), average annual temperatures are lower. Summers are typically warm to hot, with afternoon temperatures frequently reaching 40°C (104°F). Winters are usually extremely cold, with minimum temperatures

sometimes dropping below −35°C (−31°F). Many of these deserts lie in the rain shadow of an extensive mountain chain, such as the Sierra Nevada and the Cascade mountains in North America, the Himalayan Mountains in Asia, and the Andes in South America. The meager precipitation that falls comes from an occasional summer shower or a passing midlatitude cyclonic storm in winter.

Again, refer to Fig. 17.7 and notice that around the margins of the arid regions, where rainfall amounts are greater, the climate gradually changes into **semi-arid** (BS). This region, called **steppe**, typically has short bunch grass, scattered low bushes, sparse trees, or sagebrush (see ● Fig. 17.15). In North America, this climatic region includes most of the Great Plains, the southern coastal sections of California, and the northern valleys of the Great Basin. As in the arid region, northern areas experience lower winter temperatures and more frequent snowfalls. Annual precipitation is generally between 20 and 40 cm (8 and 16 in.). The climatic record for Denver, Colorado (see ● Fig. 17.16), exemplifies the semi-arid (BSk) climate.

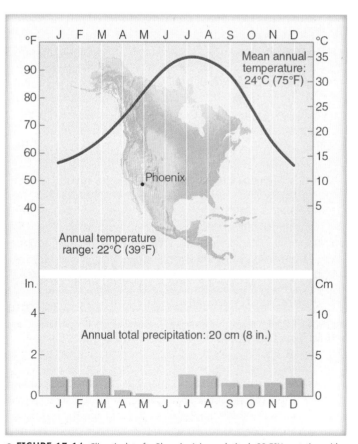

● **FIGURE 17.14** Climatic data for Phoenix, Arizona, latitude 33.5°N, a station with an arid climate (BWh).

● **FIGURE 17.15** Cumulus clouds forming over the steppe grasslands of western North America, a region with a semi-arid climate (BS).

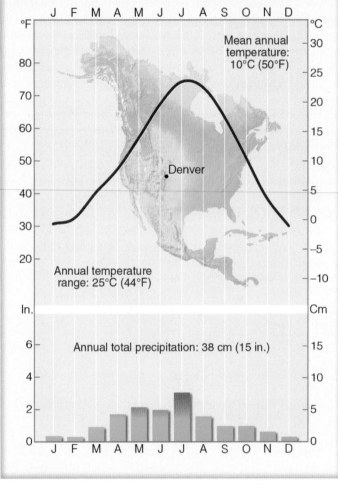

● **FIGURE 17.16** Climatic data for Denver, Colorado, latitude 40°N. A station with a semi-arid climate (BSk).

As average rainfall amounts increase, the climate gradually changes to one that is more humid. Hence, the semi-arid (steppe) climate marks the transition between the arid and the humid climatic regions. (Before reading about moist climates, you may wish to read Focus section 17.2, about deserts that experience drizzle but little rainfall.)

MOIST SUBTROPICAL MID-LATITUDE CLIMATES (GROUP C)

General characteristics: humid with mild winters (i.e., average temperature of the coldest month below 18°C, or 64°F, and above −3°C, or 27°F).

Extent: the eastern and western regions of most continents, from about 25° to 40° latitude.

Major types: humid subtropical (Cfa), marine (Cfb), and dry-summer subtropical, or Mediterranean (Cs).

The Group C climates of the middle latitudes have distinct summer and winter seasons. Additionally, they have ample precipitation to keep them from being classified as dry. Although winters can be cold, and air temperatures can change appreciably from one day to the next, no month has a mean temperature below −3°C (27°F), for if it did, it would be classified as a D climate—one with severe winters.

The first C climate we will consider is the **humid subtropical climate** (Cfa).* Notice in Fig. 17.7, pp. 482–483, that Cfa climates are found principally along the east coasts of continents, roughly between 25° and 40° latitude. They dominate the southeastern section of the United States, as well as eastern China and southern Japan. In the Southern Hemisphere, they are found in southeastern South America and along the southeastern coasts of Africa and Australia. Many large cities in China and the United States experience a humid subtropical climate.

A trademark of the humid subtropical climate is its hot, muggy summers. This sultry summer weather occurs because Cfa climates are located on the western side of subtropical highs, where maritime tropical air from lower latitudes is swept poleward into these regions. Generally, summer dew-point temperatures are high (often exceeding 23°C, or 73°F) and so is the relative humidity, even during the middle of the day. The high humidity combines with the high air temperature (usually above 32°C, or 90°F) to produce more oppressive conditions than are found in equatorial regions. Summer morning low temperatures often range between 21°C and 27°C (70°F and 81°F). Occasionally, a weak summer cool front will bring temporary relief from the sweltering conditions. However, devastating heat waves, sometimes lasting many weeks, can occur when an upper-level ridge moves over the area.

Winters tend to be relatively mild, especially in the lower latitudes, where air temperatures rarely dip much below freezing. Poleward regions experience winters that are colder and harsher. Here, frost, snow, and ice storms are more common, but heavy snowfalls are rare. Winter weather can be quite changeable, as almost summerlike conditions can give way to cold rain and wind in a matter of hours when a middle-latitude cyclonic storm and its accompanying fronts pass through the region.

Humid subtropical climates experience adequate and fairly well-distributed precipitation throughout the year, with typical annual averages between 80 and 165 cm (31 and 65 in.). In summer, when thunderstorms are common, much of the precipitation falls as afternoon showers. Tropical storms entering the

---

*In the Cfa climate, the "f" means that all seasons are wet and the "a" means that summers are long and hot. A more detailed explanation is given in Table 17.1 on p. 481.

A Desert with Clouds and Drizzle

We already know that not all deserts are hot. By the same token, not all deserts are sunny. In fact, some coastal deserts experience considerable cloudiness, especially low stratus and fog.

Amazingly, these coastal deserts are some of the driest places on Earth. They include the Atacama Desert of Chile and Peru, the coastal Sahara Desert of northwestern Africa, the Namib Desert of southwestern Africa, and a portion of the Sonoran Desert in Baja California (see ● Fig. 2). On the Atacama Desert, for example, some regions go without measurable rainfall for decades. Arica, in northern Chile, has an average annual rainfall of only 0.08 cm (0.03 in.).

This aridity is, in part, due to the fact that each region is adjacent to a large body of relatively cool water. Notice in Fig. 2 that these deserts are located along the western coastal margins of continents, where a subtropical high-pressure area causes prevailing winds to move cool water from higher latitudes along the coast. In addition, these winds help to accentuate the water's coldness by initiating *upwelling*—the rising of cold water from lower levels. The combination of these conditions tends to produce coastal water temperatures between 10°C and 15°C (50°F and 59°F), which is quite cool for such low latitudes. As surface air sweeps across the cold water, it is chilled to its dew point, often producing a blanket of fog and low clouds, from which drizzle falls. The drizzle, however, accounts for very little rainfall. In most regions, it is only

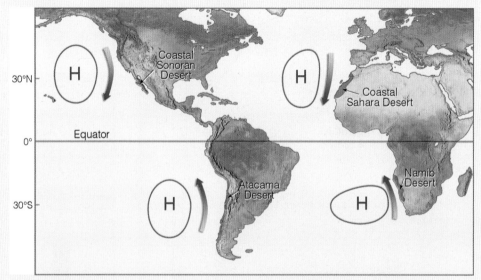

● **FIGURE 2** Location of coastal deserts (dark orange shade) that experience frequent fog, drizzle, and low clouds. (Blue arrows indicate prevailing winds and the movement of cool ocean currents.)

enough to dampen the streets with a mere trace of precipitation.

As the cool stable air moves inland, it warms, and the water droplets evaporate. Hence, most of the cloudiness and drizzle is found along the immediate coast. Although the relative humidity of this air is high, the dew-point temperature is comparatively low (often near that of the coastal surface water). Inland, further warming causes the air to rise. However, a stable subsidence inversion, associated with the subtropical highs, inhibits vertical motions by capping the rising air,

causing it to drift back toward the ocean, where it sinks, completing a rather strong sea breeze circulation. The position of the subtropical highs, which tend to remain almost stationary, plays an additional role by preventing the Intertropical Convergence Zone with its rising, unstable air from entering the region.

And so we have a desert with clouds and drizzle—a desert that owes its existence, in part, to its proximity to rather cold ocean water and, in part, to the position and air motions of a subtropical high.

United States and China can substantially add to summer and autumn rainfall totals. Winter precipitation most often occurs with eastward-trekking middle-latitude cyclonic storms. In the southeastern United States, the abundant rainfall supports a thick pine forest that becomes mixed with oak at higher latitudes. The climate data for Mobile, Alabama, a city with a Cfa climate, is given in ● Fig. 17.17.

Glance back at Fig. 17.7, pp. 482–483, and observe that C climates extend poleward along the western side of most continents from about latitude 40° to 60°. These regions are dominated by prevailing winds from the ocean that moderate the climate, keeping winters considerably milder than stations located at the same latitude farther inland. In addition to this, summers are quite cool. When the summer season is both short and cool, the climate is designated as Cfc. Equatorward, where

summers are longer (but still cool), the climate is classified as *west coast marine*, or simply **marine**, Cfb.*

Where mountains parallel the coastline, such as along the west coasts of North and South America, the marine influence is restricted to narrow belts. Over much of western Europe, prevailing westerly winds are unobstructed by high mountains, so ocean air often sweeps across the region, providing it with a **marine climate** (Cfb).

During much of the year, marine climates are characterized by low clouds, fog, and drizzle. The ocean's influence produces adequate precipitation in all months, with much of it falling as

---

*In the Cfb climate, the "b" means that summers are cooler than in those regions experiencing a Cfa climate. The temperature criteria for the various subregions are given in Table 17.1, p. 481.

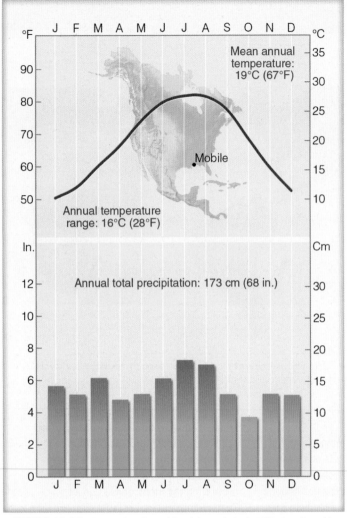

● **FIGURE 17.17** Climatic data for Mobile, Alabama, latitude 30°N. A station with a humid subtropical climate (Cfa).

● **FIGURE 17.18** Climatic data for Port Hardy, Canada, latitude 51°N. A station with a marine climate (Cfb).

light or moderate rain associated with maritime polar air masses. Snow does fall, but frequently it turns to slush after only a day or so. In some locations, topography greatly enhances precipitation totals. For example, along the west coast of North America, coastal mountains not only force air upward, enhancing precipitation, but they also slow a storm's eastward progress, which enables the storm to drop more precipitation on the area.

Along the northwest coast of North America, rainfall amounts decrease in summer. This phenomenon is caused by the northward migration of the subtropical Pacific high, which is located southwest of this region. The summer decrease in rainfall

can be seen by examining the climatic record of Port Hardy (see ● Fig. 17.18), a station situated along the coast of Canada's Vancouver Island. The data illustrate another important characteristic of marine climates: the low annual temperature range for such a high-latitude station. The ocean's influence keeps daily temperature ranges low as well. In this climate type, it rains on many days and when it is not raining, skies are usually overcast. The heavy rains produce a dense forest of Douglas fir.

Moving equatorward of marine climates, the influence of the subtropical highs becomes greater, and the summer dry period more pronounced. Gradually, the climate changes from marine to **dry-summer subtropical** (Cs), which is also called **Mediterranean** because it predominates along the coastal areas of the Mediterranean Sea. (Here the lowercase "s" stands for "summer dry.") Along the west coast of North America, Portland, Oregon—because it has rather dry summers—marks the transition between the marine climate and the dry-summer subtropical climate to the south.

The extreme summer aridity of the Mediterranean climate, which in California may exist for five months, is caused by the sinking air of the subtropical highs. In addition, these anticyclones divert summer storm systems poleward. During the winter, when the subtropical highs move equatorward, midlatitude

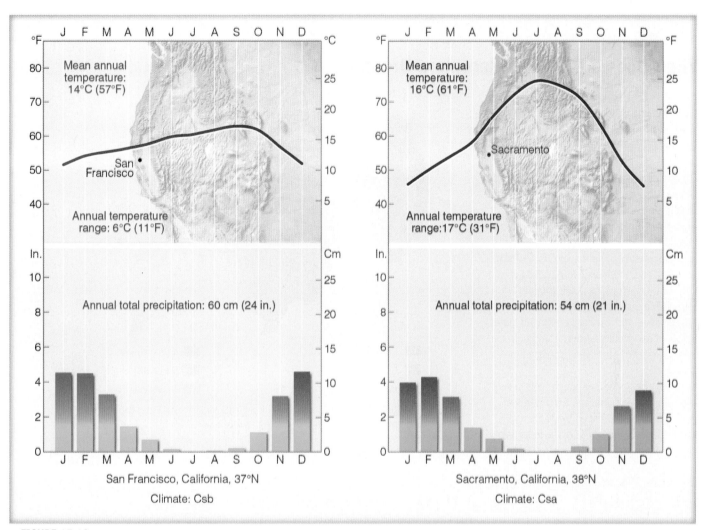

● **FIGURE 17.19** Comparison of a coastal Mediterranean climate, Csb (San Francisco, at left), with an interior Mediterranean climate, Csa (Sacramento, at right).

storms from the ocean frequent the region, bringing with them much-needed rainfall. Consequently, Mediterranean climates are characterized by mild, wet winters and mild-to-hot, dry summers.

Where surface winds parallel the coast, upwelling of cold water helps keep the water itself and the air above it cool all summer long. In these coastal areas, which are often shrouded in low clouds and fog, the climate is called *coastal Mediterranean* (Csb). Here, summer daytime maximum temperatures usually reach about 21°C (70°F), while overnight lows often drop below 15°C (59°F). Inland, away from the ocean's influence, summers are hot and winters are a little cooler than coastal areas. In this *interior Mediterranean climate* (Csa), summer afternoon temperatures usually climb above 34°C (93°F) and occasionally above 40°C (104°F).

● Figure 17.19 contrasts the coastal Mediterranean climate of San Francisco, California, with the interior Mediterranean climate of Sacramento, California. While Sacramento is only 130 km (80 mi) inland from San Francisco, Sacramento's average July temperature is 9°C (16°F) higher. As we would expect, Sacramento's annual temperature range is considerably greater, too. Although Sacramento and San Francisco both experience an occasional frost, snow in these areas is a rarity.

In Mediterranean climates, yearly precipitation amounts range between 30 and 90 cm (12 and 35 in.). However, much more precipitation falls on surrounding hillsides and mountains. Because of the summer dryness, the land supports only a scrubby mixture of low-growing woody plants and trees called *chaparral* (see ● Fig. 17.20).

● **FIGURE 17.20** In the Mediterranean-type climates of North America, typical chaparral vegetation includes chamisa, manzanita, and foothill pine.

When Does a Dry Spell Become a Drought?

When a region's average precipitation drops dramatically for an extended period of time, drought may result. The word *drought* refers to a period of abnormally dry weather that produces a number of negative consequences, such as crop damage or an adverse impact on a community's water supply. Keep in mind that drought is more than a dry spell. In the dry, summer subtropical (Csa) climate of California's Central Valley, it may not rain from May through September. This dry spell is normal for this region and, therefore, would not be considered a drought. However, if this summer dry spell were to occur in the humid subtropical (Cfa) climate of the southeastern United States, the lack of rain would constitute a drought and could be disastrous for many aspects of the community.

In an attempt to measure drought severity, Wayne Palmer, a scientist with the National Weather Service, developed the *Palmer Drought Severity Index (PDSI)*. The index takes into account average temperature and precipitation values to define drought severity. The index is most effective in assessing long-term drought that lasts several months or more. Drought conditions are indicated by a set of numbers that range from 0 (normal) to −4 (extreme drought). (See ▼ Table 1.) The index also assesses wet conditions with numbers that range from +2 (unusually moist) to +4 (extremely moist). The *Palmer Hydrological Drought Index (PHDI)* expands the PDSI by taking into account additional water (hydrological) information, such as a region's groundwater reserves and reservoir levels.

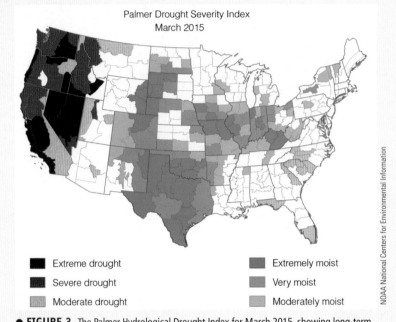

Palmer Drought Severity Index
March 2015

■ Extreme drought   ■ Extremely moist
■ Severe drought   ■ Very moist
■ Moderate drought   ■ Moderately moist

NOAA National Centers for Environmental Information

● **FIGURE 3** The Palmer Hydrological Drought Index for March 2015, showing long-term drought conditions and regions with sufficient moisture.

▼ **TABLE 1   Palmer Drought Severity Index**

VALUE	DROUGHT	VALUE	MOISTURE
−4.0 or less	Extreme	+4.0 or greater	Extremely Moist
−3.0 to −3.9	Severe	+3.0 to 3.9	Very Moist
−2.0 to −2.9	Moderate	+2.0 to 2.9	Unusually Moist
−1.9 to +1.9	Normal	−1.9 to +1.9	Normal

At this point, we should note that summers are not as dry along the Mediterranean Sea as they are along the west coast of North America. Moreover, coastal Mediterranean areas are also warmer, due to the lack of upwelling in the Mediterranean Sea.

Before leaving our discussion of C climates, note that when the dry season is in winter, the climate is classified as Cw. Over northern India and portions of China, the relatively dry winters are the result of northerly winds from continental regions circulating southward around the cold Siberian high. Many lower-latitude regions with a Cw climate would be tropical if they were not as high in elevation and, consequently, too cool to be designated as tropical.

When a moist climate turns dry, drought often results. What constitutes a drought and how is it measured? These are some of the questions addressed in Focus section 17.3.

MOIST CONTINENTAL CLIMATES (GROUP D)

General characteristics: warm-to-cool summers and cold winters (i.e., average temperature of warmest month exceeds 10°C, or 50°F, and the coldest monthly average drops below −3°C, or 27°F); winters are severe with snowstorms, blustery winds, bitter cold; climate controlled by large continent.

Extent: north of moist subtropical midlatitude climates.

A drought apparent in the PDSI can take weeks or months to be reflected in the PHDI because water supplies take time to drain down after dry conditions set in. Likewise, even after rains return to a drought-stricken region, it takes time for reservoirs and groundwater storage to be recharged, so the PHDI may lag the PSDI in showing recovery.

● Figure 3 shows the PHDI across the United States during March 2015. Notice that several parts of the West are in an extreme drought (dark red shade), including almost all of Nevada as well as the southern two-thirds of California, which at the time was in its fourth year of a destructive drought. Most of California's rains arrive in late autumn, winter, and early spring, yet drought conditions were persisting in early 2015 even after most of the rainy season had ended. Notice also in Fig. 3 that there are a few patches of green, especially over the northern Great Plains, indicating ample moisture.

Drought is not uncommon to North America. In fact, probably the worst weather-related disaster to hit the United States during the twentieth century was the great drought of the 1930s. That drought, which tragically coincided with the Great Depression, actually began in the late 1920s and continued into the late 1930s. It not only lasted a long time, but it extended over a vast area (see ● Fig. 4).

The drought, coupled with poor farming practices, left the topsoil of the Great Plains ripe for wind erosion. As a result, wind storms lifted millions of tons of soil into the air, creating vast

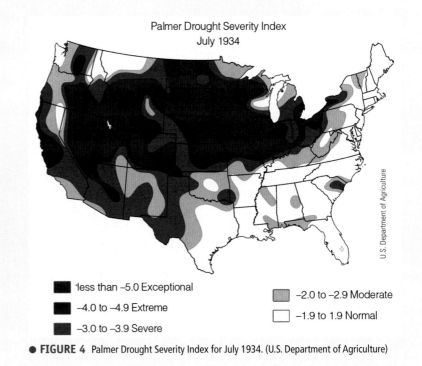

Palmer Drought Severity Index
July 1934

■ less than −5.0 Exceptional
■ −4.0 to −4.9 Extreme
■ −3.0 to −3.9 Severe
▨ −2.0 to −2.9 Moderate
□ −1.9 to 1.9 Normal

U.S. Department of Agriculture

● **FIGURE 4** Palmer Drought Severity Index for July 1934. (U.S. Department of Agriculture)

dust storms that buried whole farm houses, reduced millions of acres to an unproductive wasteland, and financially ruined thousands of families. Because of the infamous dust storms, the worst-affected region has been called the *Dust Bowl* and the 1930s are often referred to as "the Dust Bowl years." To worsen an already bad situation, the drought was accompanied by extreme summer heat that was most severe during the summers of 1934 and 1936.

How does the California drought year of 2014–2015 compare with the drought year of 1934? We can see in Fig. 4 that during 1934

most of California was experiencing some form of drought, though less extreme than in 2015. At the same time, exceptionally dry conditions existed in 1934 over a vast area of the upper Plains and far west, with many regions experiencing a Palmer Index of −4 or below. Unfortunately, this already disastrous drought became progressively worse through the mid-1930s. Millions of people were affected, many of whom eventually became destitute. Thousands migrated westward in search of employment, as depicted in the book and film *The Grapes of Wrath.*

Major types: humid continental with hot summers (Dfa), humid continental with cool summers (Dfb), and subpolar (Dfc).

The D climates are controlled by large landmasses. Therefore, they are found only in the Northern Hemisphere. Look at the climate map, Fig. 17.7, pp. 482–483, and notice that D climates extend across North America and Eurasia, from about latitude 40°N to almost 70°N. In general, they are characterized by cold winters and warm-to-cool summers.

As we know, for a station to have a D climate, the average temperature of its coldest month must dip below −3°C (27°F). This is not an arbitrary number. Köppen found that, in Europe, this temperature marked the southern limit of persistent snow

cover in winter.* Hence, D climates experience a great deal of winter snow that stays on the ground for extended periods. When the temperature drops to a point where every month has an average temperature below 10°C (50°F), the climate is classified as polar (E). Köppen found that the average monthly temperature of 10°C tended to represent the minimum temperature required for tree growth. So no matter how cold it gets in a D climate (and winters can get extremely cold), there is enough summer warmth to support the growth of trees.

---

*In North America, studies suggest that an average monthly temperature of (0°C 32°F) or below for the coldest month seems to correspond better to persistent winter snow cover.

● **FIGURE 17.21** The leaves of deciduous trees burst into brilliant color during autumn over the countryside of New York's Adirondack Park, a region with a humid continental climate.

SNEHIT/Shutterstock.com

There are two basic types of D climates: the **humid continental** (Dfa and Dfb) and the **subpolar** (Dfc). **Humid continental climates** are observed from about latitude 40°N to 50°N (60°N in Europe). In these areas, precipitation is adequate and fairly evenly distributed throughout the year, although interior stations experience maximum precipitation in summer. Annual precipitation totals usually range from 50 to 100 cm (20 to 40 in.). Native vegetation in the wetter regions includes forests of spruce, fir, pine, and oak. In autumn, nature's pageantry unveils itself as the leaves of deciduous trees turn brilliant shades of red, orange, and yellow (see ● Fig. 17.21).

Humid continental climates are subdivided on the basis of summer temperatures. Where summers are long and hot,* the climate is described as *humid continental with hot summers* (Dfa). Here, summers are often both hot and humid, especially in the southern regions. Midday temperatures often exceed 32°C (90°F) and occasionally 40°C (104°F). Summer nights are usually warm and humid, as well. The frost-free season normally lasts from five to six months, long enough to grow a wide variety of crops. Winters tend to be windy, cold, and snowy. Farther north, where summers are shorter and not as hot,** the climate is described as *humid continental with long cool summers* (Dfb). In Dfb climates, summers are not only cooler but much less humid. Temperatures

may exceed 35°C (95°F) for a time, but extended hot spells lasting many weeks are rare. The frost-free season is shorter than in the Dfa climate, and normally lasts between three and five months. Winters are long, cold, and windy. It is not uncommon for temperatures to drop below −30°C (−22°F) and stay below −18°C (0°F) for days and sometimes weeks. Autumn is short, with winter often arriving right on the heels of summer. Spring, too, is short, as late spring snowstorms are common, especially in the more northern latitudes.

● Figure 17.22 compares the Dfa climate of Des Moines, Iowa, with the Dfb climate of Winnipeg, Canada. Notice that both cities experience a large annual temperature range. This is characteristic of climates located in the northern interior of continents. In fact, as we move poleward, the annual temperature range increases. In Des Moines, it is 31°C (56°F), while 950 km (590 mi) to the north in Winnipeg, it is 38°C (68°F). The summer precipitation maximum expected for these interior continental locations shows up well in Fig. 17.22. Most of the summer rain is in the form of convective showers, although an occasional weak frontal system can produce more widespread precipitation, as can a cluster of thunderstorms—the mesoscale convective complex described in Chapter 14. The weather in both climatic types can be quite changeable, especially in winter, when a brief warm spell is replaced by blustery winds and temperatures plummeting well below −30°C (−22°F).

When winters are severe and summers short and cool, with only one to three months having a mean temperature exceeding 10°C (50°F), the climate is described as *subpolar* (Dfc). From Fig. 17.7 we can see that, in North America, this climate occurs in a broad belt across Canada and Alaska; in Eurasia, it stretches

---

*"Hot" means that the average temperature of the warmest month is above 22°C (72°F) and at least four months have a monthly mean temperature above 10°C (50°F).

**"Not as hot" means that the average temperature of the warmest month is below 22°C (72°F) and at least four months have a monthly mean temperature above 10°C (50°F).

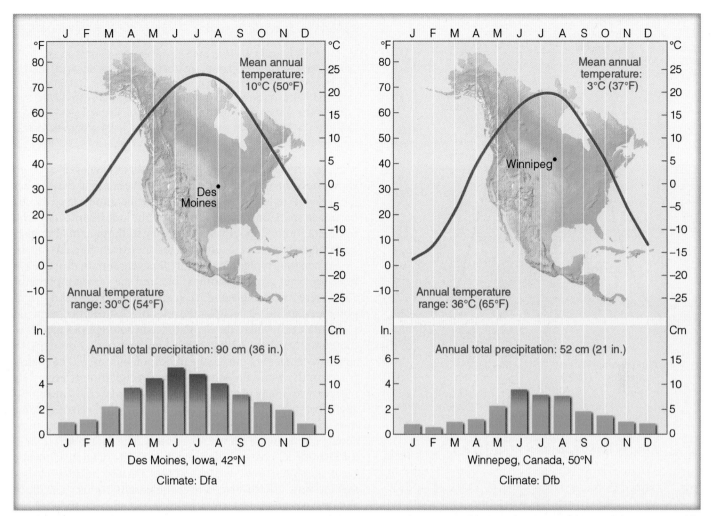

● **FIGURE 17.22** Comparison of a humid continental hot summer climate, Dfa (Des Moines, at left), with a humid continental cool summer climate, Dfb (Winnipeg, at right).

from Norway over much of Siberia. The exceedingly low temperatures of winter account for these areas being the primary source regions for continental polar and arctic air masses. Extremely cold winters coupled with cool summers produce large annual temperature ranges, as exemplified by the climate data in ● Fig. 17.23 for Fairbanks, Alaska.

Precipitation is comparatively light in the subpolar climates, especially in the interior regions, with most places receiving less than 50 cm (20 in.) annually. A good percentage of the precipitation falls when weak cyclonic storms move through the region in summer. The total snowfall is usually not large but the cold air prevents melting, so snow stays on the ground for months at a time. Because of the low temperatures, there is a low annual rate of evaporation that ensures adequate moisture to support the boreal* forests of conifers and birches known as **taiga** (see ● Fig. 17.24). Hence, the subpolar climate is known also as a *boreal climate* and as a *taiga climate*.

In the taiga region of northern Siberia and Asia, where the average temperature of the coldest month drops to a frigid 238°C (236°F) or below, the climate is designated Dfd. Where the winters are considered dry, the climate is designated Dwd.

---

*The word *boreal* comes from the ancient Greek *Boreas*, meaning "wind from the north."

POLAR CLIMATES (GROUP E)

General characteristics: year-round low temperatures (i.e., average temperature of the warmest month is below 10°C, or 50°F).

Extent: northern coastal areas of North America and Eurasia; Greenland and Antarctica.

Major types: polar tundra (ET) and polar ice caps (EF).

In the **polar tundra** (ET), the average temperature of the warmest month is below 10°C (50°F), but above freezing. (See ● Fig. 17.25, the climate data for Barrow, Alaska.) Here, the ground is permanently frozen, a condition known as **permafrost**. Permafrost can be less than 1 m (3 ft) deep or more than 1000 m (3300 ft) deep. Summer weather is usually just warm enough to thaw out the topmost part of the soil, so during the summer, the tundra turns swampy and muddy. Annual precipitation on the tundra is meager, with most stations receiving less than 20 cm (8 in.). In lower latitudes, this amount of precipitation would constitute a desert, but in the cold polar regions evaporation rates are very low and moisture remains adequate. Because of the extremely short growing season, *tundra vegetation* consists of mosses, lichens, dwarf trees, and scattered woody vegetation, typically only several centimeters tall when fully grown (see ● Fig. 17.26).

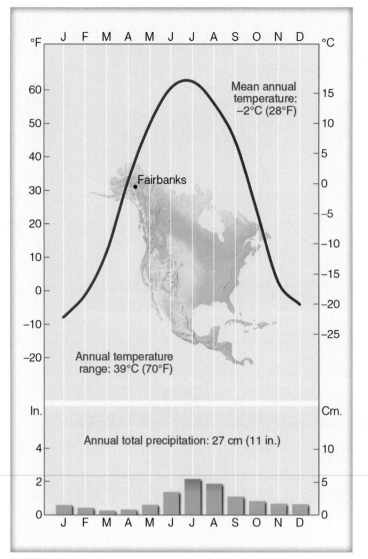

● **FIGURE 17.23** Climatic data for Fairbanks, Alaska, latitude 65°N. A station with a subpolar climate (Dfc).

● **FIGURE 17.24** Coniferous forests (taiga) such as this occur where winter temperatures are low and precipitation is abundant.

Even though summer days are long, the sun is never very high above the horizon. Additionally, some of the sunlight that reaches the surface is reflected by snow and ice, while some is used to melt the frozen soil. Consequently, in spite of the long hours of daylight, summers are quite cool. Even so, there are still large annual temperature ranges between the summers and the extremely cold winters.

When the average temperature for every month drops below freezing, plant growth is impossible, and the region is perpetually covered with snow and ice. This climatic type is known as **polar ice cap** (EF). It occupies the interior ice sheets of Greenland and Antarctica, where the depth of ice in some places measures thousands of meters. In this region, temperatures are never much above freezing, even during the middle of "summer." The coldest places in the world are located here. Precipitation is extremely meager, with many places receiving less than 10 cm (4 in.) annually. Most precipitation falls as snow during the "warmer" summer. Strong downslope katabatic winds frequently whip the snow about, adding to the climate's harshness. The data in ● Fig. 17.27 for Eismitte, Greenland, illustrate the severity of an EF climate.

In Chapter 18 we will look at how Earth's climate is changing and how it has warmed over the past 100-plus years. Such changes may have an effect on the Köppen climate boundaries, but any boundary change is often so small that it is difficult to assess. One possible way to observe these changes, however, is to see how changing temperatures are influencing regions where plants grow. In a warmer world, are the plant hardiness zones shifting poleward? Focus section 17.4 addresses this question.

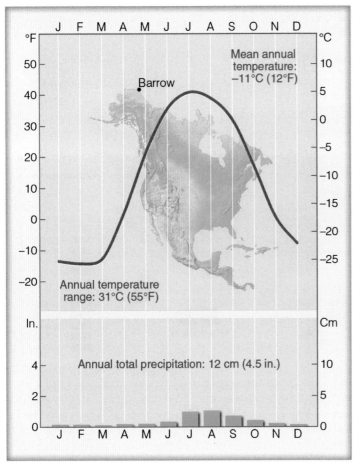

● **FIGURE 17.25** Climatic data for Barrow, Alaska, latitude 71°N. A station with a polar tundra climate (ET).

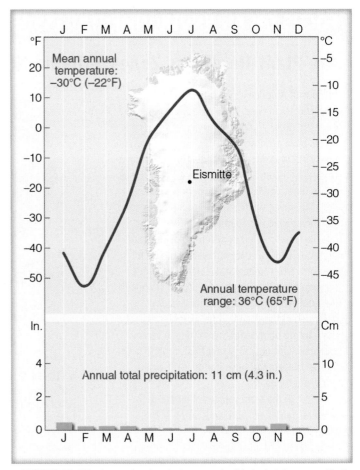

● **FIGURE 17.27** Climatic data for Eismitte, Greenland, latitude 71°N. Located in the interior of Greenland at an elevation of almost 10,000 feet above sea level. Eismitte has a polar ice cap climate (EF).

Bob Coffen/Shutterstock.com

● **FIGURE 17.26** Tundra vegetation in Alaska. This type of tundra is composed mostly of sedges and dwarfed wildflowers that bloom during the brief growing season.

CRITICAL THINKING QUESTION

What type of climate change would have to take place over the tundra in order for the tundra vegetation in Fig. 17.26 to become a coniferous forest (taiga), shown in Fig. 17.24?

Are Plant Hardiness Zones Shifting Northward?

Millions of Americans pay close attention to local climate in order to help decide what plants to grow. Temperatures across the United States have been warming in recent decades. In fact, 2012 was the warmest year in more than a century of record keeping for the 48 contiguous states, beating 1934 by a full degree Fahrenheit. Moreover, winters are warming faster than summers, which could mean that plants are less likely to be killed by severe cold. The winter of 2015–2016 was the warmest on record for the contiguous United States. These changes may not be enough to cause a drastic change to the Köppen climate classification system described in this chapter, but we might expect that climate boundaries are shifting in some locations. One way to assess this situation is to look at the temperatures that plants need to grow and thrive.

One of the main sources of guidance for farmers, gardeners, and landscapers is the plant hardiness map produced by the U.S. Department of Agriculture (USDA). This map (see ● Fig. 5) places each part of the United States, including Puerto Rico, into one of 13 numbered zones based on the average annual minimum temperature (the coldest reading one might expect to observe in a typical year). Each zone corresponds to a range of 10°F. For example, Chicago is in zone 5, which means that the average annual minimum temperature is between −20°F and −11°F. Each major zone is divided into subzones with intervals of 5°F. (For clarity, Fig. 5 does not show subzones.)

There are other important ways to measure the risk posed by cold weather to plants, such as the average dates of the first frost and freeze in autumn or the last frost and freeze in spring. However, the average annual minimum temperature serves as a useful single index for determining which plants might do well in a given location. This guideline is especially true for perennial plants such as trees and shrubs that must survive each winter's cold.

Fig. 5 shows the 2012 edition of the USDA map, based on temperatures observed from 1976 to 2005. This updated edition replaced a previous version of the map, released in 1990,

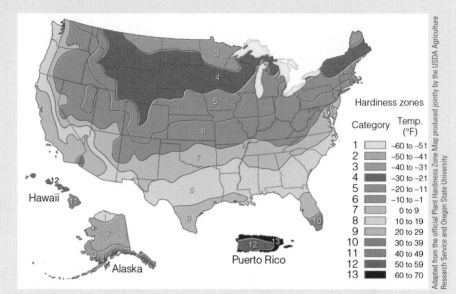

● **FIGURE 5** Plant Hardiness Zones based on lowest expected yearly temperatures. (Adapted from the official Plant Hardiness Zone Map produced jointly by the USDA Agriculture Research Service and Oregon State University.)

that was based on data from 1974 to 1986. Not surprisingly—as the planet is undergoing long-term warming related to greenhouse gases—the new map showed that many parts of the nation had shifted into warmer zones of plant hardiness. In fact, more than half of 34 cities highlighted on the 1990 map ended up in a new zone in the 2012 map.

The 1990 map was based on data spanning only 13 years, which is smaller than the 30-year period typically used in climatology. Therefore, the USDA stated that the new map was not intended to serve as an illustration of long-term climate change. However, the maps are consistent with other analyses showing an overall trend toward warmer temperatures in the United States since the 1970s. Conditions may even be leaping ahead of the revised map. One study concluded that there had been enough warming from 2005 to 2012 for about a third of the United States to shift another half zone warmer than shown on the most recent USDA map.

The winter of 2015–2016 brought many dramatic cases of mildness. December 25 was

the warmest Christmas Day on record for many East Coast cities, from Portland, Maine, which hit 62°F (17°C), to Jacksonville, Florida, which soared to 82°F (28°C). The famed cherry blossoms in Washington, D.C., reached their 2016 peak bloom on March 25, a few days ahead of average. In fact, only six years between 1995 and 2016 saw the peak bloom occur later than the long-term average of April 4.

Even in a warming climate, however, cold extremes can still occur. During the very mild winter of 2015–2016, Boston warmed to 50°F on a total of 35 days, the most recorded in any winter. Yet on February 14, the city dipped to a frigid −9°F (−23°C), its coldest reading in more than 50 years. One of the key questions in climate research is whether a warming climate will tend to make annual minimum temperatures reliably warmer, or whether the general warming will be interspersed with frigid extremes every few years—extremes that could threaten any vegetation planted with milder temperatures in mind.

● **FIGURE 17.28** Vertical view of changing vegetation and climate due to elevation in the central Sierra Nevada.

HIGHLAND CLIMATES (GROUP H) It is not necessary to visit the polar regions to experience a polar climate. Because temperature decreases with altitude, climatic changes experienced when climbing 300 m (1000 ft) in elevation are about equivalent in high latitudes to horizontal changes experienced when traveling 300 km (186 mi) northward. (This distance is equal to about 3° latitude.) Moreover, as air rises along the windward side of a mountain, precipitation amounts usually increase the higher you go. When you are ascending a high mountain, such as Mt. Everest, you can travel through many climatic regions in a relatively short distance. Thus, **highland climates** often show a great deal of variation in temperature, precipitation, and vegetation over a relatively short vertical change in elevation.

● Figure 17.28 shows how the climate and vegetation change along the western slopes of the central Sierra Nevada. (See Fig. 17.5, p. 477, for the precipitation patterns for this region.) Notice that, at the base of the mountains, the climate and vegetation represent semi-arid conditions, while in the foothills the climate becomes Mediterranean and the vegetation changes to chaparral. Higher up, thick fir and pine forests prevail. At still higher elevations, the climate is subpolar and the taiga gives way to dwarf trees and tundra vegetation. Near the summit there are permanent patches of ice and snow, with some small glaciers nestled in protected areas. Hence, in less than 13,000 vertical feet, the climate has changed from semi-arid to polar.

SUMMARY

In this chapter, we examined global temperature and precipitation patterns, as well as the various climatic regions throughout the world. Tropical climates are found in low latitudes, where the noon sun is always high, day and night are of nearly equal length, every month is warm, and no real winter season exists. Some of the rainiest places in the world exist in the tropics, especially where warm, humid air rises upslope along mountain ranges.

Dry climates prevail where potential evaporation and transpiration exceed precipitation. Some deserts, such as the Sahara, are mainly the result of sinking air associated with the subtropical highs, while others, due to the rain shadow effect, are found on the leeward side of mountains. Many deserts form in response to both of these effects.

Middle latitudes are characterized by a distinct winter and summer season. Winters tend to be milder in lower latitudes and more severe in higher latitudes. Along the east coast of some continents, summers tend to be hot and humid as moist air sweeps poleward around the subtropical highs. The air often rises and condenses into afternoon thunderstorms in this humid subtropical climate. The west coasts of many continents tend to be drier, especially in summer, as the combination of cool ocean water and sinking air of the subtropical highs largely inhibits the formation of cumuliform clouds.

In the middle of large continents, such as North America and Eurasia, summers are usually wetter than winters. Winter temperatures are generally lower than those experienced in coastal regions. As one moves northward, summers become shorter and winters longer and colder. Polar climates prevail at high latitudes, where winters are severe and there is no real summer. When ascending a high mountain, one can travel through many climatic zones in a relatively short distance.

KEY TERMS

The following terms are listed (with corresponding page numbers) in the order they appear in the text. Define each. Doing so will aid you in reviewing the material covered in this chapter.

microclimate, 474
mesoclimate, 474
macroclimate, 474
global climate, 474
climatic controls, 474

Köppen classification system, 479
P/E index, 480
tropical rain forest, 482
tropical wet climate, 482
laterite, 483
tropical monsoon climate, 484
tropical wet-and-dry climate, 484
savanna grass, 484
arid climate, 486
xerophytes, 486
semi-arid climate, 487
steppe, 487
humid subtropical climate, 488
marine climate, 489
dry-summer subtropical (Mediterranean), 490
humid continental climate, 494
subpolar climate, 494
taiga, 495
polar tundra climate, 495
permafrost, 495
polar ice cap climate, 496
highland climate, 499

QUESTIONS FOR REVIEW

1. What factors determine the global pattern of precipitation?
2. Explain why, in North America, there is typically a precipitation maximum along the West Coast in winter, a maximum on the Central Plains in summer, and a fairly even distribution between summer and winter along the East Coast.
3. Why are the lowest temperatures in polar regions observed in the interior of large landmasses?
4. What climatic information did Köppen use in classifying climates?
5. How did Köppen define tropical climate? How did he define a polar climate?
6. According to Köppen's climatic system (Fig. 17.7, pp. 482–483), what major climatic type is most abundant in each of the following areas: (a) in North America, (b) in South America, and (c) throughout the world?
7. What is the primary factor that makes a dry climate "dry"?
8. In which climatic region would each of the following be observed: tropical rain forest, xerophytes, steppe, taiga, tundra, and savanna?

9. What are the controlling factors (the climatic controls) that produce the following climatic regions: (a) tropical wet and dry, (b) Mediterranean, (c) marine, (d) humid subtropical, (e) subpolar, (f) polar ice cap

10. How do C-type climates differ from D-type climates?

11. Why are large annual temperature ranges characteristic of D-type climates?

12. Why are D climates found in the Northern Hemisphere but not in the Southern Hemisphere?

13. Explain why a tropical rainforest climate will support a tropical rain forest, while a tropical wet-and-dry climate will not.

14. Why are marine climates (Cs) usually found on the west coast of continents?

15. What is the primary distinction between a Cfa and a Dfa climate?

16. Explain how arid deserts can be found adjacent to oceans.

17. Why did Köppen use the 10°C (50°F) average temperature for July to distinguish between D and E climates?

18. What accounts for the existence of a BWk climate in the western Great Basin of North America?

19. Barrow, Alaska, receives a mere 11 cm (4.3 in.) of precipitation annually. Explain why its climate is not classified as arid or semi-arid.

20. Explain why subpolar climates are also known as boreal climates and taiga climates.

QUESTIONS FOR THOUGHT

1. Why do cities directly east of the Rockies (such as Denver, Colorado) receive much more precipitation than cities directly east of the Sierra Nevada (such as Reno and Lovelock, Nevada)?

2. What climatic controls affect the climate in your area?

3. Los Angeles, Seattle, and Boston are all coastal cities, yet Boston has a continental rather than a marine climate. Explain why.

4. Why are many structures in polar regions built on pilings?

5. Why are summer afternoon temperatures in a humid subtropical climate (Cfa) often higher than in a tropical wet climate (Af)?

6. Why are humid subtropical climates (Cfa) found in regions bounded by 20° and 40° (N or S) latitudes, and nowhere else?

7. In which of the following climate types is virga likely to occur most frequently: humid continental, arid desert, or polar tundra? Explain why.

8. As shown in Fig. 17.19, p. 491, San Francisco and Sacramento, California, have similar mean annual temperatures but different annual temperature ranges. What factors control the annual temperature ranges at these two locations?

9. Why is there a contrast in climate types on either side of the Rocky Mountains, but not on either side of the Appalachian Mountains?

10. Over the past 100 years or so, Earth has warmed by around 1.0°C (1.8°F). If additional warming should occur as expected over the next 100 years, explain how this rise in temperature might influence the boundary between C and D climates. How would the warming influence the boundary between D and E climates?

PROBLEMS AND EXERCISES

1. Suppose a city has the mean annual precipitation and temperature given in ▼ Table 17.2. Based on Köppen's climatic types, how would this climate be classified? On a map of North America, approximately where would this city be located? What type of vegetation would you expect to see there? Answer these same questions for the data in ▼ Table 17.3.

2. Compare the following climate classifications for your area: (a) ancient Greeks, (b) Köppen system, (c) Thornthwaite's system. Which classification system is best for your area's mesoclimate? Macroclimate?

3. On a blank map of the world, roughly outline where Köppen's major climatic regions are located.

 ONLINE RESOURCES

Visit www.cengagebrain.com to view additional resources, including video exercises, practice quizzes, an interactive eBook, and more.

▼ TABLE 17.2

	JAN.	FEB.	MAR.	APR.	MAY	JUNE	JULY	AUG.	SEPT.	OCT.	NOV.	DEC.	YEAR
Temperature (°F)	40	42	50	60	68	77	80	79	73	62	49	42	60
Precipitation (in.)	4.9	4.2	5.3	3.7	3.8	3.2	4.0	3.3	2.7	2.5	3.4	4.1	45

▼ TABLE 17.3

	JAN.	FEB.	MAR.	APR.	MAY	JUNE	JULY	AUG.	SEPT.	OCT.	NOV.	DEC.	YEAR
Temperature (°F)	18	18	29	42	55	65	70	68	60	48	36	23	44
Precipitation (in.)	1.9	1.5	2.2	2.6	2.9	3.6	3.8	3.0	3.1	2.9	2.8	1.9	32

Huge chunks of ice from the edge of an
Antarctic glacier plummet to the sea.

Earth's Changing Climate

A CHANGE IN OUR CLIMATE HOWEVER IS taking place very sensibly. Both heats and colds are becoming much more moderate within the memory even of the middle-aged. Snows are less frequent and less deep. They do not often lie, below the mountains, more than one, two, or three days, and very rarely a week. They are remembered to have been formerly frequent, deep, and of long continuance. The elderly inform me the earth used to be covered with snow about three months in every year. The rivers, which then seldom failed to freeze over in the course of the winter, scarcely ever do now. This change has produced an unfortunate fluctuation between heat and cold, in the spring of the year, which is very fatal to fruits. In an interval of twenty-eight years, there was no instance of fruit killed by the frost in the neighborhood of Monticello. The accumulated snows of the winter remaining to be dissolved all together in the spring, produced those overflowings of our rivers, so frequent then, and so rare now.

Thomas Jefferson, *Notes on the State of Virginia*, 1781

The opening passage of this chapter (p. 503) describes how our third American president, Thomas Jefferson, perceived the evolution of climate in the state of Virginia and in his own backyard. Some changes in climate are due to natural processes, whereas others are related to human activity. One of the great environmental concerns of our time is the **climate change** now unfolding as a result of greenhouse gases being added to our atmosphere. Glaciers are melting, sea level is rising, precipitation is becoming more intense in many areas, and global temperature is increasing each decade. Extensive research has shown that the primary cause of these changes over the last few decades is human (anthropogenic) activity—the burning of fossil fuels.

We know that climate has changed in the past, and nothing suggests that it will not continue to change, both globally and locally. As the urban environment grows, its climate differs from that of the region around it. Sometimes the difference is striking, as when city nights are warmer than the nights of the outlying rural areas. Other times, the difference is subtle, as when a layer of smoke and haze covers a city. Climate variations such as a persistent drought or a delay in the annual monsoon rains can adversely affect the lives of millions. Even small changes can be problematic when averaged over many years, as when grasslands once used for grazing gradually become uninhabited deserts. In this chapter, we will first look at the evidence for climate change in the past; then we will investigate the causes of climate change from both natural processes and human activity.

Reconstructing Past Climates

A mere 18,000 years ago, Earth was in the grip of a cold spell, with *alpine glaciers* extending their icy fingers down river valleys and huge ice sheets *(continental glaciers)* covering vast areas of North America and Europe (see ● Fig. 18.2). The ice at that time measured several kilometers thick and extended as far south as New York and the Ohio River Valley. The glaciers advanced and

● **FIGURE 18.1** Artificial light at night can create a microclimate and throw off a plant's response to the change of seasons. Notice how the leaves of this tree have their fall colors, except for those on the right-hand side directly under the streetlight.

retreated more than 20 times during the last 2.5 million years. In the warmer periods between glacier advances, average global temperatures were similar to those at present, even slightly higher in some cases. The advance and retreat of glaciers are closely related to variations in how Earth orbits the sun. Research in this area suggests we are thousands of years away from the next return of the glaciers.

Presently, glaciers cover less than 10 percent of Earth's land surface. The total volume of ice over the face of Earth amounts to about 25 million cubic kilometers. Most of this ice is in the Greenland and Antarctic ice sheets, and its accumulation over time has allowed scientists to measure past climatic changes. If global temperatures were to rise enough so that all of this ice melted, the average sea level would rise about 65 m (213 ft) (see ● Fig. 18.3). Imagine the catastrophic results: Many major cities (such as New York, Shanghai, Tokyo, and London) would be inundated. Even a rise in global temperature of several degrees Celsius may be enough to raise sea level by a meter or more this century, flooding coastal lowlands and increasing the impact of storm surges.

Geological evidence left behind by advancing and retreating glaciers suggests that global climate has undergone slow but continuous changes. To reconstruct past climates, scientists must examine and then carefully piece together all the available evidence. Unfortunately, the evidence only gives a general understanding of what past climates were like. For example, fossil pollen of a tundra plant collected in a layer of sediment in New England and dated to be 12,000 years old suggests that the climate of that region was much colder than it is today.

Other evidence of global climatic change comes from core samples taken from ocean floor sediments and ice from Greenland and Antarctica. A landmark multiuniversity research project known as CLIMAP (Climate: Long-range Investigation Mapping and Prediction) studied the past million years of global climate. Thousands of meters of ocean sediment obtained with a hollow-centered drill were analyzed. This sediment contained the remains of calcium carbonate shells of organisms that once lived near the surface. Because certain organisms can only live within a narrow range of temperature, the distribution and type of organisms within the sediment indicate the temperature of the surface water.

In addition, the oxygen-isotope* ratio of these shells provided information about the sequence of glacier advances. How? Most of the oxygen in sea water is composed of eight protons and eight neutrons in its nucleus, giving it an atomic weight of 16. However, about one out of every thousand oxygen atoms contains an extra two neutrons, giving it an atomic weight of 18. When ocean water evaporates, the heavy oxygen-18 tends to be left behind. Consequently, the oceans contain a higher concentration of oxygen-18 during periods of glacial advance, when there is more water over land in the form of ice and snow and less water in the oceans. Since the shells of marine organisms are constructed from the oxygen atoms existing in ocean water, determining the ratio of oxygen-18 to oxygen-16 within these shells yields information

---

*Isotopes are atoms whose nuclei have the same number of protons but different numbers of neutrons.

(a)

(b)

● **FIGURE 18.2** Extent of glaciation about 18,000 years ago over (a) North America and over (b) western Europe.

● **FIGURE 18.3** A guest steps out of a hotel in Miami Beach, Florida, onto a street flooded by seasonally high tides in September 2015. Such "nuisance flooding" is becoming more common in many areas as sea levels continue to rise along the U.S. East Coast. If all the ice locked up in glaciers and ice sheets were to melt, estimates are that low-lying coastal areas around the world would be under more than 61 m (200 ft) of water. Even a relatively small 1-meter rise in sea level, which is possible this century, would threaten millions of people with rising seas.

about how the climate may have varied in the past. A higher ratio of oxygen-18 to oxygen-16 in the sediment record suggests a colder climate, whereas a lower ratio suggests a warmer climate. Using data such as these, the CLIMAP project was able to reconstruct Earth's surface ocean temperature for various times during the past (see ● Fig. 18.4).

Vertical ice cores extracted from ice sheets in Antarctica and Greenland provide additional information on past temperature patterns. Glaciers form over land where temperatures are sufficiently low so that, during the course of a year, more snow falls than will melt. Successive snow accumulations over many years compact the snow, which slowly recrystallizes into ice. Since ice

is composed of hydrogen and oxygen, examining the oxygen-isotope ratio in ancient cores provides a past record of temperature trends. Generally, the colder the air when the snow fell, the richer the concentration of oxygen-16 in the core. Moreover, bubbles of ancient air trapped in the ice can be analyzed to determine the past composition of the atmosphere (illustrated later in this chapter; see Fig. 18.15, p. 515).

Ice cores also record the causes of climate changes. One such cause is deduced from layers of sulfuric acid in the ice. The sulfuric acid originally came from large volcanic explosions that injected huge quantities of sulfur into the stratosphere. The resulting sulfate aerosols eventually fell to Earth in polar regions as acid

● **FIGURE 18.4** (a) Sea surface isotherms (°C) during August 18,000 years ago and (b) during August today. During the Ice Age (diagram a) the Gulf Stream (heavy red arrow) shifted to a more easterly direction, depriving northern Europe of its warmth and producing a strong north-to-south ocean surface temperature gradient.

(a) 18,000 years ago

(b) Today

snow, which was preserved in the ice sheets. (As we discussed in Chapter 1, aerosols are tiny airborne particles or droplets.) The Greenland ice cores also provide a continuous record of sulfur from human sources. Moreover, ice cores at both poles are being analyzed for many chemicals that provide records of biological and physical changes in the climate system, such as a beryllium isotope (^{10}Be) that indicates solar activity. Various types of dust collected in the cores indicate whether the climate was arid or wet.

Still other evidence of climatic change comes from the study of annual growth rings of trees, called **dendrochronology**. As a tree grows, it produces a layer of wood cells under its bark. Each year's growth appears as a ring, which can be seen in a cross section of the trunk. The changes in thickness of the rings, especially those rings that formed later in the tree's life (the outer rings), indicate climatic changes that may have taken place from one year to the next (see ● Fig. 18.5). The presence of frost rings during particularly cold periods and the chemistry of the wood itself provide additional information about a changing climate. Tree rings are only useful in regions that experience an annual

cycle and in trees that are stressed by temperature or moisture during their growing season. The growth of tree rings has been correlated with precipitation and temperature patterns extending hundreds of years into the past in various regions of the world.

Other data that have been used to reconstruct past climates include:

1. records of natural lake-bottom sediment and soil deposits
2. the study of pollen in deep ice caves, soil deposits, and sea sediments
3. certain geologic evidence (ancient coal beds, sand dunes, and fossils), and the change in the water level of closed-basin lakes
4. documents concerning droughts, floods, crop yields, rain, snow, and dates of lakes freezing and trees blossoming
5. oxygen-isotope ratios of corals
6. the dates of calcium carbonate layers of stalactites in caves
7. borehole temperature profiles, which can be inverted to give records of past temperature change at the surface
8. deuterium (heavy hydrogen) ratios in ice cores, which indicate temperature changes

Even with all of this knowledge, our picture of past climates is still incomplete. With this shortcoming in mind, we will examine what the information gained about past climates does reveal.

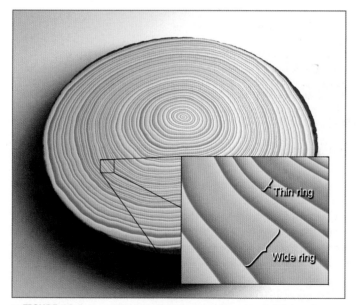

● **FIGURE 18.5** The width of tree rings represents the amount of time each year that conditions were favorable for tree growth. The width of tree rings and the ratio of the light early wood to the darker later wood is controlled by a number of factors. In dry regions, the dominant factor may be precipitation, whereas at high latitudes or high elevations, it may be temperature.

Climate Throughout the Ages

Throughout much of Earth's history, the global climate was probably much warmer than it is today. During most of this time, the polar regions were free of ice. These comparatively warm conditions, however, were interrupted by several periods of glaciation. Geologic evidence suggests that one glacial period occurred about 700 m.y.a. (which stands for "million of years ago") and another about 300 m.y.a. The most recent one—the *Pleistocene epoch* or, simply, the **Ice Age**—began about 2.5 m.y.a. Often, each advance and retreat of glaciers within the Pleistocene epoch is simply referred to as an ice age (using lowercase letters). Let's summarize the climatic conditions that led up to the Pleistocene.

About 65 m.y.a., Earth was warmer than it is now; polar ice caps did not exist. Beginning about 55 m.y.a., Earth entered a long cooling trend. After millions of years, polar ice appeared. As average

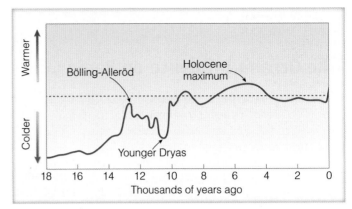

● **FIGURE 18.6** Relative air temperature variations (warmer and cooler periods) during the past 18,000 years. These data, which represent temperature records compiled from a variety of sources, only give an approximation of temperature changes. Some regions of the world experienced a cooling and other regions a warming that either preceded or lagged behind the temperature variations shown in the diagram.

temperatures continued to lower, the ice grew thicker, and by about 10 m.y.a. a deep blanket of ice covered the Antarctic. Meanwhile, snow and ice began to accumulate in high mountain valleys of the Northern Hemisphere, and alpine glaciers soon appeared.

About 2.5 m.y.a., continental glaciers appeared in the Northern Hemisphere, marking the beginning of the Pleistocene epoch—the time of the ice ages mentioned above. The Pleistocene, however, was not a period of continuous glaciation but a time when glaciers alternately advanced and retreated (melted back) over large portions of North America and Europe. Between the glacial advances were warmer periods called **interglacial periods**, which lasted for 10,000 years or more. The most recent was the *Eemian interglacial period*, which lasted from about 130,000 to about 114,000 y.a. (which stands for "years ago"). A recently obtained ice core indicates that summer temperatures over Greenland may have averaged as much as 8°C (14°F) above today's values during the peak warmth of this period.

The most recent North American glaciers reached their maximum thickness and extent between about 26,000 and 20,000 years ago (y.a.). At that time, average temperatures in Greenland were about 10°C (18°F) lower than at present and tropical average temperatures were about 4°C (7°F) lower than they are today. Because a great deal of water was in the form of ice over land, sea level was perhaps 120 m (395 ft) lower than it is now. The lower sea level exposed vast areas of land, such as the *Bering land bridge* (a strip of land that connected Siberia to Alaska as shown in Fig. 18.2a), which allowed human and animal migration from Asia to North America.

The ice began to retreat about 14,700 y.a. as surface temperatures slowly rose, producing a warm spell called the *Bölling-Alleröd* period (see ● Fig. 18.6). Then, about 13,000 y.a., the average temperature suddenly dropped and northeastern North America and northern Europe reverted back to glacial conditions. By about 10,000 y.a., the cold spell (known as the **Younger Dryas** *) had ended abruptly, and temperatures had risen rapidly in many areas. Between about 9000 and 6000 y.a., the continental ice sheets over North America disappeared, and summer temperatures in the higher latitudes of the Northern Hemisphere were, on average, several degrees Celsius above today's readings. This northern-latitude warm spell during the current interglacial period, or *Holocene epoch*, is sometimes called the **mid-Holocene maximum**. However, temperatures across much of the rest of the globe appear to have been close to, or even slightly below, their current averages. About 5000 y.a., a cooling trend set in across northern latitudes, during which extensive alpine glaciers returned, but not continental glaciers.

---

*This exceptionally cold spell is named after the *Dryas*, an arctic flower.

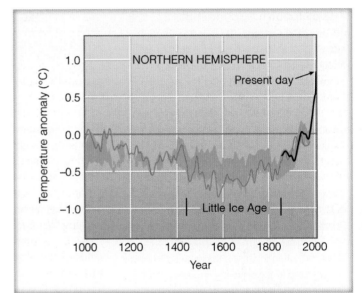

● **FIGURE 18.7** The average temperature variations over the Northern Hemisphere for the last 1000 years relative to the 1961 to 1990 average (zero line). The blue line represents air temperatures constructed from tree rings, corals, ice cores, and pollen. Yearly temperature data measured by thermometers are in black. This reconstruction has been compared to other similar reconstructions. The area shaded orange represents where these reconstructions overlap the data by 50 percent or more. (Source: Adapted from Susan Solomon (Ed.), *Climate Change 2007–The Physical Science Basis: Working Group 1 Contribution to the Fourth Assessment Report of the IPCC*, Vol. 4. Cambridge University Press, 2007. Present-day data courtesy of NOAA.)

It is interesting to note that ice core data from Greenland reveal that rapid shifts in climate (from ice age conditions to a much warmer state) took place in as little as a decade over central Greenland around the end of the Younger Dryas. The data also reveal that similar rapid shifts in climate occurred several times toward the end of the most recent ice age. What could cause such rapid changes in temperature? One possible explanation is given in Focus section 18.1.

TEMPERATURE TRENDS DURING THE PAST 1000 YEARS

● Figure 18.7 shows how the average surface air temperature changed in the Northern Hemisphere during the last

The Ocean's Influence on Rapid Climate Change

During the last glacial period, the climate around Greenland (and probably other areas of the world, such as northern Europe) underwent shifts, from ice-age temperatures to much warmer conditions in a matter of years. What could bring about such large fluctuations in temperature over such a short period of time? It now appears that a vast circulation of ocean water, known as the *conveyor belt*, plays a major role in the climate system.

● Figure 1 illustrates the movement of the ocean conveyor belt, or *thermohaline circulation*.* The conveyor-like circulation begins in the north Atlantic near Greenland and Iceland, where salty surface water is cooled through contact with cold Arctic air masses. The cold, dense water sinks and flows southward through the deep Atlantic Ocean, around Africa, and into the Indian and Pacific oceans.

In the North Atlantic, the sinking of cold water draws warm water northward from lower latitudes. As this water flows northward, evaporation increases the water's salinity (dissolved salt content) and density. When this salty, dense water reaches the far regions of the North Atlantic, it gradually sinks to great depths. This warm part of the conveyor delivers an incredible amount of tropical heat to the northern Atlantic. During the winter, this heat is transferred to the overlying atmosphere, and evaporation moistens the air. Strong westerly winds then carry this warmth and moisture into northern and western Europe, where it causes winters to be much warmer and wetter than one would normally expect for this latitude.

Ocean sediment records along with ice-core records from Greenland suggest that the giant conveyor belt has switched on and off

*Thermohaline circulations are ocean circulations produced by differences in temperature and/or salinity. Changes in ocean water temperature or salinity create changes in water density.

● **FIGURE 1** The ocean conveyor belt. In the North Atlantic, cold, salty water sinks, drawing warm water northward from lower latitudes. The warm water provides warmth and moisture for the air above, which is then swept into northern Europe by westerly winds that keep the climate of that region milder than one would normally expect. If the conveyor belt stops or slows dramatically, winters could turn much colder over northern Europe.

during the last glacial period. Such events have apparently coincided with rapid changes in climate. For example, when the conveyor belt is strong, winters in northern Europe tend to be wet and relatively mild. However, when the conveyor belt is weak or stops altogether, winters in northern Europe appear to turn much colder. This switching from a period of milder winters to one of severe cold shows up many times in the climate record. One such event—the Younger Dryas—illustrates how quickly climate can change and how western and northern Europe's climate can cool within a matter of decades, then quickly return back to milder conditions.

Apparently, one mechanism that can switch the conveyor belt off is a massive influx of freshwater. For example, about 13,000 years ago, freshwater from a huge glacial lake may have started to flow down the St. Lawrence River and into the North Atlantic. This massive inflow of freshwater might have reduced the salinity (and, hence, density) of the surface

water to the point that it stopped sinking. The conveyor may have shut down for about 1000 years during which time severe cold engulfed much of northern Europe in the Younger Dryas event. The conveyor belt possibly started up again when freshwater began to drain down the Mississippi rather than into the North Atlantic. It was during this time that milder conditions returned to northern Europe.

Will increasing levels of CO_2 have an effect on the conveyor belt? Some climate models predict that as CO_2 levels increase, more precipitation will fall over the North Atlantic. This situation reduces the density of the sea water and slows down the conveyor belt. Computer models predict that the conveyor belt could slow by about 20 to 45 percent over the twenty-first century, depending on how quickly carbon dioxide emissions increase. This slowdown of the conveyor belt would tend to keep Europe from warming as rapidly as the rest of the world. However, the conveyor belt is not expected to shut down entirely.

1000 years.* The data needed to reconstruct the temperature profile in Fig. 18.7 comes from a variety of sources, including tree rings, corals, ice cores, historical records, and thermometers.

*The National Academy of Sciences published a report comparing the work summarized in Fig. 18.7 to many other reconstructions of temperatures during the past 1000 years, and found that they all gave basically the same results.

Notice that about 1000 y.a., the Northern Hemisphere was about 1°C cooler than its average over the last several decades. However, certain regions in the Northern Hemisphere were warmer than others. For example, during this time vineyards flourished and wine was produced in England, indicating warm, dry summers and the absence of cold springs. This relatively warm, tranquil period of several hundred years over Western Europe is sometimes referred to as the *Medieval Climatic Optimum*. It was during the early part of the millennium that Vikings colonized Iceland and Greenland and traveled to North America.

Notice in Fig. 18.7 that the temperature curve indicates a relatively warm period during the eleventh to fourteenth centuries, though still cooler than the twentieth century. During this time, the relatively mild climate of Western Europe began to show large variations. For several hundred years the climate grew stormy. Both great floods and great droughts occurred. Extremely cold winters were followed by relatively warm ones. During the cold spells, the English vineyards and the Viking settlements suffered. Europe experienced several famines during the 1300s. The ice pack of Greenland began advancing, and the Viking colony in Greenland perished.*

Again look at Fig. 18.7 and observe that the Northern Hemisphere experienced a slight cooling during the fifteenth to nineteenth centuries. This cooling was significant enough in certain areas to allow alpine glaciers to increase in size and advance down river canyons. Though the cooling averaged less than 1°C across the hemisphere, it had a major impact, especially in many areas of Europe, where winters were long and severe; summers, short and wet. The vineyards in England actually vanished, and farming became impossible in the more northern latitudes. Cut off from the rest of the world by an advancing ice pack, the Viking colony in Greenland perished. While there were cold periods in other parts of the world, there is no evidence that a multi-century cold spell existed worldwide. However, this period has come to be known as the **Little Ice Age**, largely for its impacts on Europe.

TEMPERATURE TRENDS DURING THE PAST 100-PLUS YEARS
In the early 1900s, the average global surface temperature began to rise (see ● Fig. 18.8). Notice that, from about 1900 to 1945, the average temperature rose nearly 0.5°C. Following the warmer

---

*Although climate change played a role in the demise of the Viking colony in northern Greenland, it was also their inability to adapt to the climate and to learn hunting and farming techniques from the Inuit that led to their downfall.

period, Earth's temperature changed little over the next 25 years or so, with slight cooling over the Northern Hemisphere. In the late 1960s and 1970s, the cooling trend ended over most of the Northern Hemisphere, and by the late 1970s a warming trend set in that has continued into the twenty-first century. The increase in average temperature experienced over the Northern Hemisphere during the twentieth century is likely to have been the largest of any century during the past 1000 years. Globally, the year 2014 was the warmest up to that point in records going back to 1880. The next year, 2015, was even warmer, and 2016 was warmer still. Research suggests these were the three warmest years in more than 1000 years.

The average warming experienced over the globe, however, has not been uniform. The greatest warming has occurred in the Arctic and over the midlatitude continents in winter and spring, whereas a few areas have not warmed in recent decades, such as areas of the oceans in the Southern Hemisphere and parts of Antarctica. The United States has experienced less warming than some other parts of the world. Moreover, most of the warming globally has occurred at night—a situation that has lengthened the frost-free seasons in many mid- and high-latitude regions. (In recent decades, the warming has been more equally distributed between day and night.)

The changes in air temperature shown in Fig. 18.8 are derived from three main sources: air temperatures over land, air temperatures over ocean, and sea surface temperatures. There are, however, uncertainties in the temperature record. For example, during this time period recording stations have moved, and techniques for measuring temperature have varied. Also, marine

● **FIGURE 18.8** The orange and blue bars represent the annual average temperature variations over the globe (land and sea) from 1880 through 2016. Temperature changes are compared to the average surface temperature from 1951–1980. The dark solid line shows the five-year average temperature change.

observing stations are scarce. In addition, urbanization (especially in developed nations) tends to artificially raise average temperatures as cities grow (the *urban heat island* effect). Taking all of this information into account, along with improved sea-surface temperature data, the **global warming** between the early 1880s and the mid-2010s measured about $1.0°C (1.8°F)$. Look at Fig. 18.8 and observe that although the global temperature trend was relatively flat during the decade from 2000 to 2009, that decade was still warmer than the decade of the 1990s, which in turn was warmer than the decade of the 1980s. Likewise, the 2010s are almost certain to end up warmer than the 2000s.

A global increase in temperature of $1.0°C$ over a little more than 100 years may seem small, but global temperatures probably have not varied by more than $2°C$ during the past 10,000 years. Consequently, an increase of $1.0°C$ becomes quite significant when compared with temperature changes over thousands of years.

Up to this point we have examined the temperature record of Earth's surface and observed that Earth has been in a warming trend for more than 100 years. A common question is whether the warming trend is due to natural variations in the climate system, human activities, or a combination of the two. As we will see later in this chapter, climate scientists have concluded that most if not all of the warming in the last few decades is due to an enhanced greenhouse effect caused by increasing levels of greenhouse gases, such as CO_2, mainly as a result of fossil fuel burning (oil, gas, and coal).*

If human activities are primarily responsible for this global warming, what was it that led to global trends of the past, before human beings walked on the surface of this planet?

BRIEF REVIEW

Before going on to the next section, here is a brief review of some of the facts and concepts we covered so far:

- Earth's climate is constantly undergoing change. Evidence suggests that throughout much of Earth's history, Earth's climate was much warmer than it is today.

- The most recent glacial period (or Ice Age) began about 2.5 m.y.a. During this time, glacial advances were interrupted by warmer periods called interglacial periods. In North America, continental glaciers reached their maximum thickness and extent from about 26,000 to 20,000 y.a. and disappeared completely from North America by about 6000 years ago.

- The Younger Dryas event represents a time about 12,000 y.a. when northeastern North America and northern Europe reverted to glacial conditions.

- From the late 1800s to the early 2010s, Earth's surface temperature increased by about $1.0°C$. The rate of global warming has increased over the last several decades.

---

*Earth's atmospheric greenhouse effect is due mainly to the absorption and emission of infrared radiation by greenhouse gases such as water vapor, CO_2, methane, nitrous oxide, and chlorofluorocarbons. Refer back to Chapter 2 for additional information on this topic.

Climate Change Caused by Natural Events

Why does Earth's climate change? There are three "external" causes of climate change. They are:

1. changes in incoming solar radiation
2. changes in the composition of the atmosphere
3. changes in Earth's surface

Natural phenomena can cause climate to change by all three mechanisms, whereas human activities can change climate by both the second and third mechanisms. In addition to these external causes, there are "internal" causes of climate change, such as changes in the circulation patterns of the ocean and atmosphere, which redistribute energy within the climate system rather than altering the total amount of energy it holds.

Part of the complexity of the climate system is the intricate interrelationship of the elements involved. For example, if temperature changes, many other elements may be altered as well. The interactions among the atmosphere, the oceans, and the ice are extremely complex and the number of possible interactions among these systems is enormous. No climatic element within the system is isolated from the others, which is why the complete picture of Earth's changing climate is not totally understood. With this in mind, we will first investigate how feedback systems work; then we will consider some of the current theories as to why Earth's climate changes naturally.

CLIMATE CHANGE: FEEDBACK MECHANISMS In Chapter 2, we learned that the Earth-atmosphere system is in a delicate balance between incoming and outgoing energy. If this balance is upset, even slightly, global climate can undergo a series of complicated changes.

Let's assume that the Earth-atmosphere system has been disturbed to the point that Earth has entered a slow warming trend. Over the years the temperature slowly rises, and water from the oceans rapidly evaporates into the warmer air. The increased quantity of water vapor absorbs more of Earth's infrared energy, thus strengthening the atmospheric greenhouse effect.

This strengthening of the greenhouse effect raises the air temperature even more, which, in turn, allows more water vapor to evaporate into the atmosphere. The greenhouse effect becomes even stronger, and the air temperature rises even more. This situation is known as the **water vapor–greenhouse feedback**. It represents a **positive feedback mechanism** because the initial increase in temperature is reinforced by the other processes. If this feedback were left unchecked, Earth's temperature would increase until the oceans completely evaporated. Such a chain reaction is called a *runaway greenhouse effect*.

Another positive feedback mechanism is the **snow-albedo feedback**, in which an increase in global surface air temperature might cause snow and ice to melt in polar latitudes. This melting would reduce the albedo (reflectivity) of the surface, allowing more solar energy to reach the surface, which would further raise the temperature (see ● Fig. 18.9).

(a)	• High surface albedo	(b)	• Lower surface albedo	(c)	• Very low surface albedo
	• Low absorption of sunlight		• Higher absorption of sunlight		• Much higher absorption of sunlight
	• Gradual surface warming		• Surface warming increases		• Surface warming enhanced

● **FIGURE 18.9** On a warming planet, the snow-albedo positive feedback would enhance the warming. (a) In polar regions snow reflects much of the sun's energy back to space. (b) If the air temperature were to gradually increase, some of the snow would melt, less sunlight would be reflected, and more sunlight would reach the ground, warming it more quickly. (c) The warm surface would enhance the snowmelt which, in turn, would accelerate the rise in temperature.

All feedback mechanisms work simultaneously and in both directions. Consequently, the snow-albedo feedback produces a positive feedback on a cooling planet as well. Suppose, for example, Earth were in a slow cooling trend. Lower temperatures might allow for a greater snow cover in middle and high latitudes, which would increase the albedo of the surface so that much of the incoming sunlight would be reflected back to space. Lower temperatures might further increase the snow cover, causing the air temperature to lower even more. If left unchecked, the snow-albedo positive feedback would produce a *runaway ice age*, which is highly unlikely on Earth because other feedback mechanisms in the atmospheric system would be working to moderate the magnitude of the cooling.

Along with positive feedback mechanisms, there are **negative feedback mechanisms**—those that tend to weaken the interactions among the variables rather than reinforce them. For example, a warming planet emits more infrared radiation.* If Earth's climate system were in a runaway greenhouse effect, the increase in radiant energy from the surface would greatly slow the rise in temperature and help to stabilize the climate. The increase in radiant energy from the surface as the planet warms is the strongest negative feedback in the climate system, and greatly lessens the possibility of a runaway greenhouse effect. Consequently, although human-produced enhancement of the greenhouse effects could have critical impacts, there is no evidence that a runaway greenhouse effect ever occurred on Earth, and no indication that it will occur in the future.

Another negative feedback in Earth's climate system is the **chemical weathering-CO_2 feedback**. Chemical weathering is a process by which carbon dioxide is removed from the atmosphere as silicate minerals in rocks decompose in the presence of moisture. In this feedback, as chemical weathering increases, the amount of CO_2 in the atmosphere decreases. Chemical weathering (and the removal of CO_2 from the atmosphere) will generally take place more rapidly on a warmer planet, as chemical reactions speed up and greater evaporation from the oceans leads to more precipitation over the continents. As CO_2 leaves the atmosphere more quickly, CO_2 levels drop and Earth's climate begins to cool and stabilize. As temperatures dip, less water evaporates from the oceans, chemical weathering decreases, and the removal of CO_2 from the atmosphere diminishes.

In summary, the Earth-atmosphere system has a number of processes called *feedback mechanisms* that can either intensify or weaken tendencies toward climate change. Although we do not worry about a runaway greenhouse effect or an ice-covered Earth anytime in the future, there is concern that large positive feedback mechanisms may be working in the climate system to produce accelerated melting of ice in polar regions, especially in Greenland.

CLIMATE CHANGE: PLATE TECTONICS AND MOUNTAIN BUILDING

Earlier, we saw that one of the external causes of climate change is a change in the surface of Earth. During the geologic past, Earth's surface has undergone extensive modifications. One involves the slow shifting of the continents and the ocean floors. This motion is explained in the widely accepted **theory of plate tectonics**. According to this theory, Earth's outer shell is composed of huge plates that fit together like pieces of a jigsaw puzzle. The plates, which slide over a partially molten zone below them, move in relation to one another. Continents are embedded in the plates and move along like luggage riding piggyback on a conveyor belt. The rate of motion is extremely slow, only a few centimeters per year.

Besides providing insights into many geological processes, the theory of plate tectonics also helps to explain past climates. For example, we find glacial features near sea level in Africa today, suggesting that the area underwent a period of glaciation hundreds of millions of years ago. Were temperatures at low elevations near the equator ever cold enough to produce ice sheets? Probably not. The ice sheets formed when the African landmass was located at a much higher latitude. Over the many millions of

*Recall from Chapter 2, p. 40, that the outgoing infrared radiation from the surface increases at a rate proportional to the fourth power of the surface's absolute temperature. This relationship is called the Stefan-Boltzmann law. In effect, doubling the absolute temperature of Earth's surface would result in 16 times more energy emitted.

● **FIGURE 18.10** Geographical distribution of (a) landmasses about 150 million years ago, and (b) today. Arrows show the relative direction of continental movement.

years since then, the land has slowly moved to its present position. Thinking along the same lines, we can see how the fossil remains of tropical vegetation can be found under layers of ice in polar regions today.

According to the theory of plate tectonics, the now-existing continents were at one time joined together in a single huge continent, which broke apart. Its pieces slowly moved across the face of the globe, thus changing the distribution of continents and ocean basins, as illustrated in ● Fig. 18.10. When landmasses are concentrated in middle and high latitudes (as they are today), it appears that ice sheets are more likely to form. During these times, there is a greater likelihood that more sunlight will be reflected back into space from the snow that falls over the continent in winter. Less sunlight absorbed by the surface lowers the air temperature, which allows for a greater snow cover, and, over thousands of years, the formation of continental glaciers. The amplified cooling that takes place over the snow-covered land is the snow-albedo feedback mentioned earlier.

The various arrangements of the continents may also influence the path of ocean currents, which, in turn, could not only alter the transport of heat from low to high latitudes but could also change both the global wind system and the climate in middle and high latitudes. As an example, suppose that plate movement "pinches off" a rather large body of high-latitude ocean water such that the transport of warm water into the region is cut off. In winter, the surface water would eventually freeze over with ice. This freezing would, in turn, reduce the amount of sensible and latent heat given up to the atmosphere. Furthermore, the ice allows snow to accumulate on top of it, thereby setting up conditions that could lead to even lower temperatures.

There are other mechanisms by which tectonic processes may influence climate.* In ● Fig. 18.11, notice that the formation of oceanic plates (plates that lie beneath the ocean) begins at a *ridge*, where dense, molten material from inside Earth wells up to the surface, forming new sea floor material as it hardens. Spreading (on the order of several centimeters a year) takes place at the ridge center, where two oceanic plates move away from one another. When an oceanic plate encounters a lighter continental plate, it responds by diving under it, in a process called *subduction*. Heat

and pressure then melt a portion of the subducting rock, which usually consists of volcanic rock and calcium-rich ocean sediment. The molten rock may then gradually work its way to the surface, producing volcanic eruptions that spew water vapor, carbon dioxide, and minor amounts of other gases into the atmosphere. The release of these gases (called *degassing*) usually takes place at other locations as well (for instance, at ridges where new crustal rock is forming).

Some periods of climatic change, taking place over millions of years, may be related to the rate at which the plates move and, hence, related to the amount of CO_2 in the air. For example, during times of rapid movement, an increase in volcanic activity vents massive large quantities of CO_2 into the atmosphere, which enhances the atmospheric greenhouse effect, causing global temperatures to rise.

Millions of years later, when spreading rates decrease, less volcanic activity means less CO_2 is spewed into the atmosphere. If the rate at which the oceans remove CO_2 from the atmosphere does not change, CO_2 levels in the atmosphere should drop. A reduction in CO_2 levels weakens the greenhouse effect, which, in turn, causes global temperatures to decrease. The cooler Earth promotes the accumulation of ice and snow over portions of the continents, which promotes additional cooling by reflecting more sunlight back to space. (It is important to note that modern-day volcanic activity emits less than 1 percent of the CO_2 produced by human activity.)

Mountains can cause the climate to change in a variety of ways. A chain of volcanic mountains forming perpendicular to the mean wind flow may disrupt the airflow over the mountains, altering the climate both upwind and downwind of them. By the same token, mountain building that occurs when two continental plates collide (like that which presumably formed the Himalayan mountains and Tibetan highlands) can have a marked influence on global circulation patterns and, hence, on the climate of an entire hemisphere. It has also been theorized that the uplift of the Himalayas and Tibetan plateau helped to increase chemical weathering, thus pulling more carbon dioxide out of the air and lowering global temperature—perhaps enough to help trigger the onset of the Ice Age.

Up to now, we have examined how climatic variations can take place over millions of years due to the movement of continents and the associated restructuring of landmasses.

*Tectonic processes are large-scale processes that deform Earth's crust.

● **FIGURE 18.11** Earth is composed of a series of moving plates. The rate at which plates move (spread) may influence global climate. During times of rapid spreading, increased volcanic activity may promote global warming by enriching the CO_2 content of the atmosphere.

We will now turn our attention to variations in Earth's orbit that may account for climatic fluctuations that take place on a time scale of tens of thousands of years.

CLIMATE CHANGE: VARIATIONS IN EARTH'S ORBIT Another external cause of climate change involves a change in the amount of solar radiation that reaches Earth. A theory ascribing climatic changes to variations in Earth's orbit is the **Milankovitch theory**, named for the astronomer Milutin Milankovitch, who first proposed the idea in the 1930s. The basic premise of this theory is that, as Earth travels through space, three separate cyclic movements combine to produce variations in the amount of solar energy that falls on Earth.

The first cycle deals with changes in the shape (**eccentricity**) of Earth's orbit as Earth revolves about the sun. Notice in ● Fig. 18.12 that Earth's orbit changes from being elliptical (dashed line) to being nearly circular (solid line). To go from circular to elliptical and back again takes about 100,000 years. The greater the eccentricity of the orbit (that is, the more elliptical the orbit), the greater the variation in solar energy received by Earth between its closest and farthest approach to the sun.

Presently, we are in a period of low eccentricity, which means that our annual orbit around the sun is more circular. Moreover, Earth is closer to the sun in January and farther away in July (see Chapter 3, p. 60). The difference in distance (which only amounts to about 3 percent) is responsible for a nearly 7 percent increase in the solar energy received at the top of the atmosphere from July to January. When the difference in distance is 9 percent (a highly elliptical orbit), the difference in solar energy received between July and January is on the order of 20 percent. In addition, the more eccentric orbit changes the length of seasons in each hemisphere by changing the length of time between the vernal and autumnal equinoxes. Although rather large percentage changes in solar energy can occur between summer and winter, the globally and annually averaged change in solar energy received by Earth (due to orbital changes) hardly varies at all. It is the distribution of incoming solar energy that changes, not the totals.

The second cycle takes into account the fact that, as Earth rotates on its axis, it wobbles like a spinning top. This wobble, known as the **precession** of Earth's axis, occurs in a cycle of

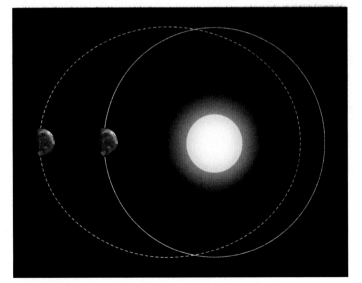

● **FIGURE 18.12** For Earth's orbit to stretch from nearly a circular (solid line) to an elliptical orbit (dashed line) and back again takes nearly 100,000 years. (Diagram is highly exaggerated and is not to scale.)

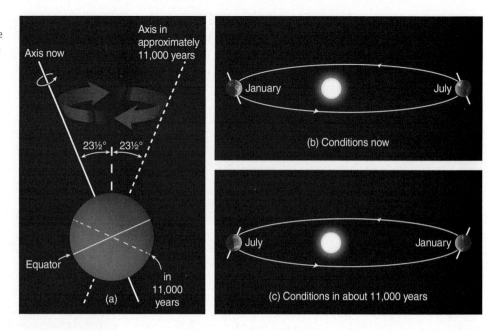

● **FIGURE 18.13** (a) Like a spinning top, Earth's axis of rotation slowly moves and traces out the path of a cone in space. (b) Presently Earth is closer to the sun in January, when the Northern Hemisphere experiences winter. (c) In about 11,000 years, due to precession, Earth will be closer to the sun in July, when the Northern Hemisphere experiences summer.

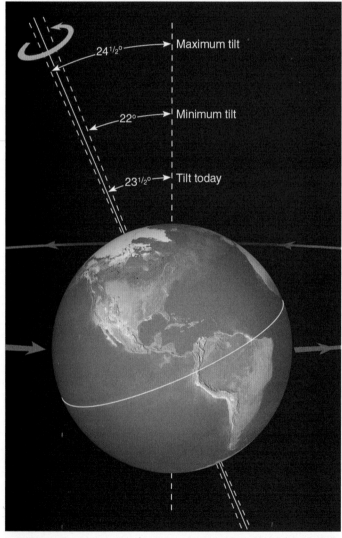

● **FIGURE 18.14** Earth currently revolves around the sun while tilted on its axis by an angle of 23 ½°. During a period of 41,000 years, this angle of tilt ranges from about 22° to 24 ½°.

about 23,000 years. Presently, Earth is closer to the sun in January and farther away in July. Due to precession, the reverse will be true in about 11,000 years (see ● Fig. 18.13). In about 23,000 years we will be back to where we are today. This means, of course, that if everything else remains the same, 11,000 years from now seasonal variations in the Northern Hemisphere should be greater than at present. The opposite would be true for the Southern Hemisphere, although the effects there in both directions would be somewhat muted because of the greater proportion of ocean to land in the Southern Hemisphere.

The third cycle takes about 41,000 years to complete and relates to the changes in tilt (**obliquity**) with respect to Earth's orbit. Presently, Earth's orbital tilt is 23 ½°, but during the 41,000-year cycle the tilt varies from about 22° to 24 ½ (see ● Fig. 18.14). The smaller the tilt, the less seasonal variation there is between summer and winter in middle and high latitudes; thus, winters tend to be milder and summers cooler. (Recall from Chapter 3 that this tilt of Earth is the true "reason for the seasons," although many people incorrectly believe that summer occurs because Earth is closer to the Sun.)

Over the extensive land areas located at high latitudes of the Northern Hemisphere, ice sheets are more likely to form when less solar radiation reaches the surface in summer. Less sunlight promotes lower summer temperatures. During the cooler summer, snow from the previous winter may not totally melt. The summer cooling predominates even though there is an increase in winter insolation, because very little sunlight reaches the far north in winter. The accumulation of snow over many years increases the albedo of the surface. Less sunlight reaches the surface, summer temperatures continue to fall, more snow accumulates, and continental ice sheets gradually form. It is interesting to note that when all of the Milankovich cycles are taken into account, the present trend should be toward *cooler summers* over high latitudes of the Northern Hemisphere.

In summary, the Milankovitch cycles that combine to produce variations in solar radiation received at Earth's surface include:

1. changes in the shape (*eccentricity*) of Earth's orbit about the sun
2. *precession* of Earth's axis of rotation, or wobbling
3. changes in the tilt (*obliquity*) of Earth's axis

In the 1970s, scientists in the CLIMAP project (described on p. 504) found strong evidence in deep-ocean sediments that variations in climate during the past several hundred thousand years were closely associated with the Milankovitch cycles. Many subsequent studies have strengthened this premise. For example, analyses show that during the past 800,000 years, the extent and depth of ice sheets have peaked about every 100,000 years. This timing corresponds to the 100,000-year cycling of the eccentricity in Earth's orbit discussed above. Superimposed on this are smaller ice advances that show up at intervals of about 41,000 years and 23,000 years; as we saw above, Earth's obliquity varies on a 41,000-year cycle. So, it does appear that the Milankovitch cycles play a role in the frequency of glaciation and the severity of natural climatic variation.

But orbital changes alone are probably not totally responsible for ice buildup and retreat. Evidence (from trapped air bubbles in the ice sheets of Greenland and Antarctica representing thousands of years of snow accumulation) reveals that CO_2 concentrations were about 30 percent lower during colder glacial periods than during warmer interglacial periods. Analysis of air bubbles in Antarctic ice cores reveals that methane follows a pattern similar to that of CO_2 (see ● Fig. 18.15). This knowledge suggests that smaller amounts of CO_2 in the atmosphere may have had the effect of amplifying the cooling initiated by the orbital changes. Likewise, increasing CO_2 levels at the end of the glacial period may have accounted for the rapid melting of the ice sheets.* Analysis of air bubbles in Antarctic ice cores reveals that methane, another important greenhouse gas, follows a pattern similar to that of CO_2 (see Fig. 18.15).

Interestingly, research shows that temperature changes thousands of years ago actually *preceded* the CO_2 changes. This observation supports the idea that CO_2 is an agent of positive feedback in the climate system, where higher temperatures lead to higher CO_2 levels and lower temperatures to lower CO_2 levels. Consequently, CO_2 can be viewed as an internal, natural part of Earth's climate system.

Just why atmospheric CO_2 levels have varied as glaciers expanded and contracted is not clear, but it may be due to changes in biological activity taking place in the oceans. Moreover, as oceans cool, they remove more CO_2 from the atmosphere. Perhaps, also, changing levels of CO_2 indicate a shift in ocean circulation patterns. Such shifts, brought on by changes in precipitation and evaporation rates, may alter the distribution of heat energy around the world. Alterations wrought in this manner could, in turn, affect the global circulation of winds, which may explain why alpine glaciers in the Southern Hemisphere expanded and contracted in tune with Northern Hemisphere glaciers during the last ice age, even though the Southern Hemisphere (according

---

*During peak CO_2 levels in preindustrial times, its concentration was less than 300 ppm, which is substantially lower than its concentration of about 400 ppm in today's atmosphere.

● **FIGURE 18.15** Variations of carbon dioxide (top, ppm), methane (middle, ppb), and temperatures (bottom, °C) during the past 650,000 years. Concentrations of gases are derived from air bubbles trapped within the ice sheets of Antarctica and extracted from ice cores. Temperatures are derived from the analysis of oxygen isotopes. The shaded orange bands indicate current and most recent interglacial warm periods. (Note: ppm represents parts per million by volume, and ppb represents parts per billion by volume.) (Source: Adapted from the Technical Summary of *Climate Change 2007: The Physical Science Basis. Contribution of Working Group I to the Fourth Assessment Report of the Intergovernmental Panel on Climate Change, 2007.* Reprinted by permission of the Intergovernmental Panel on Climate Change.)

to the Milankovitch cycles) was not in an orbital position for glaciation.

Still other factors may work in conjunction with Earth's orbital changes to explain the temperature variations between glacial and interglacial periods. Some of these are:

1. the amount of dust and other aerosols in the atmosphere
2. the reflectivity of the ice sheets
3. the concentration of other greenhouse gases
4. the changing characteristics of clouds
5. the rebounding of land, having been depressed by ice

Hence, the Milankovitch cycles, in association with other natural factors, may explain the advance and retreat of ice over periods of 10,000 to 100,000 years. But what caused the ice ages to begin in the first place? And why were periods of glaciation so infrequent during geologic time? The Milankovitch theory does not attempt to answer these questions, although it is likely that changes in Earth's geography related to plate tectonics are involved.

CLIMATE CHANGE: VARIATIONS IN SOLAR OUTPUT

Solar energy measurements made by sophisticated instruments aboard satellites show that the sun's energy output (called *brightness*) varies slightly—by a fraction of 1 percent—with sunspot activity.

Sunspots are huge magnetic storms on the sun that show up as cooler (darker) regions on the sun's surface. They occur in cycles, with the number and size reaching a maximum approximately every 11 years. During periods of maximum sunspots, the sun emits more energy (about 0.1 percent more) than during periods of sunspot minimums (see ● Fig. 18.16). Evidently, the greater number of bright areas (*faculae*) around the sunspots radiate more energy, which offsets the effect of the dark spots.

It appears that the 11-year sunspot cycle has not always prevailed. Between 1645 and 1715, during the period known as the **Maunder minimum**,* very few sunspots were observed. This minimum coincided with part of the Little Ice Age (recall that this was a cool spell in the temperature record experienced mainly over Europe). Some scientists suggest that a reduction in the sun's energy output may have played a role in exacerbating this cool spell, although the cooling started many years before the Maunder minimum began.

Fluctuations in solar output may account for small climatic changes over time scales of decades and centuries. Many theories have been proposed linking solar variations to climate change, but none has been proven. Instruments aboard satellites and solar telescopes on Earth monitor the sun to observe how its energy output may vary. To date, these measurements show that solar output has only changed a fraction of 1 percent over several decades. In fact, the solar cycle that peaked in the early 2010s was the weakest in more than 100 years, yet global temperatures continued to climb. Because high-quality data sources have only been available for several decades, it may be some time before we fully understand the relationship between solar activity and climate change on Earth.

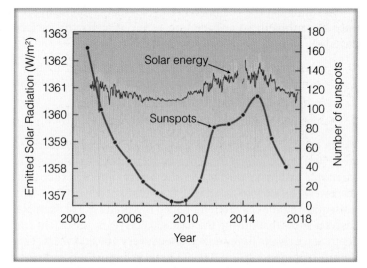

● **FIGURE 18.16** Changes in total solar irradiance (solar energy output) (red line) in watts per square meter (red line) measured by the NASA-sponsored *SORCE* satellite from 2003 to 2017. Gray line represents the number of sunspots observed each year.

CLIMATE CHANGE: ATMOSPHERIC PARTICLES

Microscopic liquid and solid particles (*aerosols*) that enter the atmosphere from both natural and human-induced sources can have an effect on climate. The effect these particles have on the climate is exceedingly complex and depends upon a number of factors, such as the particle's size, shape, color, chemical composition, and vertical distribution above the surface. In this section, we will examine those particles that enter the atmosphere through natural means.

Particles Near the Surface Particles can enter the atmosphere in a variety of natural ways. For example, wildfires can produce copious amounts of tiny smoke particles, and dust storms sweep tons of fine particles into the atmosphere. Smoldering volcanoes can release significant quantities of sulfur-rich aerosols into the lower atmosphere. And even the oceans are a major source of natural sulfur aerosols, as tiny drifting aquatic plants—phytoplankton—produce a form of sulfur (dimethylsulphide, DMS) that slowly diffuses into the atmosphere, where it combines with oxygen to form sulfur dioxide, which in turn converts to sulfate aerosols. Although the effect these particles have on the climate system is complex, the overall effect they have is to *cool the surface* by preventing sunlight from reaching the surface.

Volcanic Eruptions Volcanic eruptions can have a major impact on climate. During volcanic eruptions, fine particles of ash and dust (as well as gases) can be ejected into the atmosphere (see ● Fig 18.17). Scientists agree that the volcanic eruptions having the greatest impact on climate are those rich in sulfur gases. These gases, when ejected into the stratosphere,** combine with water vapor in the presence of sunlight to produce tiny, reflective sulfuric acid particles that grow in size, forming a dense layer of haze. The haze may reside in the stratosphere for several years,

---

*This period is named after E. W. Maunder, the British solar astronomer who first discovered the low sunspot period sometime in the late 1880s.

**You may recall from Chapter 1 that the stratosphere is a stable layer of air above the troposphere, typically about 11 to 50 km (7 to 31 mi) above Earth's surface.

● **FIGURE 18.17** Large volcanic eruptions rich in sulfur can affect climate. As sulfur gases in the stratosphere transform into tiny reflective sulfuric acid particles, they prevent a portion of the sun's energy from reaching the surface. Here, the Philippine volcano Mount Pinatubo erupts during June 1991.

absorbing and reflecting back to space a portion of the sun's incoming energy. The reflection of incoming sunlight by the haze tends to cool the air at Earth's surface, especially in the hemisphere where the eruption occurs.

Two of the largest volcanic eruptions of the twentieth century, in terms of their sulfur-rich veil, were that of El Chichón in Mexico during April 1982 and Mount Pinatubo in the Philippines during June 1991. The eruption of Mount Pinatubo in 1991 was many times greater than that of Mount St. Helens in the Pacific Northwest in 1980. In fact, the largest eruption of Mount St. Helens was a lateral explosion that pulverized a portion of the volcano's north slope. The ensuing dust and ash (and very little sulfur) had virtually no effect on global climate as the volcanic material was confined mostly to the lower atmosphere and fell out quite rapidly over a large area of the northwestern United States.

Mount Pinatubo ejected an estimated 20 million tons of sulfur dioxide (more than twice that of El Chichón), which gradually worked its way around the globe. For major eruptions such as this one, mathematical models predict that average hemispheric temperatures can drop by 0.5°C or more for one to three years after the eruption. Model predictions agreed with temperature changes brought on by the Pinatubo eruption, as in early 1992 the mean global surface temperature had decreased by about 0.5°C (see ● Fig. 18.18). The cooling might even have been greater had the eruption not coincided with a significant El Niño event in 1991–1992 (see Chapter 10, p. 279, for more information on EI Niño). In spite of El Niño, the eruption of Mount Pinatubo produced the two coolest years of the decade, 1991 and 1992. Even more modest volcanoes may have a climate impact if they erupt frequently. Scientists, for example, have found that a series of eruptions from dozens of smaller volcanoes from 2000 to 2009 may have reduced global warming over that period by 25 percent from what otherwise would have been expected.

WEATHER WATCH

The year without a summer (1816) even had its effect on literature. Inspired (or perhaps dismayed) by the cold, gloomy summer weather along the shores of Lake Geneva, Mary Shelley wrote the novel *Frankenstein*.

Climate variation and climate change are not just about temperature. In fact, precipitation may be the most important weather element in terms of the impact on humans. The Pinatubo eruption of 1991 had a large impact on the world's hydrologic cycle. It even caused drought over certain areas, due to the effects (described in the next section) that volcanic emissions can exert on cloud and precipitation formation.

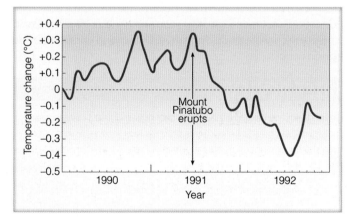

● **FIGURE 18.18** Changes in average global air temperature from 1990 to 1992. After the eruption of Mount Pinatubo in June 1991, the average global temperature by July 1992 decreased by almost 0.5°C (0.9°F) from the 1981–1990 average (dashed line). (Data courtesy of John Christy, University of Alabama in Huntsville, and Roy Spencer, NASA Marshall Space Flight Center.)

As we have just seen, volcanic eruptions rich in sulfur tend to cool Earth's surface. Such was the case following the massive eruption in 1815 of Mount Tambora in Indonesia. In addition, a smaller volcanic eruption occurred in 1809, from which the climate system may not have fully recovered when Tambora erupted in 1815. An infamous cold spell often linked to this volcanic activity occurred during the year 1816, which has come to be known as "*the year without a summer.*" In Europe that year, bad weather contributed to a poor wheat crop, and famine spread across the land. In North America, unusual blasts of cold polar air moved through Canada and the northeastern United States between May and September. Heavy snow arrived in June, with killing frosts in July and August. In the warmer days that followed each cold snap, farmers replanted, only to have another cold outbreak damage the planting. The unusually cold summer was followed by a bitterly cold winter.

In an attempt to correlate sulfur-rich volcanic eruptions with long-term trends in global climate, scientists are measuring the acidity of annual ice layers in Greenland and Antarctica. Generally, the greater the concentration of sulfuric acid particles in the atmosphere, the greater the acidity of the ice layer. Relatively acidic ice has been uncovered from about A.D. 1350 to about 1700, a time that corresponds to the early part of the *Little Ice Age.* Such findings suggest that sulfur-rich volcanic eruptions may have played an important role in triggering this comparatively cool period and, perhaps, other cool periods during the geologic past. Moreover, recent core samples taken from the northern Pacific Ocean reveal that volcanic eruptions in the northern Pacific were at least 10 times larger 2.5 m.y.a. (a time when Northern Hemisphere glaciation began) than previous volcanic events recorded elsewhere in the sediment.

Although volcanic eruptions rich in sulfur tend to cool the surface, they also have the effect of warming the lower stratosphere by absorbing radiant energy from the sun and earth. The absorption of this energy can cause the tropical stratosphere to become much warmer than the polar stratosphere. This situation produces a strong horizontal pressure gradient and strong west-to-east (zonal) stratospheric winds. These winds work their way down into the upper troposphere, where they direct milder maritime surface air from off the ocean onto the continents. The milder ocean air tends to produce warmer winters over Northern Hemisphere continents during the first or second winter after the eruption occurs.

BRIEF REVIEW

Up to this point, we have examined a number of ways Earth's climate can change by natural means. Before going onto the next section, which covers climate change brought on by human activities, here is a brief review of some of the facts and concepts we covered so far:

- The external causes of climate change include: (1) changes in incoming solar radiation; (2) changes in the composition of the atmosphere; (3) changes in the surface of Earth.

- The shifting of continents, along with volcanic activity and mountain building, are possible causes of natural climate change.

- The Milankovitch theory (in association with other natural forces) proposes that alternating glacial and interglacial episodes during the past 2.5 million years are the result of small variations in the tilt of Earth's axis and in the geometry of Earth's orbit around the sun.

- Trapped air bubbles in the ice sheets of Greenland and Antarctica reveal that CO_2 levels and methane levels were lower during colder glacial periods and higher during warmer interglacial periods. But even when the levels were higher, they still were much lower than they are today.

- Fluctuations in solar output (brightness) may account for periods of climatic change.

- Volcanic eruptions, rich in sulfur, may be responsible for cooler periods in the geologic past.

Climate Change Caused by Human (Anthropogenic) Activities

Earlier in this chapter we saw how variations in atmospheric carbon dioxide may have contributed to changes in global climate spanning thousands and even millions of years. Today, we are modifying the chemistry and characteristics of the atmosphere by injecting into it vast quantities of particles and greenhouse gases without fully understanding the long-term consequences. In this section, we will first look at how gases and particles injected into the lower atmosphere by human activities may be affecting climate. Then we will examine how CO_2 and other trace gases appear to be enhancing Earth's greenhouse effect, producing global warming.

CLIMATE CHANGE: AEROSOLS INJECTED INTO THE LOWER ATMOSPHERE In a previous section we learned that tiny solid and liquid particles (aerosols) can enter the atmosphere from both human-induced and natural sources. The human-induced sources include emissions from factories, autos, trucks, aircraft, power plants, home furnaces and fireplaces, to name a few. Many aerosols are not injected directly into the atmosphere, but form when gases convert to particles. Some particles, such as sulfates and nitrates, mainly reflect incoming sunlight, whereas others, such as soot, readily absorb sunlight. Many of the particles that reduce the amount of sunlight reaching Earth's surface tend to cause a *net cooling* of the surface air during the day.

In recent years, the effect of highly reflective **sulfate aerosols** on climate has been extensively researched. Earlier we learned that sulfate aerosols can come from natural sources, such as the oceans. However, the majority of these sulfate particles in the lower atmosphere are directly related to human activities and come primarily from the combustion of sulfur-containing fossil fuels.

Sulfur pollution, which has more than doubled globally since preindustrial times, enters the atmosphere mainly as sulfur dioxide gas. There, it transforms into tiny sulfate droplets or particles. Because these aerosols usually remain in the lower atmosphere for only a few days, they do not have time to spread around the globe. Hence, they are not well mixed and their effect is felt mostly over the Northern Hemisphere, especially over polluted regions.

Sulfate aerosols not only reflect incoming sunlight back to space, but they also serve as cloud condensation nuclei, tiny particles on which cloud droplets form. Thus, they have the potential for altering the physical characteristics of clouds. For example, if the number of sulfate aerosols and, hence, condensation nuclei inside a cloud should increase, the cloud would have to share its available moisture with the added nuclei, a situation that should produce many more (but smaller) cloud droplets. The greater number of droplets would reflect more sunlight and have the effect of *brightening* the cloud and reducing the amount of sunlight that reaches the surface. This process could also reduce the chance of precipitation falling from the cloud.

In summary, sulfate aerosols reflect incoming sunlight, which tends to lower Earth's surface temperature during the day. Sulfate aerosols may also modify clouds by increasing their reflectivity. Because sulfate pollution increased dramatically over industrialized areas of the Northern Hemisphere from the 1940s into the 1970s, it may help explain why global temperatures showed little warming during this period, and perhaps why warming intensified in the 1980s and 1990s as sulfate pollution decreased.

The effects of aerosols in the lower atmosphere on the climate system are a topic of active research. (If you are interested in a theory as to how vast quantities of particles injected into the atmosphere millions of years ago may have altered climate, and how humanity could alter climate during nuclear war, read Focus section 18.2.)

CLIMATE CHANGE: GREENHOUSE GASES   We learned in Chapter 2, p. 45, that carbon dioxide is a greenhouse gas that strongly absorbs infrared radiation and plays a major role in the warming of the lower atmosphere. Everything else being equal, the more CO_2 in the atmosphere, the warmer the surface air. We also know that CO_2 has been increasing steadily in the atmosphere, primarily due to human activities, such as the burning of fossil fuels like coal, oil, and natural gas (see Fig. 1.6, p. 8). Deforestation is also adding to this increase. Through the process of photosynthesis, the leaves of trees remove CO_2 from the atmosphere. The CO_2 is then stored in leaves, branches, and roots. When the trees are cut and burned, or left to rot, the CO_2 goes back into the atmosphere.

As of the late 2010s, the annual average of CO_2 concentration in the atmosphere was close to 410 ppm, and it was increasing by about 2 to 3 ppm per year. In recent studies, scientists have examined scenarios that show the atmospheric concentration by the end of this century as being anywhere from 421 to 1313 ppm. The actual number will depend on how much CO_2 is emitted by human activity in the coming decades and on how natural processes interact with the increase in CO_2.

To complicate the picture, increasing concentrations of other greenhouse gases—such as methane (CH_4), nitrous oxide (N_2O), and chlorofluorocarbons (CFCs)—all readily absorb infrared radiation.* Overall, these gases tend to be much less prevalent than CO_2 in the atmosphere, but many of them are more

---

*Refer back to Chapter 1 and to Table 1.1 on p. 6 for additional information on the concentration of these greenhouse gases.

powerful per molecule in their heat-trapping effect. Collectively, these gases enhance the atmospheric greenhouse effect by a substantial amount—about half of the effect now produced by CO_2.

How will increasing levels of greenhouse gases influence climate in the future? Before we address this question, we will look at how humans may be affecting climate by changing the landscape.

CLIMATE CHANGE: LAND USE CHANGES   All climate models predict that, as fossil fuels continue to spew greenhouse gases into the air, the climate will change and Earth's surface will warm. But are humans changing the climate by other activities as well? Modification of Earth's surface taking place right now could potentially be influencing the immediate climate of certain regions. For example, studies show that about half the rainfall in the Amazon River Basin is returned to the atmosphere through evaporation and through transpiration from the leaves of trees. Consequently, clearing large areas of tropical rain forests in South America to create open areas for farms and cattle ranges, as is happening now, will most likely cause a decrease in evaporative cooling. This decrease, in turn, could lead to a warming in that area of at least several degrees Celsius. In turn, the reflectivity of the deforested area will change. Similar changes in albedo result from the overgrazing and excessive cultivation of grasslands in semi-arid regions, causing an increase in desert conditions (a process known as **desertification**).

Billions of acres of the world's range and cropland, along with the welfare of millions of people, have been affected by desertification in recent decades. One of the main causes is believed to be overgrazing, although overcultivation, poor irrigation practices, and deforestation also play a role. The effect desertification will have on climate, as surface albedos increase and more dust is swept into the air, is uncertain. Some studies have proposed that allowing livestock to graze over larger areas, as their wild predecessors did, could reduce the risk of desertification by minimizing the effects of more intensive, localized grazing. (For a look at how modified land surfaces have influenced life in northern Africa, read Focus section 18.3.)

Some scientists have concluded that humans were altering climate long before modern civilizations came along. A growing amount of research indicates that preindustrial farming had been boosting methane and carbon dioxide in the atmosphere for up to 8000 years before the more rapid increase brought about by industrialization. One of the scientists involved in this research, retired professor William Ruddiman of the University of Virginia, has theorized that, without preindustrial farming, which produces methane and some carbon dioxide, we would have entered a naturally occurring ice age—an idea that remains controversial. Ruddiman even suggests that the Little Ice Age of the fourteenth through the nineteenth centuries in Europe was human-induced because plagues, which killed millions of people, caused a reduction in farming. The reasoning behind this idea goes something like this: As forests are cleared for farming, levels of CO_2 and methane increase, producing a strong greenhouse effect and a rise in surface air temperature. When catastrophic plagues strike—the bubonic plague, for instance—high mortality rates cause farms to be abandoned. As forests begin to take

Nuclear Winter, Cold Summers, and Dead Dinosaurs

A number of studies indicate that a nuclear war brought on by either human carelessness or negligence would drastically modify Earth's climate, instigating climate change unprecedented in recorded human history.

Researchers assume that a nuclear war would raise an enormous pall of thick, sooty smoke from massive fires that would burn for days, even weeks, following an attack. The smoke would drift higher into the atmosphere, where it would be caught in the upper-level westerlies and circle the middle latitudes of the Northern Hemisphere. Unlike soil dust, which mainly scatters and reflects incoming solar radiation, soot particles readily absorb sunlight. Hence, months, or perhaps years, after the war, sunlight would virtually be unable to penetrate the smoke layer, bringing darkness or, at best, twilight at midday.

Such reduction in solar energy would cause surface air temperatures over landmasses to drop below freezing, even during the summer, resulting in extensive damage to plants and crops and the death of millions (or possibly billions) of people. The dark, cold, and gloomy conditions that would be brought on by nuclear war are often referred to as *nuclear winter*.

As the lower troposphere cools, the solar energy absorbed by the smoke particles in the upper troposphere would cause this region to warm. The end result would be a strong, stable temperature inversion extending from the surface up into the higher atmosphere. A strong inversion would lead to a number of adverse effects, such as suppressing convection, altering precipitation processes, and causing major changes in the general wind patterns.

The heating of the upper part of the smoke cloud would cause it to rise upward into the stratosphere, where it would then drift around the world. Thus, about one-third of the smoke would remain in the atmosphere for up to a decade. The other two-thirds would be washed out in a month or so by precipitation. This smoke lofting, combined with persisting sea ice formed by the initial cooling, would produce climatic change that would remain for more than a decade.

Virtually all research on nuclear winter, including models and analog studies, confirms this gloomy scenario. Observations of forest fires show lower temperatures under the smoke, confirming part of the theory. The implications of nuclear winter are clear: A nuclear war would drastically alter global climate and would devastate our living environment.

Even with improved global superpower relations, and the end of the Cold War, the danger of nuclear winter remains a possibility. The current global nuclear arsenal of more than 10,000 warheads is more than is needed to produce the effects of a nuclear winter. Recent studies suggest that even a nuclear conflict limited to one region could have major effects on the protective ozone layer across the planet's stratosphere.

Could atmospheric particles and a nuclear winter–type event 65 m.y.a. have contributed to the demise of living creatures on Earth, such as dinosaurs? The dinosaurs, along with an estimated 70 percent to 75 percent of all plant and animal species on Earth, died in a mass extinction. What could cause such a catastrophe? About 65 m. y. a., a gigantic meteorite estimated to measure some 10 km (6 mi) in diameter slammed into Earth near the Yucatán Peninsula of Mexico at about 44,000 mi/hr (see ● Fig. 2). One popular theory proposes that the impact sent billions of tons of dust and debris into the upper atmosphere, where such particles circled the globe for months and greatly reduced the sunlight reaching Earth's surface. Reduced sunlight disrupted photosynthesis in plants which, in turn, led to a breakdown in the planet's food chain. Lack of food, as well as cooler conditions brought on by the dust, must have had an adverse effect on life, especially large plant-eating dinosaurs. Evidence for this catastrophic collision comes from the geologic record, which shows a thin layer of sediment deposited worldwide, about the time the dinosaurs disappeared. The sediment contains *iridium*, a rare element on Earth, but common in certain types of meteorites.

As evidence has accumulated over the last few years, most scientists now consider the Yucatán meteorite strike to be the main explanation for the extinction that killed off the dinosaurs. However, other factors may have played a role as well, including huge volcanic eruptions that occurred in India at about the same time. Have such meteorite collisions been more common in the geologic past than was once thought? And what is the likelihood of such an event occurring in the near future? These are among the questions that scientists continue to explore as they solve the detective story of past climates while looking toward our future climate.

● **FIGURE 2** Artist's interpretation of a giant meteorite striking Earth's surface 65 m.y.a., creating a nuclear winter–type event.

The Sahel—An Example of Climatic Variability and Human Existence

The Sahel is in North Africa, located between about 14° and 18°N latitude (see Fig. 3). Bounded on the north by the dry Sahara and on the south by the grasslands of the Sudan, the Sahel is a semi-arid region of variable rainfall. Precipitation totals may exceed 50 cm (20 in.) in the southern portion while in the north, rainfall is scanty. Yearly rainfall amounts are also variable as a year with adequate rainfall can be followed by a dry one.

During the winter, the Sahel is dry, but, as summer approaches, the Intertropical Convergence Zone (ITCZ) usually moves into the region, bringing rain. The inhabitants of the Sahel are mostly nomadic people who migrate to find grazing land for their cattle and goats. In the early and middle 1960s, adequate rainfall led to improved pasturelands; herds grew larger and so did the population. However, in the late 1960s, the annual rains did not reach as far north as usual, marking the beginning of a series of dry years and a severe drought.

The decrease in rainfall, along with overgrazing, turned thousands of square kilometers of pasture into barren wasteland. By 1973, when the severe drought reached its climax, rainfall totals were 50 percent of the long-term average, and perhaps 50 percent of the cattle and goats had died. The Sahara Desert had migrated southward into the northern fringes of the region, and a great famine led to the deaths of more than 100,000 people.

Although yearly rainfall has recovered substantially from the intense droughts of the 1970s and 1980s, it has not consistently reached the values observed in the 1950s and 1960s. The overall dryness of the region has caused many of the larger, shallow lakes (such as Lake Chad) to shrink in size. The wetter years of the 1950s and 1960s appear to be due to the northward displacement of the ITCZ. The drier years, however, appear to be more related to the intensity of rain that falls during the so-called rainy season. But what causes the lack of intense rain? Some scientists initially attributed

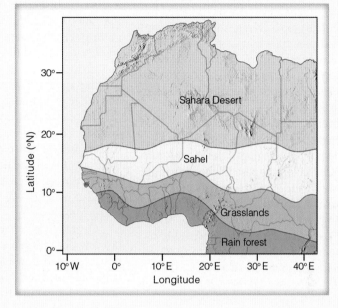

● **FIGURE 3** The semi-arid Sahel of North Africa is bounded by the Sahara Desert to the north and grasslands to the south.

this situation to a *biogeophysical feedback mechanism* wherein less rainfall and reduced vegetation cover modify the surface and promote a positive feedback relationship: Surface changes act to reduce convective activity, which in turn promotes or reinforces the dry conditions. As an example, when the vegetation is removed from the surface (perhaps through overgrazing or excessive cultivation), the surface albedo (reflectivity) increases, and the surface temperature drops. But studies show that less vegetation cover does not always result in a higher albedo.

Since the mid-1970s the Sahara Desert has not progressively migrated southward into the Sahel. In fact, during dry years, the desert does migrate southward, but in wet years, it retreats. By the same token, vegetation cover throughout the Sahel is more extensive during the wetter years. Consequently, desertification is not presently overtaking the Sahel, nor is the albedo of the region showing much year-to-year change.

So the question remains: Why did the Sahel experience such devastating drought

during the 1970s and 1980s? Recent studies suggest that the dry periods were mainly due to a cooling of the North Atlantic Ocean. This region appears to alternate between warmer and cooler conditions over a roughly 70-year cycle called the *Atlantic Multidecadal Oscillation*. Another possible factor is the presence of sulfate aerosols that enhance the formation of highly reflective clouds above the water. The increase in cloud reflectivity could cool the ocean surface and influence the circulation of the atmosphere in such a way that the ITCZ does not, on average, move as far north. The sulfate pollution that may have influenced African climate apparently originated over North America, suggesting that human activities on one continent could potentially cause climate variability on another, with the end result being a disastrous famine.*

---

*Studies also show a correlation between sulfate particles ejected into the stratosphere from volcanic eruptions and past dry spells in the Sahel.

---

over the untended land, levels of CO_2 and methane drop, causing a reduction in the greenhouse effect and a corresponding drop in air temperature. When the plague abates, the farms return, forests are cleared, levels of greenhouse gases go up, and surface air temperatures rise. (It should be emphasized that the rapid increase in greenhouse gases over the last few decades is primarily due to fossil fuel use, rather than land-use processes such as those described above.)

Climate Change: Global Warming

We have discussed several times in this chapter that Earth's atmosphere is in a warming trend that began around the turn of the twentieth century. This warming trend is real, as the average global surface air temperature since the late 1800s has risen by about 1.0°C (1.8°F). Moreover, the global average for each decade since the 1980s has been warmer than that of the preceding decade. There are many signs of increasing global warmth other than temperature readings. For example, the amount of water locked in the world's glaciers and ice sheets is steadily decreasing, and sea level is steadily rising.

Global warming might even be apparent where you live. The growing season, for example, may be getting longer (see Focus section 17.4, p. 498), or you may find the changing of the leaf color in autumn tending to happen later than in the past. Global warming in any given year, however, is small, and it only becomes significant when averaged over many years, such as decades. So, it is important not to base global warming on a specific weather event. A few facts illustrate the point. In 2014, a January cold wave across eastern North America sent temperatures plummeting. Atop Mount Mitchell in North Carolina, a low temperature of −31°C (−24°F) was the second coldest reading ever observed there. Binghamton, New York, dipped below 0°F on 10 days, the most ever recorded there in January. Yet at the same time, residents of California were experiencing the third warmest January on record, and globally, January 2014 was the fourth warmest on record. (Note that the United States only represents about 2 percent of the entire surface area of the planet, so one cannot use U.S. conditions alone as an index of how much the entire world is warming.)

It is interesting to note that 2011 was a record year for weather-related catastrophes in the United States, with floods, tornadoes, heat waves, drought, and snowstorms leading to a total of $12 billion disasters. The very next year came in close behind, with a total of $11 billion disasters. Did climate change play a role in these events? Focus section 18.4 looks at this question and the extreme weather of 2011 and 2012.

RECENT GLOBAL WARMING: PERSPECTIVE   Is this warming trend experienced over the past 100-plus years due to increasing greenhouse gases and an enhanced greenhouse effect? Before we can address this question, we need to review a few concepts we learned in Chapter 2.

WEATHER WATCH

During the summer of 2010, devastating flooding inundated almost a third of Pakistan, and a disastrous heat wave and drought hit Russia. Are those extreme weather events simply natural climate variability, or are they a consequence of global warming? Because Russia's summer climate is highly variable, global warming is not required in order to produce a heat wave as strong as the one observed in 2010. However, the odds of such heat appear to be rising. One analysis estimates that long-term climate warming has made a heat episode as strong as the one that struck Russia in July 2010 five times more likely.

Radiative Forcing Agents   We know from Chapter 2 that our world without water vapor, CO_2, and other greenhouse gases would be a colder world—about 33°C (59°F) colder than at present. With an average surface temperature of about 18°C (0°F), much of the planet would be uninhabitable. In Chapter 2, we also learned that when the rate of the incoming solar energy balances the rate of outgoing infrared energy from Earth's surface and atmosphere, the Earth-atmosphere system is in a state of *radiative equilibrium*. Increasing concentrations of greenhouse gases can disturb this equilibrium and are, therefore, referred to as **radiative forcing agents**. The **radiative forcing*** provided by extra CO_2 and other greenhouse gases has increased by about 3 W/m^2 over the past several hundred years, with the most rapid increase occurring over the last several decades. At the same time, increasing amounts of sun-blocking aerosols emitted by human activity (such as sulfates and other pollutants), together with their effects on cloudiness, have led to a decrease in radiative forcing estimated at roughly 1 W/m^2, which counteracts part of the greenhouse forcing.

Overall, it is very likely that most of the warming during the last few decades has come from increasing amounts of greenhouse gases. But what part does natural climate variability play in global warming? And with atmospheric concentrations of CO_2 increasing by more than 30 percent since the early 1900s, why might the observed increase in global temperature seem relatively small?

We already know that the climate can change due to natural events. For example, changes in the sun's energy output (called *solar irradiance*) and volcanic eruptions rich in sulfur, as we have discussed, are two major natural radiative forcing agents. Recent studies show that since the middle 1700s, changes in the sun's energy output have increased the total radiative forcing on the climate system by only a small amount, perhaps about 0.05 W/m^2. On the other hand, volcanic eruptions that inject sulfur-rich particles into the stratosphere produce a negative forcing, which lasts for a few years after the eruption. Several major eruptions occurred between 1880 and 1920, as well as between 1960 and 1991. In addition, a number of smaller eruptions took place between 2000 and 2011. The combined change in radiative forcing due to both volcanic activity and solar activity from 1998 to 2011 appears to have been slightly negative (about −0.2 W/m^2), which means that the net effect is that of *cooling* Earth's surface. Thus, natural factors may have actually reduced some of the global warming that otherwise would have been expected since 1998.

Did this cooling in combination with the cooling produced by sulfur-rich aerosols in the lower troposphere reduce the overall warming of Earth's surface during the last century? The use of climate models can help answer this question.

Climate Models and Recent Temperature Trends   We know that Earth's average surface temperature has increased by about 1.0°C since the end of the nineteenth century. How does this observed

---

*Radiative forcing is interpreted as an increase (positive) or a decrease (negative) in net radiant energy observed over an area in the middle of the tropopause. All factors being equal, an increase in *radiative forcing* may induce surface *warming*, whereas a *decrease* may induce surface *cooling*.

The Extremes of 2011 and 2012: Did Climate Change Play a Role?

Americans dealt with a succession of extreme weather events in 2011 and 2012 that seemed never-ending. Two major winter storms struck the Eastern Seaboard in January 2011, leaving New York City with its snowiest January on record. A February storm paralyzed Chicago with 21.2 inches and brought several cities their heaviest snow in history (see ●Fig. 4). The onslaught shifted in the spring, as a catastrophic four-day series of tornadoes swept through the southeast United States. The worst day by far was April 27, which brought the nation's most prolific 24-hour outbreak of twisters on record, and the deadliest in more than 80 years. At least 322 people were killed and an estimated $10 billion in damage occurred. Less than a month later, a violent tornado ripped through Joplin, Missouri, killing an estimated 159 people—the largest toll from a single tornado since 1947.

Disastrous weather continued into the late spring and summer of 2011. Record floods poured across the Missouri and Mississippi river valleys, and unprecedented drought and heat struck the Southern Plains. Texas recorded the hottest summer in history for any state, with a statewide average temperature of 86.8°F. Tropical cyclones largely bypassed the nation, but Hurricane Irene brought destructive floods to New England in August, followed by the Northeast's heaviest October snows in more than a century.

In 2012, high temperatures and dry conditions covered a larger swath of the nation's midsection. The most widespread drought in United States history since the 1930s affected more than half of the nation for most of the year, causing an estimated $30 billion in economic impact. Massive wildfires struck the western United States, with one especially intense fire in June destroying hundreds of homes near Colorado Springs, Colorado, during the city's worst heat wave on record. And in late October, Hurricane Sandy slammed into the New Jersey coast on an extremely unusual path that brought a catastrophic storm surge into the heavily populated New York and New Jersey coastlines. More than 100 people were

● **FIGURE 4** Cars are stranded on Chicago's Lake Shore Drive after a huge snowstorm during February 2011.

Scott Olson/Getty Images News/Getty Images

killed, and damages ran into the tens of billions of dollars.

Disasters such as hurricanes, tornadoes, wildfires, and floods often have people wondering whether climate change might be boosting the occurrence of such extremes. No single weather event is "caused" by a changing climate, but research shows that human-produced greenhouse gases are making some types of extreme weather more likely. For example, heat waves are expected to become more frequent and intense in coming decades. There is also evidence that precipitation is becoming increasingly concentrated in very heavy rain and snow events in many parts of the world, including the United States. A warming climate allows more water to evaporate into the atmosphere from the oceans, which could help intensify rain and snow where it is falling. However, warmer temperatures also help draw moisture from already-dry land. Computer models suggest that both precipitation and drought impacts will continue to intensify this century.

Some other kinds of extreme weather are less directly connected to increased greenhouse gases. For example, tornadoes are very localized events that develop as

various ingredients come together to create severe thunderstorms. The number of tornado reports in the United States has roughly doubled since the 1950s, but this phenomenon is mainly due to the growing number of storm spotters and chasers, as well as better post-storm surveys. There has been no significant trend in the number of the strongest tornadoes (those ranked EF3 or greater on the enhanced Fujita scale). Some research has pointed to a potential increase in severe thunderstorms across parts of the southern and eastern United States as the climate warms and average instability increases. For the most violent tornadoes to occur, this high instability must also be accompanied by substantial vertical wind shear. As the nation warms, vertical wind shear is expected to decrease on average. However, we can still expect some periods when adequate vertical wind shear is present along with high instability. Thus, it now appears that severe weather episodes may become more variable, with the possibility of fewer tornado outbreaks overall but more intense outbreaks when enough wind shear is present. In the United States, the terrible tornado year of 2011, which caused more than 500 deaths and billions of dollars in damage, was followed by the relatively tranquil tornado season of 2012, which produced the fewest tornadoes of any year in more than 60 years of record keeping.

A growing area of research focuses on *detection* and *attribution*. These studies attempt to identify a change in climate, and determine how much can be attributed to human-produced greenhouse gases. With the help of new statistical and numerical modeling tools, scientists are now beginning to estimate how the odds of a given weather event might have been boosted by our warming planet, or what part of a given event may be related to human-produced greenhouse gases. For example, one study in 2015 used models to estimate that climate change was responsible for 8 to 27 percent of the summertime soil-moisture deficit observed in California during the three years from 2012 to 2014.

temperature change compare with temperature changes derived from climate models using different forcing agents? Before we look at what climate models reveal, it is important to realize that the interactions between Earth and its atmosphere are so complex that it is difficult to unequivocally *prove* that Earth's present warming trend is due entirely to increasing concentrations of greenhouse gases. The problem is that any human-induced signal of climate change is superimposed on a background of natural climatic variations ("noise"), such as the El Niño-Southern Oscillation (ENSO) phenomenon (discussed in Chapter 10). Moreover, it can be difficult to separate a signal from the noise of natural climate variability in temperature observations. However, today's more sophisticated climate models are much better at filtering out this noise while at the same time taking into account those forcing agents that are both natural and human-induced.

● Figure 18.19a shows the predicted changes in global surface air temperature from 1900 to 2005 made by different climate models, which, as you recall, are mathematical models that simulate climate. Here the models use only natural forcing agents, such as solar energy and volcanic eruptions. Notice that the models' projected temperature (blue line) does not follow the observed trend in surface air temperature (gray line). In fact, the models project that temperatures now would be roughly similar to those in the early 1900s.

Figure 18.19b shows how the models project changes in global surface air temperature from 1900 to 2005 when *both* natural forcing agents and human forcing agents (such as greenhouse gases and sulfur aerosols) are added to the models. Notice how the projected temperature change (red line) now closely follows the observed temperature trend (gray line). It is climate studies using computer models such as these that have led scientists to conclude that *most* of the warming since the middle of the twentieth century is very likely the result of increasing levels of greenhouse gases.

The Intergovernmental Panel on Climate Change (IPCC), an activity that involves more than 2000 leading earth scientists from around the globe, has produced the world's most comprehensive reports on climate change for more than 25 years. The IPCC published in-depth climate assessments in 1990, 1995, 2001, 2007, and again in 2013. The 2013 Fifth Assessment Report of the IPCC states that:

It is extremely likely that human influence has been the dominant cause of the observed warming since the mid-twentieth century. [In the report, "extremely likely" means a probability of at least 95 percent.]

FUTURE CLIMATE CHANGE: PROJECTIONS

Climate models predict that, by the end of this century, increasing concentrations of greenhouse gases will result in an additional global warming that could be as much as several degrees Celsius. The newest, most sophisticated models take into account a number of important relationships, including the interactions between the oceans and the atmosphere, the processes by which CO_2 is removed from the atmosphere, and the cooling effect produced by sulfate aerosols in the lower atmosphere. The models also predict that, as the air warms, additional water will evaporate from the ocean surfaces and enter the atmosphere as water vapor. The added water

● **FIGURE 18.19** (a) Projected global surface air temperature changes from 1900 to 2005 using only natural forcing agents (dark blue line) compared to observed global surface air temperature changes (gray line). Light blue lines show range of model simulations. (Names and dates of major volcanic eruptions are given at the bottom of the graph.) (b) Projected global surface air temperature changes using both natural and human forcing agents (dark red line) compared to observed global surface air temperature changes (gray line). Orange lines show range of model simulations. (Temperature changes in both (a) and (b) are relative to the period 1901 to 1950.) (Adapted from the Technical Summary of *Climate Change 2007: The Physical Science Basis. Contribution of Working Group I to the Fourth Assessment Report of the Intergovernmental Panel on Climate Change,* 2007. Reprinted by permission of the Intergovernmental Panel on Climate Change.)

CRITICAL THINKING QUESTION In Fig. 18.19b, how do you think the model's projected temperature changes would appear (red line) if only human forcing agents were used in the model?

vapor (which is the most abundant greenhouse gas) will produce a feedback on the climate system by enhancing the atmospheric greenhouse effect and accelerating the temperature rise. (This phenomenon is the *water vapor-greenhouse feedback* described on p. 510.) Without this feedback produced by the added water vapor, the models predict that the warming would be much less.

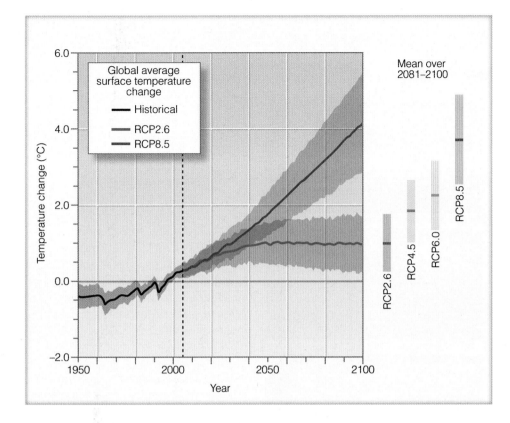

● **FIGURE 18.20** Global average projected surface air temperature changes (°C) above the 1986–2005 average (dark purple zero line) for the years 2000 to 2100. Temperature changes inside the graph and to the right of the graph are based on dozens of climate models run with different scenarios, based on representative concentration pathways (RCPs). Each scenario describes how the average temperature will change based on different concentrations of greenhouse gases and various forcing agents. The black line shows global temperature change during the twentieth century. The shaded bars on the right side of the figure indicate the likely range of temperature change for each scenario. The thick solid bar within each shaded bar gives the best estimate for temperature change by the years 2081–2100 for each scenario. (See Table 18.1 for additional information on the four RCPs.) (Source: Adapted from the Summary for Policymakers, *Climate Change 2013: The Physical Science Basis. Contribution of Working Group I to the Fifth Assessment Report of the Intergovernmental Panel on Climate Change*, 2013. Reprinted by permission of the Intergovernmental Panel on Climate Change.)

● Figure 18.20 shows climate models' projected warming during this century due to increasing levels of greenhouse gases and various forcing agents. Four sets of model simulations were produced, as shown by the four bands on the right side of the images. Each set of simulations used a different representative concentration pathway (RCP), which describes how the total radiative forcing might change over this century (see ▼ Table 18.1). To produce the low-end RCP2.6, greenhouse gas emissions would have to decline drastically over the next several decades. If emissions continue to increase rapidly, we will be closer to the high-end RCP8.5. For each RCP, several dozen climate models were used to create a range of projections. The shaded area around the blue trace (RCP2.6) and the red trace (RCP8.5) tell us that the

climate models do not all project the same amount of warming for a particular RCP. This is because each climate model handles the very complex interactions of the Earth-atmosphere system in a somewhat different way. Each model has its strengths and weaknesses, so by combining the results of several models, scientists can gain a more complete sense of what the future may hold. The other uncertainty, of course, is how much greenhouse gas will be emitted this century, as shown in the sharply different results for each of the RCPs.

In its 2013 report, the IPCC concluded that doubling the concentration of CO_2 would likely produce surface warming in the range of 0.3°C to 4.8°C by the period 2081–2100, with the actual amount depending largely on the rate of fossil fuel burning through the century. If, during this century, the surface temperature should increase by 2°C, that would be more than twice the warming experienced during the twentieth century. An increase of 4.5°C would have potentially devastating effects worldwide. Consequently, it is likely that the warming over this, the twenty-first century, will be much larger than the warming experienced during the twentieth century, and probably greater than any warming during the past 10,000 years. (Up to this point, we've looked at how climate models predict changes in future surface air temperatures. For additional information on these models, read Focus section 18.5.)

Uncertainties About Greenhouse Gases Although carbon dioxide continues to increase in the atmosphere, and model projections agree that warming can be expected this century, some uncertainties remain. At this point in time, it is unclear how water

▼ **TABLE 18.1** **Projected Average Surface Air Temperature Increases: Ranges and Best Estimates for the Period 2081–2100, Using Six Representative Concentration Pathways***

NAME OF PATHWAY	LIKELY RANGE OF TEMPERATURE RANGE, °C	MEAN ESTIMATED TEMPERATURE CHANGE, °C
RCP2.6	0.3-1.7	1.0
RCP4.5	1.1-2.6	1.8
RCP6.0	1.4-3.1	2.2
RCP8.5	2.6-4.8	3.7

*Temperature changes are relative to the average surface air temperature for the period 1986 to 2005.

Climate Models: How Do They Work?

Climate models that simulate the physical processes of the atmosphere (and the oceans) are called *general circulation models*, or GCMs for short. When an atmospheric component of a GCM is linked to an ocean component, the model is said to be "coupled" and the model is called an *atmosphere-ocean general circulation model*, or *AOGCM*. General circulation models use mathematics and the laws of physics to describe the general behavior of the atmosphere. To reduce some of the atmosphere's complexities, the models make simplified assumptions about the atmosphere and describe the atmosphere in more simplified physical terms. They also reduce many of the small-scale atmospheric processes (such as those due to clouds) into a single approximation, or parameter, which is known as *parameterization*.

The GCMs represent the atmosphere by dividing it up into grid squares, typically 100 kilometers on a side. The size of the squares has been getting steadily smaller over the years as computing power increases and model resolution improves (see ● Fig. 5). General circulation models simulate the behavior of the real atmosphere and describe the major circulation features as well as the seasonal and latitudinal temperature patterns. Today's coupled models are extremely sophisticated, taking into account land and vegetation processes, atmospheric chemistry, carbon cycles in the ocean and on land, ice and snow cover, and aerosols (see ● Fig. 6). These models are increasingly

(a) First Assessment Report, 1991

(b) Third Assessment Report, 2001

(c) Fifth Assessment Report, 2013

● **FIGURE 5** Improvements in horizontal resolution for the primary models used to predict climate change in the first, third, and fifth assessment reports conducted from the Intergovernmental Panel on Climate Change, which were released in 1991, 2001, and 2013, respectively.

referred to as *Earth system models* because they include not only an evaluation of oceans and atmosphere, but other parts of our global ecosystem.

A general circulation model is first run for a few decades to make sure that the model simulates the real atmosphere. Then, to see how some variables (such as increasing levels of CO_2) might influence the atmosphere, the model is repeatedly run with increasing concentrations of CO_2. In this manner, the GCMs reveal how the atmosphere and its circulation might change with time due to increasing levels of greenhouse gases. When the models are run with different scenarios (that is, varying concentrations of greenhouse gases and different forcing agents), the end result is usually a variation in the predicted temperature (such as those temperature projections shown in Fig. 18.19 on p. 524 and in Fig. 18.20 on p. 525).

Just as weather forecasters examine the results from several different models, each of which handles the atmosphere in a slightly different way, climate scientists make use of a variety of models from more than a dozen research centers around the globe. In addition, the researchers must consider natural climate variability. If a particular experiment is repeated using the same model but with slightly different initial conditions, the results may be somewhat different. Typically, the large-scale trends in global climate over decades are consistent from model to model and run to run, but there may be regional variations. Thus, climate researchers rely on ensembles, containing dozens of model runs, in order to see the full range of possibilities that may occur over the coming decades.

---

and land will ultimately affect increasing concentrations of CO_2. Currently, the oceans and the vegetation on land absorb about half of the CO_2 emitted by human sources, although the exact proportion varies from year to year. As a result, both oceans and landmasses play a major role in the climate system, yet the exact effect they will have on rising levels of CO_2 and global warming is not totally clear. For instance, the microscopic plants (phytoplankton) dwelling in the oceans extract CO_2 from the atmosphere during photosynthesis and store some of it below the oceans' surface, where they die. Could a warming earth trigger a large blooming of these microscopic plants, in effect reducing the rate at which atmospheric CO_2 is increasing?

Recent studies indicate that warming the planet tends to *reduce* both ocean and land intake of CO_2. Therefore, if the amount of human-induced CO_2 emissions continues to increase

at its present rate, more of that CO_2 should remain in the atmosphere, further enhancing global warming. An example of how rising temperatures can play a role in altering the way landmasses absorb and emit CO_2 is found in the Alaskan tundra. There, temperatures in recent years have risen to the point where more frozen soil melts in summer than it used to. Accordingly, during the warmer months, deep layers of exposed decaying peat moss release CO_2 into the atmosphere. Until recently, this region absorbed more CO_2 than it released. Now, however, much of the tundra acts as a producing source of CO_2. Moreover, recent analyses suggest that warmer temperatures will further increase the net amount of CO_2 released from Earth's tundra.

Deforestation accounts for about 10 to 15 percent of the observed increase in atmospheric CO_2. That percentage is lower than it was in the 1990s, in part because of progress in reducing

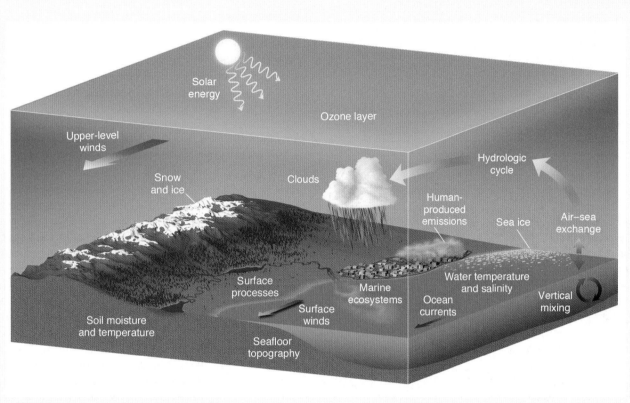

● **FIGURE 6** Components of an Earth system model.

General circulation models are not perfect: They have imperfect parameterization of all processes, and they cannot resolve processes that unfold over distances smaller than the grid spacing. However, as mentioned previously, today's models are extremely sophisticated, and they serve as the most reliable tools available for estimating climate change. As computers continue to become even more powerful, researchers are experimenting with models that blend global coverage with some aspects of higher-resolution models over specified areas. Another option is to use higher resolution when carrying out a model run for particular periods of time, such as a single decade later in the century. Simulations like these could lead to more accurate and detailed depictions of the weather and climate features of particular interest that might emerge in future climates.

deforestation. Hence, changes in land use can influence levels of CO_2 concentrations, especially if the practice of deforestation is replaced by reforestation.

Perhaps the greatest uncertainty in climate change over coming decades is the rate at which human activities will add greenhouse gases to the atmosphere. We can see in ● Fig. 18.21 the dramatic rise in CO_2 levels during the twentieth century. In the year 1990, carbon dioxide levels were increasing by about 1.5 ppm/year, whereas today they are increasing by more than 2 ppm/year. If this trend continues, CO_2 concentrations could easily exceed 550 ppm by the end of this, the twenty-first century. In Fig. 18.21 notice that the atmospheric concentration of methane has increased dramatically over the last 250 years, and it is still increasing, although with some variability in the last 20 years. Also notice that atmospheric concentrations of

nitrous oxide have risen quickly, and its concentration is still rising.

Since the mid-1990s, the atmospheric concentration of a group of greenhouse gases called *chlorofluorocarbons* (halocarbons) has been decreasing. However, the substitute compounds for chlorofluorocarbons, which are also greenhouse gases, have been increasing. Moreover, the total amount of surface ozone probably increased by more than 30 percent since 1750. However, the majority of ozone is found in the stratosphere, where its maximum concentration is typically less than 12 ppm. Although ozone is a greenhouse gas, it plays a very minor role in the enhancement of the greenhouse effect, as its concentration near Earth's surface is typically less than 0.04 ppm. The concentration of this greenhouse gas varies greatly from region to region, and depends upon the production of photochemical smog. The increase in surface

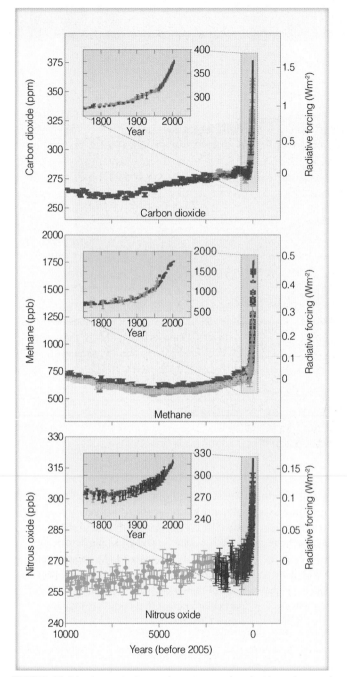

● **FIGURE 18.21** Changes in the greenhouse gases carbon dioxide, methane, and nitrous oxide indicated from ice core and modern data through 2005. Since then, carbon dioxide concentrations have risen to exceed 400 parts per million.(Source: *Climate Change 2007: The Physical Science Basis. Contribution of Working Group I to the Fourth Assessment Report of the Intergovernmental Panel on Climate Change*, 2007. Reprinted by permission of the Intergovernmental Panel on Climate Change.)

ozone has probably led to a very small increase in radiative forcing. Before going on to the next section you may wish to read Focus section 18.6, which details some of the misconceptions that have arisen about global warming, ozone, and the ozone hole.

The Question of Clouds As the atmosphere warms and more water vapor is added to the air, global cloudiness might increase as well. How, then, would clouds—which come in a variety of

shapes and sizes and form at different altitudes—affect the climate system?

Clouds reflect incoming sunlight back to space, a process that tends to cool the climate, but clouds also emit infrared radiation to Earth, which tends to warm it. Just how the climate will respond to changes in cloudiness will depend on the type of clouds that form, their height above the surface, and their physical properties, such as liquid water (or ice) content, depth, and droplet size distribution. For example, high, thin cirriform clouds (composed mostly of ice) appear to promote a net warming effect: They allow a good deal of sunlight to pass through (which warms Earth's surface), yet because they are cold, they warm the atmosphere around them by absorbing more infrared radiation from Earth's surface than they emit upward. Low stratified clouds, on the other hand, tend to promote a net cooling effect. Composed mostly of water droplets, they reflect much of the sun's incoming energy, which cools Earth's surface and, because their tops are relatively warm, they radiate to space much of the infrared energy they receive from Earth's surface. Satellite data confirm that, overall, the current global mixture of clouds has a *net cooling effect* on our planet, which means that, without any clouds, our atmosphere would be warmer.

Additional clouds in a warmer world would not necessarily have a net cooling effect, however. Their influence on the average surface air temperature would depend on their extent and on whether low or high clouds dominate the climate scene. Consequently, the feedback from clouds could potentially enhance or reduce the warming produced by increasing greenhouse gases. Most models show that as the surface air warms, there will be an increase in the typical altitude of cirrus clouds. This could cause an overall positive feedback. In other words, models indicate that warmer temperatures may change the global arrangement of clouds in a way that influences further warming.*

---

*In addition to the amount and distribution of clouds, the way in which climate models calculate the optical properties of a cloud (such as albedo) can have a large influence on the model's calculations. Also, there is much uncertainty as to how clouds will interact with aerosols and what the net effect will be.

● **FIGURE 18.22** Jet contrails can have an effect on climate by reflecting incoming sunlight and by emitting infrared energy to the surface.

Ozone and the Ozone Hole: Their Influence on Climate Change

THE IMPACT OF OZONE ON THE GREENHOUSE EFFECT AND CLIMATE CHANGE

Ozone is indeed a greenhouse gas, but its influence on the greenhouse effect is just minor. Why? Because of the following two conditions:

1. The concentration of atmospheric ozone is extremely small. Near Earth's surface, ozone averages only about 0.04 ppm, and in the stratosphere where it is more concentrated, its average value is only between 5 and 12 ppm. By comparison, the average value of carbon dioxide in our atmosphere is about 410 ppm.

2. Ozone only absorbs infrared energy in a very narrow band, near 10 μm. Look at Fig. 2.12 on p. 44 and observe that both water vapor and carbon dioxide are much more prolific absorbers of infrared energy than is ozone.

Accordingly, given these two facts, any small change in ozone concentration would have negligible impact on the greenhouse effect and on climate change.

THE IMPACT OF THE OZONE HOLE ON CLIMATE CHANGE

How does the ozone hole affect climate change? You may recall that we briefly looked at the ozone hole in Chapter 1 (see p. 10)*. There, we saw that over springtime Antarctica ozone concentrations in the stratosphere plummet, in some years leaving virtually no protective ozone above this region. Now, ozone readily absorbs incoming ultraviolet solar (UV) radiation at wavelengths below about 0.3 μm. So, does this fact mean that the formation of the ozone hole enhances global warming by allowing more UV radiation to reach the surface and warm it?

We know from Chapter 2 (p. 41) that the sun emits only a small fraction of its total

*The ozone hole is covered in more detail in Chapter 19 on p. 548.

energy output at ultraviolet wavelengths. Although UV waves do carry more energy than visible waves, there are too few of them to produce much warming. Those UV waves that do reach the surface mostly impinge upon snow and ice, ensuring that virtually no surface warming occurs.

It's interesting to note that the main temperature-related effect of ozone depletion is in the lower stratosphere, which has been *cooling*. Temperatures at this height (above about 20 km, or 12 mi) have dropped to record lows in recent years, in large part due to the loss of ozone.

Therefore, the depletion of ozone over Antarctica during its spring (that is, the ozone hole) does not enhance global warming at Earth's surface. We then must take care not to link the ozone hole with global warming. These are two distinctly different atmospheric conditions virtually unrelated—basically a case of apples and oranges.

There is another cloud-related factor for us to consider. Jet aircraft influence climate by producing contrails (condensation trails) high in the troposphere, generally above about 20,000 feet (see ● Fig. 18.22). Most contrails form as a cirrus-like trail behind the aircraft. Some disappear quickly, whereas others persist over time, occasionally stretching across the sky as streamers of cirriform clouds that coalesce into a white canopy.

Contrails can affect climate by enhancing cirriform cloudiness and by adding ice crystals to existing cirriform clouds, thus changing their albedo. Because contrails reflect sunlight and absorb infrared energy, they have the ability to alter the temperature near the ground. Overall, contrails have a net warming effect on the planet's radiative balance, so it is possible that any increase in global air travel may lead to additional warming.

The Impact of Oceans The oceans are a critical part of Earth's climate system, yet the exact effect they will have on climate change is not fully understood. For example, the oceans have a large capacity for storing heat energy. In fact, more than 90 percent of the energy trapped by increased greenhouse gases in recent decades has gone not into the atmosphere but into the ocean. Only a slight change in the rate of oceanic heat storage can thus have a big impact on atmospheric warming. Variations in this heat storage, perhaps related to ocean circulation patterns such as the Pacific Decadal Oscillation (described in Chapter 10 on p. 285), may help explain why some decades show greater atmospheric warming than others. The vast amount of heat stored by

the ocean also means that some global warming will continue to occur even if fossil-fuel emissions were to completely stop.

It appears that increased heat storage in the Pacific Ocean since the late 1990s played a role in slowing down atmospheric warming for more than a decade. However, the record global temperatures observed in 2014, 2015, and 2016, along with the strong 2014–2016 El Niño event, suggest that a larger fraction of the heat stored in the oceans has again been making its way into the atmosphere.

CONSEQUENCES OF CLIMATE CHANGE: THE POSSIBILITIES If the world continues to warm as predicted by climate models, where will most of the warming take place? Climate models predict that land areas will warm more rapidly than the global average, particularly in the northern high latitudes in winter (see ● Fig. 18.23a). We can see in Fig. 18.23b that the greatest surface warming for the period 2001 to 2006 tended to occur over landmasses in the high latitudes of the Northern Hemisphere, especially over Canada and Russia. These observations of global average temperature change suggest that climate models are on target with their warming projections.

As high-latitude regions of the Northern Hemisphere continue to warm, modification of the land may actually enhance the warming. For example, the dark green boreal forests* of the high latitudes absorb up to three times as much solar energy as does

*The boreal forest consists of woodlands (northern part) and conifers and some hardwoods (southern part). Its northern boundary is next to the tundra along the Arctic tree line.

2081–2100

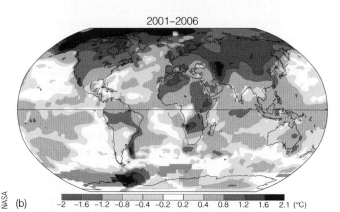

2001–2006

(a)

-2 -1.5 -1 -0.5 0 0.5 1 1.5 2 3 4 5 7 9 11 (°C)

NASA (b)

-2 -1.6 -1.2 -0.8 -0.4 -0.2 0.2 0.4 0.8 1.2 1.6 2.1 (°C)

● **FIGURE 18.23** (a) Projected surface air temperature changes averaged for the period 2081–2100 (using the RCP8.5 scenario) compared to the average surface temperature for the period 1986–2005. The largest increase in air temperature is projected to be over landmasses and in the Arctic region. (b) The average change in surface air temperature for the period 2001–2006 compared to the average for the years 1951–1980. The greatest warming was over the Arctic region and the high-latitude landmasses of the Northern Hemisphere. (Diagram [a] source: Summary for Policymakers, *Climate Change 2013: The Physical Science Basis. Contribution of Working Group I to the Fifth Assessment Report of the Intergovernmental Panel on Climate Change*, 2013. Reprinted by permission of the Intergovernmental Panel on Climate Change. Diagram [b] source: NASA)

the snow-covered tundra. Consequently, the winter temperatures in subarctic regions are, on the average, much higher than they would be without trees. If warming allows the boreal forests to expand into the tundra, the forests may accelerate the warming in that region. As the temperature rises, organic matter in the soil should decompose at a faster rate, adding more CO_2 to the air, which might accelerate the warming even more. Trees that grow in a climate zone defined by temperature may become especially hard hit as rising temperatures place them in an inhospitable environment. In a weakened state, they may become more susceptible to insects and disease. These changes in temperature will also affect people in many ways, of course, including direct effects on human health. For example, with heat waves expected to become more frequent and intense, heat-related deaths are expected to increase, although there could be some compensating decrease in cold-related illnesses.

Precipitation   Changes in precipitation and drought may be just as important as changes in temperature over the coming decades.

As with temperature, changes in precipitation will not be evenly distributed, as some areas will tend to get more precipitation and others less. Since the middle of the twentieth century, precipitation has generally increased over the middle- and high-latitude land areas of the Northern Hemisphere, while decreasing over some subtropical land areas. In many areas, there has also been an increase in the intensity of the heaviest precipitation events during the last 50 years or so.

Notice in ● Fig. 18.24 that the models project a further increase in average precipitation over high latitudes of the Northern Hemisphere and a continued decrease in precipitation over parts of the subtropics. The latter could have an adverse effect by placing added stress on agriculture. Even in places where average annual precipitation does not change, it is possible that rainfall and snowfall will be focused in more intense wet spells, with longer dry periods in between. In many parts of the world, observations show that the heaviest one-day rainfall events are already becoming heavier. In addition, warming temperatures will tend to cause soil to dry out more quickly, exacerbating the impact of drought when it occurs. In mountainous regions of western North America, where much of the precipitation falls in winter, a greater fraction of precipitation might fall as rain, causing a decrease in snowmelt runoff that fills the reservoirs during the spring. In California, the reduction in water storage could threaten the state's agriculture.

Sea Level Rise   Another major consequence of climate change is an increase in sea level, as land-based ice sheets and glaciers recede and the oceans continue to expand as they slowly warm. During the twentieth century, average global sea level rose by about 17 cm (7 in.). From 1900 to 2010, globally averaged sea level rose about 19 cm (7.5 in.), with the pace accelerating from the 1990s onward. About half of that was a result of melting glaciers and ice sheets, with the other half produced by the expansion of oceans as they warm. Globally averaged sea level has risen about twice as quickly since 1993—roughly 3.4 cm (1.3 in) per decade—as it did during the twentieth century as a whole.

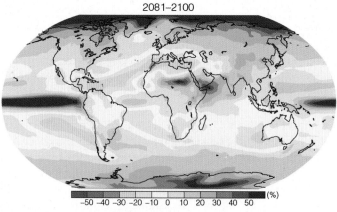

2081–2100

-50 -40 -30 -20 -10 0 10 20 30 40 50 (%)

● **FIGURE 18.24** Projected relative changes in annual mean precipitation (in percent) for the period 2081–2100 (using the RCP8.5 scenario) compared to the average for the period 1986–2005. A positive percentage represents an increase in precipitation, whereas a negative percentage represents a decrease. (Source: *Climate Change 2013: The Physical Science Basis*. Contribution of Working Group I to the Fifth Assessment Report of the Intergovernmental Panel on Climate Change, 2013. Reprinted by permission of the Intergovernmental Panel on Climate Change.)

Although we often think of sea level as being a fixed height around the world, it can actually vary by more than 30 cm (12 in.) based on natural processes. These include prevailing winds that can pile up water in the western tropical Pacific and along the east coasts of Asia and North America. In some areas, such as Scandinavia, the land is still rebounding from the weight of ice sheets thousands of years ago. Other coastal areas are plagued with subsidence, as the land slowly settles from natural processes or the withdrawal of underground fluids. Future changes in ocean circulation will lead to greater rises in sea level in some areas than in others.

The amount of additional rise in sea level this century will also depend on the future rate of greenhouse gas emissions, how much the air and water temperature increase in response to these greenhouse gases, and how quickly the vast ice sheets of Greenland and Antarctica melt. In 2013, the IPCC projected that the twenty-first-century rise in sea level will likely be somewhere between 26 and 82 cm (10 to 32 in.). The increase could be even larger if some of the ice shelves along the edge of Antarctica that hold back large expanses of ice were to fracture and collapse into the sea, which happened during previous warm periods in Earth's history. Should this happen in the coming decades, recent modeling has found that sea level could rise by as much as 150 cm (59 in.) this century, with even larger increases afterward.

Some models that take recent trends into account suggest that global average sea level could rise more by 100 cm (40 in.) or more by the year 2100. Especially concerning are recent studies showing that a large part of the West Antarctic ice sheet has entered a phase of melting that may be unstoppable. Over the next several hundred years, this melting alone could raise sea level by 3 m (10 ft) or more.

Sea level rise will be a growing issue in the coming decades for the many millions of people who live near coastlines around the world. Storm surges will occur atop a higher baseline water level. Rising ocean levels could also have a damaging influence on coastal ecosystems, such as coral reefs. In addition, coastal groundwater supplies might become contaminated with saltwater. And as we saw in Chapter 16, as sea surface temperatures increase (other factors being equal), the average intensity of hurricanes will likely increase as well. (For more information on hurricanes and global warming, read Focus section 16.4 on p. 465.)

Effects in Polar Regions   In polar regions, as elsewhere around the globe, rising temperatures produce complex interactions among temperature, precipitation, and wind patterns. Hence, in Antarctica, more snow might actually fall in the warmer (but still cold) air. This situation could allow snow to build up across the interior, although it may be counterbalanced by an increase in melting already taking place along the Antarctic coastline. Over Greenland, which is experiencing rapid melting of ice and snow, any increase in precipitation will likely be offset by rapid melting, and so the ice sheet is expected to continue to shrink.

Sea ice has been shrinking and thinning rapidly across the Arctic Ocean.* During the summer of 2007, and again in the summer of 2012, the extent of Arctic sea ice dropped dramatically to new record lows (see ● Fig. 18.25). If the warming in this region continues at its present rate, summer sea ice may, at times, shrink to cover less than 10 percent of the Arctic Ocean by the middle of this century, or even sooner.

---

*Sea ice is formed by the freezing of sea water. The melting of sea ice does not increase sea level, just as a melting ice cube does not raise the level of the water in the glass. The extent of sea ice normally expands in winter and shrinks in summer.

● **FIGURE 18.25** The extent of Arctic sea ice in (a) March 2015, when the ice cover was at or near its maximum for the year and in (b) September 2015, when the ice cover was near or at its minimum. The orange line in (a) represents the median maximum of the ice cover for the period 1979–2000. The orange line in (b) represents the median minimum extent of the ice cover for the period 1979–2000.

(a) February 2015

(b) September 2015

Around the edge of Antarctica, sea ice extent increased somewhat during the 2000s and early 2010s, reaching a record high in 2014, but it plummeted to a record low in 2016–2017. Projections consistently indicate that Antarctic sea ice will decrease later this century. (Note that most of the sea ice around the edge of Antarctica naturally disappears each summer, whereas the persistent summer ice in the center of the Arctic Ocean has played a more important role in Earth's climate balance.)

Effects on Ecosystems Increasing levels of CO_2 in a warmer world could have many other consequences. For example, greater amounts of CO_2 can be expected to act as a "fertilizer" for some plants, accelerating their growth, although this process can slow over time if water, nitrogen, and other nutrients were not plentiful enough to sustain the growth. In some ecosystems, certain some plant species could become so dominant that others are eliminated. In tropical areas, where many developing nations are located, the effects of climate change may actually decrease crop yield, whereas higher latitudes might benefit from a longer growing season and an earlier snowmelt. Extremely cold winters might become less numerous, with fewer bitter cold spells. However, wildfires may continue to become more prevalent during dry spells in forested high-latitude areas. (The city of Fort McMurray in northern Canada was engulfed by a huge wildfire in May 2016, much earlier in the year than wildfire is normally observed in the region.) Thus, while there will be some "winners" and some "losers," the most recent analyses suggest that the impact of climate change on agriculture and ecosystems may become increasingly negative by later in this century.

Following are some conclusions about global warming and its future impact on our climate system, summarized from the 2013 Fifth Assessment Report of the Intergovernmental Panel on Climate Change (IPCC):

- The primary source of the increased atmospheric concentration of carbon dioxide since the preindustrial period results from fossil fuel use, with land-use change providing another significant but smaller contribution. The atmospheric concentration of carbon dioxide in 2005 exceeds by far the natural range over the last 800,000 years (180 to 300 ppm) as determined from ice cores.

- Almost the entire globe experienced surface warming during the period from 1901 to 2002. Average Northern Hemisphere temperatures during the period from 1983 to 2012 were *very likely* higher than during any other 30-year period in at least the past 1400 years.

- It is *extremely likely* that more than half of the observed increase in global average surface temperature from 1951 to 2010 was caused by anthropogenic factors, including the increase in greenhouse gas concentrations. Temperatures of the most extreme hot days, hot nights, cold nights, and cold days are all *very likely* to have increased since 1950 due to anthropogenic forcing. It is *likely* that anthropogenic forcing has increased the risk of heat waves.

- The frequency and intensity of heavy precipitation events have *likely* increased in North America and Europe. In other continents, there are larger uncertainties due to limited observations.

- There is *high confidence* that ocean warming dominates the increase in energy stored in the climate system, accounting for more than 90 percent of the energy accumulated between 1971 and 2010. It is virtually certain that the upper ocean (02700 m) warmed from 1971 to 2010. Such warming causes seawater to expand, contributing to sea level rise.

- It is *very likely* that the mean rate of global averaged sea level rise was 1.7 [1.5 to 1.9] millimeters/year between 1901 and 2010, 2.0 [1.7 to 2.3] millimeters/year between 1971 and 2010, and 3.2 [2.8 to 3.6] millimeters/year between 1993 and 2010. The rate of sea level rise since the mid-nineteenth century has been larger than the mean rate during the previous two millennia.

- Over the last two decades, there is *high confidence* that the Greenland and Antarctic ice sheets have been losing mass, glaciers have continued to shrink almost worldwide, and Arctic sea ice and Northern Hemisphere spring snow cover have continued to decrease in extent.

- Global mean surface temperature by 2016–2035 will *likely* be 0.3°C to 0.7°C warmer than in 1986–2005. There is more uncertainty in the amount of warming by 2081–2100 compared to 1986–2005; the amount will *likely* fall between 0.3°C and 4.8°C, depending on changes in greenhouse gas emissions and other factors.

- Extreme precipitation events over most of the midlatitude landmasses and over wet tropical regions will *very likely* become more intense and more frequent by the end of this century.

- Global mean sea level will continue to rise during the twenty-first century. The rate of sea level rise will *very likely* exceed that observed during 1971–2010 due to increased ocean warming and increased loss of mass from glaciers and ice sheets. Sea level rise will not be uniform: about 30 percent of coastlines will experience at least 20 percent greater or lesser change than the global average. Sea level rise due to thermal expansion of the ocean is expected to continue for many centuries.

CLIMATE CHANGE: EFFORTS TO CURB The most obvious way to limit global warming is to reduce greenhouse gas emissions by reducing the use of fossil fuels. Burning natural gas produces less carbon dioxide than burning oil and coal, and the rapid growth of natural gas use was one factor in a drop in CO_2 emissions in the United States during the early 2010s. However, natural gas production also leads to emissions of methane (a powerful greenhouse gas) as a byproduct. Researchers are now investigating this phenomenon and the extent to which it may be counteracting the benefits of reduced CO_2 emissions.

Increasing the use of alternative energy sources can also play a major role in curbing global warming. Technologies such as solar and wind power—the two fastest-growing energy sources worldwide—produce almost no greenhouse gases other than those required to build and maintain the facilities.

Diplomatic Efforts In an attempt to mitigate the impact humans have on the climate system, representatives from 160 countries met at Kyoto, Japan, in 1997 to work out a formal agreement to limit greenhouse gas emissions in industrialized nations. The international agreement—called the *Kyoto Protocol*—was adopted in 1997 and put into force in February 2005. The Protocol set mandatory targets for reducing greenhouse gas emissions in countries that adopt the plan. Although the percent by which each country was to reduce its emissions varies, the overall goal was to reduce greenhouse gas emissions in developed countries by at least 5 percent below existing 1990 levels during the five-year period of 2008 through 2012. For the industrialized nations that participated in the Kyoto Protocol, emissions dropped by more than 22 percent. However, the United States did not ratify the protocol, and many developing nations such as China were not required to carry out emission reductions, since they had been responsible for only a small part of the accumulated CO_2 up to that point. As a result, the global total of greenhouse gas emissions actually *increased* by more than 25 percent from 1990 to 2012.

The Kyoto Protocol has been followed by the Paris Agreement, which was introduced in 2015 and adopted by virtually every one of the world's nations. Under this agreement, each nation set voluntary targets for reducing emissions and will report their progress on a regular basis. (In mid-2017, the United States announced its intention to withdraw from the Paris Agreement.) In addition, several cities and countries, including Costa Rica, Iceland, and Norway, have pledged to become *carbon neutral*—meaning that all of their greenhouse gas emissions would be offset by activities such as planting trees, so that the country ends up with no net emissions. Many global businesses are also striving to become carbon neutral.

In the United States, many cities and states have implemented their own climate change policies. For example, California has set targets for reducing greenhouse gas emissions to 1990 levels by the year 2020, with additional reductions of 40 percent below 1990 levels by 2030 and 80 percent by 2050. In addition, the mayors of more than 1000 towns and cities in the United States have pledged to reduce the levels of carbon emissions in their municipalities below 1990 levels.

Geoengineering In recent years, the idea of using technology to mitigate climate change has gained interest among researchers and policy makers. Called **geoengineering**, the idea is to use global-scale technological fixes to counter climate change by either (1) removing greenhouse gases from the atmosphere or by (2) changing the amount of sunlight that reaches Earth.

Several geoengineering ideas to remove CO_2 from the atmosphere include fertilizing the oceans with plants that absorb CO_2, sprinkling iron-rich particles over portions of the ocean to promote the growth of carbon-absorbing phytoplankton, and placing large drifting vertical pipes into the ocean so that wave activity will pull up nutrient-rich water from below to promote algae blooms. One significant drawback to all of these proposals is that they do nothing to reduce the gradual acidification of oceans, which occurs as the oceans absorb carbon dioxide. Acidification poses a serious threat to shellfish, coral, and many other forms of marine life that rely on calcification. One proposal that does take this problem into account is to extract CO_2 from the atmosphere with "synthetic trees" made of recyclable chemicals that react with CO_2 in the air.

To prevent sunlight from reaching Earth's surface, one idea proposes placing an array of reflecting mirrors in space high above Earth. Another proposal suggests injecting highly reflective sulfate aerosols into the stratosphere. In one study, scientists using climate models placed tons of sulfate aerosols—on the order of the amount lofted by Mount Pinatubo in 1991—into the stratosphere at various intervals. The study concluded that injecting these sulfate aerosols every one to four years in conjunction with reducing greenhouse gases could provide a "grace period" of up to 20 years before a major cutback in greenhouse gas emissions would be required.

All of these geoengineering proposals may have unforeseen or unwanted consequences. Injecting the stratosphere with sulfate particles, for example, might alter the temperature of the upper atmosphere and affect the fragile ozone layer. Carrying out large-scale geoengineering would be quite expensive, and it would also call for global agreement on techniques and procedures, as the results could affect the entire planet. As we have seen with the Kyoto Protocol and the Paris Agreement, such global consensus can be very difficult to obtain. In short, the science of geoengineering is intriguing, but it poses complex political, financial, and technological challenges.

Climate Change: A Final Note Cutting down on the emissions of greenhouse gases and pollutants has several potentially positive benefits. A reduction in greenhouse gas emissions could slow down the enhancement of Earth's greenhouse effect and reduce global warming while at the same time, the associated reduction in air pollutants might reduce acid rain, diminish haze, slow the production of photochemical smog, and produce significant health benefits. Even if the greenhouse warming were to end up toward the lower end of what modern climate models project, these measures would certainly benefit humanity.

SUMMARY

In this chapter, we considered some of the many ways Earth's climate can be changed. First, we saw that Earth's climate has undergone considerable change during the geologic past. Some of the evidence for a changing climate comes from tree rings (dendrochronology), chemical analysis of oxygen isotopes in ice cores and fossil shells, and geologic evidence left behind by advancing and retreating glaciers. The evidence from these suggests that, throughout much of the geologic past (long before humanity arrived on the scene), Earth was warmer than it is today. There were cooler periods, however, during which glaciers advanced over large sections of North America and Europe.

We examined some of the possible causes of climate change, noting that the problem is extremely complex, as a change in one variable in the climate system almost immediately changes other variables. Climate changes can be brought on by both natural events and by human (anthropogenic) activities. One theory of natural cause of climate change suggests that the shifting of the continents, along with volcanic activity and mountain building, may account for variations in climate that take place over millions of years.

Another theory of natural cause of climate change is the Milankovitch theory, which proposes that alternating glacial and interglacial episodes during the past 2.5 million years are the result of small variations in the tilt of Earth's axis and in the geometry of Earth's orbit around the sun. Climate change may also be brought on naturally by volcanic eruptions rich in sulfur and by variations in the sun's energy output.

Human activities, such as emitting vast quantities of greenhouse gases into the atmosphere, can produce climate changes worldwide. Global average temperatures since the late nineteenth century have risen by about 1.0°C. Many studies have found that increasing concentrations of greenhouse gases are the primary cause of this warming. Sophisticated climate models project that, as levels of CO_2 and other greenhouse gases continue to increase, Earth's surface will warm substantially by the end of this century. The models also predict that, as Earth warms, there will be a global increase in atmospheric water vapor, more extreme precipitation events, a worsening of drought impacts, a more rapid melting of sea ice, and a rise in sea level.

KEY TERMS

The following terms are listed (with corresponding page numbers) in the order they appear in the text. Define each. Doing so will aid you in reviewing the material covered in this chapter.

QUESTIONS FOR REVIEW

1. What methods do scientists use to determine climate conditions that have occurred in the past?

2. Explain how the changing climate influenced the formation of the Bering land bridge.

3. How does today's average global temperature compare with the average temperature during most of the past 1000 years?

4. What is the Younger Dryas episode? When did it occur?

5. How does a positive feedback mechanism differ from a negative feedback mechanism? Is the water vapor–greenhouse feedback considered positive or negative? Explain.

6. Explain why the chemical weathering–CO_2 feedback is a negative feedback in Earth's climate system.

7. How does the theory of plate tectonics explain climate change over periods of millions of years?

8. Describe the Milankovitch theory of climatic change by explaining how each of the three cycles alters the amount of solar energy reaching Earth.

9. Given the analysis of air bubbles trapped in polar ice during the past 800,000 years, were CO_2 levels generally higher or lower during warmer glacial periods? Were methane levels higher or lower at this time?

10. How do sulfate aerosols in the lower atmosphere affect surface air temperatures during the day?

11. Describe the scenario of nuclear winter.

12. Do volcanic eruptions rich in sulfur tend to warm or cool Earth's surface? Explain.

13. Explain how variations in the sun's energy output might influence global climate.

14. Climate models predict that increasing levels of CO_2 will cause the mean global surface temperature to rise

significantly by the year 2100. What other greenhouse gas *must* also increase in concentration for the amount of predicted temperature rise to occur?

15. Describe some of the natural and human-induced radiative forcing agents and their effect on climate.

16. List five ways natural events can cause climate change.

17. List three ways human (anthropogenic) activities can cause climate change.

18. Describe how clouds influence the climate system.

19. In Fig. 18.19a, p. 524, explain why the actual rise in surface air temperature (gray line) is much greater than the projected rise in temperature due to natural forcing agents.

20. Why have climate scientists concluded that most of the warming experienced during the last 50 years has been due to increasing concentrations of greenhouse gases?

21. List some of the potential consequences of climate change on the atmosphere and its inhabitants.

22. Is CO_2 the only greenhouse gas we should be concerned with for climate change? If not, what are the other gases?

23. Explain how the ocean's conveyor belt circulation works. How does the conveyor belt appear to influence the climate of northern Europe? (Hint: The answer is found in Focus section 18.1, p. 508.)

QUESTIONS FOR THOUGHT

1. Ice cores extracted from Greenland and Antarctica have yielded valuable information on climate changes during the past 800,000 years. What do you feel might be some of the limitations in using ice core information to evaluate past climate changes?

2. When glaciation was at a maximum (about 18,000 years ago), was global precipitation greater or less than at present? Explain your reasoning.

3. Consider the following climate change scenario. Warming global temperatures increase saturation vapor pressures over the ocean. As more water evaporates, increasing quantities of water vapor build up in the troposphere. More clouds form as the water vapor condenses. The clouds increase the albedo, resulting in decreased amounts of solar radiation reaching Earth's surface. Is this scenario plausible? What type(s) of feedback(s) is/are involved? What type of clouds (high or low)?

4. Explain two different ways that an increase in sulfate particles might lower surface air temperatures.

5. Are ice ages in the Northern Hemisphere more likely when: (a) the tilt of Earth is at a maximum or a minimum? (b) the sun is closest to Earth during summer in the Northern Hemisphere, or during winter? Explain your reasoning for both (a) and (b).

6. Most climate models show that the poles will warm faster than the tropics. What effect will this have on winter storms in midlatitudes?

7. The oceans are a major sink (absorber) of CO_2. One hypothesis states that as warming increases, less CO_2 will be dissolved in the oceans. Would you expect Earth to cool or to warm further? Why?

8. Why did periods of glacial advance in the higher latitudes of the Northern Hemisphere tend to occur with colder summers, but not necessarily with colder winters?

PROBLEMS AND EXERCISES

1. If the annual precipitation near Hudson Bay (latitude 55°N) is 38 cm (15 in.) per year, calculate how long it would take snow falling on this region to reach a thickness of 3000 m (about 10,000 ft). (Assume that all the precipitation falls as snow, that there is no melting during the summer, and that the annual precipitation remains constant. To account for compaction of the snow, use a water equivalent of 5 to 1.)

2. On a warming planet, the snow-albedo feedback produces a positive feedback. Make a diagram (or several diagrams) to illustrate this phenomenon. Now, with another diagram, show that the snow-albedo feedback produces a positive feedback on a cooling planet.

ONLINE RESOURCES

Visit www.cengagebrain.com to view additional resources, including video exercises, practice quizzes, an interactive eBook, and more.

Heavy smog clogs the air in the Guomao area of central Beijing, China, on December 25, 2015.

CHAPTER

19

Air Pollution

AIR POLLUTION MAKES THE EARTH A LESS PLEASANT place to live. It reduces the beauty of nature. This blight is particularly noticed in mountain areas. Views that once made the pulse beat faster because of the spectacular panorama of mountains and valleys are more often becoming shrouded in smoke. When once you almost always could see giant boulders sharply etched in the sky and the tapered arrowheads of spired pines, you now often see a fuzzy picture of brown and green. The polluted air acts like a translucent screen pulled down by an unhappy God.

Louis J. Battan, *The Unclean Sky*

Every deep breath fills our lungs. Most of what we inhale is gaseous nitrogen and oxygen. However, we may also inhale in minute quantities other gases and particles, some of which could be considered pollutants. These contaminants come from car exhaust, chimneys, factories, power plants, and other sources related to human activities.

Virtually every large city has to contend in some way with air pollution, which clouds the sky, injures plants, and damages property. Some pollutants merely have a noxious odor, whereas others can cause severe health problems. The cost is high. In the United States, for example, outdoor air pollution takes its toll in health care and lost work productivity at an annual expense that runs into *billions* of dollars. Estimates are that, worldwide, nearly 1 billion people in urban environments are continuously being exposed to health hazards from air pollutants. The World Health Organization estimated that in 2012, outdoor air pollution contributed to roughly 3.7 million deaths worldwide. Indoor air pollution is also a critical health issue. Emissions from indoor cookstoves, mainly in developing countries, were one of the main factors behind roughly 4.3 million deaths in 2010 that were attributed to indoor air pollution.

This chapter takes a look at this serious contemporary concern. We begin by briefly examining the history of problems in this area, and then go on to explore the types and sources of air pollution, as well as the weather that can produce an unhealthy accumulation of pollutants. Finally, we investigate how air pollution influences the urban environment and also how it brings about unwanted acid precipitation.

A Brief History of Air Pollution

Strictly speaking, air pollution is not a new problem. More than likely it began when humans invented fire whose smoke choked the inhabitants of poorly ventilated caves. In fact, very early accounts of air pollution characterized the phenomenon as "smoke problems," the major cause being people burning wood and coal to keep warm.

To alleviate the smoke problem in old England, King Edward I issued a proclamation in 1273 forbidding the use of sea coal, an impure form of coal that produced a great deal of soot and sulfur dioxide when burned. One person was reputedly executed for violating this decree. In spite of such restrictions, the use of coal grew as a heating fuel during the fifteenth and sixteenth centuries.

As industrialization increased, the smoke problem worsened. In 1661, the prominent scientist John Evelyn wrote an essay deploring London's filthy air. And by the 1850s, London had become notorious for its "pea soup" fog, a thick mixture of smoke and fog that hung over the city. These fogs could be dangerous. In 1873, one was responsible for as many as 700 deaths. Another in 1911 claimed the lives of 1150 Londoners. To describe this chronic atmospheric event, a physician, Harold Des Voeux, coined (in 1905) the word *smog*, meaning a combination of smoke and fog.

Little was done to control the burning of coal as time went by, primarily because it was extremely difficult to counter the basic attitude of the powerful industrialists: "Where there's muck, there's money." London's acute smog problem intensified. Then, during the first week of December 1952, a major disaster struck. The winds died down over London and the fog and smoke became so thick that people walking along the street literally could not see where they were going (see ● Fig. 19.1). People wore masks over their mouths and found their way along the sidewalks by feeling the walls of buildings. This particularly disastrous smog lasted 5 days and took at least 4000 lives, prompting Parliament to pass a Clean Air Act in 1956. Additional air pollution incidents occurred in England during 1956, 1957, and 1962, but due to the strong legislative measures taken against air pollution, London's air today is much cleaner, and "pea soup" fogs are a thing of the past.

● **FIGURE 19.1** The fog and smoke were so dense in London during December 1952 that visibilities were often restricted to less than 100 feet and streetlights had to be turned on during the middle of the day.

Air pollution episodes were by no means limited to Great Britain. During the winter of 1930, for instance, Belgium's highly industrialized Meuse Valley experienced an air pollution tragedy when smoke and other contaminants accumulated in a narrow steep-sided valley. The tremendous buildup of pollutants caused about 600 people to become ill, and ultimately 63 died. Not only did humans suffer, but cattle, birds, and rats also fell victim to the deplorable conditions.

The industrial revolution brought air pollution to the United States, as homes and coal-burning industries belched smoke, soot, and other undesirable emissions into the air. Soon, large industrial cities, such as St. Louis and Pittsburgh (which became known as the "Smoky City"), began to feel the effects of the ever-increasing use of coal. As early as 1911, studies documented the irritating effect of smoke particles on the human respiratory system and the "depressing and devitalizing" effects of the constant darkness brought on by giant, black clouds of smoke. By 1940, the air over some cities had become so polluted that automobile headlights had to be turned on during the day.

The first major documented air pollution disaster in the United States occurred at Donora, Pennsylvania, during October 1948, when industrial pollution became trapped in the Monongahela River Valley. During the ordeal, which lasted five days, more than 20 people died and thousands became ill.* Several times during the 1960s, air pollution levels became dangerously high over New York City. Meanwhile, on the West Coast, in cities such as Los Angeles, the ever-increasing number of automobiles, coupled with large petroleum processing plants, were instrumental in generating a different type of pollutant, *photochemical smog*—the type that forms in sunny weather and irritates the eyes. Toward the end of World War II, Los Angeles had its first (of many) smog alerts.

Air pollution episodes in Los Angeles, New York, and other large American cities led to the establishment of much stronger emission standards for industry and automobiles. The Clean Air Act of 1970, for example, empowered the federal government to set emission standards that each state was required to enforce. The Clean Air Act was revised in 1977 and updated by Congress in 1990 to include even stricter emission requirements for autos and industry. The new version of the Act also includes incentives to encourage companies to lower emissions of those pollutants contributing to the current problem of acid rain. Moreover, amendments to the Act have identified 189

*Additional information about the Donora air pollution disaster is given in Focus section 19.4, p. 558.

● **FIGURE 19.2** Pupils cover their noses after school in heavy smog on December 23, 2015, in Binzhou, China.

toxic air pollutants for regulation, and in 2001, the U.S. Supreme Court, in a unanimous ruling, made it clear that cost need not be taken into account when setting clean air standards. Meanwhile, some of the most urgent air pollution problems on Earth are now found in developing countries such as China and India, where rapidly growing populations can be subject to massive levels of emissions from coal plants, vehicles, and other sources (see ● Fig. 19.2).

Most of the pollutants we will examine in the next several sections are those more typically found outside. But as we have already mentioned, air pollution can be a major health problem inside a structure. Focus section 19.1 addresses some of the health risks of indoor air pollution.

Types and Sources of Air Pollutants

Air pollutants are airborne substances (either solids, liquids, or gases) that occur in concentrations high enough to threaten the health of people and animals, to harm vegetation and structures, or to toxify a given environment. Air pollutants come from both natural sources and human activities. Examples of natural sources include wind picking up dust and soot from Earth's surface and carrying it aloft, tons of ash and dust belched into our atmosphere by volcanoes, and vast quantities of drifting smoke produced by forest fires (see ● Fig. 19.3).

Human-induced pollution enters the atmosphere from both *fixed sources* and *mobile sources*. Fixed sources encompass industrial complexes, power plants, homes, office buildings, and so forth; mobile sources include motor vehicles, ships, and jet aircraft. Certain pollutants are called **primary air pollutants** because they enter the atmosphere directly—from smokestacks and tailpipes, for example. Other pollutants, known as **secondary air pollutants**, form only when a

● **FIGURE 19.3** Strong northeasterly Santa Ana winds on October 28, 2003, blew the smoke from massive wild fires across southern California out over the Pacific Ocean.

chemical reaction occurs between a primary pollutant and some other component of air, such as water vapor or another pollutant. ▼ Table 19.1 summarizes some of the sources of primary air pollutants.

● Figure 19.4 shows that carbon monoxide is the most abundant primary air pollutant in the United States. The primary source for all pollutants is transportation (motor vehicles and so on), with fuel combustion from stationary (fixed) sources coming in a distant second. Although hundreds of pollutants are found in our atmosphere, most fall into five groups, recognized as "criteria pollutants" under the Clean Air Act, which are summarized in the following section. Note that Fig. 19.4 also includes ammonia, a by-product of livestock operations. As of this writing, ammonia is not listed as a criteria pollutant. However, exposure to airborne ammonia can cause serious respiratory effects. The Environmental Protection Agency (EPA) began tracking ammonia emissions in 2002, and several groups have petitioned the EPA to classify ammonia as a criteria pollutant. Note in Fig. 19.4 that emissions are characterized in terms of their total weight. Because different pollutants have different lifetimes in the atmosphere, they can also be measured by the number of parts per million (ppm) or parts per billion (ppb).

▼ **TABLE 19.1** **Some of the Sources of Primary Air Pollutants**

	SOURCES		POLLUTANTS
NATURAL			
	Volcanic eruptions		Particles (dust, ash), gases (SO_2, CO_2)
	Forest fires		Smoke, unburned hydrocarbons, CO_2, nitrogen oxides, ash
	Dust storms		Suspended particulate matter
	Ocean waves		Salt particles
	Vegetation		Hydrocarbons (VOCs),* pollens
	Hot springs		Sulfurous gases
HUMAN-CAUSED			
Industrial	Paper mills		Particulate matter, sulfur oxides
	Power plants	Coal	Ash, sulfur oxides, nitrogen oxides
		Oil	Sulfur oxides, nitrogen oxides, CO
	Refineries		Hydrocarbons, sulfur oxides, CO
	Manufacturing	Sulfuric acid	SO_2, SO_3, and H_2SO_4
		Phosphate fertilizer	Particulate matter, gaseous fluoride
		Iron and steel mills	Metal oxides, smoke, fumes, dust, organic and inorganic gases
		Plastics	Gaseous resin
		Varnish/paint	Acrolein, sulfur compounds, hydrocarbons (VOCs),
Personal	Automobiles		CO, nitrogen oxides, hydrocarbons (VOCs), particulate matter
	Home furnaces/fireplaces		CO, particulate matter
	Open burning of refuse		CO, particulate matter

*VOCs are volatile organic compounds; they represent a class of organic compounds, most of which are hydrocarbons.

Indoor Air Pollution

When people think of air pollution, most think of outside air where automobiles, factories, and power plants spew countless tons of contaminants into the air. But, surprisingly, the air we breathe inside our homes and other structures can be between 5 and 100 times more polluted than the air we breathe outdoors. The most dangerous indoor pollution occurs in developing countries, where cookstoves that use wood or dung spew out large amounts of soot. Millions of deaths are caused each year by this indoor pollution. However, even homes in affluent, industrialized countries can harbor a variety of dangerous indoor pollutants (see ● Fig. 1).

The Environmental Protection Agency has identified many sources of indoor air pollution, ranging from building materials, pressed wood products, furnishings, and home cleaning products, to pesticides, adhesives, and personal care products. In addition, heating sources (such as unvented kerosene heaters, wood stoves, and fireplaces) can release a variety of pollutants into a home. The pollution impact of any heating source depends upon several factors, such as how old the source is, the level of maintenance it receives, as well as its location and access to ventilation. For example, a gas cooking stove or heating stove with improper fittings and adjustments can cause a significant emission of carbon monoxide. New carpets and padding (as well as the adhesives used in their installation) can emit volatile organic compounds.

Some pollution sources, such as building materials and foam insulation, produce a constant stream of pollutants, whereas activities such as tobacco smoking emit pollutants into the air on an intermittent basis. In certain instances, outside pollution is brought indoors. Some pollutants enter homes and other structures through cracks and holes in foundations and basements, as is the case with radon.

Radon is a colorless, odorless gas—a natural radioactive compound—that forms as the uranium in soil and rock breaks down. Radon is found everywhere on earth and only becomes a problem when it leaks out of the soil and becomes trapped inside homes and buildings. The

● **FIGURE 1** With candles burning, a fire in the fireplace, and stain-resistant material on rugs and carpets, there are probably more air pollutants in this living room than there are in a similar size volume of air outdoors.

radon gas that seeps in through cracks and other openings in a building can accumulate to levels that create a serious health threat. Studies by the EPA have shown that as many as 6 percent of the homes (roughly 7 million) in the United States may have elevated levels of radon. Radon concentrations vary greatly from structure to structure and can only be measured by devices known as *radon detectors*. Inside a home, the radon decays into *polonium,* a solid substance that attaches itself to dust in the air. The tiny dust particles can be inhaled deep into the lungs, where they attach to lung tissue. As the polonium decays, it damages the lung tissue, sometimes producing mutated cells that may develop into lung cancer.

Another major chemical pollutant found inside our homes is *formaldehyde*. It is a colorless, pungent-smelling gas, used widely to manufacture building materials, insulation, and other household products. In most homes, the significant sources of formaldehyde are urethane-formaldehyde foam insulation and pressed wood products, such as particle board and plywood paneling. Emissions of formaldehyde from new products can be greatly affected by indoor temperatures and ventilation. Exposure to formaldehyde can cause watery eyes, burning sensations in the

nose and throat, breathing difficulties, and nausea. It can also trigger attacks in individuals suffering from asthma. More than 50,000 people who lived in mobile homes provided by the federal government following hurricanes Katrina and Rita in 2005 were eligible for payments totaling more than $40 million following a class-action lawsuit related to formaldehyde emissions.

Another polluter of our indoor air is *asbestos,* a mineral fiber once used in insulation and as a fire retardant. Manufacturers in the United States have greatly reduced the use of asbestos; it is banned in many nations, but still used for construction in some developing countries. In the United States, much asbestos still remains in furnace and pipe insulation, texturing materials, and floor tiles of older buildings. The most lethal fibers of asbestos are invisible. When inhaled, these tiny particles can accumulate and remain deep in the lungs for extended periods of time, where they damage tissue and potentially cause cancer or *asbestosis,* a permanent scarring of the lung that can be fatal.

Smoking tobacco indoors can also create an extremely dangerous health situation. Environmental tobacco smoke is a complex mixture of more than 4700 different compounds. These pollutants enter the body as particles and as gases, such as carbon monoxide and hydrogen cyanide. Exposure to tobacco smoke greatly increases the risk that smokers will develop lung cancer in smokers, and the risk is increased for nonsmokers as well. The small children of smokers are also more likely to fall victim to such illnesses as bronchitis and pneumonia. Heart disease is closely linked with exposure to tobacco smoke, as is premature aging of the skin.

In summary, indoor pollutants are responsible for a wide variety of health problems. Irritation of the eyes, nose, and throat, headaches, fatigue, and dizziness are but a few of the maladies attributed to indoor air pollutants. While these symptoms can be annoying, other life-threatening diseases can occur after prolonged exposure to many of the substances mentioned earlier, such as radon, asbestos, and other toxic compounds.

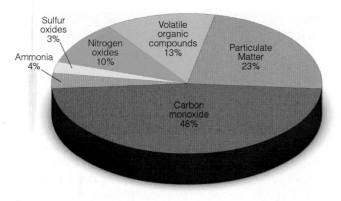

● **FIGURE 19.4** Estimates of emissions of the primary air pollutants in the United States on a per-weight basis as of 2013. (Data courtesy of United States Environmental Protection Agency.)

PRINCIPAL AIR POLLUTANTS The term **particulate matter** represents a group of solid particles and liquid droplets that are small enough to remain suspended in the air. Collectively known as *aerosols*, this grouping includes solid particles that may irritate people but are usually not poisonous, such as soot (tiny solid carbon particles), dust, smoke, and pollen. Some of the more dangerous substances include asbestos fibers and arsenic. Tiny liquid droplets of sulfuric acid, polychlorinated biphenyls (PCBs), oil, and various pesticides are also placed into this category.

Because it often dramatically reduces visibility in urban environments, particulate matter pollution is the most noticeable (see ● Fig. 19.5). Some particulate matter observed in urban air includes iron, copper, nickel, and lead. This type of pollution can immediately influence the human respiratory system. Once inside the lungs, it can make breathing difficult, particularly for those suffering from chronic respiratory disorders. Lead particles

are especially dangerous, as they tend to fall out of the atmosphere and become absorbed into the body through ingestion of contaminated food and water supplies. Lead accumulates in bone and soft tissues, and in high concentrations it can cause brain damage, convulsions, and death. Even at low doses, lead can be particularly dangerous to fetuses, infants, and children who, when exposed, may suffer central nervous system damage. Fortunately, lead emissions in the United States have been virtually eliminated (see Fig. 19.13 on p. 550). Particulate pollution, however, remains a serious problem. Researchers have found that particulate pollution poses a major risk not only to our respiratory systems but also to our cardiovascular systems. The smallest particulates—those less than 2.5 microns (0.0001 in.) in diameter—are tiny enough to pass from the lungs into the bloodstream, where they can increase the risk of heart attack, arrhythmia, congestive heart failure, and stroke. One study estimated that particulate pollution may be responsible for as many as 10,000 heart disease fatalities per year in the United States. Recent estimates are that more than half of the global deaths associated with outdoor air pollution are actually caused by cardiovascular problems, not respiratory ailments.

Roughly 18 million metric tons of particulate matter are emitted over the United States each year. One major problem is that particulate pollution may remain in the atmosphere for some time depending on the size and the amount of precipitation that occurs. Larger, heavier particles with diameters greater than about 10 μm* (0.01 mm) tend to settle to the ground in about a day or so after being emitted. Finer, lighter particles with diameters less than 1 μm (0.001 mm) can remain suspended in the lower atmosphere for several weeks.

---

*Recall that 1 micrometer (μm) is one-millionth of a meter. (The thickness of one page in the printed version of this book is about 100 micrometers.)

● **FIGURE 19.5** A thick layer of particulate matter (mostly smoke) and haze covers Santiago, Chile.

Particles with diameters smaller than 10 μm are referred to as PM_{10}. These particles pose the greatest health risk, as they are small enough to penetrate the lung's natural defense mechanisms. Moreover, winds can carry these fine particles great distances before they finally reach the surface. In fact, suspended particles from locations in Europe and Russia are believed to be responsible for the brownish cloud layer called *Arctic haze* that forms over the Arctic each spring. Strong winds over northern China can pick up dust particles and sweep them eastward, where they may settle on North America. This *Asian dust*, which has increased dramatically in recent decades, can reduce visibility, produce spectacular sunrises and sunsets, and coat everything with a thin veneer of particles (see ● Fig. 19.6).

Studies show that particulate matter with diameters less than 2.5 μm, called $PM_{2.5}$, are especially dangerous. For one thing, they can penetrate farther into the lungs. Moreover, these tiny particles frequently consist of toxic or carcinogenic (cancer-causing) combustion products. One major concern is the $PM_{2.5}$ particles found in diesel soot. Relatively high amounts of these particles have been measured inside school buses with higher amounts observed downwind of traffic corridors and truck terminals. Another important source of tiny particulates is wildfire. More than 10 percent of the total $PM_{2.5}$ emissions across the United States in recent years has been attributed to emissions from wildland fires.

Rain and snow remove many of these particles from the air; even the minute particles are removed by ice crystals and cloud droplets. In fact, numerical simulations of air pollution suggest that the predominant removal mechanism occurs when these particles act as nuclei for cloud droplets and ice crystals. Moreover, a long-lasting accumulation of suspended particles (especially those rich in sulfur) is not only aesthetically unappealing, but it also has the potential for affecting the climate, as some particles reflect incoming sunlight and prevent a portion of the sun's energy from reaching the surface. Thus, as we saw in Chapter 17, these *sulfur-rich aerosols* have the net effect of cooling Earth's surface.

Many suspended particles are hygroscopic, meaning that water vapor readily condenses onto them. As a thin film of water forms on the particles, they grow in size. When they reach a diameter between 0.1 and 1.0 μm, these *wet haze* particles effectively scatter incoming sunlight to give the sky a milky white appearance. The particles are usually sulfate or nitrate particulate matter from combustion processes, such as those produced by diesel engines and power plants. The hazy air mass may become quite thick, and on humid summer days it often becomes well defined, as illustrated in ● Fig. 19.7.

Carbon monoxide (CO), a major pollutant of city air, is a colorless, odorless, poisonous gas that forms during the incomplete combustion of carbon-containing fuels. As we saw earlier, carbon monoxide is the most plentiful of the primary pollutants (see Fig. 19.4).

The EPA estimates that more than 60 million metric tons of carbon monoxide enter the air annually over the United States through human activity alone, with about one-third of that coming from highway vehicles. Due to stricter air quality standards and the use of emission-control devices, carbon monoxide levels unrelated to wildfire decreased by about 50 percent from the early 1970s to the early 2000s, and they have dropped by more than 30 percent since then.

It is fortunate that carbon monoxide is quickly removed from the atmosphere by microorganisms in the soil, because even in small amounts, this gas is dangerous. Hence, it poses a serious problem in poorly ventilated areas, such as highway tunnels and underground parking garages. Because carbon monoxide cannot

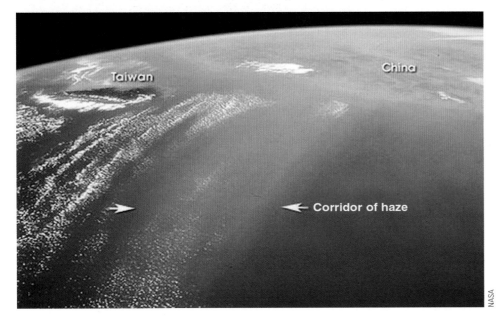

● **FIGURE 19.6** A thick haze about 200 km wide and about 600 km long covers a portion of the East China Sea on March 4, 1996. The haze is probably a mixture of industrial air pollution, dust, and smoke.

● **FIGURE 19.7** Cumulus clouds and a thunderstorm rise above the thick layer of haze that frequently covers the eastern half of the United States on humid summer days.

be seen or smelled, it can kill without warning. Here's how: Normally, your cells obtain oxygen through a blood pigment called *hemoglobin,* which picks up oxygen from the lungs, combines with it, and carries it throughout your body. Unfortunately, human hemoglobin is more receptive to carbon monoxide than to oxygen, so if there is too much carbon monoxide in the air you breathe, your brain will soon be starved of oxygen, and headache, fatigue, drowsiness, and even death may result.*

Sulfur dioxide (SO_2) is a colorless gas that comes primarily from the burning of sulfur-containing fossil fuels (such as coal and oil). Its primary sources include power plants, heating devices, smelters, petroleum refineries, and paper mills. However, it can also enter the atmosphere naturally during volcanic eruptions and as sulfate particles from ocean spray.

Sulfur dioxide readily oxidizes (combines with oxygen) to form the secondary pollutants *sulfur trioxide* (SO_3) and, in moist air, highly corrosive *sulfuric acid* (H_2SO_4). Winds can carry these particles great distances before they reach Earth as undesirable contaminants. When inhaled into the lungs, high concentrations of sulfur dioxide aggravate respiratory problems, such as asthma, bronchitis, and emphysema. Sulfur dioxide in large quantities can cause injury to certain plants, such as lettuce and spinach, sometimes producing bleached marks on their leaves and reducing their yield.

Volatile organic compounds (VOCs) represent a class of organic compounds that are mainly **hydrocarbons**—individual organic compounds composed of hydrogen and carbon. At room temperature they occur as solids, liquids, and gases. Although thousands of such compounds are known to exist, methane (which occurs naturally and poses no known dangers to health) is the most abundant. Other volatile organic compounds include benzene, formaldehyde, and some chlorofluorocarbons. The Environmental Protection Agency estimates that more than 10 million metric tons of VOCs are emitted into the air over the United States each year, with about 15 percent of the total coming

from vehicles used for transportation and about 20 percent from wildfires. The nation's VOC levels unrelated to wildfire dropped by roughly 55 percent from 1980 to 2016.

Certain VOCs, such as benzene (an industrial solvent) and benzo-a-pyrene (a product of burning wood, smoking, and barbecuing), are known to be carcinogens—cancer-causing agents. Although many VOCs are not intrinsically harmful, some will react with nitrogen oxides in the presence of sunlight to produce secondary pollutants, which are harmful to human health.

Nitrogen oxides are gases that form when some of the nitrogen in the air reacts with oxygen during the high-temperature combustion of fuel. The two primary nitrogen pollutants are **nitrogen dioxide** (NO_2) and **nitric oxide (NO)**, which, together, are commonly referred to as NO_x—or simply, *oxides of nitrogen.*

Although both nitric oxide and nitrogen dioxide are produced by natural bacterial action, their concentration in urban environments is between 10 and 100 times greater than in nonurban areas. In moist air, nitrogen dioxide reacts with water vapor to form corrosive nitric acid (HNO_3), a substance that adds to the problem of acid rain, which we will address later.

The primary sources of nitrogen oxides are motor vehicles, power plants, and waste disposal systems. High concentrations are believed to contribute to heart and lung problems, as well as to lowering the body's resistance to respiratory infections. Studies on test animals suggest that nitrogen oxides may encourage the spread of cancer. Moreover, nitrogen oxides are highly reactive gases that play a key role in producing ozone and other ingredients of photochemical smog.

As science evolves and health standards improve, the range of substances that are considered to be pollutants continues to grow. In 2007, the U.S. Supreme Court ruled that the greenhouse gas carbon dioxide (CO_2) is a pollutant covered by the Clean Air Act. Because of this ruling, the Environmental Protection Agency, in 2010, began to regulate CO_2 as a pollutant on the premise that, as a greenhouse gas, CO_2 causes a risk to public health.

OZONE IN THE TROPOSPHERE   As mentioned earlier, the word **smog** originally meant the combining of smoke and fog. Today, however, the word mainly refers to the type of smog that forms in large cities, such as Los Angeles. Because this type of smog forms when chemical reactions take place in the presence of sunlight (called *photochemical reactions*), it is termed **photochemical smog**, sometimes called *Los Angeles–type smog.* When the smog is composed of sulfurous smoke and foggy air, it is sometimes called *London-type smog.*

The main component of photochemical smog is the gas **ozone** (O_3). Ozone is an invisible but noxious substance with an unpleasant odor that irritates eyes and the mucous membranes of the respiratory system, aggravating chronic diseases, such as asthma and bronchitis. Even in healthy people, exposure to relatively low concentrations of ozone for six or seven hours during periods of moderate exercise can significantly reduce lung function. This situation often is accompanied by symptoms such as chest pain, nausea, coughing, and pulmonary congestion. Ozone also attacks rubber, retards tree growth, and damages crops. Each year, in the United States alone, ozone is responsible for crop yield losses of several billion dollars.

*Should you become trapped in your car during a snowstorm, and you have your engine and heater running to keep warm, roll down the window just a little. This action will allow the escape of any carbon monoxide that may have entered the car through leaks in the exhaust system.

We will see later that ozone forms naturally in the stratosphere through the combining of molecular oxygen and atomic oxygen. There, *stratospheric ozone* provides a protective shield against the sun's harmful ultraviolet rays. However, near the surface, in polluted air, ozone—often referred to as *tropospheric* (or *ground-level*) *ozone*—is a secondary pollutant that is not emitted directly into the air. Rather, it forms from a complex series of chemical reactions involving other pollutants, such as nitrogen oxides and volatile organic compounds (hydrocarbons). Because sunlight is required to produce ozone, concentrations of tropospheric ozone are normally higher during the afternoons (see ● Fig. 19.8) and during the summer months, when sunlight is more intense.

In polluted air, ozone production occurs along the following lines. Sunlight (with wavelengths shorter than about 0.41 μm) dissociates nitrogen dioxide into nitric oxide and atomic oxygen, which may be expressed by

$$NO_2 + \text{solar radiation} \rightarrow NO + O$$

The atomic oxygen then combines with molecular oxygen (in the presence of a third molecule, M), to form ozone, as

$$O_2 + O + m \rightarrow O_3 + M$$

The ozone is then destroyed by combining with nitric oxide; thus

$$O_3 + NO \rightarrow NO_2 + O_2$$

If sunlight is present, however, the newly formed nitrogen dioxide will break down into nitric oxide and atomic oxygen. The atomic oxygen then combines with molecular oxygen to form ozone again. Consequently, large concentrations of ozone can form in polluted air only if some of the nitric oxide reacts with other gases *without removing ozone in the process*. Under these conditions, certain hydrocarbons (emitted by autos and industrial sources) and the hydroxyl radical come into play.

The formation of the hydroxyl radical (OH) begins when ultraviolet radiation (at wavelengths of about 0.31 μm and below) dissociates some of the ozone into molecular oxygen and atomic oxygen; accordingly

$$O_3 + \text{UV radiation} \rightarrow O_2 + O$$

The atomic oxygen formed is in an excited state, which means it can react with a variety of other molecules, including water vapor, to produce two hydroxyl radical molecules; thus

$$O + H_2O \rightarrow OH + OH$$

The OH is called a "radical" because it contains an unpaired electron. This situation allows the OH molecule to react with many other atoms and molecules, including unburned or partially burned hydrocarbons (RH) released into the air by automobiles and industry, as

$$OH + RH \rightarrow R^{\cdot} + H_2O$$

The product $R^{\cdot}$ represents an organic hydrocarbon that can have a complex molecular structure. The $R^{\cdot}$ is then able to react with molecular oxygen to form $RO_2^{\cdot}$, a reactive molecule that removes nitric oxide by combining with it to form nitrogen dioxide, shown by the expression

$$RO_2^{\cdot} + NO \rightarrow NO_2 + \text{other products}$$

In this manner, nitric oxide can react with hydrocarbons to form nitrogen dioxide *without removing ozone*. Hence, the reactive hydrocarbons in polluted air allow ozone concentrations to increase by preventing nitric oxide from destroying the ozone as rapidly as it is formed.

The hydrocarbons (VOCs) also react with oxygen and nitrogen dioxide to produce other undesirable contaminants, such as *PAN* (peroxyacetyl nitrate)—a pollutant that irritates eyes and is extremely harmful to vegetation—and organic compounds. Ozone, PAN, and small amounts of other oxidating pollutants are the ingredients of photochemical smog. Instead of being specified individually, these pollutants are sometimes grouped under a single heading called *photochemical oxidants.**

In addition to the human-related sources discussed earlier, many hydrocarbons (VOCs) also occur naturally in the atmosphere, as they are given off by vegetation. Oxides of nitrogen drifting downwind from urban areas can react with these natural hydrocarbons and produce smog in relatively uninhabited areas. This phenomenon has been observed downwind of cities such as Los Angeles, London, and New York. Some regions have so much natural (background) hydrocarbon that it may be difficult to reduce ozone levels as much as desired. As a whole, though, ozone pollution has steadily decreased in recent decades across the United States, including most of the nation's large cities. When averaged across more than 200 sites nationwide, ozone concentrations on the most polluted days dropped by 32 percent between 1980 and 2015.

Up to now, we have concentrated on ozone in the troposphere, primarily in a polluted environment. The next section examines the formation and destruction of ozone in the upper atmosphere—in the stratosphere.

OZONE IN THE STRATOSPHERE   Recall from Chapter 1 that the stratosphere is a region of the atmosphere that lies above the troposphere between about 10 and 50 km (6 and 31 mi) above

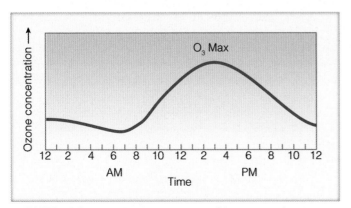

● **FIGURE 19.8** Average hourly concentrations of ozone measured at six major cities over a two-year period.

---

*An *oxidant* is a substance (such as ozone) whose oxygen combines chemically with another substance.

Earth's surface. The atmosphere is stable in the stratosphere because of a strong temperature inversion—a situation in which the air temperature increases rapidly with height (look back at Fig. 1.11, p. 13). The inversion is due, in part, to the gas ozone, which absorbs ultraviolet radiation at wavelengths of less than about 0.3 mm.

In the stratosphere, above middle latitudes, notice in ● Fig. 19.9 that ozone as a fraction of the entire atmosphere is most dense at an altitude near 25 km. Even at this altitude, its concentration is quite small, as there are only about 12 ozone molecules for every million air molecules (12 ppm).* Although thin, this layer of ozone is significant, for it shields Earth's inhabitants from harmful amounts of ultraviolet solar radiation, which at wavelengths below 0.3 μm has enough energy to cause skin cancer in humans. Also, UV radiation at 0.26 μm can destroy acids in DNA (deoxyribonucleic acid), the substance that transmits the hereditary blueprint from one generation to the next.

In recent decades, much attention has been placed on reducing the depletion of stratospheric ozone because of the many risks such a loss would pose. These include:

● an increase in the number of cases of skin cancer,

● a sharp increase in eye cataracts and sunburns,

---

*With a concentration of ozone of only 12 parts per million in the stratosphere, the composition of air here is about the same as it is near Earth's surface—mainly 78 percent nitrogen and 21 percent oxygen.

● suppression of the human immune system,

● an adverse impact on crops and animals due to an increase in ultraviolet radiation,

● a reduction in the growth of ocean phytoplankton, and

● a cooling of the stratosphere that could alter stratospheric wind patterns, possibly affecting the destruction of ozone.

Stratospheric Ozone: Production-Destruction   Ozone (O_3) forms naturally in the stratosphere when atomic oxygen (O) combines with molecular oxygen (O_2) in the presence of another molecule. Although it forms mainly above 25 km, ozone gradually drifts downward by mixing processes, producing a peak concentration in middle latitudes near 25 km. (In polar regions, its maximum concentration is found at lower levels.) Ozone is broken down into molecular and atomic oxygen when it absorbs ultraviolet (UV) radiation with wavelengths between 0.2 and 0.3 mm (see ● Fig. 19.10). Thus,

$$O_3 + UV \rightarrow O_2 + O$$

Ozone is destroyed through collisions with other atoms and molecules. For example, ozone and atomic oxygen combine to form two oxygen molecules, as

$$O_3 + O \rightarrow 2O_2$$

Likewise, the combination of two ozone molecules destroys ozone, as

$$O_3 + O_3 \rightarrow 3O_2$$

These equations also represent the net result of a number of complex chemical reactions that include trace gases of nitrogen,

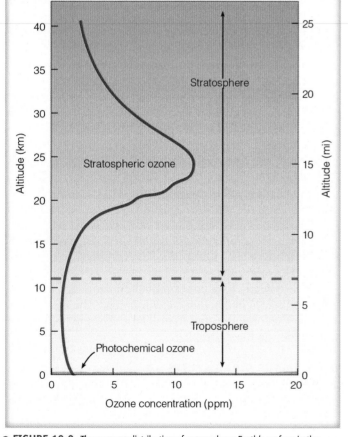

● **FIGURE 19.9**   The average distribution of ozone above Earth's surface in the middle latitudes.

● **FIGURE 19.10**   An ozone molecule absorbing ultraviolet radiation can become molecular oxygen (O_2) and atomic oxygen (O).

hydrogen, and chlorine. For example, two natural destructive gases for ozone are nitric oxide and nitrogen dioxide, which, as we have seen, are collectively known as *oxides of nitrogen.* The origin of these gases begins at Earth's surface as soil bacteria produce N_2O (nitrous oxide). This gas gradually finds its way into the stratosphere where, above about 25 km, solar energy converts some of it into ozone-destroying oxides of nitrogen. In the stratosphere, just a small amount of nitric oxide can destroy a large amount of ozone. The following sequence of chemical reactions will illustrate why. In step 1, the nitric oxide quickly combines with ozone to form nitrogen dioxide and molecular oxygen. Then, in step 2, the nitrogen dioxide combines with atomic oxygen to form nitric oxide and molecular oxygen. Thus

Step 1 $NO + O_3 \rightarrow NO_2 + O_2$

Step 2 $NO_2 + O \rightarrow NO + O_2$

The nitric oxide (NO) released in step 2 is now ready to start destroying ozone again.

Stratospheric ozone is maintained by a delicate natural balance between production and destruction. Could this balance be upset?

Stratospheric Ozone: Upsetting the Balance The concentration of ozone in the stratosphere may be changed by natural events. In the upper atmosphere, both cosmic rays* and solar particles can produce secondary electrons having sufficient energy to separate molecular nitrogen (N_2) into two nitrogen atoms (N). The nitrogen atoms combine with free atomic oxygen to form nitric oxide which, in turn, rapidly destroys ozone. Furthermore, large volcanic eruptions rich in sulfur, such as the Philippines' Mt. Pinatubo in June 1991, produce sulfate aerosols that act as a catalyst to help chlorine destroy ozone. And scientists, using measurements from satellites, discovered that even changes in ultraviolet radiation from the sun can cause small variations in the amount of stratospheric ozone.

The idea that human activities could alter the amount of stratospheric ozone was first explored in the early 1970s, at a time when Congress was pondering whether or not the United States should build a supersonic jet transport similar to the French-British Concorde. Among the gases emitted from the engines of this type of aircraft are nitrogen oxides. Although the U.S. supersonic aircraft was being designed to fly in the stratosphere below the level of maximum ozone, it was feared that the nitrogen oxides would eventually have an adverse effect on the ozone. This factor was one of many considered when Congress decided to halt the development of the U.S. version of the supersonic transport in 1971.

Concerns about ozone loss gained more prominence later in the 1970s, as scientists found that *chlorofluorocarbons* (CFCs) posed a serious threat to stratospheric ozone. At the time, chlorofluorocarbons were the most widely used propellants in spray cans, such as deodorants and hairsprays. In the troposphere, these gases are quite safe, being nonflammable, nontoxic, and unable to chemically combine with other substances. Hence,

these gases slowly diffuse upward without being destroyed. They apparently enter the stratosphere

1. near breaks in the tropopause, especially in the vicinity of jet streams, and

2. in building thunderstorms, especially those that develop in the tropics along the Intertropical Convergence Zone and penetrate the lower stratosphere.

Once chlorofluorocarbon molecules reach the middle stratosphere, ultraviolet energy that is normally absorbed by ozone breaks them up, releasing atomic *chlorine* in the process. Note in the following sequence of reactions how chlorine destroys ozone rapidly. In step 1, atomic chlorine (Cl) combines with ozone, forming chlorine monoxide (ClO)—a new substance—and molecular oxygen (O_2). Almost immediately, the chlorine monoxide combines with free atomic oxygen (step 2) to produce chlorine atoms and molecular oxygen. The free chlorine atoms are now ready to combine with and destroy more ozone molecules. Estimates are that a single chlorine atom removes as many as 100,000 ozone molecules before it is taken out of action by combining with other substances.

Step 1 $Cl + O_3 \rightarrow ClO + O_2$
Step 2 $ClO + O \rightarrow Cl + ClO_2$

Fortunately, chlorine atoms do not exist in the stratosphere forever. They are removed as chlorine monoxide combines with nitrogen dioxide to form chlorine nitrate, $ClONO_2$.

$$ClO + NO_2 \rightarrow ClONO_2$$

Free chlorine atoms may be removed by combining with methane to form hydrogen chloride (HCl) and a new substance, CH_3.

$$CH_4 + Cl \rightarrow HCl + CH_3$$

Since the average lifetime of a CFC molecule is about 50 to 100 years, any increase in the concentration of CFCs is long-lasting and a genuine threat to the concentration of ozone.

Regulation of CFCs and the Montreal Protocol In the late 1970s, the use of CFCs in aerosol cans was banned in the United States and a number of other countries. However, CFCs were still allowed to be used in refrigerators and air-conditioning units. It soon became apparent that CFCs continued to pose a threat to stratospheric ozone, especially with the discovery in the mid-1980s of major ozone depletion during the springtime above Antarctica, where several factors combine to make severe ozone loss possible. This sharp drop in ozone is known as the **ozone hole**. (More information on the ozone hole is provided in Focus section 19.2.)

The discovery of the ozone hole added urgency to the development of an international agreement called the *Montreal Protocol*, which was signed in 1987. This agreement, the first in United Nations history to be ratified by every member, established a timetable for diminishing CFC emissions and the use of bromine compounds (halons), which destroy ozone at a rate 50 times greater than do chlorine compounds.*

---

*Cosmic rays are high-energy atomic nuclei and atomic particles that travel through space at extremely high speeds. Most cosmic rays are believed to come from exploding stars (hypernovae), although some are produced in solar flares.

*There are many chemical reactions that involve chlorine and bromine and the destruction of ozone in the stratosphere. The example given so far is just one of them.

The Ozone Hole

In 1974, two chemists from the University of California, Irvine—F. Sherwood Rowland and Mario J. Molina—warned that increasing levels of CFCs would eventually deplete stratospheric ozone on a global scale. Their studies suggested that ozone depletion would occur gradually and would perhaps not be detectable for many years to come. It was surprising, then, when in 1985 British researchers identified a year-to-year decline in stratospheric ozone over Antarctica. Their findings, corroborated later by satellites and balloon-borne instruments, showed that since the late 1970s ozone concentrations had diminished each year during the months of September and October. This decrease in stratospheric ozone over springtime Antarctica is known as the *ozone hole*. In years of severe depletion, such as in 2006, the ozone hole covers almost twice the area of the Antarctic continent (see ● Fig. 2).

To understand the causes behind the ozone hole, scientists in 1986 organized the first *National Ozone Expedition*, NOZE-1, which set up a fully instrumented observing station near McMurdo Sound, Antarctica. During 1987, with the aid of instrumented aircraft, NOZE-2 got under way. The findings from these research programs helped scientists put together the pieces of the ozone puzzle.

The stratosphere above Antarctica has one of the world's highest ozone concentrations. Most of this ozone forms over the tropics and is brought to the Antarctic by stratospheric winds. During September and October (spring in the Southern Hemisphere), a belt of stratospheric winds called the *polar vortex* encircles the Antarctic region near 66°S latitude, essentially isolating the cold Antarctic stratospheric air from the warmer air of the middle latitudes. During the long, dark Antarctic winter, temperatures inside the vortex can drop to −85°C (−121°F). This frigid air allows for the formation of *polar*

● **FIGURE 2** Ozone distribution over the South Pole on September 11, 2014, as measured by ozone monitoring equipment on NASA's *Aura* satellite. Notice that the lowest ozone concentration or ozone hole (purple and blue shades) covers most of Antarctica. The color scale on the bottom of the image shows total ozone values in Dobson units (DU). A Dobson unit is the physical thickness of the ozone layer if it were brought to Earth's surface (500 DU equals 5 mm).

0 100 200 300 400 500 600 700
Total Ozone (Dobson units)

NASA

stratospheric clouds. These ice clouds are critical in facilitating chemical interactions among nitrogen, hydrogen, and chlorine atoms, the end product of which is the destruction of ozone.

In 1986, the NOZE-1 study detected unusually high levels of chlorine compounds in the stratosphere, and, in 1987, the instrumented aircraft of NOZE-2 measured enormous increases in chlorine compounds when it entered the polar vortex. These findings, in conjunction with other chemical discoveries, allowed scientists to pinpoint *chlorine* from CFCs as the main cause of the ozone hole.

Even with a decline in ozone-destroying chemicals, the largest Antarctic ozone hole observed to date occurred during September

2006 (Fig. 2), and the year 2015 brought the fourth-largest ozone hole on record. Yearly variations in the size and depth of the ozone hole are attributed mainly to changes in polar stratospheric temperatures.

Because ozone protects Earth's inhabitants from the sun's harmful UV rays, the springtime drop in ozone over the Southern Hemisphere is serious enough that government agencies in New Zealand and Australia warn their citizens to protect themselves from the sun's rays when the ozone hole forms each year.

In the Northern Hemisphere's polar Arctic, airborne instruments and satellites during the late 1980s and 1990s measured high levels of ozone-destroying chlorine compounds in the stratosphere. But observations could not detect an ozone hole like the one that forms over the Antarctic. (Compare Fig. 2 with Fig. 19.12, p. 549.)

Why would the Arctic not have an ozone hole like its southern counterpart? For one thing, the circulation of stratospheric air over the Arctic (an ocean surrounded by continents) differs from that over the Antarctic (a continent ringed by oceans). The Arctic stratosphere is normally too warm for widespread development of the polar stratospheric clouds that facilitate ozone loss. However, in 1997 and again in 2011, a very cold Arctic stratosphere teamed up with ozone-destroying chemicals to cause significant ozone depletion.

During the past 40 years or so, we have learned much about stratospheric ozone and the ozone hole. Ozone-depleting substances are now regulated, and emissions of these substances are essentially zero. The ozone hole is still there—stronger in some years, weaker in others. Based on current knowledge and trends, mid-latitude and Arctic ozone depletion is expected to end by the middle of this century, with the Antarctic ozone hole virtually extinct by later in the century.

One sign of the success of the Montreal Protocol is the decrease in atmospheric concentrations of most ozone-depleting gases since the 1990s. (See ● Fig. 19.11.) Although the use of CFCs has decreased by more than 95 percent, there are still millions of kilograms in the troposphere that will continue to slowly diffuse upward, and thus we will continue to see signs of ozone depletion for years to come.

A United Nations assessment in 2014 found that the total amount of atmospheric ozone was relatively unchanged since 2000 outside polar regions, after having declined by about 2.5 percent during the 1980s and early 1990s. Above the United States, stratospheric ozone concentrations have been running about 3 to 5 percent below their pre-1980 averages. Indications are that ozone concentrations in the stratosphere will return to

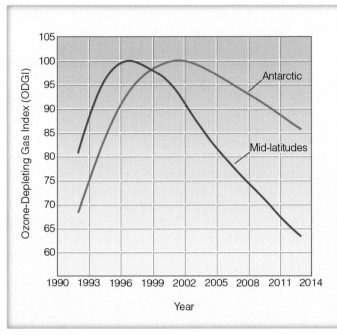

● **FIGURE 19.11** The Ozone Depleting Gas Index in the stratosphere for the Antarctic (green) and mid-latitudes (blue). The index was created to show how the concentrations of ozone-depleting gases (such as CFCs) have changed in the stratosphere since the early 1990s. A value of 100 represents the highest concentration of ozone-depleting gases, whereas a value of zero represents the point at which ozone-depleting gases will have returned to 1980 levels, and the concentration of ozone will have returned to near normal levels. (Data from NOAA)

● **FIGURE 19.12** Ozone distribution over the Arctic for March 19, 2011, as measured by the *Ozone Monitoring Instrument* (OMI) on NASA's *Aura* satellite. Total ozone amounts are in Dobson units (DU). A Dobson unit is the physical thickness of the ozone layer if it were brought to Earth's surface (500 DU equals 5 mm). Notice low readings over the Arctic (blue shade) and that these readings are not nearly as low as readings over the Antarctic as shown in Fig. 2, p. 548. (Data courtesy of Ozone Hole Watch)

CRITICAL THINKING QUESTION The ozone hole forms during September and October over the Antarctic. Why, in the Northern Hemisphere, does the "ozone hole" form over the Arctic during the months of March and April?

pre-1980 levels before the mid-twenty-first century, except over Antarctica, where the ozone concentrations will likely remain low until about 2070.

Natural processes can sometimes exacerbate the impact of ozone-depleting gases for periods of time. For example, satellite measurements in 1992 and 1993 revealed that stratospheric ozone concentrations had dropped to record low levels over much of the globe. The decrease appears to have stemmed not only from ozone-destroying chemicals but also from by the 1991 volcanic eruption of Mt. Pinatubo, which sent tons of sulfur dioxide gas into the stratosphere, where it formed tiny droplets of sulfuric acid. These droplets not only enhanced the ozone destructiveness of the human-produced chlorine chemicals but also altered the circulation of air in the stratosphere, making it more favorable for ozone depletion. A different process played out in 1997, when springtime ozone levels over the Arctic dropped dramatically. This decrease apparently was due to the interaction of ozone-destroying pollution with cold stratospheric weather patterns that favored ozone reduction.

During the early part of this century, springtime ozone concentrations over the Arctic have varied greatly from year to year. For example, 2010 had relatively high stratospheric ozone levels, whereas in 2011 ozone levels approached the lowest readings ever observed there (see ● Fig. 19.12). Although the ozone layer is in a long-term state of recovery, it remains vulnerable to large annual variations, primarily because of the year-to-year influence of stratospheric temperatures on the effects of ozone-destroying chemicals.

There are now two major substitutes for CFCs: *hydrochlorofluorocarbons* (HCFCs) and *hydrofluorocarbons* (HFCs). The HCFCs contain fewer chlorine atoms per molecule than CFCs and, therefore, pose much less danger to the ozone layer, whereas HFCs contain no chlorine at all. However, like CFCs, these two groups of substitutes are also greenhouse gases that can enhance global warming. Because of this, the Montreal Protocol now calls for HCFCs to be phased out by 2030, and an amendment to the protocol in 2016 will phase out HFCs over the next several decades. (Reducing the risk of climate change by reducing the amount of greenhouse gas added to the atmosphere has proven to be an important and beneficial side effect of the Montreal Protocol.)

AIR POLLUTION: TRENDS AND PATTERNS Over the past few decades, major strides have been made in the United States to improve the quality of the air we breathe. ● Figure 19.13 shows the estimated emission trends over the United States for the primary pollutants. Notice that since the Clean Air Act of 1970, emissions of most pollutants have fallen off substantially, with lead showing the greatest reduction, primarily due to the elimination of leaded gasoline.

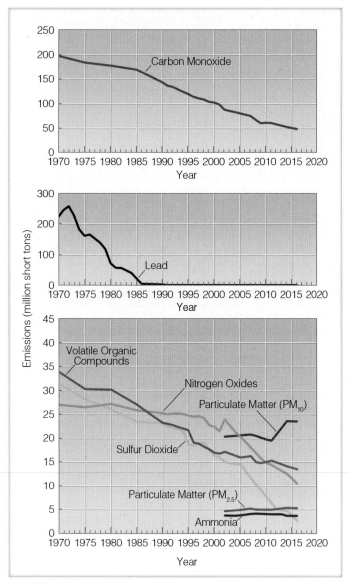

● **FIGURE 19.13** Emission estimates of six pollutants in the United States from 1970 through 2014. Emissions related to wildfires are not included. (Data courtesy of United States Environmental Protection Agency.)

WEATHER WATCH

Reduction in stratospheric ozone over North America has allowed ultraviolet (UV-B) radiation levels to increase. In fact, according to NASA scientists, the average amount of ultraviolet radiation reaching the surface was roughly 5 to 10 percent higher over the United States in 2009 than it was in 1979, with the greatest increases occurring at northern latitudes. The trend seems to have leveled off since the mid-1990s.

Although the situation has improved, we can see from Fig. 19.13 that much more needs to be done. Large quantities of pollutants are still spewed into our air. In fact, many urban areas of the United States do not conform to the standards for air quality set by the Clean Air Act of 1990. A large part of the problem of pollution control lies in the fact that even with stricter

emission laws, increasing amounts of vehicular travel and other pollution-generating activities can overwhelm control efforts. For example, vehicles in the United States drove more than 3.1 trillion miles in 2016, which was a new record total and was more than twice the amount driven in 1980. Both nationally and worldwide, an increasing fraction of the population is clustered in urban areas, which increases their potential exposure to outdoor air pollution.

Clean air standards for the United States are established by the EPA. *Primary ambient air quality standards* are set to protect human health, whereas *secondary standards* protect human welfare, as measured by the effects of air pollution on visibility, crops, and buildings. The National Ambient Air Quality Standards (NAAQS) currently in effect for six pollutants are given in ▼ Table 19.2. The table shows the concentration of each pollutant needed to exceed the NAAQS. Those areas that do not meet air quality standards are called *nonattainment areas*. The EPA also issues specific rules to address a variety of local, state, and regional air pollution concerns. For example, the Cross-State Air Pollution Rule restricts power plant emissions that are carried from one state into another. Even with stronger emission laws, estimates are that more than 160 million Americans are breathing air that at times does not meet at least one of the EPA primary standards (see ● Fig. 19.14).

To indicate the air quality in a particular region, the EPA developed the **air quality index (AQI)**. The index includes the pollutants carbon monoxide, sulfur dioxide, nitrogen dioxide, particulate matter, and ozone. On any given day, the pollutant

▼ **TABLE 19.2   The National Ambient Air Quality Standards**

POLLUTANT	AVERAGING PERIOD	PRIMARY NAAQS	SECONDARY NAAQS
Ozone (O_3)	8-hour	0.070 ppm	—
Carbon monoxide (CO)	1-hour	35 ppm	—
	8-hour	9 ppm	—
Sulfur dioxide (SO_2)	1-hour	0.075 ppm	—
	3-hour	—	0.5 ppm
Nitrogen dioxide (NO_2)	1-hour	0.100 ppm	—
	Annual	0.053 ppm	0.053 ppm
Particle pollution, 10 μg or less (PM_{10})	24-hour	150 μg / m³	150 μg / m³
Particle pollution, 2.5 μg or less ($PM_{2.5}$)	24-hour	35 μg / m³	35 μg / m³
	Annual	12 μg / m³	15 μg / m³
Lead (Pb)	Rolling 3-month average	0.15 μg/m³	0.15 μg/m³

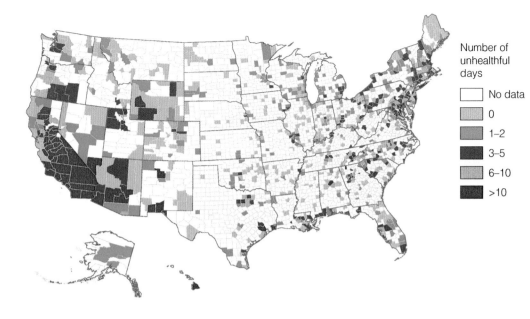

● FIGURE 19.14 The number of unhealthful days (by county) across the United States for any one of the five "criteria pollutants" (CO, SO₂, NO₂, O₃, and particulate matter) during 2013. (Data courtesy of United States Environmental Protection Agency.)

Number of unhealthful days

	No data
	0
	1–2
	3–5
	6–10
	>10

measuring the highest value is the one used in the index. The pollutant's measurement is then converted to a number that ranges from 0 to 500 (see ▼ Table 19.3). When the pollutant's value is the same as the primary ambient air quality standard, the pollutant is assigned an AQI number of 100. A pollutant is considered unhealthful when its AQI value exceeds 100. When the AQI value is between 51 and 100, the air quality is described as "moderate." Although these levels may not be harmful to humans during a 24-hour period, they may exceed long-term standards. Notice that the AQI is color-coded, with each color corresponding to an AQI level of health concern. The color green indicates "good" air quality; the color red, "unhealthy" air; and maroon, "hazardous"

▼ TABLE 19.3 **The Air Quality Index**

AQI VALUE	AIR QUALITY	GENERAL HEALTH EFFECTS	RECOMMENDED ACTIONS
0–50	Good	None	None
51–100	Moderate	There may be a moderate health concern for a very small number of individuals. People unusually sensitive to ozone may experience respiratory symptoms.	When O₃ AQI values are in this range, unusually sensitive people should consider limiting prolonged outdoor exposure.
101–150	Unhealthy for sensitive groups	Mild aggravation of symptoms in susceptible persons, including people with lung disease, older adults, and children.	Active people with respiratory or heart disease should limit prolonged outdoor exertion.
151–200	Unhealthy	Aggravation of symptoms in susceptible persons, with irritation symptoms in the healthy population.	Active children and adults with respiratory or heart disease should avoid extended outdoor activities; everyone else, especially children, should limit prolonged outdoor exertion.
201–300	Very unhealthy	Significant aggravation of symptoms and decreased exercise tolerance in persons with heart or lung disease, with widespread symptoms in the healthy population.	Active children and adults with existing heart or lung disease should avoid outdoor activities and exertion. Everyone else, especially children, should limit outdoor exertion.
301–500	Hazardous	Everyone may experience more serious health effects. Significant aggravation of symptoms. Premature onset of certain diseases. Premature death may occur in ill or elderly people.	Everyone should avoid all outdoor exertion and minimize physical outdoor activities. Older adults and persons with existing heart or lung disease should stay indoors.

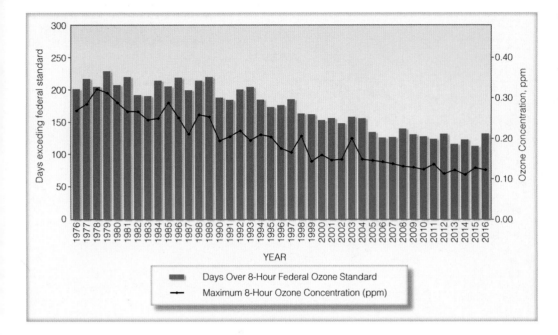

● **FIGURE 19.15** The number of days ozone exceeded the eight-hour federal standard established in 1997 (0.075 ppm) and maximum eight-hour ozone concentration (ppm) for Los Angeles and surrounding areas in the South Coast air basin. (Courtesy of South Coast Air Pollution District.)

air quality. Table 19.3 also shows the health effects and the precautions that should be taken when the AQI value reaches a certain level.

Higher emission standards, along with cleaner fuels (such as renewables and natural gas) and more energy-efficient buildings and vehicles, have made the air over our large cities cleaner today than it was years ago. In fact, total emissions of toxic chemicals spewed into the skies over the United States have been declining steadily since the EPA began its inventory of these chemicals in 1987. But particulates and ground-level ozone are still pervasive problems. Because ozone is a secondary pollutant, its formation is controlled by the concentrations of other pollutants, namely nitrogen oxides and hydrocarbons (VOCs). Moreover, weather conditions play a vital role in ozone formation, as ozone reaches its highest concentrations in hot, sunny weather when surface winds are light and a stagnant high-pressure area covers the region. As a result of these factors, year-to-year ozone trends are quite variable, although the Los Angeles area has shown a steady decline in the number of unhealthful days due to ozone (see ● Fig. 19.15). In fact, the year 2015 saw a record-low number of unhealthful days in the Los Angeles area for ozone as well as particulates.

BRIEF REVIEW

Before going on to the next several sections, here is a brief review of some of the important points presented so far.

- Near the surface, primary air pollutants (such as particulate matter, CO, SO_2, NO, NO_2, and VOCs) enter the atmosphere directly, whereas secondary air pollutants (such as O_3) form when a chemical reaction takes place between a primary pollutant and some other component of air.

- The word "smog" (coined in London in the early 1900s) originally meant the combining of smoke and fog. Today, the word mainly refers to photochemical smog—pollutants that form in the presence of sunlight.

- Stratospheric ozone forms naturally in the stratosphere and provides a protective shield against the sun's harmful ultraviolet rays. Tropospheric (ground-level) ozone, which forms in polluted air, is a health hazard and is the primary ingredient of photochemical smog.

- Human-induced chemicals, such as chlorofluorocarbons (CFCs), have been altering the amount of ozone in the stratosphere by releasing chlorine, which rapidly destroys ozone.

- Even though the emissions of most pollutants have declined across the United States since 1970, millions of Americans are breathing air that does not meet air quality standards.

Factors That Affect Air Pollution

If you live in a region that periodically experiences photochemical smog, you may have noticed that these episodes often occur with clear skies, light winds, and generally warm, sunny weather. Although this may be "typical" air pollution weather, it by no means represents the only weather conditions necessary to produce high concentrations of pollutants, as we will see in the following sections.

THE ROLE OF THE WIND Wind speed plays a role in diluting pollution. When vast quantities of pollutants are spewed into the air, the wind speed determines how quickly the pollutants mix with the surrounding air and, of course, how fast they move away from their source. Strong winds tend to lower the concentration of pollutants by spreading them apart as they move downstream. This process of spreading is called **dispersion**. Moreover, the stronger the wind, the more turbulent the air. Turbulent air produces swirling eddies that dilute the pollutants by mixing them with the cleaner surrounding air. Hence, when the wind dies down, pollutants are not readily dispersed, and they tend to become more concentrated (see ● Fig. 19.16).

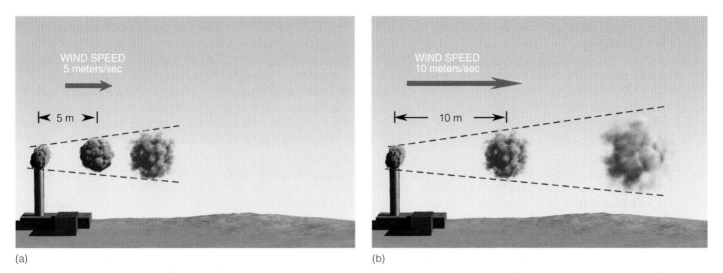

● **FIGURE 19.16** If each chimney emits a puff of smoke every second, then where the wind speed is low (a), the smoke puffs are closer together and more concentrated. Where the wind speed is greater (b), the smoke puffs are farther apart and more diluted as turbulent eddies mix the smoke with the surrounding air.

THE ROLE OF STABILITY AND INVERSIONS Recall from Chapter 6 that atmospheric stability determines the extent to which air will rise. Remember also that an unstable atmosphere favors vertical air currents, whereas a stable atmosphere strongly resists upward vertical motions. Smoke emitted into a stable atmosphere thus tends to spread horizontally rather than mixing vertically.

The stability of the atmosphere is determined by the way the air temperature changes with height (the lapse rate). When the measured air temperature decreases rapidly as we move up into the atmosphere, the atmosphere tends to be more unstable and pollutants tend to be mixed vertically as illustrated in ● Fig. 19.17a. If, however, the measured air temperature either decreases quite slowly as we ascend, or actually increases with height (remember that this is called an *inversion*), the atmosphere is stable. An inversion represents an extremely stable atmosphere where warm air lies above cool air (see Fig. 19.17b). Any air parcel that attempts to rise into the inversion will, at some point, be cooler and heavier (more dense) than the warmer air surrounding it. Hence, the inversion acts like a lid on vertical air motions.

The inversion depicted in Fig. 19.17b is called a **radiation** (or **surface) inversion**. This type of inversion typically forms during the night and early morning hours when the sky is clear and the winds are light. As we saw earlier in Chapter 3, radiation inversions also tend to be well developed during the long nights of winter.

In Figure 19.17b, notice that within the stable inversion, the smoke from the shorter stacks does not rise very high, but spreads out, contaminating the area around it. In the relatively unstable air above the inversion, smoke from the taller stack is able to rise and become dispersed. Since radiation inversions are often rather shallow, it should be apparent why taller chimneys have been replacing many shorter ones. In fact, taller chimneys disperse pollutants better than shorter ones even in the absence of a surface inversion, because the taller chimneys are able to mix pollutants throughout a greater volume of air. Although these

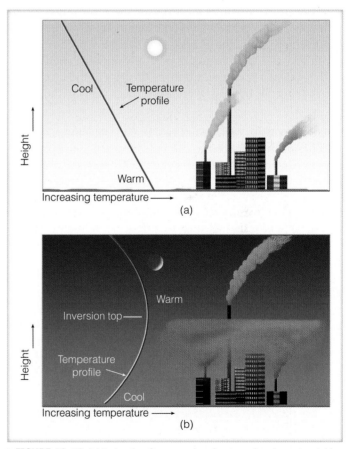

● **FIGURE 19.17** (a) During the afternoon, when the atmosphere is most unstable, pollutants rise, mix, and disperse downwind. (b) At night when a radiation inversion exists, pollutants from the shorter stacks are trapped within the inversion, while pollutants from the taller stack, above the inversion, are able to rise and disperse downwind.

taller stacks do improve the air quality in their immediate area, they may also contribute to the acid rain problem by allowing the pollutants to be swept great distances downwind.

As the sun rises and the surface warms, the radiation inversion normally weakens and disappears before noon.

By afternoon, the atmosphere is sufficiently unstable so that, with adequate winds, pollutants are able to disperse vertically (Fig. 19.17b). The changing atmospheric stability, from stable in the early morning to conditionally unstable in the afternoon, can have a profound effect on the daily concentrations of pollution in certain regions. For example, on a busy city street corner, carbon monoxide levels can be considerably higher in the early morning than in the early afternoon (with the same flow of traffic). Changes in atmospheric stability can also cause smoke plumes from chimneys to change during the course of a day. (Some of these changes are described in Focus section 19.3.)

Radiation inversions normally last just a few hours. On the other hand, **subsidence inversions** may persist for several days or longer, so these are the inversions most commonly associated with major air pollution episodes. They form as the air above a deep anticyclone slowly sinks (subsides) and warms.*

A typical temperature profile of a subsidence inversion that forms along the California coast in summer is shown in Fig. 19.18. Notice that in the relatively unstable air beneath the inversion, the pollutants are able to mix vertically up to the inversion base. The stable air of the inversion, however, inhibits vertical mixing and acts like a lid on the pollution below, preventing it from entering into the inversion.

In Fig. 19.18, the region of relatively unstable (well mixed) air that extends from the surface to the base of the inversion is referred to as the **mixing layer**. The vertical extent of the mixing layer is called the **mixing depth**. Observe that if the inversion rises, the mixing depth increases and the pollutants will be dispersed throughout a greater volume of air; if the inversion lowers, the mixing depth decreases and the pollutants will become more concentrated, sometimes reaching unhealthy levels. Since the atmosphere tends to be most unstable in the afternoon and most stable in the early morning, we typically find the greatest mixing depth in the afternoon and the shallowest one (if one exists at all) in the early morning. Consequently, during the day, the top of the mixing layer may be clearly visible (see ● Fig. 19.19). Moreover, during take-off or landing on daylight flights out of large urban areas, the top of the mixing layer may sometimes be observed.

The position of the semipermanent Pacific high off the coast of California contributes greatly to the air pollution in that region. The Pacific high promotes subsiding air, which warms the air aloft. Surface winds around the high promote upwelling of ocean water. Upwelling—the rising of cold water from below—makes the surface water cool, which, in turn, cools the air above. Warm air aloft coupled with cool, surface (marine) air together produce a strong and persistent subsidence inversion—one that exists 80 to 90 percent of the time over the city of Los Angeles between June and October, the smoggy months. The pollutants trapped within the cool marine air are occasionally swept eastward by a sea breeze. This action carries smog from the coastal regions into the interior valleys producing a *smog front* (see ● Fig. 19.20).

*Remember from Chapter 2 that sinking air always warms because it is being compressed by the surrounding air.

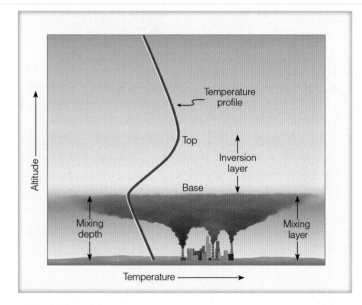

● **FIGURE 19.18** The inversion layer acts as a lid on the pollutants below. If the inversion lowers, the mixing depth decreases and the pollutants are concentrated within a smaller volume.

CRITICAL THINKING QUESTION If the surface air temperature were to increase dramatically in Fig. 19.18, would you expect the base of the inversion to raise or lower (assuming that the lapse rate in the diagram below the inversion does not change)? How would the change in the height of the inversion base influence the mixing layer and the concentration of pollutants?

WEATHER WATCH

An unusually strong and long-lived inversion covered much of the Great Basin for several weeks in January 2013. On the morning of January 23, the temperature in Salt Lake City, Utah, was a frigid 4°F in dense smoke and fog. Almost 3000 feet higher up in the mountains just to the east of town, Park City was experiencing bright sunshine and fresh air, with a temperature of 45°F.

● **FIGURE 19.19** A thick layer of polluted air is trapped in the valley. The top of the polluted air marks the base of a subsidence inversion.

Smokestack Plumes

We know that the stability of the air (especially near the surface) changes during the course of a day. These changes can influence the pollution near the ground as well as the behavior of smoke leaving a chimney. ● Figure 3 illustrates different smoke plumes that can develop with adequate wind, but different types of stability.

In Fig. 3a, it is early morning, the winds are light, and a radiation inversion extends from the surface to well above the height of the smokestack. In this stable environment, there is little up-and-down motion, so the smoke spreads horizontally rather than vertically. When viewed from above, the smoke plume resembles the shape of a fan, and it is referred to as a *fanning smoke plume.*

Later in the morning, the surface air warms quickly and destabilizes as the radiation inversion gradually disappears from the surface upward (Fig. 3b). However, the air above the chimney is still stable, as indicated by the presence of the inversion, so vertical motions are confined to the region near the surface. Because of this, the smoke mixes downwind, increasing the concentration of pollution at the surface—sometimes to dangerously high levels. This effect is called *fumigation.* Here again, we can see why a taller smokestack is preferred. A taller stack extends upward into the stable layer, producing a fanning plume that does not mix downward toward the ground.

If daytime heating of the ground continues, the depth of atmospheric instability increases. Notice in Fig. 3c that the inversion has completely disappeared. Light-to-moderate winds combine with rising and sinking air to cause the smoke to move up and down in a wavy pattern, producing a *looping smoke plume.*

The continued rising of warm air and sinking of cool air can cause the temperature profile to equal that of the dry adiabatic rate (Fig. 3d). In this neutral atmosphere, vertical and horizontal motions are about equal, and the smoke from the stack tends to take on the shape of a cone, forming a *coning smoke plume.*

● **FIGURE 3** As the vertical temperature profile changes during the course of a day (*a* through *e*), the pattern of smoke emitted from the stack changes as well.

After sunset, the ground cools rapidly and the radiation inversion reappears. When the top of the inversion extends upward to slightly above the stack, stable air is near the ground with neutral air above (Fig. 3e). Because the stable air in the inversion prevents the smoke from mixing downward, the smoke is carried upward, producing a *lofting smoke plume.* Thus, smoke plumes provide a clue to the stability of the atmosphere, and knowing the stability yields important information about the dispersion of pollutants.

Of course, other factors influence the dispersion of pollutants from a chimney, including the pollutants' temperature and exit velocity, wind speed and direction, and, as we saw in an earlier section, the chimney's height. Overall, taller chimneys, greater wind speeds, and higher exit velocities result in a lower concentration of pollutants.

● **FIGURE 19.20** The leading edge of cool, marine air carries pollutants into Riverside, California, as an advancing smog front.

Warm air

Cold air

● **FIGURE 19.21** At night, cold air and pollutants drain downhill and settle in low-lying valleys.

THE ROLE OF TOPOGRAPHY The shape of the landscape (topography) plays an important part in trapping pollutants. We know from Chapter 3 that, at night, cold air tends to drain downhill, where it settles into low-lying basins and valleys. The cold air can have several effects: It can strengthen a pre-existing surface inversion, and it can carry pollutants downhill from the surrounding hillsides (see ● Fig. 19.21).

Valleys prone to pollution are those completely enclosed by mountains and hills. The surrounding mountains tend to block the prevailing wind. With light winds, and a shallow mixing layer, the poorly ventilated cold valley air can only slosh back and forth like a murky bowl of soup.

Valleys susceptible to stagnant air exist in just about all mountainous regions. Air pollution concentrations in these valleys tend to be greatest during the colder months. During the warmer months, daytime heating can warm the sides of the valley to the point that upslope valley winds vent the pollutants upward, as if in a chimney.

The pollution problem in several large cities is, at least, partly due to topography. For example, the city of Los Angeles is surrounded on three sides by hills and mountains.* Cool marine air from off the ocean moves inland and pushes against the hills, which tend to block the air's eastward progress. Unable to rise, the cool air settles in the basin, trapping pollutants from industry and millions of autos. Baked by sunlight, the pollutants become the infamous photochemical smog (see ● Fig. 19.22). By the same token, the "mile high" city of Denver, Colorado, sits in a broad shallow basin that frequently traps both cold air and pollutants. The much larger Great Basin that spans Utah and neighboring states is also prone to trapping cold air, occasionally for weeks at a time, which can put cities in the region at risk of significant pollution episodes.

SEVERE AIR POLLUTION POTENTIAL The greatest potential for an episode of severe air pollution occurs when all of the factors mentioned in the previous sections come together simultaneously. Ingredients for a major buildup of atmospheric pollution are:

- many sources of air pollution (preferably clustered close together),
- a deep high-pressure area that becomes stationary over a region,

---

*The city of Los Angeles and its surrounding topography is shown in Fig. 9.28 on p. 245.

● **FIGURE 19.22** Thick smog, such as shown in this photo of Los Angeles, California, taken during May 2015, has become less common over the region in recent years.

- light surface winds that are unable to disperse the pollutants,
- a strong subsidence inversion produced by the sinking of air aloft,
- a shallow mixing layer with poor ventilation,
- a valley where the pollutants can accumulate,
- clear skies so that radiational cooling at night will produce a surface inversion, which can cause an even greater buildup of pollutants near the ground,
- and, for photochemical smog, adequate sunlight to produce secondary pollutants, such as ozone.

Light winds and poor vertical mixing can produce a condition known as **atmospheric stagnation**. The atmospheric conditions resulting in air stagnation usually occur during particular weather patterns. For example, a region under the domination of a surface high-pressure area or ridge often experiences clear skies, light winds, and a subsidence inversion. Moreover, where warm air rides up over cold surface air, such as ahead of an advancing warm front, stable atmospheric conditions promote the trapping of pollutants near the surface. On the other hand, the strong and gusty winds and generally less-stable air behind a cold front usually results in good dispersion (see ● Fig. 19.23).

If air stagnation conditions persist for several days or more where there are ample pollutant sources, the buildup of pollution can lead to some of the worst air pollution disasters on record, such as the one in the valley city of Donora, Pennsylvania, where in 1948 seventeen people died in 14 hours. (Additional information on the Donora disaster is found in Focus section 19.4.)

● **FIGURE 19.23** Weather patterns associated with poor, fair, and good air pollution dispersion.

Five Days in Donora—An Air Pollution Episode

On Tuesday morning, October 26, 1948, a cold surface anticyclone moved over the eastern half of the United States. There was nothing unusual about this high-pressure area; with a central pressure of only 1025 mb (30.27 in.), it was not exceptionally strong (see ● Fig. 4). Aloft, however, a large blocking-type ridge formed over the region, and the jet stream, which moves the surface pressure features along, was far to the west. Consequently, the surface anticyclone became entrenched over Pennsylvania and remained nearly stationary for five days.

The widely spaced isobars around the high-pressure system produced a weak pressure gradient and generally light winds throughout the area. These light winds, coupled with the gradual sinking of air from aloft, set the stage for a disastrous air pollution episode.

On Tuesday morning, radiation fog gradually settled over the moist ground in Donora, a small town nestled in the Monongahela Valley of western Pennsylvania. Because Donora rests on bottom land, surrounded by rolling hills, its residents were accustomed to fog, but not to what was to follow.

The strong radiational cooling that formed the fog, along with the sinking air of the anticyclone, combined to produce a strong temperature inversion. Light, downslope winds spread cool air and contaminants over Donora from the community's steel mill, zinc smelter, and sulfuric acid plant.

The fog and its burden of pollutants lingered into Wednesday. Cool drainage winds during the night strengthened the inversion and added more effluents to the already filthy air. The dense fog layer blocked sunlight from reaching the ground. With essentially no surface heating, the mixing depth lowered and the pollution became more concentrated. Unable to

● **FIGURE 4** Surface weather map that shows a stagnant anticyclone over the eastern United States on October 26, 1948. The inset map shows a schematic depiction of the town of Donora on the Monongahela River.

mix and disperse either horizontally or vertically, the dirty air became confined to a shallow, stagnant layer.

Meanwhile, the factories continued to belch impurities into the air (primarily sulfur dioxide and particulate matter) from stacks no higher than 40 m (130 ft) tall. The fog gradually thickened into a moist clot of smoke and water droplets. By Thursday, the visibility had decreased to the point where one could barely see across the street. At the same time, the air had a penetrating, almost sickening, smell of sulfur dioxide. At this point, a large percentage of the population became ill.

The episode reached a climax on Saturday, as 17 deaths were reported. As the death rate

mounted, alarm swept through the town. An emergency meeting was called between city officials and factory representatives to see what could be done to cut down on the emission of pollutants.

The light winds and unbreathable air persisted until, on Sunday, an approaching storm generated enough wind to vertically mix the air and disperse the pollutants. A welcome rain then cleaned the air further. All told, the episode had claimed the lives of 22 people. During the five-day period, about half of the area's 14,000 inhabitants experienced some ill effects from the pollution. Most of those affected were older people with a history of cardiac or respiratory disorders.

Air Pollution and the Urban Environment

For more than 100 years, it has been known that cities are generally warmer than surrounding rural areas. This region of city warmth, known as the **urban heat island**, can influence the

concentration of air pollution. However, before we look at its influence, let's see how the heat island actually forms.

The urban heat island is formed by industrial and urban development. In rural areas, a large part of the incoming solar energy evaporates water from vegetation and soil. In cities, where less vegetation and exposed soil exists, the majority of the sun's energy is absorbed by urban structures and asphalt. Hence, during warm daylight hours, less evaporative cooling in cities allows surface temperatures to rise higher than in rural areas.*

At night, the solar energy (stored as vast quantities of heat in city buildings and roads) is slowly released into the city air. Additional city heat is given off at night (and during the day) by vehicles and factories, as well as by industrial and domestic heating and cooling units. The release of heat energy is retarded by the tall city walls that do not allow infrared radiation to escape as readily as do the relatively level surfaces of the surrounding countryside. The slow release of heat tends to keep nighttime city temperatures higher than those of the faster cooling rural areas. Overall, the heat island is strongest

1. at night, when compensating sunlight is absent;
2. during the winter, when nights are longer and there is more heat generated in the city; and
3. when the region is dominated by a high-pressure area with light winds, clear skies, and less humid air.

Over time, increasing numbers (and sizes) of urban heat islands affect climatological temperature records, producing additional warming in climatic records taken in cities. Because this warming is not directly related to greenhouse gases, scientists take it into account when analyzing and interpreting climate change over the past 100-plus years.

The constant outpouring of pollutants into the environment may influence the climate of a city. Certain particles reflect solar radiation, thereby reducing the sunlight that reaches the surface. Some particles serve as nuclei upon which water and ice form. Water vapor condenses onto these particles when the relative humidity is as low as 70 percent, forming haze that greatly reduces visibility. Moreover, the added nuclei increase the frequency of city fog.**

Many studies over the last few decades indicate that precipitation can be greater in urban areas than in the surrounding countrysides. This phenomenon may be due in part to the increased roughness of city terrain, the result of large structures that cause surface air to slow and gradually converge. This pile-up of air over the city then slowly rises, much like toothpaste does when its tube is squeezed. At the same time, city heat warms the surface air, making it more unstable, which enhances rising air motions, which, in turn, aid in forming clouds and thunderstorms. This process helps explain why precipitation tends to be heavier in and near urban areas, and sometimes just downwind from them. Enhanced precipitation has been found over the years in areas near Chicago, Illinois; St. Louis, Missouri; Paris, France; Atlanta, Georgia; Beijing, China; and other locations. The effects are generally more evident with showers and thunderstorms rather than steady rains. The large concentrations of particulates in urban areas can also help boost rainfall within and

● **FIGURE 19.24** On a clear, relatively calm night, a weak country breeze carries pollutants from the outskirts into the city, where they concentrate and rise due to the warmth of the city's urban heat island. This effect may produce a pollution (or dust) dome from the suburbs to the center of town.

downwind of cities in some cases. The main exception is for cities located in dry regions, where large numbers of aerosols can actually decrease rainfall, as the particulates (nuclei) compete for the available moisture—similar to what happens when a cloud is overseeded, discussed in Chapter 7.

On clear, still nights when the heat island is pronounced, a small thermal low-pressure area may form over urban areas. Sometimes a light breeze—which may be called a **country breeze**—blows from the countryside into the city. If there are major industrial areas along the city's outskirts, pollutants are carried into the heart of town, where they become even more concentrated. Such an event is especially likely if an inversion inhibits vertical mixing and dispersion (see ● Fig. 19.24). ▼ Table 19.4 summarizes the environmental influence of cities by contrasting the urban environment with the rural.

On average, more people in the United States die from excessive heat during heat waves than from any other extreme weather event. However, many of these deaths actually occur due to exposure to high levels of air pollution. This topic is presented in more detail in Focus section 19.5.

▼ **TABLE 19.4** **Contrast of the Urban and Rural Environment (Average Conditions)***

CONSTITUENTS	URBAN AREA (CONTRASTED TO RURAL AREA)
Mean pollution level	Higher
Mean sunshine reaching the surface	Lower
Mean temperature	Higher
Mean relative humidity	Lower
Mean visibility	Lower
Mean wind speed	Lower
Mean precipitation	Higher
Mean amount of cloudiness	Higher
Mean thunderstorm frequency	Higher

*Values are omitted because they vary greatly depending upon city, size, type of industry, and season of the year.

*The cause of the urban heat island is quite involved. Depending on the location, time of year, and time of day, any or all of the following differences between cities and their surroundings can be important: albedo (reflectivity of the surface), surface roughness, emissions of heat, emissions of moisture, and emissions of particles that affect net radiation and the growth of cloud droplets.

**The impact that tiny liquid and solid particles (aerosols) may have on a larger scale is complex and depends upon many factors that were addressed in Chapter 18.

Heat Waves and Air Pollution: A Deadly Team

Some of the worst weather calamities of the twenty-first century have been triggered by intense, prolonged episodes of heat. By itself, the cumulative impact of high temperatures can be enough to kill people, especially those in poor health who have little or no way to cool down. But air pollution can also play a significant role in the deadly nature of heat waves.

The numbers involved are staggering. Mortality specialists have found that more than 40,000 people died as a result of a heat wave that gripped Europe during the summer of 2003. Likewise, the record-breaking heat that struck Moscow in 2010 (see ● Fig. 5) took more than 10,000 lives. Because a heat wave unfolds over days and weeks, death tolls such as these are calculated by taking the total number of fatalities observed and adjusting that total based on the number of deaths that would normally have been expected at a given time of year.

Like a gang of criminals, high temperatures and air pollution often appear at the same time and place, so it can be difficult to tell whether the villains acted individually or collectively. The stationary domes of upper-level high pressure that can lead to heat waves also tend to produce multiple days of sunshine and stagnant air, conditions ideal for the formation of ground-level ozone and the accumulation of small particulates in the atmosphere. These, in turn, are directly linked to a variety of dangerous, potentially fatal health conditions, including cardiovascular and respiratory problems.

Several studies of the European heat wave of 2003 examined the causes of death and the potential role of pollutants. One team found that between 13 and 30 percent of the roughly 1000 deaths in Switzerland associated with the heat wave were actually caused by ozone rather than by the heat itself. Another study attributed between 21 and 38 percent of more than 2000 heat wave fatalities in the United Kingdom to the effects of ozone and particles with diameters smaller than 10 micrometers (PM_{10}). One reason why it is difficult to narrow the ranges reported in studies like these is that researchers are not yet sure of the extent to which heat and pollution have a synergistic effect, as opposed to acting independently.

To get at this question, one analysis in the United States drew on data from the massive National Morbidity, Mortality, and Air Pollution Study (NMMAPS), which combined health- and weather-related data for about 100 urban areas from 1987 through 2000. Using NMMAPS, the researchers found that, as a rule, ozone did tend to boost the link between hot weather and deaths related to heart disease. For example, a 10°C(18°F) increase in temperature produced about 1 percent more cardiovascular deaths when ozone concentrations were at their lowest, but the increase was about 8 percent when ozone levels were highest.

The role of air pollution in heat-related health problems could be magnified in coming decades. Climate change simulations indicate that heat waves may increase in number, duration, and strength across many parts of the world. At the same time, an ever-greater proportion of the world's population is clustering in urban areas. In the absence of stringent air pollution control, this trend could increase both the amount of air pollution in a given area and the number of people vulnerable to its health effects.

June 17, 2010

Pavel L Photo and Video/Shutterstock.com

August 7, 2010

AP Images/Mikhail Metzel

● **FIGURE 5** The severe effects of Moscow's 2010 heat wave on air quality can be seen in these two photos of the same street, one taken on June 17 (before the heat wave began) and the other on August 7 (near the peak of the heat wave).

Acid Deposition

Air pollution emitted from industrial areas, especially products of combustion, such as oxides of sulfur and nitrogen, can be carried many kilometers downwind. Either these particles and gases slowly settle to the ground in dry form (*dry deposition*) or they are removed from the air during the formation of cloud particles and then carried to the ground in rain and snow (*wet deposition*). **Acid rain** and *acid precipitation* are common terms used to describe wet deposition, while **acid deposition** encompasses both dry and wet acidic substances. How, then, do these substances become acidic?

Emissions of SO_2 and oxides of nitrogen may settle on the local landscape, where they transform into acids as they interact with water, especially during the formation of dew or frost. The remaining airborne particles may transform into tiny dilute drops of sulfuric acid (H_2SO_4) and nitric acid (HNO_3) during a complex series of chemical reactions involving sunlight, water vapor, and other gases. These acid particles may then fall slowly to the surface, or they may adhere to cloud droplets or to fog droplets, producing **acid fog**. They may even act as nuclei on which the cloud droplets begin to grow. When precipitation occurs in the cloud, it carries the acids to the ground. Because of this, precipitation grew increasingly acidic during the mid- to late-twentieth century with the increase of air pollution in many parts of the world, especially downwind of major industrial areas.

High concentrations of pollutants that produce acid rain can be carried great distances from their sources. For example, scientists in one study discovered high concentrations of pollutants about 600 km off the east coast of North America. It is suspected that they came from industrial East Coast cities. Even though most pollutants are washed from the atmosphere during storms, some may be swept over the Atlantic, reaching places such as Bermuda and Ireland. Acid rain knows no national boundaries.

Although studies suggest that acid precipitation may be nearly worldwide in distribution, regions that have been noticeably affected since the mid-twentieth century include eastern North America, central Europe, and Scandinavia. More recently, as a result of rapid industrialization, China and other parts of Asia have been affected. In some places, acid precipitation occurs naturally, such as in northern Canada, where natural fires in exposed coal beds produce tremendous quantities of sulfur dioxide. By the same token, acid fog can form by natural means.

Precipitation is naturally somewhat acidic. The carbon dioxide occurring naturally in the air dissolves in precipitation, making it slightly acidic with a pH between 5.0 and 5.6. Consequently, precipitation is considered acidic when its pH is below about 5.0 (see ● Fig. 19.25). In the northeastern United States, where emissions of sulfur dioxide are primarily responsible for the acid precipitation, typical pH values range between 4.5 and 4.7 (see ● Fig. 19.26). But acid precipitation is not confined to the Northeast; it is also found in the southeastern states, too. Along the West Coast, the main cause of acid deposition appears to be the oxides of nitrogen released in automobile exhaust. In Los Angeles, acid fog is a more serious problem than acid rain, especially along the coast, where fog is most prevalent. The fog's pH is usually between 4.4 and 4.8, although pH values of 3.0 and below have been measured.

High concentrations of acid deposition can damage plants and water resources. Freshwater ecosystems seem to be particularly sensitive to changes in acidity. Concern has centered mainly on areas where interactions with alkaline soil are unable to neutralize the acidic inputs. Studies indicate that thousands of lakes in the United States and Canada became so acidified in the late twentieth century that entire fish populations may have been adversely affected.

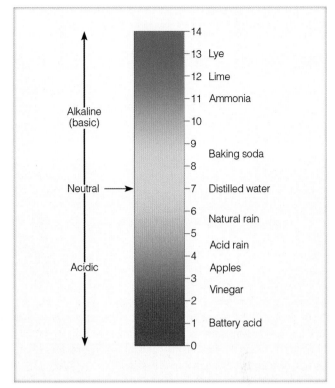

● **FIGURE 19.25** The pH scale ranges from 0 to 14, with a value of 7 considered neutral. Values greater than 7 are alkaline and below 7 are acidic. The scale is logarithmic, which means that rain with pH 3 is 10 times more acidic than rain with pH 4 and 100 times more acidic than rain with pH 5.

Acid rain may work with other factors, such as natural pathogens, to attack forests across vast areas. During the 1980s, up to half of the trees in forests across Germany showed signs of a blight that was due, in part, to acid deposition. Apparently, acidic particles raining down on the forest floor for decades caused a chemical imbalance in the soil that, in turn, causes serious deficiencies in certain elements necessary for the trees' growth. The trees were thus weakened and become susceptible to insects and drought. The impact on large forests was called "Waldsterben," or forest death. The same type of processes likely affected North American forests during this period, but at a much slower pace, with symptoms apparent in many forests at higher elevations from southeastern Canada to South Carolina (see ● Fig. 19.27). Moreover, acid precipitation has been a problem in the mountainous West, where high mountain lakes and forests seem to have been most affected.

The effects of acid rain are not only felt in rural areas. Acid deposition has eroded the foundations of structures in many cities throughout the world, as well as damaging priceless outdoor fountain sculptures and statues. The cost of this damage to building surfaces, monuments, and other structures has been estimated to run into the billions of dollars each year.

Observations and computer models in the 1980s greatly clarified the threat posed by acid precipitation and the processes behind it. As a result, the Clean Air Act of 1990 imposed a reduction in the United States' emissions of sulfur dioxide and nitrogen dioxide. Many other countries implemented similar measures. The United States also implemented a "cap and trade" system by

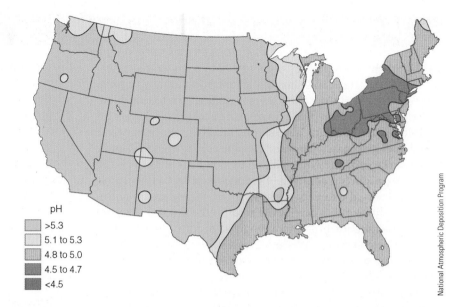

● FIGURE 19.26 Values of pH in precipitation over the United States during 2012. (National Atmospheric Deposition Program)

pH
> 5.3
5.1 to 5.3
4.8 to 5.0
4.5 to 4.7
< 4.5

National Atmospheric Deposition Program

which companies that reduced their emissions had the right to sell allowances to other companies that had yet to do so. Such measures have helped produce large decreases in emissions (in many cases, more than 50 percent). As a result, the threat of acid precipitation has lessened significantly in many areas. Some tree species in the northeast United States that were heavily affected by acid rain, including red spruce, have showed reinvigoration since 2000, and a recent study found an accelerating drop in sulfate concentrations within lakes across the northeastern United States. Because soils in certain regions are still showing evidence of acidification, it may take time before some lakes return to their previous pH levels.

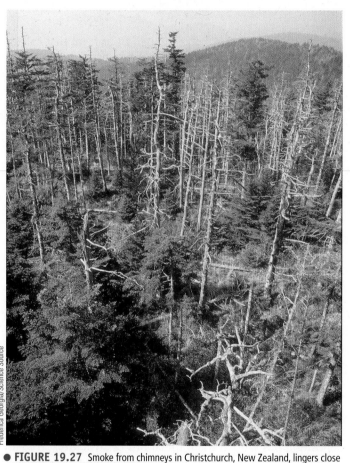

Frederica Georgia/Science Source

● FIGURE 19.27 Smoke from chimneys in Christchurch, New Zealand, lingers close to the ground on a cold winter morning.

SUMMARY

In this chapter, we found that air pollution has plagued humanity for centuries. Air pollution problems began when people tried to keep warm by burning wood and coal. These problems worsened during the industrial revolution as coal became the primary fuel for both homes and industry. Even though many American cities do not meet all of the air quality standards set by the federal Clean Air Act of 1990, the air over our large cities is cleaner today than it was years ago as a result of stricter emission standards and cleaner fuels.

We examined the types and sources of air pollution and found that primary air pollutants enter the atmosphere directly, whereas secondary pollutants form by chemical reactions that involve other pollutants. The secondary pollutant ozone is the main ingredient of photochemical smog—a smog that irritates the eyes and forms in the presence of sunlight. In polluted air near the surface, ozone forms during a series of chemical reactions involving nitrogen oxides and hydrocarbons (VOCs). In the stratosphere, ozone is a naturally occurring gas that protects us from the sun's harmful ultraviolet rays. We examined how ozone forms in the stratosphere and how it may be altered by natural means. We also learned that human-induced gases, such as chlorofluorocarbons, work their way into the stratosphere where they release atomic chlorine that rapidly destroys ozone, especially in polar regions.

We looked at the air quality index and found that a number of areas across the United States still have days considered unhealthy by the standards set by the United States Environmental Protection Agency. We also looked at the main factors affecting air pollution and found that most air pollution episodes occur when the winds are light, skies are clear, the mixing layer is shallow, the atmosphere is stable, and a strong inversion exists. These conditions usually prevail when a high-pressure area stalls over a region.

We observed that, on the average, urban environments tend to be warmer and more polluted than the rural areas that surround them. We saw that pollution from industrial areas can modify environments downwind from them, as oxides of sulfur and nitrogen are swept into the air, where they may transform into acids that fall to the surface. Many areas have seen significant improvement in acid deposition by reducing the pollutants that cause acid precipitation.

KEY TERMS

The following terms are listed (with corresponding page numbers) in the order they appear in the text. Define each. Doing so will aid you in reviewing the material covered in this chapter.

air pollutants, 539
primary air pollutants, 539
secondary air pollutants, 539
particulate matter, 542
carbon monoxide (CO), 543
sulfur dioxide (SO_2), 544
volatile organic compounds (VOCs), 544
hydrocarbons, 544
nitrogen dioxide (NO_2), 544
nitric oxide (NO), 544
smog, 544
photochemical smog, 544
ozone (O_3), 544
ozone hole, 547
air quality index (AQI), 550
dispersion (of pollutants), 552
radiation (surface) inversion, 553
subsidence inversion, 554
mixing layer, 554
mixing depth, 554
atmospheric stagnation, 557
urban heat island, 558
country breeze, 559
acid rain, 560
acid deposition, 560
acid fog, 561

QUESTIONS FOR REVIEW

1. What are some of the main sources of air pollution?
2. List a few of the substances that fall under the category of particulate matter.
3. How does PM_{10} particulate matter differ from that called $PM_{2.5}$? Which poses the greater risk to human health?
4. List two ways particulate matter is removed from the atmosphere.
5. Describe some of the sources of pollution found inside a home or building.
6. Why are high levels of radon inside a home so dangerous?
7. Describe the primary sources and some of the health problems associated with each of the following pollutants:
   (a) carbon monoxide (CO)
   (b) sulfur dioxide (SO_2)
   (c) volatile organic compounds (VOCs)
   (d) nitrogen oxides
8. How does London-type smog differ from Los Angeles–type smog?
9. What is the main component of photochemical smog? What are some of the adverse health effects of photochemical smog? Why is the production of photochemical smog more prevalent during the summer and early fall than during the middle of winter?
10. In polluted air, (a) describe the role that NO_2 plays in the production of tropospheric (ground-level) ozone, and (b) the role that NO plays in the destruction of tropospheric (ground-level) ozone. What role do hydrocarbons (VOCs) play in the production of ozone and other photochemical oxidants?
11. Describe the main processes that account for stratospheric ozone production and destruction. What natural and human-produced substances could alter the concentration of ozone in the stratosphere?
12. Why is stratospheric ozone beneficial to life on Earth, whereas tropospheric (ground-level) ozone is not?

13. (a) How are CFCs related to the destruction of stratospheric ozone?
    (b) If most of the ozone in the stratosphere were destroyed, what possible effects might this have on Earth's inhabitants?

14. Explain how scientists believe the Antarctic ozone hole forms.

15. What are primary ambient air quality standards and secondary standards intended to do? What are nonattainment areas?

16. Why is a light wind, rather than a strong wind, more conducive to high concentrations of air pollution?

17. How does atmospheric stability influence the accumulation of air pollutants near the surface?

18. What weather patterns often lead to atmospheric stagnation?

19. Why is it that polluted air and inversions seem to go hand in hand?

20. Give several reasons why taller smoke stacks are better than shorter ones at improving the air quality in their immediate area.

21. How does the mixing depth normally change during the course of a day? As the mixing depth changes, how does it affect the concentration of pollution near the surface?

22. How does topography influence the concentration of pollutants in cities such as Los Angeles and Denver? In mountainous terrain?

23. List the factors that can lead to a major buildup of atmospheric pollution.

24. What type of weather pattern is best for dispersion of pollutants?

25. What is an urban heat island? Is it more strongly developed at night or during the day? Explain.

26. What causes the "country breeze"? Why is it usually more developed at night than during the day? Would it develop more easily in summer or in winter? Explain.

27. How can pollution play a role in influencing the precipitation downwind of certain large industrial complexes?

28. Why is acid deposition considered a serious problem? How does precipitation become acidic?

QUESTIONS FOR THOUGHT

1. (a) Suppose clouds of nitrogen dioxide drift slowly from a major industrial complex over a relatively unpopulated area. If the area is essentially "free" of hydrocarbons, would you expect high levels of tropospheric (ground-level) ozone to form? Explain.
   (b) Now suppose that the clouds of nitrogen dioxide drift slowly over an area that has a high concentration of hydrocarbons (VOCs), from both natural and industrial sources. Would you expect high levels of tropospheric (ground-level) ozone to form under these conditions? Explain your reasoning.

2. For least-polluted conditions, what would be the best time of day for a farmer to burn agricultural debris? Explain why you chose that time of day.

3. Why are most severe air pollution episodes associated with subsidence inversions rather than radiation inversions?

4. Table 19.4, p. 559, shows that cloudiness is generally greater in urban areas than in rural areas. Since clouds reflect a great deal of incoming sunlight, they tend to keep daytime temperatures lower. Why then, during the day, are urban areas generally warmer than surrounding rural areas?

5. Acid snow can be a major problem. In the high mountains of the western United States, especially downwind of a major metropolitan or industrial area, explain why, for a high-mountain lake, acid snow can be a greater problem than acid rain, even when both have the same pH.

6. Why do we want to reduce high ozone concentrations at Earth's surface while, at the same time, we do not want to reduce ozone concentrations in the stratosphere?

7. Give a few reasons why, in industrial areas, nighttime pollution levels might be higher than daytime levels.

8. A large industrial smokestack located within an urban area emits vast quantities of sulfur dioxide and nitrogen dioxide. Following criticism from local residents that emissions from the stack are contributing to poor air quality in the area, the management raises the height of the stack from 10 m (33 ft) to 100 m (330 ft). Will this increase in stack height change any of the existing air quality problems? Will it create any new problems? Explain.

9. If the sulfuric acid and nitric acid in rainwater are capable of adversely affecting soil, trees, and fish, why doesn't this same acid adversely affect people when they walk in the rain?

PROBLEMS AND EXERCISES

1. Keep a log of the daily AQI readings in your area and note the pollutants listed in the index. Also, keep a record of the daily weather conditions, such as cloud cover, high temperature for the day, average wind direction and speed, etc. See if there is any relationship between these weather conditions and high AQI readings for certain pollutants.

2. Suppose the AQI reading for ozone is 300.
   (a) How would the air be described on this day?
   (b) What would be the general health effects, and what precautions should a person take under these conditions?

3. Keep a log of daily AQI readings for one urban and one rural location for days on which both areas have similar weather conditions; note the weather conditions. Compare them. How do the weather conditions influence the AQI in both locations? Do the weather conditions contribute to greater differences in AQI readings between the two locations? Explain.

ONLINE RESOURCES

Visit www.cengagebrain.com to view additional resources, including video exercises, practice quizzes, an interactive eBook, and more.

A spectacular rainbow forms in front of a light shower in the state of Minas Gerais in southeast Brazil.

CHAPTER 20

Light, Color, and Atmospheric Optics

CONTENTS

THE SKY IS CLEAR, THE WEATHER COLD, and the year, 1818. Near Baffin Island in Canada, a ship with full sails enters unknown waters. On board are the English brothers James and John Ross, who are hoping to find the elusive "Northwest Passage," the waterway linking the Atlantic and Pacific oceans. On this morning, however, their hopes would be dashed, for directly in front of the vessel, blocking their path, is a huge, towering mountain range. Disappointed, they turn back and report that the Northwest Passage does not exist.

Almost ninety years later, Rear Admiral Robert Peary of the United States Navy reported encountering the same barrier and called it "Crocker Land." What type of treasures did this mountain conceal—gold, silver, precious gems? The curiosity of explorers from all over the world had been aroused. Speculation was the rule, until, in 1913, the American Museum of Natural History commissioned Donald MacMillan to lead an expedition to solve the mystery of Crocker Land. At first, the journey was disappointing. Where Peary had seen mountains, MacMillan saw only vast stretches of open water. Finally, ahead of his ship was Crocker Land, but it was more than two hundred miles west from where Peary had encountered it. MacMillan sailed on as far as possible. Then he dropped anchor and set out on foot with a small crew of men.

As the team moved toward the mountains, the mountains seemed to move away from them. If they stood still, the mountains stood still; if they started walking, the mountains receded again. Puzzled, they trekked onward over the glittering snowfields until huge mountains surrounded them on three sides. But in the next instant the sun disappeared below the horizon and, as if by magic, the mountains dissolved into the cold arctic twilight. Dumbfounded, the men looked around only to see ice in all directions—not a mountain was in sight. There they were, the victims of one of nature's greatest practical jokes, for Crocker Land was a mirage.

The sky is full of visual events. Optical illusions (mirages) can appear as towering mountains or wet roadways. In clear weather, the sky can appear blue, while the horizon appears milky white. Sunrises and sunsets can fill the sky with brilliant shades of pink, red, orange, and purple. At night, away from city lights, the sky is black except for the light from the stars, planets, and the moon. The moon's size and color seem to vary during the night, and the stars twinkle. To understand what we see in the sky, we will take a closer look at sunlight, examining how it interacts with the atmosphere to produce an array of atmospheric visuals.

White and Colors

We know from Chapter 2 that nearly half of the solar radiation that reaches the atmosphere is in the form of visible light. As sunlight enters the atmosphere, it is either absorbed, reflected, scattered, or transmitted on through. How objects at the surface respond to this energy depends on their general nature (color, density, composition) and the wavelength of light that strikes them. How do we see? Why do we see various colors? What kinds of visual effects do we observe because of the interaction between light and matter? In particular, what can we see when light interacts with our atmosphere?

We perceive light because radiant energy from the sun travels outward in the form of electromagnetic waves. When these waves reach the human eye, they stimulate antenna-like nerve endings in the retina. These antennae are of two types—*rods* and *cones*. The rods respond to all wavelengths of visible light and give us the ability to distinguish light from dark. If people possessed rod-type receptors only, then only black and white vision would be possible. The cones respond to specific wavelengths of visible light. Radiation with a wavelength between 0.4 and 0.7 micrometers (μm) strikes the cones, which immediately fire an impulse through the nervous system to the brain, and we perceive this impulse as the sensation of color. (Color blindness is caused by missing or malfunctioning cones.) Wavelengths of radiation shorter than 0.4 μm, or longer than 0.7 μm, do not stimulate color vision in humans.

White light is perceived when all visible wavelengths strike the cones of the eye with nearly equal intensity.* Because the sun radiates almost half of its energy as visible light, all visible wavelengths from the midday sun reach the cones, and the sun usually appears nearly white. A star that is cooler than our sun radiates most of its energy at slightly longer wavelengths; therefore, it appears redder. On the other hand, a star much hotter than our sun radiates more energy at shorter wavelengths and thus appears bluer. A star whose temperature is about the same as the sun's appears white.

Objects that are not hot enough to emit radiation at visible wavelengths can still have color. Everyday objects we see as red are those that absorb all visible radiation except red. The red light is reflected from the object to our eyes. Blue objects have blue light returning from them, since they absorb all visible wavelengths except blue. Some surfaces absorb all visible wavelengths

and reflect no light at all. Since no radiation strikes the rods or cones, these surfaces appear black. Therefore, when we see colors, we know that light must be reaching our eyes.

Clouds and Scattered Light

It is interesting to watch the underside of a puffy, growing cumulus cloud change color from white to dark gray or black. When we see this change happen, our first thought is usually, "It's going to rain." Why is the cloud initially white? Why does it change color? To answer these questions, we need to investigate in more detail the concept of *scattering*.

When sunlight bounces off a surface at the same angle at which it strikes the surface, we say that it is **reflected light**, and call this phenomenon *reflection* (see ● Fig. 20.1). There are various constituents of the atmosphere, however, that tend to deflect solar radiation from its path and send it out in all directions. We know from Chapter 2 that radiation reflected in this way is said to be **scattered**. (Scattered light is also called *diffuse light*.) During the scattering process, no energy is gained or lost and, therefore, no temperature changes occur. In the atmosphere, scattering is usually caused by small objects, such as air molecules, fine particles of dust, water molecules, and some pollutants. Just as the ball in a pinball machine bounces off the pins in many directions, so solar radiation is knocked about by small particles in the atmosphere.

BLUE SKIES AND HAZY DAYS The sky appears blue because light that stimulates the sensation of blue color is reaching the retina of the eye. How does this happen?

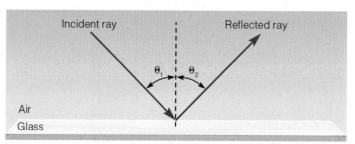

● **FIGURE 20.1** For a ray of light striking a flat, smooth surface, the angle at which the incident ray strikes the surface (the angle of incidence, or θ_1) is equal to the angle at which the reflected ray leaves the surface (the angle of reflection, or θ_2). This phenomenon is called *Snell's law*.

---

*Recall from Chapter 2 that visible white light is a combination of waves with different wavelengths. The wavelengths of visible light in decreasing order are: red (longest), orange, yellow, green, blue, and violet (shortest).

TABLE 20.1 The Various Types of Scattering of Visible Light

TYPE OF PARTICLE	PARTICLE DIAMETER (MICROMETERS, μM)	PREDOMINANT TYPE OF SCATTERING	PHENOMENA
Air molecules	0.0001 to 0.001	Rayleigh	Blue sky, red sunsets
Aerosols (pollutants)	0.01 to 1.0	Mie	Whitish haze, glare in clouds near sun
Cloud droplets	10 to 100	Geometric (nonselective)	White clouds

Individual air molecules are much smaller than cloud droplets—their diameters are small even when compared with the wavelength of visible light. Each air molecule of oxygen and nitrogen is a *selective scatterer* in that each scatters shorter waves of visible light much more effectively than longer waves. This selective scattering is also known as *Rayleigh scattering* (see ▼ Table 20.1).

As sunlight enters the atmosphere, the shorter visible wavelengths of violet, blue, and green are scattered more by atmospheric gases than are the longer wavelengths of yellow, orange, and especially red.* In fact, violet light is scattered about 16 times more than red light. Consequently, as we view the sky, the scattered waves of violet, blue, and green strike the eye from all directions. Because our eyes are more sensitive to blue light, these waves, viewed together, produce the sensation of blue coming from all around us (see ● Fig. 20.2). Therefore, when we look at the sky it appears blue (see ● Fig. 20.3). (Earth, by the way, is not the only planet with a colorful sky. On Mars, dust in the air turns the sky ruddy at midday and purple at sunset.)

The selective scattering of blue light by air molecules and very small particles can make distant mountains appear blue, such as happens with the Blue Ridge Mountains of Virginia and the Blue Mountains of Australia (see ● Fig. 20.4). In some places, a *blue haze* may cover the landscape, even in areas far removed from human contamination. Although its cause is still controversial, the blue haze appears to be the result of a particular process.

---
*The reason for this fact is that the intensity of Rayleigh scattering varies as $1/\lambda^4$, where λ is the wavelength of radiation.

Extremely tiny particles (hydrocarbons called *terpenes*) are released by vegetation to combine chemically with small amounts of ozone. This reaction produces tiny particles (about 0.2 μm in diameter) that selectively scatter blue light.

WHITE CLOUDS AND DARK BASES When somewhat larger particles are present, a phenomenon called *Mie scattering* becomes more important. Mie scattering is most noticeable when particles of dust, salt, or smoke are about the same size as the wavelengths

● **FIGURE 20.2** The sky appears blue because billions of air molecules selectively scatter the shorter wavelengths of visible light more effectively than the longer ones. This causes us to see blue light coming from all directions.

● **FIGURE 20.3** Blue skies and white clouds. The selective scattering of blue light by air molecules produces the blue sky, whereas the scattering of all wavelengths of visible light by liquid cloud droplets produces the white clouds.

CRITICAL THINKING QUESTION Why does the lower part of this developing cumulus cloud appear so much darker than the upper part?

© C. Donald Ahrens

● **FIGURE 20.4** The Blue Ridge Mountains in Virginia. The blue haze is caused by the scattering of blue light by extremely small particles—smaller than the wavelengths of visible light. Notice that the scattered blue light causes the most distant mountains to become almost indistinguishable from the sky.

Eva Hambach/AFP/Getty Images

of visible light (or between about 0.1 and 1 µm). Each wavelength is affected in roughly the same way by Mie scattering, so the result tends to be a whitish effect. When large numbers of particles of this size are present, the color of the sky begins to change from blue to milky white, and we may call the day "hazy." Mie scattering sends more light in the forward direction than sideways or backward, so its effects are most noticeable when looking in the general direction of the sun. A whitish halo around the sun, or a bright sheen visible in a nearby layer of thin clouds, is often the result of Mie scattering. If the relative humidity is high enough, soluble particles (nuclei) will "pick up" water vapor and grow into haze particles. Thus, the color of the sky gives us a hint about how much material is suspended in the air: the more particles, the more scattering, and the whiter the sky becomes. Since most

of the suspended particles are near the surface, the horizon often appears white. On top of a high mountain, when we are above many of these haze particles, the sky usually appears a deep blue.

Cloud droplets about 10 µm or so in diameter are large enough to effectively scatter all wavelengths of visible radiation more or less equally in all directions, a phenomenon we call *geometric (or nonselective) scattering* (see ● Fig. 20.5). Clouds, even small ones, are optically thick, meaning they are able to scatter vast amounts of sunlight and thus there is very little chance sunlight will pass through unscattered. These same clouds are poor absorbers of sunlight. Hence, when we look at a cloud, it appears white because countless cloud droplets scatter all wavelengths of visible sunlight in all directions (see ● Fig. 20.6).

As a cloud grows larger and taller, more sunlight is reflected from it and less light can penetrate all the way through it (see ● Fig. 20.7). In fact, relatively little light penetrates a cloud whose thickness is 1000 m (3300 ft). Since little sunlight reaches

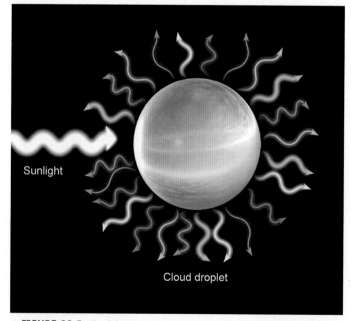

● **FIGURE 20.5** Cloud droplets scatter all wavelengths of visible white light about equally in a process called *geometric (nonselective) scattering*. The different colors represent different wavelengths of visible light.

● **FIGURE 20.6** Because tiny cloud droplets scatter visible light in all directions, light from many billions of droplets turns a cloud white.

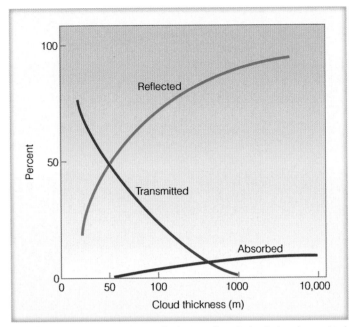

● **FIGURE 20.7** Average percent of radiation reflected, absorbed, and transmitted by clouds of various thickness.

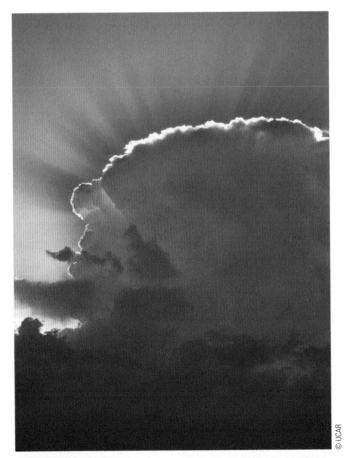

● **FIGURE 20.8** The bright light beams radiating across the sky from behind the clouds are crepuscular rays.

CRITICAL THINKING QUESTION Would you expect the crepuscular rays in Fig. 20.8 to become brighter or dimmer (less bright) if the size of particles that are scattering the light became smaller in diameter?

● **FIGURE 20.9** Anticrepuscular rays photographed near Boulder, Colorado. The rays appear to converge toward the horizon opposite from a setting sun, which is to the back of the photographer.

the underside of the cloud, little light is scattered, and the cloud base appears dark. At the same time, if droplets near the cloud base grow larger, they become less effective scatterers and better absorbers. As a result, the meager amount of visible light that does reach this part of the cloud is absorbed rather than scattered, which makes the cloud appear even darker. These same cloud droplets may even grow large and heavy enough to fall to Earth as rain. From a casual observation of clouds, we know that dark, threatening ones frequently produce rain. Now, we know why they appear so dark.

CREPUSCULAR AND ANTICREPUSCULAR RAYS Haze can scatter light from the rising or setting sun in a way that produces bright light beams, or **crepuscular rays**, radiating across the sky (see ● Fig. 20.8). When the bright light beams appear to converge toward the part of the horizon opposite from the sun, the beams of light are called **anticrepuscular rays** (see ● Fig. 20.9). A similar effect occurs when the sun shines through a break in a layer of clouds. Dust, tiny water droplets, or haze in the air beneath the clouds scatter sunlight, making that region of the sky appear bright with rays. Because these rays seem to reach downward from clouds, some people will remark that the "sun is drawing up water." In England, this same phenomenon is referred to as "Jacob's ladder." No matter what these sunbeams are called, it is the scattering of sunlight by particles in the atmosphere that makes them visible.

Red Suns and Blue Moons

At midday, the sun seems a brilliant white, while at sunset it usually appears to be yellow, orange, or red. At noon, when the sun is high in the sky, light from the sun is most intense—all wavelengths of visible light are able to reach the eye with about equal intensity, and the sun appears white. (Looking directly at the sun, especially during this time of day, can cause irreparable damage to the eye. Normally, we get only glimpses or impressions of the sun out of the corner of our eye.)

Near sunrise or sunset, however, the rays coming directly from the sun strike the atmosphere at a low angle. These rays pass through much more atmosphere than at any other time during the day. (When the sun is 4° above the horizon, sunlight must pass through an atmosphere more than 12 times thicker than when the sun is directly overhead.) By the time sunlight has penetrated this large amount of air, most of the shorter waves of visible light have been scattered away by the air molecules. Just about the only waves from a setting sun that make it on through the atmosphere on a fairly direct path are the yellow, orange, and red. Upon reaching the eye, these waves produce a bright yellow-orange sunset (see ● Fig. 20.10). Such sunsets occur only when the atmosphere is fairly clean, as it would be after a recent rain. If the atmosphere contains many fine particles with diameters that are a little larger than those of air molecules, slightly longer (yellow) waves also will be scattered away. Only orange and red waves would penetrate through to the eye, and the sun would appear red-orange. When the atmosphere becomes loaded with particles, only the longest red wavelengths are able to penetrate the atmosphere, and we see a red sun.

Natural events may produce red sunrises and sunsets over the oceans. For example, the scattering characteristics of small suspended salt particles and water molecules are responsible for the brilliant red suns that can be observed from a beach (see ● Fig. 20.11). Volcanic eruptions rich in sulfur can produce red sunsets, too. Such red sunsets are actually produced by a highly reflective cloud of sulfuric acid droplets, formed from sulfur dioxide gas injected into the stratosphere during powerful eruptions. Two examples of these eruptions are the Mexican volcano El Chichón in 1982 and the

● **FIGURE 20.11** Red sunset near the coast of Iceland. The reflection of sunlight off the slightly rough water is producing a glitter path.

Philippine volcano Mt. Pinatubo in 1991. Fine particles produced by these eruptions were carried by the winds aloft and circled the globe, producing beautiful sunrises and sunsets for months and even years after the eruptions. These same volcanic particles in the stratosphere can turn the sky red after sunset, as some of the red light from the setting sun bounces off the bottom of the particles back to Earth's surface. Generally, these volcanic red sunsets occur about an hour after the actual sunset (see ● Fig. 20.12).

Occasionally, the atmosphere becomes so laden with dust, smoke, and pollutants that even red waves are unable to pierce the filthy air. An eerie effect then occurs. Because no visible waves enter the eye, the sun literally disappears before it reaches the horizon.

● **FIGURE 20.10** Because of the selective scattering of radiant energy by a thick section of atmosphere, the sun at sunrise and sunset appears either yellow, orange, or red. The more particles in the atmosphere, the more scattering of sunlight, and the redder the sun appears.

● **FIGURE 20.12** Bright red sky over California produced by the sulfur-rich particles from the volcano Mt. Pinatubo during September 1992. The photo was taken about an hour after sunset.

The scattering of light by large quantities of atmospheric particles can cause some rather unusual sights. If the dust, smoke particles, or pollutants are roughly uniform in size, they can selectively scatter the sun's rays. Even at noon, various colored suns have appeared—orange suns, green suns, and even blue suns—although green and blue suns are extremely unusual. For blue suns to appear, the size of the suspended particles must be similar to the primary wavelengths of visible light. When these particles are present, they tend to scatter red light more than blue, which causes a bluing of the sun and a reddening of the sky. Although rare, the same phenomenon can happen to moonlight, making the moon appear blue; thus, the expression "once in a blue moon." (The term "blue moon" also refers to the second full moon in a calendar month that has two full moons).

In summary, the scattering of light by small particles in the atmosphere causes many familiar effects: white clouds, blue skies, hazy skies, crepuscular rays, and colorful sunsets. In the absence of any scattering, we would simply see a white sun against a black sky—not an attractive alternative.

Twinkling, Twilight, and the Green Flash

Light that passes through a substance is said to be *transmitted*. Upon entering a denser substance, transmitted light slows in speed. If it enters the substance at an angle, the light's path also bends. This bending is called **refraction**. The amount of refraction depends primarily on two factors: the density of the material and the angle at which the light enters the material.

Refraction can be demonstrated in a darkened room by shining a flashlight into a beaker of water (see ● Fig. 20.13). If the light is held directly above the water so that the beam strikes the surface of the water straight on, no bending occurs. But, if the light enters the water at some angle, it bends toward the *normal*, which is the dashed line in the diagram running perpendicular to the air-water boundary. (The normal is simply a line that intersects any surface at a right angle. We use it as a reference to see how much bending occurs as light enters and leaves various substances.) A small mirror on the bottom of the beaker reflects the light upward. This reflected light bends away from the normal as it re-enters the air.

● **FIGURE 20.13** The behavior of light as it enters and leaves a more-dense substance, such as water.

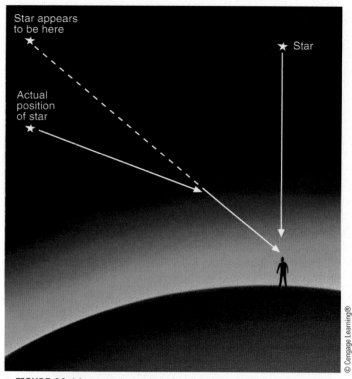

● **FIGURE 20.14** Due to the bending of starlight by the atmosphere, stars not directly overhead appear to be higher than they really are. (The angle between positions is highly exaggerated.)

▼ **TABLE 20.2** The Amount of Atmospheric Refraction (Bending) in Minutes Viewed at Sea Level Under Standard Atmospheric Conditions (60 minutes equals 1°)

ELEVATION ABOVE HORIZON (DEGREES)	REFRACTION (MINUTES)
0°	35.0
5°	10.0
20°	2.6
40°	1.2
60°	0.6
90°	0.0

We can summarize these observations as follows: *Light that travels from a less-dense to a more-dense medium loses speed and bends toward the normal, while light that enters a less-dense medium increases in speed and bends away from the normal.*

The refraction of light within the atmosphere causes a variety of visual effects. At night, for example, the light from the stars that we see directly above us is not bent, but starlight that enters Earth's atmosphere at an angle is bent. In fact, a star whose light enters the atmosphere just above the horizon has more atmosphere to penetrate and is thus refracted the most. As we can see in ● Fig. 20.14, the bending is toward the normal as the light enters the more-dense atmosphere. By the time this "bent" starlight reaches our eyes, the star appears to be higher than it actually is

because our eyes cannot detect that the light path is bent. We see light coming from a particular direction and interpret the star to be in that direction. So, the next time you take a midnight stroll, point to any star near the horizon and remember: This point is only where the star appears to be. To point to the star's true position, you would have to lower your arm just a bit (about one-half a degree, according to ▼ Table 20.2).

As starlight enters the atmosphere, it often passes through regions of differing air density. Each of these regions deflects and bends the tiny beam of starlight, constantly changing the apparent position of the star. This causes the star to appear to *twinkle* or flicker, a condition known as **scintillation**. Planets, being much closer to us, appear larger, and usually do not twinkle because their size is greater than the angle at which their light deviates as it penetrates the atmosphere. Planets sometimes twinkle, however, when they are near the horizon, where the bending of their light is greatest.

The refraction of light by the atmosphere has some other interesting consequences. For example, the atmosphere gradually bends the rays from a rising or setting sun or moon. Because light rays from the lower part of the sun (or moon) are bent more than those from the upper part, the sun appears to flatten out on the horizon, taking on an elliptical shape. (The sun in Fig. 20.16 shows this effect.) Also, since light is bent most on the horizon, the sun and moon both appear to be higher than they really are. As a result, they both rise about two minutes earlier and set about two minutes later than they would if there were no atmosphere (see ● Fig. 20.15).

● **FIGURE 20.15** The bending of sunlight by the atmosphere causes the sun to rise about two minutes earlier, and set about two minutes later, than it would otherwise. (The angle between the positions is highly exaggerated.)

● **FIGURE 20.16** The very light green on the upper rim of the sun is the green flash. Also, observe how the atmosphere makes the sun appear to flatten on the horizon into an elliptical shape.

© Pekka Parviainen

You may have noticed that on clear days the sky is often bright for some time after the sun sets. The atmosphere refracts and scatters sunlight to our eyes, even though the sun itself has disappeared from our view. (Look back at Fig. 20.12.) **Twilight** is the name given to the time after sunset (and immediately before sunrise) when the sky remains illuminated and allows outdoor activities to continue without artificial lighting. (*Civil twilight* lasts from sunset until the sun is 6° below the horizon, while *astronomical twilight* lasts until the sky is completely dark and the astronomical observation of the faintest stars is possible.)

The length of twilight depends on season and latitude. During the summer in middle latitudes, twilight adds about 30 minutes of light to each morning and evening for outdoor activities. The duration of twilight increases with increasing latitude, especially in summer. At high latitudes during the summer, morning and evening twilight may converge, producing a *white night*—a night-long twilight. At low latitudes, twilight tends to be shorter, with more abrupt transitions between daylight and darkness.

In general, without the atmosphere, there would be no refraction or scattering, and the sun would rise later and set earlier than it now does. Instead of twilight, darkness would arrive immediately at the sun's disappearance below the horizon. Imagine the number of evening softball games that would be called because of instant darkness.

Occasionally, a flash of green light—called the **green flash**—can be seen near the upper rim of a rising or setting sun (see ● Fig. 20.16). Remember from our earlier discussion that, when the sun is near the horizon, its light must penetrate a thick section of atmosphere. This thick atmosphere refracts sunlight, with purple and blue light bending the most, and red light the least. Because of this bending, more blue light should appear along the top of the sun. But because the atmosphere selectively scatters blue light, very little reaches us, and we see green light instead.

Usually, the green light is too faint for the human eye to see. However, under certain atmospheric conditions, such as when the surface air is very hot or when an upper-level inversion exists, the atmosphere magnifies the green light. When this happens, a momentary flash of green light appears, often just before the sun disappears from view.

The flash usually lasts about a second, although in polar regions it can last longer. Here, the sun slowly changes in elevation and the flash may exist for many minutes. Members of Admiral Byrd's expedition in the south polar region reported seeing the green flash for 35 minutes in September as the sun slowly rose above the horizon, marking the end of the long winter.

WEATHER WATCH

On December 14, 1890, a mirage—lasting several hours—gave the residents of Saint Vincent, Minnesota, a clear view of cattle nearly 8 miles away.

BRIEF REVIEW

Up to this point, we have examined how light can interact with our atmosphere. Before going on, here is a review of some of the important concepts and facts we have covered:

- When light is scattered, it is sent in all directions—forward, sideways, and backward.

- White clouds, blue skies, hazy skies, crepuscular rays, anticrepuscular rays, and colorful sunsets are all the result of sunlight being scattered.

- The bending of light as it travels through regions of differing density is called refraction.

- As light travels from a less-dense substance (such as outer space) and enters a more-dense substance at an angle (such as our atmosphere), the light bends downward, toward the normal. This effect causes stars, the moon, and the sun to appear just a tiny bit higher than they actually are.

The Mirage: Seeing Is Not Believing

In the atmosphere, when an object appears to be displaced from its true position, we call this phenomenon a **mirage**. A mirage is not a figment of the imagination—our minds are not playing tricks on us, but the atmosphere is.

Atmospheric mirages are created by light passing through and being bent by air layers of different densities. Such changes in air density are usually caused by sharp changes in air temperature. The greater the rate of temperature change, the more the light rays are bent. For example, on a warm, sunny day, black road surfaces absorb a great deal of solar energy and become very hot. Air in contact with these hot surfaces warms by conduction and, because air is a poor thermal conductor, we find much cooler air only a few meters higher. On hot days, these road surfaces often appear wet (see ● Fig. 20.17). Such "puddles" disappear as we approach them, and advancing cars seem to swim in them. Yet, we know the road is dry. The apparent wet pavement above a road is the result of blue sky-light refracting up into our eyes as it travels through air of different densities. A similar type of mirage occurs in deserts during the hot summer. Many thirsty travelers have been disappointed to find that what appeared to be a water hole was in actuality hot desert sand.

Sometimes, these "watery" surfaces appear to *shimmer*. The shimmering results as rising and sinking air near the ground constantly change the air density. As light moves through these regions, its path also changes, causing the shimmering effect.

When the air near the ground is much warmer than the air above, objects may not only appear to be lower than they really are, but also they often appear to be inverted. These mirages are called **inferior** (lower) **mirages**. The tree in ● Fig. 20.18 certainly doesn't grow upside down. So why does it look that way? It appears to be inverted because light reflected from the top of the tree moves outward in all directions. Rays that enter the hot, less-dense air above the sand are refracted upward, entering the eye from below. The brain is fooled into thinking that these rays come from below the ground, which makes the tree appear upside down. However, some light from the top of the tree travels directly toward the eye through air of nearly constant density and, therefore, bends very little. These rays reach the eye "straight-on," and the tree appears upright. Thus, off in the distance, we see a tree and its upside-down image beneath it. (Some of the trees in Fig. 20.17 show this effect.)

The atmosphere can play optical jokes on us in extremely cold areas, too. In polar regions, air next to a snow surface can be much colder than the air many meters above. Because the air in this cold layer is very dense, light from distant objects entering it bends toward the normal in such a way that the objects can appear to be shifted upward. This phenomenon is called a **superior** (upward) **mirage**. ● Figure 20.19 shows the atmospheric conditions favorable for a superior mirage. (A special type of mirage, the **Fata Morgana**, is described in Focus section 20.1.)

● **FIGURE 20.17** The road in the photo appears wet because blue skylight is bending up into the camera as the light passes through air of different densities.

© C. Donald Ahrens

● **FIGURE 20.18** Inferior mirage over hot desert sand.

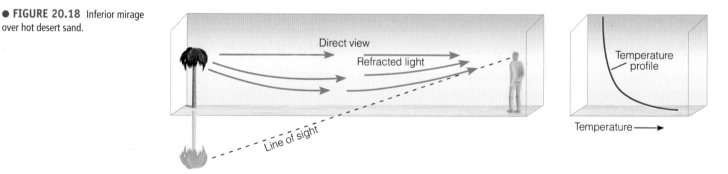

Direct view

Refracted light

Line of sight

Temperature profile

Temperature ⟶

The Fata Morgana

A special type of superior mirage is the *Fata Morgana*, a mirage that transforms a fairly uniform horizon into one of vertical walls and columns with spires (see ● Fig. 1). According to legend, Fata Morgana (Italian for "fairy Morgan") was the half-sister of King Arthur. Morgan, who was said to live in a crystal palace beneath the water, had magical powers that could build fantastic castles out of thin air. Looking across the Straits of Messina (between Italy and Sicily), residents of Reggio, Italy, on occasion would see buildings, castles, and sometimes whole cities appear, only to vanish again in minutes. The Fata Morgana is observed where the air temperature increases with height above the surface, slowly at first, then more rapidly, then slowly again. Consequently, mirages like the Fata Morgana are frequently seen where warm air rests above a cold surface, such as above large bodies of water and in polar regions.

● **FIGURE 1** The Fata Morgana mirage over water. The mirage is the result of refraction—light from small islands and ships is bent in such a way as to make them appear to rise vertically above the water.

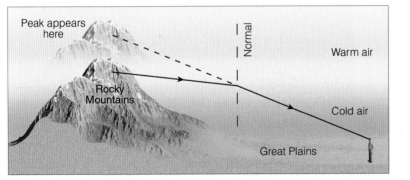

● **FIGURE 20.19** The formation of a superior mirage. When cold air lies close to the surface with warm air aloft, light from distant mountains is refracted toward the normal as it enters the cold air. This causes an observer on the ground to see mountains higher and closer than they really are.

Halos, Sundogs, and Sun Pillars

A ring of light encircling and extending outward from the sun or moon is called a **halo**. Such a display is produced when sunlight or moonlight is refracted as it passes through ice crystals. Hence, the presence of a halo indicates that *cirriform clouds* are present.

The most common type of halo is the 22° halo—a ring of light 22° from the sun or moon.* Such a halo forms when tiny suspended column-type ice crystals (with diameters less than 20 μm) become randomly oriented as air molecules constantly bump against them. The refraction of light rays through these crystals forms a halo like the one shown in ● Fig. 20.20. Less common is the 46° halo, which forms in a similar fashion to the 22° halo (see ● Fig. 20.21). With the 46° halo, however, the light is refracted through column-type ice crystals that have diameters in a narrow range between about 15 and 25 μm.

Occasionally, a bright arc of light can be seen at the top of a 22° halo (see ● Fig. 20.22). Since the arc is tangent to the halo, it is called a **tangent arc**. Apparently, the arc forms as large six-sided

---

*Extend your arm and spread your fingers apart. An angle of 22° is about the distance from the tip of the thumb to the tip of the little finger.

● **FIGURE 20.20** A 22° halo around the sun, produced by the refraction of sunlight through ice crystals.

● **FIGURE 20.22** Halo with an upper tangent arc.

● **FIGURE 20.21** The formation of a 22° and a 46° halo with column-type ice crystals.

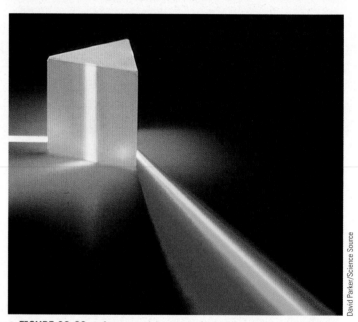

● **FIGURE 20.23** Refraction and dispersion of light through a glass prism.

(hexagonal), pencil-shaped ice crystals fall with their long axes horizontal to the ground. Refraction of sunlight through the ice crystals produces the bright arc of light. When the sun is on the horizon, the arc that forms at the top of the halo is called an *upper tangent arc.* When the sun is above the horizon, a *lower tangent arc* may form on the lower part of the halo beneath the sun. The shape of the arcs changes greatly with the position of the sun.

A halo is usually seen as a bright, white ring, but some refraction effects can add color. To understand this, we must first examine refraction more closely.

When white light passes through a glass prism, it is refracted and split into a spectrum of visible colors (see ● Fig. 20.23). Each wavelength of light is slowed by the glass, but each is slowed a little differently. Because longer wavelengths (red) slow the least and shorter wavelengths (violet) slow the most, red light bends the least, and violet light bends the most. The breaking up of white light by "selective" refraction is called **dispersion** of light. As light passes through ice crystals, dispersion causes red light to be on the inside of the halo and blue light on the outside.

When hexagonal platelike ice crystals with diameters larger than about 30 μm are present in the air, they tend to fall slowly and orient themselves horizontally (see ● Fig. 20.24). (The horizontal orientation of these ice crystals prevents a ring halo.) In this position, the ice crystals act as small prisms, refracting and dispersing sunlight that passes through them. If the sun is near the horizon in such a configuration that it, ice crystals, and observer are all in the same horizontal plane, the observer will see a pair of brightly colored spots, one on either side of the sun. These colored spots are called **sundogs,** *mock suns,* or **parhelia**—meaning "with the sun" (see ● Fig. 20.25). The colors usually grade from red

● **FIGURE 20.24** Platelike ice crystals falling with their flat surfaces parallel to Earth produce sundogs.

● **FIGURE 20.25** The bright areas on each side of the sun are sundogs.

© Mike Hollingshead/Getty Images

● FIGURE 20.26 The refraction of sunlight through cirrus clouds (ice crystals) at about 58° above the horizon produces a vibrant display of color.

● FIGURE 20.27 A brilliant red sun pillar extending upward above the sun, produced by the reflection of sunlight off ice crystals.

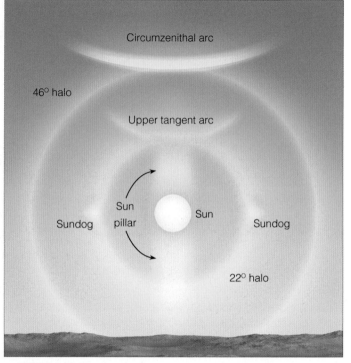

● FIGURE 20.28 Optical phenomena that form when cirriform ice crystal clouds are present. (A picture of the circumzenithal arc is in Fig. 2 on p. 583.)

As these crystals fall, they can rotate about their horizontal axes, producing many orientations that reflect sunlight. Look for sun pillars when the sun is low on the horizon and cirriform clouds are present. ● Figure 20.28 is a summary of some of the optical phenomena that form with cirriform clouds.

Rainbows

Now we come to one of the most spectacular light shows observed on Earth—the rainbow. **Rainbows** occur when rain is falling in one part of the sky, and the sun is shining in another. (Rainbows also may form by the sprays from waterfalls and water sprinklers.) To see the rainbow, we must face the falling rain with the sun at our backs. Look at ● Fig. 20.29 closely and note that, when we see a rainbow in the evening, we are facing east toward the rain shower. Behind us—in the west—it is usually clear, or at least clear enough for sunlight to reach the showers. Because clouds tend to move from west to east in middle latitudes, the clear skies in the west suggest that the showers will give way to clearing. However, when we see a rainbow in the morning, we are facing west, toward the rain shower. It is a good bet that the clouds and showers will move toward us and it will rain soon. These observations explain why the following weather rhyme became popular:

Rainbow at morning, sailor take warning;
Rainbow at night, sailor's delight.*

---

*This rhyme is often used with the words "red sky" in the place of "rainbow." The red sky makes sense when we consider that it is the result of red light from a rising or setting sun being reflected from the underside of clouds above us. In the morning, a red sky indicates that it is clear to the east and cloudy to the west. A red sky in the evening suggests the opposite.

(bent least) on the inside closest to the sun to blue (bent more) on the outside. The refraction of sunlight through ice crystals can produce a variety of optical visuals, sometimes turning white cirriform clouds into a display of vibrant colors (see ● Fig. 20.26).

Whereas sundogs, tangent arcs, and halos are caused by *refraction* of sunlight *through* ice crystals, **sun pillars** are caused by *reflection* of sunlight *off* ice crystals. Sun pillars appear most often at sunrise or sunset as a vertical shaft of light extending upward or downward from the sun (see ● Fig. 20.27). Pillars may form as hexagonal platelike ice crystals fall with their flat bases oriented horizontally. As the tiny crystals fall in still air, they tilt from side to side like a falling leaf. This motion allows sunlight to reflect off the tipped surfaces of the crystals, producing a relatively bright area in the sky above or below the sun. Pillars may also form as sunlight reflects off hexagonal pencil-shaped ice crystals that fall with their long axes oriented horizontally.

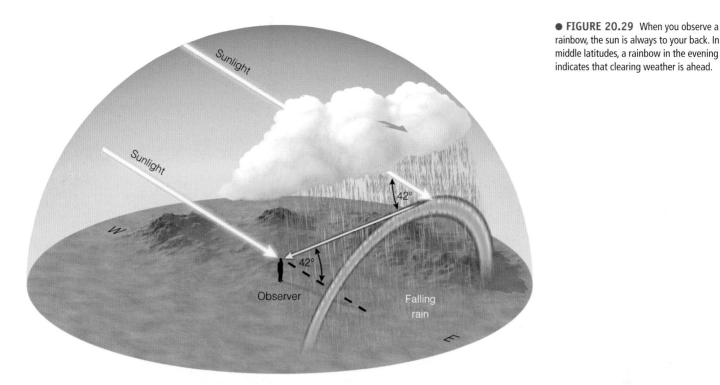

● **FIGURE 20.30** Rays of sunlight entering a raindrop bounce off the back of the drop and are redirected toward our eyes.

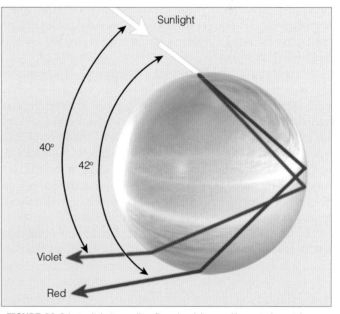

● **FIGURE 20.31** Sunlight internally reflected and dispersed by a raindrop. Light rays are internally reflected only when they strike the backside of the drop at an angle greater than the critical angle for water. Refraction of the light as it enters the drop causes the point of reflection (on the back of the drop) to be different for each color. Hence, the colors are separated from each other when the light emerges from the raindrop.

When we look at a rainbow we are looking at sunlight that has entered the falling drops, and, in effect, has been redirected back toward our eyes. Exactly how this process happens requires some discussion.

As sunlight enters a raindrop, it slows and bends (see ● Fig. 20.30). Although most of this light passes right on through the drop and we do not see it, some of it strikes the backside of the drop at such an angle that it is reflected within the drop. The angle at which this occurs is called the *critical angle*, and for water, this angle is 48°. Notice in Fig. 20.30 that light that strikes the back of a raindrop at an angle exceeding the critical angle

bounces off the back of the drop and is *internally reflected* toward our eyes. Because each light ray bends differently from the rest, each ray emerges from the drop at a slightly different angle as illustrated in Fig. 20.30.

Observe in ● Fig. 20.31 that inside a raindrop, violet light is refracted the most and red light the least. This situation causes red light to emerge from the drop at an angle of 42° from the beam of sunlight, whereas violet light emerges at an angle of 40°. The light leaving the drop is, therefore, dispersed into a spectrum of colors from red to violet. Since we see only a single color from each drop, it takes myriads of raindrops (each refracting and reflecting light back to our eyes at slightly different angles) to produce the brilliant colors of a *primary rainbow*. If raindrops are not spread widely enough across the sky, then the rainbow may not appear "complete."

Figure 20.31 might lead us erroneously to believe that red light should be at the bottom of the bow and violet at the top. A more careful observation of the behavior of light leaving two drops (see ● Fig. 20.32) shows us why the reverse is true. When violet light from the *lower drop* reaches an observer's eye, red light from the same drop is arriving further down, toward the observer's waist. Notice that red light reaches the observer's eye from the *higher drop*. Because the color red comes from higher drops and the color violet from lower drops, the colors of a primary rainbow change from red on the outside (top) to violet on the inside (bottom).

Sometimes a larger second (secondary) rainbow, with its colors reversed, can be seen above the primary bow (see ● Fig. 20.33). Usually this secondary bow is much fainter than the primary one. The *secondary bow* is caused when sunlight enters the raindrops at an angle that allows the light to make two internal reflections in each drop. Each reflection weakens the light intensity and makes the bow dimmer. ● Figure 20.34 shows that the color reversals—with red now at the bottom and violet on top—arise from the way the light emerges from each drop after going through two internal reflections.*

As you look at a rainbow, keep in mind that only one ray of light is able to enter your eye from each drop. Every time you move, whether it be up, down, or sideways, the rainbow moves with you. This is because, with every movement, light from different raindrops enters your eye. The bow you see is not exactly the same rainbow that the person standing next to you sees. In effect, each of us has a personal rainbow to ponder and enjoy! (Earlier we learned that for a rainbow to form, rain must be falling in one part of the sky. Can a rainbow actually form on a day when it is not raining? If you are unsure of the answer, read Focus section 20.2.)

---

*An even smaller amount of light is reflected a third time, which leads to a *tertiary (third-order) bow* on the opposite side of the sky, around the sun. Because these three-time reflections are so attenuated, and because the tertiary rainbow appears on the sunny side of the sky, it is almost never observed with the naked eye, but advances in digital photography confirmed its presence. In 2011, the first photograph of a *quaternary (fourth-order) bow* was captured—again surrounding the sun—and in 2012 a *quinary (fifth-order) bow* was photographed. The quinary rainbow lies between the primary and secondary bows.

● **FIGURE 20.32** The formation of a primary rainbow. The observer sees red light from the upper drop and violet light from the lower drop.

● **FIGURE 20.33** A primary and a secondary rainbow.

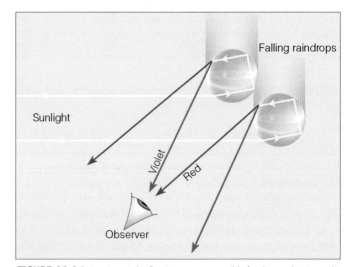

● **FIGURE 20.34** Two internal reflections are responsible for the weaker, secondary rainbow. Notice that the eye sees violet light from the upper drop and red light from the lower drop.

Can It Be a Rainbow If It Is Not Raining?

Up to this point, we have seen that the sky is full of atmospheric visuals. One that we closely examined was the rainbow. Look back at Fig. 20.33, p. 582, and then look at ● Fig. 2. Is the color display in Fig. 2 a rainbow? The colors definitely show a rainbow-like brilliance. For this reason, some people will call this phenomenon a rainbow. But remember, for a rainbow to form it must be raining, and on this day it is not.

The color display in Fig. 2 is due to the refraction of light through ice crystals. Earlier in this chapter we saw that the refraction of light produces a variety of visuals, such as halos, tangent arcs, and sundogs. (Refer back to

Fig. 20.28, p. 580.) The color display in Fig. 2 is a type of refraction phenomena called a *circumzenithal arc*. The photograph was taken in the afternoon, during late winter, when the sun was about 30° or so above the western horizon and the sky was full of ice crystal (cirrus) clouds. The arc was almost directly overhead, at the zenith; hence, its name—"circumzenithal."

The circumzenithal arc forms about 45° above the sun as platelike ice crystals fall with their flat surfaces parallel to the ground. Remember that this is similar to the formation of a sundog (see Fig. 20.24, p. 529), except that in the formation of a circumzenithal arc,

sunlight enters the top of the crystal and exits one of its sides (see ● Fig. 3). This is why the short-lived circumzenithal arc can only form when the sun is lower than 32° above the horizon. When the sun is higher than this angle, the refracted light cannot be seen by the observer.

There is a wide variety of other refraction phenomena (too many to describe here) that may at first glance appear as a rainbow. Keep in mind, however, that rainbows only form when it is raining in one part of the sky and the sun is shining in another.

© C. Donald Ahrens

● **FIGURE 2** Is this a rainbow? The photograph was taken looking almost straight up.

● **FIGURE 3** The formation of a circumzenithal arc.

Coronas, Glories, and *Heiligenschein*

When the moon is seen through a thin veil of clouds composed of tiny spherical water droplets, a bright ring of light, called a **corona** (meaning "crown"), may appear to rest on the moon (see ● Fig. 20.35). The same effect can occur with the sun, but, due to the sun's brightness, it is usually difficult to see.

The corona is a result of **diffraction** of light—the bending of light as it passes *around* objects. To understand the corona, imagine water waves moving around a small stone in a pond. As the waves spread around the stone, the trough

of one wave may meet the crest of another wave. This situation causes the waves to cancel each other, thus producing calm water. (This is known as *destructive interference*.) Where two crests come together (*constructive interference*), they produce a much larger wave. The same thing happens when light passes around tiny cloud droplets. Where light waves constructively interfere, we see bright light; where destructive interference occurs, we see darkness. Sometimes, the corona appears white, with alternating bands of light and dark. On other occasions, the rings have color (see ● Fig. 20.36).

● **FIGURE 20.35** The corona around the moon results from the diffraction of light by tiny liquid cloud droplets of uniform size.

● **FIGURE 20.36** Corona around the sun.

The colors appear when the cloud droplets (or any kind of small particles, such as volcanic ash) are of uniform size. Because the amount of bending due to diffraction depends upon the wavelength of light, the shorter wavelength blue light appears on the inside of a ring, whereas the longer wavelength red light appears on the outside. These colors may repeat over and over, becoming fainter as each ring is farther from the moon or sun. Also, the smaller the cloud droplets, the larger the ring diameter. Therefore, clouds that have recently formed (such as thin altostratus and altocumulus) are the best corona producers.

When various sizes of droplets exist within a cloud, the corona becomes distorted and irregular. Sometimes the cloud exhibits patches of color, often pastel shades of pink, blue, or green. These bright areas produced by diffraction are called **iridescence** (see ● Fig. 20.37). Cloud iridescence is most often seen within 20° of the sun, and it is often associated with clouds such as cirrocumulus and altocumulus.

Like the corona, the **glory** is also a diffraction phenomenon. When an aircraft flies above a cloud layer composed of water droplets less than 50 µm in diameter, a set of colored rings, called the *glory,* may appear around the shadow of the aircraft (see ● Fig. 20.38). The same effect can happen when you stand with your back to the sun and look into a cloud or fog

● **FIGURE 20.37** Cloud iridescence.

bank, as a bright ring of light may be seen around the shadow of your head. In this case, the glory is called the *brocken bow,* after the Brocken Mountains in Germany, where it is particularly common.

For the glory and the brocken bow to occur, the sun must be to your back, so that sunlight can be returned to your eye from the water droplets. Sunlight that enters the small water droplet along its edge is refracted, then reflected off the backside of the droplet. The light then exits at the other side of the droplet, being refracted

● **FIGURE 20.38** The series of rings surrounding the shadow of the aircraft is called the *glory*.

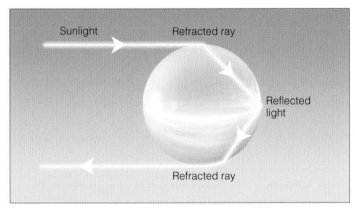

● **FIGURE 20.39** Light that produces the glory follows this path in a water droplet.

● **FIGURE 20.40** The *Heiligenschein* is the ring of light around the shadow of the observer's head.

once again (see ● Fig. 20.39). However, for the light to be returned to your eyes, the light actually clings ever so slightly to the edge of the droplet—the light actually skims along the surface of the droplet as a *surface wave* for a short distance. Diffraction of light coming from the edges of the droplets produces the ring of light we see as the glory and the brocken bow. The colorful rings may be due to the various angles at which different colors leave the droplet.

On a clear morning with dew on the grass, stand facing the dew with your back to the sun and observe that, around the shadow of your head, is a bright area—the ***Heiligenschein*** (German for "halo"). The *Heiligenschein* forms when sunlight that falls on nearly spherical dew drops is focused and reflected back toward the sun along nearly the same path that it took originally. (Light reflected in this manner is said to be *retroreflected*.) The light, however, does not travel along the exact path; it actually spreads out just enough to be seen as bright white light around the shadow of your head on a dew-covered lawn (see ● Fig. 20.40).

The scattering of sunlight in the atmosphere can produce a variety of atmospheric visuals, from hazy days and blue skies to crepuscular rays and blue moons. Refraction (bending) of light by the atmosphere causes stars near the horizon to appear higher than they really are. It also causes the sun and moon to rise earlier and set later than they otherwise would. Under certain atmospheric conditions, the amplification of green light near the upper rim of a rising or setting sun produces the illusive green flash.

Mirages form when refraction of light displaces objects from their true positions. Inferior mirages cause objects to appear lower than they really are, while superior mirages displace objects upward.

Halos and sundogs form from the refraction of light through ice crystals. Sun pillars are the result of sunlight reflecting off gently falling ice crystals. The refraction, reflection, and dispersion of light in raindrops create a rainbow. If you want to see a rainbow, the sun must be to your back, and rain must be falling in front of you. Diffraction of light produces coronas, glories, and cloud iridescence. We can see the *Heiligenschein* on a clear morning when sunlight falls on nearly spherical dew drops.

KEY TERMS

The following terms are listed (with corresponding page numbers) in the order they appear in the text. Define each. Doing so will aid you in reviewing the material covered in this chapter.

reflected light, 568	halo, 577
scattered light, 568	tangent arc, 577
crepuscular rays, 571	dispersion (of light), 578
anticrepuscular rays, 571	sundog (parhelia), 578
refraction (of light), 573	sun pillars, 580
scintillation, 574	rainbow, 580
twilight, 575	corona, 583
green flash, 575	diffraction (of light), 583
mirage, 576	iridescence, 584
inferior mirage, 576	glory, 584
superior mirage, 576	*Heiligenschein,* 585
Fata Morgana, 576	

QUESTIONS FOR REVIEW

1. (a) Why are cumulus clouds normally white?
   (b) Why do the undersides of building cumulus clouds frequently change color from white to dark gray or even black?
2. Explain why the sky is blue during the day and black at night.
3. What can make a setting (or rising) sun appear red?
4. Why do stars "twinkle"?
5. Explain why the horizon sky appears white on a hazy day.
6. How does light bend as it enters a more-dense substance at an angle? How does it bend upon leaving the more-dense substance? Make a sketch to illustrate your answer.
7. Since twilight occurs without the sun being visible, how does it tend to lengthen the day?
8. On a clear, dry, warm day, why do dark road surfaces frequently appear wet?
9. What atmospheric conditions are necessary for an inferior mirage? A superior mirage?
10. What process (refraction or scattering) produces crepuscular rays?
11. (a) Describe how a halo forms.
    (b) How is the formation of a halo different from that of a sundog?
12. At what time of day would you expect to observe the green flash?
13. Explain how sun pillars form.
14. What process (refraction, scattering, diffraction) is responsible for lengthening the day?
15. Why can a rainbow only be observed if the sun is toward the observer's back?
16. Why are secondary rainbows higher and much dimmer than primary rainbows? Explain your answer with the aid of a diagram.
17. Explain why this rhyme makes sense:Rainbow at morning, runner take warning.Rainbow at night (evening), runner's delight.
18. Suppose you look at the moon and see a bright ring of light that appears to rest on its surface.
    (a) Is this ring of light a halo or a corona?
    (b) What type of clouds (water or ice) must be present for this type of optical phenomenon to occur?
    (c) Is this ring of light produced mainly by refraction or diffraction?
19. Would you expect to see the glory when flying in an aircraft on a perfectly clear day? Explain.
20. Explain how light is able to reach your eyes when you see
    (a) a corona;
    (b) a glory;
    (c) the *Heiligenschein*
21. How would you distinguish a corona from a halo?
22. What process is primarily responsible for the formation of cloud iridescence—reflection, refraction, or diffraction of light?

QUESTIONS FOR THOUGHT

1. Explain why on a cloudless day the sky will usually appear milky white before it rains and a deep blue after it rains.

2. How long does twilight last on the moon? (Hint: The moon has no atmosphere.)

3. Why is it often difficult to see the road while driving on a foggy night with your high beam lights on?

4. What would be the color of the sky if air molecules scattered the longest wavelengths of visible light and passed the shorter wavelengths straight through? (Use a diagram to help explain your answer.)

5. Explain why the colors of the planets are not related to the temperatures of the planets, whereas the colors of the stars are related to the temperatures of the stars.

6. If there were no atmosphere surrounding Earth, what color would the sky be at sunrise? At sunset? What color would the sun be at noon? At sunrise? At sunset?

7. Why are rainbows seldom observed at noon?

8. On a cool, clear summer day, a blue haze often appears over the Great Smoky Mountains of Tennessee. Explain why the blue haze usually changes to a white haze as the relative humidity of the air increases.

9. During a lunar eclipse, Earth, the sun, and the moon are aligned as shown in ● Fig. 20.41. The earth blocks sunlight from directly reaching the moon's surface, yet the surface of the moon will often appear a pale red color during a lunar eclipse. How can you account for this phenomenon?

10. Explain why smoke rising from a cigarette often appears blue, yet appears white when blown from the mouth.

11. During Ernest Shackleton's last expedition to Antarctica, on May 8, 1915, seven days after the sun had set for the winter, he saw the sun reappear. Explain how this event—called the Novaya Zemlya effect—can occur.

12. Could a superior mirage form over land on a hot, sunny day? Explain.

13. Explain why it is easier to get sunburned on a high mountain than in the valley below. (The answer is not that you are closer to the sun on top of the mountain.)

14. Why are stars more visible on a clear night when there is no moon than on a clear night with a full moon?

15. During the day, clouds are white, and the sky is blue. Why then, during a full moon, do cumulus clouds appear faintly white, while the sky does not appear blue?

PROBLEMS AND EXERCISES

1. Choose a three-day period in which to observe the sky five times each day. Record in a notebook the number of times you see halos, crepuscular rays, coronas, cloud iridescence, sun dogs, rainbows, and other phenomena.

2. Make your own rainbow. On a sunny afternoon or morning, turn on the water sprinkler to create a spray of water drops. Stand as close to the spray as possible (without getting soaked) and observe that, as you move up, down, and sideways, the bow moves with you.
   (a) Explain why this happens.
   (b) Also, explain with the use of a diagram why the sun must be at your back in order for you to see the bow.

3. Take a large beaker or bottle and fill it with water. Add a small amount of nonfat powdered milk and stir until the water turns a faint milky white. Shine white light into the beaker, and, on the opposite side, hold a white piece of paper.
   (a) Explain why the milk has a blue cast to it and why the light shining on the paper appears ruddy.
   (b) What do you know about the size of the milk particles?
   (c) How does this demonstration relate to the color of the sky and the color of the sun—at sunrise and sunset?

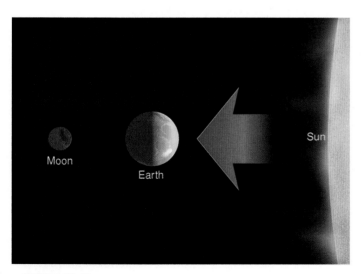 ONLINE RESOURCES

Visit www.cengagebrain.com to view additional resources, including video exercises, practice quizzes, an interactive eBook, and more.

● **FIGURE 20.41**

Units, Conversions, Abbreviations, and Equations

LENGTH

1 kilometer (km)	=	1000 m
	=	3281 ft
	=	0.62 mi
1 statute mile (mi)	=	5280 ft
	=	1609 m
	=	1.61 km
1 nautical mile (nm)	=	1.15 mi
	=	1.85 km
	=	0.87 nm
1 meter (m)	=	100 cm
	=	3.28 ft
	=	39.37 in.
1 foot (ft)	=	12 in.
	=	30.48 cm
	=	0.305 m
1 centimeter (cm)	=	0.39 in.
	=	0.01 m
	=	10 mm
1 inch (in.)	=	2.54 cm
	=	0.08 ft
1 millimeter (mm)	=	0.1 cm
	=	0.001 m
	=	0.039 in.
1 micrometer (μm)	=	0.0001 cm
	=	0.000001 m
1 degree latitude	=	111 km
	=	60 nautical mi
	=	69 statute mi

AREA

1 square centimeter (cm^2)	=	0.15 $in.^2$
1 square inch ($in.^2$)	=	6.45 cm^2
1 square meter (m^2)	=	10.76 ft^2
1 square foot (ft^2)	=	0.09 m^2

VOLUME

1 cubic centimeter (cm^3)	=	0.06 $in.^3$
1 cubic inch ($in.^3$)	=	16.39 cm^3

SPEED

1 knot	=	1 nautical mi/hr
	=	1.15 statute mi/hr
	=	0.51 m/sec
	=	1.85 km/hr
1 mile per hour (mi/hr)	=	0.87 knot
	=	0.45 m/sec
	=	1.61 km/hr
1 kilometer per hour (km/hr)	=	0.54 knot
	=	0.62 mi/hr
	=	0.28 m/sec
	=	1.94 knots
1 meter per second (m/sec)	=	2.24 mi/hr
	=	3.60 km/hr

FORCE

1 dyne	=	1 gram centimeter per second per second 1g cm/sec^2
	=	2.2481×10^{-6} pound (lb)
1 newton (N)	=	1 kilogram meter per second per second 1kg m/sec^2
	=	10^5 dynes
	=	0.2248 lb

MASS

1 gram (g)	=	0.035 ounce
	=	0.002 lb
1 kilogram (kg)	=	1000 g
	=	2.2 lb

ENERGY

1 erg	=	1 dyne per cm
	=	2.388×10^{-8} cal
1 joule (J)	=	1 newton meter
	=	0.239 cal
	=	10^7 erg
1 calorie (cal)	=	4.186 J
	=	4.186×10^7 erg

PRESSURE

1 millibar (mb) = 1000 dynes/cm^2
= 0.75 millimeter of mercury (mm Hg)
= 0.02953 inches of mercury (in. Hg)
= 0.01450 pounds per square inch (lb/in.^{2})
= 100 pascals (Pa)

1 standard atmosphere	=	1013.25 mb
	=	760 mm Hg
	=	29.92 in. Hg
	=	14.7 lb/in.^{2}
1 inch of mercury	=	33.865 mb
1 millimeter of mercury	=	1.3332 mb
1 pascal	=	0.01 mb
	=	1 N/m^2
1 hectopascal (hPa)	=	1 mb
1 kilopascal (kPa)	=	10 mb

POWER

1 watt (W)	=	1 J/sec
	=	14. 3353 cal/min
1 cal/min	=	0.06973 W
1 horsepower (hp)	=	746 W

POWERS OF TEN

PREFIX

nano	one-billionth	=	10^{-9}	=	0.000000001
micro	one-millionth	=	10^{-6}	=	0.000001
milli	one-thousandth	=	10^{-3}	=	0.001
centi	one-hundredth	=	10^{-2}	=	0.01
deci	one-tenth	=	10^{-1}	=	0.1
hecto	one hundred	=	10^2	=	100
kilo	one thousand	=	10^3	=	1000
mega	one million	=	10^6	=	1,000,000
giga	one billion	=	10^9	=	1,000,000,000

TEMPERATURE

$$°C = \tfrac{5}{9}\,(°F - 32)$$

To convert degrees Fahrenheit (°F) to degrees Celsius (°C): Subtract 32 degrees from °F, then divide by 1.8.
To convert degrees Celsius (°C) to degrees Fahrenheit (°F): Multiply °C by 1.8, then add 32 degrees.
To convert degrees Celsius (°C) to Kelvins (K): Add 273 to Celsius temperature, as

$$K = °C + 273$$

▼ TABLE A.1 **Temperature Conversions**

°F	°C	°F	°C	°F	°C	°F	°C	°F	°C	°F	°C	°F	°C	°F	°C
−40	−40	−20	−28.9	0	−17.8	20	−6.7	40	4.4	60	15.6	80	26.7	100	37.8
−39	−39.4	−19	−28.3	1	−17.2	21	−6.1	41	5	61	16.1	81	27.2	101	38.3
−38	−38.9	−18	−27.8	2	−16.7	22	−5.6	42	5.6	62	16.7	82	27.8	102	38.9
−37	−38.3	−17	−27.2	3	−16.1	23	−5	43	6.1	63	17.2	83	28.3	103	39.4
−36	−37.8	−16	−26.7	4	−15.6	24	−4.4	44	6.7	64	17.8	84	28.9	104	40
−35	−37.2	−15	−26.1	5	−15	25	−3.9	45	7.2	65	18.3	85	29.4	105	40.6
−34	−36.7	−14	−25.6	6	−14.4	26	−3.3	46	7.8	66	18.9	86	30	106	41.1
−33	−36.1	−13	−25	7	−13.9	27	−2.8	47	8.3	67	19.4	87	30.6	107	41.7
−32	−35.6	−12	−24.4	8	−13.3	28	−2.2	48	8.9	68	20	88	31.1	108	42.2
−31	−35	−11	−23.9	9	−12.8	29	−1.7	49	9.4	69	20.6	89	31.7	109	42.8
−30	−34.4	−10	−23.3	10	−12.2	30	−1.1	50	10	70	21.1	90	32.2	110	43.3
−29	−33.9	−9	−22.8	11	−11.7	31	−0.6	51	10.6	71	21.7	91	32.8	111	43.9
−28	−33.3	−8	−22.2	12	−11.1	32	0	52	11.1	72	22.2	92	33.3	112	44.4
−27	−32.8	−7	−21.7	13	−10.6	33	0.6	53	11.7	73	22.8	93	33.9	113	45
−26	−32.2	−6	−21.1	14	−10	34	1.1	54	12.2	74	23.3	94	34.4	114	45.6
−25	−31.7	−5	−20.6	15	−9.4	35	1.7	55	12.8	75	23.9	95	35	115	46.1
−24	−31.1	−4	−20	16	−8.9	36	2.2	56	13.3	76	24.4	96	35.6	116	46.7
−23	−30.6	−3	−19.4	17	−8.3	37	2.8	57	13.9	77	25	97	36.1	117	47.2
−22	−30	−2	−18.9	18	−7.8	38	3.3	58	14.4	78	25.6	98	36.7	118	47.8
−21	−29.4	−1	−18.3	19	−7.2	39	3.9	59	15	79	26.1	99	37.2	119	48.3

▼ TABLE A.2　SI Units* and Their Symbols

QUANTITY	NAME	UNIT	SYMBOL
length	meter	m	m
mass	kilogram	kg	kg
time	second	sec	sec
temperature	Kelvin	K	K
density	kilogram per cubic meter	kg/m^3	kg/m^3
speed	meter per second	m/sec	m/sec
force	newton	$m \cdot kg/sec^2$	N
pressure	pascal	N/m^2	Pa
energy	joule	$N \cdot m$	J
power	watt	J/sec	W

*SI stands for Système International, which is the international system of units and symbols.

▼ TABLE A.3　Some Useful Equations and Constants

NAME	EQUATION	CONSTANTS AND ABBREVIATIONS
Gas law (equation of state)	$p = \rho RT$	$R = 287$ J/kg $\cdot$ K (SI) or $R = 2.87 \times 10^6$ erg/g $\cdot$ K p = pressure is N/m^2 (SI) ρ = density (kg/m^3) T = temperature (K)
Stefan-Boltzmann law	$E = \sigma T^4$	$\sigma = 5.67 \times 10^{-8}$ W/m^2 $\cdot$ K^4 (SI) or $\sigma = 5.67 \times 10^{-5}$ erg/cm^2 $\cdot$ K^4 $\cdot$ sec E = radiation emitted in W/m^2 (SI)
Wien's law	$\lambda_{max} = \dfrac{w}{T}$	$w = 2897$ μm K λ_{max} = wavelength (μm)
Solar constant		1376 W/m^2 (SI) 1.97 cal/cm^2/min
Geostrophic wind equation	$Vg = \dfrac{1}{2\Omega \sin\phi\rho} \dfrac{\Delta p}{d}$	Vg = geostrophic wind (m/sec) $\Omega^g = 7.29 \times 10^{-5}$ radian*/sec ϕ = latitude d = distance (m) Δp = pressure difference (N/m^2)
Coriolis parameter	$f = 2\Omega \sin\phi$	g = force of gravity (9.8 m/sec^2)
Hydrostatic equation	$\dfrac{\Delta p}{\Delta z} = -\rho g$	Δz = change in height (m)

*2π radians equal 360°.

Weather Symbols and the Station Model

SIMPLIFIED SURFACE-STATION MODEL

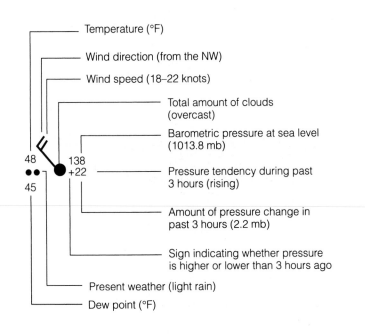

Temperature (°F)

Wind direction (from the NW)

Wind speed (18–22 knots)

Total amount of clouds (overcast)

Barometric pressure at sea level (1013.8 mb)

Pressure tendency during past 3 hours (rising)

Amount of pressure change in past 3 hours (2.2 mb)

Sign indicating whether pressure is higher or lower than 3 hours ago

Present weather (light rain)

Dew point (°F)

48
••
45

138
+22

CLOUD COVERAGE

○	Clear
⦶	1/8
◴	Scattered
◕	3/8
◑	4/8
⊖	5/8
◐	Broken
◉	7/8
●	Overcast
⊗	Obscured
Ⓜ	Missing

UPPER-AIR MODEL (500 MB)

Temperature (°C)

Height of pressure surface in meters with first 3 digits given (5640 m)

12 hour height change in meters (04 equals 40 m)

Sign indicating whether height is rising or falling

Dew point depression (difference between air temperature and dew point, °C)

Wind speed (58–62 knots)

Wind direction (from the southwest)

−15   564
5   +04

COMMON WEATHER SYMBOLS

••	Light rain	▽	Rain shower
•••	Moderate rain	✶▽	Snow shower
••••	Heavy rain	△▽	Showers of hail
✶✶	Light snow	✚	Drifting or blowing snow
✶✶	Moderate snow	ᔕ	Dust storm
✶✶✶	Heavy snow	═	Fog
,,	Light drizzle	⚯	Haze
△	Ice pellets (sleet)	⥯	Smoke
⌒•	Freezing rain	℞	Thunderstorm
⌒,	Freezing drizzle	⧫	Hurricane

WIND ENTRIES

	MILES (STATUTE) PER HOUR	KNOTS	KILOMETERS PER HOUR
Calm symbol	Calm	Calm	Calm
	1–2	1–2	1–3
	3–8	3–7	4–13
	9–14	8–12	14–19
	15–20	13–17	20–32
	21–25	18–22	33–40
	26–31	23–27	41–50
	32–37	28–32	51–60
	38–43	33–37	61–69
	44–49	38–42	70–79
	50–54	43–47	80–87
	55–60	48–52	88–96
	61–66	53–57	97–106
	67–71	58–62	107–114
	72–77	63–67	115–124
	78–83	68–72	125–134
	84–89	73–77	135–143
	119–123	103–107	144–198

PRESSURE TENDENCY

Rising, then falling

Rising, then steady; or rising, then rising more slowly

Rising steadily or unsteadily

Falling or steady, then rising; or rising, then rising more quickly

} Barometer now higher than 3 hours ago

Steady, same as 3 hours ago

Falling, then rising, same or lower than 3 hours ago

Falling, then steady; or falling, then falling more slowly

Falling steadily, or unsteadily

Steady or rising, then falling; or falling, then falling more quickly

} Barometer now lower than 3 hours ago

FRONT SYMBOLS

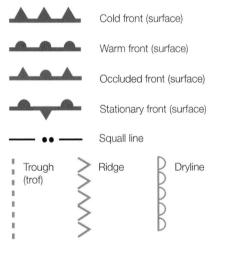

Cold front (surface)

Warm front (surface)

Occluded front (surface)

Stationary front (surface)

Squall line

Trough (trof)   Ridge   Dryline

Humidity and Dew-Point Tables (Psychrometric Tables)

To obtain the dew point (or relative humidity), simply read down the temperature column and then over to the wet-bulb depression. For example, in Table C.1, a temperature of 10°C with a wet-bulb depression of 3°C produces a dew-point temperature of 4°C. (Dew-point temperature and relative humidity readings are appropriate for pressures near 1000 mb.)

▼ TABLE C.1   Dew-Point Temperature (°C)

AIR (DRY-BULB) TEMPERATURE (°C)	WET-BULB DEPRESSION (DRY-BULB TEMPERATURE MINUS WET-BULB TEMPERATURE) (°C)															
	0.5	1.0	1.5	2.0	2.5	3.0	3.5	4.0	4.5	5.0	7.5	10.0	12.5	15.0	17.5	20.0
−20	−25	−33														
−17.5	−21	−27	−38													
−15	−19	−23	−28													
−12.5	−15	−18	−22	−29												
−10	−12	−14	−18	−21	−27	−36										
−7.5	−9	−11	−14	−17	−20	−26	−34									
−5	−7	−8	−10	−13	−16	−19	−24	−31								
−2.5	−4	−6	−7	−9	−11	−14	−17	−22	−28	−41						
0	−1	−3	−4	−6	−8	−10	−12	−15	−19	−24						
2.5	1	0	−1	−3	−4	−6	−8	−10	−13	−16						
5	4	3	2	0	−1	−3	−4	−6	−8	−10	−48					
7.5	6	6	4	3	2	1	−1	−2	−4	−6	−22					
10	9	8	7	6	5	4	2	1	0	−2	−13					
12.5	12	11	10	9	8	7	6	4	3	2	−7	−28				
15	14	13	12	12	11	10	9	8	7	5	−2	−14				
17.5	17	16	15	14	13	12	12	11	10	8	2	−7	−35			
20	19	18	18	17	16	15	14	14	13	12	6	−1	−15			
22.5	22	21	20	20	19	18	17	16	16	15	10	3	−6	−38		
25	24	24	23	22	21	21	20	19	18	18	13	7	0	−14		
27.5	27	26	26	25	24	23	23	22	21	20	16	11	5	−5	−32	
30	29	29	28	27	27	26	25	25	24	23	19	14	9	2	−11	
32.5	32	31	31	30	29	29	28	27	26	26	22	18	13	7	−2	
35	34	34	33	32	32	31	31	30	29	28	25	21	16	11	4	
37.5	37	36	36	35	34	34	33	32	32	31	28	24	20	15	9	0
40	39	39	38	38	37	36	36	35	34	34	30	27	23	18	13	6
42.5	42	41	41	40	40	39	38	38	37	36	33	30	26	22	17	11
45	44	44	43	43	42	42	41	40	40	39	36	33	29	25	21	15
47.5	47	46	46	45	45	44	44	43	42	42	39	35	32	28	24	19
50	49	49	48	48	47	47	46	45	45	44	41	38	35	31	28	23

AIR (DRY-BULB) TEMPERATURE (°C)	WET—BULB DEPRESSION (DRY—BULB TEMPERATURE MINUS WET—BULB TEMPERATURE) (°C)																	
	0.5	1.0	1.5	2.0	2.5	3.0	3.5	4.0	4.5	5.0	7.5	10.0	12.5	15.0	17.5	20.0	22.5	25.0
−20	70	41	11															
−17.5	75	51	26	2														
−15	79	58	38	18														
−12.5	82	65	47	30	13													
−10	85	69	54	39	24	10												
−7.5	87	73	60	48	35	22	10											
−5	88	77	66	54	43	32	21	11	1									
−2.5	90	80	70	60	50	42	37	22	12	3								
0	91	82	73	65	56	47	39	31	23	15								
2.5	92	84	76	68	61	53	46	38	31	24								
5	93	86	78	71	65	58	51	45	38	32	1							
7.5	93	87	80	74	68	62	56	50	44	38	11							
10	94	88	82	76	71	65	60	54	49	44	19							
12.5	94	89	84	78	73	68	63	58	53	48	25	4						
15	95	90	85	80	75	70	66	61	57	52	31	12						
17.5	95	90	86	81	77	72	68	64	60	55	36	18	2					
20	95	91	87	82	78	74	70	66	62	58	40	24	8					
22.5	96	92	87	83	80	76	72	68	64	61	44	28	14	1				
25	96	92	88	84	81	77	73	70	66	63	47	32	19	7				
27.5	96	92	89	85	82	78	75	71	68	65	50	36	23	12	1			
30	96	93	89	86	82	79	76	73	70	67	52	39	27	16	6			
32.5	97	93	90	86	83	80	77	74	71	68	54	42	30	20	11	1		
35	97	93	90	87	84	81	78	75	72	69	56	44	33	23	14	6		
37.5	97	94	91	87	85	82	79	76	73	70	58	46	36	26	18	10	3	
40	97	94	91	88	85	82	79	77	74	72	59	48	38	29	21	13	6	
42.5	97	94	91	88	86	83	80	78	75	72	61	50	40	31	23	16	9	2
45	97	94	91	89	86	83	81	78	76	73	62	51	42	33	26	18	12	6
47.5	97	94	92	89	86	84	81	79	76	74	63	53	44	35	28	21	15	9
50	97	95	92	89	87	84	82	79	77	75	64	54	45	37	30	23	17	11

▼ TABLE C.3 Dew-Point Temperature (°F)

WET-BULB DEPRESSION (DRY-BULB TEMPERATURE MINUS WET-BULB TEMPERATURE) (°F)

AIR (DRY-BULB) TEMPERATURE (°F)	1	2	3	4	5	6	7	8	9	10	11	12	13	14	15	16	17	18	19	20	25	30	35	40
0	-7	-20																						
5	-1	-9	-24																					
10	5	-2	-10	-27																				
15	11	6	0	-9	-26																			
20	16	12	8	2	-7	-21																		
25	22	19	15	10	5	-3	-15	-51																
30	27	25	21	18	14	8	2	-7	-25															
35	33	30	28	25	21	17	13	7	0	-11														
40	38	35	33	30	28	25	21	18	13	7	-1	-14												
45	43	41	38	36	34	31	28	25	22	18	13	7	-1	-14										
50	48	46	44	42	40	37	34	32	29	26	22	18	13	8	0	-13								
55	53	51	50	48	45	43	41	38	36	33	30	27	24	20	15	9	1	-12						
60	58	57	55	53	51	49	47	45	43	40	38	35	32	29	25	21	17	11	4	-8				
65	63	62	60	59	57	55	53	51	49	47	45	42	40	37	34	31	27	24	19	14				
70	69	67	65	64	62	61	59	57	55	53	51	49	47	44	42	39	36	33	30	26	-11			
75	74	72	71	69	68	66	64	63	61	59	57	55	54	51	49	47	44	42	39	36	15			
80	79	77	76	74	73	72	70	68	67	65	63	62	60	58	56	54	52	50	47	44	28	-7		
85	84	82	81	80	78	77	75	74	72	71	69	68	66	64	62	61	59	57	54	52	39	19		
90	89	87	86	85	83	82	81	79	78	76	75	73	72	70	69	67	65	63	61	59	48	32		
95	94	93	91	90	89	87	86	85	83	81	80	79	78	76	74	73	71	70	68	66	56	43	24	
100	99	98	96	95	94	93	91	90	89	87	86	85	83	82	80	79	77	76	74	72	63	52	37	12
105	104	103	101	100	99	98	96	95	94	93	91	90	89	87	86	84	83	82	80	78	70	61	48	30
110	109	108	106	105	104	103	102	100	99	98	97	95	94	93	91	90	89	87	86	84	77	68	57	43
115	114	113	112	110	109	108	107	106	104	103	102	101	99	98	97	96	94	93	92	90	83	75	65	54
120	119	118	117	115	114	113	112	111	110	108	107	106	105	104	102	101	100	98	97	96	89	81	73	63

▼ **TABLE C.4** **Relative Humidity (Percent)**

AIR (DRY-BULB) TEMPERATURE (°F)	WET–BULB DEPRESSION (DRY–BULB TEMPERATURE MINUS WET-BULB TEMPERATURE) (°F)																							
	1	2	3	4	5	6	7	8	9	10	11	12	13	14	15	16	17	18	19	20	25	30	35	40
0	67	31																						
5	73	46	20																					
10	78	56	34	13																				
15	82	64	46	29	11																			
20	85	70	55	40	26	12																		
25	87	74	62	49	37	25	13	1																
30	89	78	67	56	46	36	26	16	6															
35	91	81	72	63	54	45	36	27	19	10	2													
40	92	83	75	68	60	52	45	37	29	22	15	7												
45	93	86	78	71	64	57	51	44	38	31	25	18	12	6										
50	93	87	80	74	67	61	55	49	43	38	32	27	21	16	10	5								
55	94	88	82	76	70	65	59	54	49	43	38	33	28	23	19	14	9	5						
60	94	89	83	78	73	68	63	58	53	48	43	39	34	30	26	21	17	13	9	5				
65	95	90	85	80	75	70	66	61	56	52	48	44	39	35	31	27	24	20	16	12				
70	95	90	86	81	77	72	68	64	59	55	51	48	44	40	36	33	29	25	22	19	3			
75	96	91	86	82	78	74	70	66	62	58	54	51	47	44	40	37	34	30	27	24	7	1		
80	96	91	87	83	79	75	71	68	64	61	57	54	50	47	44	41	38	35	32	29	15	3		
85	96	92	88	84	80	76	73	69	66	62	59	56	52	49	46	43	41	38	35	32	20	8		
90	96	92	89	85	81	78	74	71	68	65	61	58	55	52	49	47	44	41	39	36	24	13	3	
95	96	93	89	85	82	79	75	72	69	66	63	60	57	54	51	49	46	43	41	38	27	17	7	1
100	96	93	89	86	83	80	77	73	70	68	65	62	59	56	54	51	49	46	44	41	30	21	12	4
105	97	93	90	87	83	80	77	74	71	69	66	63	60	58	55	53	50	48	46	43	33	23	15	7
110	97	93	90	87	84	81	78	75	73	70	67	65	62	60	57	55	52	50	48	46	36	26	18	11
115	97	94	91	88	85	82	79	76	74	71	68	66	63	61	58	56	54	52	49	47	37	28	21	13
120	97	94	91	88	85	82	80	77	74	72	69	67	65	62	60	58	55	53	51	49	40	31	23	17

Average Annual Global Precipitation

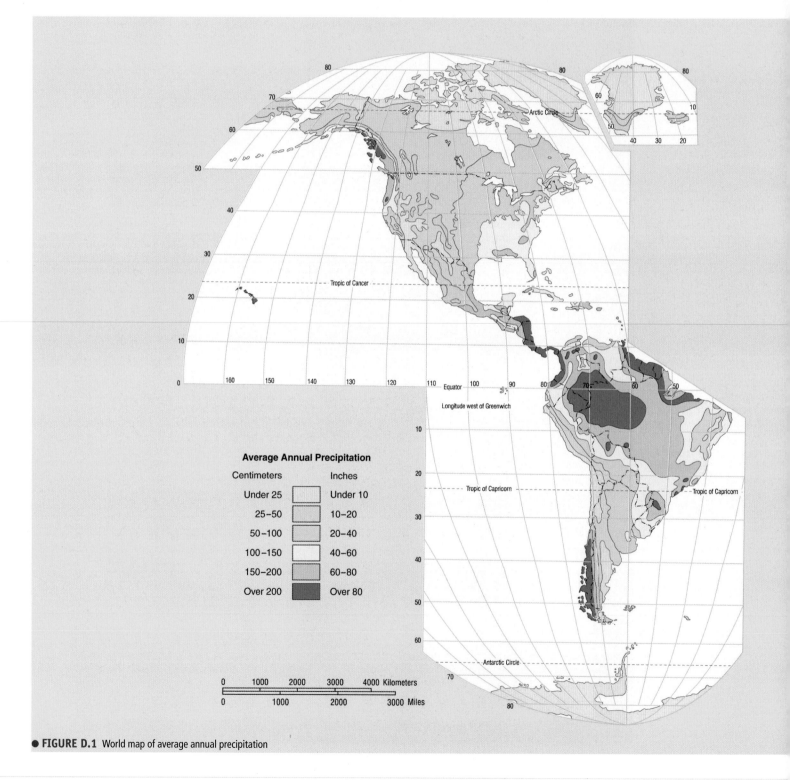

FIGURE D.1 World map of average annual precipitation

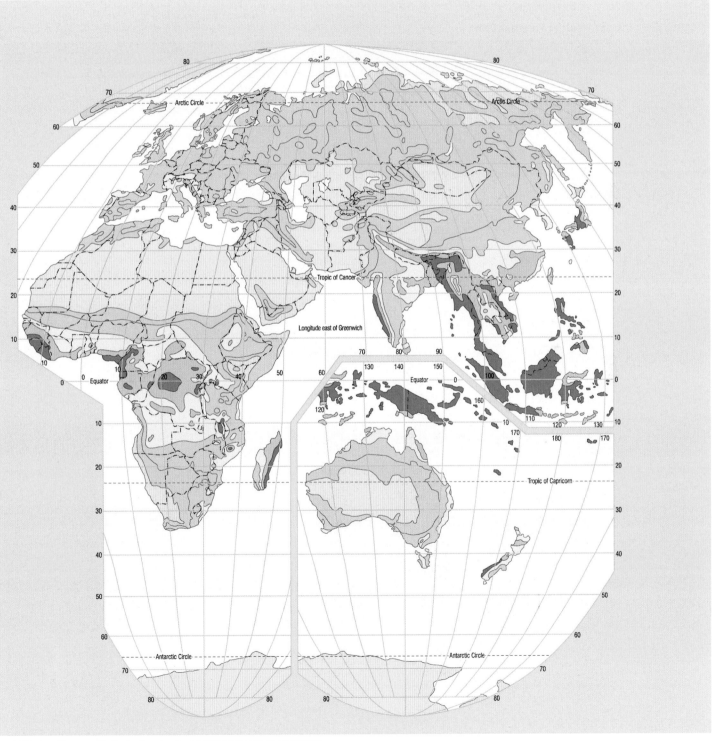

A Western Paragraphic Projection developed at Western Illinois University

Instant Weather Forecast Chart

This chart is a guide to forecasting the weather. It is applicable to most of the United States, especially the eastern two-thirds. It works best during the fall, winter, and spring, when the weather systems are most active.

▼ TABLE E.1 **Instant Weather Forecast Chart**

SEA-LEVEL PRESSURE (MB)	PRESSURE TENDENCY	SURFACE WIND DIRECTION	SKY CONDITION	24-HR WEATHER FORECAST (SEE WEATHER FORECAST CODE)
1023 or higher (30.21 in.)	rising, steady, or falling	any direction	clear, high clouds, or Cu	1, 18 (in winter, 14)
1022 to 1016 (30.20 in. to 30.00 in.)	rising or steady	SW, W, NW, N	clear, high clouds or Cu	1, 18
	falling or steady	SW, S, SE	clear, high clouds	1, 3, 17, 5
	falling	SW, S, SE	middle or low clouds	6, 17
	falling	E, NE	middle or low clouds	6, 14
	falling or steady	E, NE	clear or high clouds	3, 5, 14
1015 to 1009 (29.99 in. to 29.80 in.)	rising	SW, W, NW, N	clear	1, 14
	falling	any direction	overcast	2, 16
	falling or steady	SW, S, SE	precipitation	11, 2, 16
	falling	SW, S, SE	clear	3, 17 (dry climate summer, 1, 15)
	falling	E, NE	high clouds	3, 17, 5
	falling	SE, E, NE	middle or low clouds	7
	falling	S, SW	middle or low clouds	7, 12, 14
			overcast, precipitation	9
			overcast, precipitation	10, 13
1008 or below (29.79 in.)	rising	SW, W, NW, N	clear	1, 12
	rising	SW, W, NW, N	overcast	2, 12, 16
	rising	SW, W, NW, N	overcast with precipitation	11, 12, 16
	rising	NE	overcast	4, 12, 13, 14
	rising	NE	overcast with precipitation	11, 12, 13, 14
	rising or steady	SW, S, SE	clear	3, 6, 8, 12, 15
	falling	SW, S, SE	overcast	7, 8, 12, 13
	falling	SW, S, SE	overcast with precipitation	8, 10, 12, 13, 16
	falling	N	overcast	4, 14
	falling or steady	E, NE	overcast	7, 12, 14
	falling	E, NE	overcast with precipitation	8, 9, 12, 13

1 = clear or scattered clouds

2 = clearing

3 = increasing clouds

4 = continued overcast

5 = precipitation possible within 24 hours

6 = precipitation possible within 12 hours

7 = precipitation possible within 8 hours

8 = possible period of heavy precipitation

9 = precipitation continuing

10 = precipitation ending within 12 hours

11 = precipitation ending within 6 hours

12 = windy

13 = possible wind shift to W, NW, or N

14 = continued cool or cold

15 = continued mild or warm

16 = turning colder

17 = slowly rising temperatures

18 = little temperature change

Changing GMT and UTC to Local Time

The system of time used in meteorology is Greenwich Mean Time (GMT), which is also known as Coordinated Universal Time (UTC) and as Zulu (Z) Time. This is the time measured on the prime meridian 0° in Greenwich, England. Because Eastern Standard Time (EST) is 5 hours slower than GMT, to convert from GMT to EST simply requires subtracting 5 hours from GMT. Conversely, to change EST to GMT entails adding 5 hours to EST. Figure F.1 shows how to convert to GMT in various time zones of North America. Since in meteorology the time is given on a 24-hour clock, Table F.1 shows the relationship between the familiar two 12-hour periods of a.m. and p.m. and the 24-hour system.

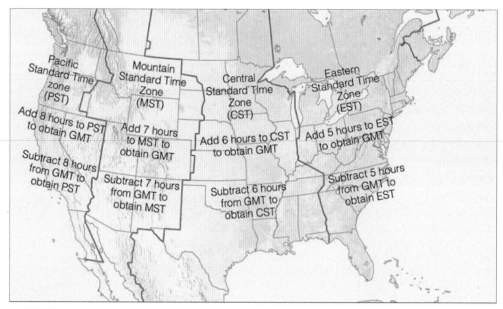

● **FIGURE F.1** The time zones of North America

▼ **TABLE F.1   Conversion of A.M. and P.M. Time System to 24-Hour System**

TIME A.M.	24-HR SYSTEM TIME	TIME A.M.	24-HR SYSTEM TIME	TIME P.M.	24-HR SYSTEM TIME	TIME P.M.	24-HR SYSTEM TIME
12:00 (midnight)	0000	6:00	0600	12:00 (noon)	1200	6:00	1800
1:00	0100	7:00	0700	1:00	1300	7:00	1900
2:00	0200	8:00	0800	2:00	1400	8:00	2000
3:00	0300	9:00	0900	3:00	1500	9:00	2100
4:00	0400	10:00	1000	4:00	1600	10:00	2200
5:00	0500	11:00	1100	5:00	1700	11:00	2300

APPENDIX G

Standard Atmosphere

ALTITUDE				PRESSURE	TEMPERATURE		DENSITY
METERS	FEET	KILOMETERS	MILES	MILLIBARS	°C	°F	kg / m³
0	0	0.0	0.0	1013.25	15.0	(59.0)	1.225
500	1,640	0.5	0.3	954.61	11.8	(53.2)	1.167
1,000	3,280	1.0	0.6	898.76	8.5	(47.3)	1.112
1,500	4,921	1.5	0.9	845.59	5.3	(41.5)	1.058
2,000	6,562	2.0	1.2	795.01	2.0	(35.6)	1.007
2,500	8,202	2.5	1.5	746.91	−1.2	(29.8)	0.957
3,000	9,842	3.0	1.9	701.21	−4.5	(23.9)	0.909
3,500	11,483	3.5	2.2	657.80	−7.7	(18.1)	0.863
4,000	13,123	4.0	2.5	616.60	−11.0	(12.2)	0.819
4,500	14,764	4.5	2.8	577.52	−14.2	(6.4)	0.777
5,000	16,404	5.0	3.1	540.48	−17.5	(0.5)	0.736
5,500	18,045	5.5	3.4	505.39	−20.7	(−5.3)	0.697
6,000	19,685	6.0	3.7	472.17	−24.0	(−11.2)	0.660
6,500	21,325	6.5	4.0	440.75	−27.2	(−17.0)	0.624
7,000	22,965	7.0	4.3	411.05	−30.4	(−22.7)	0.590
7,500	24,606	7.5	4.7	382.99	−33.7	(−28.7)	0.557
8,000	26,247	8.0	5.0	356.51	−36.9	(−34.4)	0.526
8,500	27,887	8.5	5.3	331.54	−40.2	(−40.4)	0.496
9,000	29,528	9.0	5.6	308.00	−43.4	(−46.1)	0.467
9,500	31,168	9.5	5.9	285.84	−46.6	(−51.9)	0.440
10,000	32,808	10.0	6.2	264.99	−49.9	(−57.8)	0.413
11,000	36,089	11.0	6.8	226.99	−56.4	(−69.5)	0.365
12,000	39,370	12.0	7.5	193.99	−56.5	(−69.7)	0.312
13,000	42,651	13.0	8.1	165.79	−56.5	(−69.7)	0.267
14,000	45,932	14.0	8.7	141.70	−56.5	(−69.7)	0.228
15,000	49,213	15.0	9.3	121.11	−56.5	(−69.7)	0.195
16,000	52,493	16.0	9.9	103.52	−56.5	(−69.7)	0.166
17,000	55,774	17.0	10.6	88.497	−56.5	(−69.7)	0.142
18,000	59,055	18.0	11.2	75.652	−56.5	(−69.7)	0.122
19,000	62,336	19.0	11.8	64.674	−56.5	(−69.7)	0.104
20,000	65,617	20.0	12.4	55.293	−56.5	(−69.7)	0.089
25,000	82,021	25.0	15.5	25.492	−51.6	(−60.9)	0.040
30,000	98,425	30.0	18.6	11.970	−46.6	(−51.9)	0.018
35,000	114,829	35.0	21.7	5.746	−36.6	(−33.9)	0.008
40,000	131,234	40.0	24.9	2.871	−22.8	(−9.0)	0.004
45,000	147,638	45.0	28.0	1.491	−9.0	(15.8)	0.002
50,000	164,042	50.0	31.1	0.798	−2.5	(27.5)	0.001
60,000	196,850	60.0	37.3	0.220	−26.1	(−15.0)	0.0003
70,000	229,659	70.0	43.5	0.052	−53.6	(−64.5)	0.00008
80,000	262,467	80.0	49.7	0.010	−74.5	(−102.1)	0.00002

Adiabatic Chart

● **FIGURE H.1** Adiabatic chart

Glossary

Absolute humidity The mass of water vapor in a given volume of air. It represents the density of water vapor in the air.

Absolute vorticity *See* Vorticity.

Absolute zero A temperature reading of –273°C, –460°F, or 0 K. Theoretically, there is no molecular motion at this temperature.

Absolutely stable atmosphere An atmospheric condition that exists when the environmental lapse rate is less than the moist adiabatic rate. This results in a lifted parcel of air being colder than the air around it.

Absolutely unstable atmosphere An atmospheric condition that exists when the environmental lapse rate is greater than the dry adiabatic rate. This results in a lifted parcel of air being warmer than the air around it.

Accretion The growth of a precipitation particle by the collision of an ice crystal or snowflake with a supercooled liquid droplet that freezes upon impact.

Acid deposition The depositing of acidic particles (usually sulfuric acid and nitric acid) at Earth's surface. Acid deposition occurs in dry form (*dry deposition*) or wet form (*wet deposition*). Acid rain and acid precipitation often denote wet deposition. (*See* Acid rain.)

Acid fog *See* Acid rain.

Acid rain Cloud droplets or raindrops combining with gaseous pollutants, such as oxides of sulfur and nitrogen, to make falling rain (or snow) more strongly acidic than usual—typically with a pH less than 5.0. If fog droplets combine with such pollutants it becomes *acid fog*.

Actual vapor pressure *See* Vapor pressure.

Adiabatic process A process that takes place without a transfer of heat between the system (such as an air parcel) and its surroundings. In an adiabatic process, compression always results in warming, and expansion results in cooling.

Advection The horizontal transfer of any atmospheric property by the wind.

Advection fog Occurs when warm, moist air moves over a cold surface and the air cools to below its dew point.

Advection-radiation fog Fog that forms as relatively warm moist air moves over a colder surface that is cooled mainly by radiational cooling.

Aerosols Tiny suspended solid particles (dust, smoke, etc.) or liquid droplets that enter the atmosphere from either natural or human (anthropogenic) sources, such as the burning of fossil fuels. Sulfur-containing fossil fuels, such as coal, produce *sulfate aerosols*.

Aerovane A wind instrument that indicates or records both wind speed and wind direction. Also called a *skyvane*.

Aggregation The clustering together of ice crystals to form snowflakes.

Air density *See* Density.

Air glow A faint glow of light emitted by excited gases in the upper atmosphere. Air glow is much fainter than the aurora.

Air mass A large body of air that has similar horizontal temperature and moisture characteristics.

Air-mass thunderstorm *See* Ordinary thunderstorm.

Air-mass weather A persistent type of weather that may last for several days (up to a week or more). It occurs when an area comes under the influence of a particular air mass.

Air parcel *See* Parcel of air.

Air pollutants Solid, liquid, or gaseous airborne substances that occur in concentrations high enough to threaten the health of people and animals, to harm vegetation and structures, or to toxify a given environment.

Air pressure (atmospheric pressure) The pressure exerted by the mass of air above a given point, usually expressed in millibars (mb), inches of mercury (Hg) or in hectopascals (hPa).

Air Quality Index (AQI) An index of air quality that provides daily air pollution concentrations. Intervals on the scale relate to potential health effects.

Aitken nuclei *See* Condensation nuclei.

Albedo The percent of radiation returning from a surface compared to that which strikes it.

Aleutian low The subpolar low-pressure area that is centered near the Aleutian Islands on charts that show mean sea-level pressure.

Altimeter An instrument that indicates the altitude of an object above a fixed level. Pressure altimeters use an aneroid barometer with a scale graduated in altitude instead of pressure.

Altocumulus A middle cloud, usually white or gray. Often occurs in layers or patches with wavy, rounded masses or rolls.

Altocumulus castellanus An altocumulus cloud showing vertical development. Individual cloud elements have towerlike tops, often in the shape of tiny castles.

Altostratus A middle cloud composed of gray or bluish sheets or layers of uniform appearance. In the thinner regions, the sun or moon usually appears dimly visible.

Analogue forecasting method A forecast made by comparison of past large-scale synoptic weather patterns that resemble a given (usually current) situation in its essential characteristics.

Analysis The drawing and interpretation of the patterns of various weather elements on a surface or upper-air chart.

Anemometer An instrument designed to measure wind speed.

Aneroid barometer An instrument designed to measure atmospheric pressure. It contains no liquid.

Angular momentum The product of an object's mass, speed, and radial distance of rotation.

Annual range of temperature The difference between the warmest and coldest months at any given location.

Anticyclone An area of high atmospheric pressure around which the wind blows clockwise in the Northern Hemisphere and counterclockwise in the Southern Hemisphere. Also called a *high*.

Apparent temperature What the air temperature "feels like" for various combinations of air temperature and relative humidity.

Arctic front In northern latitudes, the semi-permanent front that separates deep cold arctic air from the more shallow, less-cold polar air.

Arctic Oscillation (AO) A reversal of atmospheric pressure over the Arctic that produces changes in the upper-level westerly winds over northern latitudes. These changes in upper-level winds influence winter weather patterns over North America, Greenland, and Europe.

Arcus cloud *See* Shelf cloud.

Arid climate An extremely dry climate—drier than the semi-arid climate. Often referred to as a "true desert" climate.

Asbestos   A general name for the fibrous variety of silicate minerals that are incombustible and resist chemicals and heat.

ASOS   Acronym for *Automated Surface Observing Systems*. A system designed to provide continuous information of wind, temperature, pressure, cloud base height, and runway visibility at selected airports.

Atlantic Multidecadal Oscillation   A reversal of sea-surface temperatures every 30–40 years over the North Atlantic Ocean.

Atmosphere   The envelope of gases that surround a planet and are held to it by the planet's gravitational attraction. Earth's atmosphere is mainly nitrogen and oxygen.

Atmospheric boundary layer   The layer of air from Earth's surface usually up to about 1 km (3300 ft) where the wind is influenced by friction of Earth's surface and objects on it. Also called the *planetary boundary layer* and the *friction layer*.

Atmospheric greenhouse effect   The warming of an atmosphere by its absorbing and emitting infrared radiation while allowing shortwave radiation to pass on through. The gases mainly responsible for Earth's atmospheric greenhouse effect are water vapor and carbon dioxide. Also called the *greenhouse effect*.

Atmospheric models   Simulation of the atmosphere's behavior by mathematical equations or by physical models.

Atmospheric river   A region of upper-level flow that transports large amounts of moisture, typically from the tropics and subtropics into the midlatitudes.

Atmospheric stagnation   A condition of light winds and poor vertical mixing that can lead to a high concentration of pollutants. Air stagnations are most often associated with fair weather, an inversion, and the sinking air of a high-pressure area.

Atmospheric window   The wavelength range between 8 and 11 μm in which little absorption of infrared radiation takes place.

Attenuation   Any process in which the rate of flow of a beam of energy decreases (mainly due to absorption or scattering) with increasing distance from the energy source.

Aurora   Glowing light display in the nighttime sky caused by excited gases in the upper atmosphere giving off light. In the Northern Hemisphere it is called the *aurora borealis* (northern lights); in the Southern Hemisphere, the *aurora australis* (southern lights).

Autumnal equinox   The point during the year at which the sun passes directly over the equator following the three months of summer. Occurs around September 23 in the Northern Hemisphere.

AWIPS   Acronym for *Advanced Weather Interactive Processing System*. New computerized system that integrates and processes data received at a weather forecasting office from NEXRAD, ASOS, and analysis and guidance products prepared by NMC.

Back-door cold front   A cold front moving south or southwest along the Atlantic seaboard of the United States.

Backing wind   A wind that changes direction in a counterclockwise sense (e.g., north to northwest to west).

Ball lightning   A rare form of lightning that may consist of a reddish, luminous ball of electricity or charged air.

Banner cloud   A cloud extending downwind from an isolated mountain peak, often on an otherwise cloud-free day.

Baroclinic (atmosphere)   The state of the atmosphere where surfaces of constant pressure intersect surfaces of constant density. On an isobaric chart, isotherms cross the contour lines, and temperature advection exists.

Baroclinic instability   A type of instability arising from a meridional (north to south) temperature gradient, a strong vertical wind speed shear, temperature advection, and divergence in the flow aloft. Many mid-latitude cyclones develop as a result of this instability.

Barograph   A recording barometer.

Barometer   An instrument that measures atmospheric pressure. The two most common barometers are the *mercury barometer* and the *aneroid barometer*.

Barotropic (atmosphere)   A condition in the atmosphere where surfaces of constant density parallel surfaces of constant pressure.

Bead lightning   Lightning that appears as a series of beads tied to a string.

Bergeron process   *See* Ice-crystal process.

Bermuda high   *See* Subtropical high.

Billow clouds   Broad, nearly parallel lines of wavelike clouds oriented at right angles to the wind. Also called Kelvin–Helmholtz wave clouds.

Bimetallic thermometer   A temperature-measuring device usually consisting of two dissimilar metals that expand and contract differentially as the temperature changes.

Blackbody   A hypothetical object that absorbs all of the radiation that strikes it. It also emits radiation at a maximum rate for its given temperature.

Black ice   A thin sheet of ice that appears relatively dark and may form as supercooled droplets, drizzle, or light rain, that comes in contact with a road surface that is below freezing. Also, thin dark-appearing ice that forms on freshwater or saltwater ponds, or lakes.

Blizzard   A severe weather condition characterized by low temperatures and strong winds (greater than 35 mi/hr) bearing a great amount of snow either falling or blowing. When these conditions continue after the falling snow has ended, it is termed a *ground blizzard*.

Boulder winds   Fast-flowing, local downslope winds that may attain speeds of 100 knots or more. They are especially strong along the eastern foothills of the Rocky Mountains near Boulder, Colorado.

Boundary layer   *See* Atmospheric boundary layer.

Bow echo   A line of thunderstorms on a radar screen that appears in the shape of a bow. Bow echoes are often associated with damaging straight-line winds and small tornadoes.

Brocken bow   A bright ring of light seen around the shadow of an observer's head as the observer peers into a cloud or fog bank. Formed by *diffraction* of light.

Buoyant force (buoyancy)   The upward force exerted upon an air parcel (or any object) by virtue of the density (mainly temperature) difference between the parcel and that of the surrounding air.

Buys-Ballot's law   A law describing the relationship between the wind direction and the pressure distribution. In the Northern Hemisphere, if you stand with your back to the surface wind, then turn clockwise about 30°, lower pressure will be to your left. In the Southern Hemisphere, stand with your back to the surface wind, then turn counterclockwise about 30°; lower pressure will be to your right.

California current   The ocean current that flows southward along the west coast of the United States from about Washington to the Baja California peninsula of Mexico.

California norther   A strong, dry, northerly wind that blows in late spring, summer, and early fall in northern and central California. Its warmth and dryness are due to downslope compressional heating.

Cap   A layer of warm, dry, stable air that inhibits warm, moist, unstable air from rising through it to produce thunderstorms.

Cap cloud   *See* Pileus cloud.

CAPE   *See* Convective Available Potential Energy.

Carbon dioxide (CO_2) A colorless, odorless gas whose concentration is about 0.04 percent (400 ppm) in a volume of air near sea level. It is a selective absorber of infrared radiation and, consequently, it is important in Earth's atmospheric greenhouse effect. Solid CO_2 is called *dry ice*.

Carbon monoxide (CO)   A colorless, odorless, toxic gas that forms during the incomplete combustion of carbon-containing fuels.

Ceiling   The height of the lowest layer of clouds when the weather reports describe the sky as broken or overcast.

Ceiling balloon   A small balloon used to determine the height of the cloud base. The height is computed from the balloon's ascent rate and the time required for its disappearance into the cloud.

Ceilometer   An instrument that automatically records cloud height.

Celsius scale   A temperature scale where zero is assigned to the temperature where water freezes and 100 to the temperature where water boils (at sea level).

Centripetal acceleration   The inward-directed acceleration on a particle moving in a curved path.

Centripetal force   The radial force required to keep an object moving in a circular path. It is directed toward the center of that curved path.

Chaos   The property describing a system that exhibits erratic behavior in that very small changes in the initial state of the system rapidly lead to large and apparently unpredictable changes sometime in the future.

Chemical weathering–CO_2 feedback   A negative feedback in Earth's climate system. As chemical weathering of rocks increases, CO_2 is removed from the atmosphere more quickly, which in turn weakens the greenhouse effect, causing the atmosphere to cool.

Chinook wall cloud   A bank of clouds over the Rocky Mountains that signifies the approach of a chinook.

Chinook wind   A warm, dry wind on the eastern side of the Rocky Mountains. In the Alps, the wind is called a *Foehn*.

Chlorofluorocarbons (CFCs)   Compounds consisting of methane (CH_4) or ethane (C_2H_6) with some or all of the hydrogen replaced by chlorine or fluorine. Originally used in fire extinguishers, as refrigerants, as solvents for cleaning electronic microcircuits, and as propellants. CFCs contribute to the atmospheric greenhouse effect and destroy ozone in the stratosphere. Their use has been virtually eliminated through the Montreal Protocol.

Cirrocumulus   A high cloud that appears as a white patch of clouds without shadows. It consists of very small elements in the form of grains or ripples.

Cirrostratus   High, thin, sheetlike clouds, composed of ice crystals. They frequently cover the entire sky and often produce a halo.

Cirrus   A high cloud composed of ice crystals in the form of thin, white, featherlike clouds in patches, filaments, or narrow bands.

Clear air turbulence (CAT)   Turbulence encountered by aircraft flying through cloudless skies. Thermals, wind shear, and jet streams can each be a factor in producing CAT.

Clear ice   A layer of ice that appears transparent because of its homogeneous structure and small number and size of air pockets.

Climate   The accumulation of daily and seasonal weather events over a long period of time.

Climate change   A change in the long-term statistical average of weather elements—such as temperature and precipitation—sustained over several decades or longer. Climate change is also called *climatic change*.

Climatic controls   The relatively permanent factors that govern the general nature of the climate of a region.

Climatic optimum   *See* Mid-Holocene maximum.

Climatological forecast   A weather forecast, usually a month or more in the future, which is based upon the climate of a region rather than upon current weather conditions.

Cloud   A visible aggregate of tiny water droplets and/or ice crystals in the atmosphere above Earth's surface.

Cloudburst   Any sudden and heavy rain shower.

Cloud seeding   The introduction of artificial substances (usually silver iodide or dry ice) into a cloud for the purpose of either modifying its development or increasing its precipitation.

Cloud streets   Lines or rows of cumuliform clouds.

Coalescence   The merging of cloud droplets into a single larger droplet.

Cold advection (cold air advection)   The transport of cold air by the wind from a region of lower temperatures to a region of higher temperatures.

Cold air damming   A shallow layer of cold air that is trapped between the Atlantic coast and the Appalachian Mountains.

Cold fog   *See* Supercooled cloud.

Cold front   A transition zone where a cold air mass advances and replaces a warm air mass.

Cold occlusion   *See* Occluded front.

Cold wave   A rapid fall in temperature within 24 hours that often requires increased protection for agriculture, industry, commerce, and human activities.

Collision-coalescence process   The process of producing precipitation by liquid particles (cloud droplets and raindrops) colliding and joining (coalescing).

Comma cloud   A band of organized cumuliform clouds that looks like a comma on a satellite photograph.

Computer enhancement   A process where the temperatures of radiating surfaces are assigned different shades of gray (or different colors) on an infrared picture. This allows specific features to be more clearly delineated.

Condensation   The process by which water vapor becomes a liquid.

Condensation level   The level above the surface marking the base of a cumuliform cloud.

Condensation nuclei   Also called *cloud condensation nuclei*. Tiny particles upon whose surfaces condensation of water vapor begins in the atmosphere. Small nuclei less than 0.2 μm in radius are called *Aitken nuclei*; those with radii between 0.2 and 1 μm are *large nuclei*, while *giant nuclei* have radii larger than 1 μm.

Conditionally unstable atmosphere   An atmospheric condition that exists when the environmental lapse rate is less than the dry adiabatic rate but greater than the moist adiabatic rate. Also called *conditional instability*.

Conduction   The transfer of heat by molecular activity from one substance to another, or through a substance. Transfer is always from warmer to colder regions.

Constant height chart (constant-level chart)   A chart showing variables, such as pressure, temperature, and wind, at a specific altitude above sea level. Variation in horizontal pressure is depicted by isobars. The most common constant-height chart is the surface chart, which is also called the *sea-level chart* or *surface weather map*.

Constant pressure chart (isobaric chart)   A chart showing variables, such as temperature and wind, on a constant pressure surface. Variations in height are usually shown by lines of equal height (contour lines).

Contact freezing   The process by which contact with a nucleus such as an ice crystal causes supercooled liquid droplets to change into ice.

Continental arctic air mass   An air mass characterized by extremely low temperatures and very dry air.

Continental polar air mass   An air mass characterized by low temperatures and dry air. Not as cold as arctic air masses.

Continental tropical air mass   An air mass characterized by high temperatures and low humidity.

Contour line   A line that connects points of equal elevation above a reference level, most often sea level.

Contrail (condensation trail)   A cloudlike streamer frequently seen forming behind aircraft flying in clear, cold, humid air.

Controls of temperature   The main factors that cause variations in temperature from one place to another.

Convection   Motions in a fluid that result in the transport and mixing of the fluid's properties. In meteorology, convection usually refers to atmospheric motions that are predominantly vertical, such as rising air currents due to surface heating. The rising of heated surface air and the sinking of cooler air aloft is often called *free convection*. (Compare with *forced convection*.)

Convective Available Potential Energy (*CAPE*)   The maximum energy available to a rising air parcel. It is a measure of how fast the air parcel will rise inside a cumuliform cloud.

Convective instability   Instability arising in the atmosphere when a column of air exhibits warm, moist, nearly saturated air near the surface and cold, dry air aloft. When the lower part of the layer is lifted and saturation occurs, it becomes unstable.

Convergence   An atmospheric condition that exists when the winds cause a horizontal net inflow of air into a specified region.

Conveyor belt model   (for mid-latitude storms) A three-dimensional picture of a mid-latitude cyclone and the various air streams (called *conveyor belts*) that interact to produce the weather associated with the storm.

Cooling degree day   A form of degree day used in estimating the amount of energy necessary to reduce the effective temperature of warm air. A cooling degree day is a day on which the average temperature is one degree above a desired base temperature.

Coriolis force (effect)   An apparent force observed on any free-moving object in a rotating system. On Earth, this deflective force results from Earth's rotation and causes moving particles (including the wind) to deflect to the right in the Northern Hemisphere and to the left in the Southern Hemisphere.

Corona (optic)   A series of colored rings concentrically surrounding the disk of the sun or moon. Smaller than the halo, the corona is often caused by the diffraction of light around small water droplets of uniform size.

Country breeze   A light breeze that blows into a city from the surrounding countryside. It is best observed on clear nights when the urban heat island is most pronounced.

Crepuscular rays   Alternating light and dark bands of light that appear to fan out from the sun's position, usually at twilight.

Cumulonimbus   An exceptionally dense and vertically developed cloud, often with a top in the shape of an anvil. The cloud is frequently accompanied by heavy showers, lightning, thunder, and sometimes hail. It is also known as a *thunderstorm cloud*.

Cumulus   A cloud in the form of individual, detached domes or towers that are usually dense and well defined. It has a flat base with a bulging upper part that often resembles cauliflower. Cumulus clouds of fair weather are called *cumulus humilis*. Those that exhibit much vertical growth are called *cumulus congestus* or *towering cumulus*.

Cumulus stage   The initial stage in the development of an ordinary cell thunderstorm in which rising, warm, humid air develops into a cumulus cloud.

Curvature effect   In cloud physics, as cloud droplets decrease in size, they exhibit a greater surface curvature that causes a more rapid rate of evaporation.

Cut-off low   A cold upper-level low that has become displaced out of the basic westerly flow and lies to the south of this flow.

Cyclogenesis   The development or strengthening of middle-latitude (extratropical) cyclones.

Cyclone   An area of low pressure around which the winds blow counterclockwise in the Northern Hemisphere and clockwise in the Southern Hemisphere.

Daily range of temperature   The difference between the maximum and minimum temperatures for any given day.

Dart leader   The discharge of electrons that proceeds intermittently toward the ground along the same ionized channel taken by the initial lightning stroke.

Data assimilation   The process of incorporating various types of atmospheric observations into the initial stages of a forecast produced by a computer model.

Dendrochronology   The analysis of the annual growth rings of trees as a means of interpreting past climatic conditions.

Density   The ratio of the mass of a substance to the volume occupied by it. Air density is usually expressed as g/cm^3 or kg/m^3.

Deposition   A process that occurs in subfreezing air when water vapor changes directly to ice without becoming a liquid first.

Deposition nuclei   Tiny particles (ice nuclei) upon which an ice crystal may grow by the process of deposition.

Derecho   Strong, damaging, straight-line winds associated with a cluster of severe thunderstorms that most often form in the evening or at night.

Desertification   A general increase in the desert conditions of a region.

Desert pavement   An arrangement of pebbles and large stones that remains behind as finer dust and sand particles are blown away by the wind.

Dew   Water that has condensed onto objects near the ground when their temperatures have fallen below the dew point of the surface air.

Dew cell   An instrument used to determine the dew-point temperature.

Dew point (dew-point temperature)   The temperature to which air must be cooled (at constant pressure and constant water vapor content) for saturation to occur.

Dew point hygrometer   An instrument that determines the dew-point temperature of the air.

Diffraction   The bending of light around objects, such as cloud and fog droplets, producing fringes of light and dark or colored bands.

Dispersion (of light)   The separation of white light into its different component wavelengths.

Dispersion (of pollution)   The spreading of atmospheric pollutants.

Dissipating stage   The final stage in the development of an ordinary cell thunderstorm when downdrafts exist throughout the cumulonimbus cloud.

Divergence   An atmospheric condition that exists when the winds cause a horizontal net outflow of air from a specific region.

Doldrums   The region near the equator that is characterized by low pressure and light, shifting winds.

Doppler lidar   The use of light beams to determine the velocity of objects such as dust and falling rain by taking into account the *Doppler shift*.

Doppler radar   A radar that determines the velocity of falling precipitation either toward or away from the radar unit by taking into account the *Doppler shift*.

Doppler shift (effect)   The change in the frequency of waves that occurs when the emitter or the observer is moving toward or away from the other.

Downburst   A severe localized downdraft that can be experienced beneath a severe thunderstorm. (Compare *microburst* and *macroburst*.)

Drizzle   Small water drops between 0.2 and 0.5 mm in diameter that fall slowly and reduce visibility more than light rain.

Drought   A period of abnormally dry weather sufficiently long enough to cause serious effects on agriculture and other activities in the affected area.

Dry adiabatic rate   The rate of change of temperature in a rising or descending unsaturated air parcel. The rate of adiabatic cooling or warming is about 10°C per 1000 m (5.5°F per 1000 ft).

Dry adiabats   Lines on an adiabatic chart that show the dry adiabatic rate for rising or descending air. They represent lines of constant potential temperature.

Dry-bulb temperature   The air temperature measured by the dry-bulb thermometer of a psychrometer.

Dry climate   A climate deficient in precipitation where annual potential evaporation and transpiration exceed precipitation.

Dry haze   *See* Haze.

Dry lightning Lightning that occurs with thunderstorms that produce little, if any, appreciable precipitation that reaches the surface.

Dryline A boundary that separates warm, dry air from warm, moist air. It usually represents a zone of instability along which thunderstorms form.

Dry slot On a satellite image the dry slot represents the relatively clear region (or clear wedge) that appears just to the west of the tail of a comma cloud of a mid-latitude cyclonic storm.

Dry-summer subtropical climate A climate characterized by mild, wet winters and warm to hot, dry summers. Typically located between 30 and 45 degrees latitude on the western side of continents. Also called *Mediterranean climate*.

Dual-polarization Doppler radar A Doppler radar that sends and receives signals that are horizontally and vertically polarized. This technique allows the radar to more easily distinguish heavy rain, hail, snow, and other precipitation.

Dust devil (whirlwind) A small but rapidly rotating wind made visible by the dust, sand, and debris it picks up from the surface. It develops best on clear, dry, hot afternoons.

Earth vorticity The rotation (spin) of an object about its vertical axis brought on by the rotation of Earth on its axis. Earth's vorticity is a maximum at the poles and zero at the equator.

Easterly wave *See* Tropical wave.

Eccentricity (of Earth's orbit) The deviation of Earth's orbit from elliptical to nearly circular.

Eddy A small volume of air (or any fluid) that behaves differently from the larger flow in which it exists.

Eddy viscosity The internal friction produced by turbulent flow.

Ekman spiral An idealized description of the way the wind-driven ocean currents vary with depth. In the atmosphere it represents the way the winds vary from the surface up through the friction layer or planetary boundary layer.

Ekman transport Net surface water transport due to the Ekman spiral. In the Northern Hemisphere the transport is 90° to the right of the surface wind direction.

Electrical hygrometer *See* Hygrometer.

Electrical thermometers Thermometers that use elements that convert energy from one form to another (transducers). Common electrical thermometers include the electrical resistance thermometer, thermocouple, and thermistor.

Electromagnetic waves *See* Radiant energy.

El Niño An extensive ocean warming that typically extends westward from the coast of Peru and Ecuador across the eastern tropical Pacific Ocean, with associated atmospheric conditions. El Niño events occur roughly once every 2 to 7 years. (*See also* ENSO.)

Embryo In cloud physics, a tiny ice crystal that grows in size and becomes an ice nucleus.

Energy The property of a system that generally enables it to do work. Some forms of energy are kinetic, radiant, potential, chemical, electric, and magnetic.

Enhanced Fujita (EF) scale A modification of the original *Fujita Scale* that describes tornado intensity by observing damage caused by the tornado.

Ensemble forecasting A forecasting technique that entails running several forecast models (or different versions of a single model), each beginning with slightly different weather information. The forecaster's level of confidence is based on how well the models agree (or disagree) at the end of some specified time.

ENSO (El Niño/Southern Oscillation) A condition in the tropical Pacific whereby the reversal of surface air pressure at opposite ends of the Pacific Ocean induces westerly winds, a strengthening of the equatorial countercurrent, and extensive ocean warming.

Entrainment The mixing of environmental air into a pre-existing air current or cloud so that it becomes part of the current or cloud.

Environmental lapse rate The rate of decrease of air temperature with elevation. It is most often measured with a radiosonde.

Equilibrium vapor pressure The necessary vapor pressure around liquid water that allows the water to remain in equilibrium with its environment. Also called *saturation vapor pressure*.

Evaporation The process by which a liquid changes into a gas.

Evaporation (mixing) fog Fog produced when sufficient water vapor is added to the air by evaporation, and the moist air mixes with relatively drier air. The two common types are *steam fog*, which forms when cold air moves over warm water, and *frontal fog*, which forms as warm raindrops evaporate in a cool air mass.

Exosphere The outermost portion of the atmosphere.

Extratropical cyclone A cyclonic storm that most often forms along a front in middle and high latitudes. Also called a *middle-latitude cyclonic storm*, a *depression*, and a *low*. It is not a tropical storm or hurricane.

Eye A region in the center of a hurricane (tropical storm) where the winds are light and skies are clear to partly cloudy.

Eyewall A wall of dense thunderstorms that surrounds the eye of a hurricane.

Eyewall replacement A situation within a hurricane (tropical cyclone) where the storm's original eyewall dissipates and a new eyewall forms outward, farther away from the center of the storm.

Fahrenheit scale A temperature scale where 32 is assigned to the temperature where water freezes and 212 to the temperature where water boils (at sea level).

Fall streaks Falling ice crystals that evaporate before reaching the ground.

Fall wind A strong, cold katabatic wind that blows downslope off snow-covered plateaus.

Fata Morgana A complex mirage that is characterized by objects being distorted in such a way as to appear as castlelike features.

Feedback mechanism A process whereby an initial change in an atmospheric process will tend to either reinforce the process (*positive feedback*) or weaken the process (*negative feedback*).

Ferrel cell The name given to the middle-latitude cell in the 3-cell model of the general circulation.

Fetch The distance that the wind travels over open water.

Flash flood A flood that rises and falls quite rapidly with little or no advance warning, usually as the result of intense rainfall over a relatively small area.

Flurries of snow *See* Snow flurries.

Foehn *See* Chinook wind.

Fog A cloud with its base at Earth's surface.

Forced convection On a small scale, a form of mechanical stirring taking place when twisting eddies of air are able to mix hot surface air with the cooler air above. On a larger scale, it can be induced by the lifting of warm air along a front (*frontal uplift*) or along a topographic barrier (*orographic uplift*).

Forecast funnel A sequence of steps used by forecasters to analyze current and projected conditions, moving from larger to smaller scales during the process.

Forked lightning Cloud-to-ground lightning that exhibits downward-directed crooked branches.

Formaldehyde A colorless gaseous compound (HCHO) used in the manufacture of resins, fertilizers, and dyes. Also used as an embalming fluid, a preservative, and a disinfectant.

Free convection *See* Convection.

Freeze A condition occurring over a widespread area when the surface air temperature remains below freezing for a sufficient time to damage certain agricultural crops. A freeze most often occurs as cold air is advected into a region, causing freezing conditions to exist in a deep layer of surface air. Also called *advection frost*.

Freezing nuclei Particles that promote the freezing of supercooled liquid droplets.

Freezing rain and **freezing drizzle** Rain or drizzle that falls in liquid form and then freezes upon striking a cold object or ground. Both can produce a coating of ice on objects which is called *glaze*.

Friction layer The atmospheric layer near the surface usually extending up to about 1 km (3300 ft) where the wind is influenced by friction of Earth's surface and objects on it. Also called the *atmospheric boundary layer* and *planetary boundary layer*.

Front The transition zone between two distinct air masses.

Frontal fog *See* Evaporation fog.

Frontal inversion A temperature inversion encountered upon ascending through a sloping front, usually a warm front.

Frontal thunderstorms Thunderstorms that form in response to forced convection (forced lifting) along a front. Most go through a cycle similar to those of ordinary thunderstorms.

Frontal wave A wavelike deformation along a front in the lower levels of the atmosphere. Those that develop into storms are termed *unstable waves*, while those that do not are called *stable waves*.

Frontogenesis The formation, strengthening, or regeneration of a front.

Frontolysis The weakening or dissipation of a front.

Frost (also called **hoarfrost**) A covering of ice produced by deposition on exposed surfaces when the air temperature falls below the frost point.

Frostbite The partial freezing of exposed parts of the body, causing injury to the skin and sometimes to deeper tissues.

Frost point The temperature at which the air becomes saturated with respect to ice when cooled at constant pressure and constant water vapor content.

Frozen dew The transformation of liquid dew into tiny beads of ice when the air temperature drops below freezing.

Fujita scale A scale developed by T. Theodore Fujita for classifying tornadoes according to the damage they cause and their rotational wind speed. (*See also* Enhanced Fujita Scale.)

Fulgurite A rootlike tube (or several tubes) that forms when a lightning stroke fuses sand particles together.

Funnel cloud A funnel-shaped cloud of condensed water, usually extending from the base of a cumuliform cloud. The rapidly rotating air of the funnel is not in contact with the ground; hence, it is not a tornado.

Galaxy A huge assembly of stars (between millions and hundreds of millions) held together by gravity.

Gas law The thermodynamic law applied to a perfect gas that relates the pressure of the gas to its density and absolute temperature.

General circulation of the atmosphere Large-scale atmospheric motions over the entire Earth.

Geoengineering The use of global-scale technology fixes to mitigate climate changes.

Geostationary satellite A satellite that orbits Earth at the same rate that Earth rotates and thus remains over a fixed place above the equator.

Geostrophic wind A theoretical horizontal wind blowing in a straight path, parallel to the isobars or contours, at a constant speed. The geostrophic wind results when the Coriolis force exactly balances the horizontal pressure gradient force.

Giant nuclei *See* Condensation nuclei.

Glaciated cloud A cloud or portion of a cloud where only ice crystals exist.

Glaze A coating of ice, often clear and smooth, that forms on exposed surfaces by the freezing of a film of supercooled water deposited by rain, drizzle, or fog. As a type of aircraft icing, glaze is called *clear ice*.

Global climate Climate of the entire globe.

Global scale The largest scale of atmospheric motion. Also called the *planetary scale*.

Global warming Increasing global surface air temperatures that show up in the climate record. The term *global warming* is usually attributed to human activities, such as increasing concentrations of greenhouse gases from automobiles and industrial processes, for example.

Glory Colored rings that appear around the shadow of an object.

Gradient wind A theoretical wind that blows parallel to curved isobars or contours.

Graupel Ice particles between 2 and 5 mm in diameter that form in a cloud often by the process of accretion. Snowflakes that become rounded pellets due to riming are called *graupel* or *snow pellets*.

Green flash A small green color that occasionally appears on the upper part of the sun as it rises or sets.

Greenhouse effect *See* Atmospheric greenhouse effect.

Greenhouse gases Gases in Earth's atmosphere, such as water vapor and carbon dioxide (CO_2), that allow much of the sunlight to pass through but are strong absorbers of infrared energy emitted by Earth and the atmosphere. Other greenhouse gases include methane (CH_4), nitrous oxide (N_2O), fluorocarbons (CFCs), and ozone (O_3).

Ground fog *See* Radiation fog.

Growing degree day A form of the degree day used as a guide for crop planting and for estimating crop maturity dates.

Gulf Stream A warm, swift, narrow ocean current flowing along the east coast of the United States.

Gust front A boundary that separates a cold downdraft of a thunderstorm from warm, humid surface air. On the surface its passage resembles that of a cold front.

Gustnado A relatively weak tornado associated with a thunderstorm's outflow. It most often forms along the gust front.

Gyre A large circular, surface ocean current pattern.

Haboob A dust or sandstorm that forms as cold downdrafts from a thunderstorm turbulently lift dust and sand into the air.

Hadley cell A thermal circulation proposed by George Hadley to explain the movement of the trade winds. It consists of rising air near the equator and sinking air near 30° latitude.

Hailstones Transparent or partially opaque particles of ice that range in size from that of a pea to that of golf balls.

Hailstreak The accumulation of hail at Earth's surface along a relatively long (10 km), narrow (2 km) band.

Hair hygrometer *See* Hygrometer.

Halons A group of organic compounds used as fire retardants. In the stratosphere, these compounds release bromine atoms that rapidly destroy ozone. The production of halons is now banned by the Montreal Protocol.

Halos Rings or arcs that encircle the sun or moon when seen through an ice crystal cloud or a sky filled with falling ice crystals. Halos are produced by refraction of light.

Haze Fine dry or wet dust or salt particles dispersed through a portion of the atmosphere. Individually these are not visible but cumulatively they will diminish visibility. *Dry haze* particles are very small, on the order of 0.1 mm. *Wet haze* particles are larger.

Heat A form of energy transferred between systems by virtue of their temperature differences.

Heat burst A sudden increase in surface air temperature often accompanied by extreme drying. A heat burst is associated with the downdraft of a thunderstorm, or a cluster of thunderstorms.

Heat capacity   The ratio of the heat absorbed (or released) by a system to the corresponding temperature rise (or fall).

Heat Index (HI)   An index that combines air temperature and relative humidity to determine an apparent temperature—how hot it actually feels.

Heating degree day   A form of the degree day used as an index for fuel consumption.

Heat lightning   Distant lightning that illuminates the sky but is too far away for its thunder to be heard.

Heatstroke   A physical condition induced by a person's overexposure to high air temperatures, especially when accompanied by high humidity.

Hectopascal   Abbreviated hPa. One hectopascal is equal to 100 Newtons/m², or 1 millibar.

Heiligenschein   A faint white ring surrounding the shadow of an observer's head on a dew-covered lawn.

Helicity   A measure of the potential of the wind for helical (corkscrew-like) flow. Helicity is related to the strength of the wind flow, the amount of vertical wind shear, and the amount of turning in the wind flow, its spin or vorticity. The updraft and the rotation inside a thunderstorm combine to produce helical flow. The higher the value of helicity, the more likely for the development of rotating thunderstorms (supercells) and severe weather.

Heterosphere   The region of the atmosphere above about 85 km where the composition of the air varies with height.

High   *See* Anticyclone.

High inversion fog   A fog that lifts above the surface but does not completely dissipate because of a strong inversion (usually subsidence) that exists above the fog layer.

Highland climate   The climate of high elevations. Also called *mountain climate*. There is no single climatic type but a variety of different climate zones often characterized by a rapid change in temperature and precipitation as one ascends or descends in elevation.

Homogeneous (spontaneous) freezing   The freezing of pure water. For tiny cloud droplets, homogeneous freezing does not occur until the air temperature reaches about −40°C.

Homosphere   The region of the atmosphere below about 85 km where the composition of the air remains fairly constant.

Hook echo   The shape of an echo on a Doppler radar screen that indicates the possible presence of a tornado.

Horse latitudes   The belt of latitude at about 30° to 35° where winds are predominantly light and the weather is hot and dry.

Humid continental climate   A climate characterized by severe winters and mild to warm summers with adequate annual precipitation. Typically located over large continental areas in the Northern Hemisphere between about 40° and 70° latitude.

Humidity   A general term that refers to the air's water vapor content. (*See* Relative humidity.)

Humid subtropical climate   A climate characterized by hot muggy summers, cool to cold winters, and abundant precipitation throughout the year.

Hurricane   A tropical cyclone with sustained winds of at least 64 knots (74 mi/hr).

Hurricane hunters   A popular term for aircraft and/or personnel engaged in the reconnaissance of hurricanes (tropical cyclones).

Hurricane warning   A warning given when it is likely that a hurricane will strike an area within 24 hours.

Hurricane watch   A hurricane watch indicates that a hurricane poses a threat to an area (often within several days) and residents of the watch area should be prepared.

Hydrocarbons   Chemical compounds composed of only hydrogen and carbon—they are included under the general term volatile organic compounds (VOCs).

Hydrochlorofluorocarbons (HCFCs)   A widely used replacement for chlorofluorocarbons, with a molecular structure that includes fewer chlorine atoms and at least one hydrogen atom per molecule. HCFCs cause much less ozone damage than CFCs but are powerful greenhouse gases. Their use will be phased out in the 2020s under the Montreal Protocol.

Hydrofluorocarbons (HFCs)   A common replacement for chloro-flourocarbons, made up of hydrogen, flourine, and carbon. Because HFCs have no chlorine atoms, they cause no damage to the strato-spheric ozone layer, although some HFCs are powerful greenhouse gases.

Hydrologic cycle   A model that illustrates the movement and exchange of water among Earth, atmosphere, and oceans.

Hydrophobic   The ability to resist the condensation of water vapor. Usually used to describe "water-repelling" condensation nuclei.

Hydrostatic equation   An equation that states that the rate at which the air pressure decreases with height is equal to the air density times the acceleration of gravity. The equation relates to how quickly the air pressure decreases in a column of air.

Hydrostatic equilibrium   The state of the atmosphere when there is a balance between the vertical pressure gradient force and the downward pull of gravity.

Hygrometer   An instrument designed to measure the air's water vapor content. The sensing part of the instrument can be hair (*hair hygrometer*), a plate coated with carbon (*electrical hygrometer*), or an infrared sensor (*infrared hygrometer*).

Hygroscopic   The ability to accelerate the condensation of water vapor. Usually used to describe condensation nuclei that have an affinity for water vapor.

Hypothermia   The deterioration in one's mental and physical condition brought on by a rapid lowering of human body temperature.

Hypoxia   A condition experienced by humans when the brain does not receive sufficient oxygen.

Ice Age   *See* Pleistocene epoch.

Ice-crystal (Bergeron) process   A process that produces precipitation. The process involves tiny ice crystals in a supercooled cloud growing larger at the expense of the surrounding liquid droplets. Also called the *Bergeron process*.

Ice fog   A type of fog that forms at very low temperatures, composed of tiny suspended ice particles.

Icelandic low   The subpolar low-pressure area that is centered near Iceland on charts that show mean sea-level pressure.

Ice nuclei   Particles that act as nuclei for the formation of ice crystals in the atmosphere.

Ice pellets   *See* Sleet.

Ice storm   A winter storm characterized by a substantial amount of precipitation in the form of freezing rain, freezing drizzle, or sleet.

Indian summer   An unseasonably warm spell with clear skies near the middle of autumn. Usually follows a substantial period of cool weather.

Inferior mirage   *See* Mirage.

Infrared hygrometer   *See* Hygrometer.

Infrared radiation   Electromagnetic radiation with wavelengths between about 0.7 and 1000 μm. This radiation is longer than visible radiation but shorter than microwave radiation.

Infrared radiometer   An instrument designed to measure the intensity of infrared radiation emitted by an object. Also called *infrared sensor*.

Insolation   The *in*coming *sol*ar rad*iation* that reaches Earth and the atmosphere.

Instrument shelter   A boxlike (often wooden) structure designed to protect weather instruments from direct sunshine and precipitation.

Interglacial period   A time interval of relatively mild climate during the Ice Age when continental ice sheets were absent or limited in extent to Greenland and the Antarctic.

Intertropical Convergence Zone (ITCZ)   The boundary zone separating the northeast trade winds of the Northern Hemisphere from the southeast trade winds of the Southern Hemisphere.

Inversion   An increase in air temperature with height.

Ion   An electrically charged atom, molecule, or particle.

Ionosphere   An electrified region of the upper atmosphere where fairly large concentrations of ions and free electrons exist.

Iridescence   Brilliant spots or borders of colors, most often red and green, observed in clouds up to about 30° from the sun.

Isallobar   A line of equal change in atmospheric pressure during a specified time interval.

Isobar   A line connecting points of equal pressure.

Isobaric chart (map)   *See* Constant pressure chart.

Isobaric surface   A surface along which the atmospheric pressure is everywhere equal.

Isotach   A line connecting points of equal wind speed.

Isotherm   A line connecting points of equal temperature.

Isothermal layer   A layer where the air temperature is constant with increasing altitude. In an isothermal layer, the air temperature lapse rate is zero.

Jet maximum   *See* Jet streak.

Jet streak   A region of high wind speed that moves through the axis of a jet stream. Also called *jet maximum*.

Jet stream   Relatively strong winds concentrated within a narrow band in the atmosphere.

Katabatic (fall) wind   A wind that blow downslope under the force of gravity. It is often cold.

Kelvin   A unit of temperature. A kelvin is denoted by K and 1 K equals 1°C. Zero kelvin is absolute zero, or −273.15°C.

Kelvin scale   A temperature scale with zero degrees equal to the theoretical temperature at which all molecular motion ceases. Also called the *absolute scale*. The units are sometimes called "degrees Kelvin"; however, the correct SI terminology is "Kelvins," abbreviated K.

Kelvin wave   A large, slow-moving oceanic feature that brings warm water across the tropical Pacific Ocean.

Kinetic energy   The energy within a body that is a result of its motion.

Kirchhoff's law   A law that states: Good absorbers of a given wavelength of radiation are also good emitters of that wavelength.

Knot   A unit of speed equal to 1 nautical mile per hour. One knot equals 1.15 mi/hr.

Köppen classification system   A system for classifying climates that is based mainly on annual and monthly averages of temperature and precipitation.

Lake breeze   A wind blowing onshore from the surface of a lake.

Lake-effect snows   Localized snowstorms that form on the downwind side of a lake. Such storms are common in late fall and early winter near the Great Lakes as cold, dry air picks up moisture and warmth from the unfrozen bodies of water.

Laminar flow   A nonturbulent flow in which the fluid moves smoothly in parallel layers or sheets.

Land breeze   A coastal breeze that blows from land to sea, usually at night.

Landspout   Relatively weak nonsupercell tornado that originates with a cumiliform cloud in its growth stage and with a cloud that does not contain a mid-level mesocyclone. Its spin originates near the surface. Landspouts often look like waterspouts over land.

La Niña   A condition where the surface waters of the central and eastern tropical Pacific Ocean turn cooler than normal, with associated atmospheric conditions.

Lapse rate   The rate at which an atmospheric variable (usually temperature) decreases with height. (*See* Environmental lapse rate.)

Latent heat   The heat that is either released or absorbed by a unit mass of a substance when it undergoes a change of state, such as during evaporation, condensation, or sublimation.

Laterite   A soil formed under tropical conditions where heavy rainfall leaches soluble minerals from the soil. This leaching leaves the soil hard and poor for growing crops.

Lee-side low   Storm systems (extratropical cyclones) that form on the downwind (lee) side of a mountain chain. In the United States lee-side lows frequently form on the eastern side of the Rockies and Sierra Nevada mountains.

Lenticular cloud   A cloud in the shape of a lens.

Level of free convection   The level in the atmosphere at which a lifted air parcel becomes warmer than its surroundings in a conditionally unstable atmosphere.

Lidar   An instrument that uses a laser to generate intense pulses that are reflected from atmospheric particles of dust and smoke. Lidars have been used to determine the amount of particles in the atmosphere as well as particle movement that has been converted into wind speed. Lidar means *light detection and ranging*.

Lifting condensation level (LCL)   The level at which a parcel of air, when lifted dry adiabatically, would become saturated.

Lightning   A visible electrical discharge produced by thunderstorms.

Liquid-in-glass thermometer   *See* Thermometer.

Little Ice Age   The period from about 1350 to 1850 when average temperatures over Europe were relatively low, and alpine glaciers increased in size and advanced down mountain canyons.

Local winds   Winds that tend to blow over a relatively small area; often due to regional effects, such as mountain barriers, large bodies of water, local pressure differences, and other influences.

Long-range forecast   Generally used to describe a weather forecast that extends beyond about 8.5 days into the future.

Longwave radiation   A term most often used to describe the infrared energy emitted by Earth and the atmosphere.

Longwaves in the westerlies   A wave in the upper level of westerlies characterized by a long length (thousands of kilometers) and significant amplitude. Also called *Rossby waves*.

Low   *See* Extratropical cyclone.

Low-level jet streams   Jet streams that typically form near Earth's surface below an altitude of about 2 km and usually attain speeds of less than 60 knots.

Macroburst   A strong downdraft (*downburst*) greater than 4 km wide that can occur beneath thunderstorms. A downburst less than 4 km across is called a *microburst*.

Macroclimate   The general climate of a large area, such as a country.

Macroscale   The normal meteorological synoptic scale for obtaining weather information. It can cover an area ranging from the size of a continent to the entire globe.

Madden-Julian Oscillation   A recurring feature across the tropics that involves a large area of showers and thunderstorms propagating eastward over a period of weeks to months.

Magnetic storm   A worldwide disturbance of Earth's magnetic field caused by solar disturbances.

Magnetosphere   The region around Earth in which Earth's magnetic field plays a dominant part in controlling the physical processes that take place.

Mammatus clouds   Clouds that look like pouches hanging from the underside of a cloud.

Marine climate   A climate controlled largely by the ocean. The ocean's influence keeps winters relatively mild and summers cool.

Maritime air   Moist air whose characteristics were developed over an extensive body of water.

Maritime polar air mass   An air mass characterized by low temperatures and high humidity.

Maritime tropical air mass   An air mass characterized by high temperatures and high humidity.

Mature thunderstorm   The second stage in the three-stage cycle of an ordinary thunderstorm. This mature stage is characterized by heavy showers, lightning, thunder, and violent vertical motions inside cumulonimbus clouds.

Maunder minimum   A period from about 1645 to 1715 when few sunspots were observed.

Maximum thermometer   A thermometer with a small constriction just above the bulb. It is designed to measure the maximum air temperature.

Mean annual temperature   The average temperature at any given location for the entire year.

Mean daily temperature   The average of the highest and lowest temperature for a 24-hour period.

Mechanical turbulence   Turbulent eddy motions caused by obstructions, such as trees, buildings, mountains, and so on.

Mediterranean climate   *See* Dry-summer subtropical climate.

Medium-range forecast   Generally used to describe a weather forecast that extends from about 3 to 8.5 days into the future.

Mercury barometer   A type of barometer that uses mercury to measure atmospheric pressure. The height of the mercury column is a measure of atmospheric pressure.

Meridional flow   A type of atmospheric circulation pattern in which the north–south component of the wind is pronounced.

Mesoclimate   The climate of an area ranging in size from a few acres to several square kilometers.

Mesocyclone   A vertical column of cyclonically rotating air within a supercell thunderstorm.

Mesohigh   A relatively small area of high atmospheric pressure that forms beneath a thunderstorm.

Mesopause   The top of the mesosphere. The boundary between the mesosphere and the thermosphere, usually near 85 km.

Mesoscale   The scale of meteorological phenomena that range in size from a few km to about 100 km. It includes local winds, thunderstorms, and tornadoes.

Mesoscale convective complex (MCC)   A large, organized convective weather system comprised of a number of individual thunderstorms. An MCC can span 1000 times more area than an individual ordinary cell thunderstorm. An MCC is a particular type of mesoscale convective system.

Mesoscale convective system (MCS)   A large cloud system that represents an ensemble of thunderstorms that form by convection and produce precipitation over a wide area.

Mesoscale convective vortex   A counterclockwise circulation, usually less than 240 km (150 mi) in diameter, about an area of low pressure that forms in the mid-levels of the atmosphere in association with a mesoscale convective system.

Mesosphere   The atmospheric layer between the stratosphere and the thermosphere. Located at an average elevation between 50 and 80 km above Earth's surface.

Meteogram   A chart that shows how one or more weather variables has changed at a station over a given period of time or how the variables are likely to change with time.

Meteorology   The study of the atmosphere and atmospheric phenomena as well as the atmosphere's interaction with Earth's surface, oceans, and life in general.

Microburst   A strong localized downdraft (downburst) less than 4 km wide that occurs beneath thunderstorms. A strong downburst greater than 4 km across is called a *macroburst.*

Microclimate   The climate structure of the air space near the surface of Earth.

Micrometer (mm)   A unit of length equal to one-millionth of a meter.

Microscale   The smallest scale of atmospheric motions.

Middle latitudes   The region of the world typically described as being between 30° and 50° latitude.

Middle-latitude cyclone   *See* Extratropical cyclone.

Mid-Holocene maximum   A warm period in geologic history about 5000 to 6000 years ago that favored the development of plants.

Milankovitch theory   A theory suggesting that changes in Earth's orbit were responsible for variations in solar energy reaching Earth's surface and climatic changes.

Millibar (mb)   A unit for expressing atmospheric pressure. Sea-level pressure is normally close to 1013 mb.

Minimum thermometer   A thermometer designed to measure the minimum air temperature during a desired time period.

Mini-swirls   Small whirling eddies perhaps 30 to 100 m in diameter that form in a region of strong wind shear of a hurricane's eye-wall. They are believed to be small tornadoes.

Mirage   A refraction phenomenon that makes an object appear to be displaced from its true position. When an object appears higher than it actually is, it is called a *superior mirage.* When an object appears lower than it actually is, it is an *inferior mirage.*

Mixed cloud   A cloud containing both water drops and ice crystals.

Mixing depth   The vertical extent of the mixing layer.

Mixing layer   The unstable atmospheric layer that extends from the surface up to the base of an inversion. Within this layer, the air is well stirred.

Mixing ratio   The ratio of the mass of water vapor in a given volume of air to the mass of dry air.

Moist adiabatic rate   The rate of change of temperature in a rising or descending saturated air parcel. The rate of cooling or warming varies but a common value of 6°C per 1000 m (3.3°F per 1000 ft) is used.

Moist adiabats   Lines on an adiabatic chart that show the moist adiabatic rate for rising and descending air.

Molecular viscosity   The small-scale internal fluid friction that is due to the random motion of the molecules within a smooth-flowing fluid, such as air.

Molecule   A collection of atoms held together by chemical forces.

Monsoon   A name given to seasonal winds that typically blow from different directions during different times of the year, most often during summer and winter.

Monsoon depressions   Weak low-pressure areas that tend to form in response to divergence in an upper-level jet stream. The circulation around the low strengthens the monsoon wind system and enhances precipitation during the summer.

Monsoon wind system   A wind system that reverses direction between winter and summer. Usually the wind blows from land to sea in winter and from sea to land in summer.

Mountain and **valley breeze**   A local wind system of a mountain valley that blows downhill (*mountain breeze*) at night and uphill (*valley breeze*) during the day.

Multicell thunderstorm   A convective storm system composed of a cluster of convective cells, each one in a different stage of its life cycle.

Nacreous clouds   Clouds of unknown composition that have a soft, pearly luster and that form at altitudes about 25 to 30 km above Earth's surface. They are also called *mother-of-pearl clouds*.

Negative feedback mechanism   *See* Feedback mechanism.

Neutral stability (neutrally stable atmosphere)   An atmospheric condition that exists in dry air when the environmental lapse rate equals the dry adiabatic rate. In saturated air the environmental lapse rate equals the moist adiabatic rate.

NEXRAD   An acronym for *Nex*t Generation Weather *Rad*ar. The main component of NEXRAD is the WSR-88D Doppler radar.

Nimbostratus   A dark, gray cloud characterized by more or less continuously falling precipitation. It is rarely accompanied by lightning, thunder, or hail.

Nitric oxide (NO)   A colorless gas produced by natural bacterial action in soil and by combustion processes at high temperatures. In polluted air, nitric oxide can react with ozone and hydrocarbons to form other substances. In this manner, it acts as an agent in the production of photochemical smog.

Nitrogen (N_2)   A colorless and odorless gas that occupies about 78 percent of dry air in the lower atmosphere.

Nitrogen dioxide (NO_2)   A reddish-brown gas, produced by natural bacterial action in soil and by combustion processes at high temperatures. In the presence of sunlight, it breaks down into nitric oxide and atomic oxygen. In polluted air, nitrogen dioxide acts as an agent in the production of photochemical smog.

Nitrogen oxides (NO_X)   Gases produced by natural processes and by combustion processes at high temperatures. In polluted air, nitric oxide (NO) and nitrogen dioxide (NO_2) are the most abundant oxides of nitrogen, and both act as agents for the production of photochemical smog.

Noctilucent clouds   Wavy, thin, bluish-white clouds that are best seen at twilight in polar latitudes. They form at altitudes about 80 to 90 km above the surface.

Nocturnal inversion   *See* Radiation inversion.

Nonsupercell tornado   A tornado that occurs with a cloud that is often in its growing stage, and one that does not contain a mid-level mesocyclone, or wall cloud. Landspouts and gustnadoes are examples of nonsupercell tornadoes.

North Atlantic Oscillation (NAO)   A reversal of atmospheric pressure over the Atlantic Ocean that influences the weather over Europe and over eastern North America.

Northeaster   A name given to a strong, steady wind from the northeast that is accompanied by rain and inclement weather. It often develops when a storm system moves northeastward along the coast of North America. Also called *nor'easter*.

Northern lights   *See* Aurora.

Nowcasting   Short-term weather forecasts varying from minutes up to a few hours.

Nuclear winter   The dark, cold, and gloomy conditions that presumably would be brought on by nuclear war.

Nucleation   Any process in which the phase change of a substance to a more condensed state (such as condensation, deposition, and freezing) is initiated about a particle (nucleus).

Numerical weather prediction (NWP)   Forecasting the weather based upon the solutions of mathematical equations by high-speed computers.

Obliquity (of Earth's axis)   The tilt of Earth's axis. It represents the angle from the perpendicular to the plane of Earth's orbit.

Occluded front (occlusion)   A complex frontal system that ideally forms when a cold front overtakes a warm front. When the air behind the front is colder than the air ahead of it, the front is called a *cold occlusion*. When the air behind the front is milder than the air ahead of it, it is called a *warm occlusion*.

Ocean conveyor belt   The global circulation of ocean water that is driven by the sinking of cold, dense water near Greenland and Labrador in the North Atlantic.

Ocean-effect snow   Localized bands of snow that occur when relatively cold air flows over a warmer ocean.

Offshore wind   A breeze that blows from the land out over the water. Opposite of an onshore wind.

Omega high   A ridge in the middle or upper troposphere that has the shape of the Greek letter omega (Ω).

Onshore wind   A breeze that blows from the water onto the land. Opposite of an offshore wind.

Open wave   The stage of development of a wave cyclone (mid-latitude cyclonic storm) where a cold front and a warm front exist, but no occluded front. The center of lowest pressure in the wave is located at the junction of the two fronts.

Orchard heaters   Oil heaters placed in orchards that generate heat and promote convective circulations to protect fruit trees from damaging low temperatures. Also called *smudge pots*.

Ordinary cell thunderstorm   (also called *air-mass thunderstorm*) A thunderstorm produced by local convection within a conditionally unstable air mass. It often forms in a region of low wind shear and does not reach the intensity of a severe thunderstorm.

Orographic clouds   Clouds produced by lifting along rising terrain, usually mountains.

Orographic uplift   The lifting of air over a topographic barrier. Clouds that form in this lifting process are called *orographic clouds*.

Outflow boundary   A surface boundary formed by the horizontal spreading of cool air that originated inside a thunderstorm.

Outgassing   The release of gases dissolved in hot, molten rock.

Overrunning   A condition that occurs when air moves up and over another layer of air.

Overshooting top   A situation in a mature thunderstorm where rising air, associated with strong convection, penetrates into a stable layer (usually the stratosphere), forcing the upper part of the cloud to rise above its relatively flat anvil top.

Oxygen (O_2)   A colorless and odorless gas that occupies about 21 percent of dry air in the lower atmosphere.

Ozone (O_3)   An almost colorless gaseous form of oxygen with an odor similar to weak chlorine. The highest natural concentration is found in the stratosphere where it is known as *stratospheric ozone*. It also forms in polluted air near the surface where it is the main ingredient of photochemical smog. Here, it is called *tropospheric ozone*.

Ozone hole   A sharp drop in stratospheric ozone concentration observed over the Antarctic during the spring.

Pacific air   Cool, moist air that originates over the Pacific Ocean, moves eastward, then descends the Rocky Mountains and moves over the plains as dry, stable, relatively cool air.

Pacific Decadal Oscillation (PDO)   A reversal in ocean surface temperatures that occurs every 20 to 30 years over the northern Pacific Ocean.

Pacific high   *See* Subtropical high.

Parcel of air   An imaginary small body of air a few meters wide that is used to explain the behavior of air.

Parhelia   *See* Sundog.

Particulate matter   Solid particles or liquid droplets that are small enough to remain suspended in the air. Also called *aerosols*.

Pattern recognition   An analogue method of forecasting where the forecaster uses prior weather events (or similar weather map conditions) to make a forecast.

P/E index (precipitation–evaporation index)   An index that gives the long-range effectiveness of precipitation in promoting plant growth.

P/E ratio (precipitation–evaporation ratio)   An expression devised for the purpose of classifying climates; based on monthly totals of precipitation and evaporation.

Permafrost   A layer of soil beneath Earth's surface that remains frozen throughout the year.

Persistence forecast   A forecast that the future weather condition will be the same as the present condition.

Photochemical smog   *See* Smog.

Photodissociation   The splitting of a molecule by a photon.

Photon   A discrete quantity of energy that can be thought of as a packet of electromagnetic radiation traveling at the speed of light.

Photosphere   The visible surface of the sun from which most of its energy is emitted.

Pileus cloud   A smooth cloud in the form of a cap. Occurs above, or is attached to, the top of a cumuliform cloud. Also called a *cap cloud*.

Pilot balloon   A small balloon that rises at a constant rate and is tracked by a theodolite in order to obtain wind speed and wind direction at various levels above Earth's surface.

Planetary boundary layer   *See* Atmospheric boundary layer.

Planetary scale   The largest scale of atmospheric motion. Sometimes called the *global scale*.

Plasma   *See* Solar wind.

Plate tectonics   The theory that Earth's surface down to about 100 km is divided into a number of plates that move relative to one another across the surface of Earth. Once referred to as continental drift.

Pleistocene Epoch (or Ice Age)   The most recent period of extensive continental glaciation that saw large portions of North America and Europe covered with ice. It began about 2 million years ago and ended about 10,000 years ago.

Polar easterlies   A shallow body of easterly winds located at high latitudes poleward of the subpolar low.

Polar front   A semipermanent, semicontinuous front that separates tropical air masses from polar air masses.

Polar front jet stream (polar jet)   The jet stream that is associated with the polar front in middle and high latitudes. It is usually located at altitudes between 9 and 12 km.

Polar front theory   A theory developed by a group of Scandinavian meteorologists that explains the formation, development, and overall life history of cyclonic storms that form along the polar front.

Polar ice cap climate   A climate characterized by extreme cold, as every month has an average temperature below freezing.

Polar low   An area of low pressure that forms over polar water behind (poleward of) the main polar front.

Polar orbiting satellite   A satellite whose orbit closely parallels Earth's meridian lines and thus crosses the polar regions on each orbit.

Polar tundra climate   A climate characterized by extremely cold winters and cool summers, as the average temperature of the warmest month climbs above freezing but remains below 10°C (50°F).

Polar vortex   The semipermanent zone of upper-level low pressure found in the polar regions that is sometimes disrupted or displaced during winter.

Pollutants   Any gaseous, chemical, or organic matter that contaminates the atmosphere, soil, or water.

Positive feedback mechanism   *See* Feedback mechanism.

Potential energy   The energy that a body possesses by virtue of its position with respect to other bodies in the field of gravity.

Potential evapotranspiration (PE)   The amount of moisture that, if it were available, would be removed from a given land area by evaporation and transpiration.

Potential temperature   The temperature that a parcel of dry air would have if it were brought dry adiabatically from its original position to a pressure of 1000 mb.

Precession (of Earth's axis of rotation)   The wobble of Earth's axis of rotation that traces out the path of a cone over a period of about 23,000 years.

Precipitation   Any form of water particles—liquid or solid—that falls from the atmosphere and reaches the ground.

Pressure   The force per unit area. (*See also* Air pressure.)

Pressure gradient   The rate of decrease of pressure per unit of horizontal distance. On the same chart, when the isobars are close together, the pressure gradient is steep. When the isobars are far apart, the pressure gradient is weak.

Pressure gradient force (PGF)   The force due to differences in pressure within the atmosphere that causes air to move and, hence, the wind to blow. It is directly proportional to the pressure gradient.

Pressure tendency   The rate of change of atmospheric pressure within a specified period of time, most often three hours. Same as *barometric tendency*.

Prevailing westerlies   The dominant westerly winds that blow in middle latitudes on the poleward side of the subtropical high-pressure areas. Also called *westerlies*.

Prevailing wind   The wind direction most frequently observed during a given period.

Primary air pollutants   Air pollutants that enter the atmosphere directly.

Probability forecast   A forecast of the probability of occurrence of one or more of a mutually exclusive set of weather conditions.

Prognostic chart (prog)   A chart showing expected or forecasted conditions, such as pressure patterns, frontal positions, contour height patterns, and so on.

Prominence   *See* Solar flare.

Psychrometer   An instrument used to measure the water vapor content of the air. It consists of two thermometers (dry bulb and wet bulb). After whirling the instrument, the dew point and relative humidity can be obtained with the aid of tables.

Radar   An electronic instrument used to detect objects (such as falling precipitation) by their ability to reflect and scatter microwaves back to a receiver. (*See also* Doppler radar.)

Radiant energy (radiation)   Energy propagated in the form of electromagnetic waves. These waves do not need molecules to propagate them, and in a vacuum they travel at nearly 300,000 km per sec (186,000 mi per sec).

Radiational cooling   The process by which Earth's surface and adjacent air cool by emitting infrared radiation.

Radiation fog   Fog produced over land when radiational cooling reduces the air temperature to or below its dew point. It is also known as *ground fog* and *valley fog*.

Radiation inversion   An increase in temperature with height due to radiational cooling of Earth's surface. Also called a *nocturnal inversion*.

Radiative equilibrium temperature   The temperature achieved when an object, behaving as a blackbody, is absorbing and emitting radiation at equal rates.

Radiative forcing   An increase (positive) or a decrease (negative) in net radiant energy observed over an area at the tropopause. An increase in radiative forcing may induce surface warming, whereas a decrease may induce surface cooling.

Radiative forcing agent   Any factor (such as increasing greenhouse gases and variations in solar output) that can change the balance between incoming energy from the sun and outgoing energy from Earth and the atmosphere.

Radiometer   *See* Infrared radiometer.

Radiosonde   A balloon-borne instrument that measures and transmits pressure, temperature, and humidity to a ground-based receiving station.

Radon   A colorless, odorless, radioactive gas that forms naturally as uranium in soil and rock breaks down.

Rain   Precipitation in the form of liquid water drops that have diameters greater than that of drizzle.

Rainbow   An arc of concentric colored bands that spans a section of the sky when rain is present and the sun is positioned at the observer's back.

Rain gauge   An instrument designed to measure the amount of rain that falls during a given time interval.

Rain shadow   The region on the leeside of a mountain where the precipitation is noticeably less than on the windward side.

Rawinsonde observation   A radiosonde observation that includes wind data.

Reflected light   *See* Reflection.

Reflection   The process whereby a surface turns back a portion of the radiation that strikes it. When the radiation that is turned back (reflected) from the surface is visible light, the radiation is referred to as *reflected light.*

Refraction   The bending of light as it passes from one medium to another.

Relative humidity   The ratio of the amount of water vapor in the air compared to the amount required for saturation (at a particular temperature and pressure). The ratio of the air's actual vapor pressure to its saturation vapor pressure.

Relative vorticity   *See* Vorticity.

Return stroke   The luminous lightning stroke that propagates upward from Earth to the base of a cloud.

Ribbon lightning   Lightning that appears to spread horizontally into a ribbon of parallel luminous streaks when strong winds are blowing parallel to the observer's line of sight.

Ridge   An elongated area of high atmospheric pressure.

Rime   A white or milky granular deposit of ice formed by the rapid freezing of supercooled water drops as they come in contact with an object in below-freezing air.

Riming   *See* Accretion.

Roll cloud   A dense, roll-shaped, elongated cloud that appears to slowly spin about a horizontal axis behind the leading edge of a thunderstorm's gust front.

Rossby waves   *See* Longwaves in the westerlies.

Rotor cloud   A turbulent cumuliform type of cloud that forms on the leeward side of large mountain ranges. The air in the cloud rotates about an axis parallel to the range.

Rotors   Turbulent eddies that form downwind of a mountain chain, creating hazardous flying conditions.

Saffir–Simpson Wind Scale   A scale relating a hurricane's winds to the possible damage it is capable of inflicting.

St. Elmo's fire   A bright electric discharge that is projected from objects (usually pointed) when they are in a strong electric field, such as during a thunderstorm.

Sand dune   A hill or ridge of loose sand shaped by the winds.

Santa Ana wind   A warm, dry wind that blows into southern California from the east off the elevated desert plateau. Its warmth is derived from compressional heating.

Saturation (of air)   An atmospheric condition whereby the level of water vapor is the maximum possible at the existing temperature and pressure.

Saturation vapor pressure   The maximum amount of water vapor necessary to keep moist air in equilibrium with a surface of pure water or ice. It represents the maximum amount of water vapor that the air can hold at any given temperature and pressure. (*See* Equilibrium vapor pressure.)

Savanna   A tropical or subtropical region of grassland and drought-resistant vegetation. Typically found in tropical wet-and-dry climates.

Scales of motion   The hierarchy of atmospheric circulations from tiny gusts to giant storms.

Scattering   The process by which small particles in the atmosphere deflect radiation from its path into different directions.

Scatterometer   A specialized satellite-borne instrument that can measure surface winds above the open ocean by observing the roughness of the sea.

Scientific method   The systematic production of knowledge that involves posing a question, putting forth a hypothesis, predicting what the hypothesis would imply if it were true, and carrying out tests to see if the prediction is accurate.

Scintillation   The apparent twinkling of a star due to its light passing through regions of differing air densities in the atmosphere.

Sea breeze   A coastal local wind that blows from the ocean onto the land. The leading edge of the breeze is termed a *sea-breeze front.*

Sea breeze convergence zone   A region where sea breezes, having started in different regions, flow together and converge.

Sea breeze front   The horizontal boundary that marks the leading edge of cooler marine air associated with a sea breeze.

Sea level pressure   The atmospheric pressure at mean sea level.

Seasonal forecasts   Outlooks that extend over three-month periods ranging out to a year or more, typically showing the odds that precipitation and temperature will be above or below average.

Secondary air pollutants   Pollutants that form when a chemical reaction occurs between a primary air pollutant and some other component of air. Tropospheric ozone is a secondary air pollutant.

Secondary low   A low-pressure area (often an open wave) that forms near, or in association with, a main low-pressure area.

Seiches   Standing waves that oscillate back and forth over an open body of water.

Selective absorbers   Substances such as water vapor, carbon dioxide, clouds, and snow that absorb radiation only at particular wavelengths.

Semi-arid climate   A dry climate where potential evaporation and transpiration exceed precipitation. Not as dry as the arid climate. Typical vegetation is short grass.

Semipermanent highs and lows   Areas of high pressure (anticyclones) and low pressure (extratropical cyclones) that tend to persist at a particular latitude belt throughout the year. In the Northern Hemisphere, typically they shift slightly northward in summer and slightly southward in winter.

Sensible heat   The heat we can feel and measure with a thermometer.

Sensible temperature   The sensation of temperature that the human body feels in contrast to the actual temperature of the environment as measured with a thermometer.

Severe thunderstorms   Intense thunderstorms capable of producing heavy showers, flash floods, hail, strong and gusty surface winds, and tornadoes. The U.S. National Weather Service defines a severe thunderstorm as having at least one of the following: hail with a diameter of at least 1 in., surface wind gusts of 50 knots or greater, or a tornado.

Sferics   Radio waves produced by lightning. A contraction of *atmospherics.*

Shear   *See* Wind shear.

Sheet lightning   Occurs when the lightning flash is not seen but the flash causes the cloud (or clouds) to appear as a diffuse luminous white sheet.

Shelf cloud   A dense, arch-shaped, ominous-looking cloud that often forms along the leading edge of a thunderstorm's gust front, especially when stable air rises up and over cooler air at the surface. Also called an *arcus cloud.*

Shelterbelt   A belt of trees or shrubs arranged as a protection against strong winds.

Short-range forecast   Generally used to describe a weather forecast that extends from about 6 hours to a few days into the future.

Shortwave (in the atmosphere) A small wave that moves around longwaves in the same direction as the air flow in the middle and upper troposphere. Shortwaves are also called *shortwave troughs*.

Shortwave radiation A term most often used to describe the radiant energy emitted from the sun, in the visible and near ultraviolet wavelengths.

Shower Intermittent precipitation from a cumuliform cloud, usually of short duration but often heavy.

Siberian high A strong, shallow area of high pressure that forms over Siberia in winter.

Simoom A strong, extremely hot, dry, dust-laden wind that blows over the Sahara. Its name means "poison wind," a reference to the ill effects produced by its extremely high temperature (reportedly exceeding 120°F) and low relative humidity (often less than 10 percent).

Sleet A type of precipitation consisting of transparent pellets of ice 5 mm or less in diameter. Same as *ice pellets*.

Smog Originally *smog* meant a mixture of smoke and fog. Today, *smog* means air that has restricted visibility due to pollution, or pollution formed in the presence of sunlight—*photochemical smog*.

Smog front (also **smoke front**) The leading edge of a sea breeze that is contaminated with smoke or pollutants.

Smudge pots *See* Orchard heaters.

Snow A solid form of precipitation composed of ice crystals in complex hexagonal form.

Snow-albedo feedback A positive feedback whereby increasing surface air temperatures enhance the melting of snow and ice in polar latitudes. This reduces Earth's albedo and allows more sunlight to reach the surface, which causes the air temperature to rise even more.

Snowflake An aggregate of ice crystals that falls from a cloud.

Snow flurries Light showers of snow that fall intermittently.

Snow grains Precipitation in the form of very small, opaque grains of ice. The solid equivalent of drizzle.

Snow pellets White, opaque, approximately round ice particles between 2 and 5 mm in diameter that form in a cloud either from the sticking together of ice crystals or from the process of accretion. Also called *graupel*.

Snow rollers A cylindrical spiral of snow shaped somewhat like a child's muff and produced by the wind.

Snow squall (shower) An intermittent heavy shower of snow that greatly reduces visibility.

Solar constant The rate at which solar energy is received on a surface at the outer edge of the atmosphere perpendicular to the sun's rays when Earth is at a mean distance from the sun. The value of the solar constant is about two calories per square centimeter per minute or about 1361 W/m² in the SI system of measurement.

Solar flare A rapid eruption from the sun's surface that emits high energy radiation and energized charged particles.

Solar wind An outflow of charged particles from the sun that escapes the sun's outer atmosphere at high speed.

Solute effect The dissolving of hygroscopic particles, such as salt, in pure water, thus reducing the relative humidity required for the onset of condensation.

Sonic boom A loud explosive-like sound caused by a shock wave emanating from an aircraft (or any object) traveling at or above the speed of sound.

Sounding An upper-air observation, such as a radiosonde observation. A vertical profile of an atmospheric variable such as temperature or winds.

Source regions Regions where air masses originate and acquire their properties of temperature and moisture.

Southern Oscillation (SO) The reversal of surface air pressure at opposite ends of the tropical Pacific Ocean that occur during major El Niño events.

Specific heat The ratio of the heat absorbed (or released) by the unit mass of the system to the corresponding temperature rise (or fall).

Specific humidity The ratio of the mass of water vapor in a given parcel to the total mass of air in the parcel.

Spin-up vortices Small whirling tornadoes perhaps 30 to 100 m in diameter that form in a region of strong wind shear in a hurricane's eyewall.

Squall line A line of thunderstorms that form along a cold front or out ahead of it.

Stable air *See* Absolutely stable atmosphere.

Standard atmosphere A hypothetical vertical distribution of atmospheric temperature, pressure, and density in which the air is assumed to obey the gas law and the hydrostatic equation. The lapse rate of temperature in the troposphere is taken as 6.5°C/1000 m or 3.6°F/1000 ft.

Standard atmospheric pressure A pressure of 1013.25 millibars (mb), 29.92 inches of mercury (Hg), 760 millimeters (mm) of mercury, 14.7 pounds per square inch (lb/in.²), or 1013.25 hectopascals (hPa).

Standard rain gauge A nonrecording rain gauge with an 8-inch diameter collector funnel and a tube that amplifies rainfall by tenfold.

Stationary front A front that is nearly stationary with winds blowing almost parallel and from opposite directions on each side of the front.

Station pressure The actual air pressure computed at the observing station.

Statistical forecast A forecast based on a mathematical/statistical examination of data that represents the past observed behavior of the forecasted weather element.

Steady-state forecast A weather prediction based on the past movement of surface weather systems. It assumes that the systems will move in the same direction and at approximately the same speed as they have been moving. Also called *trend forecasting*.

Steam fog *See* Evaporation (mixing) fog.

Stefan–Boltzmann law A law of radiation which states that the amount of radiant energy emitted from a unit surface area of an object (ideally a blackbody) is proportional to the fourth power of the object's absolute temperature.

Steppe An area of grass-covered, treeless plains that has a semi-arid climate.

Stepped leader An initial discharge of electrons that proceeds intermittently toward the ground in a series of steps in a cloud-to-ground lightning stroke.

Storm surge An abnormal rise of the sea along a shore; primarily due to the winds of a storm, especially a hurricane.

Stratocumulus A low cloud, predominantly stratiform, with low, lumpy, rounded masses, often with blue sky between them.

Stratosphere The layer of the atmosphere above the troposphere and below the mesosphere (between 10 km and 50 km), generally characterized by an increase in temperature with height.

Stratospheric polar night jet A jet stream that forms near the top of the stratosphere over polar latitudes during the winter months.

Stratus A low, gray cloud layer with a rather uniform base whose precipitation is most commonly drizzle.

Streamline A line that shows the wind flow pattern.

Sublimation The process whereby ice changes directly into water vapor without melting.

Subpolar climate A climate observed in the Northern Hemisphere that borders the polar climate. It is characterized by severely cold winters and short, cool summers. Also known as *taiga climate* and *boreal climate*.

Subpolar low A belt of low pressure located between 50° and 70° latitude. In the Northern Hemisphere, this "belt" consists of the *Aleutian low* in the North Pacific and the *Icelandic low* in the North Atlantic. In the Southern Hemisphere, it exists around the periphery of the Antarctic continent.

Subsidence   The slow sinking of air, usually associated with high-pressure areas.

Subsidence inversion   A temperature inversion produced by compressional warming—the adiabatic warming of a layer of sinking air.

Subtropical front   A zone of temperature transition in the upper troposphere over subtropical latitudes, where warm air carried poleward by the Hadley cell meets the cooler air of the middle latitudes.

Subtropical high   A semipermanent high in the subtropical high-pressure belt centered near 30° latitude. The *Bermuda high* is located over the Atlantic Ocean off the east coast of North America. The *Pacific high* is located off the west coast of North America.

Subtropical jet stream   The jet stream typically found between 20° and 30° latitude at altitudes between 12 and 14 km.

Suction vortices   Small, rapidly rotating whirls perhaps 10 m in diameter that are found within large tornadoes.

Sulfate aerosols   *See* Aerosols.

Sulfur dioxide   (SO_2) A colorless gas that forms primarily in the burning of sulfur-containing fossil fuels.

Summer solstice   The point in the year when the sun is highest in the sky. Typically it occurs around June 21 in the Northern Hemisphere, with the sun directly overhead at latitude $23\frac{1}{2}°$N, the Tropic of Cancer.

Sundog   A colored luminous spot produced by refraction of light through ice crystals that appears on either side of the sun. Also called *parhelia*.

Sun pillar   A vertical streak of light extending above (or below) the sun. It is produced by the reflection of sunlight off ice crystals.

Sunspots   Relatively cooler areas on the sun's surface. They represent regions of an extremely high magnetic field.

Supercell storm   A severe thunderstorm that consists primarily of a single rotating updraft. Its organized internal structure allows the storm to maintain itself for several hours. Supercell storms can produce large hail and dangerous tornadoes.

Supercell tornadoes   Tornadoes that occur within supercell thunderstorms that contain well-developed, mid-level mesocyclones.

Supercooled cloud (or cloud droplets)   A cloud composed of liquid droplets at temperatures below 0°C (32°F). When the cloud is on the ground it is called *supercooled fog* or *cold fog*.

Superior mirage   *See* Mirage.

Supersaturation   A condition whereby the atmosphere contains more water vapor than is needed to produce saturation with respect to a flat surface of pure water or ice, and the relative humidity is greater than 100 percent.

Super typhoon   A tropical cyclone (typhoon) in the western Pacific that has sustained winds of 130 knots or greater.

Surface inversion   *See* Radiation inversion.

Surface map   A map that shows the distribution of sea-level pressure with isobars and weather phenomena. Also called a *surface chart*.

Synoptic scale   The typical weather map scale that shows features such as high- and low-pressure areas and fronts over a distance spanning a continent. Also called the *cyclonic scale*.

Taiga (boreal forest)   The open northern part of the coniferous forest. Taiga also refers to subpolar climate.

Tangent arc   An arc of light tangent to a halo. It forms by refraction of light through ice crystals.

Tcu   An abbreviation sometimes used to denote a towering cumulus cloud (cumulus congestus).

Teleconnections   A linkage between weather changes occurring in widely separated regions of the world.

Temperature   The degree of hotness or coldness of a substance as measured by a thermometer. It is also a measure of the average speed or kinetic energy of the atoms and molecules in a substance.

Temperature inversion   An increase in air temperature with height, often simply called an *inversion*.

Terminal velocity   The constant speed obtained by a falling object when the upward drag on the object balances the downward force of gravity.

Texas norther   A strong, cold wind from between the northeast and northwest associated with a cold outbreak of polar air that brings a sudden drop in temperature. Sometimes called a *blue norther*.

Theodolite   An instrument used to track the movements of a pilot balloon.

Theory of plate tectonics   *See* Plate tectonics.

Thermal   A small, rising parcel of warm air produced when Earth's surface is heated unevenly.

Thermal belts   Horizontal zones of vegetation found along hillsides that are primarily the result of vertical temperature variations.

Thermal circulations   Air flow resulting primarily from the heating and cooling of air.

Thermal lows and **thermal highs**   Areas of low and high pressure that are shallow in vertical extent and are produced primarily by surface temperatures.

Thermal tides   Atmospheric pressure variations due to the uneven heating of the atmosphere by the sun.

Thermal turbulence   Turbulent vertical motions that result from surface heating and the subsequent rising and sinking of air.

Thermograph   An instrument that measures and records air temperature.

Thermometer   An instrument for measuring temperature. The most common is liquid-in-glass, which has a sealed glass tube attached to a glass bulb filled with liquid.

Thermosphere   The atmospheric layer above the mesosphere (above about 85 km) where the temperature increases rapidly with height.

Thunder   The sound due to rapidly expanding gases along the channel of a lightning discharge.

Thunderstorm   A convective storm (cumulonimbus cloud) with lightning and thunder. Thunderstorms can be composed of an ordinary cell, multicells, or a rapidly rotating supercell.

Tipping bucket rain gauge   A rain gauge that records rainfall by collecting rain in a chamber (bucket) that tips when the chamber fills with 0.01 in. (0.025 cm) of rain.

Tornado   An intense, rotating column of air that often protrudes from a cumuliform cloud in the shape of a funnel or a rope whose circulation is present on the ground. (*See* Funnel cloud.)

Tornado alley   A region in the Great Plains of the United States extending from Texas and Oklahoma northward into Kansas and Nebraska where tornadoes are most frequent. Same as *tornado belt*.

Tornado emergency   A type of tornado warning issued when a particularly strong tornado poses the potential for major damage or loss of life.

Tornado outbreak   A series of tornadoes that forms within a particular region—a region that may include several states. Often associated with widespread damage and destruction.

Tornado vortex signature (TVS)   An image of a tornado on the Doppler radar screen that shows up as a small region of rapidly changing wind directions inside a mesocyclone.

Tornado warning   A warning issued when a tornado has been observed or is indicated on a radar screen. It may also be issued when the formation of a tornado is imminent. Tornado warnings typically cover parts of one or more counties and last from 30 to 45 minutes.

Tornado watch   A forecast issued to alert the public that tornadoes may develop within a specified area, usually a portion of one or more states, over the next few hours.

Tornadogenesis   The process by which a tornado forms.

Trace (of precipitation)   An amount of precipitation less than 0.01 in. (0.025 cm).

Trade wind inversion   A temperature inversion frequently found in the subtropics over the eastern portions of the tropical oceans.

Trade winds   The winds that occupy most of the tropics and blow from the subtropical highs to the equatorial low.

Transpiration   The process by which water in plants is transferred as water vapor to the atmosphere.

Tropical cyclone   The general term for storms (cyclones) that form over warm tropical oceans.

Tropical depression   A mass of thunderstorms and clouds generally with a cyclonic wind circulation of between 20 and 34 knots (32 and 39 mi/hr).

Tropical disturbance   An organized mass of thunderstorms with a slight cyclonic wind circulation of less than 20 knots (32 mi/hr).

Tropical easterly jet   A jet stream that forms on the equatorward side of the subtropical highs near 15 km.

Tropical monsoon climate   A tropical climate with a brief dry period of perhaps one or two months.

Tropical rain forest   A type of forest consisting mainly of lofty trees and a dense undergrowth near the ground.

Tropical storm   A region of organized thunderstorms over the tropical or subtropical ocean with a cyclonic wind circulation between 35 and 64 knots.

Tropical wave   A migratory wavelike disturbance in the tropical easterlies. Tropical waves occasionally intensify into tropical cyclones. They are also called *easterly waves.*

Tropical wet-and-dry climate   A tropical climate poleward of the tropical wet climate where a distinct dry season occurs, often lasting for two months or more.

Tropical wet climate   A tropical climate with sufficient rainfall to produce a dense tropical rain forest.

Tropopause   The boundary between the troposphere and the stratosphere.

Tropopause jets   Jet streams found near the tropopause, such as the polar front and subtropical jet streams.

Troposphere   The layer of the atmosphere extending from Earth's surface up to the tropopause (about 10 km above the ground).

Trough   An elongated area of low atmospheric pressure.

Turbulence   Any irregular or disturbed flow in the atmosphere that produces gusts and eddies.

Twilight   The time at the beginning of the day immediately before sunrise and at the end of the day after sunset when the sky remains illuminated.

Typhoon   A hurricane (tropical cyclone) that forms in the northwestern Pacific Ocean.

Ultraviolet (UV) radiation   Electromagnetic radiation with wavelengths longer than X-rays but shorter than visible light.

Unstable air   *See* Absolutely unstable atmosphere.

Upper-air front   A front that is present aloft but usually does not extend down to the ground. Also called an *upper front* and an *upper-tropospheric front.*

Upslope fog   Fog formed as moist, stable air flows upward over a topographic barrier.

Upslope precipitation   Precipitation that forms due to moist, stable air gradually rising along an elevated plain. Upslope precipitation is common over the western Great Plains, especially east of the Rocky Mountains.

Upwelling   The rising of water (usually cold) toward the surface from the deeper regions of a body of water.

Urban heat island   The increased air temperatures in urban areas as contrasted to the cooler surrounding rural areas.

Valley breeze   *See* Mountain breeze.

Valley fog   *See* Radiation fog.

Vapor pressure   The pressure exerted by the water vapor molecules in a given volume of air. (*See also* Saturation vapor pressure.)

Veering wind   The wind that changes direction in a clockwise sense—north to northeast to east, and so on.

Vernal equinox   The point during the year at which the sun passes directly over the equator following the three months of winter. Occurs around March 20 in the Northern Hemisphere.

Very short range forecast   Generally used to describe a weather forecast that is made for up to a few hours (usually less than 6 hours) into the future.

Virga   Precipitation that falls from a cloud but evaporates before reaching the ground. (*See* Fall streaks.)

Viscosity   The resistance of fluid flow. (*See* Molecular viscosity *and* Eddy viscosity.)

Visible radiation (light)   Radiation with a wavelength between 0.4 and 0.7 μm. This region of the electromagnetic spectrum is called the *visible region.*

Visible region   *See* Visible radiation.

Visibility   The greatest distance at which an observer can see and identify prominent objects.

Volatile organic compounds (VOCs)   A class of organic compounds that are released into the atmosphere from sources such as motor vehicles, paints, and solvents. VOCs (which include hydrocarbons) contribute to the production of secondary pollutants, such as ozone.

Vorticity   A measure of the spin of a fluid, usually small air parcels. *Absolute vorticity* is the combined vorticity due to Earth's rotation (*Earth's vorticity*) and the vorticity due to the air's circulation relative to Earth. *Relative vorticity* is due to the curving of the air flow and wind shear.

Vorticity advection   The transport of vorticity by the wind. *Positive vorticity advection* occurs when the wind blows from high vorticity toward low vorticity, resulting in an increase in vorticity over time at a location. *Negative vorticity advection* occurs when the wind blows from low vorticity toward high vorticity, resulting in a decrease in vorticity over time at a location.

Wall cloud   An area of rotating clouds that extends beneath a supercell thunderstorm and from which a funnel cloud may appear. Also called a *collar cloud* and *pedestal cloud.*

Warm advection (or *warm air advection*)   The transport of warm air by the wind from a region of higher temperatures to a region of lower temperatures.

Warm-core low   A low-pressure area that is warmer at its center than at its periphery. Tropical cyclones exhibit this temperature pattern.

Warm front   A front that moves in such a way that warm air replaces cold air.

Warm occlusion   *See* Occluded front.

Warm sector   The region of warm air within a wave cyclone that lies between a retreating warm front and an advancing cold front.

Water equivalent   The depth of water that would result from the melting of a snow sample. Typically about 10 inches of snow will melt to 1 inch of water, producing a water equivalent of 10 to 1.

Waterspout   A column of rotating wind over water that has characteristics of a dust devil and tornado.

Water vapor   Water in a vapor (gaseous) form. Also called *moisture.*

Water vapor–greenhouse effect feedback   A positive feedback whereby increasing surface air temperatures cause an increase in the evaporation of water from the oceans. Increasing concentrations of atmospheric water vapor enhance the greenhouse effect, which causes the surface air temperature to rise even more. Also called the *water vapor–temperature rise feedback.*

Watt (W)   The unit of power in SI units where 1 watt is equivalent to 1 joule per second.

Wave cyclone   An extratropical cyclone that forms and moves along a front. The circulation of winds about the cyclone tends to produce a wavelike deformation on the front.

Wavelength   The distance between successive crests, troughs, or identical parts of a wave.

Weather   The condition of the atmosphere at any particular time and place.

Weather elements   The elements of *air temperature, air pressure, humidity, clouds, precipitation, visibility,* and *wind* that determine the present state of the atmosphere, the weather.

Weather type forecasting   A forecasting method where weather patterns are categorized into similar groups or types.

Weather types   Certain weather patterns categorized into similar groups. Used as an aid in weather prediction.

Weather warning   A forecast indicating that hazardous weather is either imminent or actually occurring within the specified forecast area.

Weather watch   A forecast indicating that atmospheric conditions are favorable for hazardous weather to occur over a particular region during a specified time period.

Weighing rain gauge   A rain gauge that records rainfall by weighing the collected water over a given time and converting the amount of water to rainfall depth.

Westerlies   The dominant westerly winds that blow in the middle latitudes on the poleward side of the subtropical high-pressure areas.

Westerly wind burst   A persistent area of strong west winds associated with showers and thunderstorms across the tropics.

Wet-bulb depression   The difference in degrees between the air temperature (dry-bulb temperature) and the wet-bulb temperature.

Wet-bulb temperature   The lowest temperature that can be obtained by evaporating water into the air.

Wet haze   *See* Haze.

Whirlwinds   *See* Dust devils.

Wien's law   A law of radiation which states that the wavelength of maximum emitted radiation by an object (ideally a blackbody) is inversely proportional to the object's absolute temperature.

Wind   Air in motion relative to Earth's surface.

Wind chill   The cooling effect of any combination of temperature and wind, expressed as the loss of body heat. Also called *wind-chill index.*

Wind direction   The direction *from which* the wind is blowing.

Wind machines   Fans placed in orchards for the purpose of mixing cold surface air with warmer air above.

Wind profiler   A Doppler radar capable of measuring the turbulent eddies that move with the wind. Because of this, it is able to provide a vertical picture of wind speed and wind direction.

Wind rose   A diagram that shows the percent of time that the wind blows from different directions at a given location over a given time.

Wind-sculpted trees   Trees whose branches are bent, twisted, and broken off on one side by strong prevailing winds. Also called *flag trees.*

Wind shear   The rate of change of wind speed or wind direction over a given distance.

Wind sock   A tapered fabric shaped like a cone that indicates wind direction by pointing away from the wind. Also called a *wind cone.*

Wind speed   The rate at which the air moves by a stationary object, usually measured in statute miles per hour (mi/hr), nautical miles per hour (knots), kilometers per hour (km/hr), or meters per second (m/sec).

Wind vane   An instrument used to indicate wind direction.

Windward side   The side of an object facing into the wind.

Wind waves   Water waves that form due to the flow of air over the water's surface.

Winter chilling   The amount of time the air temperature during the winter must remain below a certain value so that fruit and nut trees will grow properly during the spring and summer.

Winter solstice   The point in the year when the sun is lowest in the sky. Typically it occurs around December 21 in the Northern Hemisphere. At this point, the sun is directly overhead at latitude $23\frac{1}{2}°$S, the Tropic of Capricorn, where the summer solstice is occurring at the same time.

Xerophytes   Drought-resistant vegetation.

Younger Dryas event   A cold episode that took place from about 13,000 to about 11,700 years ago, during which average temperatures dropped suddenly and portions of the Northern Hemisphere reverted back to glacial conditions.

Zonal wind flow   A wind that has a predominately west-to-east component.

Additional Reading Material

SELECTED PERIODICALS

Bulletin of the American Meteorological Society. Monthly. American Meteorological Society. http://journals.ametsoc.org/toc/bams/current

Journal of Operational Meteorology. Online only. (Deals mainly with weather forecasting.) National Weather Association. http://nwas.org/publications/journal-operational-meteorology/

Weather. Monthly. Royal Meteorological Society. http://onlinelibrary.wiley.com/journal/10.1002/%28ISSN%291477-8696

Weatherwise. Bimonthly. Taylor & Francis Group LLC. www.weatherwise.org

SELECTED TECHNICAL PERIODICALS

From the American Meteorological Society

http://journals.ametsoc.org
Earth Interactions
Journal of Applied Meteorology and Climatology
Journal of Atmospheric and Oceanic Technology
Journal of the Atmospheric Sciences
Journal of Climate
Journal of Hydrometeorology
Journal of Physical Oceanography
Monthly Weather Review
Weather and Forecasting
Weather, Climate, and Society

From the American Geophysical Union

http://agupubs.onlinelibrary.wiley.com/
Earth's Future
Earth and Space Science
EOS (https://eos.org/)
Geophysical Research Letters
Journal of Geophysical Research—Atmospheres

From other publishers, including frequent content from atmospheric science

Nature. Weekly. Macmillan Publishers Limited. www.nature.com

Nature Climate Change. Monthly. Macmillan Publishers Limited. www.nature.com/nclimate

Proceedings of the National Academy of Sciences of the United States of America. Weekly. NAS. www.pnas.org

Science. Weekly. American Association for the Advancement of Science. www.sciencemag.org

BOOKS

Many of the titles below are written at the introductory level. Those that are more advanced are marked with an asterisk.

Ahrens, C. Donald, and Robert Henson. *Essentials of Meteorology* (8th ed.), Cengage Learning, Boston, MA, 2018.

Ahrens, C. Donald, and Perry Samson. *Extreme Weather and Climate*, Cengage Learning, Boston, MA, 2011.

*Andrews, David G. *An Introduction to Atmospheric Physics* (2nd ed.), Cambridge University Press, New York, 2010.

Archer, David. *Global Warming: Understanding the Forecast* (2nd ed.), Wiley, Hoboken, NJ, 2011.

Bigg, Grant R. *The Oceans and Climate* (2nd ed.), Cambridge University Press, New York, 2003.

*Bluestein, Howard B. *Synoptic-Dynamic Meteorology in Midlatitudes. Vol. 1: Principles of Kinematics and Dynamics*, Oxford University Press, New York, 1992.

*—. *Synoptic-Dynamic Meteorology in Midlatitudes. Vol. II: Observations and Theory of Weather Systems*, Oxford University Press, New York, 1993.

—. *Tornado Alley: Monster Storms of the Great Plains*, Oxford University Press, New York, 1999.

Bohren, Craig F. *Clouds in a Glass of Beer: Simple Experiments in Atmospheric Physics*, Wiley, New York, 1987.

—. *What Light Through Yonder Window Breaks?*, Wiley, New York, 1991.

Boubel, Richard W., et al., *Fundamentals of Air Pollution* (5th ed.), Academic Press, New York, 2014.

*Bradley, Raymond S. *Paleoclimatology: Reconstructing Climates of the Quarternary* (3rd ed.), Academic Press, New York, 2014.

Brunner, Ronald D., and Amanda H. Lynch. *Adaptive Governance and Climate Change*, American Meteorological Society, Boston, MA, 2010.

Burt, Christopher C. *Extreme Weather, A Guide and Record Book*, W.W. Norton & Company, New York, 2007.

Burt, Stephen. *The Weather Observer's Handbook*, Cambridge University Press, New York, 2012.

Carlson, Toby N. *An Observer's Guide to Clouds and Weather: A Northeastern Primer on Prediction*, American Meteorological Society, Boston, MA, 2014.

Changnon, Stanley A. *Railroads and Weather: From Fogs to Floods and Heat to Hurricanes*, American Meteorological Society, Boston, MA, 2006.

Climate Change 2013. The Physical Science Basis. Working Group 1 contribution to the Fifth Assessment Report of the IPCC, Cambridge University Press, New York, 2014.

*Cotton, W. R., and R. A. Anthes. *Storm and Cloud Dynamics* (2nd ed.), Academic Press, New York, 2010.

Cotton, William R., and Roger A. Pielke. *Human Impacts on Weather and Climate* (2nd ed.), Cambridge University Press, New York, 2007.

Doswell, Charles A. III, editor. *Severe Convective Storms*, American Meteorological Society, Boston, MA, 2001.

Dow, Kirstin, and Thomas E. Downing. *The Atlas of Climate Change: Mapping the World's Greatest Challenge*. University of California Press, Oakland, CA, 2011.

Edwards, Paul. *A Vast Machine: Computer Models, Climate Data, and the Politics of Global Warming*. The MIT Press, Cambridge, MA, 2013.

Emanuel, Kerry. *Divine Wind: The History and Science of Hurricanes*, Oxford University Press, New York, 2005.

Encyclopedia of Climate and Weather (2nd ed.), Stephen H. Schneider, Terry L. Root, and Michael D. Mastrandrea, Ed., Oxford University Press, New York, 2011.

The Encyclopedia of Weather and Climate Change: A Complete Visual Guide, Juliane L. Fry, et al., University of California Press, Berkeley, Los Angeles, CA, 2010.

Fabry, Frederic. *Radar Meteorology: Principles and Practice.* Cambridge University Press, New York, 2015.

Fagan, Brian. *The Great Warming: Climate Change and the Rise and Fall of Civilizations.* Bloomsbury Press, London, England, 2009.

Fleming, James Rodger. *Fixing the Sky: The Checkered History of Weather and Climate Control.* Columbia University Press, New York, 2012.

Glossary of Meteorology. Mary M. Cairns, Ed., American Meteorological Society, Boston, MA. Online only (www.ametsoc.org/ams/index.cfm/publications/glossary-of-meteorology/)

Henson, Robert. *The Thinking Person's Guide to Climate Change,* American Meteorological Society, Boston, MA, 2014.

—. *Weather on the Air: A History of Broadcast Meteorology,* American Meteorological Society, Boston, MA, 2010.

*Hobbs, Peter V. *Basic Physical Chemistry for Atmospheric Sciences* (2nd ed.), Cambridge University Press, New York, 2000.

Hulme, Mike. *Why We Disagree about Climate Change: Understanding Controversy, Inaction and Opportunity.* Cambridge University Press, Cambridge, England, 2009.

International Cloud Atlas. World Meteorological Organization, Geneva, Switzerland, 1987. Available online (http://library.wmo.int/pmb_ged/wmo_407_en-v2.pdf).

Jacobson, Mark Z. *Air Pollution and Global Warming: History, Science, and Solutions* (2nd ed.), Cambridge University Press, New York, 2012.

*Karoly, David J., and Dayton G. Vincent, Eds. *Meteorology of the Southern Hemisphere,* American Meteorological Society, Boston, MA, 1998.

Kocin, Paul J., and Kocin L. W. Uccellini. *Northeast Snowstorms,* Vol. 1 and Vol. 2, American Meteorological Society, Boston, MA, 2004.

Lackmann, Gary. *Midlatitude Synoptic Meteorology: Dynamics, Analysis, and Forecasting,* American Meteorological Society, Boston, MA, 2012.

Lee, Raymond L., and Alistair B. Fraser. *The Rainbow Bridge: Rainbows in Art, Myth, and Science.* Penn State University Press, University Park, PA, 2001.

Lynch, David K., and William Livingston. *Color and Light in Nature* (2nd ed.), Cambridge University Press, New York, 2001.

Managing the Risks of Extreme Events and Disasters to Advance Climate Change Adaptation. Intergovernmental Panel on Climate Change, Cambridge University Press, New York, 2012.

Meinel, Aden, and Marjorie Meinel. *Sunsets, Twilights and Evening Skies,* Cambridge University Press, New York, 1991.

Mergen, Bernard. *Weather Matters: An American Cultural History since 1900.* University Press of Kansas, Lawrence, KS, 2008.

Mims, F. M. *Hawaii's Mauna Loa Observatory: Fifty Years of Monitoring the Atmosphere,* University of Hawaii Press, Honolulu, HI, 2011.

Monmonier, Mark. *Air Apparent: How Meteorologists Learned to Map, Predict, and Dramatize Weather.* University of Chicago Press, Chicago, IL, 2001.

Pretor-Pinney, Gavin. *The Cloudspotters' Guide: The Science, History, and Culture of Clouds,* TarcherPerigee, New York, 2007.

Randall, David. *Atmosphere, Clouds, and Climate.* Princeton University Press, Princeton, New Jersey, 2012.

Righter, Robert W. *Wind Energy in America: A History,* University of Oklahoma Press, Norman, OK, 2008.

*Rogers, R. R. *A Short Course in Cloud Physics* (3rd ed.), Pergamon Press, Oxford, England, 1989.

Ruddiman, William. *Earth's Climate: Past and Future* (3rd ed.), William Freeman, New York, 2013.

Schultz, David H. *Eloquent Science: A Practical Guide to Becoming a Better Writer, Speaker, and Atmospheric Scientist,* American Meteorological Society, Boston, MA, 2009.

Sheffield, Justin, and Eric F. Wood. *Drought: Past Problems and Future Scenarios.* Routledge, London, England. 2011.

Simmons, Kevin M., and Daniel Sutter. *Economic and Societal Impacts of Tornadoes,* American Meteorological Society, Boston, MA, 2011.

—. *Deadly Season: Analyzing the 2011 Tornado Outbreaks,* American Meteorological Society, Boston, MA, 2012.

Somerville, Richard C. *The Forgiving Air* (2nd ed.), American Meteorological Society, Boston, MA, 2008.

*Strangeways, Ian. *Precipitation: Theory, Measurement and Distribution.* Cambridge University Press, London, England, 2011.

*Stull, Roland B. *Practical Meteorology: An Algebra-based Survey of Atmospheric Science.* University of British Columbia, Vancouver, 2015.

Vallis, Geoffrey K. *Climate and the Oceans.* Princeton University Press, Princeton, New Jersey, 2011.

Vasquez, Tim. *Weather Forecasting and Analysis Handbook.* Weather Graphics Technologies, Austin, TX, 2015.

*Wallace, John M. and Peter V. Hobbs. *Atmospheric Science: An Introductory Survey* (2nd ed.), Academic Press, Burlington, MA, 2006.

Williams, Jack. *The AMS Weather Book: The Ultimate Guide to America's Weather,* American Meteorological Society, Boston, MA, 2009.

Index

Downdraft, 386
Downslope wind, 252
Driest regions in U.S., 478
Drizzle, 177, 188
Dropsonde, 15, 257, 447, 467n
Dropwindsonde observations, 257
Drought, 492–493
Drought severity, 492
Dry adiabat, 160–161
Dry adiabatic rate, 144
Dry air, 108
Dry-bulb temperature, 107
Dry climates (group B), 481, 486–488
Dry conveyor belt, 334
Dry deposition, 560
Dry growth regime, 187
Dry haze, 115–116
Dry/humid boundary, 481n
Dry lightning, 408
Dryline, 312–313, 398–399
Dry slot, 331, 334
Dry-summer subtropical climate (Cs),
    481, 489
DU (*see* Dobson unit (DU))
Dual-polarization radar, 191, 434, 435
Dust Bowl, 235, 493
Dust Bowl years, 493
Dust devils, 247–249
Dust storms, 247, 493, 516, 540
Dwarf planet, 5
Dwd climate, 481, 495

E
Earth
    angle of tilt, 60–61, 514
    annual energy balance, 49–51
    atmosphere (*see* Atmosphere)
    average sea-level pressure, 201
    average surface pressure, 201
    data (table), 14
    distance from sun, 60
    elliptical path around sun,
        60–61, 66
    magnetic field, 51
    magnetosphere, 51
    orbital changes, 513–516
    orbiting the sun, 60–61, 66
    radiation, 40–41
    regions of the world, 264
    tilted on its axis, 60–61, 513–516
    vorticity, 337–338
    warming of Earth's surface, 50n
Earth-atmosphere energy balance, 50

Easterly wave, 440
Eastern Pacific hurricanes, 448–450
Eccentricity, 513
Echo, 190, 257
E climates, 481, 495–496
Eddies
    air pockets, 232
    fluid friction, 229
    howling of wind, 230
    mountain wave, 231
    roll, 231
    stable/unstable air, 230
Eddy viscosity, 229, 231
Edmund Fitzgerald freighter, 326
Edward I, 538
Eemian interglacial period, 507
EF climate, 481, 496–497
EF scale, 421–422
Ehrlich, Gretel, 383
850-mb chart, 356
Eismitte, Greenland, 74, 497
Ekman spiral, 279, 452
Ekman transport, 279, 452
El Azizia, Libya, 70
El Chichón volcanic eruption,
    517, 572
Electrical hygrometer, 107
Electrical resistance thermometer, 85
Electrical thermometer, 85
Electrification of clouds, 404–405
Electromagnetic waves, 39
Electron, 108
Elephant, 43
Elevated storm, 310
El Niño, 279–285
    conditions, 281–282
    effects of, 282–284
    hurricanes, 444
    1997-1998 event, 282
    non-El Niño conditions, 281
    Oceanic Niño Index (ONI), 283
    origin of the name, 279
    predicting, 286
    regions of climactic abnormalities,
        285
    strongest El Niño events, 282
    2014-2016 event, 280, 282
    winter conditions, 284
    Winter Olympics (2010), 284
El Niño conditions, 281–282
El Niño event, 444n
El Niño-Southern Oscillation (ENSO),
    280, 286, 524

Elongated highs (ridges), 206
Elongated lows (troughs), 206
ELVES, 405
Energy
    defined, 32
    Earth's annual energy balance,
        49–51
    heat, 32
    internal, 32
    kinetic, 32
    law of conservation of energy, 32
    potential, 32
    radiant, 32, 39
    units of measurement, A-1
    wind, 255
Enhanced Fujita scale, 421–422
Enhanced infrared image, 137
Enhanced infrared imagery, 350
Ensemble forecasting, 359, 467
Ensemble 500-mb forecast chart, 358
ENSO (El Niño-Southern Oscillation),
    280, 286
Entrainment, 155, 386
Entrance region, 332
Environmental lapse rate, 145,
    149, 150
Environmental Protection Agency
    (EPA), 43, 540–541, 543, 550, 552
Environmental tobacco smoke, 540
EPA (*see* Environmental Protection
    Agency (EPA))
Equation of state (gas law), 202, A-3
Equations and constants, A-3
Equatorial low, 265
Equatorial region, 264
Equilibrium vapor pressure, 168
ET climate, 481, 495–497
European Advanced Scatterometer
    (ASCAT), 257
European heat wave (2003), 560
Europe's "Storm of the Century," 340
Evaporation, 6, 34, 92, 94
Evaporation fog, 120
Evaporation (mixing) fog, 120–122
Evaporative cooling system, 104
Evapotranspiration, 480n
Evelyn, John, 538
Exit region, 332
Exosphere, 16
Explosive cyclogenesis, 325
Extended forecast, 364
Extratropical cyclone, 19, 322 (*see also*
    Middle-latitude cyclonic storms)

CLOUD CHART

Ahrens and Henson
Meteorology
Today
and
Essentials of
Meteorology

All photos © C. Donald Ahrens

Cirrus High ice crystal clouds blown by the wind. Typically, they are observed at altitudes above about 6000 m (20,000 ft).

Altocumulus Middle clouds composed of water droplets and ice crystals at elevations usually between 2000 m and 7000 m (6500 ft and 23,000 ft) above the surface. (If you hold your hand at arm's length, the individual puffs will be about the size of your thumbnail.)

Cirrostratus A wide-spread high cloud composed of ice crystals, that is normally white and usually covers a large portion of the sky. Sometimes halos around the sun or moon are the only indication of its presence.

Cirrocumulus These high clouds are small rounded white puffs composed of ice crystals. (The hole in the center may have been produced by a jet aircraft descending through the cloud deck.)

Altostratus This gray-looking water droplet and ice crystal middle cloud often blurs the sun, making it appear watery or "dimly visible".

Nimbostratus A dark gray-looking cloud that often covers the entire sky. Steady rain or snow falls from its base. The fragmented cloud beneath the nimbostratus is *Stratus Fractus*, or *scud*.

Stratus This low, unifrom, grayish-looking cloud typically has a base below 2000 m (6500 ft). Drizzle may fall from its base. It is distinguished from altostratus in that the sun is not usually visible through stratus.

Stratocumulus A low, lumpy-looking wide spread cloud with dark and light shading. (If you hold your hand at arm's length, the individual cloud elements will appear to be about the size of your fist.)

Cumulus Small, puffy clouds with relatively flat bases and limited vertical growth.

Cumulonimbus This cumuliform cloud n develop vertically to great heights, often sprouting anvil-shaped top. It produces hea showers of rain or snow, lightning and thun and strong, gusty surface winds.

Cumulus congestus Cumulus clouds that show extensive vertical development. They can form in rows, such as these, or as individual clouds that appear as a head of cauliflower. Showery precipitation may fall from cumulus congestus.

Mammatus Downward moving air causes these clouds to hang from the base of another cloud—most often a cumulonimbus or an altostratus.

Cumulus The sprouting clouds with pronounced vertical growth are *cumulus congestus*. The clouds with much smaller vertical extent are called *cumulus humilis*. The very small, ragged-looking clouds are called *cumulus fractus*.

Lenticular These lens-shaped clouds often form in waves that develop downwind of a mountain. They normally remain in one place as the air rushes through them.